Student's Solutions Manual

Judith A. Penna

Indiana University Purdue University Indianapolis

Foundations of Mathematics

Marvin L. Bittinger

Indiana University Purdue University Indianapolis

Judith A. Penna

Indiana University Purdue University Indianapolis

PEARSON

Addison
Wesley

Boston   San Francisco   New York
London   Toronto   Sydney   Tokyo   Singapore   Madrid
Mexico City   Munich   Paris   Cape Town   Hong Kong   Montreal

Reproduced by Pearson Addison-Wesley from electronic files supplied by the author.

Copyright © 2004 Pearson Education, Inc.
Publishing as Pearson Addison-Wesley, 75 Arlington Street, Boston, MA 02116

ISBN   0-321-17421-6

1 2 3 4 5 6 QEP 06 05 04 03

Contents

Chapter 1

Whole Numbers

Exercise Set 1.1

1. 2 3 $\boxed{5}$, 8 8 8

The digit 5 means 5 thousands.

3. 1, 4 8 8, $\boxed{5}$ 2 6

The digit 5 means 5 hundreds.

5. 1, 3 7 0, 0 0 $\boxed{0}$

The last 0 tells the number of ones.

7. $\boxed{1}$, 3 7 0, 0 0 0

The digit 1 tells the number of millions.

9. 5702 = 5 thousands + 7 hundreds + 0 tens + 2 ones, or
5 thousands + 7 hundreds + 2 ones

11. 93,986 = 9 ten thousands + 3 thousands + 9 hundreds + 8 tens + 6 ones

13. 2058 = 2 thousands + 0 hundreds + 5 tens + 8 ones, or 2 thousands + 5 tens + 8 ones

15. 1268 = 1 thousand + 2 hundreds + 6 tens + 8 ones

17. 2 thousands + 4 hundreds + 7 tens + 5 ones = 2475

19. 6 ten thousands + 8 thousands + 9 hundreds + 3 tens + 9 ones = 68,939

21. 7 thousands + 3 hundreds + 0 tens + 4 ones = 7304

23. 1 thousand + 9 ones = 1009

25. A word name for 85 is eighty-five.

27.
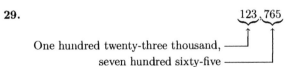

29.
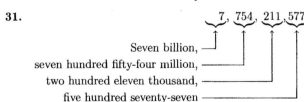

31.
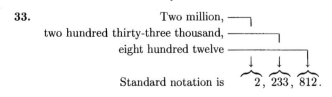

33.
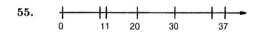

35.
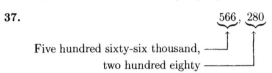

37.
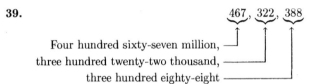

39.
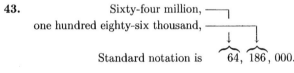

41.
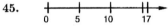
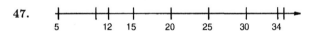

43.

45.

Since 0 is to the left of 17, 0 < 17.

47.
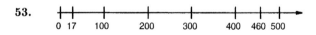

Since 34 is to the right of 12, 34 > 12.

49.

Since 1000 is to the left of 1001, 1000 < 1001.

51.

Since 133 is to the right of 132, 133 > 132.

53.

Since 460 is to the right of 17, 460 > 17.

55.

Since 37 is to the right of 11, 37 > 11.

57. Since 54 lies to the right of 47 on the number line, Thrust SCC is longer. We write the inequality 54 ft > 47 ft.

59. Since 87 lies to the right of 81 on the number line, we can write 87 yr > 81 yr. We can also write 81 yr < 87 yr since 81 lies to the left of 87 on the number line.

61. Discussion and Writing Exercise

63. First consider the whole numbers from 100 through 199. The 10 numbers 102, 112, 122, ... , 192 contain the digit 2. In addition, the 10 numbers 120, 121, 122, ... , 129 contain the digit 2. However, we do not count the number 122 in this group because it was counted in the first group of ten numbers. Thus, 19 numbers from 100 through 199 contain the digit 2. Using the same type of reasoning for the whole numbers from 300 to 400, we see that there are also 19 numbers in this group that contain the digit 2.

Finally, consider the 100 whole numbers 200 through 299. Each contains the digit 2.

Thus, there are 19 + 19 + 100, or 138 whole numbers between 100 and 400 that contain the digit 2 in their standard notation.

Exercise Set 1.2

1.
```
    3 6 4
  +   2 3
  ───────
    3 8 7
```
Add ones, add tens, then add hundreds.

3.
```
      1
    1 7 1 6
  + 3 4 8 2
  ─────────
    5 1 9 8
```
Add ones: We get 8. Add tens: We get 9 tens. Add hundreds: We get 11 hundreds, or 1 thousand + 1 hundred. Write 1 in the hundreds column and 1 above the thousands. Add thousands: We get 5 thousands.

5.
```
        1
    8 1 1 3
  +   3 9 0
  ─────────
    8 5 0 3
```
Add ones: We get 3. Add tens: We get 10 tens, or 1 hundred + 0 tens. Write 0 in the tens column and 1 above the hundreds. Add hundreds: We get 5. Add thousands: We get 8.

7.
```
      1
      3 5 6
  + 4 9 1 0
  ─────────
    5 2 6 6
```
Add ones: We get 6. Add tens: We get 6. Add hundreds: We get 12 hundreds, or 1 thousand + 2 hundreds. Write 2 in the hundreds column and 1 above the thousands. Add thousands: We get 5.

9.
```
      1
      9 9
  +     1
  ───────
    1 0 0
```
Add ones: We get 10 ones, or 1 ten + 0 ones. Write 0 in the ones column and 1 above the tens. Add tens: We get 10 tens.

11.
```
      1 1
    5 0 9 3
  + 3 2 1 7
  ─────────
    8 3 1 0
```
Add ones: We get 10 ones, or 1 ten + 0 ones. Write 0 in the ones column and 1 above the tens. Add tens: We get 11. Write 1 in the tens column and 1 above the hundreds. Add hundreds: We get 3 hundreds. Add thousands: We get 8 thousands.

13.
```
      1 1
    4 8 2 5
  + 1 7 8 3
  ─────────
    6 6 0 8
```
Add ones: We get 8. Add tens: We get 10 tens. Write 0 in the tens column and 1 above the hundreds. Add hundreds: We get 16 hundreds. Write 6 in the hundreds column and 1 above the thousands. Add thousands: We get 6 thousands.

15.
```
       1 1 1
    2 3, 4 4 3
  + 1 0, 9 8 9
  ───────────
    3 4, 4 3 2
```
Add ones: We get 12 ones, or 1 ten + 2 ones. Write 2 in the ones column and 1 above the tens. Add tens: We get 13 tens. Write 3 in the tens column and 1 above the hundreds. Add hundreds: We get 14 hundreds. Write 4 in the hundreds column and 1 above the thousands. Add thousands: We get 4 thousands. Add ten thousands: We get 3 ten thousands.

17.
```
       1 1
    1 2, 0 7 0
       2 9 5 4
  +    3 4 0 0
  ───────────
    1 8, 4 2 4
```
Add ones: We get 4. Add tens: We get 12 tens, or 1 hundred + 2 tens. Write 2 in the tens column and 1 above the hundreds. Add hundreds: We get 14 hundreds, or 1 thousand + 4 hundreds. Write 4 in the hundreds column and 1 above the hundreds. Add thousands: We get 8 thousands. Add ten thousands: We get 1 ten thousand.

19.
```
      2 4
      3 2 7
      4 2 8
      5 6 9
      7 8 7
  +   2 0 9
  ─────────
    2 3 2 0
```
Add ones: We get 40. Write 0 in the ones column and 4 above the tens. Add tens: We get 22 tens. Write 2 in the tens column and 2 above the hundreds. Add hundreds: We get 23 hundreds.

21. 14 mi + 13 mi + 8 mi + 10 mi + 47 mi + 22 mi = Perimeter
We carry out the addition.
```
    2
    1 4
    1 3
      8
    1 0
    4 7
  + 2 2
  ─────
    1 1 4
```
The perimeter of the figure is 114 mi.

23. 200 ft + 85 ft + 200 ft + 85 ft = Perimeter
We carry out the addition.
```
    1 1
    2 0 0
      8 5
    2 0 0
  +   8 5
  ───────
    5 7 0
```
The perimeter of the hockey rink is 570 ft.

25. $7 - 4 = 3$

This number gets added (after 3).

$7 = 3 + 4$

(By the commutative law of addition, $7 = 4 + 3$ is also correct.)

27. $43 - 16 = 27$

This number gets added (after 27).

$43 = 27 + 16$

(By the commutative law of addition, $43 = 16 + 27$ is also correct.)

29. $6 + 9 = 15$ ↑
This addend gets subtracted from the sum.
$6 = 15 - 9$

$6 + 9 = 15$ ↑
This addend gets subtracted from the sum.
$9 = 15 - 6$

31. $23 + 9 = 32$ ↑
This addend gets subtracted from the sum.
$23 = 32 - 9$

$23 + 9 = 32$ ↑
This addend gets subtracted from the sum.
$9 = 32 - 23$

33.
$$\begin{array}{r} 6\,5 \\ -\ 2\,1 \\ \hline 4\,4 \end{array}$$
Subtract ones, then subtract tens.

35.
$$\begin{array}{r} 4\,5\,4\,7 \\ -\ 3\,4\,2\,1 \\ \hline 1\,1\,2\,6 \end{array}$$
Subtract ones, subtract tens, subtract hundreds, then subtract thousands.

37.
$$\begin{array}{r} \overset{6\ 16}{7\,7\,\cancel{6}\,9} \\ -\ 2\,3\,8\,7 \\ \hline 5\,3\,8\,2 \end{array}$$
Subtract ones. We cannot subtract 8 tens from 6 tens. Borrow 1 hundred to get 16 tens. Subtract tens, subtract hundreds, then subtract thousands.

39.
$$\begin{array}{r} \overset{6\ 16\ \ 3\ 10}{7\,\cancel{6}\,\cancel{4}\,\cancel{0}} \\ -\ 3\,8\,0\,9 \\ \hline 3\,8\,3\,1 \end{array}$$
We cannot subtract 9 ones from 0 ones. Borrow 1 ten to get 10 ones. Subtract ones, then tens. We cannot subtract 8 hundreds from 6 hundreds. Borrow 1 thousand to get 16 hundreds. Subtract hundreds, then thousands.

41.
$$\begin{array}{r} \overset{9\ \ 9\ \ 9\ 12}{\cancel{1\,0,0\,0\,2}} \\ -\ \ \ 7\,8\,3\,4 \\ \hline 2\,1\,6\,8 \end{array}$$
We have 1 ten thousand, or 1000 tens. We borrow 1 ten to get 10 ones. We then have 999 tens. Subtract ones, then tens, then hundreds, then thousands.

43.
$$\begin{array}{r} \overset{8\ 10\ \ \ 2\ 17}{\cancel{9}\,\cancel{0},2\,\cancel{3}\,\cancel{7}} \\ -\ 4\,7,2\,0\,9 \\ \hline 4\,3,0\,2\,8 \end{array}$$

45.
$$\begin{array}{r} \overset{13}{\overset{\cancel{3}\ \ 10}{\cancel{1}\,\cancel{4}\,\cancel{0}}} \\ -\ \ \ 5\,6 \\ \hline 8\,4 \end{array}$$

47.
$$\begin{array}{r} \overset{8\ 10}{6\,\cancel{9}\,\cancel{0}} \\ -\ 2\,3\,6 \\ \hline 4\,5\,4 \end{array}$$

49.
$$\begin{array}{r} \overset{8\ 10}{\cancel{9}\,\cancel{0}\,3} \\ -\ 1\,3\,2 \\ \hline 7\,7\,1 \end{array}$$

51.
$$\begin{array}{r} \overset{8\ 12}{8\,0\,\cancel{9}\,\cancel{2}} \\ -\ 1\,0\,7\,3 \\ \hline 7\,0\,1\,9 \end{array}$$
Borrow 1 ten to get 12 ones. Subtract ones, then tens, then hundreds, then thousands.

53.
$$\begin{array}{r} \overset{13}{\overset{7\ \ \cancel{3}\ 13}{5\,\cancel{8}\,\cancel{4}\,\cancel{3}}} \\ -\ \ \ \ 9\,8 \\ \hline 5\,7\,4\,5 \end{array}$$

55.
$$\begin{array}{r} \overset{10}{\overset{1\ \ \cancel{0}\ 10\ 3\ 13}{\cancel{2}\,\cancel{1},\cancel{0}\,\cancel{4}\,\cancel{3}}} \\ -\ \ \ 8\,9\,0\,9 \\ \hline 1\,2,1\,3\,4 \end{array}$$

57.
$$\begin{array}{r} \overset{6\ 9\ 9\ 10}{\cancel{7\,0\,0\,0}} \\ -\ 2\,7\,9\,4 \\ \hline 4\,2\,0\,6 \end{array}$$
We have 7 thousands or 700 tens. We borrow 1 ten to get 10 ones. We then have 699 tens. Subtract ones, then tens, then hundreds, then thousands.

59.
$$\begin{array}{r} \overset{7\ 9\ 9\ 10}{4\,\cancel{8,0\,0\,0}} \\ -\ 3\,7,6\,9\,5 \\ \hline 1\,0,3\,0\,5 \end{array}$$
We have 8 thousands or 800 tens. We borrow 1 ten to get 10 ones. We then have 799 tens. Subtract ones, then tens, then hundreds, then thousands, then ten thousands.

61. Discussion and Writing Exercise

63. $4\,\boxed{8}\,6,2\,0\,5$

The digit 8 tells the number of ten thousands.

65. One method is described in the answer section in the text. Another method is: $1 + 100 = 101$, $2 + 99 = 101$, ..., $50 + 51 = 101$. Then $50 \cdot 101 = 5050$.

4

Chapter 1: Whole Numbers

Exercise Set 1.3

1.
$$\begin{array}{r} \overset{2}{9}4 \\ \times\ 6 \\ \hline 564 \end{array}$$ Multiplying by 6

3.
$$\begin{array}{r} 2340 \\ \times 1000 \\ \hline 2,340,000 \end{array}$$ Multiplying by 1 thousand (We write 000 and then multiply 2340 by 1.)

5.
$$\begin{array}{r} \overset{2}{5}09 \\ \times\ 3 \\ \hline 1527 \end{array}$$ Multiplying by 3

7.
$$\begin{array}{r} \overset{1}{9}\overset{2}{2}\overset{6}{2}9 \\ \times\ 7 \\ \hline 64,603 \end{array}$$ Multiplying by 7

9.
$$\begin{array}{r} \overset{2}{5}3 \\ \times\ 90 \\ \hline 4770 \end{array}$$ Multiplying by 9 tens (We write 0 and then multiply 53 by 9.)

11.
$$\begin{array}{r} \overset{2}{\overset{3}{8}}5 \\ \times 47 \\ \hline 595 \\ 3400 \\ \hline 3995 \end{array}$$ Multiplying by 7 / Multiplying by 40 / Adding

13.
$$\begin{array}{r} \overset{2}{6}40 \\ \times 72 \\ \hline 1280 \\ 44800 \\ \hline 46,080 \end{array}$$ Multiplying by 2 / Multiplying by 70 / Adding

15.
$$\begin{array}{r} \overset{1}{\overset{1}{4}}\overset{1}{\overset{1}{4}}4 \\ \times 33 \\ \hline 1332 \\ 13320 \\ \hline 14,652 \end{array}$$ Multiplying by 3 / Multiplying by 30 / Adding

17.
$$\begin{array}{r} \overset{3}{\overset{7}{5}}09 \\ \times 408 \\ \hline 4072 \\ 203600 \\ \hline 207,672 \end{array}$$ Multiplying by 8 / Multiplying by 4 hundreds (We write 00 and then multiply 509 by 4.)

19.
$$\begin{array}{r} \overset{4}{\overset{1}{\overset{3}{8}}}\overset{2}{\overset{1}{5}}3 \\ \times 936 \\ \hline 5118 \\ 25590 \\ 767700 \\ \hline 798,408 \end{array}$$ Multiplying by 6 / Multiplying by 30 / Multiplying by 900 / Adding

21.
$$\begin{array}{r} \overset{1}{\overset{1}{\overset{1}{\overset{1}{6}}}}\overset{2}{\overset{1}{\overset{3}{4}}}28 \\ \times 3224 \\ \hline 25712 \\ 128560 \\ 1285600 \\ 19284000 \\ \hline 20,723,872 \end{array}$$ Multiplying by 4 / Multiplying by 20 / Multiplying by 200 / Multiplying by 3000 / Adding

23.
$$\begin{array}{r} \overset{1}{3}\overset{3}{4}82 \\ \times 104 \\ \hline 13928 \\ 348200 \\ \hline 362,128 \end{array}$$ Multiplying by 4 / Multiplying by 1 hundred (We write 00 and then multiply 3482 by 1.)

25.
$$\begin{array}{r} \overset{2}{\overset{4}{5}}006 \\ \times 4008 \\ \hline 40048 \\ 20024000 \\ \hline 20,064,048 \end{array}$$ Multiplying by 8 / Multiplying by 4 thousands (We write 000 and then multiply 5006 by 4.)

27.
$$\begin{array}{r} \overset{2}{\overset{3}{5}}\overset{3}{\overset{4}{6}}08 \\ \times 4500 \\ \hline 2804000 \\ 22432000 \\ \hline 25,236,000 \end{array}$$ Multiplying by 5 hundreds (We write 00 and then multiply 5608 by 5.) / Multiplying by 4000 / Adding

29. $A = 728 \text{ mi} \times 728 \text{ mi} = 529,984$ square miles

31. $A = l \times w = 90 \text{ ft} \times 90 \text{ ft} = 8100$ square feet

33. $18 \div 3 = 6$ The 3 moves to the right. A related multiplication sentence is $18 = 6 \cdot 3$. (By the commutative law of multiplication, there is also another multiplication sentence: $18 = 3 \cdot 6$.)

35. $22 \div 22 = 1$ The 22 on the right of the $\div$ symbol moves to the right. A related multiplication sentence is $22 = 1 \cdot 22$. (By the commutative law of multiplication, there is also another multiplication sentence: $22 = 22 \cdot 1$.)

37. $9 \times 5 = 45$

Move a factor to the other side and then write a division.

$9 \times 5 = 45$ $9 \times 5 = 45$

$9 = 45 \div 5$ $5 = 45 \div 9$

39. Two related division sentences for $37 \cdot 1 = 37$ are:

$37 = 37 \div 1$ $(37 \cdot 1 = 37 \quad)$

and

$1 = 37 \div 37$ $(37 \cdot 1 = 37 \quad)$

41.
```
    1 2
  6 ) 7 2
    6 0
    ----
    1 2
    1 2
    ----
      0
```
Think: 7 tens $\div$ 6. Estimate 1 ten.
Think: 12 ones $\div$ 6. Estimate 2 ones.

The answer is 12.

43. $\dfrac{23}{23} = 1$ Any nonzero number divided by itself is 1.

45. $22 \div 1 = 22$ Any number divided by 1 is that same number.

47. $\dfrac{16}{0}$ is not defined, because division by 0 is not defined.

49.
```
      5 5
  5 ) 2 7 7
    2 5 0
    -----
      2 7
      2 5
      -----
        2
```
Think: 2 hundreds $\div$ 5. There are no hundreds in the quotient.
Think: 27 tens $\div$ 5. Estimate 5 tens.
Think: 27 ones $\div$ 5. Estimate 5 ones.

The answer is 55 R 2.

51.
```
      1 0 8
  8 ) 8 6 4
    8 0 0
    -----
      6 4
      6 4
      -----
        0
```
Think: 8 hundreds $\div$ 8. Estimate 1 hundred.
Think: 6 tens $\div$ 8. There are no tens in the quotient (other than the tens in 100). Write a 0 to show this.
Think: 64 ones $\div$ 8. Estimate 8 ones.

The answer is 108.

53.
```
        3 0 7
  4 ) 1 2 2 8
    1 2 0 0
    -------
        2 8
        2 8
        -------
          0
```
Think: 12 hundreds $\div$ 4. Estimate 3 hundreds.
Think: 2 tens $\div$ 4. There are no tens in the quotient (other than the tens in 300). Write a 0 to show this.
Think: 28 ones $\div$ 4. Estimate 7 ones.

The answer is 307.

55.
```
      9 2
  8 ) 7 3 8
    7 2 0
    -----
      1 8
      1 6
      -----
        2
```
Think: 73 tens $\div$ 8. Estimate 9 tens.
Think: 18 ones $\div$ 8. Estimate 2 ones.

The answer is 92 R 2.

57.
```
        1 7 0 3
  5 ) 8 5 1 5
    5 0 0 0
    -------
    3 5 1 5
    3 5 0 0
    -------
        1 5
        1 5
        -------
          0
```
Think: 8 thousands $\div$ 5. Estimate 1 thousand.
Think: 35 hundreds $\div$ 5. Estimate 7 hundreds.
Think: 1 ten $\div$ 5. There are no tens in the quotient (other than the tens in 1700). Write a 0 to show this.
Think: 15 ones $\div$ 5. Estimate 3 ones.

The answer is 1703.

59.
```
                 1 2 7
  1 0 0 0 ) 1 2 7,0 0 0
        1 0 0,0 0 0
        -----------
          2 7,0 0 0
          2 0,0 0 0
          -----------
            7 0 0 0
            7 0 0 0
            -----------
                  0
```
Think: 1270 hundreds $\div$ 1000. Estimate 1 hundred.
Think: 2700 tens $\div$ 1000. Estimate 2 tens.
Think: 7000 ones $\div$ 1000. Estimate 7 ones.

The answer is 127.

61.
```
          2 9
  3 0 ) 8 7 5
      6 0 0
      -----
      2 7 5
      2 7 0
      -----
          5
```
Think: 87 tens $\div$ 30. Estimate 2 tens.
Think: 275 ones $\div$ 30. Estimate 9 ones.

The answer is 29 R 5.

63.
```
          4 0
  2 1 ) 8 5 2
      8 4 0
      -----
        1 2
```
Round 21 to 20.
Think: 85 tens $\div$ 20. Estimate 4 tens.
Think: 12 ones $\div$ 20. There are no ones in the quotient (other than the ones in 40). Write a 0 to show this.

The answer is 40 R 12.

65.
```
              8
  8 5 ) 7 6 7 2
      6 8 0 0
      -------
      [8 7] 2
```
Round 85 to 90.
Think: 767 tens $\div$ 90. Estimate 8 tens.

Since 87 is larger than the divisor, the estimate is too low.

```
          9 0
  8 5 ) 7 6 7 2
      7 6 5 0
      -------
          2 2
```
Think: 767 tens $\div$ 90. Estimate 9 tens.
Think: 22 ones $\div$ 90. There are no ones in the quotient (other than the ones in 90). Write a 0 to show this.

The answer is 90 R 22.

67.
```
            3
  1 1 1 [ 3 2 1 9
          3 3 3 0
```
Round 111 to 100.
Think: 321 tens ÷ 100. Estimate 3 tens.

Since we cannot subtract 3330 from 3219, the estimate is too high.

```
            2 9
  1 1 1 [ 3 2 1 9
          2 2 2 0
          ─────
            9 9 9
            9 9 9
          ─────
                0
```
Think: 321 tens ÷ 100. Estimate 2 tens.
Think: 999 ones ÷ 100. Estimate 9 ones.

The answer is 29.

69.
```
        1 0 0 7
  5 [ 5 0 3 6
      5 0 0 0
      ─────
          3 6
          3 5
          ───
            1
```
Think: 5 thousands ÷ 5. Estimate 1 thousand.
Think: 0 hundreds ÷ 5. There are no hundreds in the quotient (other than the hundreds in 1000). Write a 0 to show this.
Think: 3 tens ÷ 5. There are no tens in the quotient (other than the tens in 1000). Write a 0 to show this.
Think: 36 ones ÷ 5. Estimate 7 ones.

The answer is 1007 R 1.

71.
```
          2 2
  4 6 [ 1 0 5 8
        9 2 0
        ─────
        1 3 8
          9 2
          ───
          4 6
```
Round 46 to 50.
Think: 105 tens ÷ 50. Estimate 2 tens.
Think: 138 ones ÷ 50. Estimate 2 ones.

Since 46 is not smaller than the divisor, 46, the estimate is too low.

```
          2 3
  4 6 [ 1 0 5 8
        9 2 0
        ─────
        1 3 8
        1 3 8
        ─────
            0
```
Think: 138 ones ÷ 50. Estimate 3 ones.

The answer is 23.

73.
```
          4
  2 4 [ 8 8 8 0
        9 6 0 0
```
Round 24 to 20.
Think: 88 hundreds ÷ 20. Estimate 4 hundreds.

Since we cannot subtract 9600 from 8880, the estimate is too high.

```
          3 8
  2 4 [ 8 8 8 0
        7 2 0 0
        ─────
        1 6 8 0
        1 9 2 0
```
Think: 88 hundreds ÷ 20. Estimate 3 hundreds.
Think: 168 tens ÷ 20. Estimate 8 tens.

Since we cannot subtract 1920 from 1680, the estimate is too high.

```
          3 7 0
  2 4 [ 8 8 8 0
        7 2 0 0
        ─────
        1 6 8 0
        1 6 8 0
        ─────
              0
```
Think: 168 tens ÷ 20. Estimate 7 tens.
Think: 0 ones ÷ 20. There are no ones in the quotient (other than the ones in 370). Write a 0 to show this.

The answer is 370.

75.
```
            5
  2 8 [ 1 7 , 0 6 7
        1 4 0 0 0
        ─────────
        [3 0] 6 7
```
Round 28 to 30.
Think: 170 hundreds ÷ 30. Estimate 5 hundreds.

Since 30 is larger than the divisor, 28, the estimate is too low.

```
            6 0 8
  2 8 [ 1 7 , 0 6 7
        1 6 8 0 0
        ─────────
            2 6 7
            2 2 4
            ─────
            [4 3]
```
Think: 170 hundreds ÷ 30. Estimate 6 hundreds.
Think: 26 tens ÷ 30. There are no tens in the quotient (other than the tens in 600.) Write a zero to show this.
Think: 267 ones ÷ 30. Estimate 8 ones.

Since 43 is larger than the divisor, 28, the estimate is too low.

```
            6 0 9
  2 8 [ 1 7 , 0 6 7
        1 6 8 0 0
        ─────────
            2 6 7
            2 5 2
            ─────
            1 5
```
Think: 267 ones ÷ 30. Estimate 9 ones.

The answer is 609 R 15.

77.
```
              3 5 0 8
  2 8 5 [ 9 9 9 , 9 9 9
          8 5 5 0 0 0
          ───────────
          1 4 4 9 9 9
          1 4 2 5 0 0
          ───────────
              2 4 9 9
              2 2 8 0
              ───────
                2 1 9
```

The answer is 3508 R 219.

79. Round 48 to the nearest ten.

4 [8]
↑

The digit 4 is in the tens place. Consider the next digit to the right. Since the digit, 8, is 5 or higher, round 4 tens up to 5 tens. Then change the digit to the right of the tens digit to zero.

The answer is 50.

81. Round 467 to the nearest ten.

$$4 \; 6 \; \boxed{7}$$
$$\uparrow$$

The digit 6 is in the tens place. Consider the next digit to the right. Since the digit, 7, is 5 or higher, round 6 tens up to 7 tens. Then change the digit to the right of the tens digit to zero.

The answer is 470.

83. Round 731 to the nearest ten.

$$7 \; 3 \; \boxed{1}$$
$$\uparrow$$

The digit 3 is in the tens place. Consider the next digit to the right. Since the digit, 1, is 4 or lower, round down, meaning that 3 tens stays as 3 tens. Then change the digit to the right of the tens digit to zero.

The answer is 730.

85. Round 895 to the nearest ten.

$$8 \; 9 \; \boxed{5}$$
$$\uparrow$$

The digit 9 is in the tens place. Consider the next digit to the right. Since the digit, 5, is 5 or higher, we round up. The 89 tens become 90 tens. Then change the digit to the right of the tens digit to zero.

The answer is 900.

87. Round 146 to the nearest hundred.

$$1 \; \boxed{4} \; 6$$
$$\uparrow$$

The digit 1 is in the hundreds place. Consider the next digit to the right. Since the digit, 4, is 4 or lower, round down, meaning that 1 hundred stays as 1 hundred. Then change all digits to the right of the hundreds digit to zeros.

The answer is 100.

89. Round 957 to the nearest hundred.

$$9 \; \boxed{5} \; 7$$
$$\uparrow$$

The digit 9 is in the hundreds place. Consider the next digit to the right. Since the digit, 5, is 5 or higher, round up. The 9 hundreds become 10 hundreds. Then change all digits to the right of the hundreds digit to zeros.

The answer is 1000.

91. Round 9079 to the nearest hundred.

$$9 \; 0 \; \boxed{7} \; 9$$
$$\uparrow$$

The digit 0 is in the hundreds place. Consider the next digit to the right. Since the digit, 7, is 5 or higher, round 0 hundreds up to 1 hundred. Then change all digits to the right of the hundreds digit to zeros.

The answer is 9100.

93. Round 32,850 to the nearest hundred.

$$3 \; 2, 8 \; \boxed{5} \; 0$$
$$\uparrow$$

The digit 8 is in the hundreds place. Consider the next digit to the right. Since the digit, 5, is 5 or higher, round 8 hundreds up to 9 hundreds. Then change all digits to the right of the hundreds digit to zero.

The answer is 32,900.

95. Round 5876 to the nearest thousand.

$$5 \; \boxed{8} \; 7 \; 6$$
$$\uparrow$$

The digit 5 is in the thousands place. Consider the next digit to the right. Since the digit, 8, is 5 or higher, round 5 thousands up to 6 thousands. Then change all digits to the right of the thousands digit to zeros.

The answer is 6000.

97. Round 7500 to the nearest thousand.

$$7 \; \boxed{5} \; 0 \; 0$$
$$\uparrow$$

The digit 7 is in the thousands place. Consider the next digit to the right. Since the digit, 5, is 5 or higher, round 7 thousands up to 8 thousands. Then change all the digits to the right of the thousands digit to zeros.

The answer is 8000.

99. Round 45,340 to the nearest thousand.

$$4 \; 5, \boxed{3} \; 4 \; 0$$
$$\uparrow$$

The digit 5 is in the thousands place. Consider the next digit to the right. Since the digit, 3, is 4 or lower, round down, meaning that 5 thousands stays as 5 thousands. Then change all the digits to the right of the thousands digit to zeros.

The answer is 45,000.

101. Round 373,405 to the nearest thousand.

$$3 \; 7 \; 3, \boxed{4} \; 0 \; 5$$
$$\uparrow$$

The digit 3 is in the thousands place. Consider the next digit to the right. Since the digit, 4, is 4 or lower, round down, meaning that 3 thousands stays as 3 thousands. Then change all the digits to the right of the thousands digit to zeros.

The answer is 373,000.

103.

	Rounded to the nearest ten
7 8	8 0
+ 9 7	+ 1 0 0
	1 8 0 ← Estimated answer

105.

	Rounded to the nearest ten
8 0 7 4	8 0 7 0
− 2 3 4 7	− 2 3 5 0
	5 7 2 0 ← Estimated answer

107.

	Rounded to the nearest hundred
7 3 4 8	7 3 0 0
+ 9 2 4 7	+ 9 2 0 0
	1 6, 5 0 0 ← Estimated answer

109.

	Rounded to the nearest hundred
6 8 5 2	6 9 0 0
− 1 7 4 8	− 1 7 0 0
	5 2 0 0 ← Estimated answer

111.

	Rounded to the nearest thousand
9 6 4 3	1 0, 0 0 0
4 8 2 1	5 0 0 0
8 9 4 3	9 0 0 0
+ 7 0 0 4	+ 7 0 0 0
	3 1, 0 0 0 ← Estimated answer

113.

	Rounded to the nearest thousand
9 2, 1 4 9	9 2, 0 0 0
− 2 2, 5 5 5	− 2 3, 0 0 0
	6 9, 0 0 0 ← Estimated answer

115.

	Rounded to the nearest ten
4 5	5 0
× 6 7	× 7 0
	3 5 0 0 ← Estimated answer

117.

	Rounded to the nearest ten
3 4	3 0
× 2 9	× 3 0
	9 0 0 ← Estimated answer

119.

	Rounded to the nearest hundred
8 7 6	9 0 0
× 3 4 5	× 3 0 0
	2 7 0, 0 0 0 ← Estimated answer

121.

	Rounded to the nearest hundred
4 3 2	4 0 0
× 1 9 9	× 2 0 0
	8 0, 0 0 0 ← Estimated answer

123. Discussion and Writing Exercise

125. $7882 = 7$ thousands $+ 8$ hundreds $+ 8$ tens $+ 2$ ones

127. $21 - 16 = 5$
$\uparrow$
This number gets added (after 5).
$\downarrow$
$21 = 5 + 16$

(By the commutative law of addition, $21 = 16 + 5$ is also correct.)

129. $47 + 9 = 56$ $\qquad$ $47 + 9 = 56$
$\quad\uparrow$ $\qquad\qquad\qquad\quad\uparrow$
This addend gets subtracted from the sum. $\qquad$ This addend gets subtracted from the sum.
$\qquad\qquad\downarrow$ $\qquad\qquad\qquad\qquad\downarrow$
$47 = 56 - 9$ $\qquad\qquad$ $9 = 56 - 47$

131.

a	b	$a \cdot b$	$a + b$
	68	3672	
84			117
		32	12
		304	35

To find a in the first row we divide $a \cdot b$ by b:

$\quad 3672 \div 68 = 54$

Then we add to find $a + b$:

$\quad 54 + 68 = 122$

To find b in the second row we subtract a from $a + b$:

$\quad 117 - 84 = 33$

Then we multiply to find $a \cdot b$:

$\quad 84 \cdot 33 = 2772$

To find a and b in the third row we find a pair of numbers whose product is 32 and whose sum is 12. Pairs of numbers whose product is 32 are 1 and 32, 2 and 16, 4 and 8. Since $4 + 8 = 12$, the numbers we want are 4 and 8. We will let $a = 4$ and $b = 8$. (We could also let $a = 8$ and $b = 4$).

To find a and b in the last row we find a pair of numbers whose product is 304 and whose sum is 35. Pairs of numbers whose product is 304 are 1 and 304, 2 and 152, 4 and 76, 8 and 38, 16 and 19. Since $16 + 19 = 35$, the numbers we want are 16 and 19. We will let $a = 16$ and $b = 19$. (We could also let $a = 19$ and $b = 16$.)

The completed table is shown below.

a	b	$a \cdot b$	$a + b$
54	68	3672	122
84	33	2772	117
4	8	32	12
16	19	304	35

133. We divide 1231 by 42:

```
        2 9
  4 2 ) 1 2 3 1
        8 4 0
        ─────
        3 9 1
        3 7 8
        ─────
          1 3
```

The answer is 29 R 13. Since 13 students will be left after 29 buses are filled, then 30 buses are needed.

135. Use a calculator to perform the computations in this exercise.

First find the total area of each floor:

$$A = l \times w = 172 \times 84 = 14,448 \text{ square feet}$$

Find the area lost to the elevator and the stairwell:

$$A = l \times w = 35 \times 20 = 700 \text{ square feet}$$

Subtract to find the area available as office space on each floor:

$$14,448 - 700 = 13,748 \text{ square feet}$$

Finally, multiply by the number of floors, 18, to find the total area available as office space:

$$18 \times 13,748 = 247,464 \text{ square feet}$$

Exercise Set 1.4

1. $x + 0 = 14$

We replace x by different numbers until we get a true equation. If we replace x by 14, we get a true equation: $14 + 0 = 14$. No other replacement makes the equation true, so the solution is 14.

3. $y \cdot 17 = 0$

We replace y by different numbers until we get a true equation. If we replace y by 0, we get a true equation: $0 \cdot 17 = 0$. No other replacement makes the equation true, so the solution is 0.

5.
$$13 + x = 42$$
$$13 + x - 13 = 42 - 13 \qquad \text{Subtracting 13 on both sides}$$
$$0 + x = 29 \qquad \text{13 plus } x \text{ minus 13 is } 0 + x.$$
$$x = 29$$

Check: $13 + x = 42$

$13 + 29 \; ? \; 42$

$42 \;\big|\;$ TRUE

The solution is 29.

7.
$$12 = 12 + m$$
$$12 - 12 = 12 + m - 12 \qquad \text{Subtracting 12 on both sides}$$
$$0 = 0 + m \qquad \text{12 plus } m \text{ minus 12 is } 0 + m.$$
$$0 = m$$

Check: $12 = 12 + m$

$12 \; ? \; 12 + 0$

$12 \;\big|\;$ TRUE

The solution is 0.

9. $3 \cdot x = 24$
$$\frac{3 \cdot x}{3} = \frac{24}{3} \qquad \text{Dividing by 3 on both sides}$$
$$x = 8 \qquad \text{3 times } x \text{ divided by 3 is } x.$$

Check: $3 \cdot x = 24$

$3 \cdot 8 \; ? \; 24$

$24 \;\big|\;$ TRUE

The solution is 8.

11. $112 = n \cdot 8$
$$\frac{112}{8} = \frac{n \cdot 8}{8} \qquad \text{Dividing by 8 on both sides}$$
$$14 = n$$

Check: $112 = n \cdot 8$

$112 \; ? \; 14 \cdot 8$

$112 \;\big|\;$ TRUE

The solution is 14.

13. $45 \times 23 = x$

To solve the equation we carry out the calculation.

$$\begin{array}{r} 4\,5 \\ \times\,2\,3 \\ \hline 1\,3\,5 \\ 9\,0\,0 \\ \hline 1\,0\,3\,5 \end{array}$$

We can check by repeating the calculation. The solution is 1035.

15. $t = 125 \div 5$

To solve the equation we carry out the calculation.

$$\begin{array}{r} 2\,5 \\ 5\,\overline{)1\,2\,5} \\ 1\,0\,0 \\ \hline 2\,5 \\ 2\,5 \\ \hline 0 \end{array}$$

We can check by repeating the calculation. The solution is 25.

17. $p = 908 - 458$

To solve the equation we carry out the calculation.

$$\begin{array}{r} 9\,0\,8 \\ -\,4\,5\,8 \\ \hline 4\,5\,0 \end{array}$$

We can check by repeating the calculation. The solution is 450.

19. $x = 12,345 + 78,555$

To solve the equation we carry out the calculation.

$$\begin{array}{r} 1\,2,3\,4\,5 \\ +\,7\,8,5\,5\,5 \\ \hline 9\,0,9\,0\,0 \end{array}$$

We can check by repeating the calculation. The solution is 90,900.

21. $3 \cdot m = 96$
$$\frac{3 \cdot m}{3} = \frac{96}{3} \qquad \text{Dividing by 3 on both sides}$$
$$m = 32$$

Check: $3 \cdot m = 96$

$3 \cdot 32 \; ? \; 96$

$96 \;\big|\;$ TRUE

The solution is 32.

23. $715 = 5 \cdot z$

$\dfrac{715}{5} = \dfrac{5 \cdot z}{5}$ Dividing by 5 on both sides

$143 = z$

Check: $\overline{715 = 5 \cdot x}$

$715\ ?\ 5 \cdot 143$

$\big|\ 715$ TRUE

The solution is 143.

25. $10 + x = 89$

$10 + x - 10 = 89 - 10$

$x = 79$

Check: $\overline{10 + x = 89}$

$10 + 79\ ?\ 89$

$89\ \big|$ TRUE

The solution is 79.

27. $61 = 16 + y$

$61 - 16 = 16 + y - 16$

$45 = y$

Check: $\overline{61 = 16 + y}$

$61\ ?\ 16 + 45$

$\big|\ 61$ TRUE

The solution is 45.

29. $6 \cdot p = 1944$

$\dfrac{6 \cdot p}{6} = \dfrac{1944}{6}$

$p = 324$

Check: $\overline{6 \cdot p = 1944}$

$6 \cdot 324\ ?\ 1944$

$1944\ \big|$ TRUE

The solution is 324.

31. $5 \cdot x = 3715$

$\dfrac{5 \cdot x}{5} = \dfrac{3715}{5}$

$x = 743$

The number 743 checks. It is the solution.

33. $47 + n = 84$

$47 + n - 47 = 84 - 47$

$n = 37$

The number 37 checks. It is the solution.

35. $x + 78 = 144$

$x + 78 - 78 = 144 - 78$

$x = 66$

The number 66 checks. It is the solution.

37. $165 = 11 \cdot n$

$\dfrac{165}{11} = \dfrac{11 \cdot n}{11}$

$15 = n$

The number 15 checks. It is the solution.

39. $624 = t \cdot 13$

$\dfrac{624}{13} = \dfrac{t \cdot 13}{13}$

$48 = t$

The number 48 checks. It is the solution.

41. $x + 214 = 389$

$x + 214 - 214 = 389 - 214$

$x = 175$

The number 175 checks. It is the solution.

43. $567 + x = 902$

$567 + x - 567 = 902 - 567$

$x = 335$

The number 335 checks. It is the solution.

45. $18 \cdot x = 1872$

$\dfrac{18 \cdot x}{18} = \dfrac{1872}{18}$

$x = 104$

The number 104 checks. It is the solution.

47. $40 \cdot x = 1800$

$\dfrac{40 \cdot x}{40} = \dfrac{1800}{40}$

$x = 45$

The number 45 checks. It is the solution.

49. $2344 + y = 6400$

$2344 + y - 2344 = 6400 - 2344$

$y = 4056$

The number 4056 checks. It is the solution.

51. $8322 + 9281 = x$

$17,603 = x$ Doing the addition

The number 17,603 checks. It is the solution.

53. $234 \cdot 78 = y$

$18,252 = y$ Doing the multiplication

The number 18,252 checks. It is the solution.

55. $58 \cdot m = 11,890$

$\dfrac{58 \cdot m}{58} = \dfrac{11,890}{58}$

$m = 205$

The number 205 checks. It is the solution.

57. Discussion and Writing Exercise

59. $7 + 8 = 15$ $7 + 8 = 15$

$\uparrow$ $\uparrow$

This number gets   This number gets
subtracted from the   subtracted from the
sum. $\downarrow$ sum. $\downarrow$

$7 = 15 - 8$ $8 = 15 - 7$

61. Since 123 is to the left of 789 on the number line, $123 < 789$.

63. Since 688 is to the right of 0 on the number line, $688 > 0$.

65.
$$\begin{array}{r} 1\,4\,2 \\ 9\overline{\smash{)}1\,2\,8\,3} \\ 9\,0\,0 \\ \hline 3\,8\,3 \\ 3\,6\,0 \\ \hline 2\,3 \\ 1\,8 \\ \hline 5 \end{array}$$
Think: 12 hundreds ÷ 9. Estimate 1 hundred.
Think: 38 tens ÷ 9. Estimate 4 tens.
Think: 23 ones ÷ 9. Estimate 2 ones.

The answer is 142 R 5.

67.
$$\begin{array}{r} 3\,3\,4 \\ 1\,7\overline{\smash{)}5\,6\,7\,8} \\ 5\,1\,0\,0 \\ \hline 5\,7\,8 \\ 5\,1\,0 \\ \hline 6\,8 \\ 6\,8 \\ \hline 0 \end{array}$$
Think 56 hundreds ÷ 17. Estimate 3 hundreds.
Think 57 tens ÷ 17. Estimate 3 tens.
Think 68 ones ÷ 17. Estimate 4 ones.

The answer is 334.

69. $23,465 \cdot x = 8,142,355$
$$\frac{23,465 \cdot x}{23,465} = \frac{8,142,355}{23,465}$$
$x = 347$ Using a calculator to divide
The number 347 checks. It is the solution.

Exercise Set 1.5

---

1. *Familiarize*. We visualize the situation. Since we are combining quantities, addition can be used.

$$\boxed{57{,}601} + \boxed{45{,}488} + \boxed{44{,}007} + \boxed{31{,}572}$$
to   to   to   to
AOL   Yahoo! Microsoft Lycos

Let n = the total number of visits to all the sites, in millions.

Translate. We translate to an equation.

$$57,601 + 45,488 + 44,007 + 31,572 = n$$

***Solve*.** We carry out the addition.

$$\begin{array}{r} {\scriptstyle 1\ \ 1\ \ 1\,1} \\ 5\,7,6\,0\,1 \\ 4\,5,4\,8\,8 \\ 4\,4,0\,0\,7 \\ +\ 3\,1,5\,7\,2 \\ \hline 1\,7\,8,6\,6\,8 \end{array}$$

Thus, $178,668 = n$.

***Check*.** We can repeat the calculation. We can also estimate by rounding, say to the nearest thousand.

$$57,601 + 45,488 + 44,007 + 31,572$$
$$\approx 58,000 + 45,000 + 44,000 + 32,000$$
$$\approx 179,000$$

Since the estimated answer is close to the calculated answer, our result is probably correct.

***State*.** There were 178,668 million, or 178,668,000,000 visits to the web sites.

3. *Familiarize*. We visualize the situation. Let n = the number by which the number of visits to the AOL network site exceeded the number of visits to the Yahoo! sites, in millions.

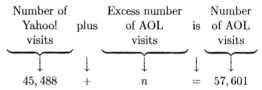

Yahoo! visits	Excess AOL visits
45,488	n
Number of AOL visits	
57,601	

***Translate*.** We see this as a "missing addend" situation.

Number of Yahoo! visits	plus	Excess number of AOL visits	is	Number of AOL visits
↓	↓	↓	↓	↓
45,488	+	n	=	57,601

***Solve*.** We subtract 45,488 on both sides of the equation.

$$45,488 + n = 57,601$$
$$45,488 + n - 45,488 = 57,601 - 45,488$$
$$n = 12,113$$

***Check*.** We can add the difference 12,113, to the subtrahend, 45,488: $45,488 + 12,113 = 57,601$. We can also estimate:

$$57,601 - 45,488 \approx 58,000 - 45,000$$
$$\approx 13,000 \approx 12,113$$

The answer checks.

***State*.** The AOL Network site had 12,113 million, or 12,113,000,000 more visits than the Yahoo! sites.

5. *Familiarize*. We visualize the situation. We are combining quantities, so addition can be used.

$$\boxed{1976} + \boxed{24\ \text{years}}$$

Let y = the year in which the Concorde had its first crash.

***Translate*.** We translate to an equation.

$$1976 + 24 = y$$

***Solve*.** We carry out the addition.

$$\begin{array}{r} {\scriptstyle 1\ \ 1\ \ 1} \\ 1\,9\,7\,6 \\ +\ \ \ \ 2\,4 \\ \hline 2\,0\,0\,0 \end{array}$$

Thus, $2000 = y$.

***Check*.** We can repeat the calculation. We can also estimate: $1976 + 24 \approx 1980 + 20 \approx 2000$. The answer checks.

***State*.** The Concorde had its first crash in 2000.

7. *Familiarize*. We visualize the situation. We are combining quantities, so addition can be used.

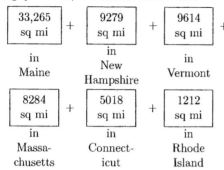

Let a = the total area of New England, in square miles.

***Translate*.** We translate to an equation.

$$33,265 + 9279 + 9614 + 8284 + 5018 + 1212 = a$$

***Solve*.** We carry out the addition.

$$
\begin{array}{r}
{\scriptstyle 3\ 1\ 2\ 3} \\
3\,3,2\,6\,5 \\
9\,2\,7\,9 \\
9\,6\,1\,4 \\
8\,2\,8\,4 \\
5\,0\,1\,8 \\
+\ 1\,2\,1\,2 \\
\hline
6\,6,6\,7\,2
\end{array}
$$

Thus, $66,672 = a$.

***Check*.** We can repeat the calculation. We can also estimate: $33,265 + 9279 + 9614 + 8284 + 5018 + 1212 \approx 33,000 + 9000 + 10,000 + 8000 + 5000 + 1000 \approx 66,000 \approx 66,672$. The answer checks.

***State*.** The total area of New England is 66,672 sq mi.

9. *Familiarize*. We visualize the situation. We are combining quantities, so addition can be used.

$\boxed{\$70,000} + \boxed{\$86,235} + \boxed{\$80,725} +$

in Maine	in New Hampshire	in Vermont

$\boxed{\$75,000} + \boxed{\$78,000} + \boxed{\$69,900}$

in Massa- chusetts	in Connect- icut	in Rhode Island

Let s = the total amount paid in salaries to the governors of the New England states.

***Translate*.** We translate to an equation.

$$70,000 + 86,235 + 80,725 + 75,000 + 78,000 + 69,900 = s$$

***Solve*.** We carry out the addition.

$$
\begin{array}{r}
{\scriptstyle 2\ 1\quad 1} \\
7\,0,0\,0\,0 \\
8\,6,2\,3\,5 \\
8\,0,7\,2\,5 \\
7\,5,0\,0\,0 \\
7\,8,0\,0\,0 \\
+\ 6\,9,9\,0\,0 \\
\hline
4\,5\,9,8\,6\,0
\end{array}
$$

Thus, $459,860 = s$.

***Check*.** We can repeat the calculation. We can also estimate: $70,000 + 86,235 + 80,725 + 75,000 + 78,000 + 69,900 \approx 70,000 + 86,000 + 81,000 + 75,000 + 78,000 + 70,000 \approx 460,000 \approx 459,860$. The answer checks.

***State*.** A total of \$459,860 was paid in salaries to the governors of the New England states.

11. *Familiarize*. We visualize the situation. Let n = the number by which the people in the Navy in 1990 exceeded the number of people in the Navy in 2000.

Number in Navy in 2000 372,000	Excess number in 1990 n
Number in Navy in 1990 583,000	

***Translate*.** We see this as a "missing addend" situation.

Number in Navy in 2000	plus	Excess number in 1990	is	Number in Navy in 1990
$372,000$	$+$	n	$=$	$583,000$

***Solve*.** We subtract 372,000 on both sides of the equation.

$$372,000 + n = 583,000$$
$$372,000 + n - 372,000 = 583,000 - 372,000$$
$$n = 211,000$$

***Check*.** We can repeat the calculation. We can also add the difference, 211,000, to the subtrahend, 372,000: $372,000 + 211,000 = 583,000$. The answer checks.

***State*.** In 1990 there were 211,000 more people in the Navy than in 2000.

13. *Familiarize*. We visualize the situation. Let l = the excess length of the Nile River, in miles.

Length of Missouri-Mississippi 3860 miles	Excess length of Nile l
Length of Nile 4100 miles	

***Translate*.** This is a "missing addend" situation. We translate to an equation.

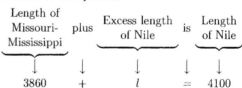

Length of Missouri- Mississippi	plus	Excess length of Nile	is	Length of Nile
3860	$+$	l	$=$	4100

***Solve*.** We subtract 3860 on both sides of the equation.

$$3860 + l = 4100$$
$$3860 + l - 3860 = 4100 - 3860$$
$$l = 240$$

***Check*.** We can check by adding the difference, 240, to the subtrahend, 3860: $3860 + 240 = 4100$. Our answer checks.

***State*.** The Nile River is 240 mi longer than the Missouri-Mississippi River.

15. *Familiarize*. We first draw a picture of the situation. Let g = the number of gallons that will be used in 6144 mi of city driving.

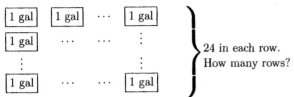

24 in each row. How many rows?

Translate. We translate to an equation.

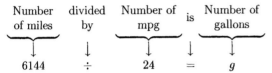

Number of miles	divided by	Number of mpg	is	Number of gallons
↓	↓	↓	↓	↓
6144	÷	24	=	g

Solve. We carry out the division.

```
      2 5 6
2 4 ⟌ 6 1 4 4
      4 8 0 0
      ───────
      1 3 4 4
      1 2 0 0
      ───────
        1 4 4
        1 4 4
        ─────
            0
```

Thus, $256 = g$.

Check. We can check by multiplying the number of gallons by the number of miles per gallon: $24 \cdot 256 = 6144$. The answer checks.

State. The Beetle will use 256 gal of gasoline in 6144 mi of city driving.

17. *Familiarize*. We draw a picture of the situation. Let p = the number of pixels on the computer screen. Repeated addition works well here.

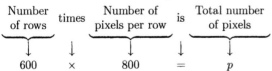

600 addends

Translate. We translate to an equation.

Number of rows	times	Number of pixels per row	is	Total number of pixels
↓	↓	↓	↓	↓
600	×	800	=	p

Solve. We carry out the multiplication.

```
      8 0 0
    × 6 0 0
  ─────────
  4 8 0,0 0 0
```

Thus, $480,000 = p$.

Check. We can repeat the calculation. The answer checks.

State. There are 480,000 pixels on the computer screen.

19. *Familiarize*. We draw a picture of the situation. Let c = the total cost of the purchase. Repeated addition works well here.

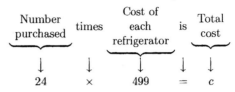

24 addends

Translate. We translate to an equation.

Number purchased	times	Cost of each refrigerator	is	Total cost
↓	↓	↓	↓	↓
24	×	499	=	c

Solve. We carry out the multiplication.

```
      1 1
      3 3
      4 9 9
    ×   2 4
  ─────────
    1 9 9 6
    9 9 8 0
  ─────────
  1 1,9 7 6
```

Thus, $11,976 = c$.

Check. We can repeat the calculation. We can also estimate: $24 \times 499 \approx 25 \times 500 \approx 12,500 \approx 11,976$. The answer checks.

State. The total cost of the purchase is $11,976.

21. *Familiarize*. We visualize the situation. Let a = the amount by which the number of CDs sold in 1999 exceeded the amount sold in 1995, in millions.

CDs sold in 1995	Excess sold in 1999
723	a
CDs sold in 1999	
939	

Translate. We see this as a "missing addend" situation.

Number sold in 1995	plus	Excess number sold in 1999	is	Number sold in 1999
↓	↓	↓	↓	↓
723	+	a	=	939

Solve. We subtract 723 on both sides of the equation.

$$723 + a = 939$$
$$723 + a - 723 = 939 - 723$$
$$a = 216$$

Check. We can add the difference, 216, to the subtrahend, 723: $723 + 216 = 939$. The answer checks.

State. There were 216 million, or 216,000,000, more CDs sold in 1999 than in 1995.

23. *Familiarize*. We visualize the situation. Since we are combining quantities, addition can be used.

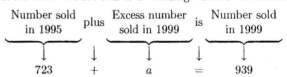

759	+	847	+	939
in 1997		in 1998		in 1999

Let n = the total number of CDs sold from 1997 through 1999, in millions.

Translate. We translate to an equation.

$$759 + 847 + 939 = n$$

Solve. We carry out the addition.

$$\begin{array}{r} {}^{1\ 2} \\ 7\ 5\ 9 \\ 8\ 4\ 7 \\ +\ 9\ 3\ 9 \\ \hline 2\ 5\ 4\ 5 \end{array}$$

Thus, $2545 = n$.

Check. We can repeat the calculation. We can also estimate: $759 + 847 + 939 \approx 800 + 800 + 900 \approx 2500 \approx 2545$. The answer checks.

State. From 1997 through 1999, 2545 million, or 2,545,000,000 CDs were sold.

25. **Familiarize**. We first draw a picture. Let $w =$ the number of full weeks the episodes can run.

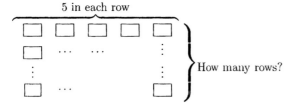

Translate. We translate to an equation.

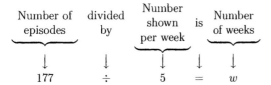

Solve. We carry out the division.

$$\begin{array}{r} 3\ 5 \\ 5\ \overline{)1\ 7\ 7} \\ 1\ 5\ 0 \\ \hline 2\ 7 \\ 2\ 5 \\ \hline 2 \end{array}$$

Check. We can check by multiplying the number of weeks by 5 and adding the remainder, 2:

$$5 \cdot 35 = 175, \qquad 175 + 2 = 177$$

State. 35 full weeks will pass before the station must start over. There will be 2 episodes left over.

27. **Familiarize**. We first draw a picture. Let $h =$ the number of hours in a week. Repeated addition works well here.

$$\underbrace{\boxed{24\ \text{hours}} + \boxed{24\ \text{hours}} + \cdots + \boxed{24\ \text{hours}}}_{7\ \text{addends}}$$

Translate. We translate to an equation.

$$\begin{array}{ccccccc} \underbrace{\begin{array}{c}\text{Number of} \\ \text{hours in} \\ \text{a day}\end{array}} & \text{times} & \underbrace{\begin{array}{c}\text{Number of} \\ \text{days in} \\ \text{a week}\end{array}} & \text{is} & \underbrace{\begin{array}{c}\text{Number of} \\ \text{hours in} \\ \text{a week}\end{array}} \\ \downarrow & \downarrow & \downarrow & \downarrow & \downarrow \\ 24 & \times & 7 & = & h \end{array}$$

Solve. We carry out the multiplication.

$$\begin{array}{r} 2\ 4 \\ \times\ \ \ 7 \\ \hline 1\ 6\ 8 \end{array}$$

Thus, $168 = h$, or $h = 168$.

Check. We can repeat the calculation. We an also estimate:

$$24 \times 7 \approx 20 \times 10 = 200 \approx 168$$

Our answer checks.

State. There are 168 hours in a week.

29. **Familiarize**. We first draw a picture. We let $x =$ the amount of each payment.

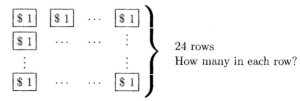

Translate. We translate to an equation.

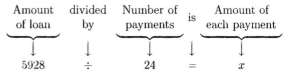

Solve. We carry out the division.

$$\begin{array}{r} 2\ 4\ 7 \\ 2\ 4\ \overline{)5\ 9\ 2\ 8} \\ 4\ 8\ 0\ 0 \\ \hline 1\ 1\ 2\ 8 \\ 9\ 6\ 0 \\ \hline 1\ 6\ 8 \\ 1\ 6\ 8 \\ \hline 0 \end{array}$$

Thus, $247 = x$, or $x = 247$.

Check. We can check by multiplying 247 by 24: $24 \cdot 247 = 5928$. The answer checks.

State. Each payment is \$247.

31. **Familiarize**. We visualize the situation. Let $p =$ the population of Atlanta in 1990.

Population in 1990	Increase
p	897,597
Population in 1999	
3,857,097	

Translate. We see this as a "missing addend" situation.

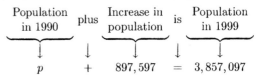

Solve. We subtract 897,597 on both sides of the equation.

$$p + 897,597 = 3,857,097$$
$$p + 897,597 - 897,597 = 3,857,097 - 897,597$$
$$p = 2,959,500$$

Check. We can add the difference, 2,959,500, to the subtrahend, 897,597: $897,597 + 2,959,500 = 3,857,097$. The answer checks.

State. In 1990 the population of Atlanta was 2,959,500.

33. Familiarize. First we draw a picture. Let $c =$ the number of columns. The number of columns is the same as the number of squares in each row.

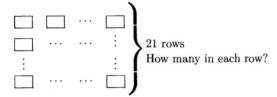

Translate. We translate to an equation.

Number of squares divided by Number of rows is Number of columns.

$$441 \div 21 = c$$

Solve. We carry out the division.

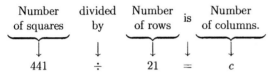

Thus, $21 = c$.

Check. We can check by multiplying the number of rows by the number of columns: $24 \cdot 21 = 441$. The answer checks.

State. The puzzle has 21 columns.

35. Familiarize. We draw a picture of the situation. Let $n =$ the number of 20-bar packages that can be filled.

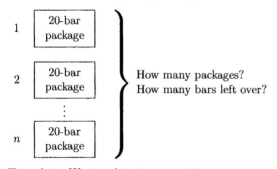

Translate. We translate to an equation.

Number of bars divided by Number per package is Number of packages

$$11,267 \div 20 = n$$

Solve. We carry out the division.

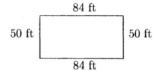

Thus, $n = 563$ R 7.

Check. We can check by multiplying the number of packages by 20 and then adding the remainder, 7:

$$20 \cdot 563 = 11,260 \qquad 11,260 + 7 = 11,267$$

The answer checks.

State. 563 packages can be filled. There will be 7 bars left over.

37. Familiarize. We first draw a picture. Let $A =$ the area and $P =$ the perimeter of the court, in feet.

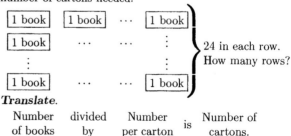

Translate. We write one equation to find the area and another to find the perimeter.

a) Using the formula for the area of a rectangle, we have

$$A = l \cdot w = 84 \cdot 50$$

b) Recall that the perimeter is the distance around the court.

$$P = 84 + 50 + 84 + 50$$

Solve. We carry out the calculations.

a)
$$\begin{array}{r} 5\,0 \\ \times\,8\,4 \\ \hline 2\,0\,0 \\ 4\,0\,0\,0 \\ \hline 4\,2\,0\,0 \end{array}$$

Thus, $A = 4200$.

b) $P = 84 + 50 + 84 + 50 = 268$

Check. We can repeat the calculation. The answers check.

State. a) The area of the court is 4200 square feet.

b) The perimeter of the court is 268 ft.

39. Familiarize. We first draw a picture. We let $c =$ the number of cartons needed.

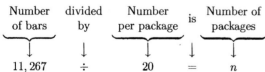

Translate.

Number of books divided by Number per carton is Number of cartons.

$$840 \div 24 = c$$

Solve. We carry out the division.

$$
\begin{array}{r}
3\,5 \\
2\,4\,\overline{\smash{)}\,8\,4\,0} \\
7\,2\,0 \\
\hline
1\,2\,0 \\
1\,2\,0 \\
\hline
0
\end{array}
$$

Thus, $35 = c$, or $c = 35$.

Check. We can check by multiplying: $24 \cdot 35 = 840$. The answer checks.

State. It will take 35 cartons to ship 840 books.

41. *Familiarize*. We visualize the situation as we did in Exercise 39. Let $c =$ the number of cartons that can be filled.

Translate.

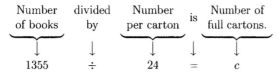

Number of books	divided by	Number per carton	is	Number of full cartons.
1355	÷	24	=	c

Solve. We carry out the division.

$$
\begin{array}{r}
5\,6 \\
2\,4\,\overline{\smash{)}\,1\,3\,5\,5} \\
1\,2\,0\,0 \\
\hline
1\,5\,5 \\
1\,4\,4 \\
\hline
1\,1
\end{array}
$$

Check. We can check by multiplying the number of cartons by 24 and adding the remainder, 11:

$$24 \cdot 56 = 1344, \qquad 1344 + 11 = 1355$$

Our answer checks.

State. 56 cartons can be filled. There will be 11 books left over. If 1355 books are to be shipped, it will take 57 cartons.

43. *Familiarize*. First we find the distance in reality between two cities that are 6 in. apart on the map. We make a drawing. Let $d =$ the distance between the cities, in miles. Repeated addition works well here.

$$\boxed{64 \text{ miles}} + \boxed{64 \text{ miles}} + \cdots + \boxed{64 \text{ miles}}$$

$$\underbrace{\qquad\qquad\qquad\qquad\qquad}_{\text{6 addends}}$$

Translate.

Number of miles per inch	times	Number of inches	is	Distance, in miles.
64	×	6	=	d

Solve. We carry out the multiplication.

$$
\begin{array}{r}
6\,4 \\
\times\quad 6 \\
\hline
3\,8\,4
\end{array}
$$

Thus, $384 = d$.

Check. We can repeat the calculation or estimate the product. Our answer checks.

State. Two cities that are 6 in. apart on the map are 384 miles apart in reality.

Next we find distance on the map between two cities that, in reality, are 1728 mi apart.

Familiarize. We visualize the situation. Let $m =$ the distance between the cities on the map.

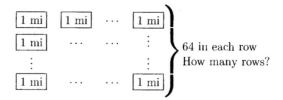

Translate.

Number of miles	divided by	Number of miles per inch	is	Distance, in inches.
1728	÷	64	=	m

Solve. We carry out the division.

$$
\begin{array}{r}
2\,7 \\
6\,4\,\overline{\smash{)}\,1\,7\,2\,8} \\
1\,2\,8\,0 \\
\hline
4\,4\,8 \\
4\,4\,8 \\
\hline
0
\end{array}
$$

Thus, $27 = m$, or $m = 27$.

Check. We can check by multiplying: $64 \cdot 27 = 1728$. Our answer checks.

State. The cities are 27 in. apart on the map.

45. *Familiarize*. We first make a drawing. Let $r =$ the number of rows.

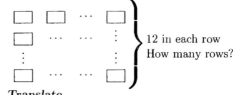

Translate.

Number of holes	divided by	Number per row	is	Number of rows.
216	÷	12	=	r

Solve. We carry out the division.

$$
\begin{array}{r}
1\,8 \\
1\,2\,\overline{\smash{)}\,2\,1\,6} \\
1\,2\,0 \\
\hline
9\,6 \\
9\,6 \\
\hline
0
\end{array}
$$

Thus, $18 = r$, or $r = 18$.

Check. We can check by multiplying: $12 \cdot 18 = 216$. Our answer checks.

State. There are 18 rows.

47. *Familiarize*. This is a multistep problem.

We must find the total price of the 5 video games. Then we must find how many 10's there are in the total price. Let p = the total price of the games.

To find the total price of the 5 video games we can use repeated addition.

$$\underbrace{\boxed{\$64} + \boxed{\$64} + \boxed{\$64} + \boxed{\$64} + \boxed{\$64}}_{\text{5 addends}}$$

Translate.

Price per game	times	Number of games	is	Total price of games
↓	↓	↓	↓	↓
64	·	5	=	p

Solve. First we carry out the multiplication.
$$64 \cdot 5 = p$$
$$320 = p$$

The total price of the 5 video games is $320. Repeated addition can be used again to find how many 10's there are in $320. We let x = the number of $10 bills required.

$320			
$10	$10	$\cdots$	$10

Translate to an equation and solve.
$$10 \cdot x = 320$$
$$\frac{10 \cdot x}{10} = \frac{320}{10}$$
$$x = 32$$

Check. We repeat the calculations. The answer checks.

State. It took 32 ten dollar bills.

49. *Familiarize*. This is a multistep problem. We must find the total amount of the checks written. Then we subtract this amount from the original balance and add the amount of the deposit. Let a = the total amount of the checks written. To find this we can add.

Translate.

First check	plus	Second check	plus	Third check	is	Total amount
↓	↓	↓	↓	↓	↓	↓
46	+	87	+	129	=	a

Solve. First we carry out the addition.

```
  1 2
    4 6
    8 7
+ 1 2 9
-------
  2 6 2
```

Thus, $262 = a$.

Now let b = the amount left in the account after the checks are written.

Amount left	is	Original amount	minus	Amount of checks
↓	↓	↓	↓	↓
b	=	568	−	262

We solve this equation by carrying out the subtraction.

```
  5 6 8
− 2 6 2
-------
  3 0 6
```

Thus, $b = 306$.

Finally, let f = the final amount in the account after the deposit is made.

Final amount	is	Amount after checks	plus	Amount of deposit
↓	↓	↓	↓	↓
f	=	306	+	94

We solve this equation by carrying out the addition.

```
  1 1
  3 0 6
+   9 4
-------
  4 0 0
```

Thus, $f = 400$.

Check. We repeat the calculations. The answer checks.

State. There is $400 left in the account.

51. *Familiarize*. This is a multistep problem. We begin by visualizing the situation.

One pound 3500 calories			
100 cal 8 min	100 cal 8 min	$\cdots$	100 cal 8 min

Let x = the number of hundreds in 3500. Repeated addition applies here.

Translate. We translate to an equation.

100 calories	times	How many 100's	is	3500?
↓	↓	↓	↓	↓
100	·	x	=	3500

Solve. We divide by 100 on both sides of the equation.
$$100 \cdot x = 3500$$
$$\frac{100 \cdot x}{100} = \frac{3500}{100}$$
$$x = 35$$

We know that running for 8 min will burn 100 calories. This must be done 35 times in order to lose one pound. Let t = the time it takes to lose one pound. We have:
$$t = 35 \times 8$$
$$t = 280$$

Check. $280 \div 8 = 35$, so there are 35 8's in 280 min, and $35 \cdot 100 = 3500$, the number of calories that must be burned in order to lose one pound. The answer checks.

State. You must run for 280 min, or 4 hr, 40 min, at a brisk pace in order to lose one pound.

53. *Familiarize*. This is a multistep problem. We begin by visualizing the situation.

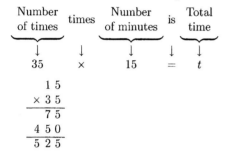

One pound 3500 calories			
100 cal 15 min	100 cal 15 min	...	100 cal 15 min

From Exercise 51 we know that there are 35 100's in 3500. From the chart we know that doing aerobic exercise for 15 min burns 100 calories. Thus we must do 15 min of exercise 35 times in order to lose one pound. Let $t =$ the number of minutes of aerobic exercise required to lose one pound.

Translate. We translate to an equation.

Number of times	times	Number of minutes	is	Total time
↓	↓	↓	↓	↓
35	×	15	=	t

$$\begin{array}{r} 1\,5 \\ \times\,3\,5 \\ \hline 7\,5 \\ 4\,5\,0 \\ \hline 5\,2\,5 \end{array}$$

Thus, $525 = t$.

Check. $525 \div 15 = 35$, so there are 35 15's in 525 min, and $35 \cdot 100 = 3500$, the number of calories that must be burned in order to lose one pound. The answer checks.

State. You must do aerobic exercise for 525 min, or 8 hr, 45 min, in order to lose one pound.

55. *Familiarize*. This is a multistep problem. We will find the number of bones in both hands and the number in both feet and then the total of these two numbers. Let $h =$ the number of bones in two human hands, $f =$ the number of bones in two human feet, and $t =$ the total number of bones in two hands and two feet.

Translate. We translate to three equations.

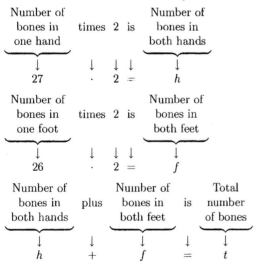

Number of bones in one hand	times	2	is	Number of bones in both hands
↓	↓	↓	↓	↓
27	·	2	=	h

Number of bones in one foot	times	2	is	Number of bones in both feet
↓	↓	↓	↓	↓
26	·	2	=	f

Number of bones in both hands	plus	Number of bones in both feet	is	Total number of bones
↓	↓	↓	↓	↓
h	+	f	=	t

Solve. We solve each equation.

$$27 \cdot 2 = h \qquad 26 \cdot 2 = f$$
$$54 = h \qquad 52 = f$$

$$h + f = t$$
$$54 + 52 = t$$
$$106 = t$$

Check. We repeat the calculations. The answer checks.

State. In all, a human has 106 bones in both hands and both feet.

57. *Familiarize*. This is a multistep problem.

We must find the total cost of the 4 shirts and the total cost of the 6 pairs of pants. The total cost of the clothing is the sum of these two totals.

Repeated addition works well in finding the total cost of the 4 shirts and the total cost of the 6 pairs of pants. We let $x =$ the total cost of the shirts and $y =$ the total cost of the pants.

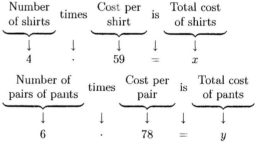

$$\underbrace{\boxed{\$59} + \boxed{\$59} + \boxed{\$59} + \boxed{\$59}}_{\text{4 addends}}$$

$$\underbrace{\boxed{\$78} + \boxed{\$78} + \cdots \boxed{\$78}}_{\text{6 addends}}$$

Translate. We translate to two equations.

Number of shirts	times	Cost per shirt	is	Total cost of shirts
↓	↓	↓	↓	↓
4	·	59	=	x

Number of pairs of pants	times	Cost per pair	is	Total cost of pants
↓	↓	↓	↓	↓
6	·	78	=	y

Solve. To solve these equations, we carry out the multiplications.

$$\begin{array}{r} 5\,9 \\ \times\quad 4 \\ \hline 2\,3\,6 \end{array} \qquad \text{Thus, } x = \$236.$$

$$\begin{array}{r} 7\,8 \\ \times\quad 6 \\ \hline 4\,6\,8 \end{array} \qquad \text{Thus } y = \$468.$$

We let $a =$ the total amount spent.

Total cost of shirts	plus	Total cost of pants	is	Amount spent
↓	↓	↓	↓	↓
236	+	468	=	a

To solve the equation, carry out the addition.

$$\begin{array}{r} 2\,3\,6 \\ +\,4\,6\,8 \\ \hline 7\,0\,4 \end{array}$$

Check. We repeat the calculations. The answer checks.

State. The total cost of the clothing is $704.

59. Discussion and Writing Exercise

61. Round 234,562 to the nearest hundred.

$$2\,3\,4,\,5\,\boxed{6}\,2$$
$$\uparrow$$

The digit 5 is in the hundreds place. Consider the next digit to the right. Since the digit, 6, is 5 or higher, round 5 hundreds up to 6 hundreds. Then change all digits to the right of the hundreds place to zeros.

The answer is 234,600.

63. Round 234,562 to the nearest thousand.

$$2\,3\,4,\,\boxed{5}\,6\,2$$
$$\uparrow$$

The digit 4 is in the thousands place. Consider the next digit to the right. Since the digit, 5, is 5 or higher, round 4 thousands up to 5 thousands. Then change all digits to the right of the thousands place to zeros.

The answer is 235,000.

65.

	Rounded to the nearest thousand
$2\,8,\,4\,3\,0$	$2\,8,\,0\,0\,0$
$-\,1\,1,\,9\,7\,7$	$-\,1\,2,\,0\,0\,0$
	$1\,6,\,0\,0\,0 \leftarrow$ Estimated answer

67.

	Rounded to the nearest thousand
$5\,8\,0\,0$	$6\,0\,0\,0$
$-\,2\,1\,0\,0$	$-\,2\,0\,0\,0$
	$4\,0\,0\,0 \leftarrow$ Estimated answer

69.

	Rounded to the nearest hundred
$7\,9\,9$	$8\,0\,0$
$\times\,8\,8\,7$	$\times\,\quad 9\,0\,0$
	$7\,2\,0,\,0\,0\,0 \leftarrow$ Estimated answer

71. *Familiarize*. This is a multistep problem. First we will find the differences in the distances traveled in 1 second. Then we will find the differences for 18 seconds. Let $d =$ the difference in the number of miles light would travel per second in a vacuum and in ice. Let $g =$ the difference in the number of miles light would travel per second in a vacuum and in glass.

Translate. Each is a "how much more" situation.

Distance in ice	plus	Additional distance	is	Distance in a vacuum.
↓	↓	↓	↓	↓
142,000	+	d	=	186,000

Distance in glass	plus	Additional distance	is	Distance in a vacuum.
↓	↓	↓	↓	↓
109,000	+	g	=	186,000

Solve. We begin by solving each equation.

$$142,000 + d = 186,000$$
$$142,000 + d - 142,000 = 186,000 - 142,000$$
$$d = 44,000$$

$$109,000 + g = 186,000$$
$$109,000 + g - 109,000 = 186,000 - 109,000$$
$$g = 77,000$$

Now to find the differences in the distances in 18 seconds, we multiply each solution by 18.

For ice: $18 \cdot 44,000 = 792,000$

For glass: $18 \cdot 77,000 = 1,386,000$

Check. We repeat the calculations. Our answers check.

State. In 18 seconds light travels 792,000 miles farther in ice and 1,386,000 miles farther in glass than in a vacuum.

Exercise Set 1.6

1. Exponential notation for $3 \cdot 3 \cdot 3 \cdot 3$ is 3^4.

3. Exponential notation for $5 \cdot 5$ is 5^2.

5. Exponential notation for $7 \cdot 7 \cdot 7 \cdot 7 \cdot 7$ is 7^5.

7. Exponential notation for $10 \cdot 10 \cdot 10$ is 10^3.

9. $7^2 = 7 \cdot 7 = 49$

11. $9^3 = 9 \cdot 9 \cdot 9 = 729$

13. $12^4 = 12 \cdot 12 \cdot 12 \cdot 12 = 20,736$

15. $11^2 = 11 \cdot 11 = 121$

17. $12 + (6 + 4) = 12 + 10$ Doing the calculation inside the parentheses
           $= 22$ Adding

19. $52 - (40 - 8) = 52 - 32$ Doing the calculation inside the parentheses
           $= 20$ Subtracting

21. $1000 \div (100 \div 10)$
    $= 1000 \div 10$ Doing the calculation inside the parentheses
    $= 100$ Dividing

23. $(256 \div 64) \div 4 = 4 \div 4$ Doing the calculation inside the parentheses
           $= 1$ Dividing

25. $(2 + 5)^2 = 7^2$ Doing the calculation inside the parentheses
           $= 49$ Evaluating the exponential expression

27. $(11 - 8)^2 - (18 - 16)^2$
    $= 3^2 - 2^2$ Doing the calculations inside the parentheses
    $= 9 - 4$ Evaluating the exponential expressions
    $= 5$ Subtracting

29. $16 \cdot 24 + 50 = 384 + 50$ Doing all multiplications and divisions in order from left to right

$\qquad\qquad = 434$ Doing all additions and subtractions in order from left to right

31. $83 - 7 \cdot 6 = 83 - 42$ Doing all multiplications and divisions in order from left to right

$\qquad\qquad = 41$ Doing all additions and subtractions in order from left to right

33. $10 \cdot 10 - 3 \times 4$

$\quad = 100 - 12$ Doing all multiplications and divisions in order from left to right

$\quad = 88$ Doing all additions and subtractions in order from left to right

35. $4^3 \div 8 - 4$

$\quad = 64 \div 8 - 4$ Evaluating the exponential expression

$\quad = 8 - 4$ Doing all multiplications and divisions in order from left to right

$\quad = 4$ Doing all additions and subtractions in order from left to right

37. $17 \cdot 20 - (17 + 20)$

$\quad = 17 \cdot 20 - 37$ Carrying out the operation inside parentheses

$\quad = 340 - 37$ Doing all multiplications and divisions in order from left to right

$\quad = 303$ Doing all additions and subtractions in order from left to right

39. $6 \cdot 10 - 4 \cdot 10$

$\quad = 60 - 40$ Doing all multiplications and divisions in order from left to right

$\quad = 20$ Doing all additions and subtractions in order from left to right

41. $300 \div 5 + 10$

$\quad = 60 + 10$ Doing all multiplications and divisions in order from left to right

$\quad = 70$ Doing all additions and subtractions in order from left to right

43. $3 \cdot (2 + 8)^2 - 5 \cdot (4 - 3)^2$

$\quad = 3 \cdot 10^2 - 5 \cdot 1^2$ Carrying out operations inside parentheses

$\quad = 3 \cdot 100 - 5 \cdot 1$ Evaluating the exponential expressions

$\quad = 300 - 5$ Doing all multiplications and divisions in order from left to right

$\quad = 295$ Doing all additions and subtractions in order from left to right

45. $4^2 + 8^2 \div 2^2 = 16 + 64 \div 4$

$\qquad\qquad\quad = 16 + 16$

$\qquad\qquad\quad = 32$

47. $10^3 - 10 \cdot 6 - (4 + 5 \cdot 6) = 10^3 - 10 \cdot 6 - (4 + 30)$

$\qquad\qquad\qquad\qquad\quad = 10^3 - 10 \cdot 6 - 34$

$\qquad\qquad\qquad\qquad\quad = 1000 - 10 \cdot 6 - 34$

$\qquad\qquad\qquad\qquad\quad = 1000 - 60 - 34$

$\qquad\qquad\qquad\qquad\quad = 940 - 34$

$\qquad\qquad\qquad\qquad\quad = 906$

49. $6 \times 11 - (7 + 3) \div 5 - (6 - 4) = 6 \times 11 - 10 \div 5 - 2$

$\qquad\qquad\qquad\qquad\qquad\qquad = 66 - 2 - 2$

$\qquad\qquad\qquad\qquad\qquad\qquad = 64 - 2$

$\qquad\qquad\qquad\qquad\qquad\qquad = 62$

51. $\qquad 120 - 3^3 \cdot 4 \div (5 \cdot 6 - 6 \cdot 4)$

$\quad = 120 - 3^3 \cdot 4 \div (30 - 24)$

$\quad = 120 - 3^3 \cdot 4 \div 6$

$\quad = 120 - 27 \cdot 4 \div 6$

$\quad = 120 - 108 \div 6$

$\quad = 120 - 18$

$\quad = 102$

53. $2^9 \cdot 2^6 \div 2^7 = 512 \cdot 64 \div 128$

$\qquad\qquad\quad = 32,768 \div 128$

$\qquad\qquad\quad = 256$

55. We add the numbers and then divide by the number of addends.

$$\frac{\$64 + \$97 + \$121}{3} = \frac{\$282}{3}$$
$$= \$94$$

57. $8 \times 13 + \{42 \div [18 - (6 + 5)]\}$

$\quad = 8 \times 13 + \{42 \div [18 - 11]\}$

$\quad = 8 \times 13 + \{42 \div 7\}$

$\quad = 8 \times 13 + 6$

$\quad = 104 + 6$

$\quad = 110$

59. $[14 - (3 + 5) \div 2] - [18 \div (8 - 2)]$

$\quad = [14 - 8 \div 2] - [18 \div 6]$

$\quad = [14 - 4] - 3$

$\quad = 10 - 3$

$\quad = 7$

61. $(82 - 14) \times [(10 + 45 \div 5) - (6 \cdot 6 - 5 \cdot 5)]$

$\quad = (82 - 14) \times [(10 + 9) - (36 - 25)]$

$\quad = (82 - 14) \times [19 - 11]$

$\quad = 68 \times 8$

$\quad = 544$

63. $4 \times \{(200 - 50 \div 5) - [(35 \div 7) \cdot (35 \div 7) - 4 \times 3]\}$

$\quad = 4 \times \{(200 - 10) - [5 \cdot 5 - 4 \times 3]\}$

$\quad = 4 \times \{190 - [25 - 12]\}$

$\quad = 4 \times \{190 - 13\}$

$\quad = 4 \times 177$

$\quad = 708$

65. $\{[18 - 2 \cdot 6] - [40 \div (17 - 9)]\}+$
$\qquad \{48 - 13 \times 3 + [(50 - 7 \cdot 5) + 2]\}$
$= \{[18 - 12] - [40 \div 8]\}+$
$\qquad \{48 - 13 \times 3 + [(50 - 35) + 2]\}$
$= \{6 - 5\} + \{48 - 13 \times 3 + [15 + 2]\}$
$= 1 + \{48 - 13 \times 3 + 17\}$
$= 1 + \{48 - 39 + 17\}$
$= 1 + 26$
$= 27$

67. Discussion and Writing Exercise

69. $\qquad x + 341 = 793$
$x + 341 - 341 = 793 - 341$
$\qquad\qquad x = 452$

The solution is 452.

71. $\qquad 7 \cdot x = 91$
$\qquad \dfrac{7 \cdot x}{7} = \dfrac{91}{7}$
$\qquad\qquad x = 13$

The solution is 13.

73. $\qquad 3240 = y + 898$
$3240 - 898 = y + 898 - 898$
$\qquad 2342 = y$

The solution is 2342.

75. $\qquad 25 \cdot t = 625$
$\qquad \dfrac{25 \cdot t}{25} = \dfrac{625}{25}$
$\qquad\qquad t = 25$

The solution is 25.

77. *Familiarize*. We first make a drawing.

380 mi

Translate. We use the formula for the area of a rectangle.
$$A = l \cdot w = 380 \cdot 270$$

Solve. We carry out the multiplication.
$$A = 380 \cdot 270 = 102,600$$

Check. We repeat the calculation. The answer checks.

State. The area is 102,600 square miles.

79. $1 + 5 \cdot 4 + 3 = 1 + 20 + 3$
$\qquad\qquad = 24 \qquad$ Correct answer

To make the incorrect answer correct we add parentheses:
$$1 + 5 \cdot (4 + 3) = 36$$

81. $12 \div 4 + 2 \cdot 3 - 2 = 3 + 6 - 2$
$\qquad\qquad\qquad = 7 \qquad$ Correct answer

To make the incorrect answer correct we add parentheses:
$$12 \div (4 + 2) \cdot 3 - 2 = 4$$

Exercise Set 1.7

1. We divide the first number by the second.

$$\begin{array}{r} 3 \\ 14 \overline{)52} \\ 42 \\ \hline 10 \end{array}$$

The remainder is not 0, so 14 is not a factor of 52.

3. We divide the first number by the second.

$$\begin{array}{r} 25 \\ 25 \overline{)625} \\ 500 \\ \hline 125 \\ 125 \\ \hline 0 \end{array}$$

The remainder is 0, so 25 is a factor of 625.

5. We first find some factorizations:
$18 = 1 \cdot 18 \qquad 18 = 3 \cdot 6$
$18 = 2 \cdot 9$

Factors: 1, 2, 3, 6, 9, 18

7. We first find some factorizations:
$54 = 1 \cdot 54 \qquad 54 = 3 \cdot 18$
$54 = 2 \cdot 27 \qquad 54 = 6 \cdot 9$

Factors: 1, 2, 3, 6, 9, 18, 27, 54

9. We first find some factorizations:
$4 = 1 \cdot 4 \qquad 4 = 2 \cdot 2$

Factors: 1, 2, 4

11. The only factorization is $7 = 1 \cdot 7$.

Factors: 1, 7

13. The only factorization is $1 = 1 \cdot 1$.

Factor: 1

15. We first find some factorizations:
$98 = 1 \cdot 98 \qquad 98 = 7 \cdot 14$
$98 = 2 \cdot 49$

Factors: 1, 2, 7, 14, 49, 98

17.
$1 \cdot 4 = 4$	$6 \cdot 4 = 24$
$2 \cdot 4 = 8$	$7 \cdot 4 = 28$
$3 \cdot 4 = 12$	$8 \cdot 4 = 32$
$4 \cdot 4 = 16$	$9 \cdot 4 = 36$
$5 \cdot 4 = 20$	$10 \cdot 4 = 40$

19.
$1 \cdot 20 = 20$	$6 \cdot 20 = 120$
$2 \cdot 20 = 40$	$7 \cdot 20 = 140$
$3 \cdot 20 = 60$	$8 \cdot 20 = 160$
$4 \cdot 20 = 80$	$9 \cdot 20 = 180$
$5 \cdot 20 = 100$	$10 \cdot 20 = 200$

21.
$1 \cdot 3 = 3$	$6 \cdot 3 = 18$
$2 \cdot 3 = 6$	$7 \cdot 3 = 21$
$3 \cdot 3 = 9$	$8 \cdot 3 = 24$
$4 \cdot 3 = 12$	$9 \cdot 3 = 27$
$5 \cdot 3 = 15$	$10 \cdot 3 = 30$

23. $1 \cdot 12 = 12$ $6 \cdot 12 = 72$
$2 \cdot 12 = 24$ $7 \cdot 12 = 84$
$3 \cdot 12 = 36$ $8 \cdot 12 = 96$
$4 \cdot 12 = 48$ $9 \cdot 12 = 108$
$5 \cdot 12 = 60$ $10 \cdot 12 = 120$

25. $1 \cdot 10 = 10$ $6 \cdot 10 = 60$
$2 \cdot 10 = 20$ $7 \cdot 10 = 70$
$3 \cdot 10 = 30$ $8 \cdot 10 = 80$
$4 \cdot 10 = 40$ $9 \cdot 10 = 90$
$5 \cdot 10 = 50$ $10 \cdot 10 = 100$

27. $1 \cdot 9 = 9$ $6 \cdot 9 = 54$
$2 \cdot 9 = 18$ $7 \cdot 9 = 63$
$3 \cdot 9 = 27$ $8 \cdot 9 = 72$
$4 \cdot 9 = 36$ $9 \cdot 9 = 81$
$5 \cdot 9 = 45$ $10 \cdot 9 = 90$

29. We divide 26 by 6.

$$
\begin{array}{r}
4 \\
6 \overline{\smash{\big)}\, 26} \\
2\,4 \\
\hline
2
\end{array}
$$

Since the remainder is not 0, 26 is not divisible by 6.

31. We divide 1880 by 8.

$$
\begin{array}{r}
235 \\
8 \overline{\smash{\big)}\, 1880} \\
1600 \\
\hline
280 \\
240 \\
\hline
40 \\
40 \\
\hline
0
\end{array}
$$

Since the remainder is 0, 1880 is divisible by 8.

33. We divide 256 by 16.

$$
\begin{array}{r}
16 \\
16 \overline{\smash{\big)}\, 256} \\
160 \\
\hline
96 \\
96 \\
\hline
0
\end{array}
$$

Since the remainder is 0, 256 is divisible by 16.

35. We divide 4227 by 9.

$$
\begin{array}{r}
469 \\
9 \overline{\smash{\big)}\, 4227} \\
3600 \\
\hline
627 \\
540 \\
\hline
87 \\
81 \\
\hline
6
\end{array}
$$

Since the remainder is not 0, 4227 is not divisible by 9.

37. We divide 8650 by 16.

$$
\begin{array}{r}
540 \\
16 \overline{\smash{\big)}\, 8650} \\
8000 \\
\hline
650 \\
640 \\
\hline
10
\end{array}
$$

Since the remainder is not 0, 8650 is not divisible by 16.

39. 1 is neither prime nor composite.

41. The number 9 has factors 1, 3, and 9.

Since 9 is not 1 and not prime, it is composite.

43. The number 11 is prime. It has only the factors 1 and 11.

45. The number 29 is prime. It has only the factors 1 and 29.

47.
$$
\begin{array}{r}
2 \quad \leftarrow \quad \text{2 is prime.} \\
2 \overline{\smash{\big)}\, 4} \\
2 \overline{\smash{\big)}\, 8}
\end{array}
$$
$8 = 2 \cdot 2 \cdot 2$

49.
$$
\begin{array}{r}
7 \quad \leftarrow \quad \text{7 is prime.} \\
2 \overline{\smash{\big)}\, 14}
\end{array}
$$
$14 = 2 \cdot 7$

51.
$$
\begin{array}{r}
7 \quad \leftarrow \quad \text{7 is prime.} \\
3 \overline{\smash{\big)}\, 21} \\
2 \overline{\smash{\big)}\, 42}
\end{array}
$$
$42 = 2 \cdot 3 \cdot 7$

53.
$$
\begin{array}{r}
5 \quad \leftarrow \quad \text{5 is prime.} \\
5 \overline{\smash{\big)}\, 25}
\end{array}
$$
(25 is not divisible by 2 or 3. We move to 5.)
$25 = 5 \cdot 5$

55.
$$
\begin{array}{r}
5 \quad \leftarrow \quad \text{5 is prime.} \\
5 \overline{\smash{\big)}\, 25} \\
2 \overline{\smash{\big)}\, 50}
\end{array}
$$
(25 is not divisible by 2 or 3. We move to 5.)
$50 = 2 \cdot 5 \cdot 5$

57.
$$
\begin{array}{r}
13 \quad \leftarrow \quad \text{13 is prime.} \\
13 \overline{\smash{\big)}\, 169}
\end{array}
$$
(169 is not divisible by 2, 3, 5, 7 or 11. We move to 13.)
$169 = 13 \cdot 13$

59.
$$
\begin{array}{r}
5 \quad \leftarrow \quad \text{5 is prime.} \\
5 \overline{\smash{\big)}\, 25} \\
2 \overline{\smash{\big)}\, 50} \\
2 \overline{\smash{\big)}\, 100}
\end{array}
$$
(25 is not divisible by 2 or 3. We move to 5.)
$100 = 2 \cdot 2 \cdot 5 \cdot 5$

61.
$$
\begin{array}{r}
7 \quad \leftarrow \quad \text{7 is prime.} \\
5 \overline{\smash{\big)}\, 35}
\end{array}
$$
(35 is not divisible by 2 or 3. We move to 5.)
$35 = 5 \cdot 7$

63.
```
       3    ← 3 is prime.
    3 ⟌ 9      (9 is not divisible by 2. We move
    2 ⟌ 1 8        to 3.)
    2 ⟌ 3 6
    2 ⟌ 7 2
```

$72 = 2 \cdot 2 \cdot 2 \cdot 3 \cdot 3$

65.
```
      1 1    ← 11 is prime.
    7 ⟌ 7 7     (77 is not divisible by 2, 3, or 5.
                   We move to 7.)
```

$77 = 7 \cdot 11$

67.
```
        1 0 3    ← 103 is prime.
    7 ⟌ 7 2 1
    2 ⟌ 1 4 4 2
    2 ⟌ 2 8 8 4
```

$2884 = 2 \cdot 2 \cdot 7 \cdot 103$

69.
```
      1 7    ← 17 is prime.
    3 ⟌ 5 1     (51 is not divisible by 2. We move
                   to 3.)
```

$51 = 3 \cdot 17$

71. Discussion and Writing Exercise

73.
```
      1 3
    ×   2
    ─────
      2 6
```

75.
```
        3
      2 5
    ×  1 7
    ─────
      1 7 5    Multiplying by 7
      2 5 0    Multiplying by 10
    ─────
      4 2 5    Adding
```

77.
```
          0
    2 2 ⟌ 0
          0
        ───
          0
```

The answer is 0.

79.
```
          1
    2 2 ⟌ 2 2
          2 2
        ─────
            0
```

The answer is 1.

81. *Familiarize*. This is a multistep problem. Find the total cost of the shirts and the total cost of the pants and then find the sum of the two.

We let p = the total cost of the shirts and p = the total cost of the pants.

Translate. We write two equations.

Number of shirts	times	Cost of one shirt	is	Total cost of shirts
↓	↓	↓	↓	↓
7	·	48	=	s

Number of pairs of pants	times	Cost of one pair	is	Total cost of pants
↓	↓	↓	↓	↓
4	·	69	=	p

Solve. We carry out the multiplication.

$$7 \cdot 48 = s$$
$$336 = s \quad \text{Doing the multiplication}$$

The total cost of the 7 shirts is $336.

$$4 \cdot 69 = p$$
$$276 = p \quad \text{Doing the multiplication}$$

The total cost of the 4 pairs of pants is $276.

Now we find the total amount spent. We let t = this amount.

Total cost of shirts	plus	Total cost of pants	is	Total amount spent
↓	↓	↓	↓	↓
336	+	276	=	t

To solve the equation, carry out the addition.

```
      3 3 6
    + 2 7 6
    ───────
      6 1 2
```

Check. We can repeat the calculations. The answer checks.

State. The total cost is $612.

83. Row 1: 48, 90, 432, 63; row 2: 7, 2, 2, 10, 8, 6, 21, 10; row 3: 9, 18, 36, 14, 12, 11, 21; row 4: 29, 19, 42

Exercise Set 1.8

1. A number is divisible by 2 if its <u>ones digit</u> is even.

4<u>6</u> is divisible by 2 because <u>6</u> is even.
22<u>4</u> is divisible by 2 because <u>4</u> is even.
1<u>9</u> is not divisible by 2 because <u>9</u> is not even.
55<u>5</u> is not divisible by 2 because <u>5</u> is not even.
30<u>0</u> is divisible by 2 because <u>0</u> is even.
3<u>6</u> is divisible by 2 because <u>6</u> is even.
45,27<u>0</u> is divisible by 2 because <u>0</u> is even.
444<u>4</u> is divisible by 2 because <u>4</u> is even.
8<u>5</u> is not divisible by 2 because <u>5</u> is not even.
71<u>1</u> is not divisible by 2 because <u>1</u> is not even.
13,25<u>1</u> is not divisible by 2 because <u>1</u> is not even.
254,76<u>5</u> is not divisible by 2 because <u>5</u> is not even.
25<u>6</u> is divisible by 2 because <u>6</u> is even.
806<u>4</u> is divisible by 2 because <u>4</u> is even.
186<u>7</u> is not divisible by 2 because <u>7</u> is not even.
21,56<u>8</u> is divisible by 2 because <u>8</u> is even.

3. A number is divisible by 4 if the <u>number</u> named by the last <u>two</u> digits is divisible by 4.

<u>46</u> is not divisible by 4 because <u>46</u> is not divisible by 4.

2<u>24</u> is divisible by 4 because <u>24</u> is divisible by 4.

<u>19</u> is not divisible by 4 because <u>19</u> is not divisible by 4.

5<u>55</u> is not divisible by 4 because <u>55</u> is not divisible by 4.

3<u>00</u> is divisible by 4 because <u>00</u> is divisible by 4.

<u>36</u> is divisible by 4 because <u>36</u> is divisible by 4.

45,2<u>70</u> is not divisible by 4 because <u>70</u> is not divisible by 4.

44<u>44</u> is divisible by 4 because <u>44</u> is divisible by 4.

<u>85</u> is not divisible by 4 because <u>85</u> is not divisible by 4.

7<u>11</u> is not divisible by 4 because <u>11</u> is not divisible by 4.

13,2<u>51</u> is not divisible by 4 because <u>51</u> is not divisible by 4.

254,7<u>65</u> is not divisible by 4 because <u>65</u> is not divisible by 4.

2<u>56</u> is divisible by 4 because <u>56</u> is divisible by 4.

80<u>64</u> is divisible by 4 because <u>64</u> is divisible by 4.

18<u>67</u> is not divisible by 4 because <u>67</u> is not divisible by 4.

21,5<u>68</u> is divisible by 4 because <u>68</u> is divisible by 4.

5. For a number to be divisible by 6, the sum of the digits must be divisible by 3 and the ones digit must be 0, 2, 4, 6 or 8 (even). It is most efficient to determine if the ones digit is even first and then, if so, to determine if the sum of the digits is divisible by 3.

46 is not divisible by 6 because 46 is not divisible by 3.

$$4 + 6 = 10$$
↑
Not divisible by 3

224 is not divisible by 6 because 224 is not divisible by 3.

$$2 + 2 + 4 = 8$$
↑
Not divisible by 3

19 is not divisible by 6 because 19 is not even.

19
↑
Not even

555 is not divisible by 6 because 555 is not even.

555
↑
Not even

300 is divisible by 6.

300          $3 + 0 + 0 = 3$
↑   ↑
Even   Divisible by 3

36 is divisible by 6.

36          $3 + 6 = 9$
↑   ↑
Even   Divisible by 3

45,270 is divisible by 6.

45,270 $4 + 5 + 2 + 7 + 0 = 18$
↑   ↑
Even   Divisible by 3

4444 is not divisible by 6 because 4444 is not divisible by 3.

$$4 + 4 + 4 + 4 = 16$$
↑
Not divisible by 3

85 is not divisible by 6 because 85 is not even.

85
↑
Not even

711 is not divisible by 6 because 711 is not even.

711
↑
Not even

13,251 is not divisible by 6 because 13,251 is not even.

13,251
↑
Not even

254,765 is not divisible by 6 because 254,765 is not even.

254,765
↑
Not even

256 is not divisible by 6 because 256 is not divisible by 3.

$$2 + 5 + 6 = 13$$
↑
Not divisible by 3

8064 is divisible by 6.

8064 $8 + 0 + 6 + 4 = 18$
↑   ↑
Even   Divisible by 3

1867 is not divisible by 6 because 1867 is not even.

1867
↑
Not even

21,568 is not divisible by 6 because 21,568 is not divisible by 3.

$$2+1+5+6+8=22$$
↑
Not divisible by 3

7. A number is divisible by 9 if the sum of the digits is divisible by 9.

46 is not divisible by 9 because $4 + 6 = 10$ and 10 is not divisible by 9.

224 is not divisible by 9 because $2 + 2 + 4 = 8$ and 8 is not divisible by 9.

19 is not divisible by 9 because $1 + 9 = 10$ and 10 is not divisible by 9.

555 is not divisible by 9 because $5 + 5 + 5 = 15$ and 15 is not divisible by 9.

300 is not divisible by 9 because $3 + 0 + 0 = 3$ and 3 is not divisible by 9.

36 is divisible by 9 because $3 + 6 = 9$ and 9 is divisible by 9.

45,270 is divisible by 9 because $4 + 5 + 2 + 7 + 0 = 18$ and 18 is divisible by 9.

4444 is not divisible by 9 because $4 + 4 + 4 + 4 = 16$ and 16 is not divisible by 9.

85 is not divisible by 9 because $8 + 5 = 13$ and 13 is not divisible by 9.

711 is divisible by 9 because $7 + 1 + 1 = 9$ and 9 is divisible by 9.

13,251 is not divisible by 9 because $1 + 3 + 2 + 5 + 1 = 12$ and 12 is not divisible by 9.

254,765 is not divisible by 9 because $2+5+4+7+6+5 = 29$ and 29 is not divisible by 9.

256 is not divisible by 9 because $2 + 5 + 6 = 13$ and 13 is not divisible by 9.

8064 is divisible by 9 because $8 + 0 + 6 + 4 = 18$ and 18 is divisible by 9.

1867 is not divisible by 9 because $1 + 8 + 6 + 7 = 22$ and 22 is not divisible by 9.

21,568 is not divisible by 9 because $2 + 1 + 5 + 6 + 8 = 22$ and 22 is not divisible by 9.

9. A number is divisible by 3 if the sum of the digits is divisible by 3.

56 is not divisible by 3 because $5 + 6 = 11$ and 11 is not divisible by 3.

324 is divisible by 3 because $3 + 2 + 4 = 9$ and 9 is divisible by 3.

784 is not divisible by 3 because $7 + 8 + 4 = 19$ and 19 is not divisible by 3.

55,555 is not divisible by 3 because $5 + 5 + 5 + 5 + 5 = 25$ and 25 is not divisible by 3.

200 is not divisible by 3 because $2 + 0 + 0 = 2$ and 2 is not divisible by 3.

42 is divisible by 3 because $4 + 2 = 6$ and 6 is divisible by 3.

501 is divisible by 3 because $5 + 0 + 1 = 6$ and 6 is divisible by 3.

3009 is divisible by 3 because $3 + 0 + 0 + 9 = 12$ and 12 is divisible by 3.

75 is divisible by 3 because $7 + 5 = 12$ and 12 is divisible by 3.

812 is not divisible by 3 because $8 + 1 + 2 = 11$ and 11 is not divisible by 3.

2345 is not divisible by 3 because $2 + 3 + 4 + 5 = 14$ and 14 is not divisible by 3.

2001 is divisible by 3 because $2 + 0 + 0 + 1 = 3$ and 3 is divisible by 3.

35 is not divisible by 3 because $3 + 5 = 8$ and 8 is not divisible by 3.

402 is divisible by 3 because $4 + 0 + 2 = 6$ and 6 is divisible by 3.

111,111 is divisible by 3 because $1 + 1 + 1 + 1 + 1 + 1 = 6$ and 6 is divisible by 3.

1005 is divisible by 3 because $1 + 0 + 0 + 5 = 6$ and 6 is divisible by 3.

11. A number is divisible by 5 if the ones digit is 0 or 5.

5<u>6</u> is not divisible by 5 because the ones digit (6) is not 0 or 5.

32<u>4</u> is not divisible by 5 because the ones digit (4) is not 0 or 5.

78<u>4</u> is not divisible by 5 because the ones digit (4) is not 0 or 5.

55,55<u>5</u> is divisible by 5 because the ones digit is 5.

20<u>0</u> is divisible by 5 because the ones digit is 0.

4<u>2</u> is not divisible by 5 because the ones digit (2) is not 0 or 5.

50<u>1</u> is not divisible by 5 because the ones digit (1) is not 0 or 5.

300<u>9</u> is not divisible by 5 because the ones digit (9) is not 0 or 5.

7<u>5</u> is divisible by 5 because the ones digit is 5.

81<u>2</u> is not divisible by 5 because the ones digit (2) is not 0 or 5.

234<u>5</u> is divisible by 5 because the ones digit is 5.

200<u>1</u> is not divisible by 5 because the ones digit (1) is not 0 or 5.

3<u>5</u> is divisible by 5 because the ones digit is 5.

40<u>2</u> is not divisible by 5 because the ones digit (2) is not 0 or 5.

111,11<u>1</u> is not divisible by 5 because the ones digit (1) is not 0 or 5.

100<u>5</u> is divisible by 5 because the ones digit is 5.

13. A number is divisible by 9 if the sum of the digits is divisible by 9.

56 is not divisible by 9 because $5 + 6 = 11$ and 11 is not divisible by 9.

324 is divisible by 9 because $3 + 2 + 4 = 9$ and 9 is divisible by 9.

784 is not divisible by 9 because $7 + 8 + 4 = 19$ and 19 is not divisible by 9.

55,555 is not divisible by 9 because $5 + 5 + 5 + 5 + 5 = 25$ and 25 is not divisible by 9.

200 is not divisible by 9 because $2 + 0 + 0 = 2$ and 2 is not divisible by 9.

42 is not divisible by 9 because $4 + 2 = 6$ and 6 is not divisible by 9.

501 is not divisible by 9 because $5 + 0 + 1 = 6$ and 6 is not divisible by 9.

3009 is not divisible by 9 because $3 + 0 + 0 + 9 = 12$ and 12 is not divisible by 9.

75 is not divisible by 9 because $7 + 5 = 12$ and 12 is not divisible by 9.

812 is not divisible by 9 because $8 + 1 + 2 = 11$ and 11 is not divisible by 9.

2345 is not divisible by 9 because $2 + 3 + 4 + 5 = 14$ and 14 is not divisible by 9.

2001 is not divisible by 9 because $2 + 0 + 0 + 1 = 3$ and 3 is not divisible by 9.

35 is not divisible by 9 because $3 + 5 = 8$ and 8 is not divisible by 9.

402 is not divisible by 9 because $4 + 0 + 2 = 6$ and 6 is not divisible by 9.

111,111 is not divisible by 9 because $1 + 1 + 1 + 1 + 1 + 1 = 6$ and 6 is not divisible by 9.

1005 is not divisible by 9 because $1 + 0 + 0 + 5 = 6$ and 6 is not divisible by 9.

15. A number is divisible by 10 if the ones digit is 0.

Of the numbers under consideration, the only one whose ones digit is 0 is 200. Therefore, 200 is divisible by 10. None of the other numbers is divisible by 10.

17. Discussion and Writing Exercise

19.
$$56 + x = 194$$
$$56 + x - 56 = 194 - 56 \quad \text{Subtracting 56 on both sides}$$
$$x = 138$$

The solution is 138.

21.
$$3008 = x + 2134$$
$$3008 - 2134 = x + 2134 - 2134 \quad \text{Subtracting 2134 on both sides}$$
$$874 = x$$

The solution is 874.

23.
$$24 \cdot m = 624$$
$$\frac{24 \cdot m}{24} = \frac{624}{24} \quad \text{Dividing by 24 on both sides}$$
$$m = 26$$

The solution is 26.

25.
```
      2 3 4
9 | 2 1 0 6
    1 8 0 0
    -------
      3 0 6
      2 7 0
    -------
        3 6
        3 6
      -----
         0
```

The answer is 234.

27. *Familiarize.* We visualize the situation. Let $g =$ the number of gallons of gasoline the automobile will use to travel 1485 mi.

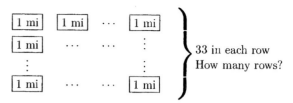

33 in each row
How many rows?

Translate. We translate to an equation.

Number of miles	divided by	Miles per gallon	is	Number of gallons
↓	↓	↓	↓	↓
1485	÷	33	=	g

Solve. We carry out the division.

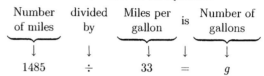

Thus $45 = g$, or $g = 45$.

Check. We can repeat the calculation. The answer checks.

State. The automobile will use 45 gallons of gasoline to travel 1485 mi.

29. 780**0** is divisible by 2 because the ones digit (0) is even.

$7800 \div 2 = 3900$ so $7800 = 2 \cdot 3900$.

390**0** is divisible by 2 because the ones digit (0) is even.

$3900 \div 2 = 1950$ so $3900 = 2 \cdot 1950$ and $7800 = 2 \cdot 2 \cdot 1950$.

195**0** is divisible by 2 because the ones digit (0) is even.

$1950 \div 2 = 975$ so $1950 = 2 \cdot 975$ and $7800 = 2 \cdot 2 \cdot 2 \cdot 975$.

97**5** is not divisible by 2 because the ones digit (5) is not even. Move on to 3.

975 is divisible by 3 because the sum of the digits $(9 + 7 + 5 = 21)$ is divisible by 3.

$975 \div 3 = 325$ so $975 = 3 \cdot 325$ and $7800 = 2 \cdot 2 \cdot 2 \cdot 3 \cdot 325$.

Since 975 is not divisible by 2, none of its factors is divisible by 2. Therefore, we no longer need to check for divisibility by 2.

325 is not divisible by 3 because the sum of the digits $(3 + 2 + 5 = 10)$ is not divisible by 3. Move on to 5.

32**5** is divisible by 5 because the ones digit is 5.

$325 \div 5 = 65$ so $325 = 5 \cdot 65$ and $7800 = 2 \cdot 2 \cdot 2 \cdot 3 \cdot 5 \cdot 65$.

Since 325 is not divisible by 3, none of its factors is divisible by 3. Therefore, we no longer need to check for divisibility by 3.

6**5** is divisible by 5 because the ones digit is 5.

$65 \div 5 = 13$ so $65 = 5 \cdot 13$ and $7800 = 2 \cdot 2 \cdot 2 \cdot 3 \cdot 5 \cdot 5 \cdot 13$.

13 is prime so the prime factorization of 7800 is $2 \cdot 2 \cdot 2 \cdot 3 \cdot 5 \cdot 5 \cdot 13$.

31. 277<u>2</u> is divisible by 2 because the ones digit (2) is even.

$2772 \div 2 = 1386$ so $2772 = 2 \cdot 1386$.

138<u>6</u> is divisible by 2 because the ones digit (6) is even.

$1386 \div 2 = 693$ so $1386 = 2 \cdot 693$ and $2772 = 2 \cdot 2 \cdot 693$.

69<u>3</u> is not divisible by 2 because the ones digit (3) is not even. We move to 3.

693 is divisible by 3 because the sum of the digits ($6 + 9 + 3 = 18$) is divisible by 3.

$693 \div 3 = 231$ so $693 = 3 \cdot 231$ and $2772 = 2 \cdot 2 \cdot 3 \cdot 231$.

Since 693 is not divisible by 2, none of its factors is divisible by 2. Therefore, we no longer need to check divisibility by 2.

231 is divisible by 3 because the sum of the digits ($2 + 3 + 1 = 6$) is divisible by 3.

$231 \div 3 = 77$ so $231 = 3 \cdot 77$ and $2772 = 2 \cdot 2 \cdot 3 \cdot 3 \cdot 77$.

77 is not divisible by 3 since the sum of the digits ($7 + 7 = 14$) is not divisible by 3. We move to 5.

7<u>7</u> is not divisible by 5 because the ones digit (7) is not 0 or 5. We move to 7.

We have not stated a test for divisibility by 7 so we will just try dividing by 7.

$$
\begin{array}{r}
1\,1 \\
7\overline{)\,7\,7}
\end{array} \quad \leftarrow \text{11 is prime}
$$

$77 \div 7 = 11$ so $77 = 7 \cdot 11$ and the prime factorization of 2772 is $2 \cdot 2 \cdot 3 \cdot 3 \cdot 7 \cdot 11$.

33. The sum of the given digits is $9 + 5 + 8$, or 22. If the number is divisible by 99, it is also divisible by 9 since 99 is divisible by 9. The smallest number that is divisible by 9 and also greater than 22 is 27. Then the sum of the two missing digits must be at least $27 - 22$, or 5. We try various combinations of two digits whose sum is 5, using a calculator to divide the resulting number by 99:

95,058 is not divisible by 99.

95,148 is not divisible by 99.

95,238 is divisible by 99.

Thus, the missing digits are 2 and 3 and the number is 95,238.

Exercise Set 1.9

In this section we will find the LCM using the list of multiples method in Exercises 1 - 19 and the prime factorization method in Exercises 21 - 43.

1. a) 4 is a multiple of 2, so it is the LCM.

   c) The LCM = 4.

3. a) 25 is not a multiple of 10.

   b) Check multiples:

       $2 \cdot 25 = 50$ A multiple of 10

   c) The LCM = 50.

5. a) 40 is a multiple of 20, so it is the LCM.

   c) The LCM = 40.

7. a) 27 is not a multiple of 18.

   b) Check multiples:

       $2 \cdot 27 = 54$ A multiple of 18

   c) The LCM = 54.

9. a) 50 is not a multiple of 30.

   b) Check multiples:

       $2 \cdot 50 = 100$ Not a multiple of 30

       $3 \cdot 50 = 150$ A multiple of 30

   c) The LCM = 150.

11. a) 40 is not a multiple of 30.

   b) Check multiples:

       $2 \cdot 40 = 80$ Not a multiple of 30

       $3 \cdot 40 = 120$ A multiple of 30

   c) The LCM = 120.

13. a) 24 is not a multiple of 18.

   b) Check multiples:

       $2 \cdot 24 = 48$ Not a multiple of 18

       $3 \cdot 24 = 72$ A multiple of 18

   c) The LCM = 72.

15. a) 70 is not a multiple of 60.

   b) Check multiples:

       $2 \cdot 70 = 140$ Not a multiple of 60

       $3 \cdot 70 = 210$ Not a multiple of 60

       $4 \cdot 70 = 280$ Not a multiple of 60

       $5 \cdot 70 = 350$ Not a multiple of 60

       $6 \cdot 70 = 420$ A multiple of 60

   c) The LCM = 420.

17. a) 36 is not a multiple of 16.

   b) Check multiples:

       $2 \cdot 36 = 72$ Not a multiple of 16

       $3 \cdot 36 = 108$ Not a multiple of 16

       $4 \cdot 36 = 144$ A multiple of 16

   c) The LCM = 144.

19. a) 36 is not a multiple of 32.

   b) Check multiples:

       $2 \cdot 36 = 72$ Not a multiple of 32

       $3 \cdot 36 = 108$ Not a multiple of 32

       $4 \cdot 36 = 144$ Not a multiple of 32

       $5 \cdot 36 = 180$ Not a multiple of 32

       $6 \cdot 36 = 216$ Not a multiple of 32

       $7 \cdot 36 = 252$ Not a multiple of 32

       $8 \cdot 36 = 288$ A multiple of 32

   c) The LCM = 288.

21. Note that each of the numbers 2, 3, and 5 is prime. They have no common prime factor. When this happens, the LCM is just the product of the numbers.

The LCM is $2 \cdot 3 \cdot 5$, or 30.

23. Note that each of the numbers 3, 5, and 7 is prime. They have no common prime factor. When this happens, the LCM is just the product of the numbers.

The LCM is $3 \cdot 5 \cdot 7$, or 105.

25. a) Find the prime factorization of each number.

$$24 = 2 \cdot 2 \cdot 2 \cdot 3$$
$$36 = 2 \cdot 2 \cdot 3 \cdot 3$$
$$12 = 2 \cdot 2 \cdot 3$$

b) Create a product by writing factors, using each the greatest number of times it occurs in any one factorization.

Consider the factor 2. The greatest number of times 2 occurs in any one factorization is three. We write 2 as a factor three times.

$$2 \cdot 2 \cdot 2 \cdot ?$$

Consider the factor 3. The greatest number of times 3 occurs in any one factorization is two. We write 3 as a factor two times.

$$2 \cdot 2 \cdot 2 \cdot 3 \cdot 3 \cdot ?$$

Since there are no other prime factors in any of the factorizations, the LCM is $2 \cdot 2 \cdot 2 \cdot 3 \cdot 3$, or 72.

27. a) Find the prime factorization of each number.

$$5 = 5 \qquad (5 \text{ is prime.})$$
$$12 = 2 \cdot 2 \cdot 3$$
$$15 = 3 \cdot 5$$

b) Create a product by writing each factor the greatest number of times it occurs in any one factorization.

The greatest number of times 2 occurs in any one factorization is two times.

The greatest number of times 3 occurs in any one factorization is one time.

The greatest number of times 5 occurs in any one factorization is one time.

Since there are no other prime factors in any of the factorizations, the LCM is $2 \cdot 2 \cdot 3 \cdot 5$, or 60.

29. a) Find the prime factorization of each number.

$$9 = 3 \cdot 3$$
$$12 = 2 \cdot 2 \cdot 3$$
$$6 = 2 \cdot 3$$

b) Create a product by writing each factor the greatest number of times it occurs in any one factorization.

The greatest number of times 2 occurs in any one factorization is two times.

The greatest number of times 3 occurs in any one factorization is two times.

Since there are no other prime factors in any of the factorizations, the LCM is $2 \cdot 2 \cdot 3 \cdot 3$, or 36.

31. a) Find the prime factorization of each number.

$$180 = 2 \cdot 2 \cdot 3 \cdot 3 \cdot 5$$
$$100 = 2 \cdot 2 \cdot 5 \cdot 5$$
$$450 = 2 \cdot 3 \cdot 3 \cdot 5 \cdot 5$$

b) Create a product by writing each factor the greatest number of times it occurs in any one factorization.

The greatest number of times 2 occurs in any one factorization is two times.

The greatest number of times 3 occurs in any one factorization is two times.

The greatest number of times 5 occurs in any one factorization is two times.

Since there are no other prime factors in any of the factorizations, the LCM is $2 \cdot 2 \cdot 3 \cdot 3 \cdot 5 \cdot 5$, or 900.

We can also find the LCM using exponents.

$$180 = 2^2 \cdot 3^2 \cdot 5^1$$
$$100 = 2^2 \cdot 5^2$$
$$450 = 2^1 \cdot 3^2 \cdot 5^2$$

The largest exponents of 2, 3, 5 in any of the factorizations are each 2. Thus, the LCM $= 2^2 \cdot 3^2 \cdot 5^2$, or 900.

33. Note that 8 is a factor of 48. If one number is a factor of another, the LCM is the greater number.

The LCM is 48.

The factorization method will also work here if you do not recognize at the outset that 8 is a factor of 48.

35. Note that 5 is a factor of 50. If one number is a factor of another, the LCM is the greater number.

The LCM is 50.

37. Note that 11 and 13 are prime. They have no common prime factor. When this happens, the LCM is just the product of the numbers.

The LCM is $11 \cdot 13$, or 143.

39. a) Find the prime factorization of each number.

$$12 = 2 \cdot 2 \cdot 3$$
$$35 = 5 \cdot 7$$

b) Note that the two numbers have no common prime factor. When this happens, the LCM is just the product of the numbers.

The LCM is $12 \cdot 35$, or 420.

41. a) Find the prime factorization of each number.

$$54 = 3 \cdot 3 \cdot 3 \cdot 2$$
$$63 = 3 \cdot 3 \cdot 7$$

b) Create a product by writing each factor the greatest number of times it occurs in any one factorization.

The greatest number of times 2 occurs in any one factorization is one time.

The greatest number of times 3 occurs in any one factorization is three times.

The greatest number of times 7 occurs in any one factorization is one time.

Since there are no other prime factors in any of the factorizations, the LCM is $2 \cdot 3 \cdot 3 \cdot 3 \cdot 7$, or 378.

43. a) Find the prime factorization of each number.

$$81 = 3 \cdot 3 \cdot 3 \cdot 3$$
$$90 = 2 \cdot 3 \cdot 3 \cdot 5$$

b) Create a product by writing each factor the greatest number of times it occurs in any one factorization.

The greatest number of times 2 occurs in any one factorization is one time.

The greatest number of times 3 occurs in any one factorization is four times.

The greatest number of times 5 occurs in any one factorization is one time.

Since there are no other prime factors in any of the factorizations, the LCM is $2 \cdot 3 \cdot 3 \cdot 3 \cdot 3 \cdot 5$, or 810.

45. We find the LCM of the number of years it takes Jupiter and Saturn to make a complete revolution around the sun.

Jupiter: $12 = 2 \cdot 2 \cdot 3$

Saturn: $30 = 2 \cdot 3 \cdot 5$

The LCM $= 2 \cdot 2 \cdot 3 \cdot 5$, or 60. Thus, Jupiter and Saturn will appear in the same direction in the night sky once every 60 years.

47. We find the LCM of the number of years it takes Saturn and Uranus to make a complete revolution around the sun.

Saturn: $30 = 2 \cdot 3 \cdot 5$

Uranus: $84 = 2 \cdot 2 \cdot 3 \cdot 7$

The LCM is $2 \cdot 2 \cdot 3 \cdot 5 \cdot 7$, or 420. Thus, Saturn and Uranus will appear in the same direction in the night sky once every 420 years.

49. Discussion and Writing Exercise

51. 9001 is to the left of 10,001 on the number line, so $9001 < 10,001$

53.
$$
\begin{array}{r}
{\scriptstyle 1\ \ 1\ \ 1\ \ 1} \\
2\ 3,4\ 5\ 6 \\
5\ 6\ 7\ 7 \\
+\quad 4\ 0\ 0\ 2 \\
\hline
3\ 3,1\ 3\ 5
\end{array}
$$

55. 2 ten thousands + 4 thousands + 6 hundreds + 0 tens + 5 ones, or 2 ten thousands + 4 thousands + 6 hundreds + 5 ones

57. From Example 8 we know that the LCM of 27, 90, and 84 is $2 \cdot 3 \cdot 3 \cdot 5 \cdot 3 \cdot 2 \cdot 7$, so the LCM of 27, 90, 84, 210, 108, and 50 must contain at least these factors. We write the prime factorizations of 210, 108, and 50:

$$210 = 2 \cdot 3 \cdot 5 \cdot 7$$
$$108 = 2 \cdot 2 \cdot 3 \cdot 3 \cdot 3$$
$$50 = 2 \cdot 5 \cdot 5$$

Neither of the four factorizations above contains the other three.

Begin with the LCM of 27, 90, and 84, $2 \cdot 3 \cdot 3 \cdot 5 \cdot 3 \cdot 2 \cdot 7$. Neither 210 nor 108 contains any factors that are missing in this factorization. Next we look for factors of 50 that

are missing. Since 50 contains a second factor of 5, we multiply by 5:

$$2 \cdot 3 \cdot 3 \cdot 5 \cdot 3 \cdot 2 \cdot 7 \cdot 5$$

The LCM is $2 \cdot 3 \cdot 3 \cdot 5 \cdot 3 \cdot 2 \cdot 7 \cdot 5$, or 18,900.

59. The width of the carton will be the common width, 5 in. The length of the carton must be a multiple of both 6 and 8. The shortest length carton will be the least common multiple of 6 and 8.

$$6 = 2 \cdot 3$$
$$8 = 2 \cdot 2 \cdot 2$$

LCM is $2 \cdot 2 \cdot 2 \cdot 3$, or 24.

The shortest carton is 24 in. long.

Chapter 2

Fraction Notation

Exercise Set 2.1

---

1. The top number is the numerator, and the bottom number is the denominator.

$$\frac{3}{4} \quad \begin{array}{l} \leftarrow \text{Numerator} \\ \leftarrow \text{Denominator} \end{array}$$

3. The top number is the numerator, and the bottom number is the denominator.

$$\frac{1\,1}{2\,0} \quad \begin{array}{l} \leftarrow \text{Numerator} \\ \leftarrow \text{Denominator} \end{array}$$

5. The dollar is divided into 4 parts of the same size, and 2 of them are shaded. This is $2 \cdot \frac{1}{4}$ or $\frac{2}{4}$. Thus, $\frac{2}{4}$ (two-fourths) of the dollar is shaded.

7. The yard is divided into 8 parts of the same size, and 1 of them is shaded. Thus, $\frac{1}{8}$ (one-eighth) of the yard is shaded.

9. We have 2 quarts, each divided into 3 parts. We take 4 of those parts. This is $4 \cdot \frac{1}{3}$ or $\frac{4}{3}$. Thus, $\frac{4}{3}$ of a quart is shaded.

11. Each inch on the ruler is divided into 16 equal parts. The shading extends to the 12th mark, so $\frac{12}{16}$ is shaded.

13. The triangle is divided into 4 triangles of the same size, and 3 of them are shaded. This is $3 \cdot \frac{1}{4}$ or $\frac{3}{4}$. Thus, $\frac{3}{4}$ (three-fourths) of the triangle is shaded.

15. The pie is divided into 8 parts of the same size, and 4 of them are shaded. This is $4 \cdot \frac{1}{8}$, or $\frac{4}{8}$. Thus, $\frac{4}{8}$ (four-eighths) of the pie is shaded.

17. The acre is divided into 12 parts of the same size, and 6 of them are shaded. This is $6 \cdot \frac{1}{12}$, or $\frac{6}{12}$. Thus, $\frac{6}{12}$ (six-twelfths) of the acre is shaded.

19. The gas gauge is divided into 8 equal parts.

a) The needle is 2 marks from the E (empty) mark, so the amount of gas in the tank is $\frac{2}{8}$ of a full tank.

b) The needle is 6 marks from the F (full) mark, so $\frac{6}{8}$ of a full tank of gas has been burned.

21. The gas gauge is divided into 8 equal parts.

a) The needle is 3 marks from the E (empty) mark, so the amount of gas in the tank is $\frac{3}{8}$ of a full tank.

b) The needle is 5 marks from the F (full) mark, so $\frac{5}{8}$ of a full tank of gas has been burned.

23. Remember: $\frac{n}{1} = n$

$$\frac{18}{1} = 18$$

Think of taking 18 objects and dividing them into 1 part. (We do not divide them.) We have 18 objects.

25. Remember: $\frac{0}{n} = 0$, for n that is not 0.

$$\frac{0}{8} = 0$$

Think of dividing an object into 8 parts and taking none of them. We get 0.

27. Remember: $\frac{n}{n} = 1$, for n that is not 0.

$$\frac{20}{20} = 1$$

If we divide an object into 20 parts and take 20 of them, we get all of the object (1 whole object).

29. $\dfrac{5}{6-6} = \dfrac{5}{0}$

Remember: $\frac{n}{0}$ is not defined for any whole number n.

Thus, $\dfrac{5}{6-6}$ is not defined.

31. Remember: $\frac{n}{0}$ is not defined for any whole number n.

$$\frac{729}{0} \text{ is not defined.}$$

33. Remember: $\frac{n}{n} = 1$, for n that is not 0.

$$\frac{87}{87} = 1$$

If we divide an object into 87 parts and take 87 of them, we get all of the object (1 whole object).

35. $\dfrac{1}{2} \cdot \dfrac{1}{3} = \dfrac{1 \cdot 1}{2 \cdot 3} = \dfrac{1}{6}$

37. $5 \times \dfrac{1}{8} = \dfrac{5 \times 1}{8} = \dfrac{5}{8}$

39. $\dfrac{2}{3} \times \dfrac{1}{5} = \dfrac{2 \times 1}{3 \times 5} = \dfrac{2}{15}$

41. $\dfrac{2}{5} \cdot \dfrac{2}{3} = \dfrac{2 \cdot 2}{5 \cdot 3} = \dfrac{4}{15}$

43. $\dfrac{3}{4} \cdot \dfrac{3}{4} = \dfrac{3 \cdot 3}{4 \cdot 4} = \dfrac{9}{16}$

45. $\dfrac{2}{3} \cdot \dfrac{7}{13} = \dfrac{2 \cdot 7}{3 \cdot 13} = \dfrac{14}{39}$

47. $7 \cdot \dfrac{3}{4} = \dfrac{7 \cdot 3}{4} = \dfrac{21}{4}$

49. $\dfrac{7}{8} \cdot \dfrac{7}{8} = \dfrac{7 \cdot 7}{8 \cdot 8} = \dfrac{49}{64}$

51. Since $2 \cdot 5 = 10$, we multiply by $\dfrac{5}{5}$.

$$\dfrac{1}{2} = \dfrac{1}{2} \cdot \dfrac{5}{5} = \dfrac{1 \cdot 5}{2 \cdot 5} = \dfrac{5}{10}$$

53. Since $8 \cdot 4 = 32$, we multiply by $\dfrac{4}{4}$.

$$\dfrac{5}{8} = \dfrac{5}{8} \cdot \dfrac{4}{4} = \dfrac{5 \cdot 4}{8 \cdot 4} = \dfrac{20}{32}$$

55. Since $3 \cdot 15 = 45$, we multiply by $\dfrac{15}{15}$.

$$\dfrac{5}{3} = \dfrac{5}{3} \cdot \dfrac{15}{15} = \dfrac{5 \cdot 15}{3 \cdot 15} = \dfrac{75}{45}$$

57. Since $22 \cdot 6 = 132$, we multiply by $\dfrac{6}{6}$.

$$\dfrac{7}{22} = \dfrac{7}{22} \cdot \dfrac{6}{6} = \dfrac{7 \cdot 6}{22 \cdot 6} = \dfrac{42}{132}$$

59.
$$\begin{aligned}
\dfrac{6}{8} &= \dfrac{3 \cdot 2}{4 \cdot 2} \quad \longleftarrow \text{ Factor the numerator} \\
&\phantom{= \dfrac{3 \cdot 2}{4 \cdot 2}} \longleftarrow \text{ Factor the denominator} \\
&= \dfrac{3}{4} \cdot \dfrac{2}{2} \quad \longleftarrow \text{ Factor the fraction} \\
&= \dfrac{3}{4} \cdot 1 \quad \longleftarrow \dfrac{2}{2} = 1 \\
&= \dfrac{3}{4} \quad \longleftarrow \text{ Removing a factor of 1}
\end{aligned}$$

61.
$$\begin{aligned}
\dfrac{2}{15} &= \dfrac{1 \cdot 3}{5 \cdot 3} \quad \longleftarrow \text{ Factor the numerator} \\
&\phantom{= \dfrac{1 \cdot 3}{5 \cdot 3}} \longleftarrow \text{ Factor the denominator} \\
&= \dfrac{1}{5} \cdot \dfrac{3}{3} \quad \longleftarrow \text{ Factor the fraction} \\
&= \dfrac{1}{5} \cdot 1 \quad \longleftarrow \dfrac{3}{3} = 1 \\
&= \dfrac{1}{5} \quad \longleftarrow \text{ Removing a factor of 1}
\end{aligned}$$

63. $\dfrac{24}{8} = \dfrac{3 \cdot 8}{1 \cdot 8} = \dfrac{3}{1} \cdot \dfrac{8}{8} = \dfrac{3}{1} \cdot 1 = \dfrac{3}{1} = 3$

65. $\dfrac{18}{24} = \dfrac{3 \cdot 6}{4 \cdot 6} = \dfrac{3}{4} \cdot \dfrac{6}{6} = \dfrac{3}{4} \cdot 1 = \dfrac{3}{4}$

67. $\dfrac{14}{16} = \dfrac{7 \cdot 2}{8 \cdot 2} = \dfrac{7}{8} \cdot \dfrac{2}{2} = \dfrac{7}{8} \cdot 1 = \dfrac{7}{8}$

69. $\dfrac{150}{25} = \dfrac{6 \cdot 25}{1 \cdot 25} = \dfrac{6}{1} \cdot \dfrac{25}{25} = \dfrac{6}{1} \cdot 1 = \dfrac{6}{1} = 6$

We could also simplify $\dfrac{150}{25}$ by doing the division $150 \div 25$.
That is, $\dfrac{150}{25} = 150 \div 25 = 6$.

71. $\dfrac{17}{51} = \dfrac{1 \cdot 17}{3 \cdot 17} = \dfrac{1}{3} \cdot \dfrac{17}{17} = \dfrac{1}{3} \cdot 1 = \dfrac{1}{3}$

73. Discussion and Writing Exercise

75. We can think of the object as being divided into 6 sections, each the size of the area shaded. Thus, $\dfrac{1}{6}$ of the object is shaded.

77. We can think of the object as being divided into 16 sections, each the size of one of the shaded sections. Since 2 sections are shaded, $\dfrac{2}{16}$ of the object is shaded. We could also express this as $\dfrac{1}{8}$.

Exercise Set 2.2

1. $\dfrac{2}{3} \cdot \dfrac{1}{2} = \dfrac{2 \cdot 1}{3 \cdot 2} = \dfrac{2}{2} \cdot \dfrac{1}{3} = 1 \cdot \dfrac{1}{3} = \dfrac{1}{3}$

3. $\dfrac{1}{4} \cdot \dfrac{2}{3} = \dfrac{1 \cdot 2}{4 \cdot 3} = \dfrac{1 \cdot 2}{2 \cdot 2 \cdot 3} = \dfrac{2}{2} \cdot \dfrac{1}{2 \cdot 3} = \dfrac{1}{2 \cdot 3} = \dfrac{1}{6}$

5. $\dfrac{12}{5} \cdot \dfrac{9}{8} = \dfrac{12 \cdot 9}{5 \cdot 8} = \dfrac{4 \cdot 3 \cdot 9}{5 \cdot 2 \cdot 4} = \dfrac{4}{4} \cdot \dfrac{3 \cdot 9}{5 \cdot 2} = \dfrac{3 \cdot 9}{5 \cdot 2} = \dfrac{27}{10}$

7. $\dfrac{10}{9} \cdot \dfrac{7}{5} = \dfrac{10 \cdot 7}{9 \cdot 5} = \dfrac{5 \cdot 2 \cdot 7}{9 \cdot 5} = \dfrac{5}{5} \cdot \dfrac{2 \cdot 7}{9} = \dfrac{2 \cdot 7}{9} = \dfrac{14}{9}$

9. $9 \cdot \dfrac{1}{9} = \dfrac{9 \cdot 1}{9} = \dfrac{9 \cdot 1}{9 \cdot 1} = 1$

11. $\dfrac{7}{5} \cdot \dfrac{5}{7} = \dfrac{7 \cdot 5}{5 \cdot 7} = \dfrac{7 \cdot 5}{7 \cdot 5} = 1$

13. $24 \cdot \dfrac{1}{6} = \dfrac{24 \cdot 1}{6} = \dfrac{24}{6} = \dfrac{4 \cdot 6}{1 \cdot 6} = \dfrac{4}{1} \cdot \dfrac{6}{6} = \dfrac{4}{1} = 4$

15. $12 \cdot \dfrac{3}{4} = \dfrac{12 \cdot 3}{4} = \dfrac{4 \cdot 3 \cdot 3}{4 \cdot 1} = \dfrac{4}{4} \cdot \dfrac{3 \cdot 3}{1} = \dfrac{3 \cdot 3}{1} = 9$

17. $\dfrac{7}{10} \cdot 28 = \dfrac{7 \cdot 28}{10} = \dfrac{7 \cdot 2 \cdot 14}{2 \cdot 5} = \dfrac{2}{2} \cdot \dfrac{7 \cdot 14}{5} = \dfrac{7 \cdot 14}{5} = \dfrac{98}{5}$

19. $240 \cdot \dfrac{1}{8} = \dfrac{240 \cdot 1}{8} = \dfrac{240}{8} = \dfrac{8 \cdot 30}{8 \cdot 1} = \dfrac{8}{8} \cdot \dfrac{30}{1} = \dfrac{30}{1} = 30$

21. $\dfrac{4}{10} \cdot \dfrac{5}{10} = \dfrac{4 \cdot 5}{10 \cdot 10} = \dfrac{2 \cdot 2 \cdot 5 \cdot 1}{2 \cdot 5 \cdot 2 \cdot 5} = \dfrac{2 \cdot 2 \cdot 5}{2 \cdot 2 \cdot 5} \cdot \dfrac{1}{5} = \dfrac{1}{5}$

23. $\dfrac{8}{10} \cdot \dfrac{45}{100} = \dfrac{8 \cdot 45}{10 \cdot 100} = \dfrac{2 \cdot 2 \cdot 2 \cdot 5 \cdot 9}{2 \cdot 5 \cdot 2 \cdot 5 \cdot 2 \cdot 5}$

$$= \dfrac{2 \cdot 2 \cdot 2 \cdot 5}{2 \cdot 2 \cdot 2 \cdot 5} \cdot \dfrac{9}{5 \cdot 5} = \dfrac{9}{5 \cdot 5} = \dfrac{9}{25}$$

25. $\dfrac{11}{24} \cdot \dfrac{3}{5} = \dfrac{11 \cdot 3}{24 \cdot 5} = \dfrac{11 \cdot 3}{3 \cdot 8 \cdot 5} = \dfrac{3}{3} \cdot \dfrac{11}{8 \cdot 5} = \dfrac{11}{8 \cdot 5} = \dfrac{11}{40}$

27. $\dfrac{10}{21} \cdot \dfrac{3}{4} = \dfrac{10 \cdot 3}{21 \cdot 4} = \dfrac{2 \cdot 5 \cdot 3}{3 \cdot 7 \cdot 2 \cdot 2}$

$$= \dfrac{2 \cdot 3}{2 \cdot 3} \cdot \dfrac{5}{7 \cdot 2} = \dfrac{5}{7 \cdot 2} = \dfrac{5}{14}$$

29. $\dfrac{5}{6}$ Interchange the numerator and denominator.

The reciprocal of $\dfrac{5}{6}$ is $\dfrac{6}{5}$. $\left(\dfrac{5}{6} \cdot \dfrac{6}{5} = \dfrac{30}{30} = 1 \right)$

31. Think of 6 as $\dfrac{6}{1}$.

$\dfrac{6}{1}$ Interchange the numerator and denominator.

The reciprocal of $\dfrac{6}{1}$ is $\dfrac{1}{6}$. $\left(\dfrac{6}{1}\cdot\dfrac{1}{6}=\dfrac{6}{6}=1\right)$

33. $\dfrac{1}{6}$ Interchange the numerator and denominator.

The reciprocal of $\dfrac{1}{6}$ is 6. $\left(\dfrac{6}{1}=6;\ \dfrac{1}{6}\cdot\dfrac{6}{1}=\dfrac{6}{6}=1\right)$

(Note that we also found that 6 and $\dfrac{1}{6}$ are reciprocals in Exercise 31.)

35. $\dfrac{10}{3}$ Interchange the numerator and denominator.

The reciprocal of $\dfrac{10}{3}$ is $\dfrac{3}{10}$. $\left(\dfrac{10}{3}\cdot\dfrac{3}{10}=\dfrac{30}{30}=1\right)$

37. $\dfrac{3}{5}\div\dfrac{3}{4}=\dfrac{3}{5}\cdot\dfrac{4}{3}$ Multiplying the dividend $\left(\dfrac{3}{5}\right)$ by the reciprocal of the divisor $\left(\text{The reciprocal of }\dfrac{3}{4}\text{ is }\dfrac{4}{3}.\right)$

$=\dfrac{3\cdot4}{5\cdot3}$ Multiplying numerators and denominators

$=\dfrac{3}{3}\cdot\dfrac{4}{5}=\dfrac{4}{5}$ Simplifying

39. $\dfrac{3}{5}\div\dfrac{9}{4}=\dfrac{3}{5}\cdot\dfrac{4}{9}$ Multiplying the dividend $\left(\dfrac{3}{5}\right)$ by the reciprocal of the divisor $\left(\text{The reciprocal of }\dfrac{9}{4}\text{ is }\dfrac{4}{9}.\right)$

$=\dfrac{3\cdot4}{5\cdot9}$ Multiplying numerators and denominators

$=\dfrac{3\cdot4}{5\cdot3\cdot3}$

$=\dfrac{3}{3}\cdot\dfrac{4}{5\cdot3}$ Simplifying

$=\dfrac{4}{5\cdot3}=\dfrac{4}{15}$

41. $\dfrac{4}{3}\div\dfrac{1}{3}=\dfrac{4}{3}\cdot3=\dfrac{4\cdot3}{3}=\dfrac{3}{3}\cdot4=4$

43. $\dfrac{1}{3}\div\dfrac{1}{6}=\dfrac{1}{3}\cdot6=\dfrac{1\cdot6}{3}=\dfrac{1\cdot2\cdot3}{1\cdot3}=\dfrac{1\cdot3}{1\cdot3}\cdot2=2$

45. $\dfrac{3}{8}\div3=\dfrac{3}{8}\cdot\dfrac{1}{3}=\dfrac{3\cdot1}{8\cdot3}=\dfrac{3}{3}\cdot\dfrac{1}{8}=\dfrac{1}{8}$

47. $\dfrac{12}{7}\div4=\dfrac{12}{7}\cdot\dfrac{1}{4}=\dfrac{12\cdot1}{7\cdot4}=\dfrac{4\cdot3\cdot1}{7\cdot4}=\dfrac{4}{4}\cdot\dfrac{3\cdot1}{7}=$

$\dfrac{3\cdot1}{7}=\dfrac{3}{7}$

49. $12\div\dfrac{3}{2}=12\cdot\dfrac{2}{3}=\dfrac{12\cdot2}{3}=\dfrac{3\cdot4\cdot2}{3\cdot1}=\dfrac{3}{3}\cdot\dfrac{4\cdot2}{1}$

$=\dfrac{4\cdot2}{1}=\dfrac{8}{1}=8$

51. $28\div\dfrac{4}{5}=28\cdot\dfrac{5}{4}=\dfrac{28\cdot5}{4}=\dfrac{4\cdot7\cdot5}{4\cdot1}=\dfrac{4}{4}\cdot\dfrac{7\cdot5}{1}$

$=\dfrac{7\cdot5}{1}=35$

53. $\dfrac{5}{8}\div\dfrac{5}{8}=\dfrac{5}{8}\cdot\dfrac{8}{5}=\dfrac{5\cdot8}{8\cdot5}=\dfrac{5\cdot8}{5\cdot8}=1$

55. $\dfrac{8}{15}\div\dfrac{4}{5}=\dfrac{8}{15}\cdot\dfrac{5}{4}=\dfrac{8\cdot5}{15\cdot4}=\dfrac{2\cdot4\cdot5}{3\cdot5\cdot4}=\dfrac{4\cdot5}{4\cdot5}\cdot\dfrac{2}{3}=\dfrac{2}{3}$

57. $\dfrac{9}{5}\div\dfrac{4}{5}=\dfrac{9}{5}\cdot\dfrac{5}{4}=\dfrac{9\cdot5}{5\cdot4}=\dfrac{5}{5}\cdot\dfrac{9}{4}=\dfrac{9}{4}$

59. $120\div\dfrac{5}{6}=120\cdot\dfrac{6}{5}=\dfrac{120\cdot6}{5}=\dfrac{5\cdot24\cdot6}{5\cdot1}=\dfrac{5}{5}\cdot\dfrac{24\cdot6}{1}$

$=\dfrac{24\cdot6}{1}=144$

61. $\dfrac{4}{5}\cdot x=60$

$x=60\div\dfrac{4}{5}$ Dividing on both sides by $\dfrac{4}{5}$

$x=60\cdot\dfrac{5}{4}$ Multiplying by the reciprocal

$=\dfrac{60\cdot5}{4}=\dfrac{4\cdot15\cdot5}{4\cdot1}=\dfrac{4}{4}\cdot\dfrac{15\cdot5}{1}=\dfrac{15\cdot5}{1}=75$

63. $\dfrac{5}{3}\cdot y=\dfrac{10}{3}$

$y=\dfrac{10}{3}\div\dfrac{5}{3}$ Dividing on both sides by $\dfrac{5}{3}$

$y=\dfrac{10}{3}\cdot\dfrac{3}{5}$ Multiplying by the reciprocal

$=\dfrac{10\cdot3}{3\cdot5}=\dfrac{2\cdot5\cdot3}{3\cdot5\cdot1}=\dfrac{5\cdot3}{5\cdot3}\cdot\dfrac{2}{1}=\dfrac{2}{1}=2$

65. $x\cdot\dfrac{25}{36}=\dfrac{5}{12}$

$x=\dfrac{5}{12}\div\dfrac{25}{36}=\dfrac{5}{12}\cdot\dfrac{36}{25}=\dfrac{5\cdot36}{12\cdot25}=\dfrac{5\cdot3\cdot12}{12\cdot5\cdot5}$

$=\dfrac{5\cdot12}{5\cdot12}\cdot\dfrac{3}{5}=\dfrac{3}{5}$

67. $n\cdot\dfrac{8}{7}=360$

$n=360\div\dfrac{8}{7}=360\cdot\dfrac{7}{8}=\dfrac{360\cdot7}{8}=\dfrac{8\cdot45\cdot7}{8\cdot1}$

$=\dfrac{8}{8}\cdot\dfrac{45\cdot7}{1}=\dfrac{45\cdot7}{1}=315$

69. Discussion and Writing Exercise

71.
$$4\,\overline{)\begin{array}{c}6\ 7\\2\ 6\ 8\end{array}}$$
$$\begin{array}{r}2\ 4\ 0\\\hline 2\ 8\\2\ 8\\\hline 0\end{array}$$
The answer is 67.

73.
$$2\ 4\,\overline{)\begin{array}{c}2\ 8\ 5\\6\ 8\ 4\ 2\end{array}}$$
$$\begin{array}{r}4\ 8\ 0\ 0\\\hline 2\ 0\ 4\ 2\\1\ 9\ 2\ 0\\\hline 1\ 2\ 2\\1\ 2\ 0\\\hline 2\end{array}$$
The answer is 285 R 2.

75. $4 \cdot x = 268$

$\dfrac{4 \cdot x}{4} = \dfrac{268}{4}$ Dividing by 4 on both sides

$x = 67$

The solution is 67.

77. $y + 502 = 9001$

$y + 502 - 502 = 9001 - 502$ Subtracting 502 on both sides

$y = 8499$

The solution is 8499.

79. Let n = the number.

$\dfrac{1}{3} \cdot n = \dfrac{1}{4}$

$n = \dfrac{1}{4} \div \dfrac{1}{3} = \dfrac{1}{4} \cdot \dfrac{3}{1} = \dfrac{1 \cdot 3}{4 \cdot 1} = \dfrac{3}{4}$

The number is $\dfrac{3}{4}$. Now we find $\dfrac{1}{2}$ of $\dfrac{3}{4}$.

$\dfrac{1}{2} \cdot \dfrac{3}{4} = \dfrac{1 \cdot 3}{2 \cdot 4} = \dfrac{3}{8}$

One-half of the number is $\dfrac{3}{8}$.

Exercise Set 2.3

1. $\dfrac{7}{8} + \dfrac{1}{8} = \dfrac{7+1}{8} = \dfrac{8}{8} = 1$

3. $\dfrac{1}{8} + \dfrac{5}{8} = \dfrac{1+5}{8} = \dfrac{6}{8} = \dfrac{3 \cdot 2}{4 \cdot 2} = \dfrac{3}{4} \cdot \dfrac{2}{2} = \dfrac{3}{4} \cdot 1 = \dfrac{3}{4}$

5. $\dfrac{2}{3} + \dfrac{5}{6}$ 3 is a factor of 6, so the LCD is 6.

$= \underbrace{\dfrac{2}{3} \cdot \dfrac{2}{2}}_{} + \dfrac{5}{6}$ ← This fraction already has the LCD as denominator.

⌐ Think: $3 \times \square = 6$. The answer is 2, so we multiply by 1, using $\dfrac{2}{2}$.

$= \dfrac{4}{6} + \dfrac{5}{6} = \dfrac{9}{6}$

$= \dfrac{3}{2}$ Simplifying

7. $\dfrac{1}{8} + \dfrac{1}{6}$ $8 = 2 \cdot 2 \cdot 2$ and $6 = 2 \cdot 3$, so the LCD is $2 \cdot 2 \cdot 2 \cdot 3$, or 24

$= \underbrace{\dfrac{1}{8} \cdot \dfrac{3}{3}}_{} + \underbrace{\dfrac{1}{6} \cdot \dfrac{4}{4}}_{}$

⌐ Think: $6 \times \square = 24$. The answer is 4, so we multiply by 1, using $\dfrac{4}{4}$.

└ Think: $8 \times \square = 24$. The answer is 3, so we multiply by 1, using $\dfrac{3}{3}$.

$= \dfrac{3}{24} + \dfrac{4}{24}$

$= \dfrac{7}{24}$

9. $\dfrac{4}{5} + \dfrac{7}{10}$ 5 is a factor of 10, so the LCD is 10.

$= \underbrace{\dfrac{4}{5} \cdot \dfrac{2}{2}}_{} + \dfrac{7}{10}$ ← This fraction already has the LCD as denominator.

⌐ Think: $5 \times \square = 10$. The answer is 2, so we multiply by 1, using $\dfrac{2}{2}$.

$= \dfrac{8}{10} + \dfrac{7}{10} = \dfrac{15}{10}$

$= \dfrac{3}{2}$ Simplifying

11. $\dfrac{5}{12} + \dfrac{3}{8}$ $12 = 2 \cdot 2 \cdot 3$ and $8 = 2 \cdot 2 \cdot 2$, so the LCD is $2 \cdot 2 \cdot 2 \cdot 3$, or 24.

$= \underbrace{\dfrac{5}{12} \cdot \dfrac{2}{2}}_{} + \underbrace{\dfrac{3}{8} \cdot \dfrac{3}{3}}_{}$

⌐ Think: $8 \times \square = 24$. The answer is 3, so we multiply by 1, using $\dfrac{3}{3}$.

└ Think: $12 \times \square = 24$. The answer is 2, so we multiply by 1, using $\dfrac{2}{2}$.

$= \dfrac{10}{24} + \dfrac{9}{24} = \dfrac{19}{24}$

13. $\dfrac{3}{20} + \dfrac{3}{4}$ 4 is a factor of 20, so the LCD is 20.

$= \dfrac{3}{20} + \dfrac{3}{4} \cdot \dfrac{5}{5}$ Multiplying by 1

$= \dfrac{3}{20} + \dfrac{15}{20} = \dfrac{18}{20} = \dfrac{9}{10}$

15. $\dfrac{5}{6} + \dfrac{7}{9}$ $6 = 2 \cdot 3$ and $9 = 3 \cdot 3$, so the LCD is $2 \cdot 3 \cdot 3$, or 18.

$= \dfrac{5}{6} \cdot \dfrac{3}{3} + \dfrac{7}{9} \cdot \dfrac{2}{2}$ Multiplying by 1

$= \dfrac{15}{18} + \dfrac{14}{18} = \dfrac{29}{18}$

17. $\dfrac{3}{10} + \dfrac{1}{100}$ 10 is a factor of 100, so the LCD is 100.

$= \dfrac{3}{10} \cdot \dfrac{10}{10} + \dfrac{1}{100}$

$= \dfrac{30}{100} + \dfrac{1}{100} = \dfrac{31}{100}$

19. $\dfrac{5}{12} + \dfrac{4}{15}$ $12 = 2 \cdot 2 \cdot 3$ and $15 = 3 \cdot 5$, so the LCD is $2 \cdot 2 \cdot 3 \cdot 5$, or 60.

$= \dfrac{5}{12} \cdot \dfrac{5}{5} + \dfrac{4}{15} \cdot \dfrac{4}{4}$

$= \dfrac{25}{60} + \dfrac{16}{60} = \dfrac{41}{60}$

21. $\dfrac{9}{10} + \dfrac{99}{100}$ 10 is a factor of 100, so the LCD is 100.

$= \dfrac{9}{10} \cdot \dfrac{10}{10} + \dfrac{99}{100}$

$= \dfrac{90}{100} + \dfrac{99}{100} = \dfrac{189}{100}$

23. $\dfrac{7}{8} + \dfrac{0}{1}$ 1 is a factor of 8, so the LCD is 8.

$= \dfrac{7}{8} + \dfrac{0}{1} \cdot \dfrac{8}{8}$

$= \dfrac{7}{8} + \dfrac{0}{8} = \dfrac{7}{8}$

Note that if we had observed at the outset that $\dfrac{0}{1} = 0$, the computation becomes $\dfrac{7}{8} + 0 = \dfrac{7}{8}$.

25. $\dfrac{3}{8} + \dfrac{1}{6}$ $8 = 2 \cdot 2 \cdot 2$ and $6 = 2 \cdot 3$, so the LCD is $2 \cdot 2 \cdot 2 \cdot 3$, or 24.

$= \dfrac{3}{8} \cdot \dfrac{3}{3} + \dfrac{1}{6} \cdot \dfrac{4}{4}$

$= \dfrac{9}{24} + \dfrac{4}{24} = \dfrac{13}{24}$

27. $\dfrac{5}{12} + \dfrac{7}{24}$ 12 is a factor of 24, so the LCD is 24.

$= \dfrac{5}{12} \cdot \dfrac{2}{2} + \dfrac{7}{24}$

$= \dfrac{10}{24} + \dfrac{7}{24} = \dfrac{17}{24}$

29. $\dfrac{3}{16} + \dfrac{5}{16} + \dfrac{4}{16} = \dfrac{3+5+4}{16} = \dfrac{12}{16} = \dfrac{3}{4}$

31. $\dfrac{8}{10} + \dfrac{7}{100} + \dfrac{4}{1000}$ 10 and 100 are factors of 1000, so the LCD is 1000.

$= \dfrac{8}{10} \cdot \dfrac{100}{100} + \dfrac{7}{100} \cdot \dfrac{10}{10} + \dfrac{4}{1000}$

$= \dfrac{800}{1000} + \dfrac{70}{1000} + \dfrac{4}{1000} = \dfrac{874}{1000}$

$= \dfrac{437}{500}$

33. $\dfrac{3}{8} + \dfrac{5}{12} + \dfrac{8}{15}$

$= \dfrac{3}{2 \cdot 2 \cdot 2} + \dfrac{5}{2 \cdot 2 \cdot 3} + \dfrac{8}{3 \cdot 5}$ Factoring the denominators

The LCM is $2 \cdot 2 \cdot 2 \cdot 3 \cdot 5$, or 120.

$= \dfrac{3}{2 \cdot 2 \cdot 2} \cdot \dfrac{3 \cdot 5}{3 \cdot 5} + \dfrac{5}{2 \cdot 2 \cdot 3} \cdot \dfrac{2 \cdot 5}{2 \cdot 5} + \dfrac{8}{3 \cdot 5} \cdot \dfrac{2 \cdot 2 \cdot 2}{2 \cdot 2 \cdot 2}$

In each case we multiply by 1 to obtain the LCD in the denominator.

$= \dfrac{3 \cdot 3 \cdot 5}{2 \cdot 2 \cdot 2 \cdot 3 \cdot 5} + \dfrac{5 \cdot 2 \cdot 5}{2 \cdot 2 \cdot 3 \cdot 2 \cdot 5} + \dfrac{8 \cdot 2 \cdot 2 \cdot 2}{3 \cdot 5 \cdot 2 \cdot 2 \cdot 2}$

$= \dfrac{45}{120} + \dfrac{50}{120} + \dfrac{64}{120}$

$= \dfrac{159}{120} = \dfrac{53}{40}$

35. $\dfrac{15}{24} + \dfrac{7}{36} + \dfrac{91}{48}$

$= \dfrac{15}{2 \cdot 2 \cdot 2 \cdot 3} + \dfrac{7}{2 \cdot 2 \cdot 3 \cdot 3} + \dfrac{91}{2 \cdot 2 \cdot 2 \cdot 2 \cdot 3}$

Factoring the denominators.

The LCM is $2 \cdot 2 \cdot 2 \cdot 2 \cdot 3 \cdot 3$, or 144.

$= \dfrac{15}{2 \cdot 2 \cdot 2 \cdot 3} \cdot \dfrac{2 \cdot 3}{2 \cdot 3} + \dfrac{7}{2 \cdot 2 \cdot 3 \cdot 3} \cdot \dfrac{2 \cdot 2}{2 \cdot 2} +$

$\dfrac{91}{2 \cdot 2 \cdot 2 \cdot 2 \cdot 3} \cdot \dfrac{3}{3}$

In each case we multiply by 1 to obtain the LCD in the denominator.

$= \dfrac{15 \cdot 2 \cdot 3}{2 \cdot 2 \cdot 2 \cdot 3 \cdot 2 \cdot 3} + \dfrac{7 \cdot 2 \cdot 2}{2 \cdot 2 \cdot 3 \cdot 3 \cdot 2 \cdot 2} + \dfrac{91 \cdot 3}{2 \cdot 2 \cdot 2 \cdot 2 \cdot 3 \cdot 3}$

$= \dfrac{90}{144} + \dfrac{28}{144} + \dfrac{273}{144} = \dfrac{391}{144}$

37. When denominators are the same, subtract the numerators and keep the denominator.

$\dfrac{5}{6} - \dfrac{1}{6} = \dfrac{5-1}{6} = \dfrac{4}{6} = \dfrac{2 \cdot 2}{2 \cdot 3} = \dfrac{2}{2} \cdot \dfrac{2}{3} = \dfrac{2}{3}$

39. When denominators are the same, subtract the numerators and keep the denominator.

$$\frac{11}{12} - \frac{2}{12} = \frac{11-2}{12} = \frac{9}{12} = \frac{3 \cdot 3}{3 \cdot 4} = \frac{3}{3} \cdot \frac{3}{4} = \frac{3}{4}$$

41. The LCM of 4 and 8 is 8.

$$\frac{3}{4} - \frac{1}{8} = \frac{3}{4} \cdot \frac{2}{2} - \frac{1}{8} \leftarrow \text{This fraction already has the LCM as the denominator.}$$

Think: $4 \times \square = 8$. The answer is 2, so we multiply by 1, using $\frac{2}{2}$.

$$= \frac{6}{8} - \frac{1}{8} = \frac{5}{8}$$

43. The LCM of 8 and 12 is 24.

$$\frac{1}{8} - \frac{1}{12} = \frac{1}{8} \cdot \frac{3}{3} - \frac{1}{12} \cdot \frac{2}{2}$$

Think: $12 \times \square = 24$. The answer is 2, so we multiply by 1, using $\frac{2}{2}$.

Think: $8 \times \square = 24$. The answer is 3, so we multiply by 1, using $\frac{3}{3}$.

$$= \frac{3}{24} - \frac{2}{24} = \frac{1}{24}$$

45. The LCM of 3 and 6 is 6.

$$\frac{4}{3} - \frac{5}{6} = \frac{4}{3} \cdot \frac{2}{2} - \frac{5}{6}$$
$$= \frac{8}{6} - \frac{5}{6} = \frac{3}{6}$$
$$= \frac{1 \cdot 3}{2 \cdot 3} = \frac{1}{2} \cdot \frac{3}{3}$$
$$= \frac{1}{2}$$

47. The LCM of 4 and 28 is 28.

$$\frac{3}{4} - \frac{3}{28} = \frac{3}{4} \cdot \frac{7}{7} - \frac{3}{28}$$
$$= \frac{21}{28} - \frac{3}{28}$$
$$= \frac{18}{28} = \frac{9 \cdot 2}{14 \cdot 2}$$
$$= \frac{9}{14} \cdot \frac{2}{2} = \frac{9}{14}$$

49. The LCM of 4 and 20 is 20.

$$\frac{3}{4} - \frac{3}{20} = \frac{3}{4} \cdot \frac{5}{5} - \frac{3}{20}$$
$$= \frac{15}{20} - \frac{3}{20} = \frac{12}{20}$$
$$= \frac{3 \cdot 4}{5 \cdot 4} = \frac{3}{5} \cdot \frac{4}{4}$$
$$= \frac{3}{5}$$

51. The LCM of 4 and 20 is 20.

$$\frac{3}{4} - \frac{1}{20} = \frac{3}{4} \cdot \frac{5}{5} - \frac{1}{20}$$
$$= \frac{15}{20} - \frac{1}{20} = \frac{14}{20}$$
$$= \frac{2 \cdot 7}{2 \cdot 10} = \frac{2}{2} \cdot \frac{7}{10}$$
$$= \frac{7}{10}$$

53. The LCM of 12 and 15 is 60.

$$\frac{5}{12} - \frac{2}{15} = \frac{5}{12} \cdot \frac{5}{5} - \frac{2}{15} \cdot \frac{4}{4}$$
$$= \frac{25}{60} - \frac{8}{60} = \frac{17}{60}$$

55. The LCM of 10 and 100 is 100.

$$\frac{6}{10} - \frac{7}{100} = \frac{6}{10} \cdot \frac{10}{10} - \frac{7}{100}$$
$$= \frac{60}{100} - \frac{7}{100} = \frac{53}{100}$$

57. The LCM of 15 and 25 is 75.

$$\frac{7}{15} - \frac{3}{25} = \frac{7}{15} \cdot \frac{5}{5} - \frac{3}{25} \cdot \frac{3}{3}$$
$$= \frac{35}{75} - \frac{9}{75} = \frac{26}{75}$$

59. The LCM of 10 and 100 is 100.

$$\frac{99}{100} - \frac{9}{10} = \frac{99}{100} - \frac{9}{10} \cdot \frac{10}{10}$$
$$= \frac{99}{100} - \frac{90}{100} = \frac{9}{100}$$

61. The LCM of 3 and 8 is 24.

$$\frac{2}{3} - \frac{1}{8} = \frac{2}{3} \cdot \frac{8}{8} - \frac{1}{8} \cdot \frac{3}{3}$$
$$= \frac{16}{24} - \frac{3}{24}$$
$$= \frac{13}{24}$$

63. The LCM of 5 and 2 is 10.

$$\frac{3}{5} - \frac{1}{2} = \frac{3}{5} \cdot \frac{2}{2} - \frac{1}{2} \cdot \frac{5}{5}$$

$$= \frac{6}{10} - \frac{5}{10}$$

$$= \frac{1}{10}$$

65. The LCM of 12 and 8 is 24.

$$\frac{5}{12} - \frac{3}{8} = \frac{5}{12} \cdot \frac{2}{2} - \frac{3}{8} \cdot \frac{3}{3}$$

$$= \frac{10}{24} - \frac{9}{24}$$

$$= \frac{1}{24}$$

67. The LCM of 8 and 16 is 16.

$$\frac{7}{8} - \frac{1}{16} = \frac{7}{8} \cdot \frac{2}{2} - \frac{1}{16}$$

$$= \frac{14}{16} - \frac{1}{16}$$

$$= \frac{13}{16}$$

69. The LCM of 25 and 15 is 75.

$$\frac{17}{25} - \frac{4}{15} = \frac{17}{25} \cdot \frac{3}{3} - \frac{4}{15} \cdot \frac{5}{5}$$

$$= \frac{51}{75} - \frac{20}{75}$$

$$= \frac{31}{75}$$

71. The LCM of 25 and 150 is 150.

$$\frac{23}{25} - \frac{112}{150} = \frac{23}{25} \cdot \frac{6}{6} - \frac{112}{150}$$

$$= \frac{138}{150} - \frac{112}{150} = \frac{26}{150}$$

$$= \frac{2 \cdot 13}{2 \cdot 75} = \frac{2}{2} \cdot \frac{13}{75}$$

$$= \frac{13}{75}$$

73. Since there is a common denominator, compare the numerators.

$$5 < 6, \text{ so } \frac{5}{8} < \frac{6}{8}.$$

75. The LCD is 12.

$$\frac{1}{3} \cdot \frac{4}{4} = \frac{4}{12} \quad \text{We multiply by 1 to get the LCD.}$$

$$\frac{1}{4} \cdot \frac{3}{3} = \frac{3}{12} \quad \text{We multiply by 1 to get the LCD.}$$

Since $4 > 3$, it follows that $\frac{4}{12} > \frac{3}{12}$, so $\frac{1}{3} > \frac{1}{4}$.

77. The LCD is 21.

$$\frac{2}{3} \cdot \frac{7}{7} = \frac{14}{21} \quad \text{We multiply by 1 to get the LCD.}$$

$$\frac{5}{7} \cdot \frac{3}{3} = \frac{15}{21} \quad \text{We multiply by 1 to get the LCD.}$$

Since $14 < 15$, it follows that $\frac{14}{21} < \frac{15}{21}$, so $\frac{2}{3} < \frac{5}{7}$.

79. The LCD is 30.

$$\frac{4}{5} \cdot \frac{6}{6} = \frac{24}{30}$$

$$\frac{5}{6} \cdot \frac{5}{5} = \frac{25}{30}$$

Since $24 < 25$, it follows that $\frac{24}{30} < \frac{25}{30}$, so $\frac{4}{5} < \frac{5}{6}$.

81. The LCD is 20.

The denominator of $\frac{19}{20}$ is the LCD.

$$\frac{4}{5} \cdot \frac{4}{4} = \frac{16}{20}$$

Since $19 > 16$, it follows that $\frac{19}{20} > \frac{16}{20}$, so $\frac{19}{20} > \frac{4}{5}$.

83. The LCD is 20.

The denominator of $\frac{19}{20}$ is the LCD.

$$\frac{9}{10} \cdot \frac{2}{2} = \frac{18}{20}$$

Since $19 > 18$, it follows that $\frac{19}{20} > \frac{18}{20}$, so $\frac{19}{20} > \frac{9}{10}$.

85. The LCD is $21 \cdot 13$, or 273.

$$\frac{31}{21} \cdot \frac{13}{13} = \frac{403}{273}$$

$$\frac{41}{13} \cdot \frac{21}{21} = \frac{861}{273}$$

Since $403 < 861$, it follows that $\frac{403}{273} < \frac{861}{273}$, so $\frac{31}{21} < \frac{41}{13}$.

87.
$$x + \frac{1}{30} = \frac{1}{10}$$

$$x + \frac{1}{30} - \frac{1}{30} = \frac{1}{10} - \frac{1}{30} \quad \text{Subtracting } \frac{1}{30} \text{ on both sides}$$

$$x + 0 = \frac{1}{10} \cdot \frac{3}{3} - \frac{1}{30} \quad \text{The LCD is 30. We multiply by 1 to get the LCD.}$$

$$x = \frac{3}{30} - \frac{1}{30} = \frac{2}{30}$$

$$x = \frac{1 \cdot 2}{2 \cdot 15} = \frac{1}{15} \cdot \frac{2}{2}$$

$$x = \frac{1}{15}$$

The solution is $\frac{1}{15}$.

89.
$$\frac{2}{3} + t = \frac{4}{5}$$

$$\frac{2}{3} + t - \frac{2}{3} = \frac{4}{5} - \frac{2}{3} \quad \text{Subtracting } \frac{2}{3} \text{ on both sides.}$$

$$t + 0 = \frac{4}{5} \cdot \frac{3}{3} - \frac{2}{3} \cdot \frac{5}{5} \quad \begin{array}{l}\text{The LCD is 15. We} \\ \text{multiply by 1 to get} \\ \text{the LCD.}\end{array}$$

$$t = \frac{12}{15} - \frac{10}{15} = \frac{2}{15}$$

The solution is $\frac{2}{15}$.

91.
$$x + \frac{1}{3} = \frac{5}{6}$$

$$x + \frac{1}{3} - \frac{1}{3} = \frac{5}{6} - \frac{1}{3}$$

$$x + 0 = \frac{5}{6} - \frac{1}{3} \cdot \frac{2}{2}$$

$$x = \frac{5}{6} - \frac{2}{6} = \frac{3}{6}$$

$$x = \frac{1 \cdot 3}{2 \cdot 3} = \frac{1}{2} \cdot \frac{3}{3}$$

$$x = \frac{1}{2}$$

The solution is $\frac{1}{2}$.

93. Discussion and Writing Exercise

95. Remember: $\frac{n}{n} = 1$, for n that is not 0.

$$\frac{38}{38} = 1$$

97. Remember: $\frac{n}{0}$ is not defined for any whole number n.

$\frac{124}{0}$ is not defined.

99. *Familiarize*. Let x = the amount by which the spending per person on gifts in 2000 exceeded the spending per person on gifts in 1999.

***Translate*.** We consider this to be a "missing addend" situation.

Amount spent per person in 1999	plus	How much more	is	Amount spent per person in 2000?
↓	↓	↓	↓	↓
1088	+	x	=	1161

***Solve*.** We subtract 1088 on both sides of the equation.
$$1088 + x = 1161$$
$$1088 + x - 1088 = 1161 - 1088$$
$$x = 73$$

***Carry out*.** We can repeat the calculation. We can also round and estimate: $1161 - 1088 \approx 1200 - 1100 \approx 100 \approx 73$. the answer checks.

***State*.** In 2000 consumers spent an average of $73 more per person on gifts than in 1999.

101. *Familiarize*. Let t = the amount by which the spending per person on travel in 2000 exceeded the spending per person on travel in 1999.

***Translate*.** We consider this to be a "missing addend" situation.

Amount spent per person in 1999	plus	How much more	is	Amount spent per person in 2000?
↓	↓	↓	↓	↓
151	+	t	=	154

***Solve*.** We subtract 151 on both sides of the equation.
$$151 + t = 154$$
$$151 + t - 151 = 154 - 151$$
$$t = 3$$

***Check*.** We can repeat the calculation. The answer checks.

***State*.** In 2000 consumers spent an average of $3 more per person on travel than in 1999.

103. *Familiarize*. Let t = the total expenditure per person in 1999. We are combining amounts, so we add.

***Translate*.** We translate to an equation.
$$1088 + 188 + 151 + 77 + 54 = t$$

***Solve*.** We carry out the addition.

$$\begin{array}{r} {\scriptstyle 3\ 2} \\ 1\,0\,8\,8 \\ 1\,8\,8 \\ 1\,5\,1 \\ 7\,7 \\ +\quad 5\,4 \\ \hline 1\,5\,5\,8 \end{array}$$

***Check*.** We can repeat the calculation. The answer checks.

***State*.** The total expenditure per person in 1999 was $1558.

105.
$$\frac{7}{8} - \frac{1}{10} \times \frac{5}{6} = \frac{7}{8} - \frac{1 \times 5}{10 \times 6} = \frac{7}{8} - \frac{1 \times 5}{2 \times 5 \times 6}$$

$$= \frac{7}{8} - \frac{5}{5} \times \frac{1}{2 \times 6} = \frac{7}{8} - \frac{1}{2 \times 6} = \frac{7}{8} - \frac{1}{12}$$

$$= \frac{7}{8} \cdot \frac{3}{3} - \frac{1}{12} \cdot \frac{2}{2} = \frac{21}{24} - \frac{2}{24} = \frac{19}{24}$$

107.
$$\left(\frac{2}{3}\right)^2 + \left(\frac{3}{4}\right)^2 = \frac{4}{9} + \frac{9}{16} = \frac{4}{9} \cdot \frac{16}{16} + \frac{9}{16} \cdot \frac{9}{9} =$$

$$\frac{64}{144} + \frac{81}{144} = \frac{145}{144}$$

109. *Familiarize*. We visualize the situation. We let h = the elevation at which the climber finished.

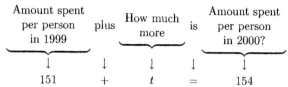

Translate.

First climb   minus   First descent   plus   Second climb   minus

$$\frac{3}{5} \quad - \quad \frac{1}{4} \quad + \quad \frac{1}{3} \quad -$$

Second descent   is   Final elevation

$$\frac{1}{7} \quad = \quad h$$

Solve. We carry out the calculation. The LCD is $5 \cdot 4 \cdot 3 \cdot 7$, or 420.

$$\frac{3}{5} - \frac{1}{4} + \frac{1}{3} - \frac{1}{7} = h$$

$$\frac{3}{5} \cdot \frac{4 \cdot 3 \cdot 7}{4 \cdot 3 \cdot 7} - \frac{1}{4} \cdot \frac{5 \cdot 3 \cdot 7}{5 \cdot 3 \cdot 7} + \frac{1}{3} \cdot \frac{5 \cdot 4 \cdot 7}{5 \cdot 4 \cdot 7} -$$

$$\frac{1}{7} \cdot \frac{5 \cdot 4 \cdot 3}{5 \cdot 4 \cdot 3} = h$$

$$\frac{252}{5 \cdot 4 \cdot 3 \cdot 7} - \frac{105}{5 \cdot 4 \cdot 3 \cdot 7} + \frac{140}{5 \cdot 4 \cdot 3 \cdot 7} - \frac{60}{5 \cdot 4 \cdot 3 \cdot 7} = h$$

$$\frac{252 - 105 + 140 - 60}{5 \cdot 4 \cdot 3 \cdot 7} = h$$

$$\frac{227}{5 \cdot 4 \cdot 3 \cdot 7} = h$$

$$\frac{227}{420} = h$$

Check. We repeat the calculation.

State. The climber's final elevation is $\frac{227}{420}$ km.

Exercise Set 2.4

1. [b] [a] Multiply: $8 \cdot 3 = 24$.

$3\frac{5}{8} = \frac{29}{8}$ [b] Add: $24 + 5 = 29$.

[a] [c] Keep the denominator.

[b] [a] Multiply: $4 \cdot 2 = 8$.

$2\frac{3}{4} = \frac{11}{4}$ [b] Add: $8 + 3 = 11$.

[a] [c] Keep the denominator.

3. [b] [a] Multiply: $5 \cdot 3 = 15$.

$5\frac{2}{3} = \frac{17}{3}$ [b] Add: $15 + 2 = 17$.

[a] [c] Keep the denominator.

5. $9\frac{5}{6} = \frac{59}{6}$ $(9 \cdot 6 = 54, \; 54 + 5 = 59)$

7. $12\frac{3}{4} = \frac{51}{4}$ $(12 \cdot 4 = 48, \; 48 + 3 = 51)$

9. To convert $\frac{18}{5}$ to a mixed numeral, we divide.

$$5\overline{\smash{)}18} \qquad \frac{18}{5} = 3\frac{3}{5}$$
$$\underline{15}$$
$$\;3$$

(quotient 3)

11.
$$10\overline{\smash{)}57} \qquad \frac{57}{10} = 5\frac{7}{10}$$
$$\underline{50}$$
$$\;7$$

(quotient 5)

13.
$$8\overline{\smash{)}345} \qquad \frac{345}{8} = 43\frac{1}{8}$$
$$\underline{320}$$
$$\;25$$

(quotient 43)

15.
$$2\frac{7}{8}$$
$$+3\frac{5}{8}$$
$$\overline{5\frac{12}{8}} = 5 + \frac{12}{8}$$
$$= 5 + 1\frac{1}{2}$$
$$= 6\frac{1}{2}$$

To find a mixed numeral for $\frac{12}{8}$ we divide:

$$8\overline{\smash{)}12} \qquad \frac{12}{8} = 1\frac{4}{8} = 1\frac{1}{2}$$
$$\underline{8}$$
$$4$$

17. The LCD is 12.

$$1\,\frac{1}{4} \cdot \frac{3}{3} = 1\frac{3}{12}$$
$$+1\,\frac{2}{3} \cdot \frac{4}{4} = +1\frac{8}{12}$$
$$\overline{2\frac{11}{12}}$$

19. The LCD is 12.

$$8\,\frac{3}{4} \cdot \frac{3}{3} = 8\frac{9}{12}$$
$$+5\,\frac{5}{6} \cdot \frac{2}{2} = +5\frac{10}{12}$$
$$\overline{13\frac{19}{12}} = 13 + \frac{19}{12}$$
$$= 13 + 1\frac{7}{12}$$
$$= 14\frac{7}{12}$$

21. The LCD is 10.

$$12\;\boxed{\dfrac{4}{5}\cdot\dfrac{2}{2}} = 12\;\dfrac{8}{10}$$
$$+8\;\dfrac{7}{10} = +8\;\dfrac{7}{10}$$
$$\overline{\hspace{3cm}}$$
$$20\;\dfrac{15}{10} = 20 + \dfrac{15}{10}$$
$$= 20 + 1\dfrac{5}{10}$$
$$= 21\dfrac{5}{10}$$
$$= 21\dfrac{1}{2}$$

23. The LCD is 8.

$$14\;\dfrac{5}{8} = 14\;\dfrac{5}{8}$$
$$+13\;\boxed{\dfrac{1}{4}\cdot\dfrac{2}{2}} = +13\;\dfrac{2}{8}$$
$$\overline{\hspace{3cm}}$$
$$27\;\dfrac{7}{8}$$

25. The LCD is 24.

$$7\;\boxed{\dfrac{1}{8}\cdot\dfrac{3}{3}} = 7\;\dfrac{3}{24}$$
$$9\;\boxed{\dfrac{2}{3}\cdot\dfrac{8}{8}} = 9\;\dfrac{16}{24}$$
$$+10\;\boxed{\dfrac{3}{4}\cdot\dfrac{6}{6}} = +10\;\dfrac{18}{24}$$
$$\overline{\hspace{3cm}}$$
$$26\;\dfrac{37}{24} = 26 + \dfrac{37}{24}$$
$$= 26 + 1\dfrac{13}{24}$$
$$= 27\dfrac{13}{24}$$

27.
$$4\dfrac{1}{5} = 3\dfrac{6}{5}$$
$$-2\dfrac{3}{5} = -2\dfrac{3}{5}$$
$$\overline{\hspace{2cm}}$$
$$1\dfrac{3}{5}$$

Since $\frac{1}{5}$ is smaller than $\frac{3}{5}$, we cannot subtract until we borrow: $4\dfrac{1}{5} = 3 + \dfrac{5}{5} + \dfrac{1}{5} = 3 + \dfrac{6}{5} = 3\dfrac{6}{5}$

29. The LCD is 10.

$$6\;\boxed{\dfrac{3}{5}\cdot\dfrac{2}{2}} = 6\;\dfrac{6}{10}$$
$$-2\;\boxed{\dfrac{1}{2}\cdot\dfrac{5}{5}} = -2\;\dfrac{5}{10}$$
$$\overline{\hspace{3cm}}$$
$$4\;\dfrac{1}{10}$$

31.
$$34 = 33\dfrac{8}{8} \quad \left(34 = 33 + 1 = 33 + \dfrac{8}{8} = 33\dfrac{8}{8}\right)$$
$$-18\dfrac{5}{8} = -18\dfrac{5}{8}$$
$$\overline{\hspace{3cm}}$$
$$15\dfrac{3}{8}$$

33. The LCD is 12.

$$21\;\boxed{\dfrac{1}{6}\cdot\dfrac{2}{2}} = 21\;\dfrac{2}{12} = 20\;\dfrac{14}{12}$$
$$-13\;\boxed{\dfrac{3}{4}\cdot\dfrac{3}{3}} = -13\;\dfrac{9}{12} = -13\;\dfrac{9}{12}$$
$$\overline{\hspace{5cm}}$$
$$7\;\dfrac{5}{12}$$

$\left(\text{Since } \dfrac{2}{12} \text{ is smaller than } \dfrac{9}{12}, \text{ we cannot subtract until we borrow: } 21\dfrac{2}{12} = 20 + \dfrac{12}{12} + \dfrac{2}{12} = 20 + \dfrac{14}{12} = 20\dfrac{14}{12}.\right)$

35. The LCD is 8.

$$14\;\dfrac{1}{8} = 14\;\dfrac{1}{8} = 13\;\dfrac{9}{8}$$
$$-\;\boxed{\dfrac{3}{4}\cdot\dfrac{2}{2}} = -\dfrac{6}{8} = -\dfrac{6}{8}$$
$$\overline{\hspace{5cm}}$$
$$13\;\dfrac{3}{8}$$

$\left(\text{Since } \dfrac{1}{8} \text{ is smaller than } \dfrac{6}{8}, \text{ we cannot subtract until we borrow: } 14\dfrac{1}{8} = 13 + \dfrac{8}{8} + \dfrac{1}{8} = 13 + \dfrac{9}{8} = 13\dfrac{9}{8}.\right)$

37. The LCD is 18.

$$25\;\boxed{\dfrac{1}{9}\cdot\dfrac{2}{2}} = 25\;\dfrac{2}{18} = 24\;\dfrac{20}{18}$$
$$-13\;\boxed{\dfrac{5}{6}\cdot\dfrac{3}{3}} = -13\;\dfrac{15}{18} = -13\;\dfrac{15}{18}$$
$$\overline{\hspace{5cm}}$$
$$11\;\dfrac{5}{18}$$

$\left(\text{Since } \dfrac{2}{18} \text{ is smaller than } \dfrac{15}{18}, \text{ we cannot subtract until we borrow: } 25\dfrac{2}{18} = 24 + \dfrac{18}{18} + \dfrac{2}{18} = 24 + \dfrac{20}{18} = 24\dfrac{20}{18}.\right)$

39. $8 \cdot 2\dfrac{5}{6}$

$$= \dfrac{8}{1}\cdot\dfrac{17}{6} \quad \text{Writing fractional notation}$$
$$= \dfrac{8\cdot 17}{1\cdot 6} = \dfrac{2\cdot 4\cdot 17}{1\cdot 2\cdot 3} = \dfrac{2}{2}\cdot\dfrac{4\cdot 17}{1\cdot 3} = \dfrac{68}{3} = 22\dfrac{2}{3}$$

41. $3\dfrac{5}{8}\cdot\dfrac{2}{3}$

$$= \dfrac{29}{8}\cdot\dfrac{2}{3} \quad \text{Writing fractional notation}$$
$$= \dfrac{29\cdot 2}{8\cdot 3} = \dfrac{29\cdot 2}{2\cdot 4\cdot 3} = \dfrac{2}{2}\cdot\dfrac{29}{4\cdot 3} = \dfrac{29}{12} = 2\dfrac{5}{12}$$

43. $3\frac{1}{2} \cdot 2\frac{1}{3} = \frac{7}{2} \cdot \frac{7}{3} = \frac{49}{6} = 8\frac{1}{6}$

45. $3\frac{2}{5} \cdot 2\frac{7}{8} = \frac{17}{5} \cdot \frac{23}{8} = \frac{391}{40} = 9\frac{31}{40}$

47. $4\frac{7}{10} \cdot 5\frac{3}{10} = \frac{47}{10} \cdot \frac{53}{10} = \frac{2491}{100} = 24\frac{91}{100}$

49. $20\frac{1}{2} \cdot 10\frac{1}{5} \cdot 4\frac{2}{3} = \frac{41}{2} \cdot \frac{51}{5} \cdot \frac{14}{3} = \frac{41 \cdot 51 \cdot 14}{2 \cdot 5 \cdot 3} =$

$\frac{41 \cdot 3 \cdot 17 \cdot 2 \cdot 7}{2 \cdot 5 \cdot 3} = \frac{2 \cdot 3}{2 \cdot 3} \cdot \frac{41 \cdot 17 \cdot 7}{5} = \frac{4879}{5} = 975\frac{4}{5}$

51.

$20 \div 3\frac{1}{5}$

$= 20 \div \frac{16}{5}$ Writing fractional notation

$= 20 \cdot \frac{5}{16}$ Multiplying by the reciprocal

$= \frac{20 \cdot 5}{16} = \frac{4 \cdot 5 \cdot 5}{4 \cdot 4} = \frac{4}{4} \cdot \frac{5 \cdot 5}{4} = \frac{25}{4} = 6\frac{1}{4}$

53. $8\frac{2}{5} \div 7$

$= \frac{42}{5} \div 7$ Writing fractional notation

$= \frac{42}{5} \cdot \frac{1}{7}$ Multiplying by the reciprocal

$= \frac{42 \cdot 1}{5 \cdot 7} = \frac{6 \cdot 7}{5 \cdot 7} = \frac{7}{7} \cdot \frac{6}{5} = \frac{6}{5} = 1\frac{1}{5}$

55. $4\frac{3}{4} \div 1\frac{1}{3} = \frac{19}{4} \div \frac{4}{3} = \frac{19}{4} \cdot \frac{3}{4} = \frac{19 \cdot 3}{4 \cdot 4} = \frac{57}{16} = 3\frac{9}{16}$

57. $1\frac{7}{8} \div 1\frac{2}{3} = \frac{15}{8} \div \frac{5}{3} = \frac{15}{8} \cdot \frac{3}{5} = \frac{15 \cdot 3}{8 \cdot 5} = \frac{5 \cdot 3 \cdot 3}{8 \cdot 5}$

$= \frac{5}{5} \cdot \frac{3 \cdot 3}{8} = \frac{3 \cdot 3}{8} = \frac{9}{8} = 1\frac{1}{8}$

59. $5\frac{1}{10} \div 4\frac{3}{10} = \frac{51}{10} \div \frac{43}{10} = \frac{51}{10} \cdot \frac{10}{43} = \frac{51 \cdot 10}{10 \cdot 43}$

$= \frac{10}{10} \cdot \frac{51}{43} = \frac{51}{43} = 1\frac{8}{43}$

61. $20\frac{1}{4} \div 90 = \frac{81}{4} \div 90 = \frac{81}{4} \cdot \frac{1}{90} = \frac{81 \cdot 1}{4 \cdot 90} = \frac{9 \cdot 9 \cdot 1}{4 \cdot 9 \cdot 10}$

$= \frac{9}{9} \cdot \frac{9 \cdot 1}{4 \cdot 10} = \frac{9}{40}$

63. Discussion and Writing Exercise

65. 45,765

We found up since the tens digit, 6, is 5 or higher. We have 45,800.

67. The sum of the digits is $9 + 9 + 9 + 3 = 30$. Since 30 is divisible by 3, then 9993 is divisible by 3.

69. The sum of the digits is $2 + 3 + 4 + 5 = 14$. Since 14 is not divisible by 9, then 2345 is not divisible by 9.

71. The ones digit of 2335 is not 0, so 2335 is not divisible by 10.

73. The last three digits of 18,888 are divisible by 8, so 18,888 is divisible by 8.

75.

$$\begin{array}{r} {\scriptstyle 2\ \ 9\ \ 9\ \ 14} \\ \cancel{3\ 0\ 0\ 4} \\ -\ 2\ 9\ 5\ 7 \\ \hline 4\ 7 \end{array}$$

77.

$$\begin{array}{r} {\scriptstyle 10} \\ {\scriptstyle 9\ \ 9\ \ \cancel{0}\ \ 13} \\ \cancel{1\,0,0\,1\,3} \\ -\ \ 9,9\,8\,8 \\ \hline 2\ 5 \end{array}$$

79. Use a calculator.

$15\frac{2}{11} \cdot 23\frac{31}{43} = \frac{167}{11} \cdot \frac{1020}{43} = \frac{167 \cdot 1020}{11 \cdot 43} =$

$\frac{170,340}{473} = 360\frac{60}{473}$

81. $8 \div \frac{1}{2} + \frac{3}{4} + \left(5 - \frac{5}{8}\right)^2$

$= 8 \div \frac{1}{2} + \frac{3}{4} + \left(\frac{40}{8} - \frac{5}{8}\right)^2$

$= 8 \div \frac{1}{2} + \frac{3}{4} + \left(\frac{35}{8}\right)^2$

$= 8 \div \frac{1}{2} + \frac{3}{4} + \frac{1225}{64}$

$= 8 \cdot 2 + \frac{3}{4} + \frac{1225}{64}$

$= 16 + \frac{3}{4} + \frac{1225}{64}$

$= \frac{1024}{64} + \frac{48}{64} + \frac{1225}{64}$

$= \frac{2297}{64} = 35\frac{57}{64}$

Exercise Set 2.5

1. *Familiarize*. A picture of the situation appears in the text. Let f = the fraction of the floor that has been tiled.

Translate. The multiplication sentence $\frac{3}{5} \cdot \frac{3}{4} = f$ corresponds to the situation.

Solve. We multiply.

$$\frac{3}{5} \cdot \frac{3}{4} = \frac{3 \cdot 3}{5 \cdot 4} = \frac{9}{20}$$

Check. We repeat the calculation. The answer checks.

State. $\frac{9}{20}$ of the floor has been tiled.

3. **Familiarize**. Recall that area is length times width. We draw a picture. We will let A = the area of the table top.

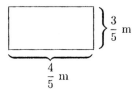

$$\left.\begin{array}{c} \\ \\ \\ \end{array}\right\} \frac{3}{5} \text{ m}$$

$$\underbrace{}_{\frac{4}{5} \text{ m}}$$

Translate. Then we translate.

$$\begin{array}{ccccc} \text{Area} & \text{is} & \text{length} & \text{times} & \text{width} \\ \downarrow & \downarrow & \downarrow & \downarrow & \downarrow \\ A & = & \dfrac{4}{5} & \times & \dfrac{3}{5} \end{array}$$

Solve. The sentence tells us what to do. We multiply.

$$\frac{4}{5} \times \frac{3}{5} = \frac{4 \times 3}{5 \times 5} = \frac{12}{25}$$

Check. We repeat the calculation. The answer checks.

State. The area is $\frac{12}{25}$ m².

5. **Familiarize**. Let n = the number of inches the screw will go into the piece of oak when it is turned 10 complete rotations.

Translate. We write an equation.

$$\begin{array}{ccccc} \underbrace{\begin{array}{c}\text{Total}\\\text{distance}\end{array}} & \text{is} & \underbrace{\begin{array}{c}\text{Distance}\\\text{for one}\\\text{revolution}\end{array}} & \text{times} & \underbrace{\begin{array}{c}\text{Number of}\\\text{revolutions}\end{array}} \\ \downarrow & \downarrow & \downarrow & \downarrow & \downarrow \\ n & = & \dfrac{1}{16} & \cdot & 10 \end{array}$$

Solve. We carry out the multiplication.

$$\begin{aligned} n &= \frac{1}{16} \cdot 10 = \frac{1 \cdot 10}{16} \\ &= \frac{1 \cdot 2 \cdot 5}{2 \cdot 8} = \frac{2}{2} \cdot \frac{1 \cdot 5}{8} \\ &= \frac{5}{8} \end{aligned}$$

Check. We can repeat the calculation. We can also determine that the answer seems reasonable since we multiplied 10 by a number less than 10 and the result is less than 10. The answer checks.

State. The screw will go $\frac{5}{8}$ in. into the piece of oak when it is turned 10 completed rotations.

7. **Familiarize**. Let p = the pitch of the screw, in inches. The distance the screw has traveled into the wallboard is found by multiplying the pitch by the number of complete rotations.

Translate. We translate to an equation.

$$\begin{array}{ccccc} \underbrace{\begin{array}{c}\text{Pitch}\\\text{of screw}\end{array}} & \text{times} & \underbrace{\begin{array}{c}\text{Number of}\\\text{rotations}\end{array}} & \text{is} & \underbrace{\begin{array}{c}\text{Distance}\\\text{traveled}\end{array}} \\ \downarrow & \downarrow & \downarrow & \downarrow & \downarrow \\ p & \cdot & 8 & = & \dfrac{1}{2} \end{array}$$

Solve. We divide on both sides of the equation by 8 and carry out the division.

$$p = \frac{1}{2} \div 8 = \frac{1}{2} \cdot \frac{1}{8} = \frac{1 \cdot 1}{2 \cdot 8} = \frac{1}{16}$$

Check. We repeat the calculation. The answer checks.

State. The pitch of the screw is $\frac{1}{16}$ in.

9. **Familiarize**. We know that 1 of 39 high school football players plays college football. That is, $\frac{1}{39}$ of high school football players play college football. In addition, we know that 1 of 39 college players plays professional football. That is, $\frac{1}{39}$ of college football players play professional football or $\frac{1}{39}$ of the $\frac{1}{39}$ of high school players who play college football also play professionally. We let f = the fractional part of high school football players that plays professional football.

Translate. The multiplication sentence $\frac{1}{39} \cdot \frac{1}{39}$ corresponds to this situation.

Solve. We carry out the multiplication.

$$\frac{1}{39} \cdot \frac{1}{39} = \frac{1 \cdot 1}{39 \cdot 39} = \frac{1}{1521}$$

Check. We repeat the calculation. The answer checks.

State. The fractional part of high school players that plays professional football is $\frac{1}{1521}$.

11. **Familiarize**. We visualize the situation. We let n = the number of addresses that will be incorrect after one year.

Mailing list 2500 addresses			
1/4 of the addresses n			

Translate.

$$\begin{array}{ccccc} \underbrace{\begin{array}{c}\text{Number incorrect}\end{array}} & \text{is} & \frac{1}{4} & \text{of} & \underbrace{\begin{array}{c}\text{Number of addresses}\end{array}} \\ \downarrow & \downarrow \downarrow \downarrow & & & \downarrow \\ n & = & \dfrac{1}{4} & \cdot & 2500 \end{array}$$

Solve. We carry out the multiplication.

$$\begin{aligned} n &= \frac{1}{4} \cdot 2500 = \frac{1 \cdot 2500}{4} = \frac{2500}{4} \\ &= \frac{4 \cdot 625}{4 \cdot 1} = \frac{4}{4} \cdot \frac{625}{1} \\ &= 625 \end{aligned}$$

Check. We can repeat the calculation. We can also determine that the answer seems reasonable since we multiplied 2500 by a number less than 1 and the result is less than 2500. The answer checks.

State. After one year 625 addresses will be incorrect.

13. *Familiarize.* We draw a picture.

1 in.
240 miles

We let n = the number of miles represented by $\frac{2}{3}$ in.

Translate. The multiplication sentence

$$n = \frac{2}{3} \cdot 240$$

corresponds to the situation.

Solve. We multiply and simplify:

$$n = \frac{2}{3} \cdot 240 = \frac{2 \cdot 240}{3} = \frac{2 \cdot 3 \cdot 80}{1 \cdot 3}$$
$$= \frac{3}{3} \cdot \frac{2 \cdot 80}{1} = \frac{2 \cdot 80}{1}$$
$$= 160$$

Check. We can repeat the calculation. We can also determine that the answer seems reasonable since we multiplied 240 by a number less than 1 and the result is less than 240.

State. $\frac{2}{3}$ in. on the map represents 160 miles.

15. *Familiarize.* We draw a picture.

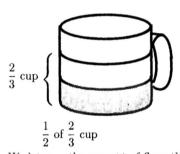

$\frac{2}{3}$ cup

$\frac{1}{2}$ of $\frac{2}{3}$ cup

We let n = the amount of flour the chef should use.

Translate. The multiplication sentence

$$\frac{1}{2} \cdot \frac{2}{3} = n$$

corresponds to the situation.

Solve. We multiply and simplify:

$$n = \frac{1}{2} \cdot \frac{2}{3} = \frac{1 \cdot 2}{2 \cdot 3} = \frac{2}{2} \cdot \frac{1}{3} = \frac{1}{3}$$

Check. We can repeat the calculation. We can also determine that the answer seems reasonable since we multiplied $\frac{2}{3}$ by a number less than 1 and the result is less than $\frac{2}{3}$. The answer checks.

State. The chef should use $\frac{1}{3}$ cup of flour.

17. *Familiarize.* We draw a picture. Let t = the number of times Benny will be able to brush his teeth.

$$\boxed{\frac{2}{5}\text{ g}} \quad \boxed{\frac{2}{5}\text{ g}} \cdots \boxed{\frac{2}{5}\text{ g}}$$

t brushings

Translate. The multiplication that corresponds to the situation is

$$\frac{2}{5} \cdot t = 30.$$

Solve. We solve the equation by dividing on both sides by $\frac{2}{5}$ and carrying out the division:

$$t = 30 \div \frac{2}{5} = 30 \cdot \frac{5}{2} = \frac{30 \cdot 5}{2} = \frac{2 \cdot 15 \cdot 5}{2 \cdot 1} = \frac{2}{2} \cdot \frac{15 \cdot 5}{1}$$
$$= \frac{15 \cdot 5}{1} = 75$$

Check. We repeat the calculation. The answer checks.

State. Benny can brush his teeth 75 times with a 30-g tube of toothpaste.

19. *Familiarize.* This is a multistep problem. First we find the total weight of the cubic meter of concrete mix. We visualize the situation, letting w = the total weight.

420 kg	150 kg	120 kg
w		

Translate. An addition sentence corresponds to this situation.

Weight of cement	plus	Weight of stone	plus	Weight of sand	is	Total weight
↓	↓	↓	↓	↓	↓	↓
420	+	150	+	120	=	w

Solve. We carry out the addition.

$$420 + 150 + 120 = w$$
$$690 = w$$

Since the mix contains 420 kg of cement, the part that is cement is $\frac{420}{690} = \frac{14 \cdot 30}{23 \cdot 30} = \frac{14}{23} \cdot \frac{30}{30} = \frac{14}{23}$.

Since the mix contains 150 kg of stone, the part that is stone is $\frac{150}{690} = \frac{5 \cdot 30}{23 \cdot 30} = \frac{5}{23} \cdot \frac{30}{30} = \frac{5}{23}$.

Since the mix contains 120 kg of sand, the part that is sand is $\frac{120}{690} = \frac{4 \cdot 30}{23 \cdot 30} = \frac{4}{23} \cdot \frac{30}{30} = \frac{4}{23}$.

We add these amounts: $\frac{14}{23} + \frac{5}{23} + \frac{4}{23} = \frac{14 + 5 + 4}{23} = \frac{23}{23} = 1.$

Check. We repeat the calculations. We also note that since the total of the fractional parts is 1, the answer is probably correct.

State. The total weight of the cubic meter of concrete mix is 690 kg. Of this, $\frac{14}{23}$ is cement, $\frac{5}{23}$ is stone, and $\frac{4}{23}$ is sand. The result when we add these amounts is 1.

21. Familiarize. We draw a picture. We let t = the total thickness.

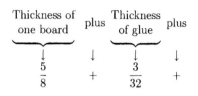

Translate. We translate to an equation.

$$\underbrace{\text{Thickness of one board}}_{\downarrow} \quad \text{plus} \quad \underbrace{\text{Thickness of glue}}_{\downarrow} \quad \text{plus}$$
$$\qquad\quad \frac{5}{8} \qquad\qquad + \qquad\qquad \frac{3}{32} \qquad\qquad +$$

$$\underbrace{\text{Thickness of second board}}_{\downarrow} \quad \text{is} \quad \underbrace{\text{Total thickness}}_{\downarrow}$$
$$\qquad\quad \frac{7}{8} \qquad\qquad = \qquad\qquad t$$

Solve. We carry out the addition. The LCD is 32 since 8 is a factor of 32.

$$\frac{5}{8} + \frac{3}{32} + \frac{7}{8} = t$$
$$\frac{5}{8} \cdot \frac{4}{4} + \frac{3}{32} + \frac{7}{8} \cdot \frac{4}{4} = t$$
$$\frac{20}{32} + \frac{3}{32} + \frac{28}{32} = t$$
$$\frac{51}{32} = t$$

Check. We repeat the calculation. We also note that the sum is larger than any of the individual thicknesses, as expected.

State. The result is $\frac{51}{32}$ in. thick.

23. Familiarize. We draw a picture. We let n = the number of pairs of basketball shorts that can be made.

$$\boxed{\frac{3}{4}\ \text{yd}} \quad \boxed{\frac{3}{4}\ \text{yd}} \quad \cdots \quad \boxed{\frac{3}{4}\ \text{yd}}$$
$$\underbrace{\hphantom{\boxed{\frac{3}{4}\ \text{yd}} \quad \boxed{\frac{3}{4}\ \text{yd}} \quad \cdots \quad \boxed{\frac{3}{4}\ \text{yd}}}}_{n \text{ pairs of shorts}}$$

Translate. The multiplication that corresponds to the situation is
$$\frac{3}{4} \cdot n = 24.$$

Solve. We solve the equation by dividing on both sides by $\frac{3}{4}$ and carrying out the division:
$$n = 24 \div \frac{3}{4} = 24 \cdot \frac{4}{3} = \frac{24 \cdot 4}{3} = \frac{3 \cdot 8 \cdot 4}{3 \cdot 1} = \frac{3}{3} \cdot \frac{8 \cdot 4}{1}$$
$$= \frac{8 \cdot 4}{1} = 32$$

Check. We repeat the calculation. The answer checks.

State. 32 pairs of basketball shorts can be made from 24 yd of fabric.

25. Familiarize. We draw a picture. We let n = the number of sugar bowls that can be filled.

$$\boxed{\frac{2}{3}\ \text{cup}} \quad \boxed{\frac{2}{3}\ \text{cup}} \quad \cdots \quad \boxed{\frac{2}{3}\ \text{cup}}$$
$$\underbrace{\hphantom{\boxed{\frac{2}{3}\ \text{cup}} \quad \boxed{\frac{2}{3}\ \text{cup}} \quad \cdots \quad \boxed{\frac{2}{3}\ \text{cup}}}}_{n \text{ bowls}}$$

Translate. We write a multiplication sentence:
$$\frac{2}{3} \cdot n = 16$$

Solve. Solve the equation as follows:
$$\frac{2}{3} \cdot n = 16$$
$$n = 16 \div \frac{2}{3} = 16 \cdot \frac{3}{2} = \frac{16 \cdot 3}{2} = \frac{2 \cdot 8 \cdot 3}{2 \cdot 1}$$
$$= \frac{2}{2} \cdot \frac{8 \cdot 3}{1} = \frac{8 \cdot 3}{1} = 24$$

Check. We repeat the calculation. The answer checks.

State. 24 sugar bowls can be filled.

27. Familiarize. We draw a picture. We let n = the amount the bucket could hold.

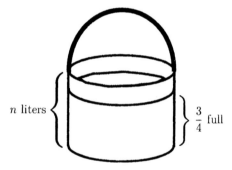

Translate. We write a multiplication sentence:
$$\frac{3}{4} \cdot n = 12$$

Solve. Solve the equation as follows:
$$\frac{3}{4} \cdot n = 12$$
$$n = 12 \div \frac{3}{4} = 12 \cdot \frac{4}{3} = \frac{12 \cdot 4}{3} = \frac{3 \cdot 4 \cdot 4}{3 \cdot 1}$$
$$= \frac{3}{3} \cdot \frac{4 \cdot 4}{1} = \frac{4 \cdot 4}{1} = 16$$

Check. We repeat the calculation. The answer checks.

State. The bucket could hold 16 L.

29. Familiarize. We draw a picture. We let p = the number of pounds of tea Rene bought.

$\frac{1}{3}$ lb	$\frac{1}{2}$ lb
p	

Translate. An addition sentence corresponds to this situation.

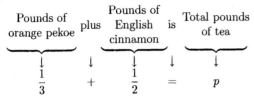

Solve. We carry out the addition. Since 3 and 2 are both prime numbers, the LCM of the denominators is their product $3 \cdot 2$, or 6.

$$\frac{1}{3} \cdot \frac{2}{2} + \frac{1}{2} \cdot \frac{3}{3} = p$$
$$\frac{2}{6} + \frac{3}{6} = p$$
$$\frac{5}{6} = p$$

Check. We check by repeating the calculation. We also note that the sum is larger than either of the individual weights, so the answer seems reasonable.

State. Rene bought $\frac{5}{6}$ lb of tea.

31. *Familiarize*. We draw a picture. We let $D = $ the total distance Russ walked.

$$\frac{7}{6}\text{ mi} \qquad \frac{3}{4}\text{ mi}$$
$$\underbrace{\qquad\qquad\qquad}_{D}$$

Translate. An addition sentence corresponds to this situation.

Distance to friend's house	plus	Distance to class	is	Total distance
$\frac{7}{6}$	$+$	$\frac{3}{4}$	$=$	D

Solve. To solve the equation, carry out the addition. Since $6 = 2 \cdot 3$ and $4 = 2 \cdot 2$, the LCM of the denominators is $2 \cdot 2 \cdot 3$, or 12.

$$\frac{7}{6} \cdot \frac{2}{2} + \frac{3}{4} \cdot \frac{3}{3} = D$$
$$\frac{14}{12} + \frac{9}{12} = D$$
$$\frac{23}{12} = D$$

Check. We repeat the calculation. We also note that the sum is larger than either of the original distances, so the answer seems reasonable.

State. Russ walked $\frac{23}{12}$ mi.

33. *Familiarize*. We visualize the situation. Let $d = $ the number of inches by which the new tread depth exceeds the more typical depth.

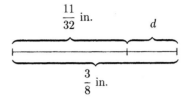

Translate. This is a "missing addend" situation.

Typical tread depth	plus	Excess depth	is	New tread depth
$\frac{11}{32}$	$+$	d	$=$	$\frac{3}{8}$

Solve. We subtract $\frac{11}{32}$ on both sides of the equation.

$$\frac{11}{32} + d - \frac{11}{32} = \frac{3}{8} - \frac{11}{32}$$
$$d + 0 = \frac{3}{8} \cdot \frac{4}{4} - \frac{11}{32} \qquad \text{The LCD is 32. We}$$
$$\qquad\qquad\qquad\qquad\qquad \text{multiply by 1 to}$$
$$d = \frac{12}{32} - \frac{11}{32} \qquad \text{get the LCD.}$$
$$d = \frac{1}{32}$$

Check. We return to the original problem and add.
$$\frac{11}{32} + \frac{1}{32} = \frac{12}{32} = \frac{4 \cdot 3}{4 \cdot 8} = \frac{4}{4} \cdot \frac{3}{8} = \frac{3}{8}$$
The answer checks.

State. The new tread depth is $\frac{1}{32}$ in. deeper than the more typical depth of $\frac{11}{32}$ in.

35. *Familiarize*. We visualize the situation. Let $x = $ the portion of the business owned by the third person.

$$\vdash\!\!\!-\!\!\frac{7}{12}\!\!-\!\!+\!\!\frac{1}{6}\!\!+\!\!x\!\!-\!\!\dashv$$
$$\vdash\!\!\!-\!\!-\text{1 entire business}-\!\!-\!\!\dashv$$

Translate. This is a "missing addend" situation.

First owner's portion	plus	Second owner's portion	plus	Third owner's portion	is	Entire business
$\frac{7}{12}$	$+$	$\frac{1}{6}$	$+$	x	$=$	1

Solve. We begin by adding the fractions on the left side of the equation.

$$\frac{7}{12} + \frac{1}{6} \cdot \frac{2}{2} + x = 1 \qquad \text{The LCD is 12.}$$

$$\frac{7}{12} + \frac{2}{12} + x = 1$$

$$\frac{9}{12} + x = 1$$

$$\frac{3}{4} + x = 1 \qquad \text{Simplifying } \frac{9}{12}$$

$$\frac{3}{4} + x - \frac{3}{4} = 1 - \frac{3}{4} \qquad \text{Subtracting } \frac{3}{4} \text{ on both sides}$$

$$x + 0 = 1 \cdot \frac{4}{4} - \frac{3}{4} \qquad \text{The LCD is 4.}$$

$$x = \frac{4}{4} - \frac{3}{4}$$

$$x = \frac{1}{4}$$

Check. We return to the original problem and add.

$$\frac{7}{12} + \frac{1}{6} + \frac{1}{4} = \frac{7}{12} + \frac{1}{6} \cdot \frac{2}{2} + \frac{1}{4} \cdot \frac{3}{3} =$$

$$\frac{7}{12} + \frac{2}{12} + \frac{3}{12} = \frac{12}{12} = 1$$

State. The third person owned $\frac{1}{4}$ of the business.

37. Familiarize. We visualize the situation. Let a = the amount of olive oil left in the bottle, in cups.

Translate. This is a "missing addend" situation.

Amount served	plus	Amount left	is	Original amount
↓	↓	↓	↓	↓
$\frac{1}{4}$	+	a	=	$\frac{11}{12}$

Solve. We subtract $\frac{1}{4}$ on both sides of the equation.

$$\frac{1}{4} + a - \frac{1}{4} = \frac{11}{12} - \frac{1}{4}$$

$$a + 0 = \frac{11}{12} - \frac{1}{4} \cdot \frac{3}{3} \qquad \text{The LCD is 12. We multiply by 1 to get the LCD.}$$

$$a = \frac{11}{12} - \frac{3}{12}$$

$$a = \frac{8}{12} = \frac{2}{3} \qquad \text{Simplifying } \frac{8}{12}$$

Check. We return to the original problem and add.

$$\frac{1}{4} + \frac{2}{3} = \frac{1}{4} \cdot \frac{3}{3} + \frac{2}{3} \cdot \frac{4}{4} = \frac{3}{12} + \frac{8}{12} = \frac{11}{12}$$

The answer checks.

State. There is $\frac{2}{3}$ cup of olive oil left in the bottle.

39. Familiarize. Let f = the number of yards of fabric needed to make the outfit.

Translate. We write an equation.

Fabric for dress	+	Fabric for band	+	Fabric for jacket	is	Total fabric
↓	↓	↓	↓	↓	↓	↓
$1\frac{3}{8}$	+	$\frac{5}{8}$	+	$3\frac{3}{8}$	=	f

Solve. We add.

$$\begin{array}{r} 1\frac{3}{8} \\ \frac{5}{8} \\ + 3\frac{3}{8} \\ \hline 4\frac{11}{8} \end{array} = 4 + \frac{11}{8}$$

$$= 4 + 1\frac{3}{8}$$

$$= 5\frac{3}{8}$$

Check. We can repeat the calculation. Also note that the answer is reasonable since it is larger than any of the individual amounts of fabric.

State. The outfit requires $5\frac{3}{8}$ yd of fabric.

41. Familiarize. We let w = the total weight of the meat.

Translate. We write an equation.

Weight of one package	plus	Weight of second package	is	Total weight
↓	↓	↓	↓	↓
$1\frac{2}{3}$	+	$5\frac{3}{4}$	=	w

Solve. We carry out the addition. The LCD is 12.

$$\begin{array}{r} 1\,\boxed{\frac{2}{3} \cdot \frac{4}{4}} = 1\frac{8}{12} \\ +5\,\boxed{\frac{3}{4} \cdot \frac{3}{3}} = +5\frac{9}{12} \\ \hline 6\frac{17}{12} = 6 + \frac{17}{12} \end{array}$$

$$= 6 + 1\frac{5}{12}$$

$$= 7\frac{5}{12}$$

Check. We repeat the calculation. We also note that the answer is larger than either of the individual weights, so the answer seems reasonable.

State. The total weight of the meat was $7\frac{5}{12}$ lb.

43. Familiarize. We draw a picture. We let D = the distance from Los Angeles at the end of the second day.

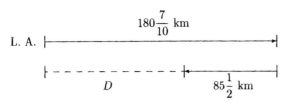

L. A.

$180\frac{7}{10}$ km

D $85\frac{1}{2}$ km

Translate. We write an equation.

Distance away from Los Angeles	minus	Distance toward Los Angeles	is	Distance from Los Angeles
↓	↓	↓	↓	↓
$180\frac{7}{10}$	$-$	$85\frac{1}{2}$	$=$	D

Solve. To solve the equation we carry out the subtraction. The LCD is 10.

$$
\begin{array}{rl}
180\ \dfrac{7}{10} & = \quad 180\ \dfrac{7}{10} \\[2mm]
-\ 85\ \boxed{\dfrac{1}{2}\cdot\dfrac{5}{5}} & = -\ 85\ \dfrac{5}{10} \\[2mm]
\hline
 & \quad 95\ \dfrac{2}{10} = 95\dfrac{1}{5}
\end{array}
$$

Check. We add the distance from Los Angeles to the distance the person drove toward Los Angeles:

$$95\frac{1}{5} + 85\frac{1}{2} = 95\frac{2}{10} + 85\frac{5}{10} = 180\frac{7}{10}$$

This checks.

State. Kim Park was $95\frac{1}{5}$ mi from Los Angeles.

45. Familiarize. We make a drawing. We let t = the number of hours the designer worked on the third day.

$\vdash\!\!-\!2\frac{1}{2}$ hr $-\!\!\vdash\!\!-\ 4\frac{1}{5}$ hr $-\!\!-\!\!\vdash\!-t-\!\dashv$

$\vdash\!\!-\!\!-\!\!-\!\!-\!\!-\ 10\frac{1}{2}$ hr $-\!\!-\!\!-\!\!-\!\!-\!\dashv$

Translate. We write an addition sentence.

$$2\frac{1}{2} + 4\frac{1}{5} + t = 10\frac{1}{2}$$

Solve. This is a two-step problem.

First we add $2\frac{1}{2} + 4\frac{1}{5}$ to find the time worked on the first two days. The LCD is 10.

$$
\begin{array}{rl}
2\ \boxed{\dfrac{1}{2}\cdot\dfrac{5}{5}} & = \quad 2\ \dfrac{5}{10} \\[2mm]
+\ 4\ \boxed{\dfrac{1}{5}\cdot\dfrac{2}{2}} & = +\ 4\ \dfrac{2}{10} \\[2mm]
\hline
 & \quad 6\ \dfrac{7}{10}
\end{array}
$$

Then we subtract $6\frac{7}{10}$ from $10\frac{1}{2}$ to find the time worked

on the third day. The LCD is 10.

$$6\frac{7}{10} + t = 10\frac{1}{2}$$

$$t = 10\frac{1}{2} - 6\frac{7}{10}$$

$$
\begin{array}{rl}
10\ \boxed{\dfrac{1}{2}\cdot\dfrac{5}{5}} = & 10\ \dfrac{5}{10} = \quad 9\ \dfrac{15}{10} \\[2mm]
-\ 6\ \dfrac{7}{10} = & -\ 6\ \dfrac{7}{10} = -\ 6\ \dfrac{7}{10} \\[2mm]
\hline
 & \qquad\qquad\qquad 3\ \dfrac{8}{10} = 3\dfrac{4}{5}
\end{array}
$$

Check. We repeat the calculations.

State. Sue worked $3\frac{4}{5}$ hr the third day.

47. Familiarize. We let m = the number of miles per gallon the car got.

Translate. We write an equation.

Total number of miles traveled	÷	Number of gallons of gas used	=	Miles per gallon
↓	↓	↓	↓	↓
213	÷	$14\frac{2}{10}$	=	m

Solve. To solve the equation we carry out the division.

$$m = 213 \div 14\frac{2}{10} = 213 \div \frac{142}{10}$$

$$= 213 \cdot \frac{10}{142} = \frac{3\cdot 71\cdot 2\cdot 5}{2\cdot 71\cdot 1}$$

$$= \frac{2\cdot 71}{2\cdot 71}\cdot\frac{3\cdot 5}{1} = 15$$

Check. We repeat the calculation.

State. The car got 15 miles per gallon of gas.

49. The length of each of the five sides is $5\frac{3}{4}$ yd. We add to find the distance around the figure.

$$5\frac{3}{4} + 5\frac{3}{4} + 5\frac{3}{4} + 5\frac{3}{4} + 5\frac{3}{4} = 25\frac{15}{4} = 25 + 3\frac{3}{4} = 28\frac{3}{4}$$

The distance is $28\frac{3}{4}$ yd.

51. Familiarize. We let t = the Fahrenheit temperature.

Translate.

Celsius temperature	times $1\frac{4}{5}$ plus 32° is					Fahrenheit temperature
↓	↓	↓	↓	↓	↓	↓
20	·	$1\frac{4}{5}$	+	32	=	t

Solve. We multiply and then add, according to the rules for order of operations.

$$t = 20 \cdot 1\frac{4}{5} + 32 = \frac{20}{1} \cdot \frac{9}{5} + 32 = \frac{20 \cdot 9}{1 \cdot 5} + 32 =$$

$$\frac{4 \cdot 5 \cdot 9}{1 \cdot 5} + 32 = \frac{5}{5} \cdot \frac{4 \cdot 9}{1} + 32 = 36 + 32 = 68$$

Check. We repeat the calculation.

State. 68° Fahrenheit corresponds to 20° Celsius.

53. We see that d and the two smallest distances combined are the same as the largest distance. We translate and solve.

$$2\frac{3}{4} + d + 2\frac{3}{4} = 12\frac{7}{8}$$

$$d = 12\frac{7}{8} - 2\frac{3}{4} - 2\frac{3}{4}$$

$$= 10\frac{1}{8} - 2\frac{3}{4} \quad \text{Subtracting } 2\frac{3}{4} \text{ from } 12\frac{7}{8}$$

$$= 7\frac{3}{8} \quad \text{Subtracting } 2\frac{3}{4} \text{ from } 10\frac{1}{8}$$

The length of d is $7\frac{3}{8}$ ft.

55. Familiarize. Let s = the number of teaspoons of sodium 10 average American women consume in one day.

Translate. A multiplication corresponds to this situation.

$$s = 10 \cdot 1\frac{1}{3}$$

Solve. We carry out the multiplication.

$$s = 10 \cdot 1\frac{1}{3} = 10 \cdot \frac{4}{3} = \frac{10 \cdot 4}{3} = \frac{40}{3} = 13\frac{1}{3}$$

Check. We repeat the calculation. The answer checks.

State. In one day 10 average American women consume $13\frac{1}{3}$ tsp of sodium.

57. Familiarize. We draw a picture.

$\frac{1}{3}$ lb	$\frac{1}{3}$ lb	$\cdots$	$\frac{1}{3}$ lb

$$\longleftarrow \quad 5\frac{1}{2} \text{ lb} \quad \longrightarrow$$

We let s = the number of servings that can be prepared from $5\frac{1}{2}$ lb of flounder fillet.

Translate. The situation corresponds to a division sentence.

$$s = 5\frac{1}{2} \div \frac{1}{3}$$

Solve. We carry out the division.

$$s = 5\frac{1}{2} \div \frac{1}{3} = \frac{11}{2} \div \frac{1}{3}$$

$$= \frac{11}{2} \cdot \frac{3}{1} = \frac{33}{2}$$

$$= 16\frac{1}{2}$$

Check. We check by multiplying. If $16\frac{1}{2}$ servings are prepared, then

$$16\frac{1}{2} \cdot \frac{1}{3} = \frac{33}{2} \cdot \frac{1}{3} = \frac{3 \cdot 11 \cdot 1}{2 \cdot 3} = \frac{3}{3} \cdot \frac{11 \cdot 1}{2} = \frac{11}{2} = 5\frac{1}{2} \text{ lb}$$

of flounder is used. Our answer checks.

State. $16\frac{1}{2}$ servings can be prepared from $5\frac{1}{2}$ lb of flounder fillet.

59. Familiarize. We let w = the weight of $5\frac{1}{2}$ cubic feet of water.

Translate. We write an equation.

Weight per cubic foot	$\cdot$	Number of cubic feet	$=$	Total weight
$\downarrow$	$\downarrow$	$\downarrow$	$\downarrow$	$\downarrow$
$62\frac{1}{2}$	$\cdot$	$5\frac{1}{2}$	$=$	w

Solve. To solve the equation we carry out the multiplication.

$$w = 62\frac{1}{2} \cdot 5\frac{1}{2}$$

$$= \frac{125}{2} \cdot \frac{11}{2} = \frac{125 \cdot 11}{2 \cdot 2}$$

$$= \frac{1375}{4} = 343\frac{3}{4}$$

Check. We repeat the calculation. We also note that $62\frac{1}{2} \approx 60$ and $5\frac{1}{2} \approx 5$. Then the product is about 300. Our answer seems reasonable.

State. The weight of $5\frac{1}{2}$ cubic feet of water is $343\frac{3}{4}$ lb.

61. Familiarize. The figure is composed of two rectangles. One has dimensions s by $\frac{1}{2} \cdot s$, or $6\frac{7}{8}$ in. by $\frac{1}{2} \cdot 6\frac{7}{8}$ in. The other has dimensions $\frac{1}{2} \cdot s$ by $\frac{1}{2} \cdot s$, or $\frac{1}{2} \cdot 6\frac{7}{8}$ in. by $\frac{1}{2} \cdot 6\frac{7}{8}$ in. The total area is the sum of the areas of these two rectangles. We let A = the total area.

Translate. We write an equation.

$$A = \left(6\frac{7}{8}\right) \cdot \left(\frac{1}{2} \cdot 6\frac{7}{8}\right) + \left(\frac{1}{2} \cdot 6\frac{7}{8}\right) \cdot \left(\frac{1}{2} \cdot 6\frac{7}{8}\right)$$

Solve. We carry out each multiplication and then add.

$$A = \left(6\frac{7}{8}\right) \cdot \left(\frac{1}{2} \cdot 6\frac{7}{8}\right) + \left(\frac{1}{2} \cdot 6\frac{7}{8}\right) \cdot \left(\frac{1}{2} \cdot 6\frac{7}{8}\right)$$

$$= \frac{55}{8} \cdot \left(\frac{1}{2} \cdot \frac{55}{8}\right) + \left(\frac{1}{2} \cdot \frac{55}{8}\right) \cdot \left(\frac{1}{2} \cdot \frac{55}{8}\right)$$

$$= \frac{55}{8} \cdot \frac{55}{16} + \frac{55}{16} \cdot \frac{55}{16}$$

$$= \frac{3025}{128} + \frac{3025}{256} = \frac{3025}{128} \cdot \frac{2}{2} + \frac{3025}{256}$$

$$= \frac{6050}{256} + \frac{3025}{256} = \frac{9075}{256}$$

$$= 35\frac{115}{256}$$

Check. We repeat the calculation.

State. The area is $35\frac{115}{256}$ sq in.

63. *Familiarize*. We make a drawing.

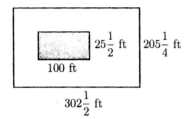

Translate. We let A = the area of the lot not covered by the building.

$$\underbrace{\text{Area left over}}_{} \;\;\underset{\text{is}}{\downarrow}\;\; \underbrace{\text{Area of lot}}_{} \;\;\underset{\text{minus}}{\downarrow}\;\; \underbrace{\text{Area of building}}_{}$$

$$A \;\;=\;\; \left(302\frac{1}{2}\right)\cdot\left(205\frac{1}{4}\right) \;\;-\;\; (100)\cdot\left(25\frac{1}{2}\right)$$

Solve. We do each multiplication and then find the difference.

$$A = \left(302\frac{1}{2}\right)\cdot\left(205\frac{1}{4}\right) - (100)\cdot\left(25\frac{1}{2}\right)$$

$$= \frac{605}{2}\cdot\frac{821}{4} - \frac{100}{1}\cdot\frac{51}{2}$$

$$= \frac{605\cdot821}{2\cdot4} - \frac{100\cdot51}{1\cdot2}$$

$$= \frac{605\cdot821}{2\cdot4} - \frac{2\cdot50\cdot51}{1\cdot2} = \frac{605\cdot821}{2\cdot4} - \frac{2}{2}\cdot\frac{50\cdot51}{1}$$

$$= \frac{496,705}{8} - 2550 = 62,088\frac{1}{8} - 2550$$

$$= 59,538\frac{1}{8}$$

Check. We repeat the calculation.

State. The area left over is $59,538\frac{1}{8}$ sq ft.

65. Discussion and Writing Exercise

67. *Familiarize*. Let a = the amount by which the annual earnings of a person with a bachelor's degree exceed the earnings of a person with no high school education. We visualize the situation.

No high school $16,053	Excess earnings a
Bachelor's degree $43,782	

Translate. We consider this to be a "missing addend" situation.

$$\underbrace{\text{Earnings with no high school}}_{} \;\;\underset{\text{plus}}{\downarrow}\;\; \underbrace{\text{Excess earnings}}_{} \;\;\underset{\text{is}}{\downarrow}\;\; \underbrace{\text{Bachelor's degree earnings}}_{}$$

$$16,053 \;\;+\;\; a \;\;=\;\; 43,782$$

Solve. We subtract 16,053 on both sides of the equation.

$$16,053 + a = 43,782$$
$$16,053 + a - 16,053 = 43,782 - 16,053$$
$$a = 27,729$$

Check. We can add the difference, 27,729, to the subtrahend, 16,053: $16,053 + 27,729 = 43,782$. The answer checks.

State. The annual earnings of a person with a bachelor's degree are $27,729 more than the annual earnings of a person with no high school education.

69. *Familiarize*. We make a drawing. We let A = the area.

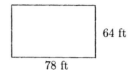

Translate. Using the formula for area, we have

$$A = l \cdot w = 78 \cdot 64.$$

Using the formula for perimeter, we have

$$P = 2l + 2w = 2\cdot78 + 2\cdot64$$

Solve. We carry out the computations.

$$\begin{array}{r} 7\,8 \\ \times\ 6\,4 \\ \hline 3\,1\,2 \\ 4\,6\,8\,0 \\ \hline 4\,9\,9\,2 \end{array}$$

Thus, $A = 4992$.

$$P = 2\cdot78 + 2\cdot64 = 156 + 128 = 284$$

Check. We repeat the calculations. The answers check.

State. The area is 4992 ft². The perimeter is 284 ft.

71.
$$\begin{array}{r} \overset{8\ \ 9\ \ 9\ \ 11}{\cancel{9\,0\,0\,1}} \\ -\ 6\,7\,9\,8 \\ \hline 2\,2\,0\,3 \end{array}$$

73.
$$\begin{array}{r} 2\,0\,4 \\ 3\,5\,\overline{\smash{)}\,7\,1\,4\,0} \\ 7\,0\,0\,0 \\ \hline 1\,4\,0 \\ 1\,4\,0 \\ \hline 0 \end{array}$$

The answer is 204.

75.
$$30\cdot x = 150$$
$$\frac{30\cdot x}{30} = \frac{150}{30} \quad \text{Dividing by 30 on both sides}$$
$$x = 5$$

The solution is 5.

77.
$$5280 = 1760 + t$$
$$5280 - 1760 = 1760 + t - 1760 \quad \text{Subtracting 1760}$$
$$\hspace{6cm}\text{on both sides}$$
$$3520 = t$$

The solution is 3520.

79.

$$8 \cdot 12 - (63 \div 9 + 13 \cdot 3)$$
$$= 8 \cdot 12 - (7 + 13 \cdot 3) \quad \text{Dividing inside the parentheses}$$
$$= 8 \cdot 12 - (7 + 39) \quad \text{Multiplying inside the parentheses}$$
$$\doteq 8 \cdot 12 - 46 \quad \text{Adding inside the parentheses}$$
$$= 96 - 46 \quad \text{Multiplying}$$
$$= 50 \quad \text{Subtracting}$$

81. *Familiarize.* First we find the fractional part of the band's pay that the guitarist received. We let $f =$ this fraction.

Translate. We translate to an equation.

One-third	of	one-half	plus	one-fifth	of	one-half	is	fractional part
$\frac{1}{3}$	$\cdot$	$\frac{1}{2}$	$+$	$\frac{1}{5}$	$\cdot$	$\frac{1}{2}$	$=$	f

Solve. We carry out the calculation.

$$\frac{1}{3} \cdot \frac{1}{2} + \frac{1}{5} \cdot \frac{1}{2} = f$$

$$\frac{1}{6} + \frac{1}{10} = f \quad \text{LCD is 30.}$$

$$\frac{1}{6} \cdot \frac{5}{5} + \frac{1}{10} \cdot \frac{3}{3} = f$$

$$\frac{5}{30} + \frac{3}{30} = f$$

$$\frac{8}{30} = f$$

$$\frac{4}{15} = f$$

Now we find how much of the $1200 received by the band was paid to the guitarist. We let $p =$ the amount.

Four-fifteenths	of	$1200	=	guitarist's pay
$\frac{4}{15}$	$\cdot$	1200	=	p

We solve the equation.

$$\frac{4}{15} \cdot 1200 = p$$

$$\frac{4 \cdot 1200}{15} = p$$

$$\frac{4 \cdot 3 \cdot 5 \cdot 80}{3 \cdot 5} = p$$

$$320 = p$$

Check. We repeat the calculations.

State. The guitarist received $\frac{4}{15}$ of the band's pay. This was $320.

83. *Familiarize.* We are told that $\frac{2}{3}$ of $\frac{7}{8}$ of the students are high school graduates who are older than 20, and $\frac{1}{7}$ of this fraction are left-handed. Thus, we want to find $\frac{1}{7}$ of $\frac{2}{3}$ of $\frac{7}{8}$. We let f represent this fraction.

Translate. The multiplication sentence

$$f = \frac{1}{7} \cdot \frac{2}{3} \cdot \frac{7}{8}$$

corresponds to this situation.

Solve. We multiply and simplify.

$$f = \frac{1}{7} \cdot \frac{2}{3} \cdot \frac{7}{8} = \frac{1 \cdot 2}{7 \cdot 3} \cdot \frac{7}{8} = \frac{1 \cdot 2 \cdot 7}{7 \cdot 3 \cdot 8} = \frac{1 \cdot 2 \cdot 7}{7 \cdot 3 \cdot 2 \cdot 4} =$$
$$\frac{2 \cdot 7}{2 \cdot 7} \cdot \frac{1}{3 \cdot 4} = \frac{1}{3 \cdot 4} = \frac{1}{12}$$

Check. We repeat the calculation. The result checks.

State. $\frac{1}{12}$ of the students are left-handed high school graduates over the age of 20.

Exercise Set 2.6

1. $\frac{1}{2} \cdot \frac{1}{3} \cdot \frac{1}{4}$

$$= \frac{1}{6} \cdot \frac{1}{4} \quad \text{Doing the multiplications in}$$

$$= \frac{1}{24} \quad \text{order from left to right}$$

3. $6 \div 3 \div 5$

$$= 2 \div 5 \quad \text{Doing the divisions in}$$

$$= \frac{2}{5} \quad \text{order from left to right}$$

5. $\frac{2}{3} \div \frac{4}{3} \div \frac{7}{8}$

$$= \frac{2}{3} \cdot \frac{3}{4} \div \frac{7}{8} \quad \text{Doing the first division; multiplying by the reciprocal of } \frac{4}{3}$$

$$= \frac{2 \cdot 3 \cdot 1}{3 \cdot 2 \cdot 2} \div \frac{7}{8}$$

$$= \frac{1}{2} \div \frac{7}{8} \quad \text{Removing a factor of 1}$$

$$= \frac{1}{2} \cdot \frac{8}{7} \quad \text{Dividing; multiplying by the reciprocal of } \frac{7}{8}$$

$$= \frac{1 \cdot 2 \cdot 4}{2 \cdot 7}$$

$$= \frac{4}{7} \quad \text{Removing a factor of 1}$$

7. $\dfrac{5}{8} \div \dfrac{1}{4} - \dfrac{2}{3} \cdot \dfrac{4}{5}$

$= \dfrac{5}{8} \cdot \dfrac{4}{1} - \dfrac{2}{3} \cdot \dfrac{4}{5}$ Dividing

$= \dfrac{5 \cdot 4}{2 \cdot 4 \cdot 1} - \dfrac{2}{3} \cdot \dfrac{4}{5}$

$= \dfrac{5}{2} - \dfrac{2}{3} \cdot \dfrac{4}{5}$ Removing a factor of 1

$= \dfrac{5}{2} - \dfrac{2 \cdot 4}{3 \cdot 5}$ Multiplying

$= \dfrac{5}{2} - \dfrac{8}{15}$

$= \dfrac{5}{2} \cdot \dfrac{15}{15} - \dfrac{8}{15} \cdot \dfrac{2}{2}$ Multiplying by 1 to obtain the LCD

$= \dfrac{75}{30} - \dfrac{16}{30}$

$= \dfrac{59}{30}$, or $1\dfrac{29}{30}$ Subtracting

9. $\dfrac{3}{4} - \dfrac{2}{3} \cdot \left(\dfrac{1}{2} + \dfrac{2}{5} \right)$

$= \dfrac{3}{4} - \dfrac{2}{3} \cdot \left(\dfrac{5}{10} + \dfrac{4}{10} \right)$ Adding inside

$= \dfrac{3}{4} - \dfrac{2}{3} \cdot \dfrac{9}{10}$ the parentheses

$= \dfrac{3}{4} - \dfrac{2 \cdot 9}{3 \cdot 10}$ Multiplying

$= \dfrac{3}{4} - \dfrac{2 \cdot 3 \cdot 3}{3 \cdot 2 \cdot 5}$

$= \dfrac{3}{4} - \dfrac{2 \cdot 3}{2 \cdot 3} \cdot \dfrac{3}{5}$

$= \dfrac{3}{4} - \dfrac{3}{5}$

$= \dfrac{15}{20} - \dfrac{12}{20}$

$= \dfrac{3}{20}$ Subtracting

11. $28\dfrac{1}{8} - 5\dfrac{1}{4} + 3\dfrac{1}{2}$

$= 28\dfrac{1}{8} - 5\dfrac{2}{8} + 3\dfrac{1}{2}$ Doing the additions and

$= 27\dfrac{9}{8} - 5\dfrac{2}{8} + 3\dfrac{1}{2}$ subtractions in order

$= 22\dfrac{7}{8} + 3\dfrac{1}{2}$ from left to right

$= 22\dfrac{7}{8} + 3\dfrac{4}{8}$

$= 25\dfrac{11}{8}$

$= 26\dfrac{3}{8}$, or $\dfrac{211}{8}$

13. $\dfrac{7}{8} \div \dfrac{1}{2} \cdot \dfrac{1}{4}$

$= \dfrac{7}{8} \cdot \dfrac{2}{1} \cdot \dfrac{1}{4}$ Dividing

$= \dfrac{7 \cdot 2}{2 \cdot 4 \cdot 1} \cdot \dfrac{1}{4}$

$= \dfrac{7}{4} \cdot \dfrac{1}{4}$ Removing a factor of 1

$= \dfrac{7}{16}$ Multiplying

15. $\left(\dfrac{2}{3} \right)^2 - \dfrac{1}{3} \cdot 1\dfrac{1}{4}$

$= \dfrac{4}{9} - \dfrac{1}{3} \cdot 1\dfrac{1}{4}$ Evaluating the exponental expression

$= \dfrac{4}{9} - \dfrac{1}{3} \cdot \dfrac{5}{4}$

$= \dfrac{4}{9} - \dfrac{5}{12}$ Multiplying

$= \dfrac{4}{9} \cdot \dfrac{4}{4} - \dfrac{5}{12} \cdot \dfrac{3}{3}$

$= \dfrac{16}{36} - \dfrac{15}{36}$

$= \dfrac{1}{36}$ Subtracting

17. $\dfrac{1}{2} - \left(\dfrac{1}{2} \right)^2 + \left(\dfrac{1}{2} \right)^3$

$= \dfrac{1}{2} - \dfrac{1}{4} + \dfrac{1}{8}$ Evaluating the exponental expressions

$= \dfrac{2}{4} - \dfrac{1}{4} + \dfrac{1}{8}$ Doing the additions and

$= \dfrac{1}{4} + \dfrac{1}{8}$ subtractions in order

$= \dfrac{2}{8} + \dfrac{1}{8}$ from left to right

$= \dfrac{3}{8}$

19. Add the numbers and divide by the number of addends.

$\left(\dfrac{2}{3} + \dfrac{7}{8} \right) \div 2$

$= \left(\dfrac{16}{24} + \dfrac{21}{24} \right) \div 2$ Doing the operations inside parentheses first; finding the LCD

$= \dfrac{37}{24} \div 2$ Adding

$= \dfrac{37}{24} \cdot \dfrac{1}{2}$ Dividing

$= \dfrac{37}{48}$

21. Add the numbers and divide by the number of addends.

$$\left(\frac{1}{6} + \frac{1}{8} + \frac{3}{4}\right) \div 3$$

$$= \left(\frac{4}{24} + \frac{3}{24} + \frac{18}{24}\right) \div 3$$

$$= \frac{25}{24} \div 3$$

$$= \frac{25}{24} \cdot \frac{1}{3}$$

$$= \frac{25}{72}$$

23. Add the numbers and divide by the number of addends.

$$\left(3\frac{1}{2} + 9\frac{3}{8}\right) \div 2$$

$$= \left(3\frac{4}{8} + 9\frac{3}{8}\right) \div 2$$

$$= 12\frac{7}{8} \div 2$$

$$= \frac{103}{8} \div 2$$

$$= \frac{103}{8} \cdot \frac{1}{2}$$

$$= \frac{103}{16}, \text{ or } 6\frac{7}{16}$$

25. *Familiarize.* To compute an average, we add the values and then divide the sum by the number of values. Let $w =$ the average birthweight, in pounds.

Translate. We have
$$w = \frac{2\frac{9}{16} + 2\frac{9}{32} + 2\frac{1}{8} + 2\frac{5}{16}}{4}.$$

Solve. First we add.
$$2\frac{9}{16} + 2\frac{9}{32} + 2\frac{1}{8} + 2\frac{5}{16}$$
$$= 2\frac{18}{32} + 2\frac{9}{32} + 2\frac{4}{32} + 2\frac{10}{32}$$
$$= 8\frac{41}{32} = 9\frac{9}{32}$$

Then we divide:
$$9\frac{9}{32} \div 4 = \frac{297}{32} \div 4$$
$$= \frac{297}{32} \cdot \frac{1}{4}$$
$$= \frac{297}{128} = 2\frac{41}{128}$$

Check. As a partial check we note that the average weight is larger than the smallest weight and smaller than the largest weight. We could also repeat the calculations.

State. The average birthweight of the quadruplets was $2\frac{41}{128}$ lb.

27. *Familiarize.* We will add the values and then divide the sum by the number of values. Let $d =$ the number of days the bulbs burned, on average.

Translate. We have
$$d = \frac{17 + 16\frac{1}{2} + 20 + 18\frac{1}{2} + 21}{5}.$$

Solve. First we add.
$$17 + 16\frac{1}{2} + 20 + 18\frac{1}{2} + 21$$
$$= 92\frac{2}{2} = 93$$

Then we divide.
$$\frac{93}{5} = 18\frac{3}{5}$$

Check. As a partial check we note that the average is larger than the smallest value and smaller than the largest value. We could also repeat the calculations.

State. On average, the bulbs burned $18\frac{3}{5}$ days.

29. $\left(\frac{2}{3} + \frac{3}{4}\right) \div \left(\frac{5}{6} - \frac{1}{3}\right)$

$$= \left(\frac{8}{12} + \frac{9}{12}\right) \div \left(\frac{5}{6} - \frac{2}{6}\right) \quad \begin{array}{l}\text{Doing the operations} \\ \text{inside parentheses}\end{array}$$

$$= \frac{17}{12} \div \frac{3}{6}$$

$$= \frac{17}{12} \cdot \frac{6}{3}$$

$$= \frac{17 \cdot \cancel{6}}{2 \cdot \cancel{6} \cdot 3}$$

$$= \frac{17}{6}, \text{ or } 2\frac{5}{6}$$

31. $\left(\dfrac{1}{2} + \dfrac{1}{3}\right)^2 \cdot 144 - \dfrac{5}{8} \div 10\dfrac{1}{2}$

$= \left(\dfrac{3}{6} + \dfrac{2}{6}\right)^2 \cdot 144 - \dfrac{5}{8} \div 10\dfrac{1}{2}$

$= \left(\dfrac{5}{6}\right)^2 \cdot 144 - \dfrac{5}{8} \div 10\dfrac{1}{2}$

$= \dfrac{25}{36} \cdot 144 - \dfrac{5}{8} \div 10\dfrac{1}{2}$

$= \dfrac{25 \cdot 36 \cdot 4}{36 \cdot 1} - \dfrac{5}{8} \div 10\dfrac{1}{2}$

$= 100 - \dfrac{5}{8} \div 10\dfrac{1}{2}$

$= 100 - \dfrac{5}{8} \div \dfrac{21}{2}$

$= 100 - \dfrac{5}{8} \cdot \dfrac{2}{21}$

$= 100 - \dfrac{5 \cdot 2}{2 \cdot 4 \cdot 21}$

$= 100 - \dfrac{5}{84}$

$= 99\dfrac{79}{84}, \text{ or } \dfrac{8395}{84}$

33. $\dfrac{2}{47}$

Because 2 is very small compared to 47, $\dfrac{2}{47} \approx 0$.

35. $\dfrac{1}{13}$

Because 1 is very small compared to 13, $\dfrac{1}{13} \approx 0$.

37. $\dfrac{6}{11}$

Because $2 \cdot 6 = 12$ and 12 is close to 11, the denominator is about twice the numerator. Thus, $\dfrac{6}{11} \approx \dfrac{1}{2}$.

39. $\dfrac{7}{15}$

Because $2 \cdot 7 = 14$ and 14 is close to 15, the denominator is about twice the numerator. Thus, $\dfrac{7}{15} \approx \dfrac{1}{2}$.

41. $\dfrac{7}{100}$

Because 7 is very small compared to 100, $\dfrac{7}{100} \approx 0$.

43. $\dfrac{19}{20}$

Because 19 is very close to 20, $\dfrac{19}{20} \approx 1$.

45. $\dfrac{\square}{11}$

A fraction is close to $\dfrac{1}{2}$ when the denominator is about

twice the numerator. Since $2 \cdot 5 = 10$ and $2 \cdot 6 = 12$ and both 10 and 12 are close to 11, we know that $\dfrac{5}{11}$ and $\dfrac{6}{11}$ are both close to $\dfrac{1}{2}$. We also want a fraction that is greater than $\dfrac{1}{2}$, so we choose 6 for the numerator and obtain $\dfrac{6}{11}$. Answers may vary.

47. $\dfrac{\square}{23}$

A fraction is close to $\dfrac{1}{2}$ when the denominator is about twice the numerator. Since $2 \cdot 11 = 22$ and $2 \cdot 12 = 24$, and both 22 and 24 are close to 23, we know that $\dfrac{11}{23}$ and $\dfrac{12}{23}$ are both close to $\dfrac{1}{2}$. We also want a fraction that is greater than $\dfrac{1}{2}$, so we choose 12 for the numerator and obtain $\dfrac{12}{23}$. Answers may vary.

49. $\dfrac{10}{\square}$

A fraction is close to $\dfrac{1}{2}$ when the denominator is about twice the numerator. Since $2 \cdot 10 = 20$, we know a number close to 20 will yield a fraction close to $\dfrac{1}{2}$. We also want a fraction that is greater than $\dfrac{1}{2}$. We can choose 19 for the denominator and obtain $\dfrac{10}{19}$. Answers may vary.

51. $\dfrac{7}{\square}$

If the denominator were 7, the fraction would be equivalent to 1. Then a denominator of 6 will make the fraction close to but greater than 1. Answers may vary.

53. $\dfrac{13}{\square}$

If the denominator were 13, the fraction would be equivalent to 1. Then a denominator of 12 will make the fraction close to but greater than 1. Answers may vary.

55. $\dfrac{\square}{15}$

If the numerator were 15, the fraction would be equivalent to 1. Then a numerator of 16 will make the fraction close to but greater than 1. Answers may vary.

57. $2\dfrac{7}{8}$

Since $\dfrac{7}{8} \approx 1$, we have $2\dfrac{7}{8} = 2 + \dfrac{7}{8} \approx 2 + 1$, or 3.

59. $12\dfrac{5}{6}$

Since $\dfrac{5}{6} \approx 1$, we have $12\dfrac{5}{6} = 12 + \dfrac{5}{6} \approx 12 + 1$, or 13.

61. $\dfrac{4}{5} + \dfrac{7}{8} \approx 1 + 1 = 2$

63. $\dfrac{2}{3} + \dfrac{7}{13} + \dfrac{5}{9} \approx 1 + \dfrac{1}{2} + \dfrac{1}{2} = 2$

65. $\frac{43}{100} + \frac{1}{10} - \frac{11}{1000} \approx \frac{1}{2} + 0 - 0 = \frac{1}{2}$

67. $7\frac{29}{80} + 10\frac{12}{13} \cdot 24\frac{2}{17} \approx 7\frac{1}{2} + 11 \cdot 24 =$

$7\frac{1}{2} + 264 = 271\frac{1}{2}$

69. $24 \div 7\frac{8}{9} \approx 24 \div 8 = 3$

71. $76\frac{3}{14} + 23\frac{19}{20} \approx 76 + 24 = 100$

73. $16\frac{1}{5} \div 2\frac{1}{11} + 25\frac{9}{10} - 4\frac{11}{23} \approx$

$16 \div 2 + 26 - 4\frac{1}{2} = 8 + 26 - 4\frac{1}{2} =$

$34 - 4\frac{1}{2} = 29\frac{1}{2}$

75. Discussion and Writing Exercise

77.
```
      1 2 6
  ×     2 7
      8 8 2
    2 5 2 0
    3 4 0 2
```

79.
```
              5 9
    1 3 2 ) 7 8 6 5
            6 6 0 0
            1 2 6 5
            1 1 8 8
                7 7
```

The answer is 59 R 77.

81. $\frac{3}{2} \cdot 522 = \frac{3 \cdot 522}{2} = \frac{3 \cdot 2 \cdot 261}{2 \cdot 1} = \frac{2}{2} \cdot \frac{3 \cdot 261}{1} = 783$

83. $\frac{3}{10} \div \frac{4}{5} = \frac{3}{10} \cdot \frac{5}{4} = \frac{3 \cdot 5}{10 \cdot 4} = \frac{3 \cdot 5}{2 \cdot 5 \cdot 4} = \frac{5}{5} \cdot \frac{3}{2 \cdot 4} = \frac{3}{8}$

85. *Familiarize*. We make a drawing.

$\underbrace{\bigcirc \; \bigcirc \; \bigcirc \; \cdots \; \bigcirc}_{\frac{3}{8} \text{ lb per person}}$ } 6 lb feeds how many people?

We let $p =$ the number of people who can attend the luncheon.

Translate. The problem translates to the following equation:

$$p = 6 \div \frac{3}{8}$$

Solve. We carry out the division.

$p = 6 \div \frac{3}{8}$

$= 6 \cdot \frac{3}{8} = \frac{6 \cdot 8}{3}$

$= \frac{2 \cdot 3 \cdot 8}{3 \cdot 1} = \frac{3}{3} \cdot \frac{2 \cdot 8}{1}$

$= 16$

Check. If each of 16 people is allotted $\frac{3}{8}$ lb of cold cuts, a total of

$$16 \cdot \frac{3}{8} = \frac{16 \cdot 3}{8} = \frac{2 \cdot 8 \cdot 3}{8} = 2 \cdot 3,$$

or 6 lb of cold cuts are used. Our answer checks.

State. 16 people can attend the luncheon.

87. a) The area of the larger rectangle is $13 \cdot 9\frac{1}{4}$, and the area of the smaller rectangle is $8\frac{1}{4} \cdot 7\frac{1}{4}$. We express the sum of their areas as $13 \cdot 9\frac{1}{4} + 8\frac{1}{4} \cdot 7\frac{1}{4}$.

b) $13 \cdot 9\frac{1}{4} + 8\frac{1}{4} \cdot 7\frac{1}{4}$

$= 13 \cdot \frac{37}{4} + \frac{33}{4} \cdot \frac{29}{4}$

$= \frac{481}{4} + \frac{957}{16}$

$= \frac{1924}{16} + \frac{957}{16}$

$= \frac{2881}{16}$, or $180\frac{1}{16}$ in^2

c) In order to simplify the expression, we must multiply before adding.

89.
$$\frac{a}{17} + \frac{1b}{23} = \frac{35a}{391}$$

$$\frac{a}{17} \cdot \frac{23}{23} + \frac{1b}{23} \cdot \frac{17}{17} = \frac{35a}{391}$$

$$\frac{a \cdot 23 + 1b \cdot 17}{391} = \frac{35a}{391}$$

Equating numerators, we have $a \cdot 23 + 1b \cdot 17 = 35a$. Try $a = 1$. We have:

$1 \cdot 23 + 1b \cdot 17 = 351$

$23 + 1b \cdot 17 = 351$

$1b \cdot 17 = 351 - 23$

$1b \cdot 17 = 328$

$1b = \frac{328}{17} = 19\frac{5}{17}$

Since $328/17$ is not a whole number, $a \neq 1$. Try $a = 2$. We have:

$2 \cdot 23 + 1b \cdot 17 = 352$

$46 + 1b \cdot 17 = 352$

$1b \cdot 17 = 352 - 46$

$1b \cdot 17 = 306$

$1b = \frac{306}{17}$

$1b = 18$

Thus, $a = 2$ and $b = 8$.

91. The largest sum will occur when the largest numbers, 4 and 5, are used for the numerators. Since $\frac{4}{3} + \frac{5}{2} > \frac{4}{2} + \frac{5}{3}$, the largest possible sum is $\frac{4}{3} + \frac{5}{2} = \frac{23}{6}$.

Chapter 3

Decimal Notation

Exercise Set 3.1

1. 249.94

  a) Write a word name for the whole number. | Two hundred forty-nine |

  b) Write "and" for the decimal point.

    Two hundred forty-nine

    | and |

  c) Write a word name for the number to the right of the decimal point, followed by the place value of the last digit.

    Two hundred forty-nine

    and

    | ninety-four hundredths |

A word name for 249.94 is two hundred forty-nine and ninety-four hundredths.

3. 96.4375

  a) Write a word name for the whole number. | Ninety-six |

  b) Write "and" for the decimal point.

    Ninety-six

    | and |

  c) Write a word name for the number to the right of the decimal point, followed by the place value of the last digit.

    Ninety-six

    and

    | four thousand three hundred seventy-five ten-thousandths |

A word name for 96.4375 is ninety-six and four thousand three hundred seventy-five ten-thousandths.

5.

  Thirty-four

  and

  eight hundred ninety-one thousandths

     34 . 891

7. Write "and 48 cents" as "and $\frac{48}{100}$ dollars." A word name for \$326.48 is three hundred twenty-six and $\frac{48}{100}$ dollars.

9. Write "and 72 cents" as "and $\frac{72}{100}$ dollars." A word name for \$36.72 is thirty-six and $\frac{72}{100}$ dollars.

11. 8.<u>3</u>   8.3. $\frac{83}{10}$

  1 place Move 1 place. 1 zero

  $8.3 = \frac{83}{10}$

13. 3.<u>56</u>   3.56. $\frac{356}{100}$

  2 places Move 2 places. 2 zeros

  $3.56 = \frac{356}{100}$

15. 46.<u>03</u>   46.03. $\frac{4603}{100}$

  2 places Move 2 places. 2 zeros

  $46.03 = \frac{4603}{100}$

17. 0.<u>00013</u>   0.00013. $\frac{13}{100,000}$

  5 places Move 5 places. 5 zeros

  $0.00013 = \frac{13}{100,000}$

19. 1.<u>0008</u>   1.0008. $\frac{10,008}{10,000}$

  4 places Move 4 places. 4 zeros

  $1.0008 = \frac{10,008}{10,000}$

21. 20.<u>003</u>   20.003. $\frac{20,003}{1000}$

  3 places Move 3 places. 3 zeros

  $20.003 = \frac{20,003}{1000}$

23. $\frac{8}{1\underline{0}}$ 0.8.

  1 zero Move 1 place.

  $\frac{8}{10} = 0.8$

25. $\frac{889}{1\underline{00}}$ 8.89.

  2 zeros Move 2 places.

  $\frac{889}{100} = 8.89$

27. $\dfrac{3798}{1000}$ $3\,\underset{\curvearrowright}{.\,798.}$

3 zeros   Move 3 places.

$\dfrac{3798}{1000} = 3.798$

29. $\dfrac{78}{10,000}$ $0.\underset{\curvearrowright}{0078.}$

4 zeros   Move 4 places.

$\dfrac{78}{10,000} = 0.0078$

31. $\dfrac{19}{100,000}$ $0.\underset{\curvearrowright}{00019.}$

5 zeros   Move 5 places.

$\dfrac{19}{100,000} = 0.00019$

33. $\dfrac{376,193}{1,000,000}$ $0.\underset{\curvearrowright}{376193.}$

6 zeros   Move 6 places.

$\dfrac{376,193}{1,000,000} = 0.376193$

35. $99\dfrac{44}{100} = 99 + \dfrac{44}{100} = 99 \text{ and } \dfrac{44}{100} = 99.44$

37. $3\dfrac{798}{1000} = 3 + \dfrac{798}{1000} = 3 \text{ and } \dfrac{798}{1000} = 3.798$

39. $2\dfrac{1739}{10,000} = 2 + \dfrac{1739}{10,000} = 2 \text{ and } \dfrac{1739}{10,000} = 2.1739$

41. $8\dfrac{953,073}{1,000,000} = 8 + \dfrac{953,073}{1,000,000} =$
$8 \text{ and } \dfrac{953,073}{1,000,000} = 8.953073$

43. To compare two numbers in decimal notation, start at the left and compare corresponding digits moving from left to right. When two digits differ, the number with the larger digit is the larger of the two numbers.

0.06
↑ Different; 5 is larger than 0.
0.58

Thus, 0.58 is larger.

45. 0.905
↑ Starting at the left, these digits are the first to differ; 1 is larger than 0.
0.91

Thus, 0.91 is larger.

47. 0.0009
↑ Starting at the left, these digits are the first to differ, and 1 is larger than 0.
0.001

Thus, 0.001 is larger.

49. 234.07
↑ Starting at the left, these digits are the first to differ, and 5 is larger than 4.
235.07

Thus, 235.07 is larger.

51. $\dfrac{4}{100} = 0.04$ so we compare 0.004 and 0.04.

0.004
↑ Starting at the left, these digits are the first to differ, and 4 is larger than 0.
0.04

Thus, 0.04 or $\dfrac{4}{100}$ is larger.

53. 0.4320
↑ Starting at the left, these digits are the first to differ, and 5 is larger than 0.
0.4325

Thus, 0.4325 is larger.

55.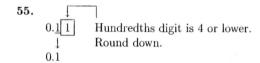
0.1⌐1⌐   Hundredths digit is 4 or lower.
↓   Round down.
0.1

57.
0.4⌐9⌐   Hundredths digit is 5 or higher.
↓   Round up.
0.5

59.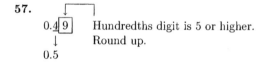
2.7⌐4⌐49   Hundredths digit is 4 or lower.
↓   Round down.
2.7

61.
123.6⌐5⌐   Hundredths digit is 5 or higher.
↓   Round up.
123.7

63.
0.89⌐3⌐   Thousandths digit is 4 or lower.
↓   Round down.
0.89

65.
0.66⌐6⌐6   Thousandths digit is 5 or higher.
↓   Round up.
0.67

67.
0.99⌐5⌐   Thousandths digit is 5 or higher.
↓   Round up.
1.00

(When we make the hundredths digit a 10, we carry 1 to the tenths place. This then requires us to carry 1 to the ones place.)

69.

0.09⟦4⟧   Thousandths digit is 4 or lower.
      Round down.
0.09

71.

0.324⟦6⟧   Ten-thousandths digit is 5 or higher.
      Round up.
0.325

73.

17.001⟦5⟧   Ten-thousandths digit is 5 or higher.
      Round up.
17.002

75.

10.101⟦1⟧   Ten-thousandths digit is 4 or lower.
      Round down.
10.101

77.

9.998⟦9⟧   Ten-thousandths digit is 5 or higher.
      Round up.
9.999

79.

8⟦0⟧9.4732   Tens digit is 4 or lower.
      Round down.
800

81.

809.473⟦2⟧   Ten-thousandths digit is 4 or lower.
      Round down.
809.473

83.

809.⟦4⟧732   Tenths digit is 4 or lower.
      Round down.
809

85.

34.5438⟦9⟧   Hundred-thousandths digit is 5 or higher.
      Round up.
34.5439

87.

34.54⟦3⟧89   Thousandths digit is 4 or lower.
      Round down.
34.54

89.

34.⟦5⟧4389   Tenths digit is 5 or higher.
      Round up.
35

91. Discussion and Writing Exercise

93. Round 617⟦2⟧ to the nearest ten.

The digit 7 is in the tens place. Since the next digit to the right, 2, is 4 or lower, round down, meaning that 7 tens stays as 7 tens. Then change the digit to the right of the tens digit to zero.

The answer is 6170.

95. Round 6⟦1⟧72 to the nearest thousand.

The digit 6 is in the thousands place. Since the next digit to the right, 1, is 4 or lower, round down, meaning that 6 thousands stays as 6 thousands. Then change all digits to the right of the thousands digit to zeros.

The answer is 6000.

97.
$$24 = 23\frac{5}{5}$$
$$-17\frac{2}{5} = -17\frac{2}{5}$$
$$6\frac{3}{5}$$

99. We use a string of successive divisions.

$$
\begin{array}{r}
1\,7 \\
5\,\overline{)\,8\,5} \\
3\,\overline{)\,2\,5\,5} \\
3\,\overline{)\,7\,6\,5} \\
2\,\overline{)\,1\,5\,3\,0}
\end{array}
$$

$1530 = 2 \cdot 3 \cdot 3 \cdot 5 \cdot 17$, or $2 \cdot 3^2 \cdot 5 \cdot 17$

101. We use a string of successive divisions.

$$
\begin{array}{r}
1\,1 \\
7\,\overline{)\,7\,7} \\
7\,\overline{)\,5\,3\,9} \\
2\,\overline{)\,1\,0\,7\,8} \\
2\,\overline{)\,2\,1\,5\,6} \\
2\,\overline{)\,4\,3\,1\,2}
\end{array}
$$

$4312 = 2 \cdot 2 \cdot 2 \cdot 7 \cdot 7 \cdot 11$, or $2^3 \cdot 7^2 \cdot 11$

103. The greatest number of decimals places occurring in any of the numbers is 6, so we add extra zeros to the first six numbers so that each number has 6 decimal places. Then we start at the left and compare corresponding digits, moving from left to right. The numbers, given from smallest to largest are 2.000001, 2.0119, 2.018, 2.0302, 2.1, 2.108, 2.109.

105. 6.78346⟦1902⟧ ←Drop all decimal places
          past the fifth place.
6.78346

107. 0.03030⟦3030303⟧ ←Drop all decimal places
          past the fifth place.
0.03030

Exercise Set 3.2

1.

$$
\begin{array}{r}
{\scriptstyle 1} \\
3\,1\,6.2\,5 \\
+\ \ 1\,8.1\,2 \\
\hline
3\,3\,4.3\,7
\end{array}
$$

Add hundredths.
Add tenths.
Write a decimal point in the answer.
Add ones.
Add tens.
Add hundreds.

3.
```
     1 1
   6 5 9.4 0 3
 + 9 1 6.8 1 2
 ─────────────
 1 5 7 6.2 1 5
```
Add thousandths.
Add hundredths.
Add tenths.
Write a decimal point in the answer.
Add ones.
Add tens.
Add hundreds.

5.
```
    1       1
      9.1 0 4
 + 1 2 3.4 5 6
 ─────────────
 1 3 2.5 6 0
```

7.
```
       1
   8 1.0 0 8
 +   3.4 0 9
 ───────────
   8 4.4 1 7
```

9.
```
   2 0.0 1 2 4
 + 3 0.0 1 2 4
 ─────────────
   5 0.0 2 4 8
```

11. Line up the decimal points.
```
   1
   3 9.0 0 0     Writing 2 extra zeros
 +   1.0 0 7
 ───────────
   4 0.0 0 7
```

13. Line up the decimal points.
```
         1
       0.3 4 0     Writing an extra zero
       3.5 0 0     Writing 2 extra zeros
       0.1 2 7
 + 7 6 8.0 0 0     Writing in the decimal point
 ─────────────       and 3 extra zeros
   7 7 1.9 6 7     Adding
```

15.
```
   1     1 1
   1 7.0 0 0 0     Writing in the decimal point.
    3.2 4 0 0     You may find it helpful to
    0.2 5 6 0     write extra zeros.
 +   0.3 6 8 9
 ─────────────
   2 0.8 6 4 9
```

17.
```
   1 2 1     1
       2.7 0 3 0
     7 8.3 3 0 0
     2 8.0 0 0 9
 + 1 1 8.4 3 4 1
 ───────────────
   2 2 7.4 6 8 0
```

19.
```
   1 2 1   1
       9 9.6 0 0 1
     7 2 8 5.1 8 0 0
       5 0 0.0 4 2 0
 +     8 7 0.0 0 0 0
 ─────────────────
   8 7 5 4.8 2 2 1
```

21.
```
   4 12
   5̶.2̶
 − 3.9
 ─────
   1.3
```
Borrow ones to subtract tenths.
Subtract tenths.
Write a decimal point in the answer.
Subtract ones.

23.
```
   4 11 2 11
   5̶ 1̶.3̶ 1̶
 −    2.2 9
 ──────────
   4 9.0 2
```
Borrow tenths to subtract hundredths.
Subtract hundredths.
Subtract tenths.
Write a decimal point in the answer.
Borrow tens to subtract ones.
Subtract ones.
Subtract tens.

25.
```
   4 8.7 6
 −  3.1 5
 ─────────
   4 5.6 1
```

27.
```
           11
   8 1̶ 13
   9̶ 2̶.3̶ 4 1
 −    6.4 2
 ──────────
   8 5.9 2 1
```

29.
```
   4 9 9 10
   2.5̶-0̶-0̶-0̶     Writing 3 extra zeros
 − 0.0 0 2 5
 ───────────
   2.4 9 7 5
```

31.
```
   3 9 10
   3.4̶-0̶-0̶     Writing 2 extra zeros
 − 0.0 0 3
 ─────────
   3.3 9 7
```

33. Line up the decimal points. Write an extra zero if desired.
```
       17 11
   1 7 1̶ 10
   2̶ 8̶.2̶ 0̶
 − 1 9.3 5
 ─────────
     8.8 5
```

35.
```
   3 10
   3 4̶.0̶ 7
 − 3 0.7
 ───────
     3.3 7
```

37.
```
     4 10
   8.4 5̶ 0̶
 − 7.4 0 5
 ─────────
   1.0 4 5
```

39.
```
   5 10
   6̶.0̶ 0 3
 − 2.3
 ─────────
   3.7 0 3
```

41.
```
   9 9 9 10
   1̶-0̶-0̶-0̶ 0̶     Writing in the decimal point
 − 0. 0 0 9 8       and 4 extra zeros
 ────────────
   0. 9 9 0 2     Subtracting
```

43.
```
   9 9 . 9 10
   1̶-0̶-0̶.0̶ 0̶
 −      0. 3 4
 ────────────
   9 9. 6 6
```

45.
```
   6 14
   7̶. 4̶ 8
 − 2. 6
 ───────
   4. 8 8
```

47.
```
   2 9 9 10
   3̶.-0̶-0̶ 0̶
 − 2.0 0 6
 ─────────
   0.9 9 4
```

49.
```
   8 9 9 10
   1 9̶.-0̶-0̶ 0̶
 −    1.1 9 8
 ───────────
   1 7.8 0 2
```

51.
$$\begin{array}{r} \overset{4}{}\overset{9}{}\overset{10}{} \\ 6\,\cancel{5}.\,\cancel{0}\,\cancel{0} \\ -\ 1\,3.\,8\,7 \\ \hline 5\,1.\,1\,3 \end{array}$$

53.
$$\begin{array}{r} \overset{8}{}\overset{10}{} \\ 3.\,\cancel{9}\,\cancel{0}\,7 \\ -\ 1.\,4\,1\,6 \\ \hline 2.\,4\,9\,1 \end{array}$$

55.
$$\begin{array}{r} \overset{8}{}\overset{17}{} \\ 3\,2.\,7\,\cancel{9}\,\cancel{7}\,8 \\ -\ \ \ 0.\,0\,5\,9\,2 \\ \hline 3\,2.\,7\,3\,8\,6 \end{array}$$

57.
$$\begin{array}{r} \overset{2}{}\overset{9}{}\overset{10}{}\overset{6}{}\overset{14}{} \\ \cancel{3}.\,\cancel{0}\,\cancel{0}\,\cancel{7}\,\cancel{4} \\ -\ 1.\,3\,4\,0\,8 \\ \hline 1.\,6\,6\,6\,6 \end{array}$$

59.
$$\begin{array}{r} \overset{18}{} \\ \overset{4}{}\overset{8}{}\ \overset{9}{}\overset{17}{} \\ 2\,3\,4\,\cancel{5}.\,\cancel{9}\,\cancel{0}\,\cancel{7}\,8\,6 \\ -\ \ \ \ \ \ \ 0.\,9\,9\,9 \\ \hline 2\,3\,4\,4.\,9\,0\,8\,8\,6 \end{array}$$

61.
$$x + 17.5 = 29.15$$
$$x + 17.5 - 17.5 = 29.15 - 17.5 \qquad \text{Subtracting 17.5}$$
$$\text{on both sides}$$
$$x = 11.65$$
$$\begin{array}{r} \overset{8}{}\overset{11}{} \\ 2\,\cancel{9}.\,\cancel{1}\,5 \\ -\ 1\,7.\,5 \\ \hline 1\,1.\,6\,5 \end{array}$$

63.
$$3.205 + m = 22.456$$
$$3.205 + m - 3.205 = 22.456 - 3.205$$
$$\text{Subtracting 3.205}$$
$$\text{on both sides}$$
$$m = 19.251$$
$$\begin{array}{r} \overset{1}{}\overset{12}{} \\ \cancel{2}\,\cancel{2}.\,4\,5\,6 \\ -\ \ \ 3.\,2\,0\,5 \\ \hline 1\,9.\,2\,5\,1 \end{array}$$

65.
$$17.95 + p = 402.63$$
$$17.95 + p - 17.95 = 402.63 - 17.95$$
$$\text{Subtracting 17.95}$$
$$\text{on both sides}$$
$$p = 384.68$$
$$\begin{array}{r} \overset{11}{}\overset{15}{} \\ 3\ 9\ \overset{1}{\cancel{1}}\ \overset{5}{\cancel{6}}\ 13 \\ \cancel{4}\,\cancel{0}\,\cancel{2}.\,\cancel{6}\,\cancel{3} \\ -\ \ \ \ \ 1\,7.\,9\,5 \\ \hline 3\,8\,4.\,6\,8 \end{array}$$

67.
$$13,083.3 = x + 12,500.33$$
$$13,083.3 - 12,500.33 = x + 12,500.33 - 12,500.33$$
$$\text{Subtracting 12,500.33}$$
$$\text{on both sides}$$
$$582.97 = x$$
$$\begin{array}{r} \overset{12}{} \\ \overset{2}{}\ \overset{10}{}\ \ \overset{2}{}\ \overset{2}{\cancel{}}\ \overset{10}{} \\ 1\,\cancel{3},\,\cancel{0}\,8\,\cancel{3}.\,\cancel{3}\,\cancel{0} \\ -\ 1\,2,\,5\,0\,0.\,3\,3 \\ \hline 5\,8\,2.\,9\,7 \end{array}$$

69.
$$x + 2349 = 17,684.3$$
$$x + 2349 - 2349 = 17,684.3 - 2349$$
$$\text{Subtracting 2349}$$
$$\text{on both sides}$$
$$x = 15,335.3$$
$$\begin{array}{r} \overset{7}{}\overset{14}{} \\ 1\,7,\,6\,\cancel{8}\,\cancel{4}.\,3 \\ -\ \ \ \ 2\,3\,4\,9.\,0 \\ \hline 1\,5,\,3\,3\,5.\,3 \end{array}$$

71. First we add the payments/debits:

$27.44 + 123.95 + 124.02 + 12.43 + 137.78 + 2800.00 = 3225.62$

Then we add the deposits/credits:

$1000.00 + 2500.00 + 18.88 = 3518.88$

We add the total of the deposits to the balance brought forward:

$9704.56 + 3518.88 = 13,223.44$

Now we subtract the debit total:

$13,223.44 - 3225.62 = 9997.82$

The result should be the ending balance, 10,483.66. Since $9997.82 \neq 10,483.66$, an error has been made. Now we successively add or subtract deposits/credits and payments/debits and check the result in the balance forward column.

$$9704.56 - 27.44 = 9677.12$$

This subtraction was done correctly.

$$9677.12 + 1000.00 = 10,677.12$$

This addition was done correctly.

$$10,677.12 - 123.95 = 10,553.17$$

This subtract was done correctly.

$$10,553.17 - 124.02 = 10,429.15$$

This subtraction was done incorrectly. It appears that 124.02 was added rather than subtracted. We correct the balance line and continue.

$$10,429.15 - 12.43 = 10,416.72$$

If the previous checkbook balance had been correct, this subtraction would have been correct. We work with the corrected balance and continue.

$$10,416.72 + 2500.00 = 12,916.72$$
$$12,916.72 - 137.78 = 12,778.94$$
$$12,778.94 + 18.88 = 12,797.82$$
$$12,797.82 - 2800.00 = 9997.82$$

The correct checkbook balance is \$9997.82.

73. Discussion and Writing Exercise

75.
$$32 = 2 \cdot 2 \cdot 2 \cdot 2 \cdot 2$$
$$85 = 5 \cdot 17$$
$$\text{LCM} = 2 \cdot 2 \cdot 2 \cdot 2 \cdot 2 \cdot 5 \cdot 17 \text{ or } 2720$$

77.
$$\frac{13}{24} - \frac{3}{8} = \frac{13}{24} - \frac{3}{8} \cdot \frac{3}{3}$$
$$= \frac{13}{24} - \frac{9}{24}$$
$$= \frac{13 - 9}{24} = \frac{4}{24}$$
$$= \frac{4 \cdot 1}{4 \cdot 6} = \frac{4}{4} \cdot \frac{1}{6}$$
$$= \frac{1}{6}$$

79.
$$\begin{array}{r} \overset{7}{}\overset{9}{}\overset{15}{} \\ 8\,\cancel{8}\,\cancel{0}\,\cancel{5} \\ -\ 2\,6\,3\,9 \\ \hline 6\,1\,6\,6 \end{array}$$

81. *Familiarize*. We draw a picture.

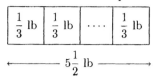

We let s = the number of servings that can be prepared from $5\frac{1}{2}$ lb of flounder fillet.

Translate. The situation corresponds to a division sentence.

$$s = 5\frac{1}{2} \div \frac{1}{3}$$

Solve. We carry out the division.

$$s = 5\frac{1}{2} \div \frac{1}{3} = \frac{11}{2} \div \frac{1}{3}$$
$$= \frac{11}{2} \cdot \frac{3}{1} = \frac{33}{2}$$
$$= 16\frac{1}{2}$$

Check. We check by multiplying. If $16\frac{1}{2}$ servings are prepared, then

$$16\frac{1}{2} \cdot \frac{1}{3} = \frac{33}{2} \cdot \frac{1}{3} = \frac{3 \cdot 11 \cdot 1}{2 \cdot 3} = \frac{3}{3} \cdot \frac{11 \cdot 1}{2} = \frac{11}{2} = 5\frac{1}{2} \text{ lb}$$

of flounder is used. Our answer checks.

State. $16\frac{1}{2}$ servings can be prepared from $5\frac{1}{2}$ lb of flounder fillet.

83. First, "undo" the incorrect addition by subtracting 235.7 from the incorrect answer:

$$\begin{array}{r} 8\,1\,7.\,2 \\ -\ 2\,3\,5.\,7 \\ \hline 5\,8\,1.\,5 \end{array}$$

The original minuend was 581.5. Now subtract 235.7 from this as the student originally intended:

$$\begin{array}{r} 5\,8\,1.\,5 \\ -\ 2\,3\,5.\,7 \\ \hline 3\,4\,5.\,8 \end{array}$$

The correct answer is 345.8.

Exercise Set 3.3

1.
$$\begin{array}{r} 8.\,6 \\ \times\quad 7 \\ \hline 6\,0.\,2 \end{array}$$
(1 decimal place)
(0 decimal places)
(1 decimal place)

3.
$$\begin{array}{r} 0.\,8\,4 \\ \times\quad 8 \\ \hline 6.\,7\,2 \end{array}$$
(2 decimal places)
(0 decimal places)
(2 decimal places)

5.
$$\begin{array}{r} 6.\,3 \\ \times\ 0.\,0\,4 \\ \hline 0.\,2\,5\,2 \end{array}$$
(1 decimal place)
(2 decimal places)
(3 decimal places)

7.
$$\begin{array}{r} 8\,7 \\ \times\ 0.\,0\,0\,6 \\ \hline 0.\,5\,2\,2 \end{array}$$
(0 decimal places)
(3 decimal places)
(3 decimal places)

9. $1\underline{0} \times 23.76$ $\qquad$ 23.7.6

1 zero $\qquad$ Move 1 place to the right.

$10 \times 23.76 = 237.6$

11. $1\underline{000} \times 583.686852$ $\qquad$ 583.686.852

3 zeros $\qquad$ Move 3 places to the right.

$1000 \times 583.686852 = 583{,}686.852$

13. $7.8 \times 1\underline{00}$ $\qquad$ 7.80.

2 zeros $\qquad$ Move 2 places to the right.

$7.8 \times 100 = 780$

15. $0.\underline{1} \times 89.23$ $\qquad$ 8.9.23

1 decimal place $\qquad$ Move 1 place to the left.

$0.1 \times 89.23 = 8.923$

17. $0.\underline{001} \times 97.68$ $\qquad$ 0.097.68

3 decimal places $\qquad$ Move 3 places to the left.

$0.001 \times 97.68 = 0.09768$

19. $78.2 \times 0.\underline{01}$ $\qquad$ 0.78.2

2 decimal places $\qquad$ Move 2 places to the left.

$78.2 \times 0.01 = 0.782$

21.
$$\begin{array}{r} 3\,2.\,6 \\ \times\quad 1\,6 \\ \hline 1\,9\,5\,6 \\ 3\,2\,6\,0 \\ \hline 5\,2\,1.\,6 \end{array}$$
(1 decimal place)
(0 decimal places)

(1 decimal place)

23.
$$\begin{array}{r} 0.\,9\,8\,4 \\ \times\quad 3.\,3 \\ \hline 2\,9\,5\,2 \\ 2\,9\,5\,2\,0 \\ \hline 3.\,2\,4\,7\,2 \end{array}$$
(3 decimal places)
(1 decimal place)

(4 decimal places)

25.
$$\begin{array}{r} 3\,7\,4 \\ \times\quad 2.\,4 \\ \hline 1\,4\,9\,6 \\ 7\,4\,8\,0 \\ \hline 8\,9\,7.\,6 \end{array}$$
(0 decimal places)
(1 decimal place)

(1 decimal place)

27.
$$\begin{array}{r} 7\,4\,9 \\ \times\ 0.\,4\,3 \\ \hline 2\,2\,4\,7 \\ 2\,9\,9\,6\,0 \\ \hline 3\,2\,2.\,0\,7 \end{array}$$
(0 decimal places)
(2 decimal places)

(2 decimal places)

29.
$$\begin{array}{r} 0.\,8\,7 \\ \times\quad 6\,4 \\ \hline 3\,4\,8 \\ 5\,2\,2\,0 \\ \hline 5\,5.\,6\,8 \end{array}$$
(2 decimal places)
(0 decimal places)

(2 decimal places)

31.
```
    4 6 . 5 0     (2 decimal places)
  ×       7 5     (0 decimal places)
  ─────────────
    2 3 2 5 0
  3 2 5 5 0 0
  ─────────────
  3 4 8 7 . 5 0   (2 decimal places)
```
Since the last decimal place is 0, we could also write this answer as 3487.5.

33.
```
      8 1 . 7     (1 decimal place)
  × 0 . 6 1 2     (3 decimal places)
  ─────────────
    1 6 3 4
    8 1 7 0
  4 9 0 2 0 0
  ─────────────
  5 0 . 0 0 0 4   (4 decimal places)
```

35.
```
      1 0 . 1 0 5     (3 decimal places)
  × 1 1 . 3 2 4       (3 decimal places)
  ───────────────────
        4 0 4 2 0
      2 0 2 1 0 0
    3 0 3 1 5 0 0
  1 0 1 0 5 0 0 0
  1 0 1 0 5 0 0 0 0
  ───────────────────
  1 1 4 . 4 2 9 0 2 0   (6 decimal places)
```
or 114.42902

37.
```
      1 2 . 3     (1 decimal place)
  × 1 . 0 8       (2 decimal places)
  ───────────────
      9 8 4
  1 2 3 0 0
  ───────────────
  1 3 . 2 8 4     (3 decimal places)
```

39.
```
      3 2 . 4     (1 decimal place)
  ×     2 . 8     (1 decimal place)
  ─────────────
    2 5 9 2
    6 4 8 0
  ─────────────
    9 0 . 7 2     (2 decimal places)
```

41.
```
      0 . 0 0 3 4 2     (5 decimal places)
  ×         0 . 8 4     (2 decimal places)
  ───────────────────
        1 3 6 8
      2 7 3 6 0
  ───────────────────
  0 . 0 0 2 8 7 2 8     (7 decimal places)
```

43.
```
      0 . 3 4 7     (3 decimal places)
  ×     2 . 0 9     (2 decimal places)
  ─────────────────
      3 1 2 3
    6 9 4 0 0
  ─────────────────
  0 . 7 2 5 2 3     (5 decimal places)
```

45.
```
      3 . 0 0 5     (3 decimal places)
  × 0 . 6 2 3       (3 decimal places)
  ─────────────────
      9 0 1 5
    6 0 1 0 0
  1 8 0 3 0 0 0
  ─────────────────
  1 . 8 7 2 1 1 5   (6 decimal places)
```

47. $1\underline{000} \times 45.678$ 45.678.

3 zeros   Move 3 places to the right.

$1000 \times 45.678 = 45{,}678$

49. Move 2 places to the right.

$28.88.¢

Change from $ sign in front to ¢ sign at end.

$28.88 = 2888¢

51. Move 2 places to the right.

$0.66.¢

Change from $ sign in front to ¢ sign at end.

$0.66 = 66¢

53. Move 2 places to the left.

$0.34.¢

Change from ¢ sign at end to $ sign in front.

34¢ = $0.34

55. Move 2 places to the left.

$34.45.¢

Change from ¢ sign at end to $ sign in front.

3345¢ = $34.45

57. 93 million $= 93 \times 1,\underbrace{000,000}_{6\ zeros}$

93.000000.

Move 6 places to the right.

93 million $= 93{,}000{,}000$

59. $7.2 billion $= \$7.2 \times 1,\underbrace{000,000,000}_{9\ zeros}$

$7.200000000.

Move 9 places to the right.

$7.2 billion $= \$7{,}200{,}000{,}000$

61. Discussion and Writing Exercise

63. $2\dfrac{1}{3} \cdot 4\dfrac{4}{5} = \dfrac{7}{3} \cdot \dfrac{24}{5} = \dfrac{7 \cdot 3 \cdot 8}{3 \cdot 5}$

$\qquad = \dfrac{3}{3} \cdot \dfrac{7 \cdot 8}{5} = \dfrac{56}{5}$

$\qquad = 11\dfrac{1}{5}$

65.
$$4\frac{4}{5} = 4\frac{12}{15}$$
$$-2\frac{1}{3} = -2\frac{5}{15}$$
$$\overline{\phantom{-2\frac{1}{3} = }2\frac{7}{15}}$$

67.
```
        3 4 2
  2 4 ⟌8 2 0 8
        7 2 0 0
        ───────
        1 0 0 8
          9 6 0
          ─────
            4 8
            4 8
            ───
              0
```
The answer is 342.

69.
```
          4 5 6 6
    7 ⟌3 1,9 6 2
        2 8 0 0 0
        ─────────
          3 9 6 2
          3 5 0 0
          ───────
            4 6 2
            4 2 0
            ─────
              4 2
              4 2
              ───
                0
```
The answer is 4566.

71.
```
          8 7
  4 0 ⟌3 4 8 0
        3 2 0 0
        ───────
          2 8 0
          2 8 0
          ─────
              0
```
The answer is 87.

73. (1 trillion) · (1 billion)

$= 1,\underbrace{000,000,000,000}_{\text{12 zeros}} \times 1,\underbrace{000,000,000}_{\text{9 zeros}}$

$= 1,\underbrace{000,000,000,000,000,000,000}_{\text{21 zeros}}$

$= 10^{21} = 1 \text{ sextillion}$

75. (1 trillion) · (1 trillion)

$= 1,\underbrace{000,000,000,000}_{\text{12 zeros}} \times 1,\underbrace{000,000,000,000}_{\text{12 zeros}}$

$= 1,\underbrace{000,000,000,000,000,000,000,000}_{\text{24 zeros}}$

$= 10^{24} = 1 \text{ septillion}$

Exercise Set 3.4

1.
```
      2. 9 9
  2 ⟌5. 9 8
      4 0 0
      ─────
      1 9 8
      1 8 0
      ─────
        1 8
        1 8
        ───
          0
```
Divide as though dividing whole numbers. Place the decimal point directly above the decimal point in the dividend.

3.
```
      2 3. 7 8
  4 ⟌9 5. 1 2
      8 0 0 0
      ───────
      1 5 1 2
      1 2 0 0
      ───────
        3 1 2
        2 8 0
        ─────
          3 2
          3 2
          ───
            0
```
Divide as though dividing whole numbers. Place the decimal point directly above the decimal point in the dividend.

5.
```
        7. 4 8
  1 2 ⟌8 9. 7 6
        8 4 0 0
        ───────
          5 7 6
          4 8 0
          ─────
            9 6
            9 6
            ───
              0
```

7.
```
          7. 2
  3 3 ⟌2 3 7. 6
        2 3 1 0
        ───────
            6 6
            6 6
            ───
              0
```

9.
```
      1. 1 4 3
  8 ⟌9. 1 4 4
      8 0 0 0
      ───────
      1 1 4 4
        8 0 0
        ─────
        3 4 4
        3 2 0
        ─────
          2 4
          2 4
          ───
            0
```

11.
```
      4. 0 4 1
  3 ⟌1 2. 1 2 3
      1 2 0 0 0
      ─────────
          1 2 3
          1 2 0
          ─────
              3
              3
              ─
              0
```

13.
```
      0. 0 7
  5 ⟌0. 3 5
      3 5
      ───
        0
```

15.
$$0.1\,2_{\wedge}\overline{\smash)8.4\,0_{\wedge}}$$ quotient $70.$
$$\underline{8\,4\,0}$$
$$0$$

Multiply the divisor by 100 (move the decimal point 2 places). Multiply the same way in the dividend (move 2 places). Then divide.

17.
$$3.4_{\wedge}\overline{\smash)6\,8.0_{\wedge}}$$ quotient $20.$
$$\underline{6\,8\,0}$$
$$0$$

Put a decimal point at the end of the whole number. Multiply the divisor by 10 (move the decimal point 1 place). Multiply the same way in the dividend (move 1 place), adding an extra 0. Then divide.

19.
$$1\,5\,\overline{\smash)6.0}$$ quotient 0.4
$$\underline{6\,0}$$
$$0$$

Put a decimal point at the end of the whole number. Write an extra 0 to the right of the decimal point. Then divide.

21.
$$3\,6\,\overline{\smash)1\,4.7\,6}$$ quotient $0.4\,1$
$$\underline{1\,4\,4\,0}$$
$$3\,6$$
$$\underline{3\,6}$$
$$0$$

23.
$$3.2_{\wedge}\overline{\smash)2\,7.2_{\wedge}0}$$ quotient 8.5
$$\underline{2\,5\,6}$$
$$1\,6\;\;0$$ Write an extra 0.
$$\underline{1\,6\;\;0}$$
$$0$$

25.
$$4.2_{\wedge}\overline{\smash)3\,9.0_{\wedge}6}$$ quotient 9.3
$$\underline{3\,7\,8\;\;0}$$
$$1\,2\;\;6$$
$$\underline{1\,2\;\;6}$$
$$0$$

27.
$$8\,\overline{\smash)5\,.0\,0\,0}$$ quotient $0\,.6\,2\,5$
$$\underline{4\;\;8}$$
$$2\;\;0$$ Write an extra 0.
$$\underline{1\;\;6}$$
$$4\;\;0$$ Write an extra 0.
$$\underline{4\;\;0}$$
$$0$$

29.
$$0.4\,7_{\wedge}\overline{\smash)0.\,1\,2_{\wedge}2\,2}$$ quotient $0.2\,6$
$$\underline{9\,4\;\;0}$$
$$2\;\;8\,2$$
$$\underline{2\;\;8\,2}$$
$$0$$

31.
$$4.8_{\wedge}\overline{\smash)7\,5.0_{\wedge}0\,0\,0}$$ quotient $1\,5.6\,2\,5$
$$\underline{4\,8\,0}$$
$$2\,7\,0$$
$$\underline{2\,4\,0}$$
$$3\,0\;\;0$$
$$\underline{2\,8\;\;8}$$
$$1\;\;2\,0$$
$$\underline{9\;\;6}$$
$$2\;\;4\,0$$
$$\underline{2\;\;4\,0}$$
$$0$$

33.
$$0.0\,3\,2_{\wedge}\overline{\smash)0.\,0\,7\,4_{\wedge}8\,8}$$ quotient $2.3\,4$
$$\underline{6\,4\,0\,0}$$
$$1\,0\;\;8\,8$$
$$\underline{9\;\;6\,0}$$
$$1\;\;2\,8$$
$$\underline{1\;\;2\,8}$$
$$0$$

35.
$$8\,2\,\overline{\smash)3\,8.5\,4}$$ quotient $0.4\,7$
$$\underline{3\,2\,8\,0}$$
$$5\;\;7\,4$$
$$\underline{5\;\;7\,4}$$
$$0$$

37. $\dfrac{213.4567}{1\underline{000}}$

$0.\underset{\uparrow\_\_\rfloor}{213}.4567$

3 zeros   Move 3 places to the left.

$\dfrac{213.4567}{1000} = 0.2134567$

39. $\dfrac{213.4567}{1\underline{0}}$

$21.\underset{\uparrow\rfloor}{3}.4567$

1 zero   Move 1 place to the left.

$\dfrac{213.4567}{10} = 21.34567$

41. $\dfrac{1.0237}{0.\underline{001}}$

$1.023.\underset{\lfloor\_\_\uparrow}{7}$

3 decimal places   Move 3 places to the right.

$\dfrac{1.0237}{0.001} = 1023.7$

43. $4.2 \cdot x = 39.06$

$\dfrac{4.2 \cdot x}{4.2} = \dfrac{39.06}{4.2}$ Dividing on both sides by 4.2

$x = 9.3$

$$4.2_{\wedge}\overline{\smash)3\,9.0_{\wedge}6}$$ quotient $0\,9.3$
$$\underline{3\,7\,8\;\;0}$$
$$1\,2\;\;6$$
$$\underline{1\,2\;\;6}$$
$$0$$

The solution is 9.3.

45. $1000 \cdot y = 9.0678$

$\dfrac{1000 \cdot y}{1000} = \dfrac{9.0678}{1000}$ Dividing on both sides by 1000

$y = 0.0090678$ Moving the decimal point 3 places to the left

The solution is 0.0090678.

47. $1048.8 = 23 \cdot t$

$\dfrac{1048.8}{23} = \dfrac{23 \cdot t}{23}$ Dividing on both sides by 23

$45.6 = t$

```
              4 5. 6
    2 3 | 1 0 4 8. 8
          9 2 0 0
          1 2 8 8
          1 1 5 0
            1 3 8
            1 3 8
                0
```

The solution is 45.6.

49. $14 \times (82.6 + 67.9) = 14 \times (150.5)$ Doing the calculation inside the parentheses

$= 2107$ Multiplying

51. $0.003 + 3.03 \div 0.01 = 0.003 + 303$ Dividing first
$= 303.003$ Adding

53. $42 \times (10.6 + 0.024)$

$= 42 \times 10.624$ Doing the calculation inside the parentheses
$= 446.208$ Multiplying

55. $4.2 \times 5.7 + 0.7 \div 3.5$

$= 23.94 + 0.2$ Doing the multiplications and divisions in order from left to right
$= 24.14$ Adding

57. $9.0072 + 0.04 \div 0.1^2$

$= 9.0072 + 0.04 \div 0.01$ Evaluating the exponential expression
$= 9.0072 + 4$ Dividing
$= 13.0072$ Adding

59. $(8 - 0.04)^2 \div 4 + 8.7 \times 0.4$

$= (7.96)^2 \div 4 + 8.7 \times 0.4$ Doing the calculation inside the parentheses
$= 63.3616 \div 4 + 8.7 \times 0.4$ Evaluating the exponential expression
$= 15.8404 + 3.48$ Doing the multiplications and divisions in order from left to right
$= 19.3204$ Adding

61. $86.7 + 4.22 \times (9.6 - 0.03)^2$

$= 86.7 + 4.22 \times (9.57)^2$ Doing the calculation inside the parentheses
$= 86.7 + 4.22 \times 91.5849$ Evaluating the exponential expression
$= 86.7 + 386.488278$ Multiplying
$= 473.188278$ Adding

63. $4 \div 0.4 + 0.1 \times 5 - 0.1^2$

$= 4 \div 0.4 + 0.1 \times 5 - 0.01$ Evaluating the exponential expression
$= 10 + 0.5 - 0.01$ Doing the multiplications and divisions in order from left to right
$= 10.49$ Adding and subtracting in order from left to right

65. $5.5^2 \times [(6 - 4.2) \div 0.06 + 0.12]$

$= 5.5^2 \times [1.8 \div 0.06 + 0.12]$ Doing the calculation in the innermost parentheses first
$= 5.5^2 \times [30 + 0.12]$ Doing the calculation inside the parentheses
$= 5.5^2 \times 30.12$
$= 30.25 \times 30.12$ Evaluating the exponential expression
$= 911.13$ Multiplying

67. $200 \times \{[(4 - 0.25) \div 2.5] - (4.5 - 4.025)\}$

$= 200 \times \{[3.75 \div 2.5] - 0.475\}$ Doing the calculations in the innermost parentheses first
$= 200 \times \{1.5 - 0.475\}$ Again, doing the calculations in the innermost parentheses
$= 200 \times 1.025$ Subtracting inside the parentheses
$= 205$ Multiplying

69. We add the numbers and then divide by the number of addends.

$(\$1276.59 + \$1350.49 + \$1123.78 + \$1402.56) \div 4$
$= \$5153.42 \div 4$
$= \$1288.355$
$\approx \$1288.36$

71. We add the sales amounts and divide by the number of addends, 5.

$\dfrac{7152 + 12{,}980 + 17{,}239 + 20{,}877 + 23{,}000}{5}$
$= \dfrac{81{,}248}{5} = 16{,}249.6$

The average number of sales per year over the five-year period was 16,249.6.

73. Discussion and Writing exercise

75. $\dfrac{36}{42} = \dfrac{6 \cdot 6}{6 \cdot 7} = \dfrac{6}{6} \cdot \dfrac{6}{7} = \dfrac{6}{7}$

77. $\dfrac{38}{146} = \dfrac{2 \cdot 19}{2 \cdot 73} = \dfrac{2}{2} \cdot \dfrac{19}{73} = \dfrac{19}{73}$

79.
```
          1 9
      3 | 5 7
    3 | 1 7 1
  2 | 3 4 2
2 | 6 8 4
```

$684 = 2 \cdot 2 \cdot 3 \cdot 3 \cdot 19$, or $2^2 \cdot 3^2 \cdot 19$

81.
$$\begin{array}{r} 2\,2\,3 \\ 3\sqrt{6\,6\,9} \\ 3\sqrt{2\,0\,0\,7} \end{array}$$
$2007 = 3 \cdot 3 \cdot 223$, or $3^2 \cdot 223$

83. $10\dfrac{1}{2} + 4\dfrac{5}{8} = 10\dfrac{4}{8} + 4\dfrac{5}{8}$

$\qquad\qquad = 14\dfrac{9}{8} = 15\dfrac{1}{8}$

85. Use a calculator.

$\qquad 9.0534 - 2.041^2 \times 0.731 \div 1.043^2$

$\qquad = 9.0534 - 4.165681 \times 0.731 \div 1.087849$

$\qquad\qquad$ Evaluating the exponential expressions

$\qquad = 9.0534 - 3.045112811 \div 1.087849$

$\qquad\qquad\qquad$ Multiplying and dividing

$\qquad = 9.0534 - 2.799205415$ in order from left to right

$\qquad = 6.254194585$

87.
$\qquad 439.57 \times 0.01 \div 1000 \times \underline{\quad} = 4.3957$
$\qquad\quad 4.3957 \div 1000 \times \underline{\quad} = 4.3957$
$\qquad\quad 0.0043957 \times \underline{\quad} = 4.3957$

We need to multiply 0.0043957 by a number that moves the decimal point 3 places to the right. Thus, we need to multiply by 1000. This is the missing value.

89.
$\qquad 0.0329 \div 0.001 \times 10^4 \div \underline{\quad} = 3290$
$\qquad 0.0329 \div 0.001 \times 10{,}000 \div \underline{\quad} = 3290$
$\qquad\qquad 32.9 \times 10{,}000 \div \underline{\quad} = 3290$
$\qquad\qquad\qquad 329{,}000 \div \underline{\quad} = 3290$

We need to divide 329,000 by a number that moves the decimal point 2 places to the left. Thus, we need to divide by 100. This is the missing value.

Exercise Set 3.5

1. $\dfrac{23}{100} = 0.23$

3. $\dfrac{3}{5} = \dfrac{3}{5} \cdot \dfrac{2}{2}$ $\qquad$ We use $\dfrac{2}{2}$ for 1 to get a denominator of 10.

$\qquad = \dfrac{6}{10} = 0.6$

5. $\dfrac{13}{40} = \dfrac{13}{40} \cdot \dfrac{25}{25}$ $\qquad$ We use $\dfrac{25}{25}$ for 1 to get a denominator of 1000.

$\qquad = \dfrac{325}{1000} = 0.325$

7. $\dfrac{1}{5} = \dfrac{1}{5} \cdot \dfrac{2}{2} = \dfrac{2}{10} = 0.2$

9. $\dfrac{17}{20} = \dfrac{17}{20} \cdot \dfrac{5}{5} = \dfrac{85}{100} = 0.85$

11. $\dfrac{3}{8} = 3 \div 8$

$$\begin{array}{r} 0.3\,7\,5 \\ 8\sqrt{3.0\,0\,0} \\ \underline{2\,4} \\ 6\,0 \\ \underline{5\,6} \\ 4\,0 \\ \underline{4\,0} \\ 0 \end{array}$$

$\dfrac{3}{8} = 0.375$

13. $\dfrac{39}{40} = \dfrac{39}{40} \cdot \dfrac{25}{25} = \dfrac{975}{1000} = 0.975$

15. $\dfrac{13}{25} = \dfrac{13}{25} \cdot \dfrac{4}{4} = \dfrac{52}{100} = 0.52$

17. $\dfrac{2502}{125} = \dfrac{2502}{125} \cdot \dfrac{8}{8} = \dfrac{20{,}016}{1000} = 20.016$

19. $\dfrac{1}{4} = \dfrac{1}{4} \cdot \dfrac{25}{25} = \dfrac{25}{100} = 0.25$

21. $\dfrac{29}{25} = \dfrac{29}{25} \cdot \dfrac{4}{4} = \dfrac{116}{100} = 1.16$

23. $\dfrac{19}{16} = \dfrac{19}{16} \cdot \dfrac{625}{625} = \dfrac{11{,}875}{10{,}000} = 1.1875$

25. $\dfrac{4}{15} = 4 \div 15$

$$\begin{array}{r} 0.\,2\,6\,6 \\ 15\sqrt{4.\,0\,0\,0} \\ \underline{3\,0} \\ 1\,0\,0 \\ \underline{9\,0} \\ 1\,0\,0 \\ \underline{9\,0} \\ 1\,0 \end{array}$$

Since 10 keeps reappearing as a remainder, the digits repeat and

$\dfrac{4}{15} = 0.2666\ldots$ or $0.2\overline{6}$.

27. $\dfrac{1}{3} = 1 \div 3$

$$\begin{array}{r} 0.\,3\,3\,3 \\ 3\sqrt{1.\,0\,0\,0} \\ \underline{9} \\ 1\,0 \\ \underline{9} \\ 1\,0 \\ \underline{9} \\ 1 \end{array}$$

Since 1 keeps reappearing as a remainder, the digits repeat and

$\dfrac{1}{3} = 0.333\ldots$ or $0.\overline{3}$.

29. $\dfrac{4}{3} = 4 \div 3$

```
        1. 3 3
    3 | 4. 0 0
        3
        ‾‾‾
        1 0
          9
        ‾‾‾
          1 0
             9
          ‾‾‾
             1
```

Since 1 keeps reappearing as a remainder, the digits repeat and

$\dfrac{4}{3} = 1.333\ldots$ or $1.\overline{3}$.

31. $\dfrac{7}{6} = 7 \div 6$

```
        1. 1 6 6
    6 | 7. 0 0 0
        6
        ‾‾‾
        1 0
          6
        ‾‾‾
          4 0
          3 6
          ‾‾‾
            4 0
            3 6
            ‾‾‾
              4
```

Since 4 keeps reappearing as a remainder, the digits repeat and

$\dfrac{7}{6} = 1.166\ldots$ or $1.1\overline{6}$.

33. $\dfrac{4}{7} = 4 \div 7$

```
        0. 5 7 1 4 2 8 5
    7 | 4. 0 0 0 0 0 0 0
        3 5
        ‾‾‾
          5 0
          4 9
          ‾‾‾
            1 0
               7
            ‾‾‾
               3 0
               2 8
               ‾‾‾
                 2 0
                 1 4
                 ‾‾‾
                   6 0
                   5 6
                   ‾‾‾
                     4 0
                     3 5
                     ‾‾‾
                       5
```

Since 5 reappears as a remainder, the sequence repeats and

$\dfrac{4}{7} = 0.571428571428\ldots$ or $0.\overline{571428}$.

35. $\dfrac{11}{12} = 11 \div 12$

```
           0. 9 1 6 6
    1 2 | 1 1. 0 0 0 0
          1 0 8
          ‾‾‾‾‾
              2 0
              1 2
              ‾‾‾
                8 0
                7 2
                ‾‾‾
                  8 0
                  7 2
                  ‾‾‾
                    8
```

Since 8 keeps reappearing as a remainder, the digits repeat and $\dfrac{11}{12} = 0.91666\ldots$ or $0.91\overline{6}$.

37. Round $0.\,2\,\boxed{6}\,6\,6\ldots$ to the nearest tenth.

      Hundredths digit is 5 or higher.

 $0.\,3$ Round up.

Round $0.\,2\,\underline{6}\,\boxed{6}\,6\ldots$ to the nearest hundredth.

      Thousandths digit is 5 or higher.

 $0.\,2\,7$ Round up.

Round $0.\,2\,6\,6\,\underline{6}\,\boxed{6}\ldots$ to the nearest thousandth.

      Ten-thousandths digit is 5 or higher.

 $0.\,2\,6\,7$ Round up.

39. Round $0.\,3\,\boxed{3}\,3\,3\ldots$ to the nearest tenth.

      Hundredths digit is 4 or lower.

 $0.\,3$ Round down.

Round $0.\,3\,\underline{3}\,\boxed{3}\,3\ldots$ to the nearest hundredth.

      Thousandths digit is 4 or lower.

 $0.\,3\,3$ Round down.

Round $0.\,3\,3\,3\,\underline{3}\,\boxed{3}\ldots$ to the nearest thousandth.

      Ten-thousandths digit is 4 or lower.

 $0.\,3\,3\,3$ Round down.

41. Round $1.\,3\,\boxed{3}\,3\,3\ldots$ to the nearest tenth.

      Hundredths digit is 4 or lower.

 $1.\,3$ Round down.

Round $1.\,3\,\underline{3}\,\boxed{3}\,3\ldots$ to the nearest hundredth.

      Thousandths digit is 4 or lower.

 $1.\,3\,3$ Round down.

Round $1.\,3\,3\,3\,\underline{3}\,\boxed{3}\ldots$ to the nearest thousandth.

      Ten-thousandths digit is 4 or lower.

 $1.\,3\,3\,3$ Round down.

43. Round $1.\,1\,\boxed{6}\,6\,6\ldots$ to the nearest tenth.

      Hundredths digit is 5 or higher.

 $1.\,2$ Round up.

Round 1. 1 6 ⬚6⬚ 6 ... to the nearest hundredth.

     ↓ ↑⎿———— Thousandths digit is 5 or higher.

    1. 1 7   Round up.

Round 1. 1 6 6 ⬚6⬚ ... to the nearest thousandth.

      ↓ ↑⎿———— Ten-thousandths digit is 5 or higher.

    1. 1 6 7   Round up.

45. $0.\overline{571428}$

Round to the nearest tenth.

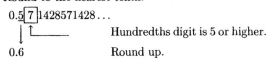

0.6   Round up.

Round to the nearest hundredth.

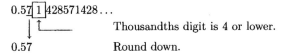

0.57   Round down.

Round to the nearest thousandth.

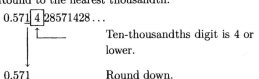

0.571   Round down.

47. Round 0. 9 ⬚1⬚ 6 6 ... to the nearest tenth.

      ↓ ↑⎿———— Hundredths digit is 4 or lower.

    0. 9   Round down.

Round 0. 9 1 ⬚6⬚ 6 ... to the nearest hundredth.

      ↓ ↑⎿———— Thousandths digit is 5 or higher.

    0. 9 2   Round up.

Round 0. 9 1 6 ⬚6⬚ ... to the nearest thousandth.

      ↓ ↑⎿———— Ten-thousandths digit is 5 or higher.

    0. 9 1 7   Round up.

49. Round 0. 1 ⬚8⬚ 1 8 ... to the nearest tenth.

      ↓ ↑⎿———— Hundredths digit is 5 or higher.

    0. 2   Round up.

Round 0. 1 8 ⬚1⬚ 8 ... to the nearest hundredth.

      ↓ ↑⎿———— Thousandths digit is 4 or lower.

    0. 1 8   Round down.

Round 0. 1 8 1 ⬚8⬚ ... to the nearest thousandth.

      ↓ ↑⎿———— Ten-thousandths digit is 5 or higher.

    0. 1 8 2   Round up.

51. Round 0. 2 ⬚7⬚ 7 7 ... to the nearest tenth.

      ↓ ↑⎿———— Hundredths digit is 5 or higher.

    0. 3   Round up.

Round 0. 2 7 ⬚7⬚ 7 ... to the nearest hundredth.

      ↓ ↑⎿———— Thousandths digit is 5 or higher.

    0. 2 8   Round up.

Round 0. 2 7 7 ⬚7⬚ ... to the nearest thousandth.

      ↓ ↑⎿———— Ten-thousandths digit is 5 or higher.

    0. 2 7 8   Round up.

53. Note that there are 5 women and 4 men, so there are $5+4$, or 9, people.

(a) $\dfrac{\text{Women}}{\text{Number of people}} = \dfrac{5}{9} = 0.55555\ldots \approx 0.556$

(b) $\dfrac{\text{Women}}{\text{Men}} = \dfrac{5}{4} = 1.25$

(c) $\dfrac{\text{Men}}{\text{Number of people}} = \dfrac{4}{9} = 0.44444\ldots \approx 0.444$

(d) $\dfrac{\text{Men}}{\text{Women}} = \dfrac{4}{5} = 0.8$

55. $\dfrac{\text{Miles driven}}{\text{Gasoline used}} = \dfrac{285}{18} = 15.833\ldots \approx 15.8$

The gasoline mileage was about 15.8 miles to the gallon.

57. $\dfrac{\text{Miles driven}}{\text{Gasoline used}} = \dfrac{324.8}{18.2} \approx 17.8$

The gasoline mileage was about 17.8 miles to the gallon.

59. We add the wind speeds and divide by the number of addends, 6.

$$\frac{35.3 + 12.5 + 11.3 + 10.7 + 10.7 + 10.4}{6}$$

$$= \frac{90.9}{6} = 15.15 \approx 15.2$$

The average of the wind speeds is about 15.2 mph.

61. $41\dfrac{11}{16} = 41 + \dfrac{11}{16}$

We convert $\dfrac{11}{16}$ to decimal notation.

```
      0. 6 8 7 5
1 6 ⟌ 1 1. 0 0 0 0
      9 6
      ─────
      1 4 0
      1 2 8
      ─────
        1 2 0
        1 1 2
        ─────
          8 0
          8 0
          ───
            0
```

We have $\dfrac{11}{16} = 0.6875 \approx 0.69$, so $\$41\dfrac{11}{16} = \$41.6875 \approx \$41.69$.

63. $25\frac{7}{8} = 25 + \frac{7}{8}$

We convert $\frac{7}{8}$ to decimal notation.

$$
\begin{array}{r}
0.875 \\
8\overline{)7.000} \\
\underline{6\,4} \\
6\,0 \\
\underline{5\,6} \\
4\,0 \\
\underline{4\,0} \\
0
\end{array}
$$

We have $\frac{7}{8} = 0.875 \approx 0.88$, so $\$25\frac{7}{8} = \$25.875 \approx \$25.88$.

65. $19\frac{3}{64} = 19 + \frac{3}{64}$

We convert $\frac{3}{64}$ to decimal notation.

$$
\begin{array}{r}
0.046875 \\
64\overline{)3.000000} \\
\underline{2\,5\,6} \\
4\,4\,0 \\
\underline{3\,8\,4} \\
5\,6\,0 \\
\underline{5\,1\,2} \\
4\,8\,0 \\
\underline{4\,4\,8} \\
3\,2\,0 \\
\underline{3\,2\,0} \\
0
\end{array}
$$

We have $\frac{3}{64} = 0.046875 \approx 0.05$, so
$\$19\frac{3}{64} = \$19.046875 \approx \$19.05$.

67. We will use the second method discussed in the text.

$$
\begin{aligned}
\frac{7}{8} \times 12.64 &= \frac{7}{8} \times \frac{1264}{100} = \frac{7 \cdot 1264}{8 \cdot 100} \\
&= \frac{7 \cdot 2 \cdot 2 \cdot 2 \cdot 2 \cdot 79}{2 \cdot 2 \cdot 2 \cdot 2 \cdot 2 \cdot 5 \cdot 5} \\
&= \frac{2 \cdot 2 \cdot 2 \cdot 2}{2 \cdot 2 \cdot 2 \cdot 2} \cdot \frac{7 \cdot 79}{2 \cdot 5 \cdot 5} \\
&= 1 \cdot \frac{7 \cdot 79}{2 \cdot 5 \cdot 5} \\
&= \frac{7 \cdot 79}{2 \cdot 5 \cdot 5} = \frac{553}{50}, \text{ or } 11.06
\end{aligned}
$$

69. $2\frac{3}{4} + 5.65 = 2.75 + 5.65$ Writing $2\frac{3}{4}$ using decimal notation

$\phantom{2\frac{3}{4} + 5.65} = 8.4$ Adding

71. We will use the first method discussed in the text.

$$
\begin{aligned}
\frac{47}{9} \times 79.95 &= 5.\overline{2} \times 79.95 \\
&\approx 5.222 \times 79.95 = 417.4989
\end{aligned}
$$

Note that this answer is not as accurate as those found using either of the other methods, due to rounding. The result using the other methods is $417.51\overline{6}$.

73. $\frac{1}{2} - 0.5 = 0.5 - 0.5$ Writing $\frac{1}{2}$ using decimal notation

$\phantom{\frac{1}{2} - 0.5} = 0$

75. $4.875 - 2\frac{1}{16} = 4.875 - 2.0625$ Writing $2\frac{1}{16}$ using decimal notation

$\phantom{4.875 - 2\frac{1}{16}} = 2.8125$

77. We will use the third method discussed in the text.

$$
\begin{aligned}
\frac{5}{6} \times 0.0765 + \frac{5}{4} \times 0.1124 &= \frac{5}{6} \times \frac{0.0765}{1} + \frac{5}{4} \times \frac{0.1124}{1} \\
&= \frac{5 \times 0.0765}{6 \times 1} + \frac{5 \times 0.1124}{4 \times 1} \\
&= \frac{0.3825}{6} + \frac{0.562}{4} \\
&= 0.06375 + 0.1405 \\
&= 0.20425
\end{aligned}
$$

79. We use the rules for order of operations, doing the multiplication first and then the division. Then we add.

$$
\begin{aligned}
\frac{4}{5} \times 384.8 + 24.8 \div \frac{8}{3} &= 307.84 + 24.8 \cdot \frac{3}{8} \\
&= 307.84 + 9.3 \\
&= 317.14
\end{aligned}
$$

81. We do the multiplications in order from left to right. Then we subtract.

$$
\begin{aligned}
\frac{7}{8} \times 0.86 - 0.76 \times \frac{3}{4} &= 0.7525 - 0.76 \times \frac{3}{4} \\
&= 0.7525 - 0.57 \\
&= 0.1825
\end{aligned}
$$

83. $3.375 \times 5\frac{1}{3} = 3.375 \times \frac{16}{3}$ Writing $5\frac{1}{3}$ using fractional notation

$\phantom{3.375 \times 5\frac{1}{3}} = 18$ Multiplying

85. $6.84 \div 2\frac{1}{2} = 6.84 \div 2.5$ Writing $2\frac{1}{2}$ using decimal notation

$\phantom{6.84 \div 2\frac{1}{2}} = 2.736$ Dividing

87. Discussion and Writing Exercise

89. $9 \cdot 2\frac{1}{3} = \frac{9}{1} \cdot \frac{7}{3} = \frac{9 \cdot 7}{1 \cdot 3} = \frac{3 \cdot 3 \cdot 7}{1 \cdot 3} = \frac{3}{3} \cdot \frac{3 \cdot 7}{1} = 21$

91. $84 \div 8\frac{2}{5} = 84 \div \frac{42}{5} = \frac{84}{1} \cdot \frac{5}{42} = \frac{84 \cdot 5}{42} = $

$\frac{42 \cdot 2 \cdot 5}{42 \cdot 1} = \frac{42}{42} \cdot \frac{2 \cdot 5}{1} = 10$

93. $17\frac{5}{6} + 32\frac{3}{8} = 17\frac{20}{24} + 32\frac{9}{24} = 49\frac{29}{24} = 50\frac{5}{24}$

95. $16\frac{1}{10} - 14\frac{3}{5} = 16\frac{1}{10} - 14\frac{6}{10} = 15\frac{11}{10} - 14\frac{6}{10} = $

$1\frac{5}{10} = 1\frac{1}{2}$

97. *Familiarize.* We draw a picture and let $c = $ the total number of cups of liquid ingredients.

$\frac{2}{3}$ cup	$\frac{1}{4}$ cup	$\frac{1}{8}$ cup
c		

Translate. The problem can be translated to an equation as follows:

Amount of water	plus	Amount of milk	plus	Amount of oil	is	Amount of liquid
↓	↓	↓	↓	↓	↓	↓
$\frac{2}{3}$	$+$	$\frac{1}{4}$	$+$	$\frac{1}{8}$	$=$	c

Solve. We carry out the addition. Since $3 = 3$, $4 = 2 \cdot 2$, and $8 = 2 \cdot 2 \cdot 2$, the LCM of the denominators is $3 \cdot 2 \cdot 2 \cdot 2$, or 24.

$$\frac{2}{3} + \frac{1}{4} + \frac{1}{8} = c$$

$$\frac{2}{3} \cdot \frac{8}{8} + \frac{1}{4} \cdot \frac{6}{6} + \frac{1}{8} \cdot \frac{3}{3} = c$$

$$\frac{16}{24} + \frac{6}{24} + \frac{3}{24} = c$$

$$\frac{25}{24} = c$$

Check. We repeat the calculation. We also note that the sum is larger than any of the individual amounts, as expected.

State. The recipe calls for $\frac{25}{24}$ cups, or $1\frac{1}{24}$ cups, of liquid ingredients.

99. $15 = 3 \cdot 5$

$27 = 3 \cdot 3 \cdot 3$

$30 = 2 \cdot 3 \cdot 5$

$\text{LCM} = 2 \cdot 3 \cdot 3 \cdot 3 \cdot 5$, or 270

101. Using a calculator we find that
$\frac{1}{7} = 1 \div 7 = 0.\overline{142857}$.

103. Using a calculator we find that
$\frac{3}{7} = 3 \div 7 = 0.\overline{428571}$.

105. Using a calculator we find that
$\frac{5}{7} = 5 \div 7 = 0.\overline{714285}$.

107. Using a calculator we find that
$\frac{1}{9} = 1 \div 9 = 0.\overline{1}$.

109. Using a calculator we find that
$\frac{1}{999} = 0.\overline{001}$.

Exercise Set 3.6

1. We are estimating the sum

$$\$109.95 + \$249.95.$$

We round both numbers to the nearest ten. The estimate is

$$\$110 + \$250 = \$360.$$

Answer (d) is correct.

3. We are estimating the difference

$$\$299 - \$249.95.$$

We round both numbers to the nearest ten. The estimate is

$$\$300 - \$250 = \$50.$$

Answer (c) is correct.

5. We are estimating the product

$$9 \times \$299.$$

We round $299 to the nearest ten. The estimate is

$$9 \times \$300 = \$2700.$$

Answer (a) is correct.

7. We are estimating the quotient

$$\$1700 \div \$299.$$

Rounding $299, we get $300. Since $1700 is close to $1800, which is a multiple of $300, we estimate

$$\$1800 \div \$300,$$

so the answer is about 6.

Answer (c) is correct.

9. This is about $0.0 + 1.3 + 0.3$, so the answer is about 1.6.

11. This is about $6 + 0 + 0$, so the answer is about 6.

13. This is about $52 + 1 + 7$, so the answer is about 60.

15. This is about $2.7 - 0.4$, so the answer is about 2.3.

17. This is about $200 - 20$, so the answer is about 180.

19. This is about 50×8, rounding 49 to the nearest ten and 7.89 to the nearest one, so the answer is about 400. Answer (a) is correct.

21. This is about 100×0.08, rounding 98.4 to the nearest ten and 0.083 to the nearest hundredth, so the answer is about 8. Answer (c) is correct.

23. This is about $4 \div 4$, so the answer is about 1. Answer (b) is correct.

25. This is about $75 \div 25$, so the answer is about 3. Answer (b) is correct.

27. We estimate the quotient $1454 \text{ ft} \div 0.39166 \text{ ft}$. We round as follows:

$$1500 \text{ ft} \div 0.5 \text{ ft} = 3000$$

It would take about 3000 PDAs to reach the top of the Sears Tower. Answers will vary depending on the estimate chosen.

29. Discussion and Writing Exercise

31.
$$
\begin{array}{r}
3 \\
3\overline{\smash{)}9} \\
3\overline{\smash{)}27} \\
2\overline{\smash{)}54} \\
2\overline{\smash{)}108}
\end{array}
$$
$108 = 2 \cdot 2 \cdot 3 \cdot 3 \cdot 3$, or $2^2 \cdot 3^3$

33.
$$
\begin{array}{r}
13 \\
5\overline{\smash{)}65} \\
5\overline{\smash{)}325}
\end{array}
$$
$325 = 5 \cdot 5 \cdot 13$, or $5^2 \cdot 13$

35.
$$
\begin{array}{r}
3 \\
3\overline{\smash{)}9} \\
3\overline{\smash{)}27} \\
2\overline{\smash{)}54} \\
2\overline{\smash{)}108} \\
2\overline{\smash{)}216} \\
2\overline{\smash{)}432} \\
2\overline{\smash{)}864} \\
2\overline{\smash{)}1728}
\end{array}
$$
$1728 = 2 \cdot 2 \cdot 2 \cdot 2 \cdot 2 \cdot 2 \cdot 3 \cdot 3 \cdot 3$, or $2^6 \cdot 3^3$

37. $\dfrac{3225}{6275} = \dfrac{25 \cdot 129}{25 \cdot 251} = \dfrac{25}{25} \cdot \dfrac{129}{251} = \dfrac{129}{251}$

39. $\dfrac{325}{625} = \dfrac{25 \cdot 13}{25 \cdot 25} = \dfrac{25}{25} \cdot \dfrac{13}{25} = \dfrac{13}{25}$

41. We round each factor to the nearest ten. The estimate is $180 \times 60 = 10,800$. The estimate is close to the result given, so the decimal point was placed correctly.

43. We round each number on the left to the nearest one. The estimate is $19 - 1 \times 4 = 19 - 4 = 15$. The estimate is not close to the result given, so the decimal point was not placed correctly.

45. a) Observe that $2^{13} = 8192 \approx 8000$, $156,876.8 \approx 160,000$, and $8000 \times 20 = 160,000$. Thus, we want to find the product of 2^{13} and a number that is approximately 20. Since $0.37 + 18.78 = 19.15 \approx 20$, we add inside the parentheses and then multiply:
$$(0.37 + 18.78) \times 2^{13} = 156,876.8$$
We can use a calculator to confirm this result.

b) Observe that $312.84 \approx 6 \cdot 50$. We start by multiplying 6.4 and 51.2, getting 327.68. Then we can use a calculator to find that if we add 2.56 to this product and then subtract 17.4, we have the desired result. Thus, we have
$$2.56 + 6.4 \times 51.2 - 17.4 = 312.84.$$

Exercise Set 3.7

1. Familiarize. We let $a =$ the amount by which the high value exceeded the low value, in dollars.

Translate. This is a missing addend situation.

Low value	plus	Amount of increase	is	High value
↓	↓	↓	↓	↓
17.13	+	a	=	27.63

Solve. We solve the equation, subtracting 17.13 on both sides.
$$17.13 + a = 27.63$$
$$17.13 + a - 17.13 = 27.63 - 17.13$$
$$a = 10.50$$

Check. We can check by adding 10.50 to 17.13 to get 27.63. The answer checks.

State. The high value differed from the low value by $10.50.

3. Familiarize. This is a two-step problem. First we find the total cost of the purchase. Then we find the amount of change Roberto received. Let $t =$ the total cost of the purchase, in dollars.

Translate and Solve.

Purchase price	plus	Sales tax	is	Total cost
↓	↓	↓	↓	↓
14.99	+	1.14	=	t

To solve the equation we carry out the addition.
$$
\begin{array}{r}
\overset{1\ \ 1}{1\ 4.9\ 9} \\
+\ \ 1.1\ 4 \\
\hline
1\ 6.1\ 3
\end{array}
$$

Thus, $t = 16.13$.

Now we find the amount of the change.

We visualize the situation. We let $c =$ the amount of change.

$20	
$16.13	c

This is a "take-away" situation.

Amount paid	minus	Amount of purchase	is	Amount of change
↓	↓	↓	↓	↓
$20	−	$16.13	=	c

To solve the equation we carry out the subtraction.
$$
\begin{array}{r}
\overset{1\ \ 9\ \ 9\ \ 10}{2\ 0.\ 0\ 0} \\
-\ 1\ 6.\ 1\ 3 \\
\hline
3.\ 8\ 7
\end{array}
$$

Thus, $c = \$3.87$.

Check. We check by adding 3.87 to 16.13 to get 20. This checks.

State. The change was $3.87.

5. Familiarize. We visualize the situation. We let $n =$ the new temperature.

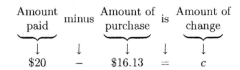

98.6°	4.2°
n	

Translate. We are combining amounts.

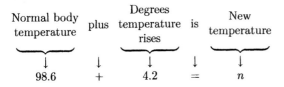

Solve. To solve the equation we carry out the addition.

$$\begin{array}{r} 1 \\ 9\,8.6 \\ +4.2 \\ \hline 1\,0\,2.8 \end{array}$$

Thus, $n = 102.8$.

Check. We can check by repeating the addition. We can also check by rounding:

$$98.6 + 4.2 \approx 99 + 4 = 103 \approx 102.8$$

State. The new temperature was $102.8°$F.

7. *Familiarize*. We visualize the situation. Let $w =$ each winner's share.

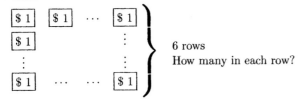

Translate.

Total prize $\div$ Number of winners $=$ Each winner's share

$$127{,}315 \div 6 = w$$

Solve. We carry out the division.

$$\begin{array}{r} 2\,1,2\,1\,9.\,1\,6\,6 \\ 6\,\overline{)\,1\,2\,7,3\,1\,5.\,0\,0\,0} \\ 1\,2\,0\,0\,0\,0 \\ \hline 7\,3\,1\,5 \\ 6\,0\,0\,0 \\ \hline 1\,3\,1\,5 \\ 1\,2\,0\,0 \\ \hline 1\,1\,5 \\ 6\,0 \\ \hline 5\,5 \\ 5\,4 \\ \hline 1\,0 \\ 6 \\ \hline 4\,0 \\ 3\,6 \\ \hline 4\,0 \\ 3\,6 \\ \hline 4 \end{array}$$

Rounding to the nearest cent, or hundredth, we get $w = 21{,}219.17$.

Check. We can repeat the calculation. The answer checks.

State. Each winner's share is \$21,219.17.

9. *Familiarize*. Let $A =$ the area, in sq cm, and $P =$ the perimeter, in cm.

Translate. We use the formulas $A = l \cdot w$ and $P = l + w + l + w$ and substitute 3.25 for l and 2.5 for w.

$$A = l \cdot w = (3.25) \cdot (2.5)$$
$$P = l + w + l + w = 3.25 + 2.5 + 3.25 + 2.5$$

Solve. To find the area we carry out the multiplication.

$$\begin{array}{r} 3.\,2\,5 \\ \times\,2.\,5 \\ \hline 1\,6\,2\,5 \\ 6\,5\,0\,0 \\ \hline 8.\,1\,2\,5 \end{array}$$

Thus, $A = 8.125$

To find the perimeter we carry out the addition.

$$\begin{array}{r} 3.\,2\,5 \\ 2.\,5 \\ 3.\,2\,5 \\ +\,2.\,5 \\ \hline 1\,1.\,5\,0 \end{array}$$

Then $P = 11.5$.

Check. We can obtain partial checks by estimating.

$$(3.25) \times (2.5) \approx 3 \times 3 \approx 9 \approx 8.125$$
$$3.25 + 2.5 + 3.25 + 2.5 \approx 3 + 3 + 3 + 3 = 12 \approx 11.5$$

The answers check.

State. The area of the stamp is 8.125 sq cm, and the perimeter is 11.5 cm.

11. *Familiarize*. We visualize the situation. We let $m =$ the odometer reading at the end of the trip.

22,456.8 mi	234.7 mi
m	

Translate. We are combining amounts.

Reading before trip plus Miles driven is Reading at end of trip

$$22{,}456.8 + 234.7 = m$$

Solve. To solve the equation we carry out the addition.

$$\begin{array}{r} 1\ \ 1 \\ 2\,2{,}4\,5\,6.8 \\ +2\,3\,4.7 \\ \hline 2\,2{,}6\,9\,1.5 \end{array}$$

Thus, $m = 22{,}691.5$.

Check. We can check by repeating the addition. We can also check by rounding:

$$22{,}456.8 + 234.7 \approx 22{,}460 + 230 = 22{,}690 \approx 22{,}691.5$$

State. The odometer reading at the end of the trip was 22,691.5.

13. *Familiarize*. This is a two-step problem. First, we find the number of miles that have been driven between fillups. This is a "how-much-more" situation. We let $n =$ the number of miles driven.

Translate and Solve.

First odometer reading	plus	Number of miles driven	is	Second odometer reading
↓	↓	↓	↓	↓
26,342.8	+	n	=	26,736.7

To solve the equation we subtract 26,342.8 on both sides.

$n = 26,736.7 - 26,342.8$
$n = 393.9$

$$\begin{array}{r} 2\,6,7\,3\,6.7 \\ -\ 2\,6,3\,4\,2.8 \\ \hline 3\,9\,3.9 \end{array}$$

Second, we divide the total number of miles driven by the number of gallons. This gives us $m =$ the number of miles per gallon.

$$393.9 \div 19.5 = m$$

To find the number m, we divide.

$$\begin{array}{r} 2\,0.2 \\ 19.5_\wedge \overline{)3\,9\,3.\,9_\wedge 0} \\ 3\,9\,0\,0 \\ \hline 3\,9\,0 \\ 3\,9\,0 \\ \hline 0 \end{array}$$

Thus, $m = 20.2$.

Check. To check, we first multiply the number of miles per gallon times the number of gallons:

$$19.5 \times 20.2 = 393.9$$

Then we add 393.9 to 26,342.8:

$$26,342.8 + 393.9 = 26,736.7$$

The number 20.2 checks.

State. The driver gets 20.2 miles per gallon.

15. Familiarize. This is a two-step problem. First, we find the number of games that can be played in one hour. Think of an array containing 60 minutes (1 hour = 60 minutes) with 1.5 minutes in each row. We want to find how many rows there are. We let g represent this number.

Translate and Solve. We think (Number of minutes) ÷ (Number of minutes per game) = (Number of games).

$$60 \div 1.5 = g$$

To solve the equation we carry out the division.

$$\begin{array}{r} 4\,0. \\ 1.5_\wedge \overline{)6\,0.\,0_\wedge} \\ 6\,0\,0 \\ \hline 0 \\ 0 \\ \hline 0 \end{array}$$

Thus, $g = 40$.

Second, we find the cost t of playing 40 video games. Repeated addition fits this situation. (We express 25¢ as $0.25.)

Cost of one game	times	Number of games played	is	Total cost
↓	↓	↓	↓	↓
0.25	×	40	=	t

To solve the equation we carry out the multiplication.

$$\begin{array}{r} 0.2\,5 \\ \times\ \ \ 4\,0 \\ \hline 1\,0.0\,0 \end{array}$$

Thus, $t = 10$.

Check. To check, we first divide the total cost by the cost per game to find the number of games played:

$$10 \div 0.25 = 40$$

Then we multiply 40 by 1.5 to find the total time:

$$1.5 \times 40 = 60$$

The number 10 checks.

State. It costs $10 to play video games for one hour.

17. Familiarize. We visualize a rectangular array consisting of 748.45 objects with 62.5 objects in each row. We want to find n, the number of rows.

Translate. We think (Total number of pounds) ÷ (Pounds per cubic foot) = (Number of cubic feet).

$$748.45 \div 62.5 = n$$

Solve. We carry out the division.

$$\begin{array}{r} 1\,1.9\,7\,5\,2 \\ 6\,2.5_\wedge \overline{)7\,4\,8.\,4_\wedge 5\,0\,0\,0} \\ 6\,2\,5\,0\,0 \\ \hline 1\,2\,3\,4\,5 \\ 6\,2\,5\,0 \\ \hline 6\,0\,9\,5 \\ 5\,6\,2\,5 \\ \hline 4\,7\,0\,0 \\ 4\,3\,7\,5 \\ \hline 3\,2\,5\,0 \\ 3\,1\,2\,5 \\ \hline 1\,2\,5\,0 \\ 1\,2\,5\,0 \\ \hline 0 \end{array}$$

Thus, $n = 11.9752$.

Check. We obtain a partial check by rounding and estimating:

$$748.45 \div 62.5 \approx 700 \div 70 = 10 \approx 11.9752$$

State. The tank holds 11.9752 cubic feet of water.

19. Familiarize. We let $d =$ the distance around the figure, in cm.

Translate. We are combining lengths.

The sum of the lengths of the 5 sides	is	the distance around the figure
↓	↓	↓
8.9 + 23.8 + 4.7 + 22.1 + 18.6	=	d

Solve. To solve we carry out the addition.

$$\begin{array}{r} {\scriptstyle 2\ \ 3} \\ 8.9 \\ 2\,3.8 \\ 4.7 \\ 2\,2.1 \\ +\ 1\,8.6 \\ \hline 7\,8.1 \end{array}$$

Thus, $d = 78.1$.

Check. To check we can repeat the addition. We can also check by rounding:

$8.9+23.8+4.7+22.1+18.6 \approx 9+24+5+22+19 = 79 \approx 78.1$

State. The distance around the figure is 78.1 cm.

21. Familiarize. Let $d =$ the distance around the figure, in cm. The figure consists of 6 vertical sides, each with length 2.5 cm, and 6 horizontal sides, each with length 2.25 cm.

Translate.

Vertical distances	plus	Horizontal distances	is	Distance around figure
↓	↓	↓	↓	↓
$6 \times (2.5)$	$+$	$6 \times (2.25)$	$=$	d

Solve. We carry out the computation.

$6 \times (2.5) + 6 \times (2.25) = 15 + 13.5 = 28.5$

Thus, $d = 28.5$

Check. We can obtain a partial check by estimating:

$6 \times (2.5) + 6 \times (2.25) \approx 6 \times 3 + 6 \times 2 \approx 18 + 12 \approx 30 \approx 28.5$

The answer checks.

State. The perimeter of the figure is 28.5 cm.

23. Familiarize. This is a multistep problem. First we find the sum s of the two 0.8 cm segments. Then we use this length to find d.

Translate and Solve.

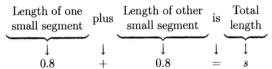

Length of one small segment	plus	Length of other small segment	is	Total length
↓	↓	↓	↓	↓
0.8	$+$	0.8	$=$	s

To solve we carry out the addition.

$$\begin{array}{r} {\scriptstyle 1} \\ 0.8 \\ +\ 0.8 \\ \hline 1.6 \end{array}$$

Thus, $s = 1.6$.

Now we find d.

Total length of smaller segments	plus	length of d	is 3.91 cm
↓	↓	↓	↓ ↓
1.6	$+$	d	$=$ 3.91

To solve we subtract 1.6 on both sides of the equation.

$d = 3.91 - 1.6$ $$\begin{array}{r} 3.9\ 1 \\ -\ 1.6\ 0 \\ \hline 2.3\ 1 \end{array}$$
$d = 2.31$

Check. We repeat the calculations.

State. The length d is 2.31 cm.

25. Familiarize. This is a two-step problem. First, we find how many minutes there are in 2 hr. We let m represent this number. Repeated addition fits this situation (Remember that 1 hr = 60 min.)

Translate and Solve.

Number of minutes in 1 hour	times	Number of hours	is	Total number of minutes
↓	↓	↓	↓	↓
60	$\cdot$	2	$=$	m

To solve the equation we carry out the multiplication.

$$\begin{array}{r} 6\ 0 \\ \times\ \ 2 \\ \hline 1\ 2\ 0 \end{array}$$

Thus, $m = 120$.

Next, we find how many calories are burned in 120 minutes. We let t represent this number. Repeated addition fits this situation also.

Number of calories burned in 1 minute	times	Number of minutes	is	Total number of calories burned
↓	↓	↓	↓	↓
7.3	$\times$	120	$=$	t

To solve the equation we carry out the multiplication.

$$\begin{array}{r} 1\ 2\ 0 \\ \times\ \ 7.\,3 \\ \hline 3\ 6\ 0 \\ 8\ 4\ 0\ 0 \\ \hline 8\ 7\ 6.\,0 \end{array}$$

Thus, $t = 876$.

Check. To check, we first divide the total number of calories by the number of calories burned in one minute to find the total number of minutes the person mowed:

$876 \div 7.3 = 120$

Then we divide 120 by 60 to find the number of hours:

$120 \div 60 = 2$

The number 876 checks.

State. In 2 hr of mowing, 876 calories would be burned.

27. Familiarize. This is a multistep problem. We will first find the total amount of the debits. Then we will find how much is left in the account after the debits are deducted. Finally, we will use this amount and the amount of the deposit to find the balance in the account after all the changes. We will let $d =$ the total amount of the debits.

Translate and Solve. We are combining amounts.

First debit	plus	Second debit	plus	Third debit	is	Total amount of debits
↓	↓	↓	↓	↓	↓	↓
23.82	$+$	507.88	$+$	98.32	$=$	d

To solve the equation we carry out the addition.

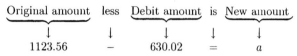

$$\begin{array}{r} {\scriptstyle 1\ 2\ \ 2\ 1} \\ 2\ 3.8\ 2 \\ 5\ 0\ 7.8\ 8 \\ +\quad 9\ 8.3\ 2 \\ \hline 6\ 3\ 0.0\ 2 \end{array}$$

Thus, $d = 630.02$.

Now we let $a =$ the amount in the account after the debits are deducted.

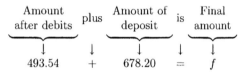

Original amount	less	Debit amount	is	New amount
↓	↓	↓	↓	↓
1123.56	−	630.02	=	a

To solve the equation we carry out the subtraction.

$$\begin{array}{r} {\scriptstyle 10} \\ {\scriptstyle \not{0}\ \ 12} \\ \not{1}\ \not{1}\ \not{2}\ 3.5\ 6 \\ -\quad 6\ 3\ 0.0\ 2 \\ \hline 4\ 9\ 3.5\ 4 \end{array}$$

Thus, $a = 493.54$.

Finally, we let $f =$ the amount in the account after the check is deposited.

Amount after debits	plus	Amount of deposit	is	Final amount
↓	↓	↓	↓	↓
493.54	+	678.20	=	f

We carry out the addition.

$$\begin{array}{r} {\scriptstyle 1\ \ 1} \\ 4\ 9\ 3.5\ 4 \\ +\ 6\ 7\ 8.2\ 0 \\ \hline 1\ 1\ 7\ 1.7\ 4 \end{array}$$

Thus, $f = 1171.74$.

Check. We repeat the calculations.

State. There is $1171.74 in the account after the changes.

29. **Familiarize.** We make and label a drawing. The question deals with a rectangle and a square, so we also list the relevant area formulas. We let $g =$ the area covered by grass.

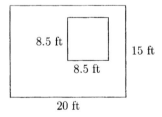

Area of a rectangle with length l and width w: $A = l \times w$

Area of a square with side s: $A = s^2$

Translate. We subtract the area of the square from the area of the rectangle.

Area of rectangle	minus	Area of square	is	Area covered by grass
↓	↓	↓	↓	↓
20×15	−	$(8.5)^2$	=	g

Solve. We carry out the computations.

$$\begin{aligned} 20 \times 15 - (8.5)^2 &= g \\ 20 \times 15 - 72.25 &= g \\ 300 - 72.25 &= g \\ 227.75 &= g \end{aligned}$$

Check. We can repeat the calculations. Also note that 227.75 is less than the area of the yard but more than the area of the flower garden. This agrees with the impression given by our drawing.

State. Grass covers 227.75 ft^2 of the yard.

31. **Familiarize.** The part of the at-bats that were hits is a fraction whose numerator is the number of hits and whose denominator is the number of at-bats. We let $a =$ the part of the at-bats that were hits.

Translate. We think (Number of hits) ÷ (Number of at-bats) = (Part of at-bats that were hits).

$$216 \div 580 = a$$

Solve. We carry out the division.

$$\begin{array}{r} 0.\ 3\ 7\ 2\ 4 \\ 5\ 8\ 0\ \overline{)\ 2\ 1\ 6.\ 0\ 0\ 0\ 0} \\ \underline{1\ 7\ 4\ 0} \\ 4\ 2\ 0\ 0 \\ \underline{4\ 0\ 6\ 0} \\ 1\ 4\ 0\ 0 \\ \underline{1\ 1\ 6\ 0} \\ 2\ 4\ 0\ 0 \\ \underline{2\ 3\ 2\ 0} \\ 8\ 0 \end{array}$$

We stop dividing at this point, because we will round to the nearest thousandth. Thus, $a \approx 0.372$.

Check. We can obtain a partial check by rounding and estimating:

$$216 \div 580 \approx 200 \div 600 = 0.\overline{3} \approx 0.372.$$

State. 0.372 of the at-bats were hits.

33. **Familiarize.** This is a multistep problem. First we find the number of minutes in excess of 400. Then we find the charge for the excess minutes. Finally we add this charge to the monthly charge for 400 minutes to find the total cost for the month. Let $m =$ the number of minutes in excess of 400.

Translate and Solve. First we have a missing addend situation.

First 400 minutes	plus	Excess minutes	is	Total minutes
↓	↓	↓	↓	↓
400	+	m	=	517

We subtract 400 on both sides of the equation.

$$\begin{aligned} 400 + m &= 517 \\ 400 + m - 400 &= 517 - 400 \\ m &= 117 \end{aligned}$$

We see that 117 minutes are charged at the rate of $0.25 per minute. We multiply to find c, the cost of these minutes.

```
        1 1 7
    × 0. 2 5
    ─────────
        5 8 5
    2 3 4 0
    ─────────
    2 9. 2 5
```

Thus, $c = 29.25$.

Finally we add the cost of the first 400 minutes and the cost of the additional 117 minutes to find t, the total cost for the month.

```
    3 9. 9 9
  + 2 9. 2 5
  ───────────
    6 9. 2 4
```

Thus, $t = 69.24$.

Check. We can repeat the calculations. The answer checks.

State. The total cost for the month was $69.24.

35. Familiarize. This is a multistep problem. First we find the number of overtime hours worked. Then we find the pay for the first 40 hours as well as the pay for the overtime hours. Finally we add these amounts to find the total pay. Let h = the number of overtime hours worked.

Translate and Solve. First we have a missing addend situation.

First 40 hours	plus	Overtime hours	is	Total hours
↓	↓	↓	↓	↓
40	+	h	=	46

We subtract 40 on both sides of the equation.

$$40 + h = 46$$
$$40 + h - 40 = 46 - 40$$
$$h = 6$$

Now we multiply to find p, the amount of pay for the first 40 hours.

```
      1 8. 5 0
    ×       4 0
    ───────────
    7 4 0. 0 0
```

Thus, $p = 740$.

Next we multiply to find a, the additional pay for the overtime hours.

```
      2 7. 7 5
    ×         6
    ───────────
    1 6 6. 5 0
```

Then $a = 166.50$.

Finally we add to find t, the total pay.

```
      7 4 0. 0 0
    + 1 6 6. 5 0
    ─────────────
      9 0 6. 5 0
```

We have $t = 906.50$.

Check. We repeat the calculations. The answer checks.

State. The construction worker's pay was $906.50.

37. Familiarize. This is a two-step problem. First we find the number of eggs in 20 dozen (1 dozen = 12). We let n represent this number.

Translate and Solve. We think (Number of dozens) · (Number in a dozen) = (Number of eggs).

$$20 \cdot 12 = n$$
$$240 = n$$

Second, we find the cost c of one egg. We think (Total cost) ÷ (Number of eggs) = (Cost of one egg).

$$\$13.80 \div 240 = c$$

We carry out the division.

```
              0.0 5 7 5
    2 4 0 ⟌ 1 3.8 0 0 0
            1 2 0 0
            ─────────
              1 8 0 0
              1 6 8 0
              ─────────
                1 2 0 0
                1 2 0 0
                ─────────
                      0
```

Thus, $c = 0.0575 \approx 0.058$ (rounded to the nearest tenth of a cent).

Check. We repeat the calculations.

State. Each egg cost about $0.058, or 5.8¢.

39. Familiarize. This is a three-step problem. We will find the area S of a standard soccer field and the area F of a standard football field using the formula Area = $l \cdot w$. Then we will find E, the amount by which the area of a soccer field exceeds the area of a football field.

Translate and Solve.

$$S = l \cdot w = 114.9 \times 74.4 = 8548.56$$
$$F = l \cdot w = 120 \times 53.3 = 6396$$

Area of football field	plus	Excess area of soccer field	is	Area of soccer field
↓	↓	↓	↓	↓
6396	+	E	=	8548.56

To solve the equation we subtract 6396 on both sides.

$$E = 8548.56 - 6396$$
$$E = 2152.56$$

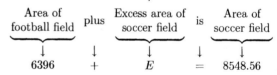

Check. We can obtain a partial check by rounding and estimating:

$$114.9 \times 74.4 \approx 110 \times 75 = 8250 \approx 8548.56$$
$$120 \times 53.3 \approx 120 \times 50 = 6000 \approx 6396$$
$$8250 - 6000 = 2250 \approx 2152.56$$

State. The area of a soccer field is 2152.56 sq yd greater than the area of a football field.

41. We add the population figures, keeping in mind that they are given in billions, and divide by the number of addends, 6.

$$\frac{2.565 + 3.050 + 3.721 + 4.477 + 5.320 + 6.241}{6}$$
$$= \frac{25.374}{6} = 4.229$$

The average population of the world for the years 1950 through 2000 was 4.229 billion.

43. Familiarize. We visualize the situation. We let $t =$ the number of degrees by which the temperature of the bath water exceeds normal body temperature.

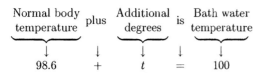

Translate. We have a missing addend situation.

$$\underbrace{\text{Normal body temperature}}_{98.6} \underset{\downarrow}{\text{ plus }} \underbrace{\text{Additional degrees}}_{t} \underset{\downarrow}{\text{ is }} \underbrace{\text{Bath water temperature}}_{100}$$

$$98.6 + t = 100$$

Solve. To solve we subtract 98.6 on both sides of the equation.

$$t = 100 - 98.6$$
$$t = 1.4$$

$$\begin{array}{r} {}^{9\;9\;10} \\ \cancel{1}\,\cancel{0}\,\cancel{0}.\,\cancel{0} \\ -\;\;9\;8.\;6 \\ \hline 1.\;4 \end{array}$$

Check. To check we add 1.4 to 98.6 to get 100. This checks.

State. The temperature of the bath water is 1.4°F above normal body temperature.

45. Familiarize. We let $C =$ the cost of the home in San Francisco. Using the table in Example 8, find the indexes of Hollywood Hills and San Francisco.

Translate. Using the formula given in Example 8, we translate to an equation.

$$C = \$125,000 \div 271 \times 310$$

Solve. We carry out the computation.

$$C = \$125,000 \div 271 \times 310$$
$$\approx \$461.2546125 \times 310 \quad \text{Dividing}$$
$$\approx \$142,989 \qquad \begin{array}{l}\text{Multiplying and rounding} \\ \text{to the nearest one}\end{array}$$

Check. We can repeat the computations. We can also estimate:

$$C = \$125,000 \div 271 \times 310$$
$$\approx \$125,000 \div 300 \times 300$$
$$\approx \$125,000 \approx \$142,989$$

The answer checks.

State. A home selling for $125,000 in Hollywood Hills would cost about $142,989 in San Francisco.

47. Familiarize. We let $C =$ the cost of the home in Tampa. Using the table in Example 8, find the indexes of Indianapolis and Tampa.

Translate. Using the formula given in Example 8, we translate to an equation.

$$C = \$96,000 \div 63 \times 74$$

Solve. We carry out the computation.

$$C = \$96,000 \div 63 \times 74$$
$$\approx \$1523.809524 \times 74 \quad \text{Dividing}$$
$$\approx \$112,762 \qquad \begin{array}{l}\text{Multiplying and rounding} \\ \text{to the nearest one}\end{array}$$

Check. We can repeat the computations. We can also estimate:

$$C = \$96,000 \div 63 \times 74$$
$$\approx \$96,000 \div 60 \times 70$$
$$\approx \$112,000 \approx \$112,762$$

The answer checks.

State. A home selling for $96,000 in Indianapolis would cost about $112,762 in Tampa.

49. Familiarize. We let $C =$ the cost of the home in Atlanta. Using the table in Example 8, find the indexes of San Francisco and Atlanta.

Translate. Using the formula given in Example 8, we translate to an equation.

$$C = \$240,000 \div 310 \times 97$$

Solve. We carry out the computation.

$$C = \$240,000 \div 310 \times 97$$
$$\approx \$774.1935484 \times 97 \quad \text{Dividing}$$
$$\approx \$75,097 \qquad \begin{array}{l}\text{Multiplying and rounding} \\ \text{to the nearest one}\end{array}$$

Check. We can repeat the computations. We can also estimate:

$$C = \$240,000 \div 310 \times 97$$
$$\approx \$240,000 \div 300 \times 100$$
$$\approx \$80,000 \approx \$75,097$$

The answer checks.

State. A home selling for $240,000 in San Francisco would cost about $75,097 in Atlanta.

51. Discussion and Writing Exercise

53.
$$\begin{array}{r} {}^{1\;1\;1} \\ 4\;5\;6\;9 \\ +\;1\;7\;6\;6 \\ \hline 6\;3\;3\;5 \end{array}$$

55.
$$4\,\boxed{\dfrac{1}{3} \cdot \dfrac{2}{2}} = \quad 4\dfrac{2}{6}$$
$$+\,2\,\boxed{\dfrac{1}{2} \cdot \dfrac{3}{3}} = +\,2\dfrac{3}{6}$$
$$\overline{\qquad\qquad\qquad 6\dfrac{5}{6}}$$

57.
$$\dfrac{2}{3} - \dfrac{5}{8} = \dfrac{2}{3} \cdot \dfrac{8}{8} - \dfrac{5}{8} \cdot \dfrac{3}{3}$$
$$= \dfrac{16}{24} - \dfrac{15}{24} = \dfrac{16 - 15}{24}$$
$$= \dfrac{1}{24}$$

59.
$$2\dfrac{2}{7} \cdot 3\dfrac{1}{2} = \dfrac{16}{7} \cdot \dfrac{7}{2}$$
$$= \dfrac{16 \cdot 7}{7 \cdot 2}$$
$$= \dfrac{2 \cdot 8 \cdot 7}{7 \cdot 2 \cdot 1}$$
$$= \dfrac{2 \cdot 7}{2 \cdot 7} \cdot \dfrac{8}{1}$$
$$= 8$$

61. $6\frac{4}{5} \cdot \frac{1}{2} = \frac{34}{5} \cdot \frac{1}{2}$

$\qquad = \frac{34 \cdot 1}{5 \cdot 2}$

$\qquad = \frac{2 \cdot 17 \cdot 1}{5 \cdot 2}$

$\qquad = \frac{2}{2} \cdot \frac{17 \cdot 1}{5}$

$\qquad = \frac{17}{5}$

$\qquad = 3\frac{2}{5}$

63. $8\frac{2}{3} \div 3 = \frac{26}{3} \div 3$

$\qquad = \frac{26}{3} \cdot \frac{1}{3}$

$\qquad = \frac{26 \cdot 1}{3 \cdot 3}$

$\qquad = \frac{26}{9}$

$\qquad = 2\frac{8}{9}$

65. Familiarize. Visualize the situation as a rectangular array containing 469 revolutions with $16\frac{3}{4}$ revolutions in each row. We must determine how many rows the array has. (The last row may be incomplete.) We let t = the time the wheel rotates.

Translate. The division that corresponds to the situation is

$$469 \div 16\frac{3}{4} = t.$$

Solve. We carry out the division.

$$t = 469 \div 16\frac{3}{4} = 469 \div \frac{67}{4} = 469 \cdot \frac{4}{67} =$$

$$\frac{67 \cdot 7 \cdot 4}{67 \cdot 1} = \frac{67}{67} \cdot \frac{7 \cdot 4}{1} = 28$$

Check. We check by multiplying the time by the number of revolutions per minute.

$$16\frac{3}{4} \cdot 28 = \frac{67}{4} \cdot 28 = \frac{67 \cdot 7 \cdot 4}{4 \cdot 1} = \frac{4}{4} \cdot \frac{67 \cdot 7}{1} = 469$$

The answer checks.

State. The water wheel rotated for 28 min.

67. Familiarize. This is a multistep problem. First we will find the total number of cards purchased. Then we will find the number of half-dozens in this number. Finally, we will find the purchase price. Let t = the total number of cards, h = the number of half-dozens, and p = the purchase price. Recall that 1 dozen = 12, so $\frac{1}{2}$ dozen = $\frac{1}{2} \cdot 12 = 6$.

Translate. We write an equation to find the total number of cards purchased.

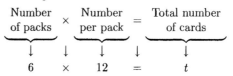

$$6 \times 12 = t$$

Solve. We carry out the multiplication.

$$t = 6 \times 12 = 72$$

Next we divide by 6 to find the number of half-dozens in 72:

$$h = 72 \div 6 = 12$$

Finally, we multiply the number of half-dozens by the price per half dozen to find the purchase price. Twelve dozen cents = $12 \cdot 12\cancel{c} = 144\cancel{c} = \$1.44$.

$$p = 12 \times \$1.44 = \$17.28$$

Check. We can repeat the calculations. The answer checks.

State. The purchase price of the cards is \$17.28.

Chapter 4

Percent Notation

1. The ratio of 4 to 5 is $\frac{4}{5}$.

3. The ratio of 56.78 to 98.35 is $\frac{56.78}{98.35}$.

5. If four of every five fatal accidents involving a Corvette do not involve another vehicle, then $5 - 4$, or 1, involves a Corvette and at least one other vehicle. Thus, the ratio of fatal accidents involving just a Corvette to those involving a Corvette and at least one other vehicle is $\frac{4}{1}$.

7. The ratio of 18 to 24 is $\frac{18}{24} = \frac{3 \cdot 6}{4 \cdot 6} = \frac{3}{4} \cdot \frac{6}{6} = \frac{3}{4}$.

9. The ratio of 2.8 to 3.6 is $\frac{2.8}{3.6} = \frac{2.8}{3.6} \cdot \frac{10}{10} = \frac{28}{36} = \frac{4 \cdot 7}{4 \cdot 9} = \frac{4}{4} \cdot \frac{7}{9} = \frac{7}{9}$.

11. The ratio of length to width is $\frac{478}{213}$.

    The ratio of width to length is $\frac{213}{478}$.

13. $\frac{120 \text{ km}}{3 \text{ hr}} = 40 \frac{\text{km}}{\text{hr}}$

15. $\frac{217 \text{ mi}}{29 \text{ sec}} \approx 7.48 \frac{\text{mi}}{\text{sec}}$

17. $\frac{434 \text{ mi}}{15.5 \text{ gal}} = 28 \text{ mpg}$

19. $\frac{623 \text{ gal}}{1000 \text{ sq ft}} = 0.623 \text{ gal/ft}^2$

21. $\frac{310 \text{ km}}{2.5 \text{ hr}} = 124 \frac{\text{km}}{\text{hr}}$

23. $\frac{1500 \text{ beats}}{60 \text{ min}} = 25 \frac{\text{beats}}{\text{min}}$

25. We can use cross-products:

$5 \cdot 9 = 45 \qquad 6 \cdot 7 = 42$

Since the cross-products are not the same, $45 \neq 42$, we know that the numbers are not proportional.

27. We can use cross-products:

$1 \cdot 20 = 20 \qquad 2 \cdot 10 = 20$

Since the cross-products are the same, $20 = 20$, we know that $\frac{1}{2} = \frac{10}{20}$, so the numbers are proportional.

29. We can use cross-products:

$2.4 \cdot 2.7 = 6.48 \qquad 3.6 \cdot 1.8 = 6.48$

Since the cross-products are the same, $6.48 = 6.48$, we know that $\frac{2.4}{3.6} = \frac{1.8}{2.7}$, so the numbers are proportional.

31. We can use cross-products:

$5\frac{1}{3} \cdot 9\frac{1}{2} = 50\frac{2}{3} \qquad 8\frac{1}{4} \cdot 2\frac{1}{5} = 18\frac{3}{20}$

Since the cross-products are not the same, $50\frac{2}{3} \neq 18\frac{3}{20}$, we know that the numbers are not proportional.

33.
$$\frac{18}{4} = \frac{x}{10}$$
$18 \cdot 10 = 4 \cdot x \qquad$ Equating cross-products
$\dfrac{18 \cdot 10}{4} = \dfrac{4 \cdot x}{4} \qquad$ Dividing by 4
$\dfrac{18 \cdot 10}{4} = x$
$\dfrac{180}{4} = x \qquad$ Multiplying
$45 = x \qquad$ Dividing

35.
$$\frac{t}{12} = \frac{5}{6}$$
$6 \cdot t = 12 \cdot 5$
$\dfrac{6 \cdot t}{6} = \dfrac{12 \cdot 5}{6}$
$t = \dfrac{12 \cdot 5}{6}$
$t = \dfrac{60}{6}$
$t = 10$

37.
$$\frac{2}{5} = \frac{8}{n}$$
$2 \cdot n = 5 \cdot 8$
$\dfrac{2 \cdot n}{2} = \dfrac{5 \cdot 8}{2}$
$n = \dfrac{5 \cdot 8}{2}$
$n = \dfrac{40}{2}$
$n = 20$

39.
$$\frac{16}{12} = \frac{24}{x}$$
$$16 \cdot x = 12 \cdot 24$$
$$\frac{16 \cdot x}{16} = \frac{12 \cdot 24}{6}$$
$$x = \frac{12 \cdot 24}{16}$$
$$x = \frac{288}{16}$$
$$x = 18$$

41.
$$\frac{t}{0.16} = \frac{0.15}{0.40}$$
$$0.40 \times t = 0.16 \times 0.15$$
$$\frac{0.40 \times t}{0.40} = \frac{0.16 \times 0.15}{0.40}$$
$$t = \frac{0.16 \times 0.15}{0.40}$$
$$t = \frac{0.024}{0.40}$$
$$t = 0.06$$

43.
$$\frac{100}{25} = \frac{20}{n}$$
$$100 \cdot n = 25 \cdot 20$$
$$\frac{100 \cdot n}{100} = \frac{25 \cdot 20}{100}$$
$$n = \frac{25 \cdot 20}{100}$$
$$n = \frac{500}{100}$$
$$n = 5$$

45.
$$\frac{\frac{1}{4}}{\frac{1}{2}} = \frac{\frac{1}{2}}{x}$$
$$\frac{1}{4} \cdot x = \frac{1}{2} \cdot \frac{1}{2}$$
$$\frac{\frac{1}{4} \cdot x}{\frac{1}{4}} = \frac{\frac{1}{2} \cdot \frac{1}{2}}{\frac{1}{4}}$$
$$x = \frac{\frac{1}{2} \cdot \frac{1}{2}}{\frac{1}{4}}$$
$$x = \frac{\frac{1}{4}}{\frac{1}{4}}$$
$$x = 1$$

47.
$$\frac{1.28}{3.76} = \frac{4.28}{y}$$
$$1.28 \times y = 3.76 \times 4.28$$
$$\frac{1.28 \times y}{1.28} = \frac{3.76 \times 4.28}{1.28}$$
$$y = \frac{3.76 \times 4.28}{1.28}$$
$$y = \frac{16.0928}{1.28}$$
$$y = 12.5725$$

49. ***Familiarize.*** Let n = the number of Americans who would be considered overweight.

Translate. We translate to a proportion.

$$\text{Overweight} \rightarrow \frac{60}{100} = \frac{n}{281,000,000} \begin{array}{l} \leftarrow \text{Overweight} \\ \leftarrow \quad \text{Total} \end{array}$$
$$\text{Total} \quad \rightarrow$$

Solve. We solve the proportion.

$$60 \cdot 281,000,000 = 100 \cdot n \quad \text{Equating cross products}$$
$$\frac{60 \cdot 281,000,000}{100} = \frac{100 \cdot n}{100}$$
$$\frac{60 \cdot 281,000,000}{100} = n$$
$$168,600,000 = n$$

Check. We substitute in the proportion and check cross products.

$$\frac{60}{100} = \frac{168,600,000}{281,000,000}$$
$$60 \cdot 281,000,000 = 16,860,000,000$$
$$100 \cdot 168,600,000 = 16,860,000,000$$

The cross products are the same.

State. 168,600,000, or 168.6 million, Americans would be considered overweight.

51. ***Familiarize.*** Let g = the number of gallons of gasoline needed to travel 126 mi.

Translate. We translate to a proportion.

$$\begin{array}{l} \text{Miles} \rightarrow \\ \text{Gallons} \rightarrow \end{array} \frac{84}{6.5} = \frac{126}{g} \begin{array}{l} \leftarrow \text{Miles} \\ \leftarrow \text{Gallons} \end{array}$$

Solve.

$$84 \cdot g = 6.5 \cdot 126 \quad \text{Equating cross products}$$
$$g = \frac{6.5 \cdot 126}{84} \quad \text{Dividing by 84}$$
$$g = \frac{819}{84}$$
$$g = 9.75$$

Check. We substitute in the proportion and check cross products.

$$\frac{84}{6.5} = \frac{126}{9.75}$$
$$84 \cdot 9.75 = 819; \ 6.5 \cdot 126 = 819$$

The cross products are the same.

State. 9.75 gallons of gasoline are needed to travel 126 mi.

53. ***Familiarize.*** Let d = the number of defective bulbs in a lot of 2500.

Translate. We translate to a proportion.

$$\begin{array}{l} \text{Defective bulbs} \rightarrow \\ \text{Bulbs in lot} \quad \rightarrow \end{array} \frac{7}{100} = \frac{d}{2500} \begin{array}{l} \leftarrow \text{Defective bulbs} \\ \leftarrow \quad \text{Bulbs in lot} \end{array}$$

Solve.
$$7 \cdot 2500 = 100 \cdot d$$
$$\frac{7 \cdot 2500}{100} = d$$
$$\frac{7 \cdot 25 \cdot 100}{100} = d$$
$$7 \cdot 25 = d$$
$$175 = d$$

Check. We substitute in the proportion and check cross products.
$$\frac{7}{100} = \frac{175}{2500}$$
$$7 \cdot 2500 = 17,500; \ 100 \cdot 175 = 17,500$$

State. There will be 175 defective bulbs in a lot of 2500.

55. Familiarize. Let s = the number of square feet of siding that Fred can paint with 7 gal of paint.

Translate. We translate to a proportion.
$$\text{Gallons} \rightarrow \ \frac{3}{1275} = \frac{7}{s} \ \leftarrow \text{Gallons}$$
$$\text{Siding} \rightarrow \qquad \qquad \leftarrow \text{Siding}$$

Solve.
$$3 \cdot s = 1275 \cdot 7$$
$$s = \frac{1275 \cdot 7}{3}$$
$$s = \frac{3 \cdot 425 \cdot 7}{3}$$
$$s = 425 \cdot 7$$
$$s = 2975$$

Check. We find the number of square feet covered by 1 gallon of paint and then multiply that number by 7.
$$1275 \div 3 = 425 \text{ and } 425 \cdot 7 = 2975$$
The answer checks.

State. Fred can paint 2975 ft^2 of siding with 7 gal of paint.

57. a) Familiarize. Let a = the number of Australian dollars equivalent to 250 U.S. dollars.

Translate. We translate to a proportion.
$$\begin{array}{c} \text{U.S.} \ \rightarrow \\ \text{Australian} \rightarrow \end{array} \frac{1}{1.80} = \frac{250}{a} \begin{array}{c} \leftarrow \ \text{U.S.} \\ \leftarrow \text{Australian} \end{array}$$

Solve.
$$1 \cdot a = 1.80 \cdot 250 \quad \text{Equating cross products}$$
$$a = 450$$

Check. We substitute in the proportion and check cross products.
$$\frac{1}{1.80} = \frac{250}{450}$$
$$1 \cdot 450 = 450; \ 1.80 \cdot 250 = 450$$
The cross products are the same.

State. 250 U.S. dollars would be worth 450 Australian dollars.

b) Familiarize. Let c = the cost of the sweatshirt in U.S. dollars.

Translate. We translate to a proportion.
$$\begin{array}{c} \text{U.S.} \ \rightarrow \\ \text{Australian} \rightarrow \end{array} \frac{1}{180} = \frac{c}{50} \begin{array}{c} \leftarrow \ \text{U.S.} \\ \leftarrow \text{Australian} \end{array}$$

Solve.
$$1 \cdot 50 = 1.80 \cdot c \quad \text{Equating cross products}$$
$$\frac{1 \cdot 50}{1.80} = c$$
$$27.78 \approx c$$

Check. We substitute in the proportion and check cross products.
$$\frac{1}{1.80} = \frac{27.78}{50}$$
$$1 \cdot 50 = 50; \ 1.80 \cdot 27.78 = 50.004 \approx 50$$
The cross products are about the same. Remember that we rounded the value of c.

State. The sweatshirt cost $27.78 in U.S. dollars.

59. a) Familiarize. Let g = the number of gallons of gasoline needed to drive 2690 mi.

Translate. We translate to a proportion.
$$\begin{array}{c} \text{Gallons} \rightarrow \\ \text{Miles} \ \rightarrow \end{array} \frac{15.5}{434} = \frac{g}{2690} \begin{array}{c} \leftarrow \text{Gallons} \\ \leftarrow \ \text{Miles} \end{array}$$

Solve.
$$15.5 \cdot 2690 = 434 \cdot g \quad \text{Equating cross products}$$
$$\frac{15.5 \cdot 2690}{434} = g$$
$$96.07 \approx g$$

Check. We find how far the car can be driven on 1 gallon of gasoline and then divide to find the number of gallons required for a 2690-mi trip.
$$434 \div 15.5 = 28 \text{ and } 2690 \div 28 \approx 96$$
The answer checks.

State. It will take about 96 gal of gasoline to drive 2690 mi.

b) Familiarize. Let d = the number of miles the car can be driven on 140 gal of gasoline.

Translate. We translate to a proportion.
$$\begin{array}{c} \text{Gallons} \rightarrow \\ \text{Miles} \ \rightarrow \end{array} \frac{15.5}{434} = \frac{140}{d} \begin{array}{c} \leftarrow \text{Gallons} \\ \leftarrow \ \text{Miles} \end{array}$$

Solve.
$$15.5 \cdot d = 434 \cdot 140 \quad \text{Equating cross products}$$
$$d = \frac{434 \cdot 140}{15.5}$$
$$d = 3920$$

Check. From the check in part (a) we know that the car can be driven 28 mi on 1 gal of gasoline. We multiply to find how far it can be driven on 140 gal.
$$140 \cdot 28 = 3920$$

The answer checks.

State. The car can be driven 3920 mi on 140 gal of gasoline.

61. **Familiarize**. Let p = the number of gallons of paint Helen should buy.

Translate. We translate to a proportion.

$$\text{Area} \to \frac{950}{2} = \frac{30,000}{p} \gets \text{Area} \atop \gets \text{Paint}$$

Area → 950   30,000 ← Area
Paint → 2 = p ← Paint

Solve.

$$950 \cdot p = 2 \cdot 30,000$$
$$p = \frac{2 \cdot 30,000}{950}$$
$$p = \frac{2 \cdot 50 \cdot 600}{19 \cdot 50}$$
$$p = \frac{2 \cdot 600}{19}$$
$$p = \frac{1200}{19}, \text{ or } 63\frac{3}{19}$$

Check. We find the area covered by 1 gal of paint and then divide to find the number of gallons needed to paint a 30,000-ft^2 wall.

$$950 \div 2 = 475 \text{ and } 30,000 \div 475 = 63\frac{3}{39}$$

The answer checks.

State. Since Helen is buying paint in one gallon cans, she will have to buy 64 cans of paint.

63. **Familiarize**. Let D = the number of deer in the game preserve.

Translate. We translate to a proportion.

Deer tagged originally → 318   56 ← Tagged deer caught later

Deer in game preserve → D = 168 ← Deer caught later

$$\frac{318}{D} = \frac{56}{168}$$

Solve.

$$318 \cdot 168 = 56 \cdot D$$
$$\frac{318 \cdot 168}{56} = D$$
$$954 = D$$

Check. We substitute in the proportion and check cross-products.

$$\frac{318}{954} = \frac{56}{168}; \ 318 \cdot 168 = 53,424; \ 954 \cdot 56 = 53,424$$

Since the cross-products are the same, the answer checks.

State. We estimate that there are 954 deer in the game preserve.

65. **Familiarize**. Let d = the actual distance between the cities.

Translate. We translate to a proportion.

Map distance → $\frac{1}{16.6}$ 3.5 ← Map distance

Actual distance → 16.6 = d ← Actual distance

Solve.

$$1 \cdot d = 16.6 \cdot 3.5$$
$$d = 58.1$$

Check. We use a different approach. Since 1 in. represents 16.6 mi, we multiply 16.6 by 3.5:

$$3.5(16.6) = 58.1$$

The answer checks.

State. The cities are 58.1 mi apart.

67. **Familiarize**. Let z = the number of pounds of zinc in the alloy.

Translate. We translate to a proportion.

Zinc → $\frac{3}{13}$ z ← Zinc

Copper → 13 = 520 ← Copper

Solve.

$$3 \cdot 520 = 13 \cdot z$$
$$\frac{3 \cdot 520}{13} = z$$
$$\frac{3 \cdot 13 \cdot 40}{13} = z$$
$$3 \cdot 40 = z$$
$$120 = z$$

Check. We substitute in the proportion and check cross products.

$$\frac{3}{13} = \frac{120}{520}$$
$$3 \cdot 520 = 1560; \ 13 \cdot 120 = 1560$$

The cross products are the same.

State. There are 120 lb of zinc in the alloy.

69. Discussion and Writing Exercise

71.
```
      6 5
  4 ) 2 6 0
      2 4 0
      -----
        2 0
        2 0
        ---
         0
```
The answer is 65.

73.
```
       2 9 0. 5
  1 6 ) 4 6 4 8. 0
       3 2 0 0
       -------
       1 4 4 8
       1 4 4 0
       -------
           8 0
           8 0
           ---
            0
```
The answer is 290.5.

75. **Familiarize**. Let f = the number of faculty positions required to maintain the current student-to-faculty ratio after the university expands.

Translate. We translate to a proportion.

Students → 2700   2900 ← Students

Faculty → 217 = f ← Faculty

Solve.

$$2700 \cdot f = 217 \cdot 2900$$
$$f = \frac{217 \cdot 2900}{2700}$$
$$f = \frac{6293}{27}, \text{ or } 233\frac{2}{27}$$

Since it is impossible to create a fractional part of a position, we round up to the nearest whole position. Thus, 234 positions will be required after the university expands. We subtract to find how many new positions should be created:

$$234 - 217 = 17$$

Check. We substitute in the proportion and check cross-products.

$$\frac{2700}{217} = \frac{2900}{6293/27}; \quad 2700 \cdot \frac{6293}{27} = 629,300;$$

$$217 \cdot 2900 = 629,300$$

State. 17 new faculty positions should be created.

Exercise Set 4.2

1. $90\% = \dfrac{90}{100}$ A ratio of 90 to 100

$90\% = 90 \times \dfrac{1}{100}$ Replacing % with $\times \dfrac{1}{100}$

$90\% = 90 \times 0.01$ Replacing % with $\times 0.01$

3. $12.5\% = \dfrac{12.5}{100}$ A ratio of 12.5 to 100

$12.5\% = 12.5 \times \dfrac{1}{100}$ Replacing % with $\times \dfrac{1}{100}$

$12.5\% = 12.5 \times 0.01$ Replacing % with $\times 0.01$

5. 67%

a) Replace the percent symbol with $\times 0.01$.

    67×0.01

b) Move the decimal point two places to the left.

    0.67.

Thus, 67% = 0.67.

7. 45.6%

a) Replace the percent symbol with $\times 0.01$.

    45.6×0.01

b) Move the decimal point two places to the left.

    0.45.6

Thus, 45.6% = 0.456.

9. 59.01%

a) Replace the percent symbol with $\times 0.01$.

    59.01×0.01

b) Move the decimal point two places to the left.

    0.59.01

Thus, 59.01% = 0.5901.

11. 10%

a) Replace the percent symbol with $\times 0.01$.

    10×0.01

b) Move the decimal point two places to the left.

    0.10.

Thus, 10% = 0.1.

13. 1%

a) Replace the percent symbol with $\times 0.01$.

    1×0.01

b) Move the decimal point two places to the left.

    0.01.

Thus, 1% = 0.01.

15. 200%

a) Replace the percent symbol with $\times 0.01$.

    200×0.01

b) Move the decimal point two places to the left.

    2.00.

Thus, 200% = 2.

17. 0.1%

a) Replace the percent symbol with $\times 0.01$.

    0.1×0.01

b) Move the decimal point two places to the left.

    0.00.1

Thus, 0.1% = 0.001.

19. 0.09%

a) Replace the percent symbol with $\times 0.01$.

    0.09×0.01

b) Move the decimal point two places to the left.

    0.00.09

Thus, 0.09% = 0.0009.

21. 0.18%

a) Replace the percent symbol with $\times 0.01$.

    0.18×0.01

b) Move the decimal point two places to the left.

    0.00.18

Thus, 0.18% = 0.0018.

23. 23.19%

a) Replace the percent symbol with ×0.01.

23.19 × 0.01

b) Move the decimal point two places to the left.

0.23.19

Thus, 23.19% = 0.2319.

25. $14\frac{7}{8}\%$

a) Convert $14\frac{7}{8}$ to decimal notation and replace the percent symbol with ×0.01.

14.875 × 0.01

b) Move the decimal point two places to the left.

0.14.875

Thus, $14\frac{7}{8}\% = 0.14875$.

27. $56\frac{1}{2}\%$

a) Convert $56\frac{1}{2}$ to decimal notation and replace the percent symbol with ×0.01.

56.5 × 0.01

b) Move the decimal point two places to the left.

0.56.5

Thus, $56\frac{1}{2}\% = 0.565$.

29. 40%

a) Replace the percent symbol with ×0.01.

40 × 0.01

b) Move the decimal point two places to the left.

0.40.

Thus, 40% = 0.4.

31. 18.6%

a) Replace the percent symbol with ×0.01.

18.6 × 0.01

b) Move the decimal point two places to the left.

0.18.6

Thus, 18.6% = 0.186.

33. 29%

a) Replace the percent symbol with ×0.01.

29 × 0.01

b) Move the decimal point two places to the left.

0.29.

Thus, 29% = 0.29.

35. 0.47

a) Move the decimal point two places to the right.

0.47.

b) Write a percent symbol: 47%

Thus, 0.47 = 47%.

37. 0.03

a) Move the decimal point two places to the right.

0.03.

b) Write a percent symbol: 3%

Thus, 0.03 = 3%.

39. 8.7

a) Move the decimal point two places to the right.

8.70.

b) Write a percent symbol: 870%

Thus, 8.7 = 870%.

41. 0.334

a) Move the decimal point two places to the right.

0.33.4

b) Write a percent symbol: 33.4%

Thus, 0.334 = 33.4%.

43. 0.75

a) Move the decimal point two places to the right.

0.75.

b) Write a percent symbol: 75%

Thus, 0.75 = 75%.

45. 0.4

a) Move the decimal point two places to the right.

0.40.

b) Write a percent symbol: 40%

Thus, 0.4 = 40%.

47. 0.006

a) Move the decimal point two places to the right.

0.00.6

b) Write a percent symbol: 0.6%

Thus, 0.006 = 0.6%.

49. 0.017

a) Move the decimal point two places to the right.

0.01.7

b) Write a percent symbol: 1.7%

Thus, 0.017 = 1.7%.

51. 0.2718

a) Move the decimal point two places to the right.

0.27.18

b) Write a percent symbol: 27.18%

Thus, 0.2718 = 27.18%.

53. 0.0239

a) Move the decimal point two places to the right.

0.02.39

b) Write a percent symbol: 2.39%

Thus, 0.0239 = 2.39%.

55. 0.526

a) Move the decimal point two places to the right.

0.52.6

b) Write a percent symbol: 52.6%

Thus, 0.526 = 52.6%.

57. 0.17

a) Move the decimal point two places to the right.

0.17.

b) Write a percent symbol: 17%

Thus, 0.17 = 17%.

59. 0.411

a) Move the decimal point two places to the right.

0.41.1

b) Write a percent symbol: 41.1%

Thus, 0.411 = 41.1%.

61. Discussion and Writing Exercise

63. To convert $\dfrac{100}{3}$ to a mixed numeral, we divide.

$$
\begin{array}{r}
3\ 3 \\
3\,\overline{\smash{\big)}\,1\ 0\ 0} \\
9\ 0 \\
\hline
1\ 0 \\
9 \\
\hline
1
\end{array}
\qquad \frac{100}{3} = 33\frac{1}{3}
$$

65. To convert $\dfrac{75}{8}$ to a mixed numeral, we divide.

$$
\begin{array}{r}
9 \\
8\,\overline{\smash{\big)}\,7\ 5} \\
7\ 2 \\
\hline
3
\end{array}
\qquad \frac{75}{8} = 9\frac{3}{8}
$$

67. To convert $\dfrac{567}{98}$ to a mixed numeral, we divide.

$$
\begin{array}{r}
5 \\
9\ 8\,\overline{\smash{\big)}\,5\ 6\ 7} \\
4\ 9\ 0 \\
\hline
7\ 7
\end{array}
\qquad \frac{567}{98} = 5\frac{77}{98} = 5\frac{11}{14}
$$

69. To convert $\dfrac{2}{3}$ to decimal notation, we divide.

$$
\begin{array}{r}
0.6\ 6 \\
3\,\overline{\smash{\big)}\,2.0\ 0} \\
1\ 8 \\
\hline
2\ 0 \\
1\ 8 \\
\hline
2
\end{array}
$$

Since 2 keeps reappearing as a remainder, the digits repeat and

$$\frac{2}{3} = 0.66\ldots \quad \text{or} \quad 0.\overline{6}.$$

71. To convert $\dfrac{5}{6}$ to decimal notation, we divide.

$$
\begin{array}{r}
0.8\ 3 \\
6\,\overline{\smash{\big)}\,5.0\ 0} \\
4\ 8 \\
\hline
2\ 0 \\
1\ 8 \\
\hline
2
\end{array}
$$

Since 2 keeps reappearing as a remainder, the digits repeat and

$$\frac{5}{6} = 0.833\ldots \quad \text{or} \quad 0.8\overline{3}.$$

73. To convert $\dfrac{8}{3}$ to decimal notation, we divide.

$$
\begin{array}{r}
2.6\ 6 \\
3\,\overline{\smash{\big)}\,8.0\ 0} \\
6 \\
\hline
2\ 0 \\
1\ 8 \\
\hline
2\ 0
\end{array}
$$

Since 2 keeps reappearing as a remainder, the digits repeat and

$$\frac{8}{3} = 2.66\ldots \text{ or } 2.\overline{6}.$$

Exercise Set 4.3

1. We use the definition of percent as a ratio.

$$\frac{41}{100} = 41\%$$

3. We use the definition of percent as a ratio.

$$\frac{5}{100} = 5\%$$

5. We multiply by 1 to get 100 in the denominator.
$$\frac{2}{10} = \frac{2}{10} \cdot \frac{10}{10} = \frac{20}{100} = 20\%$$

7. We multiply by 1 to get 100 in the denominator.
$$\frac{3}{10} = \frac{3}{10} \cdot \frac{10}{10} = \frac{30}{100} = 30\%$$

9. $\frac{1}{2} = \frac{1}{2} \cdot \frac{50}{50} = \frac{50}{100} = 50\%$

11. Find decimal notation by division.

```
        0.8 7 5
    8 ┌7.0 0 0
        6 4
        ───
          6 0
          5 6
          ───
            4 0
            4 0
            ───
              0
```

$$\frac{7}{8} = 0.875$$

Convert to percent notation.

```
    0.87.5
      └─↑
```

$$\frac{7}{8} = 87.5\%, \text{ or } 87\frac{1}{2}\%$$

13. $\frac{4}{5} = \frac{4}{5} \cdot \frac{20}{20} = \frac{80}{100} = 80\%$

15. Find decimal notation by division.

```
        0.6 6 6
    3 ┌2.0 0 0
        1 8
        ───
          2 0
          1 8
          ───
            2 0
            1 8
            ───
              2
```

We get a repeating decimal: $\frac{2}{3} = 0.66\overline{6}$

Convert to percent notation.

```
    0.66.6̄
      └─↑
```

$$\frac{2}{3} = 66.\overline{6}\%, \text{ or } 66\frac{2}{3}\%$$

17.
```
        0.1 6 6
    6 ┌1.0 0 0
        6
        ───
          4 0
          3 6
          ───
            4 0
            3 6
            ───
              4
```

We get a repeating decimal: $\frac{1}{6} = 0.16\overline{6}$

Convert to percent notation.

```
    0.16.6̄
      └─↑
```

$$\frac{1}{6} = 16.\overline{6}\%, \text{ or } 16\frac{2}{3}\%$$

19.
```
        0.1 8 7 5
   1 6 ┌3.0 0 0 0
        1 6
        ───
        1 4 0
        1 2 8
        ─────
          1 2 0
          1 1 2
          ─────
              8 0
              8 0
              ───
                0
```

$$\frac{3}{16} = 0.1875$$

Convert to percent notation.

```
    0.18.75
      └─↑
```

$$\frac{3}{16} = 18.75\%, \text{ or } 18\frac{3}{4}\%$$

21.
```
        0.8 1 2 5
   1 6 ┌1 3.0 0 0 0
        1 2 8
        ─────
            2 0
            1 6
            ───
              4 0
              3 2
              ───
                8 0
                8 0
                ───
                  0
```

$$\frac{13}{16} = 0.8125$$

Convert to percent notation.

```
    0.81.25
      └─↑
```

$$\frac{13}{16} = 81.25\%, \text{ or } 81\frac{1}{4}\%$$

23. $\frac{4}{25} = \frac{4}{25} \cdot \frac{4}{4} = \frac{16}{100} = 16\%$

25. $\frac{1}{20} = \frac{1}{20} \cdot \frac{5}{5} = \frac{5}{100} = 5\%$

27. $\frac{17}{50} = \frac{17}{50} \cdot \frac{2}{2} = \frac{34}{100} = 34\%$

29. $\frac{2}{25} = \frac{2}{25} \cdot \frac{4}{4} = \frac{8}{100} = 8\%$

31. $\frac{21}{100} = 21\%$ Using the definition of percent as a ratio

33. $\frac{6}{25} = \frac{6}{25} \cdot \frac{4}{4} = \frac{24}{100} = 24\%$

35. $85\% = \frac{85}{100}$ Definition of percent

$$= \frac{5 \cdot 17}{5 \cdot 20}$$
$$= \frac{5}{5} \cdot \frac{17}{20} \Bigg\} \text{ Simplifying}$$
$$= \frac{17}{20}$$

37. $62.5\% = \dfrac{62.5}{100}$ Definition of percent

$= \dfrac{62.5}{100} \cdot \dfrac{10}{10}$ Multiplying by 1 to eliminate the decimal point in the numerator

$= \dfrac{625}{1000}$

$= \dfrac{5 \cdot 125}{8 \cdot 125}$

$= \dfrac{5}{8} \cdot \dfrac{125}{125}$ Simplifying

$= \dfrac{5}{8}$

39. $33\dfrac{1}{3}\% = \dfrac{100}{3}\%$ Converting from mixed numeral to fractional notation

$= \dfrac{100}{3} \times \dfrac{1}{100}$ Definition of percent

$= \dfrac{100 \cdot 1}{3 \cdot 100}$ Multiplying

$= \dfrac{1}{3} \cdot \dfrac{100}{100}$ Simplifying

$= \dfrac{1}{3}$

41. $16.\overline{6}\% = 16\dfrac{2}{3}\%$ $\left(16.\overline{6} = 16\dfrac{2}{3}\right)$

$= \dfrac{50}{3}\%$ Converting from mixed numeral to fractional notation

$= \dfrac{50}{3} \times \dfrac{1}{100}$ Definition of percent

$= \dfrac{50 \cdot 1}{3 \cdot 50 \cdot 2}$ Multiplying

$= \dfrac{1}{2 \cdot 3} \cdot \dfrac{50}{50}$ Simplifying

$= \dfrac{1}{6}$

43. $7.25\% = \dfrac{7.25}{100} = \dfrac{7.25}{100} \cdot \dfrac{100}{100}$

$= \dfrac{725}{10,000} = \dfrac{29 \cdot 25}{400 \cdot 25} = \dfrac{29}{400} \cdot \dfrac{25}{25}$

$= \dfrac{29}{400}$

45. $0.8\% = \dfrac{0.8}{100} = \dfrac{0.8}{100} \cdot \dfrac{10}{10}$

$= \dfrac{8}{1000} = \dfrac{1 \cdot 8}{125 \cdot 8} = \dfrac{1}{125} \cdot \dfrac{8}{8}$

$= \dfrac{1}{125}$

47. $25\dfrac{3}{8}\% = \dfrac{203}{8}\%$

$= \dfrac{203}{8} \times \dfrac{1}{100}$ Definition of percent

$= \dfrac{203}{800}$

49. $78\dfrac{2}{9}\% = \dfrac{704}{9}\%$

$= \dfrac{704}{9} \times \dfrac{1}{100}$ Definition of percent

$= \dfrac{4 \cdot 176 \cdot 1}{9 \cdot 4 \cdot 25}$

$= \dfrac{4}{4} \cdot \dfrac{176 \cdot 1}{9 \cdot 25}$

$= \dfrac{176}{225}$

51. $64\dfrac{7}{11}\% = \dfrac{711}{11}\%$

$= \dfrac{711}{11} \times \dfrac{1}{100}$

$= \dfrac{711}{1100}$

53. $150\% = \dfrac{150}{100} = \dfrac{3 \cdot 50}{2 \cdot 50} = \dfrac{3}{2} \cdot \dfrac{50}{50} = \dfrac{3}{2}$

55. $0.0325\% = \dfrac{0.0325}{100} = \dfrac{0.0325}{100} \cdot \dfrac{10,000}{10,000} = \dfrac{325}{1,000,000} = \dfrac{25 \cdot 13}{25 \cdot 40,000} = \dfrac{25}{25} \cdot \dfrac{13}{40,000} = \dfrac{13}{40,000}$

57. Note that $33.\overline{3}\% = 33\dfrac{1}{3}\%$ and proceed as in Exercise 39; $33.\overline{3}\% = \dfrac{1}{3}$.

59. $26\% = \dfrac{26}{100}$

$= \dfrac{13 \cdot 2}{50 \cdot 2} = \dfrac{13}{50} \cdot \dfrac{2}{2}$

$= \dfrac{13}{50}$

61. $5\% = \dfrac{5}{100}$

$= \dfrac{1 \cdot 5}{20 \cdot 5} = \dfrac{1}{20} \cdot \dfrac{5}{5}$

$= \dfrac{1}{20}$

63. $6\% = \dfrac{6}{100}$

$= \dfrac{3 \cdot 2}{50 \cdot 2} = \dfrac{3}{50} \cdot \dfrac{2}{2}$

$= \dfrac{3}{50}$

65. $45\% = \dfrac{45}{100}$

$= \dfrac{9 \cdot 5}{20 \cdot 5} = \dfrac{9}{20} \cdot \dfrac{5}{5}$

$= \dfrac{9}{20}$

67. $47\% = \dfrac{47}{100}$

69. $\dfrac{1}{8} = 1 \div 8$

$$
\begin{array}{r}
0.1\,2\,5 \\
8\,\overline{)1.0\,0\,0} \\
\underline{8} \\
2\,0 \\
\underline{1\,6} \\
4\,0 \\
\underline{4\,0} \\
0
\end{array}
$$

$\dfrac{1}{8} = 0.125 = 12\dfrac{1}{2}\%, \text{ or } 12.5\%$

$\dfrac{1}{6} = 1 \div 6$

$$
\begin{array}{r}
0.1\,6\,6 \\
6\,\overline{)1.0\,0\,0} \\
\underline{6} \\
4\,0 \\
\underline{3\,6} \\
4\,0 \\
\underline{3\,6} \\
4
\end{array}
$$

We get a repeating decimal: $0.1\overline{6}$

$0.16.\overline{6}$ $\qquad$ $0.1\overline{6} = 16.\overline{6}\%$

$\dfrac{1}{6} = 0.1\overline{6} = 16.\overline{6}\%, \text{ or } 16\dfrac{2}{3}\%$

$20\% = \dfrac{20}{100} = \dfrac{1}{5} \cdot \dfrac{20}{20} = \dfrac{1}{5}$

$0.20.$ $\qquad$ $20\% = 0.2$

$\dfrac{1}{5} = 0.2 = 20\%$

$0.25.$ $\qquad$ $0.25 = 25\%$

$25\% = \dfrac{25}{100} = \dfrac{1}{4} \cdot \dfrac{25}{25} = \dfrac{1}{4}$

$\dfrac{1}{4} = 0.25 = 25\%$

$33\dfrac{1}{3}\% = \dfrac{100}{3}\% = \dfrac{100}{3} \times \dfrac{1}{100} = \dfrac{100}{300} = \dfrac{1}{3} \cdot \dfrac{100}{100} = \dfrac{1}{3}$

$0.33.\overline{3}$ $\qquad$ $33.\overline{3}\% = 0.33\overline{3}, \text{ or } 0.\overline{3}$

$\dfrac{1}{3} = 0.\overline{3} = 33\dfrac{1}{3}\%, \text{ or } 33.\overline{3}\%$

$37.5\% = \dfrac{37.5}{100} = \dfrac{37.5}{100} \cdot \dfrac{10}{10} = \dfrac{375}{1000} = \dfrac{3}{8} \cdot \dfrac{125}{125} = \dfrac{3}{8}$

$0.37.5$ $\qquad$ $37.5\% = 0.375$

$\dfrac{3}{8} = 0.375 = 37\dfrac{1}{2}\%, \text{ or } 37.5\%$

$40\% = \dfrac{40}{100} = \dfrac{2}{5} \cdot \dfrac{20}{20} = \dfrac{2}{5}$

$0.40.$ $\qquad$ $40\% = 0.4$

$\dfrac{2}{5} = 0.4 = 40\%$

71. $0.50.$ $\qquad$ $0.5 = 50\%$

$50\% = \dfrac{50}{100} = \dfrac{1}{2} \cdot \dfrac{50}{50} = \dfrac{1}{2}$

$\dfrac{1}{2} = 0.5 = 50\%$

$\dfrac{1}{3} = 1 \div 3$

$$
\begin{array}{r}
0.3 \\
3\,\overline{)1.0} \\
\underline{9} \\
1
\end{array}
$$

We get a repeating decimal: $0.\overline{3}$

$0.33.\overline{3}$ $\qquad$ $0.\overline{3} = 33.\overline{3}\%$

$\dfrac{1}{3} = 0.\overline{3} = 33.\overline{3}\%, \text{ or } 33\dfrac{1}{3}\%$

$25\% = \dfrac{25}{100} = \dfrac{25}{25} \cdot \dfrac{1}{4} = \dfrac{1}{4}$

$0.25.$ $\qquad$ $25\% = 0.25$

$\dfrac{1}{4} = 0.25 = 25\%$

$16\dfrac{2}{3}\% = \dfrac{50}{3}\% = \dfrac{50}{3} \times \dfrac{1}{100} = \dfrac{50 \cdot 1}{3 \cdot 2 \cdot 50} = \dfrac{50}{50} \cdot \dfrac{1}{6} = \dfrac{1}{6}$

$\dfrac{1}{6} = 1 \div 6$

$$
\begin{array}{r}
0.1\,6 \\
6\,\overline{)1.0\,0} \\
\underline{6} \\
4\,0 \\
\underline{3\,6} \\
4
\end{array}
$$

We get a repeating decimal: $0.1\overline{6}$

$\dfrac{1}{6} = 0.1\overline{6} = 16\dfrac{2}{3}\%, \text{ or } 16.\overline{6}\%$

$0.12.5$ $\qquad$ $0.125 = 12.5\%$

$$12.5\% = \frac{12.5}{100} = \frac{12.5}{100} \cdot \frac{10}{10} = \frac{125}{1000} = \frac{125}{125} \cdot \frac{1}{8} = \frac{1}{8}$$

$$\frac{1}{8} = 0.125 = 12.5\%, \text{ or } 12\frac{1}{2}\%$$

$$\frac{3}{4} = \frac{3}{4} \cdot \frac{25}{25} = \frac{75}{100} = 75\%$$

$$0.75. \qquad\qquad 75\% = 0.75$$

$$\frac{3}{4} = 0.75 = 75\%$$

$$0.8\overline{3} = 0.83.\overline{3} \qquad 0.8\overline{3} = 83.\overline{3}\%$$

$$83.\overline{3}\% = 83\frac{1}{3}\% = \frac{250}{3}\% = \frac{250}{3} \times \frac{1}{100} = \frac{5 \cdot 50}{3 \cdot 2 \cdot 50} =$$
$$\frac{5}{6} \cdot \frac{50}{50} = \frac{5}{6}$$

$$\frac{5}{6} = 0.8\overline{3} = 83.\overline{3}\%, \text{ or } 83\frac{1}{3}\%$$

$$\frac{3}{8} = 3 \div 8$$

$$\begin{array}{r} 0.3\,7\,5 \\ 8\overline{\smash{)}3.0\,0\,0} \\ \underline{2\,4} \\ 6\,0 \\ \underline{5\,6} \\ 4\,0 \\ \underline{4\,0} \\ 0 \end{array}$$

$$\frac{3}{8} = 0.375$$

$$0.37.5 \qquad\qquad 0.375 = 37.5\%$$

$$\frac{3}{8} = 0.375 = 37.5\%, \text{ or } 37\frac{1}{2}\%$$

73. Discussion and Writing Exercise

75. $13 \cdot x = 910$

$$\frac{13 \cdot x}{13} = \frac{910}{13}$$

$$x = 70$$

77. $0.05 \times b = 20$

$$\frac{0.05 \times b}{0.05} = \frac{20}{0.05}$$

$$b = 400$$

79. $\frac{1}{2} \cdot x = 2$

$$2 \cdot \frac{1}{2} \cdot x = 2 \cdot 2 \quad \text{Multiplying by 2 on both sides}$$

$$x = 4$$

The solution is 4.

81.
$$\begin{array}{r} 3\,3 \\ 3\overline{\smash{)}1\,0\,0} \\ \underline{9\,0} \\ 1\,0 \\ \underline{9} \\ 1 \end{array}$$

$$\frac{100}{3} = 33\frac{1}{3}$$

83.
$$\begin{array}{r} 8\,3 \\ 3\overline{\smash{)}2\,5\,0} \\ \underline{2\,4\,0} \\ 1\,0 \\ \underline{9} \\ 1 \end{array}$$

$$\frac{250}{3} = 83\frac{1}{3}$$

85.
$$\begin{array}{r} 4\,3 \\ 8\overline{\smash{)}3\,4\,5} \\ \underline{3\,2\,0} \\ 2\,5 \\ \underline{2\,4} \\ 1 \end{array}$$

$$\frac{345}{8} = 43\frac{1}{8}$$

87.
$$\begin{array}{r} 1\,8 \\ 4\overline{\smash{)}7\,5} \\ \underline{4\,0} \\ 3\,5 \\ \underline{3\,2} \\ 3 \end{array}$$

$$\frac{75}{4} = 18\frac{3}{4}$$

89. Use a calculator.

$$\frac{41}{369} = 0.11.\overline{1} = 11.\overline{1}\%$$

91. $2.5\overline{74631} = 2.57.\overline{46317} = 257.\overline{46317}\%$

93. $\frac{14}{9}\% = \frac{14}{9} \times \frac{1}{100} = \frac{2 \cdot 7 \cdot 1}{9 \cdot 2 \cdot 50} = \frac{2}{2} \cdot \frac{7}{450} = \frac{7}{450}$

To find decimal notation for $\frac{7}{450}$ we divide.

$$\begin{array}{r} 0.0\,1\,5\,5 \\ 4\,5\,0\overline{\smash{)}7.0\,0\,0\,0} \\ \underline{4\,5\,0} \\ 2\,5\,0\,0 \\ \underline{2\,2\,5\,0} \\ 2\,5\,0\,0 \\ \underline{2\,2\,5\,0} \\ 2\,5\,0 \end{array}$$

We get a repeating decimal: $\frac{14}{9}\% = 0.01\overline{5}$

95. $\dfrac{729}{7}\% = \dfrac{729}{7} \times \dfrac{1}{100} = \dfrac{729}{700}$

To find decimal notation for $\dfrac{729}{700}$ we divide.

```
              1.0 4 1 4 2 8 5 7
  7 0 0 [ 7 2 9.0 0 0 0 0 0 0 0
          7 0 0
          ─────
          2 9 0 0
          2 8 0 0
          ───────
            1 0 0 0
              7 0 0
            ─────
            3 0 0 0
            2 8 0 0
            ───────
              2 0 0 0
              1 4 0 0
              ───────
                6 0 0 0
                5 6 0 0
                ───────
                  4 0 0 0
                  3 5 0 0
                  ───────
                    5 0 0 0
                    4 9 0 0
                    ───────
                      1 0 0
```

We get a repeating decimal: $\dfrac{729}{7}\% = 1.04\overline{142857}$.

Exercise Set 4.4

1. What is 32% of 78?

$\quad a\ \ =\ 32\%\ \times\ 78$

3. 89 is what percent of 99?

$\quad 89 =\quad n\quad \times\ 99$

5. 13 is 25% of what?

$\quad 13 = 25\%\ \times\quad b$

7. What is 85% of 276?

Translate: $a = 85\% \cdot 276$

Solve: The letter is by itself. To solve the equation we convert 85% to decimal notation and multiply.

```
        2 7 6
      × 0. 8 5        (85% = 0.85)
      ───────
      1 3 8 0
    2 2 0 8 0
  a = 2 3 4. 6 0
```

234.6 is 85% of 276. The answer is 234.6.

9. 150% of 30 is what?

Translate: $150\% \times 30 = a$

Solve: Convert 150% to decimal notation and multiply.

```
        3 0
      × 1. 5        (150% = 1.5)
      ─────
      1 5 0
    3 0 0
  a = 4 5. 0
```

150% of 30 is 45. The answer is 45.

11. What is 6% of $300?

Translate: $a = 6\% \cdot \$300$

Solve: Convert 6% to decimal notation and multiply.

```
        $ 3 0 0
      × 0. 0 6        (6% = 0.06)
  a = $ 1 8. 0 0
```

$18 is 6% of $300. The answer is $18.

13. 3.8% of 50 is what?

Translate: $3.8\% \cdot 50 = a$

Solve: Convert 3.8% to decimal notation and multiply.

```
          5 0
      ×  0. 0 3 8     (3.8% = 0.038)
      ─────────
          4 0 0
        1 5 0 0
  a = 1. 9 0 0
```

3.8% of 50 is 1.9. The answer is 1.9.

15. $39 is what percent of $50?

Translate: $39 = n \times 50$

Solve: To solve the equation we divide on both sides by 50 and convert the answer to percent notation.

$$n \cdot 50 = 39$$
$$\frac{n \cdot 50}{50} = \frac{39}{50}$$
$$n = 0.78 = 78\%$$

$39 is 78% of $50. The answer is 78%.

17. 20 is what percent of 10?

Translate: $20 = n \times 10$

Solve: To solve the equation we divide on both sides by 10 and convert the answer to percent notation.

$$n \cdot 10 = 20$$
$$\frac{n \cdot 10}{10} = \frac{20}{10}$$
$$n = 2 = 200\%$$

20 is 200% of 10. The answer is 200%.

19. What percent of $300 is $150?

Translate: $n \times 300 = 150$

Solve: $n \cdot 300 = 150$

$$\frac{n \cdot 300}{300} = \frac{150}{300}$$
$$n = 0.5 = 50\%$$

50% of $300 is $150. The answer is 50%.

21. What percent of 80 is 100?

Translate: $n \times 80 = 100$

Solve: $n \cdot 80 = 100$

$$\frac{n \cdot 80}{80} = \frac{100}{80}$$
$$n = 1.25 = 125\%$$

125% of 80 is 100. The answer is 125%.

23. 20 is 50% of what?

Translate: $20 = 50\% \times b$

Solve: To solve the equation we divide on both sides by 50%:

$$\frac{20}{50\%} = \frac{50\% \times b}{50\%}$$

$$\frac{20}{0.5} = b \quad (50\% = 0.5)$$

$$40 = b$$

$$\begin{array}{r} 4\ 0\ . \\ 0.\ 5_\wedge \overline{)2\ 0.\ 0_\wedge} \\ \underline{2\ 0\ 0} \\ 0 \\ \underline{0} \\ 0 \end{array}$$

20 is 50% of 40. The answer is 40.

25. 40% of what is $16?

Translate: $40\% \times b = 16$

Solve: To solve the equation we divide on both sides by 40%:

$$\frac{40\% \times b}{40\%} = \frac{16}{40\%}$$

$$b = \frac{16}{0.4} \quad (40\% = 0.4)$$

$$b = 40$$

$$\begin{array}{r} 4\ 0\ . \\ 0.\ 4_\wedge \overline{)1\ 6.\ 0_\wedge} \\ \underline{1\ 6\ 0} \\ 0 \\ \underline{0} \\ 0 \end{array}$$

40% of $40 is $16. The answer is $40.

27. 56.32 is 64% of what?

Translate: $56.32 = 64\% \times b$

Solve: $\dfrac{56.32}{64\%} = \dfrac{64\% \times b}{64\%}$

$$\frac{56.32}{0.64} = b$$

$$88 = b$$

$$\begin{array}{r} 8\ 8\ . \\ 0.\ 6\ 4_\wedge \overline{)5\ 6.\ 3\ 2_\wedge} \\ \underline{5\ 1\ 2\ 0} \\ 5\ 1\ 2 \\ \underline{5\ 1\ 2} \\ 0 \end{array}$$

56.32 is 64% of 88. The answer is 88.

29. 70% of what is 14?

Translate: $70\% \times b = 14$

Solve: $\dfrac{70\% \times b}{70\%} = \dfrac{14}{70\%}$

$$b = \frac{14}{0.7}$$

$$b = 20$$

$$\begin{array}{r} 2\ 0\ . \\ 0.\ 7_\wedge \overline{)1\ 4.\ 0_\wedge} \\ \underline{1\ 4\ 0} \\ 0 \\ \underline{0} \\ 0 \end{array}$$

70% of 20 is 14. The answer is 20.

31. What is $62\frac{1}{2}\%$ of 10?

Translate: $a = 62\frac{1}{2}\% \times 10$

Solve: $a = 0.625 \times 10 \quad (62\frac{1}{2}\% = 0.625)$

$ a = 6.25 \qquad$ Multiplying

6.25 is $62\frac{1}{2}\%$ of 10. The answer is 6.25.

33. What is 8.3% of $10,200?

Translate: $a = 8.3\% \times 10,200$

Solve: $a = 8.3\% \times 10,200$

$ a = 0.083 \times 10,200 \quad (8.3\% = 0.083)$

$ a = 846.6 \qquad$ Multiplying

$846.60 is 8.3% of $10,200. The answer is $846.60.

35. Discussion and Writing Exercise

37. $0.\underline{09} = \dfrac{9}{1\underline{00}}$

2 decimal places $\qquad$ 2 zeros

39. $0.\underline{875} = \dfrac{875}{1\underline{000}}$

3 decimal places $\qquad$ 3 zeros

$$\frac{875}{1000} = \frac{7 \cdot 125}{8 \cdot 125} = \frac{7}{8} \cdot \frac{125}{125} = \frac{7}{8}$$

Thus, $0.875 = \dfrac{875}{1000}$, or $\dfrac{7}{8}$.

41. $0.\underline{9375} = \dfrac{9375}{1\underline{0,000}}$

4 decimal places $\qquad$ 4 zeros

$$\frac{9375}{10,000} = \frac{15 \cdot 625}{16 \cdot 625} = \frac{15}{16} \cdot \frac{625}{625} = \frac{15}{16}$$

Thus, $0.9375 = \dfrac{9375}{10,000}$, or $\dfrac{15}{16}$.

43. $\dfrac{89}{1\underline{00}} \qquad 0.\underline{89}.$

2 zeros $\qquad$ Move 2 places

$$\frac{89}{100} = 0.89$$

45. $\dfrac{3}{1\underline{0}} \qquad 0.\underline{3}.$

1 zero $\qquad$ Move 1 place

$$\frac{3}{10} = 0.3$$

47. Estimate: Round 7.75% to 8% and $10,880 to $11,000. Then translate:

$$\begin{array}{ccccc} \text{What} & \text{is} & 8\% & \text{of} & \$11,000? \\ \downarrow & \downarrow & \downarrow & \downarrow & \downarrow \\ a & = & 8\% & \times & 11,000 \end{array}$$

We convert 8% to decimal notation and multiply.

$$\begin{array}{r} 1\,1,0\,0\,0 \\ \times\quad 0.0\,8 \\ \hline 8\,8\,0.0\,0 \end{array} \quad (8\% = 0.08)$$

$880 is about 7.75% of $10,880. (Answers may vary.)

Calculate: First we translate.
   What is 7.75% of $10,880?
   ↓   ↓   ↓   ↓   ↓
   a $= 7.75\% \times$ 10,880

Use a calculator to multiply:

$0.0775 \times 10,880 = 843.2$

$843.20 is 7.75% of $10,880.

49. Estimate: Round $2496 to $2500 and 24% to 25%. Then translate:
   $2500 is 25% of what?
   ↓   ↓   ↓   ↓   ↓
   2500 $= 25\% \times$ b

We convert 25% to decimal notation and divide.

$$\frac{2500}{0.25} = \frac{0.25 \times b}{0.25}$$
$$10,000 = b$$

$2496 is 24% of about $10,000. (Answers may vary.)

Calculate: First we translate.
   $2496 is 24% of what?
   ↓   ↓   ↓   ↓   ↓
   2496 $= 0.24 \times$ b

Use a calculator to divide:

$$\frac{2496}{0.24} = 10,400$$

$2496 is 24% of $10,400.

51. We translate:

40% of $18\frac{3}{4}\%$ of $25,000 is what?
   ↓ ↓ ↓ ↓ ↓ ↓ ↓
   $40\% \times 18\frac{3}{4}\% \times 25,000 = a$

We convert 40% and $18\frac{3}{4}\%$ to decimal notation and multiply.

$0.4 \times 0.1875 \times 25,000 = a$

$$\begin{array}{r} 0.1\,8\,7\,5 \\ \times\quad\quad 0.4 \\ \hline 0.0\,7\,5\,0\,0 \end{array}$$

$$\begin{array}{r} 2\,5,0\,0\,0 \\ \times\quad\ 0.0\,7\,5 \\ \hline 1\,2\,5\,0\,0\,0 \\ 1\,7\,5\,0\,0\,0\,0 \\ \hline 1\,8\,7\,5.0\,0\,0 \end{array}$$

40% of $18\frac{3}{4}\%$ of $25,000 is $1875.

Exercise Set 4.5

1. What is   37%   of 74?
   ↓   ↓   ↓
   amount   number of   base
            hundredths

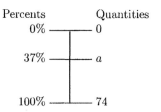

$$\frac{37}{100} = \frac{a}{74}$$

3. 4.3 is   what   percent of 5.9?
   ↓   ↓   ↓
   amount   number of   base
            hundredths

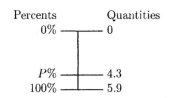

$$\frac{P}{100} = \frac{4.3}{5.9}$$

5. 14 is   25%   of what?
   ↓   ↓   ↓
   amount   number of   base
            hundredths

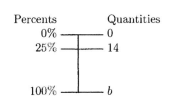

$$\frac{25}{100} = \frac{14}{b}$$

7. What is   76%   of 90?
   ↓   ↓   ↓
   amount   number of   base
            hundredths

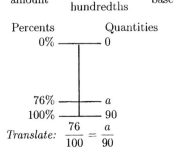

Translate: $\dfrac{76}{100} = \dfrac{a}{90}$

Solve: $76 \cdot 90 = 100 \cdot a$ Equating cross-products

$$\frac{76 \cdot 90}{100} = \frac{100 \cdot a}{100}$$ Dividing by 100

$$\frac{6840}{100} = a$$

$$68.4 = a$$ Simplifying

68.4 is 76% of 90. The answer is 68.4.

9.   70%   of 660 is what?

number of hundredths   base   amount

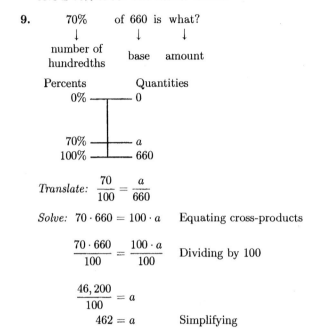

Translate: $\dfrac{70}{100} = \dfrac{a}{660}$

Solve: $70 \cdot 660 = 100 \cdot a$ Equating cross-products

$$\frac{70 \cdot 660}{100} = \frac{100 \cdot a}{100}$$ Dividing by 100

$$\frac{46,200}{100} = a$$

$$462 = a$$ Simplifying

70% of 660 is 462. The answer is 462.

11.   What is   4%   of 1000?

amount   number of hundredths   base

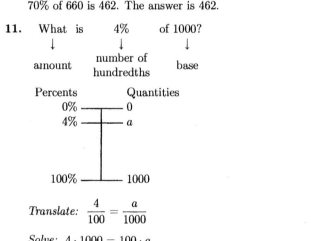

Translate: $\dfrac{4}{100} = \dfrac{a}{1000}$

Solve: $4 \cdot 1000 = 100 \cdot a$

$$\frac{4 \cdot 1000}{100} = \frac{100 \cdot a}{100}$$

$$\frac{4000}{100} = a$$

$$40 = a$$

40 is 4% of 1000. The answer is 40.

13.   4.8%   of 60 is what?

number of hundredths   base   amount

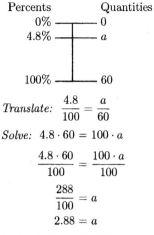

Translate: $\dfrac{4.8}{100} = \dfrac{a}{60}$

Solve: $4.8 \cdot 60 = 100 \cdot a$

$$\frac{4.8 \cdot 60}{100} = \frac{100 \cdot a}{100}$$

$$\frac{288}{100} = a$$

$$2.88 = a$$

4.8% of 60 is 2.88. The answer is 2.88.

15. $24   is   what   percent of $96?

amount   number of hundredths   base

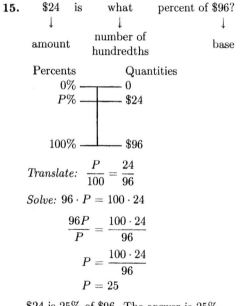

Translate: $\dfrac{P}{100} = \dfrac{24}{96}$

Solve: $96 \cdot P = 100 \cdot 24$

$$\frac{96P}{P} = \frac{100 \cdot 24}{96}$$

$$P = \frac{100 \cdot 24}{96}$$

$$P = 25$$

$24 is 25% of $96. The answer is 25%.

17.   102   is   what   percent of 100?

amount   number of hundredths   base

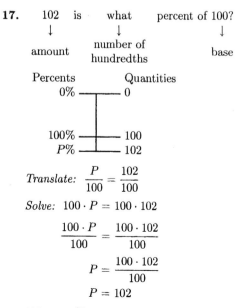

Translate: $\dfrac{P}{100} = \dfrac{102}{100}$

Solve: $100 \cdot P = 100 \cdot 102$

$$\frac{100 \cdot P}{100} = \frac{100 \cdot 102}{100}$$

$$P = \frac{100 \cdot 102}{100}$$

$$P = 102$$

102 is 102% of 100. The answer is 102%.

19.

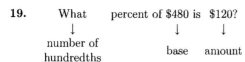

What   percent of $480 is $120?
↓   ↓   ↓
number of   base   amount
hundredths

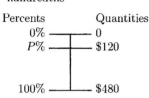

Percents   Quantities
  0% ———————— 0
  P% ———————— $120

 100% ———————— $480

Translate: $\dfrac{P}{100} = \dfrac{120}{480}$

Solve: $480 \cdot P = 100 \cdot 120$

$$\frac{480 \cdot P}{480} = \frac{100 \cdot 120}{480}$$

$$P = \frac{100 \cdot 120}{480}$$

$$P = 25$$

25% of $480 is $120. The answer is 25%.

21.

What   percent of 160 is   150?
↓   ↓   ↓
number of   base   amount
hundredths

Percents   Quantities
  0% ———————— 0

  P% ———————— 150
 100% ———————— 160

Translate: $\dfrac{P}{100} = \dfrac{150}{160}$

Solve: $160 \cdot P = 100 \cdot 150$

$$\frac{160 \cdot P}{160} = \frac{100 \cdot 150}{160}$$

$$P = \frac{100 \cdot 150}{160}$$

$$P = 93.75$$

93.75% of 160 is 150. The answer is 93.75%.

23. $18   is   25%   of what?
         ↓   ↓   ↓
       amount   number of   base
                     hundredths

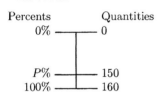

Percents   Quantities
  0% ———————— 0
 25% ———————— $18

 100% ———————— b

Translate: $\dfrac{25}{100} = \dfrac{18}{b}$

Solve: $25 \cdot b = 100 \cdot 18$

$$\frac{25 \cdot b}{b} = \frac{100 \cdot 18}{25}$$

$$b = \frac{100 \cdot 18}{25}$$

$$b = 72$$

$18 is 25% of $72. The answer is $72.

25.   60%   of what is $54?
             ↓   ↓   ↓
         number of   base   amount
         hundredths

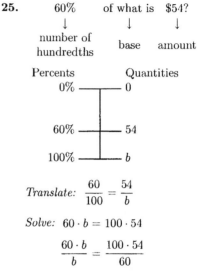

Percents   Quantities
  0% ———————— 0

 60% ———————— 54
 100% ———————— b

Translate: $\dfrac{60}{100} = \dfrac{54}{b}$

Solve: $60 \cdot b = 100 \cdot 54$

$$\frac{60 \cdot b}{b} = \frac{100 \cdot 54}{60}$$

$$b = \frac{100 \cdot 54}{60}$$

$$b = 90$$

60% of 90 is 54. The answer is 90.

27.   65.12   is   74%   of what?
          ↓   ↓   ↓
        amount   number of   base
                       hundredths

Percents   Quantities
  0% ———————— 0

 74% ———————— 65.12
 100% ———————— b

Translate: $\dfrac{74}{100} = \dfrac{65.12}{b}$

Solve: $74 \cdot b = 100 \cdot 65.12$

$$\frac{74 \cdot b}{74} = \frac{100 \cdot 65.12}{74}$$

$$b = \frac{100 \cdot 65.12}{74}$$

$$b = 88$$

65.12 is 74% of 88. The answer is 88.

29.   80%   of what is   16?
             ↓   ↓   ↓
         number of   base   amount
         hundredths

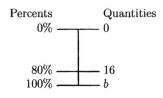

Translate: $\dfrac{80}{100} = \dfrac{16}{b}$

Solve: $80 \cdot b = 100 \cdot 16$

$$\dfrac{80 \cdot b}{80} = \dfrac{100 \cdot 16}{80}$$

$$b = \dfrac{100 \cdot 16}{80}$$

$$b = 20$$

80% of 20 is 16. The answer is 20.

31.
What is	$62\frac{1}{2}\%$	of 40?
$\downarrow$	$\downarrow$	$\downarrow$
amount	number of hundredths	base

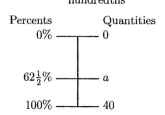

Translate: $\dfrac{62\frac{1}{2}}{100} = \dfrac{a}{40}$

Solve: $62\frac{1}{2} \cdot 40 = 100 \cdot a$

$$\dfrac{125}{2} \cdot \dfrac{40}{1} = 100 \cdot a$$

$$2500 = 100 \cdot a$$

$$\dfrac{2500}{100} = \dfrac{100 \cdot a}{100}$$

$$25 = a$$

25 is $62\frac{1}{2}\%$ of 40. The answer is 25.

33.
What is	9.4%	of $8300?
$\downarrow$	$\downarrow$	$\downarrow$
amount	number of hundredths	base

Percents   Quantities
0% ———————— 0
9.4% ——————— a

100% ———————— 8300

Translate: $\dfrac{9.4}{100} = \dfrac{a}{8300}$

Solve: $9.4 \cdot 8300 = 100 \cdot a$

$$\dfrac{9.4 \cdot 8300}{100} = \dfrac{100 \cdot a}{100}$$

$$\dfrac{78,020}{100} = a$$

$$780.2 = a$$

$780.20 is 9.4% of $8300. The answer is $780.20.

35. Discussion and Writing Exercise

37. $\dfrac{x}{188} = \dfrac{2}{47}$

$$47 \cdot x = 188 \cdot 2$$

$$x = \dfrac{188 \cdot 2}{47}$$

$$x = \dfrac{4 \cdot 47 \cdot 2}{47}$$

$$x = 8$$

39. $\dfrac{4}{7} = \dfrac{x}{14}$

$$4 \cdot 14 = 7 \cdot x$$

$$\dfrac{4 \cdot 14}{7} = x$$

$$\dfrac{4 \cdot 2 \cdot 7}{7} = x$$

$$8 = x$$

41. $\dfrac{5000}{t} = \dfrac{3000}{60}$

$$5000 \cdot 60 = 3000 \cdot t$$

$$\dfrac{5000 \cdot 60}{3000} = t$$

$$\dfrac{5 \cdot 1000 \cdot 3 \cdot 20}{3 \cdot 1000} = t$$

$$100 = t$$

43. $\dfrac{x}{1.2} = \dfrac{36.2}{5.4}$

$$5.4 \cdot x = 1.2(36.2)$$

$$x = \dfrac{1.2(36.2)}{5.4}$$

$$x = 8.0\overline{4}$$

45. ***Familiarize.*** Let q = the number of quarts of liquid ingredients the recipe calls for.

Translate.

Butter-milk	plus	Skim milk	plus	Oil	is	Total liquid ingredients
$\downarrow$	$\downarrow$	$\downarrow$	$\downarrow$	$\downarrow$	$\downarrow$	$\downarrow$
$\frac{1}{2}$	$+$	$\frac{1}{3}$	$+$	$\frac{1}{16}$	$=$	q

Solve. We carry out the addition. The LCM of the denominators is 48, so the LCD is 48.

$$\dfrac{1}{2} \cdot \dfrac{24}{24} + \dfrac{1}{3} \cdot \dfrac{16}{16} + \dfrac{1}{16} \cdot \dfrac{3}{3} = q$$

$$\dfrac{24}{48} + \dfrac{16}{48} + \dfrac{3}{48} = q$$

$$\dfrac{43}{48} = q$$

Check. We repeat the calculation. The answer checks.

State. The recipe calls for $\frac{43}{48}$ qt of liquid ingredients.

47. Estimate: Round 8.85% to 9%, and \$12,640 to \$12,600.

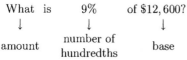

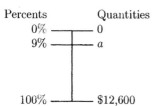

Translate: $\dfrac{9}{100} = \dfrac{a}{12,600}$

Solve: $9 \cdot 12,600 = 100 \cdot a$

$$\frac{9 \cdot 12,600}{100} = \frac{100 \cdot a}{100}$$

$$\frac{113,400}{100} = a$$

$$1134 = a$$

\$1134 is about 8.85% of \$12,640. (Answers may vary.)

Calculate:

What is   8.85%   of \$12,640?
   ↓   ↓   ↓
amount   number of   base
        hundredths

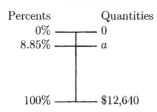

Translate: $\dfrac{8.85}{100} = \dfrac{a}{12,640}$

Solve: $8.85 \cdot 12,640 = 100 \cdot a$

$$\frac{8.85 \cdot 12,640}{100} = \frac{100 \cdot a}{100}$$

$$\frac{111,864}{100} = a \qquad \text{Use a calculator to}$$
$$\text{multiply and divide.}$$

$$1118.64 = a$$

\$1118.64 is 8.85% of \$12,640.

Exercise Set 4.6

---

1. a) *Familiarize.* The question asks for a percent. Note that $\dfrac{90}{133} \approx \dfrac{90}{135} = \dfrac{2}{3} = 66.\overline{6}\%$, so the answer is close to $66.\overline{6}\%$. Let $p =$ the percent of panda cubs born in captivity that lived to be a month old.

Translate. We will translate to a percent equation.

90 is what percent of 133?
   ↓ ↓   ↓   ↓ ↓
90 = p × 133

Solve.

$$90 = p \cdot 133$$

$$\frac{90}{133} = \frac{p \cdot 133}{133}$$

$$0.677 \approx p$$

$$67.7\% \approx p$$

Check. Note that the result, 67.7%, is close to the estimate in the Familiarize step, $66.\overline{6}\%$. The answer checks.

State. 67.7% of pandas born in captivity lived to be a month old.

b) *Familiarize.* Let $P =$ the percent of panda cubs that received the special formula and lived to be a month old.

Translate. We translate to a proportion.

$$\frac{P}{100} = \frac{18}{20}$$

$$P \cdot 20 = 100 \cdot 18$$

$$\frac{P \cdot 20}{20} = \frac{1800}{20}$$

$$P = 90$$

Check. We can repeat the calculation. The answer checks.

State. 90% of panda cubs that received the special formula lived to be a month old.

3. *Familiarize.* Let $b =$ the number of passes Trent Dilfer completed in the 2000 season.

Translate. We translate to a percent equation.

What number is 59.3% of 226?
     ↓   ↓ ↓ ↓ ↓
     b = 59.3% of 226

Solve. We convert 59.3% to decimal notation and multiply.

$$b = 0.593 \times 226 = 134.018 \approx 134$$

Check. We can repeat the calculations. We can also estimate: $59.3\% \times 226 \approx 60\% \times 225 = 135$. Since 134 is close to 135, the answer checks.

State. Trent Dilfer completed 134 passes in the 2000 season.

5. *Familiarize.* Let $x =$ the number of people in the U.S. who are overweight and $y =$ the number who are obese, in millions.

Translate. We translate to percent equations.

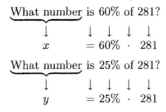

Solve.

$$x = 0.6 \cdot 281 = 168.6$$
$$y = 0.25 \cdot 281 = 70.25$$

Check. We can repeat the calculations. Also note that $60\% \cdot 281 \approx 60\% \cdot 300 = 180$ and $25\% \cdot 281 \approx 25 \cdot 300 = 75$. Since 168.6 is close to 180 and 70.25 is close to 75, the answer checks.

State. 168.6 million, or 168,600,000 people in the U.S. are overweight and 70.25 million, or 70,250,000 are obese.

7. **Familiarize.** First we find the amount of the solution that is acid. We let $a =$ this amount.

Translate. We translate to a percent equation.

What is 3% of 680?
$$\downarrow \quad \downarrow \quad \downarrow \quad \downarrow \quad \downarrow$$
$$a \quad = \quad 3\% \text{ of } 680$$

Solve. We convert 3% to decimal notation and multiply.

$$a = 3\% \times 680 = 0.03 \times 680 = 20.4$$

Now we find the amount that is water. We let $w =$ this amount.

Total amount	minus	Amount of acid	is	Amount of water
$\downarrow$	$\downarrow$	$\downarrow$	$\downarrow$	$\downarrow$
680	−	20.4	=	w

To solve the equation we carry out the subtraction.

$$w = 680 - 20.4 = 659.6$$

Check. We can repeat the calculations. Also, observe that, since 3% of the solution is acid, 97% is water. Because 97% of $680 = 0.97 \times 680 = 659.6$, our answer checks.

State. The solution contains 20.4 mL of acid and 659.6 mL of water.

9. **Familiarize.** Let $a =$ the number of field goals Vince Carter attempted in the 2000-2001 season.

Translate. We translate to a proportion.

$$\frac{45.2}{100} = \frac{288}{a}$$

Solve.

$$\frac{45.2}{100} = \frac{288}{a}$$
$$45.2 \cdot a = 100 \cdot 288$$
$$\frac{45.2 \cdot a}{45.2} = \frac{100 \cdot 288}{45.2}$$
$$a \approx 637$$

Check. We can repeat the calculations. Also note that $\frac{288}{637} \approx \frac{300}{600} = \frac{1}{2} = 50\% \approx 45.2\%$. The answer checks.

State. Carter attempted 637 field goals.

11. **Familiarize.** First we find the percent of the test items Antonio got correct. Let P represent this number.

Translate. We translate to a proportion.

$$\frac{P}{100} = \frac{76}{80}$$

Solve.

$$\frac{P}{100} = \frac{76}{80}$$
$$P \cdot 80 = 100 \cdot 76$$
$$\frac{P \cdot 80}{80} = \frac{100 \cdot 76}{80}$$
$$P = 95$$

If Antonio got 95% of the items correct, then he got $100\% - 95\%$, or 5% incorrect.

Check. We can repeat the calculations. Also note that Antonio got $80 - 76$, or 4, items incorrect and $\frac{4}{80} = \frac{1}{20} = 5\%$. The answer checks.

State. Antonio got 95% of the test items correct and 5% incorrect.

13. **Familiarize.** First we find the number of items Christina got correct. Let b represent this number.

Translate. We translate to a percent equation.

What number is 91% of 40?
$$\downarrow \qquad \quad \downarrow \quad \downarrow \quad \downarrow \quad \downarrow$$
$$b \qquad \quad = 91\% \quad \cdot \quad 40$$

Solve. We convert 91% to decimal notation and multiply.

$$b = 0.91 \cdot 40 = 36.4$$

We subtract to find the number of items Christina got incorrect:

$$40 - 36.4 = 3.6$$

Check. We can repeat the calculation. Also note that $91\% \cdot 40 \approx 90\% \cdot 40 = 36 \approx 36.4$. The answer checks.

State. Christina got 36.4 items correct and 3.6 items incorrect.

15. **Familiarize.** Let $a =$ the number of items on the test.

Translate. We translate to a proportion.

$$\frac{86}{100} = \frac{81.7}{a}$$

Solve.

$$\frac{86}{100} = \frac{81.7}{a}$$
$$86 \cdot a = 100 \cdot 81.7$$
$$\frac{86 \cdot a}{86} = \frac{100 \cdot 81.7}{86}$$
$$a = 95$$

Check. We can repeat the calculation. Also note that $\frac{81.7}{95} \approx \frac{82}{100} = 82\% \approx 86\%$. The answer checks.

State. There were 95 items on the test.

17. **Familiarize.** We let $n =$ the percent of time that television sets are on.

Translate. We translate to a percent equation.

2190 is what percent of 8760?
$$\downarrow \quad \downarrow \qquad \downarrow \qquad \downarrow \quad \downarrow$$
$$2190 = \qquad n \qquad \times 8760$$

Solve. We divide on both sides by 8760 and convert the result to percent notation.

$$2190 = n \times 8760$$

$$\frac{2190}{8760} = \frac{n \times 8760}{8760}$$

$$0.25 = n$$

$$25\% = n$$

Check. To check we find 25% of 8760:

$25\% \times 8760 = 0.25 \times 8760 = 2190$. The answer checks.

State. Television sets are on for 25% of the year.

19. First we find the maximum heart rate for a 25 year old person.

Familiarize. Note that $220 - 25 = 195$. We let $x =$ the maximum heart rate for a 25 year old person.

Translate. We translate to a percent equation.

What is 85% of 195?

$x = 85\% \times 195$

Solve. We convert 85% to a decimal and simplify.

$$x = 0.85 \times 195 = 165.75 \approx 166$$

Check. We can repeat the calculations. Also, 85% of $195 \approx 0.85 \times 200 = 170 \approx 166$. The answer checks.

State. The maximum heart rate for a 25 year old person is 166 beats per minute.

Next we find the maximum heart rate for a 36 year old person.

Familiarize. Note that $220 - 36 = 184$. We let $x =$ maximum heart rate for a 36 year old person.

Translate. We translate to a percent equation.

What is 85% of 184?

$x = 85\% \times 184$

Solve. We convert 85% to a decimal and simplify.

$$x = 0.85 \times 184 = 156.4 \approx 156$$

Check. We can repeat the calculations. Also, 85% of $184 \approx 0.9 \times 180 = 162 \approx 156$. The answer checks.

State. The maximum heart rate for a 36 year old person is 156 beats per minute.

Next we find the maximum heart rate for a 48 year old person.

Familiarize. Note that $220 - 48 = 172$. We let $x =$ the maximum heart rate for a 48 year old person.

Translate. We translate to a percent equation.

What is 85% of 172?

$x = 85\% \times 172$

Solve. We convert 85% to a decimal and simplify.

$$x = 0.85 \times 172 = 146.2 \approx 146$$

Check. We can repeat the calculations. Also, 85% of $172 \approx 0.9 \times 170 = 153 \approx 146$. The answer checks.

State. The maximum heart rate for a 48 year old person is 146 beats per minute.

We find the maximum heart rate for a 55 year old person.

Familiarize. Note that $220 - 55 = 165$. We let $x =$ the maximum heart rate for a 55 year old person.

Translate. We translate to a percent equation.

What is 85% of 165?

$x = 85\% \times 165$

Solve. We convert 85% to a decimal and simplify.

$$x = 0.85 \times 165 = 140.25 \approx 140$$

Check. We can repeat the calculations. Also, 85% of $165 \approx 0.9 \times 160 = 144 \approx 140$. The answer checks.

State. The maximum heart rate for a 55 year old person is 140 beats per minute.

Finally we find the maximum heart rate for a 76 year old person.

Familiarize. Note that $220 - 76 = 144$. We let $x =$ the maximum heart rate for a 76 year old person.

Translate. We translate to a percent equation.

What is 85% of 144?

$x = 85\% \times 144$

Solve. We convert 85% to a decimal and simplify.

$$x = 0.85 \times 144 = 122.4 \approx 122$$

Check. We can repeat the calculations. Also, 85% of $144 \approx 0.9 \times 140 = 126 \approx 122$. The answer checks.

State. The maximum heart rate for a 76 year old person is 122 beats per minute.

21. Familiarize. Use the drawing in the text to visualize the situation. Note that the increase in the amount was $16.

Let $n =$ the percent of increase.

Translate. We translate to a percent equation.

$16 is what percent of $200?

$16 = n \times 200$

Solve. We divide by 200 on both sides and convert the result to percent notation.

$$16 = n \times 200$$

$$\frac{16}{200} = \frac{n \times 200}{200}$$

$$0.08 = n$$

$$8\% = n$$

Check. Find 8% of 200: $8\% \times 200 = 0.08 \times 200 = 16$. Since this is the amount of the increase, the answer checks.

State. The percent of increase was 8%.

23. Familiarize. We use the drawing in the text to visualize the situation. Note that the reduction is $18.

We let $n =$ the percent of decrease.

Translate. We translate to a percent equation.

$18 is what percent of $90?

$18 = n \times 90$

Solve. To solve the equation, we divide on both sides by 90 and convert the result to percent notation.

$$\frac{18}{90} = \frac{n \times 90}{90}$$
$$0.2 = n$$
$$20\% = n$$

Check. We find 20% of 90: $20\% \times 90 = 0.2 \times 90 = 18$. Since this is the price decrease, the answer checks.

State. The percent of decrease was 20%.

25. Familiarize. First we find the amount of increase.

$$\begin{array}{r} 4,3\,0\,1,2\,6\,1 \\ -3,2\,9\,4,3\,9\,4 \\ \hline 1,0\,0\,6,8\,6\,7 \end{array}$$

Let P = the percent of increase.

Translate. We translate to a proportion.

$$\frac{P}{100} = \frac{1,006,867}{3,294,394}$$

Solve.

$$\frac{P}{100} = \frac{1,006,867}{3,294,394}$$
$$P \cdot 3,294,394 = 100 \cdot 1,006,867$$
$$\frac{P \cdot 3,294,394}{3,294,394} = \frac{100 \cdot 1,006,867}{3,294,394}$$
$$P \approx 30.6$$

Check. We can repeat the calculation. Also note that $30.6\% \cdot 3,294,394 \approx 30\% \cdot 3,300,000 = 990,000 \approx 1,006,867$. The answer checks.

State. The percent of increase was about 30.6%.

27. Familiarize. We note that the amount of the raise can be found and then added to the old salary. A drawing helps us visualize the situation.

$28,600	$?
100%	5%

We let x = the new salary.

Translate. We translate to a percent equation.

What is	the old salary	plus 5% of	the old salary?
↓ ↓	↓	↓ ↓ ↓	↓
x =	28,600	+ 5% ×	28,600

Solve. We convert 5% to a decimal and simplify.

$$x = 28,600 + 0.05 \times 28,600$$
$$= 28,600 + 1430 \qquad \text{The raise is } \$1430.$$
$$= 30,030$$

Check. To check, we note that the new salary is 100% of the old salary plus 5% of the old salary, or 105% of the old salary. Since $1.05 \times 28,600 = 30,030$, our answer checks.

State. The new salary is $30,030.

29. Familiarize. Let d = the amount of depreciation the first year.

Translate. We translate to a proportion.

$$\frac{25}{100} = \frac{d}{21,566}$$

Solve.

$$\frac{25}{100} = \frac{d}{21,566}$$
$$25 \cdot 21,566 = 100 \cdot d$$
$$\frac{25 \cdot 21,566}{100} = d$$
$$5391.50 = d$$

Now we subtract to find the depreciated value after 1 year.

$$\begin{array}{r} 2\,1,5\,6\,6.0\,0 \\ -\quad 5\,3\,9\,1.5\,0 \\ \hline 1\,6,1\,7\,4.5\,0 \end{array}$$

The second year the car depreciates 25% of the value after 1 year. We use a proportion to find this amount, a.

$$\frac{25}{100} = \frac{a}{16,174.50}$$
$$25 \cdot 16,174.50 = 100 \cdot a$$
$$\frac{25 \cdot 16,174.50}{100} = a$$
$$4043.63 \approx a$$

Now we subtract to find the value of the car after 2 years.

$$\begin{array}{r} 1\,6,1\,7\,4.5\,0 \\ -\quad 4\,0\,4\,3.6\,3 \\ \hline 1\,2,1\,3\,0.8\,7 \end{array}$$

Check. We can repeat the calculations. Also note that after 1 year the value of the car will be $100\% - 25\%$, or 75%, of the original value:

$$75\% \times \$21,566 = \$16,174.50$$

After 2 years the value of the car will be $100\% - 25\%$, or 75%, of the value after 1 year:

$$75\% \times \$16,174.50 \approx \$12,130.88$$

The slight discrepancy in this amount is due to rounding. The answers check.

State. After 1 year the value of the car will be $16,174.50. After 2 years, its value will be $12,130.87.

31. Familiarize. First we find the amount of the decrease.

$$\begin{array}{r} 1\,9\,9.9\,8 \\ -1\,4\,9.9\,9 \\ \hline 4\,9.9\,9 \end{array}$$

Let P = the percent of decrease.

Translate. We translate to a proportion.

$$\frac{P}{100} = \frac{49.99}{199.98}$$

Solve.

$$\frac{P}{100} = \frac{49.99}{199.98}$$
$$P \cdot 199.98 = 100 \cdot 49.99$$
$$\frac{P \cdot 199.98}{199.98} = \frac{100 \cdot 49.99}{199.98}$$
$$P \approx 25$$

Check. We can repeat the calculations. Also note that $25\% \cdot \$199.98 \approx 25\% \cdot \$200 = \$50 \approx \$49.99$. The answer checks.

State. The percent of decrease is 25%.

33. **Familiarize**. This is a multistep problem. First we find the area of a cross-section of a finished board and of a rough board using the formula $A = l \cdot w$. Then we find the amount of wood removed in planing and drying and finally we find the percent of wood removed. Let $f =$ the area of a cross-section of a finished board and let $r =$ the area of a cross-section of a rough board.

Translate. We find the areas.

$$f = 3\frac{1}{2} \cdot 1\frac{1}{2}$$
$$r = 4 \cdot 2$$

Solve. We carry out the multiplications.

$$f = 3\frac{1}{2} \cdot 1\frac{1}{2} = \frac{7}{2} \cdot \frac{3}{2} = \frac{21}{4}$$
$$r = 4 \cdot 2 = 8$$

Now we subtract to find the amount of wood removed in planing and drying.

$$8 - \frac{21}{4} = \frac{32}{4} - \frac{21}{4} = \frac{11}{4}$$

Finally we find p, the percent of wood removed in planing and drying.

$$\underset{\downarrow}{\frac{11}{4}} \quad \underset{\downarrow}{\text{is}} \quad \underset{\downarrow}{\overbrace{\text{what percent}}} \quad \underset{\downarrow \;\; \downarrow}{\text{of 8?}}$$
$$\frac{11}{4} \quad = \qquad p \qquad \cdot \quad 8$$

We solve the equation.

$$\frac{11}{4} = p \cdot 8$$
$$\frac{1}{8} \cdot \frac{11}{4} = p$$
$$\frac{11}{32} = p$$
$$0.34375 = p$$
$$34.375\% = p, \text{ or}$$
$$34\frac{3}{8}\% = p$$

Check. We repeat the calculations. The answer checks.

State. 34.375%, or $34\frac{3}{8}\%$, of the wood is removed in planing and drying.

35. a) **Familiarize**. First we find the amount of the decrease.

$$\begin{array}{r} 6\,0\,6,9\,0\,0 \\ -\,5\,7\,2,0\,5\,9 \\ \hline 3\,4,8\,4\,1 \end{array}$$

Let $P =$ the percent of decrease.

Translate. We translate to a proportion.

$$\frac{P}{100} = \frac{34,841}{606,900}$$

Solve.

$$\frac{P}{100} = \frac{34,841}{606,900}$$
$$P \cdot 606,900 = 100 \cdot 34,841$$
$$\frac{P \cdot 606,900}{606,900} = \frac{100 \cdot 34,841}{606,900}$$
$$P \approx 5.7$$

Check. We can repeat the calculations. Also note that $5.7\% \cdot 606,900 \approx 6\% \cdot 600,000 = 36,000 \approx 34,841$. The answer checks.

State. The percent of decrease was about 5.7%

b) **Familiarize**. First we find the amount of the decrease in the next decade. Let b represent this number.

Translate. We translate to a proportion.

$$\frac{5.7}{100} = \frac{b}{572,059}$$

Solve.

$$\frac{5.7}{100} = \frac{b}{572,059}$$
$$5.7 \cdot 572,059 = 100 \cdot b$$
$$\frac{5.7 \cdot 572,059}{100} = \frac{100 \cdot b}{100}$$
$$32,607 \approx b$$

We subtract to find the population in 2010:

$$572,059 - 32,607 = 539,452$$

Check. We can repeat the calculations. Also note that the population in 2010 will be $100\% - 5.7\%$, or 94.3%, of the 2000 population and $94.3\% \cdot 572,059 \approx 539,452$. The answer checks.

State. In 2010 the population will be 539,452.

37. **Familiarize**. First we subtract to find the amount of the increase.

$$\begin{array}{r} 7\,3\,5 \\ -\,4\,3\,0 \\ \hline 3\,0\,5 \end{array}$$

Now let $p =$ the percent of increase.

Translate. We translate to an equation.

$$\underset{\downarrow \;\; \downarrow}{305 \text{ is}} \quad \underset{\downarrow}{\overbrace{\text{what percent}}} \quad \underset{\downarrow \;\; \downarrow}{\text{of 430?}}$$
$$305 = \qquad p \qquad \cdot \quad 430$$

Solve.

$$305 = p \cdot 430$$
$$\frac{305}{430} = \frac{p \cdot 430}{430}$$
$$0.71 \approx p$$
$$71\% \approx p$$

Check. We can repeat the calculations. Also note that $171\% \cdot 430 = 735.3 \approx 735$. The answer checks.

State. The percent of increase is about 71%.

39. **Familiarize**. Let $a =$ the amount of the increase.

Translate. We translate to a proportion.

$$\frac{100}{100} = \frac{a}{780}$$

Solve.

$$\frac{100}{100} = \frac{a}{780}$$

$$100 \cdot 780 = 100 \cdot a$$

$$\frac{100 \cdot 780}{100} = a$$

$$780 = a$$

Now we add to find the higher rate:

$$\begin{array}{r} 7\,8\,0 \\ +\ \ 7\,8\,0 \\ \hline 1\,5\,6\,0 \end{array}$$

Check. We can repeat the calculations. Also note that $200\% \cdot \$780 = \$1560$. The answer checks.

State. The rate for smokers is $1560.

41. Familiarize. First we subtract to find the amount of the increase.

$$\begin{array}{r} 2\,9\,5\,5 \\ -1\,6\,4\,5 \\ \hline 1\,3\,1\,0 \end{array}$$

Now let $p =$ the percent of increase.

Translate. We translate to an equation.

1310 is what percent of 1645?

$$1310 = p \cdot 1645$$

Solve.

$$1310 = p \cdot 1645$$

$$\frac{1310}{1645} = p$$

$$0.80 \approx p$$

$$80\% \approx p$$

Check. We can repeat the calculations. Also note that $180\% \cdot 1645 = 2961 \approx 2955$. The answer checks.

State. The percent of increase is about 80%.

43. Familiarize. First we subtract to find the amount of change.

$$\begin{array}{r} 6\,0\,8,8\,2\,7 \\ -5\,6\,2,7\,5\,8 \\ \hline 4\,6,0\,6\,9 \end{array}$$

Now let $p =$ the percent of change.

Translate. We translate to a proportion.

$$\frac{p}{100} = \frac{46,069}{562,758}$$

Solve.

$$\frac{p}{100} = \frac{46,069}{562,768}$$

$$p \cdot 562,768 = 100 \cdot 46,069$$

$$p = \frac{100 \cdot 46,069}{562,768}$$

$$p \approx 8.2$$

Check. We can repeat the calculations. Also note that $108.2\% \cdot 562,758 \approx 608,904 \approx 608,827$. The answer checks.

State. The population of Vermont increased by 46,069. This was an 8.2% increase.

45. Familiarize. First we subtract to find the amount of the change.

$$\begin{array}{r} 5,8\,9\,4,1\,2\,1 \\ -4,8\,6\,6,6\,9\,2 \\ \hline 1,0\,2\,7,4\,2\,9 \end{array}$$

Now let $p =$ the percent of change.

Translate. We translate to an equation.

1,027,429 is what percent of 4,866,692?

$$1,027,429 = p \cdot 4,866,692$$

Solve.

$$1,027,429 = p \cdot 4,866,692$$

$$\frac{1,027,429}{4,866,692} = p$$

$$0.211 \approx p$$

$$21.1\% \approx p$$

Check. We can repeat the calculations. Also note that $121.1\% \cdot 4,866,692 \approx 5,893,564 \approx 5,894,121$. The answer checks.

State. The population of Washington increased by 1,027,429. This was a 21.1% increase.

47. Familiarize. First we subtract to find the population in 1990.

$$\begin{array}{r} 5,3\,6\,3,6\,7\,5 \\ -\ \ \ 4\,7\,1,9\,0\,6 \\ \hline 4,8\,9\,1,7\,6\,9 \end{array}$$

Now let $p =$ the percent of change.

Translate. We translate to a proportion.

$$\frac{p}{100} = \frac{471,906}{4,891,769}$$

Solve.

$$\frac{p}{100} = \frac{471,906}{4,891,769}$$

$$p \cdot 4,891,769 = 100 \cdot 471,906$$

$$p = \frac{100 \cdot 471,906}{4,891,769}$$

$$p \approx 9.6$$

Check. We can repeat the calculations. Also note that $109.6\% \cdot 4,891,769 \approx 5,361,379 \approx 5,363,675$. The answer checks.

State. The population of Wisconsin in 1990 was 4,891,769. The population had increased by 9.6% in 2000.

49. Familiarize. Since the car depreciates 25% in the first year, its value after the first year is $100\% - 25\%$, or 75%, of the original value. To find the decrease in value, we ask:

$27,300 is 75% of what?

Let $b =$ the original cost.

Translate. We translate to an equation.

$27,300 is 75% of what?

$$\$27,300 = 75\% \times b$$

Solve.
$$27,300 = 75\% \times b$$
$$\frac{27,300}{75\%} = \frac{75\% \times b}{75\%}$$
$$\frac{27,300}{0.75} = b$$
$$36,400 = b$$

Check. We find 25% of 36,400 and then subtract this amount from 36,400:

$$0.25 \times 36,400 = 9100 \text{ and}$$
$$36,400 - 9100 = 27,300$$

The answer checks.

State. The original cost was $36,400.

51. Familiarize. First we use the formula $A = l \times w$ to find the area of the strike zone:

$$A = 30 \times 17 = 510 \text{ in}^2$$

When a 2-in. border is added to the outside of the strike zone, the dimensions of the larger zone are 19 in. by 34 in. The area of this zone is

$$A = 34 \times 21 = 714 \text{ in}^2$$

We subtract to find the increase in area:

$$714 \text{ in}^2 - 510 \text{ in}^2 = 204 \text{ in}^2$$

We let $p =$ the percent of increase in the area.

Translate. We translate to a proportion.

204 is what percent of 510?

$$204 = P \times 510$$

Solve. We divide by 510 on both sides and convert to percent notation.

$$\frac{204}{510} = \frac{p \times 510}{510}$$
$$0.4 = p$$
$$40\% = p$$

Check. We repeat the calculations.

State. The area of the strike zone is increased by 40%.

53. Discussion and Writing Exercise

55. $\frac{25}{11} = 25 \div 11$

$$\begin{array}{r} 2.27 \\ 11\overline{)25.00} \\ \underline{22} \\ 30 \\ \underline{22} \\ 80 \\ \underline{77} \\ 3 \end{array}$$

Since the remainders begin to repeat, we have a repeating decimal.

$$\frac{25}{11} = 2.\overline{27}$$

57. $\frac{27}{8} = 27 \div 8$

$$\begin{array}{r} 3.375 \\ 8\overline{)27.000} \\ \underline{24} \\ 30 \\ \underline{24} \\ 60 \\ \underline{56} \\ 40 \\ \underline{40} \\ 0 \end{array}$$

$$\frac{27}{8} = 3.375$$

We could also do this conversion as follows:

$$\frac{27}{8} = \frac{27}{8} \cdot \frac{125}{125} = \frac{3375}{1000} = 3.375$$

59. $\frac{23}{25} = \frac{23}{25} \cdot \frac{4}{4} = \frac{92}{100} = 0.92$

61. $\frac{14}{32} = 14 \div 32$

$$\begin{array}{r} 0.4375 \\ 32\overline{)14.0000} \\ \underline{128} \\ 120 \\ \underline{96} \\ 240 \\ \underline{224} \\ 160 \\ \underline{160} \\ 0 \end{array}$$

$$\frac{14}{32} = 0.4375$$

(Note that we could have simplified the fraction first, getting $\frac{7}{16}$ and then found the quotient $7 \div 16$.)

63. Since 10,000 has 4 zeros, we move the decimal point in the number in the numerator 4 places to the left.

$$\frac{34,809}{10,000} = 3.4809$$

65. Familiarize. We will express 4 ft, 8 in. as 56 in. (4 ft + 8 in. = $4 \cdot 12$ in. + 8 in. = 48 in. + 8 in. = 56 in.) We let $h =$ Cynthia's final adult height.

Translate. We translate to an equation.

56 in. is 84.4% of what?

$$56 = 84.4\% \times h$$

Solve. First we convert 84.4% to a decimal.

$$56 = 0.844 \times h$$
$$\frac{56}{0.844} = \frac{0.844 \times h}{0.844}$$
$$66 \approx h$$

Check. We find 84.4% of 66: $0.844 \times 66 \approx 56$. The answer checks.

State. Cynthia's final adult height will be about 66 in., or 5 ft, 6 in.

67. *Familiarize*. If p is 120% of q, then $p = 1.2q$. Let $n =$ the percent of p that q represents.

Translate. We translate to an equation. We use $1.2q$ for p.

$$q \text{ is what percent of } p?$$
$$q = \quad n \quad \times \, 1.2q$$

Solve.

$$q = n \times 1.2q$$
$$\frac{q}{1.2q} = \frac{n \times 1.2q}{1.2q}$$
$$\frac{1}{1.2} = n$$
$$0.8\overline{3} = n$$
$$83.\overline{3}\%, \text{ or } 83\frac{1}{3}\% = n$$

Check. We find $83\frac{1}{3}\%$ of $1.2q$:

$$0.8\overline{3} \times 1.2q = q$$

The answer checks.

State. q is $83.\overline{3}\%$, or $83\frac{1}{3}\%$, of p.

Exercise Set 4.7

1. The sales tax on an item costing $25.95 is

$$\text{Sales tax rate} \times \text{Purchase price}$$
$$8.25\% \quad \times \quad \$25.95,$$

or 0.0825×25.95, or about 2.14. Thus the tax is $2.14.

3. *Rephrase*:

$$\text{Sales tax is what percent of purchase price?}$$

Translate: $48 = r \times 960$

To solve the equation, we divide on both sides by 960.

$$\frac{48}{960} = \frac{r \times 960}{960}$$
$$0.05 = r$$
$$5\% = r$$

The sales tax rate is 5%.

5. a) We first find the cost of the telephones. It is

$$5 \times \$53 = \$265.$$

b) The sales tax on items costing $265 is

$$\text{Sales tax rate} \times \text{Purchase price}$$
$$7\% \quad \times \quad \$265,$$

or 0.07×265, or 18.55. Thus the tax is $18.55.

c) The total price is given by the purchase price plus the sales tax:

$$\$265 + \$18.55, \text{ or } \$283.55.$$

To check, note that the total price is the purchase price plus 7% of the purchase price. Thus the total price is 107% of the purchase price. Since $1.07 \times \$265 = \$283.55$, we have a check. The total price is $283.55.

7. *Rephrase*:

$$\text{Sales tax is what percent of purchase price?}$$

Translate: $35.80 = r \times 895$

To solve the equation, we divide on both sides by 895.

$$\frac{35.80}{895} = \frac{r \times 895}{895}$$
$$0.04 = r$$
$$4\% = r$$

The sales tax rate is 4%.

9. *Rephrase*:   Sales tax is 5% of what?

Translate:
$$100 = 5\% \times b, \text{ or}$$
$$100 = 0.05 \times b$$

To solve the equation, we divide on both sides by 0.05.

$$\frac{100}{0.05} = \frac{0.05 \times b}{0.05}$$
$$2000 = b$$

$$\begin{array}{r} 2\ 0\ 0\ 0\ . \\ 0.0\ 5\overline{)1\ 0\ 0.0\ 0}_{\wedge} \\ \underline{1\ 0\ 0\ 0\ 0} \\ 0 \end{array}$$

The purchase price is $2000.

11. *Rephrase*:   Sales tax is 3.5% of what?

Translate:
$$28 = 3.5\% \times b, \text{ or}$$
$$28 = 0.035 \times b$$

To solve the equation, we divide on both sides by 0.035.

$$\frac{28}{0.035} = \frac{0.035 \times b}{0.035}$$
$$800 = b$$

$$\begin{array}{r} 8\ 0\ 0\ . \\ 0.0\ 3\ 5\overline{)2\ 8.0\ 0\ 0}_{\wedge} \\ \underline{2\ 8\ 0\ 0\ 0} \\ 0 \end{array}$$

The purchase price is $800.

13. a) We first find the cost of the shower units. It is

$$2 \times \$332.50 = \$665.$$

b) The total tax rate is the city tax rate plus the state tax rate, or $2\% + 6.25\% = 8.25\%$. The sales tax paid on items costing $665 is

$$\text{Sales tax rate} \times \text{Purchase price}$$
$$8.25\% \quad \times \quad \$665,$$

or 0.0825×665, or about 54.86. Thus the tax is $54.86.

c) The total price is given by the purchase price plus the sales tax:

$$\$665 + \$54.86 = \$719.86.$$

To check, note that the total price is the purchase price plus 8.25% of the purchase price. Thus the total price is 108.25% of the purchase price. Since $1.0825 \times 665 \approx 719.86$, we have a check. The total amount paid for the 2 shower units is $719.86.

15. *Rephrase:*

Sales tax	is	what percent	of	purchase price?
↓	↓	↓	↓	↓

Translate: $1030.40 = r \times 18,400$

To solve the equation, we divide on both sides by 18,400.

$$\frac{1030.40}{18,400} = \frac{r \times 18,400}{18,400}$$
$$0.056 = r$$
$$5.6\% = r$$

The sales tax rate is 5.6%.

17. Commission = Commission rate $\times$ Sales
$$C \quad = \quad 6\% \quad \times 45,000$$

This tells us what to do. We multiply.

$$\begin{array}{r} 4\,5,0\,0\,0 \\ \times \quad 0.\,0\,6 \\ \hline 2\,7\,0\,0.\,0\,0 \end{array} \quad (6\% = 0.06)$$

The commission is $2700.

19. Commission = Commission rate $\times$ Sales
$$120 \quad = \quad r \quad \times 2400$$

To solve this equation we divide on both sides by 2400:

$$\frac{120}{2400} = \frac{r \times 2400}{2400}$$

We can divide, but this time we simplify by removing a factor of 1:

$$r = \frac{120}{2400} = \frac{1}{20} \cdot \frac{120}{120} = \frac{1}{20} = 0.05 = 5\%$$

The commission rate is 5%.

21. Commission = Commission rate $\times$ Sales
$$392 \quad = \quad 40\% \quad \times \quad S$$

To solve this equation we divide on both sides by 0.4:

$$\frac{392}{0.4} = \frac{0.4 \times S}{0.4}$$
$$980 = S$$

$$\begin{array}{r} 9\,8\,0. \\ 0.4_{\wedge}\overline{)3\,9\,2.0_{\wedge}} \\ 3\,6\,0\,0 \\ \hline 3\,2\,0 \\ 3\,2\,0 \\ \hline 0 \\ 0 \\ \hline 0 \end{array}$$

$980 worth of artwork was sold.

23. Commission = Commission rate $\times$ Sales
$$C \quad = \quad 6\% \quad \times 98,000$$

This tells us what to do. We multiply.

$$\begin{array}{r} 9\,8,0\,0\,0 \\ \times \quad 0.\,0\,6 \\ \hline 5\,8\,8\,0.\,0\,0 \end{array} \quad (6\% = 0.06)$$

The commission is $5880.

25. Commission = Commission rate $\times$ Sales
$$280.80 \quad = \quad r \quad \times 2340$$

To solve this equation we divide on both sides by 2340.

$$\frac{280.80}{2340} = \frac{r \times 2340}{2340}$$
$$0.12 = r$$
$$12\% = r$$

$$\begin{array}{r} 0.1\,2 \\ 2\,3\,4\,0\,\overline{)2\,8\,0.8\,0} \\ 2\,3\,4\,0 \\ \hline 4\,6\,8\,0 \\ 4\,6\,8\,0 \\ \hline 0 \end{array}$$

The commission is 12%.

27. First we find the commission on the first $2000 of sales.
Commission = Commission rate $\times$ Sales
$$C \quad = \quad 5\% \quad \times 2000$$

This tells us what to do. We multiply.

$$\begin{array}{r} 2\,0\,0\,0 \\ \times \quad 0.\,0\,5 \\ \hline 1\,0\,0.0\,0 \end{array}$$

The commission on the first $2000 of sales is $100.

Next we subtract to find the amount of sales over $2000.

$$\$6000 - \$2000 = \$4000$$

Miguel had $4000 in sales over $2000.

Then we find the commission on the sales over $2000.
Commission = Commission rate $\times$ Sales
$$C \quad = \quad 8\% \quad \times 4000$$

This tells us what to do. We multiply.

$$\begin{array}{r} 4\,0\,0\,0 \\ \times \quad 0.\,0\,8 \\ \hline 3\,2\,0.0\,0 \end{array}$$

The commission on the sales over $2000 is $320.

Finally we add to find the total commission.

$$\$100 + \$320 = \$420$$

The total commission is $420.

29. Discount = Marked price $-$ Sale price
$$D \quad = \quad 1275 \quad - \quad 888$$

We subtract:
$$\begin{array}{r} 1\,2\,7\,5 \\ - \quad 8\,8\,8 \\ \hline 3\,8\,7 \end{array}$$

The discount is $387.

Discount = Rate of discount $\times$ Marked price
$$387 \quad = \quad R \quad \times \quad 1275$$

To solve the equation we divide on both sides by 1275.

$$\frac{387}{1275} = \frac{R \times 1275}{1275}$$
$$0.30353 \approx R$$
$$30.353\% \approx R, \text{ or}$$
$$30\frac{6}{17}\% = R$$

To check, note that a discount rate of $30\frac{6}{17}\%$ means that $69\frac{11}{17}\%$, or about 69.647%, of the marked price is paid: $0.69647 \times 1275 = 887.99925 \approx 888$. Since that is the sale price, the answer checks.

The rate of discount is $30\frac{6}{17}\%$, or about 30.4%.

31. Discount = Rate of discount × Marked price
$$D \quad = \quad 10\% \quad \times \quad \$300$$

Convert 10% to decimal notation and multiply.

$$\begin{array}{r} 3\,0\,0 \\ \times \quad 0.\,1 \\ \hline 3\,0.\,0 \end{array} \qquad (10\% = 0.10 = 0.1)$$

The discount is $30.

Sale price = Marked price − Discount
$$S \quad = \quad 300 \quad - \quad 30$$

We subtract:
$$\begin{array}{r} 3\,0\,0 \\ -\quad 3\,0 \\ \hline 2\,7\,0 \end{array}$$

To check, note that the sale price is 90% of the marked price: $0.9 \times 300 = 270$.

The sale price is $270.

33. Discount = Rate of discount × Marked price
$$D \quad = \quad 15\% \quad \times \quad \$17$$

Convert 15% to decimal notation and multiply.

$$\begin{array}{r} 1\,7 \\ \times\, 0.\,1\,5 \\ \hline 8\,5 \\ 1\,7\,0 \\ \hline 2.\,5\,5 \end{array} \qquad (15\% = 0.15)$$

The discount is $2.55.

Sale price = Marked price − Discount
$$S \quad = \quad 17 \quad - \quad 2.55$$

We subtract:
$$\begin{array}{r} 1\,7.\,0\,0 \\ -\quad 2.\,5\,5 \\ \hline 1\,4.\,4\,5 \end{array}$$

To check, note that the sale price is 85% of the marked price: $0.85 \times 17 = 14.45$.

The sale price is $14.45.

35. Discount = Rate of discount × Marked price
$$12.50 \quad = \quad 10\% \quad \times \quad M$$

To solve the equation we divide on both sides by 0.1.
$$\frac{12.50}{0.1} = \frac{0.1 \times M}{0.1}$$
$$125 = M$$

The marked price is $125.

Sale price = Marked price − Discount
$$S \quad = \quad 125.00 \quad - \quad 12.50$$

We subtract:
$$\begin{array}{r} 1\,2\,5.\,0\,0 \\ -\quad 1\,2.\,5\,0 \\ \hline 1\,1\,2.\,5\,0 \end{array}$$

To check, note that the sale price is 90% of the marked price: $0.9 \times 125 = 112.50$.

The sale price is $112.50.

37. Discount = Rate of discount × Marked price
$$240 \quad = \quad r \quad \times \quad 600$$

To solve the equation we divide on both sides by 600.
$$\frac{240}{600} = \frac{r \times 600}{600}$$

We can simplify by removing a factor of 1:
$$r = \frac{240}{600} = \frac{2}{5} \cdot \frac{120}{120} = \frac{2}{5} = 0.4 = 40\%$$

The rate of discount is 40%.

Sale price = Marked price − Discount
$$S \quad = \quad 600 \quad - \quad 240$$

We subtract:
$$\begin{array}{r} 6\,0\,0 \\ -\,2\,4\,0 \\ \hline 3\,6\,0 \end{array}$$

To check, note that a 40% discount rate means that 60% of the marked price is paid. Since $\frac{360}{600} = 0.6$, or 60%, we have a check.

The sale price is $360.

39. $I = P \cdot r \cdot t$
$$= \$200 \times 8\% \times 1$$
$$= \$200 \times 0.08$$
$$= \$16$$

41. $I = P \cdot r \cdot t$
$$= \$2000 \times 8.4\% \times \frac{1}{2}$$
$$= \frac{\$2000 \times 0.084}{2}$$
$$= \$84$$

43. $I = P \cdot r \cdot t$
$$= \$4300 \times 10.56\% \times \frac{1}{4}$$
$$= \frac{\$4300 \times 0.1056}{4}$$
$$= \$113.52$$

45. $I = P \cdot r \cdot t$
$$= \$20,000 \times 7\frac{5}{8}\% \times 1$$
$$= \$20,000 \times 0.07625$$
$$= \$1525$$

47. $I = P \cdot r \cdot t$
$$= \$50,000 \times 5\frac{3}{8}\% \times \frac{1}{4}$$
$$= \frac{\$50,000 \times 0.05375}{4}$$
$$\approx \$671.88$$

49. a) We express 90 days as a fractional part of a year and find the interest.

$$I = P \cdot r \cdot t$$
$$= \$6500 \times 8\% \times \frac{90}{365}$$
$$= \$6500 \times 0.08 \times \frac{90}{365}$$
$$\approx \$128.22 \quad \text{Using a calculator}$$

The interest due for 90 days is \$128.22.

b) The total amount that must be paid after 90 days is the principal plus the interest.

$$6500 + 128.22 = 6628.22$$

The total amount due is \$6628.22.

51. a) After 1 year, the account will contain 110% of \$400.

$$1.1 \times \$400 = \$440$$

```
    4 0 0
  × 1. 1
  -------
    4 0 0
  4 0 0 0
  -------
  4 4 0.0
```

b) At the end of the second year, the account will contain 110% of \$440.

$$1.1 \times \$440 = \$484$$

```
    4 4 0
  × 1. 1
  -------
    4 4 0
  4 4 0 0
  -------
  4 8 4.0
```

The amount in the account after 2 years is \$484.

(Note that we could have used the formula $A = P \cdot \left(1 + \frac{r}{n}\right)^{n \cdot t}$, substituting \$400 for P, 10% for r, 1 for n, and 2 for t.)

53. We use the compound interest formula, substituting \$2000 for P, 8.8% for r, 1 for n, and 4 for t.

$$A = P \cdot \left(1 + \frac{r}{n}\right)^{n \cdot t}$$
$$= \$2000 \cdot \left(1 + \frac{8.8\%}{1}\right)^{1 \cdot 4}$$
$$= \$2000 \cdot (1 + 0.088)^4$$
$$= \$2000 \cdot (1.088)^4$$
$$\approx \$2802.50$$

The amount in the account after 4 years is \$2802.50.

55. We use the compound interest formula, substituting \$4300 for P, 10.56% for r, 1 for n, and 6 for t.

$$A = P \cdot \left(1 + \frac{r}{n}\right)^{n \cdot t}$$
$$= \$4300 \cdot \left(1 + \frac{10.56\%}{1}\right)^{1 \cdot 6}$$
$$= \$4300 \cdot (1 + 0.1056)^6$$
$$= \$4300 \cdot (1.1056)^6$$
$$\approx \$7853.38$$

The amount in the account after 4 years is \$7853.38.

57. We use the compound interest formula, substituting \$20,000 for P, $7\frac{5}{8}\%$ for r, 1 for n, and 25 for t.

$$A = P \cdot \left(1 + \frac{r}{n}\right)^{n \cdot t}$$
$$= \$20,000 \cdot \left(1 + \frac{7\frac{5}{8}\%}{1}\right)^{1 \cdot 25}$$
$$= \$20,000 \cdot (1 + 0.07625)^{25}$$
$$= \$20,000 \cdot (1.07625)^{25}$$
$$\approx \$125,562.26$$

The amount in the account after 25 years is \$125,562.26.

59. We use the compound interest formula, substituting \$4000 for P, 7% for r, 2 for n, and 1 for t.

$$A = P \cdot \left(1 + \frac{r}{n}\right)^{n \cdot t}$$
$$= \$4000 \cdot \left(1 + \frac{7\%}{2}\right)^{2 \cdot 1}$$
$$= \$4000 \cdot \left(1 + \frac{0.07}{2}\right)^2$$
$$= \$4000 \cdot (1.035)^2$$
$$= \$4284.90$$

The amount in the account after 1 year is \$4284.90.

61. We use the compound interest formula, substituting \$20,000 for P, 8.8% for r, 2 for n, and 4 for t.

$$A = P \cdot \left(1 + \frac{r}{n}\right)^{n \cdot t}$$
$$= \$20,000 \cdot \left(1 + \frac{8.8\%}{2}\right)^{2 \cdot 4}$$
$$= \$20,000 \cdot \left(1 + \frac{0.088}{2}\right)^8$$
$$= \$20,000 \cdot (1.044)^8$$
$$\approx \$28,225.00$$

The amount in the account after 4 years is \$28,225.00.

63. We use the compound interest formula, substituting \$5000 for P, 10.56% for r, 2 for n, and 6 for t.

$$A = P \cdot \left(1 + \frac{r}{n}\right)^{n \cdot t}$$
$$= \$5000 \cdot \left(1 + \frac{10.56\%}{2}\right)^{2 \cdot 6}$$
$$= \$5000 \cdot \left(1 + \frac{0.1056}{2}\right)^{12}$$
$$= \$5000 \cdot (1.0528)^{12}$$
$$\approx \$9270.87$$

The amount in the account after 6 years is \$9270.87.

65. We use the compound interest formula, substituting \$20,000 for P, $7\frac{5}{8}\%$ for r, 2 for n, and 25 for t.

$$A = P \cdot \left(1 + \frac{r}{n}\right)^{n \cdot t}$$

$$= \$20,000 \cdot \left(1 + \frac{7\frac{5}{8}\%}{2}\right)^{2 \cdot 25}$$

$$= \$20,000 \cdot \left(1 + \frac{0.07625}{2}\right)^{50}$$

$$= \$20,000 \cdot (1.038125)^{50}$$

$$\approx \$129,871.09$$

The amount in the account after 25 years is \$129,871.09.

67. We use the compound interest formula, substituting \$4000 for P, 6% for r, 12 for n, and $\frac{5}{12}$ for t.

$$A = P \cdot \left(1 + \frac{r}{n}\right)^{n \cdot t}$$

$$= \$4000 \cdot \left(1 + \frac{6\%}{2}\right)^{12 \cdot \frac{5}{12}}$$

$$= \$4000 \cdot \left(1 + \frac{0.06}{12}\right)^{5}$$

$$= \$4000 \cdot (1.005)^{5}$$

$$\approx \$4101.01$$

The amount in the account after 5 months is \$4101.01.

69. Discussion and Writing Exercise

71.   0.<u>93</u>   0.93. $\frac{93}{100}$

    2 places   Move 2 places.   2 zeros

$$0.93 = \frac{93}{100}$$

73. $\frac{13}{11} = 13 \div 11$

```
        1. 1 8 1 8
  1 1 ⌐1 3. 0 0 0 0
        1 1
        ‾‾‾
          2 0
          1 1
          ‾‾‾
            9 0
            8 8
            ‾‾‾
              2 0
              1 1
              ‾‾‾
                9 0
                8 8
                ‾‾‾
                  2
```

We get a repeating decimal.

$$\frac{13}{11} = 1.\overline{18}$$

Exercise Set 4.8

1. a) We multiply the balance by 2%:

$$0.02 \times \$4876.54 = \$97.5308.$$

Antonio's minimum payment, rounded to the nearest dollar, is \$98.

b) We find the amount of interest on \$4876.54 at 21.3% for one month.

$$I = P \cdot r \cdot t$$

$$= \$4876.54 \times 0.213 \times \frac{1}{12}$$

$$\approx \$86.56$$

We subtract to find the amount applied to decrease the principal in the first payment.

$$\$98 - \$86.56 = \$11.44$$

The principal is decreased by \$11.44 with the first payment.

c) We find the amount of interest on \$4876.54 at 12.6% for one month.

$$I = P \cdot r \cdot t$$

$$= \$4876.54 \times 0.126 \times \frac{1}{12}$$

$$\approx \$51.20$$

We subtract to find the amount applied to decrease the principal in the first payment.

$$\$98 - \$51.20 = \$46.80.$$

The principal is decreased by \$46.80 with the first payment.

d) With the 12.6% rate the principal was decreased by \$46.80 − \$11.44, or \$35.36 more than at the 21.3% rate. This also means that the interest at 12.6% is \$35.36 less than at 21.3%.

3. a) We find the interest on \$44,560 at 6.5% for one month.

$$I = P \cdot r \cdot t$$

$$= \$44,560 \times 0.065 \times \frac{1}{12}$$

$$\approx \$241.37$$

The amount of interest in the first payment is \$241.37.

We subtract to find amount applied to the principal.

$$\$505.97 - \$241.37 = \$264.60$$

With the first payment the principal will decrease by \$264.60.

b) We find the interest on \$44,560 at 8.5% for one month.

$$I = P \cdot r \cdot t$$

$$= \$44,560 \times 0.085 \times \frac{1}{12}$$

$$\approx \$315.63$$

At 8.5% the additional interest in the first payment is \$315.63 − \$241.37 = \$74.26.

c) For the 6.5% loan there will be 120 payments of \$505.97:

$$120 \times \$505.97 = \$60,716.40$$

The total interest at this rate is

$$\$60,716.40 - \$44,560 = \$16,156.40.$$

For the 8.5% loan there will be 120 payments of $552.48:

$$120 \times \$552.48 = \$66,297.60$$

The total interest at this rate is

$$\$66,297.60 - \$44,560 = \$21,737.60$$

At 8.5% Grace would pay

$$\$21,737.60 - \$16,156.40 = \$5581.20$$

more in interest than at 6.5%.

5. a) We find the interest on $150,000 at 6.98% for one month.

$$I = P \cdot r \cdot t$$
$$= \$150,000 \times 0.0698 \times \frac{1}{12}$$
$$\approx \$872.50$$

The amount of interest applied to the principal is

$$\$995.94 - \$872.50 = \$123.44.$$

b) The total paid will be

$$360 \times \$995.94 = \$358,538.40.$$

Then the total amount of interest paid is

$$\$358,538.40 - \$150,000 = \$208,538.40.$$

c) We subtract to find the new principal after the first payment.

$$\$150,000 - \$123.44 = \$149,876.56$$

Now we find the interest on $149,876.56 at 6.98% for one month.

$$I = P \cdot r \cdot t$$
$$= \$149,876.56 \times 0.0698 \times \frac{1}{12}$$
$$\approx \$871.78$$

We subtract to find the amount applied to the principal.

$$\$995.94 - \$871.78 = \$124.16$$

7. a) From Exercise 5(a) we know that the amount of interest in the first payment is $872.50. The amount applied to the principal is

$$\$1346.57 - \$872.50 = \$474.07$$

b) The total paid will be

$$180 \times \$1346.57 = \$242,382.60.$$

Then the total amount of interest paid is

$$\$242,382.60 - \$150,000 = \$92,382.60.$$

c) On the 15-yr loan the Martinez family will pay

$$\$208,538.40 - \$92,382.60 = \$116,155.80$$

less than on the 30-yr loan.

9. Interest in first payment:

$$I = P \cdot r \cdot t$$
$$= \$100,000 \times 0.0698 \times \frac{1}{12}$$
$$\approx \$581.67$$

Amount of principal in first payment:

$$\$663.96 - \$581.67 = \$82.29$$

Principal after first payment:

$$\$100,000 - \$82.29 = \$99,917.71$$

Interest in second payment:

$$I = P \cdot r \cdot t$$
$$= \$99,917.71 \times 0.0698 \times \frac{1}{12}$$
$$\approx \$581.19$$

Amount of principal in second payment:

$$\$663.96 - \$581.19 = \$82.77$$

Principal after second payment:

$$\$99,917.71 - \$82.77 = \$99,834.94$$

11. Interest in first payment:

$$I = P \cdot r \cdot t$$
$$= \$100,000 \times 0.0804 \times \frac{1}{12}$$
$$\approx \$670.00$$

Amount of principal in first payment:

$$\$957.96 - \$670.00 = \$287.96$$

Principal after first payment:

$$\$100,000 - \$287.96 = \$99,712.04$$

Interest in second payment:

$$I = P \cdot r \cdot t$$
$$= \$99,712.04 \times 0.0804 \times \frac{1}{12}$$
$$\approx \$668.07$$

Amount of principal in second payment:

$$\$957.96 - \$668.07 = \$289.89$$

Principal after second payment:

$$\$99,712.04 - \$289.89 = \$99,422.15$$

13. Interest in first payment:

$$I = P \cdot r \cdot t$$
$$= \$150,000 \times 0.0724 \times \frac{1}{12}$$
$$= \$905.00$$

Amount of principal in first payment:

$$\$1022.25 - \$905.00 = \$117.25$$

Principal after first payment:

$$\$150,000 - \$117.25 = \$149,882.75$$

Interest in second payment:

$$I = P \cdot r \cdot t$$
$$= \$149,882.75 \times 0.0724 \times \frac{1}{12}$$
$$\approx \$904.29$$

Amount of principal in second payment:

$$\$1022.25 - \$904.29 = \$117.96$$

Principal after second payment:

$$\$149,882.75 - \$117.96 = \$149,764.79$$

15. Interest in first payment:

$$I = P \cdot r \cdot t$$

$$= \$200,000 \times 0.0724 \times \frac{1}{12}$$

$$\approx \$1206.67$$

Amount of principal in first payment:

$$\$1824.60 - \$1206.67 = \$617.93$$

Principal after first payment:

$$\$200,000 - \$617.93 = \$199,382.07$$

Interest in second payment:

$$I = P \cdot r \cdot t$$

$$= \$199,382.07 \times 0.0724 \times \frac{1}{12}$$

$$\approx \$1202.94$$

Amount of principal in second payment:

$$\$1824.60 - \$1202.94 = \$621.66$$

Principal after second payment:

$$\$199,382.07 - \$621.66 = \$198,760.41$$

17. a) We find the interest on \$15,000 at 8.99% for one month.

$$I = P \cdot r \cdot t$$

$$= \$15,000 \times 0.0899 \times \frac{1}{12}$$

$$\approx \$112.38$$

We subtract to find the amount applied to reduce the principal.

$$\$373.20 - \$112.38 = \$260.82$$

b) We subtract the find the principal at the beginning of the second month.

$$\$15,000 - \$260.82 = \$14,739.18$$

Now we find the interest paid the second month.

$$I = P \cdot r \cdot t$$

$$= \$14,739.18 \times 0.0899 \times \frac{1}{12}$$

$$\approx \$110.42$$

In the second month Janice will pay

$$\$112.38 - \$110.42 = \$1.96$$

less interest than in the first month.

c) There will be 48 payments of \$373.20:

$$48 \times \$373.20 = \$17,913.60$$

The total interest paid will be

$$\$17,913.60 - \$15,000 = \$2913.60.$$

19. a) The down payment is 10% of \$7900:

$$0.1 \times \$7900 = \$790$$

We subtract to find the amount borrowed:

$$\$7900 - \$790 = \$7110$$

b) We find the interest on \$7110 at 12.49% for one month.

$$I = P \cdot r \cdot t$$

$$= \$7110 \times 0.1249 \times \frac{1}{12}$$

$$\approx \$74.00$$

We subtract to find the amount applied to reduce the principal:

$$\$237.82 - \$74.00 = \$163.82$$

c) There will be 36 payments of \$237.82:

$$36 \times \$237.82 = \$8561.52$$

The total interest paid will be

$$\$8561.52 - \$7110 = \$1451.52$$

21. Discussion and Writing Exercise

23. Discussion and Writing Exercise

25.
$$\frac{x}{12} = \frac{24}{16}$$

$$16 \cdot x = 12 \cdot 24 \qquad \text{Equating cross products}$$

$$x = \frac{12 \cdot 24}{16} \qquad \text{Dividing by 16 on both sides}$$

$$x = \frac{288}{16}$$

$$x = 18$$

The solution is 18.

27.
$$0.64 \times x = 170$$

$$\frac{0.64 \cdot x}{0.64} = \frac{170}{0.64} \qquad \text{Dividing by 0.64 on both sides}$$

$$x = 265.625$$

The solution is 265.625.

29. $\dfrac{5}{9} = 5 \div 9$

$$
\begin{array}{r}
0.\,5\,5 \\
9\,\overline{\smash{\big)}\,5.\,0\,0} \\
\underline{4\,5} \\
5\,0 \\
\underline{4\,5} \\
5
\end{array}
$$

We get a repeating decimal.

$$\frac{5}{9} = 0.\bar{5}$$

31. $\dfrac{11}{12} = 11 \div 12$

$$
\begin{array}{r}
0.\,9\,1\,6\,6 \\
1\,2\,\overline{\smash{\big)}\,1\,1.\,0\,0\,0\,0} \\
\underline{1\,0\,8} \\
2\,0 \\
\underline{1\,2} \\
8\,0 \\
\underline{7\,2} \\
8\,0 \\
\underline{7\,2} \\
8
\end{array}
$$

We get a repeating decimal.

$$\frac{11}{12} = 0.91\bar{6}$$

33. $\dfrac{15}{7} = 15 \div 7$

$$
\begin{array}{r}
2.142857 \\
7\,\overline{\smash)15.000000} \\
\underline{14} \\
10 \\
\underline{7} \\
30 \\
\underline{28} \\
20 \\
\underline{14} \\
60 \\
\underline{56} \\
40 \\
\underline{35} \\
50 \\
\underline{49} \\
1
\end{array}
$$

We get a repeating decimal.

$\dfrac{15}{7} = 2.\overline{142857}$

35. $4.03 trillion = \$4.03 \times 1$ trillion
$ = \$4.03 \times 1,000,000,000,000$
$ = \$4,030,000,000,000$

37.   42.7 million $= 42.7 \times 1$ million
$ = 42.7 \times 1,000,000$
$ = 42,700,000$

Chapter 5

Geometry

Exercise Set 5.1

1. The segment consists of the endpoints G and H and all points between them.

It can be named $\overline{GH}$ or $\overline{HG}$.

3. The ray with endpoint Q extends forever in the direction of point D.

In naming a ray, the endpoint is always given first. This ray is named $\overrightarrow{QD}$.

5.

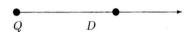

The line can be named with the small letter l, or it can be named by any two points on it. This line can be named

$$l, \overleftrightarrow{DE}, \overleftrightarrow{ED}, \overleftrightarrow{DF}, \overleftrightarrow{FD}, \overleftrightarrow{EF}, \text{ or } \overleftrightarrow{FE}.$$

7. The angle can be named in five different ways:

angle GHI, angle IHG, $\angle\, GHI$, $\angle\, IHG$, or $\angle\, H$.

9. Place the $\triangle$ of the protractor at the vertex of the angle, and line up one of the sides at 0°. We choose the horizontal side. Since 0° is on the inside scale, we check where the other side of the angle crosses the inside scale. It crosses at 10°. Thus, the measure of the angle is 10°.

11. Place the $\triangle$ of the protractor at the vertex of the angle, point B. Line up one of the sides at 0°. We choose the side that contains point A. Since 0° is on the outside scale, we check where the other side crosses the outside scale. It crosses at 180°. Thus, the measure of the angle is 180°.

13. Place the $\triangle$ of the protractor at the vertex of the angle, and line up one of the sides at 0°. We choose the horizontal side. Since 0° is on the inside scale, we check where the other side crosses the inside scale. It crosses at 130°. Thus, the measure of the angle is 130°.

15. Using a protractor, we find that the measure of the angle in Exercise 7 is 148°. Since its measure is greater than 90°and less than 180°, it is an obtuse angle.

17. The measure of the angle in Exercise 9 is 10°. Since its measure is greater than 0°and less than 90°, it is an acute angle.

19. The measure of the angle in Exercise 11 is 180°. It is a straight angle.

21. The measure of the angle in Exercise 13 is 130°. Since its measure is greater than 90°and less than 180°, it is an obtuse angle.

23. The measure of the angle in Margin Exercise 12 is 30°. Since its measure is greater than 0°and less than 90°, it is an acute angle.

25. The measure of the angle in Margin Exercise 14 is 126°. Since its measure is greater than 90°and less than 180°, it is an obtuse angle.

27. Using a protractor, we find that the lines do not intersect to form a right angle. They are not perpendicular.

29. Using a protractor, we find that the lines intersect to form a right angle. They are perpendicular.

31. All the sides are of different lengths. The triangle is a scalene triangle.

One angle is an obtuse angle. The triangle is an obtuse triangle.

33. All the sides are of different lengths. The triangle is a scalene triangle.

One angle is a right angle. The triangle is a right triangle.

35. All the sides are the same length. The triangle is an equilateral triangle.

All three angles are acute. The triangle is an acute triangle.

37. All the sides are of different lengths. The triangle is a scalene triangle.

One angle is an obtuse angle. The triangle is an obtuse triangle.

39. The polygon has 4 sides. It is a quadrilateral.

41. The polygon has 5 sides. It is a pentagon.

43. The polygon has 3 sides. It is a triangle.

45. The polygon has 5 sides. It is a pentagon.

47. The polygon has 6 sides. It is a hexagon.

49. If a polygon has n sides, the sum of its angle measures is $(n-2)\cdot 180°$. A decagon has 10 sides. Substituting 10 for n in the formula, we get

$$(n-2)\cdot 180° = (10-2)\cdot 180°$$
$$= 8\cdot 180°$$
$$= 1440°.$$

51. If a polygon has n sides, the sum of its angle measures is $(n-2)\cdot 180°$. A heptagon has 7 sides. Substituting 7 for n in the formula, we get
$$(n-2)\cdot 180° = (7-2)\cdot 180°$$
$$= 5\cdot 180°$$
$$= 900°.$$

53. If a polygon has n sides, the sum of its angle measures is $(n-2)\cdot 180°$. To find the sum of the angle measures for a 14-sided polygon, substitute 14 for n in the formula.
$$(n-2)\cdot 180° = (14-2)\cdot 180°$$
$$= 12\cdot 180°$$
$$= 2160°$$

55. If a polygon has n sides, the sum of its angle measures is $(n-2)\cdot 180°$. To find the sum of the angle measures for a 20-sided polygon, substitute 20 for n in the formula.
$$(n-2)\cdot 180° = (20-2)\cdot 180°$$
$$= 18\cdot 180°$$
$$= 3240°$$

57. $$m(\angle A) + m(\angle B) + m(\angle C) = 180°$$
$$42° + 92° + x = 180°$$
$$134° + x = 180°$$
$$x = 180° - 134°$$
$$x = 46°$$

59. $$31° + 29° + x = 180°$$
$$60° + x = 180°$$
$$x = 180° - 60°$$
$$x = 120°$$

61. $$m(\angle R) + m(\angle S) + m(\angle T) = 180°$$
$$x + 58° + 79° = 180°$$
$$x + 137° = 180°$$
$$x = 180° - 137°$$
$$x = 43°$$

63. Discussion and Writing Exercise

65. $$I = P\cdot r\cdot t$$
$$= \$2000\cdot 8\%\cdot 1$$
$$= \$2000\cdot 0.08\cdot 1$$
$$= \$160$$

67. $$I = P\cdot r\cdot t$$
$$= \$4000\cdot 7.4\%\cdot \frac{1}{2}$$
$$= \$4000\cdot 0.074\cdot \frac{1}{2}$$
$$= \$148$$

69. $$A = P\cdot \left(1+\frac{r}{n}\right)^{n\cdot t}$$
$$= \$25,000\cdot \left(1+\frac{6\%}{2}\right)^{2\cdot 5}$$
$$= \$25,000\cdot \left(1+\frac{0.06}{2}\right)^{10}$$
$$= \$25,000(1.03)^{10}$$
$$\approx \$33,597.91$$

71. $$A = P\cdot \left(1+\frac{r}{n}\right)^{n\cdot t}$$
$$= \$150,000\cdot \left(1+\frac{7.4\%}{2}\right)^{2\cdot 20}$$
$$= \$150,000\cdot \left(1+\frac{0.074}{2}\right)^{40}$$
$$= \$150,000(1.037)^{40}$$
$$\approx \$641,566.26$$

73. We find $m\angle 2$:
$$m\angle 6 + m\angle 1 + m\angle 2 = 180°$$
$$33.07° + 79.8° + m\angle 2 = 180°$$
$$112.87° + m\angle 2 = 180°$$
$$m\angle 2 = 180° - 112.87°$$
$$m\angle 2 = 67.13°$$
The measure of angle 2 is $67.13°$.

We find $m\angle 3$:
$$m\angle 1 + m\angle 2 + m\angle 3 = 180°$$
$$79.8° + 67.13° + m\angle 3 = 180°$$
$$146.93° + m\angle 3 = 180°$$
$$m\angle 3 = 180° - 146.93°$$
$$m\angle 3 = 33.07°$$
The measure of angle 3 is $33.07°$.

We find $m\angle 4$:
$$m\angle 2 + m\angle 3 + m\angle 4 = 180°$$
$$67.13° + 33.07° + m\angle 4 = 180°$$
$$100.2° + m\angle 4 = 180°$$
$$m\angle 4 = 180° - 100.2°$$
$$m\angle 4 = 79.8°$$
The measure of angle 4 is $79.8°$.

To find $m\angle 5$, note that $m\angle 6 + m\angle 1 + m\angle 5 = 180°$. Then to find $m\angle 5$ we follow the same procedure we used to find $m\angle 2$. Thus, the measure of angle 5 is $67.13°$.

75. $\angle ACB$ and $\angle ACD$ are complementary angles. Since $m\angle ACD = 40°$ and $90° - 40° = 50°$, we have $m\angle ACB = 50°$.

Now consider triangle ABC. We know that the sum of the measures of the angles is $180°$. Then
$$m\angle ABC + m\angle BCA + m\angle CAB = 180°$$
$$50° + 90° + m\angle CAB = 180°$$
$$140° + m\angle CAB = 180°$$
$$m\angle CAB = 180° - 140°$$
$$m\angle CAB = 40°,$$
so $m\angle CAB = 40°$.

To find $m \angle EBC$ we first find $m \angle CEB$. We note that $\angle DEC$ and $\angle CEB$ are supplementary angles. Since $m \angle DEC = 100°$ and $180° - 100° = 80°$, we have $m \angle CEB = 80°$. Now consider triangle BCE. We know that the sum of the measures of the angles is 180°. Note that $\angle ACB$ can also be named $\angle BCE$. Then

$$m \angle BCE + m \angle CEB + m \angle EBC = 180°$$
$$50° + 80° + m \angle EBC = 180°$$
$$130° + m \angle EBC = 180°$$
$$m \angle EBC = 180° - 130°$$
$$m \angle EBC = 50°,$$

so $m \angle EBC = 50°$.

$\angle EBA$ and $\angle EBC$ are complementary angles. Since $m \angle EBC = 50°$ and $90° - 50° = 40°$, we have $m \angle EBA = 40°$.

Now consider triangle ABE. We know that the sum of the measures of the angles is 180°. Then

$$m \angle CAB + m \angle EBA + m \angle AEB = 180°$$
$$40° + 40° + m \angle AEB = 180°$$
$$80° + m \angle AEB = 180°$$
$$m \angle AEB = 180° - 80°$$
$$m \angle AEB = 100°,$$

so $m \angle AEB = 100°$.

To find $m \angle ADB$ we first find $m \angle EDC$. Consider triangle CDE. We know that the sum of the measures of the angles is 180°. Then

$$m \angle DEC + m \angle ECD + m \angle EDC = 180°$$
$$100° + 40° + m \angle EDC = 180°$$
$$140° + m \angle EDC = 180°$$
$$m \angle EDC = 180° - 140°$$
$$m \angle EDC = 40°,$$

so $m \angle EDC = 40°$. We now note that $\angle ADB$ and $\angle EDC$ are complementary angles. Since $m \angle EDC = 40°$ and $90° - 40° = 50°$, we have $m \angle ADB = 50°$.

Exercise Set 5.2

1. Perimeter $= 4 \text{ mm} + 6 \text{ mm} + 7 \text{ mm}$
$= (4 + 6 + 7) \text{ mm}$
$= 17 \text{ mm}$

3. Perimeter $= 3.5 \text{ in.} + 3.5 \text{ in.} + 4.25 \text{ in.} +$
$0.5 \text{ in.} + 3.5 \text{ in.}$
$= (3.5 + 3.5 + 4.25 + 0.5 + 3.5) \text{ in.}$
$= 15.25 \text{ in.}$

5. $P = 2 \cdot (l + w)$ Perimeter of a rectangle
$P = 2 \cdot (5.6 \text{ km} + 3.4 \text{ km})$
$P = 2 \cdot (9 \text{ km})$
$P = 18 \text{ km}$

7. $P = 2 \cdot (l + w)$ Perimeter of a rectangle
$P = 2 \cdot (5 \text{ ft} + 10 \text{ ft})$
$P = 2 \cdot (15 \text{ ft})$
$P = 30 \text{ ft}$

9. $P = 2 \cdot (l + w)$ Perimeter of a rectangle
$P = 2 \cdot (34.67 \text{ cm} + 4.9 \text{ cm})$
$P = 2 \cdot (39.57 \text{ cm})$
$P = 79.14 \text{ cm}$

11. $P = 4 \cdot s$ Perimeter of a square
$P = 4 \cdot 22 \text{ ft}$
$P = 88 \text{ ft}$

13. $P = 4 \cdot s$ Perimeter of a square
$P = 4 \cdot 45.5 \text{ mm}$
$P = 182 \text{ mm}$

15. Familiarize. First we find the perimeter of the field. Then we multiply to find the cost of the fence wire. We make a drawing.

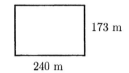

Translate. The perimeter of the field is given by

$$P = 2 \cdot (l + w) = 2 \cdot (240 \text{ m} + 173 \text{ m}).$$

Solve. We calculate the perimeter.

$$P = 2 \cdot (240 \text{ m} + 173 \text{ m}) = 2 \cdot (413 \text{ m}) = 826 \text{ m}$$

Then we multiply to find the cost of the fence wire.

$$\begin{aligned} \text{Cost} &= \$1.45/\text{m} \times \text{Perimeter} \\ &= \$1.45/\text{m} \times 826 \text{ m} \\ &= \$1197.70 \end{aligned}$$

Check. Repeat the calculations.

State. The perimeter of the field is 826 m. The fencing will cost \$1197.70.

17. Familiarize. We make a drawing and let $P =$ the perimeter.

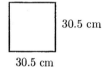

Translate. The perimeter of the square is given by

$$P = 4 \cdot s = 4 \cdot (30.5 \text{ cm}).$$

Solve. We do the calculation.

$$P = 4 \cdot (30.5 \text{ cm}) = 122 \text{ cm}.$$

Check. Repeat the calculation.

State. The perimeter of the tile is 122 cm.

19. *Familiarize.* We label the missing lengths on the drawing and let P = the perimeter.

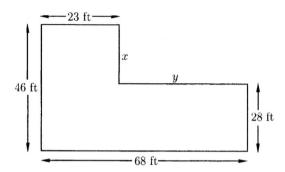

Translate. First we find the missing lengths x and y.

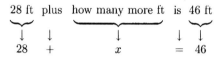

28 ft	plus	how many more ft	is	46 ft
↓	↓	↓	↓	↓
28	+	x	=	46

23 ft	plus	how many more ft	is	68 ft
↓	↓	↓	↓	↓
23	+	y	=	68

Solve. We solve for x and y.

$$28 + x = 46 \qquad 23 + y = 68$$
$$x = 46 - 28 \qquad y = 68 - 23$$
$$x = 18 \qquad y = 45$$

a) To find the perimeter we add the lengths of the sides of the house.

$$P = 23 \text{ ft} + 18 \text{ ft} + 45 \text{ ft} + 28 \text{ ft} + 68 \text{ ft} + 46 \text{ ft}$$
$$= (23 + 18 + 45 + 28 + 68 + 46) \text{ ft}$$
$$= 228 \text{ ft}$$

b) Next we find t, the total cost of the gutter.

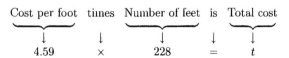

Cost per foot	times	Number of feet	is	Total cost
↓	↓	↓	↓	↓
4.59	×	228	=	t

We carry out the multiplication.

```
      2 2 8
    × 4 .5 9
    ---------
    2 0 5 2
  1 1 4 0 0
  9 1 2 0 0
  ---------
1 0 4 6 .5 2
```

Thus, $t = 1046.52$.

Check. We can repeat the calculations.

State. (a) The perimeter of the house is 228 ft. (b) The total cost of the gutter is $1046.52.

21. Discussion and Writing Exercise

23. 56.1%

a) Replace the percent symbol with × 0.01.

56.1×0.01

b) Move the decimal point two places to the left.

0.56.1

Thus, $56.1\% = 0.561$.

25. a) First find decimal notation by division.

```
        1. 1 2 5
    8 ⟌ 9. 0 0 0
        8
        ---
        1 0
          8
        ---
          2 0
          1 6
          ---
            4 0
            4 0
            ---
              0
```

$$\frac{9}{8} = 1.125$$

b) Convert the decimal notation to percent notation. Move the decimal point two places to the right and write a % symbol.

1.12.5

$$\frac{9}{8} = 112.5\%, \text{ or } 112\frac{1}{2}\%$$

27. $10^2 = 10 \cdot 10 = 100$

29. 4.7 million $= 4.7 \times 1,\underbrace{000,000}_{\text{6 zeros}}$

4.700000.

Move 6 places to the right.

4.7 million $= 4,700,000$

31. 18 in. $= 18 \text{ in.} \times \dfrac{1 \text{ ft}}{12 \text{ in.}} = \dfrac{18}{12} \times 1 \text{ ft} = \dfrac{3}{2} \text{ ft}$

$$P = 2 \cdot (l + w)$$
$$P = 2 \cdot \left(3 \text{ ft} + \frac{3}{2} \text{ ft}\right)$$
$$P = 2 \cdot \left(\frac{9}{2} \text{ ft}\right)$$
$$P = 9 \text{ ft}$$

Exercise Set 5.3

1. $A = l \cdot w$ Area of a rectangular region
$A = (5 \text{ km}) \cdot (3 \text{ km})$
$A = 5 \cdot 3 \cdot \text{ km} \cdot \text{ km}$
$A = 15 \text{ km}^2$

3. $A = l \cdot w$ Area of a rectangular region
$A = (2 \text{ in.}) \cdot (0.7 \text{ in.})$
$A = 2 \cdot 0.7 \cdot \text{ in.} \cdot \text{ in.}$
$A = 1.4 \text{ in}^2$

5. $A = s \cdot s$ Area of a square
$A = \left(2\frac{1}{2} \text{ yd}\right) \cdot \left(2\frac{1}{2} \text{ yd}\right)$
$A = \left(\frac{5}{2} \text{ yd}\right) \cdot \left(\frac{5}{2} \text{ yd}\right)$
$A = \frac{5}{2} \cdot \frac{5}{2} \cdot \text{ yd} \cdot \text{ yd}$
$A = \frac{25}{4} \text{ yd}^2, \text{ or } 6\frac{1}{4} \text{ yd}^2$

7. $A = s \cdot s$ Area of a square
$A = (90 \text{ ft}) \cdot (90 \text{ ft})$
$A = 90 \cdot 90 \cdot \text{ ft} \cdot \text{ ft}$
$A = 8100 \text{ ft}^2$

9. $A = l \cdot w$ Area of a rectangular region
$A = (10 \text{ ft}) \cdot (5 \text{ ft})$
$A = 10 \cdot 5 \cdot \text{ ft} \cdot \text{ ft}$
$A = 50 \text{ ft}^2$

11. $A = l \cdot w$ Area of a rectangular region
$A = (34.67 \text{ cm}) \cdot (4.9 \text{ cm})$
$A = 34.67 \cdot 4.9 \cdot \text{ cm} \cdot \text{ cm}$
$A = 169.883 \text{ cm}^2$

13. $A = l \cdot w$ Area of a rectangular region
$A = \left(4\frac{2}{3} \text{ in.}\right) \cdot \left(8\frac{5}{6} \text{ in.}\right)$
$A = \left(\frac{14}{3} \text{ in.}\right) \cdot \left(\frac{53}{6} \text{ in.}\right)$
$A = \frac{14}{3} \cdot \frac{53}{6} \cdot \text{ in.} \cdot \text{ in.}$
$A = \frac{2 \cdot 7 \cdot 53}{3 \cdot 2 \cdot 3} \text{ in}^2$
$A = \frac{2}{2} \cdot \frac{7 \cdot 53}{3 \cdot 3} \text{ in}^2$
$A = \frac{371}{9} \text{ in}^2, \text{ or } 41\frac{2}{9} \text{ in}^2$

15. $A = s \cdot s$ Area of a square
$A = (22 \text{ ft}) \cdot (22 \text{ ft})$
$A = 22 \cdot 22 \cdot \text{ ft} \cdot \text{ ft}$
$A = 484 \text{ ft}^2$

17. $A = s \cdot s$ Area of a square
$A = (56.9 \text{ km}) \cdot (56.9 \text{ km})$
$A = 56.9 \cdot 56.9 \cdot \text{ km} \cdot \text{ km}$
$A = 3237.61 \text{ km}^2$

19. $A = s \cdot s$ Area of a square
$A = \left(5\frac{3}{8} \text{ yd}\right) \cdot \left(5\frac{3}{8} \text{ yd}\right)$
$A = \left(\frac{43}{8} \text{ yd}\right) \cdot \left(\frac{43}{8} \text{ yd}\right)$
$A = \frac{43}{8} \cdot \frac{43}{8} \cdot \text{ yd} \cdot \text{ yd}$
$A = \frac{1849}{64} \text{ yd}^2, \text{ or } 28\frac{57}{64} \text{ yd}^2$

21. $A = b \cdot h$ Area of a parallelogram
$A = 8 \text{ cm} \cdot 4 \text{ cm}$ Substituting 8 cm for b and 4 cm for h
$A = 32 \text{ cm}^2$

23. $A = \frac{1}{2} \cdot b \cdot h$ Area of a triangle
$A = \frac{1}{2} \cdot 15 \text{ in.} \cdot 8 \text{ in.}$ Substituting 15 in. for b and 8 in. for h
$A = 60 \text{ in}^2$

25. $A = \frac{1}{2} \cdot h \cdot (a + b)$ Area of a trapezoid
$A = \frac{1}{2} \cdot 8 \text{ ft} \cdot (6 + 20) \text{ ft}$ Substituting 8 ft for h, 6 ft for a, and 20 ft for b
$A = \frac{8 \cdot 26}{2} \text{ ft}^2$
$A = 104 \text{ ft}^2$

27. $A = \frac{1}{2} \cdot h \cdot (a + b)$ Area of a trapezoid
$A = \frac{1}{2} \cdot 7 \text{ in.} \cdot (4.5 + 8.5) \text{ in.}$ Substituting 7 in. for h, 4.5 in. for a, and 8.5 in. for b
$A = \frac{7 \cdot 13}{2} \text{ in}^2$
$A = \frac{91}{2} \text{ in}^2$
$A = 45.5 \text{ in}^2$

29. $A = b \cdot h$ Area of a parallelogram
$A = 2.3 \text{ cm} \cdot 3.5 \text{ cm}$ Substituting 2.3 cm for b and 3.5 cm for h
$A = 8.05 \text{ cm}^2$

31. $A = \frac{1}{2} \cdot h \cdot (a + b)$ Area of a trapezoid
$A = \frac{1}{2} \cdot 18 \text{ cm} \cdot (9 + 24) \text{ cm}$ Substituting 18 cm for h, 9 cm for a, and 24 cm for b
$A = \frac{18 \cdot 33}{2} \text{ cm}^2$
$A = 297 \text{ cm}^2$

33. $A = \frac{1}{2} \cdot b \cdot h$ Area of a triangle
$A = \frac{1}{2} \cdot 4 \text{ m} \cdot 3.5 \text{ m}$ Substituting 4 m for b and 3.5 m for h
$A = \frac{4 \cdot 3.5}{2} \text{ m}^2$
$A = 7 \text{ m}^2$

35. *Familiarize.* We draw a picture.

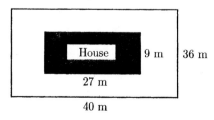

Translate. We let A = the area left over.

Area left over	is	Area of lot	minus	Area of house
$\downarrow$	$\downarrow$	$\downarrow$	$\downarrow$	$\downarrow$
A	$=$	$(40 \text{ m}) \cdot (36 \text{ m})$	$-$	$(27 \text{ m}) \cdot (9 \text{ m})$

Solve. The area of the lot is

$$(40 \text{ m}) \cdot (36 \text{ m}) = 40 \cdot 36 \cdot \text{ m} \cdot \text{ m} = 1440 \text{ m}^2.$$

The area of the house is

$$(27 \text{ m}) \cdot (9 \text{ m}) = 27 \cdot 9 \cdot \text{ m} \cdot \text{ m} = 243 \text{ m}^2.$$

The area left over is

$$A = 1440 \text{ m}^2 - 243 \text{ m}^2 = 1197 \text{ m}^2.$$

Check. Repeat the calculations.

State. The area left over for the lawn is 1197 m^2.

37. a) First find the area of the entire yard, including the sandbox:

$$A = l \cdot w = \left(93\frac{2}{3} \text{ ft}\right) \cdot (60 \text{ ft})$$
$$= \left(\frac{281}{3} \text{ ft}\right) \cdot (60 \text{ ft})$$
$$= 5620 \text{ ft}^2$$

Now find the area of the sandbox:

$$A = s \cdot s = (4.5 \text{ ft}) \cdot (4.5 \text{ ft}) = 20.25 \text{ ft}^2$$

Finally, subtract to find the area of the lawn:

$$5620 \text{ ft}^2 - 20.25 \text{ ft}^2 = 5599.75 \text{ ft}^2$$

b) Let c = the cost of mowing the law. We translate to an equation.

The cost of mowing	is	$0.008	times	the area of the lawn.
$\downarrow$	$\downarrow$	$\downarrow$	$\downarrow$	$\downarrow$
c	$=$	0.008	$\cdot$	5599.75

We multiply to solve the equation.

$$c = 0.008 \cdot 5599.75 = \$44.798 \approx \$45$$

The total cost of the mowing is about $45.

39. Familiarize. We use the drawing in the text.

Translate. We let A = the area of the sidewalk.

Area of sidewalk	is	Total area	minus	Area of building
$\downarrow$	$\downarrow$	$\downarrow$	$\downarrow$	$\downarrow$
A	$=$	$(113.4 \text{ m}) \times (75.4 \text{ m})$	$-$	$(110 \text{ m}) \times (72 \text{ m})$

Solve. The total area is

$$(113.4 \text{ m}) \times (75.4 \text{ m}) = 113.4 \times 75.4 \times \text{ m} \times \text{ m} = 8550.36 \text{ m}^2.$$

The area of the building is

$$(110 \text{ m}) \times (72 \text{ m}) = 110 \times 72 \times \text{ m} \times \text{ m} = 7920 \text{ m}^2.$$

The area of the sidewalk is

$$A = 8550.36 \text{ m}^2 - 7920 \text{ m}^2 = 630.36 \text{ m}^2.$$

Check. Repeat the calculations.

State. The area of the sidewalk is 630.36 m^2.

41. Familiarize. The dimensions are as follows:

Two walls are 15 ft by 8 ft.

Two walls are 20 ft by 8 ft.

The ceiling is 15 ft by 20 ft.

The total area of the walls and ceiling is the total area of the rectangles described above less the area of the windows and the door.

Translate. a) We let A = the total area of the walls and ceiling. The total area of the two 15 ft by 8 ft walls is

$$2 \cdot (15 \text{ ft}) \cdot (8 \text{ ft}) = 2 \cdot 15 \cdot 8 \cdot \text{ ft} \cdot \text{ ft} = 240 \text{ ft}^2$$

The total area of the two 20 ft by 8 ft walls is

$$2 \cdot (20 \text{ ft}) \cdot (8 \text{ ft}) = 2 \cdot 20 \cdot 8 \cdot \text{ ft} \cdot \text{ ft} = 320 \text{ ft}^2$$

The area of the ceiling is

$$(15 \text{ ft}) \cdot (20 \text{ ft}) = 15 \cdot 20 \cdot \text{ ft} \cdot \text{ ft} = 300 \text{ ft}^2$$

The area of the two windows is

$$2 \cdot (3 \text{ ft}) \cdot (4 \text{ ft}) = 2 \cdot 3 \cdot 4 \cdot \text{ ft} \cdot \text{ ft} = 24 \text{ ft}^2$$

The area of the door is

$$\left(2\frac{1}{2} \text{ ft}\right) \cdot \left(6\frac{1}{2} \text{ ft}\right) = \left(\frac{5}{2} \text{ ft}\right) \cdot \left(\frac{13}{2} \text{ ft}\right)$$
$$= \frac{5}{2} \cdot \frac{13}{2} \cdot \text{ ft} \cdot \text{ ft}$$
$$= \frac{65}{4} \text{ ft}^2, \text{ or } 16\frac{1}{4} \text{ ft}^2$$

Thus

$$A = 240 \text{ ft}^2 + 320 \text{ ft}^2 + 300 \text{ ft}^2 - 24 \text{ ft}^2 - 16\frac{1}{4} \text{ ft}^2$$
$$= 819\frac{3}{4} \text{ ft}^2, \text{ or } 819.75 \text{ ft}^2$$

b) We divide to find how many gallons of paint are needed.

$$819.75 \div 86.625 \approx 9.46$$

It will be necessary to buy 10 gallons of paint in order to have the required 9.46 gallons.

c) We multiply to find the cost of the paint.

$$10 \times \$17.95 = \$179.50$$

Check. We repeat the calculations.

State. (a) The total area of the walls and ceiling is 819.75 ft^2. (b) 10 gallons of paint are needed. (c) It will cost $179.50 to paint the room.

43.

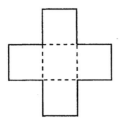

Each side is 4 cm.

The region is composed of 5 squares, each with sides of length 4 cm. The area is

$$A = 5 \cdot (s \cdot s) = 5 \cdot (4 \text{ cm} \cdot 4 \text{ cm}) = 5 \cdot 4 \cdot 4 \text{ cm} \cdot \text{cm} = 80 \text{ cm}^2$$

45. *Familiarize.* We look for the kinds of figures whose areas we can calculate using area formulas that we already know.

Translate. The shaded region consists of a square region with a triangular region removed from it. The sides of the square are 30 cm, and the triangle has base 30 cm and height 15 cm. We find the area of the square using the formula $A = s \cdot s$, and the area of the triangle using $A = \frac{1}{2} \cdot b \cdot h$. Then we subtract.

Solve. Area of the square: $A = 30 \text{ cm} \cdot 30 \text{ cm} = 900 \text{ cm}^2$.

Area of the triangle: $A = \frac{1}{2} \cdot 30 \text{ cm} \cdot 15 \text{ cm} = 225 \text{ cm}^2$.

Area of the shaded region: $A = 900 \text{ cm}^2 - 225 \text{ cm}^2 = 675 \text{ cm}^2$.

Check. We repeat the calculations.

State. The area of the shaded region is 675 cm².

47. *Familiarize.* We have one large triangle with height and base each 6 cm. We also have 6 small triangles, each with height and base 1 cm.

Translate. We will find the area of each type of triangle using the formula $A = \frac{1}{2} \cdot b \cdot h$. Next we will multiply the area of the smaller triangle by 6. And, finally, we will add this product to the area of the larger triangle to find the total area.

Solve.

For the large triangle: $A = \frac{1}{2} \cdot 6 \text{ cm} \cdot 6 \text{ cm} = 18 \text{ cm}^2$

For one small triangle: $A = \frac{1}{2} \cdot 1 \text{ cm} \cdot 1 \text{ cm} = \frac{1}{2} \text{ cm}^2$

Find the area of the 6 small triangles: $6 \cdot \frac{1}{2} \text{ cm}^2 = 3 \text{ cm}^2$

Add to find the total area: $18 \text{ cm}^2 + 3 \text{ cm}^2 = 21 \text{ cm}^2$

Check. We repeat the calculations.

State. The area of the shaded region is 21 cm².

49. *Familiarize.* We make a drawing, shading the area left over after the triangular piece is cut from the sailcloth.

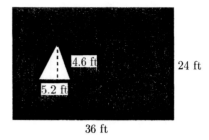

24 ft

36 ft

Translate. The shaded region consists of a rectangular region with a triangular region removed from it. The rectangular region has dimensions 36 ft by 24 ft, and the triangular region has base 5.2 ft and height 4.6 ft. We will find the area of the rectangular region using the formula $A = b \cdot h$, and the area of the triangle using $A = \frac{1}{2} \cdot b \cdot h$. Then we will subtract to find the area of the shaded region.

Solve. Area of the rectangle: $A = 36 \text{ ft} \cdot 24 \text{ ft} = 864 \text{ ft}^2$.

Area of the triangle: $A = \frac{1}{2} \cdot 5.2 \text{ ft} \cdot 4.6 \text{ ft} = 11.96 \text{ ft}^2$.

Area of the shaded region: $A = 864 \text{ ft}^2 - 11.96 \text{ ft}^2 = 852.04 \text{ ft}^2$.

Check. We repeat the calculation.

State. The area left over is 852.04 ft².

51. Discussion and Writing Exercise

53. $35\% = \dfrac{35}{100} = \dfrac{5 \cdot 7}{5 \cdot 20} = \dfrac{5}{5} \cdot \dfrac{7}{20} = \dfrac{7}{20}$

55. $37\frac{1}{2}\% = \dfrac{75}{2}\% = \dfrac{75}{2} \times \dfrac{1}{100} = \dfrac{75}{2 \cdot 100} = \dfrac{25 \cdot 3}{2 \cdot 25 \cdot 4} = \dfrac{25}{25} \cdot \dfrac{3}{2 \cdot 4} = \dfrac{3}{8}$

57. $83.\overline{3}\% = 83\frac{1}{3}\% = \dfrac{250}{3}\% = \dfrac{250}{3} \times \dfrac{1}{100} = \dfrac{250 \cdot 1}{3 \cdot 100} = \dfrac{5 \cdot 50 \cdot 1}{3 \cdot 2 \cdot 50} = \dfrac{50}{50} \cdot \dfrac{5 \cdot 1}{3 \cdot 2} = \dfrac{5}{6}$

59. *Familiarize.* Let $s =$ the number of sheets in 15 reams of paper. Repeated addition works well here.

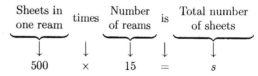

15 addends

Translate.

Sheets in one ream	times	Number of reams	is	Total number of sheets
↓	↓	↓	↓	↓
500	×	15	=	s

Solve. We multiply.

$500 \times 15 = 7500$, so $7500 = s$, or $s = 7500$.

Check. We can repeat the calculation. The answer checks.

State. There are 7500 sheets in 15 reams of paper.

61.

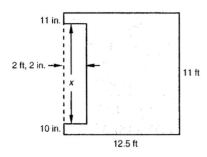

2 ft = 2 × 1 ft = 2 × 12 in. = 24 in., so 2 ft, 2 in. = 2 ft + 2 in. = 24 in. + 2 in. = 26 in.

11 ft = 11 × 1 ft = 11 × 12 in. = 132 in.

12.5 ft = 12.5 × 1 ft = 12.5 × 12 in. = 150 in.

We solve an equation to find x, in inches:

$$11 + x + 10 = 132$$
$$21 + x = 132$$
$$21 + x - 21 = 132 - 21$$
$$x = 111$$

Then the area of the shaded region is the area of a 150 in. by 132 in. rectangle less the area of a 111 in. by 26 in. rectangle.

$$A = (150 \text{ in.}) \cdot (132 \text{ in.}) - (111 \text{ in.}) \cdot (26 \text{ in.})$$
$$A = 19,800 \text{ in}^2 - 2886 \text{ in}^2$$
$$A = 16,914 \text{ in}^2$$

Exercise Set 5.4

1. $d = 2 \cdot r$

$d = 2 \cdot 7 \text{ cm} = 14 \text{ cm}$

$C = 2 \cdot \pi \cdot r$

$C \approx 2 \cdot \dfrac{22}{7} \cdot 7 \text{ cm} = \dfrac{2 \cdot 22 \cdot 7}{7} \text{ cm} = 44 \text{ cm}$

$A = \pi \cdot r \cdot r$

$A \approx \dfrac{22}{7} \cdot 7 \text{ cm} \cdot 7 \text{ cm} = \dfrac{22}{7} \cdot 49 \text{ cm}^2 = 154 \text{ cm}^2$

3. $d = 2 \cdot r$

$d = 2 \cdot \dfrac{3}{4} \text{ in.} = \dfrac{6}{4} \text{ in.} = \dfrac{3}{2} \text{ in., or } 1\dfrac{1}{2} \text{ in.}$

$C = 2 \cdot \pi \cdot r$

$C \approx 2 \cdot \dfrac{22}{7} \cdot \dfrac{3}{4} \text{ in.} = \dfrac{2 \cdot 22 \cdot 3}{7 \cdot 4} \text{ in.} = \dfrac{132}{28} \text{ in.} = \dfrac{33}{7} \text{ in.,}$

or $4\dfrac{5}{7}$ in.

$A = \pi \cdot r \cdot r$

$A \approx \dfrac{22}{7} \cdot \dfrac{3}{4} \text{ in.} \cdot \dfrac{3}{4} \text{ in.} = \dfrac{22 \cdot 3 \cdot 3}{7 \cdot 4 \cdot 4} \text{ in}^2 = \dfrac{99}{56} \text{ in}^2, \text{ or } 1\dfrac{43}{56} \text{ in}^2$

5. $r = \dfrac{d}{2}$

$r = \dfrac{32 \text{ ft}}{2} = 16 \text{ ft}$

$C = \pi \cdot d$

$C \approx 3.14 \cdot 32 \text{ ft} = 100.48 \text{ ft}$

$A = \pi \cdot r \cdot r$

$A \approx 3.14 \cdot 16 \text{ ft} \cdot 16 \text{ ft} \qquad (r = \dfrac{d}{2}; r = \dfrac{32 \text{ ft}}{2} = 16 \text{ ft})$

$A = 3.14 \cdot 256 \text{ ft}^2$

$A = 803.84 \text{ ft}^2$

7. $r = \dfrac{d}{2}$

$r = \dfrac{1.4 \text{ cm}}{2} = 0.7 \text{ cm}$

$C = \pi \cdot d$

$C \approx 3.14 \cdot 1.4 \text{ cm} = 4.396 \text{ cm}$

$A = \pi \cdot r \cdot r$

$A \approx 3.14 \cdot 0.7 \text{ cm} \cdot 0.7 \text{ cm}$
$\qquad (r = \dfrac{d}{2}; r = \dfrac{1.4 \text{ cm}}{2} = 0.7 \text{ cm})$

$A = 3.14 \cdot 0.49 \text{ cm}^2 = 1.5386 \text{ cm}^2$

9. $r = \dfrac{d}{2}$

$r = \dfrac{6 \text{ cm}}{2} = 3 \text{ cm}$

The radius is 3 cm.

$C = \pi \cdot d$

$C \approx 3.14 \cdot 6 \text{ cm} = 18.84 \text{ cm}$

The circumference is about 18.84 cm.

$A = \pi \cdot r \cdot r$

$A \approx 3.14 \cdot 3 \text{ cm} \cdot 3 \text{ cm} = 28.26 \text{ cm}^2$

The area is about 28.26 cm^2.

11. $r = \dfrac{d}{2}$

$r = \dfrac{14 \text{ ft}}{2} = 7 \text{ ft}$

$A = \pi \cdot r \cdot r$

$A \approx 3.14 \cdot 7 \text{ ft} \cdot 7 \text{ ft} = 153.86 \text{ ft}^2$

The area of the trampoline is about 153.86 ft^2.

13. $\qquad C = \pi \cdot d$

$7.85 \text{ cm} \approx 3.14 \cdot d \qquad$ Substituting 7.85 cm for C and 3.14 for π

$\dfrac{7.85 \text{ cm}}{3.14} = d \qquad$ Dividing on both sides by 3.14

$2.5 \text{ cm} = d$

The diameter is about 2.5 cm.

$r = \dfrac{d}{2}$

$r = \dfrac{2.5 \text{ cm}}{2} = 1.25 \text{ cm}$

The radius is about 1.25 cm.

$A = \pi \cdot r \cdot r$

$A \approx 3.14 \cdot 1.25 \text{ cm} \cdot 1.25 \text{ cm} = 4.90625 \text{ cm}^2$

The area is about 4.90625 cm^2.

15. $C = \pi \cdot d$

$C \approx 3.14 \cdot 1.1 \text{ ft} = 3.454 \text{ ft}$

The circumference of the elm tree is about 3.454 ft.

17. Find the area of the larger circle (pool plus walk). Its diameter is 1 yd + 20 yd + 1 yd, or 22 yd. Thus its radius is $\frac{22}{2}$ yd, or 11 yd.

$$A = \pi \cdot r \cdot r$$
$$A \approx 3.14 \cdot 11 \text{ yd} \cdot 11 \text{ yd} = 379.94 \text{ yd}^2$$

Find the area of the pool. Its diameter is 20 yd. Thus its radius is $\frac{20}{2}$ yd, or 10 yd.

$$A = \pi \cdot r \cdot r$$
$$A \approx 3.14 \cdot 10 \text{ yd} \cdot 10 \text{ yd} = 314 \text{ yd}^2$$

We subtract to find the area of the walk:
$$A = 379.94 \text{ yd}^2 - 314 \text{ yd}^2$$
$$A = 65.94 \text{ yd}^2$$

The area of the walk is 65.94 yd^2.

19. The perimeter consists of the circumferences of three semicircles, each with diameter 8 ft, and one side of a square of length 8 ft. We first find the circumference of one semicircle. This is one-half the circumference of a circle with diameter 8 ft:

$$\frac{1}{2} \cdot \pi \cdot d \approx \frac{1}{2} \cdot 3.14 \cdot 8 \text{ ft} = 12.56 \text{ ft}$$

Then we multiply by 3:

$$3 \cdot (12.56 \text{ ft}) = 37.68 \text{ ft}$$

Finally we add the circumferences of the semicircles and the length of the side of the square:

$$37.68 \text{ ft} + 8 \text{ ft} = 45.68 \text{ ft}$$

The perimeter is 45.68 ft.

21. The perimeter consists of three-fourths of the circumference of a circle with radius 4 yd and two sides of a square with sides of length 4 yd. We first find three-fourths of the circumference of the circle:

$$\frac{3}{4} \cdot 2 \cdot \pi \cdot r \approx 0.75 \cdot 2 \cdot 3.14 \cdot 4 \text{ yd} = 18.84 \text{ yd}$$

Then we add this length to the lengths of two sides of the square:
$$18.84 \text{ yd} + 4 \text{ yd} + 4 \text{ yd} = 26.84 \text{ yd}$$

The perimeter is 26.84 yd.

23. The perimeter consists of three-fourths of the perimeter of a square with side of length 10 yd and the circumference of a semicircle with diameter 10 yd. First we find three-fourths of the perimeter of the square:

$$\frac{3}{4} \cdot 4 \cdot s = \frac{3}{4} \cdot 4 \cdot 10 \text{ yd} = 30 \text{ yd}$$

Then we find one-half of the circumference of a circle with diameter 10 yd:

$$\frac{1}{2} \cdot \pi \cdot d \approx \frac{1}{2} \cdot 3.14 \cdot 10 \text{ yd} = 15.7 \text{ yd}$$

Then we add:

$$30 \text{ yd} + 15.7 \text{ yd} = 45.7 \text{ yd}$$

The perimeter is 45.7 yd.

25. The shaded region consists of a circle of radius 8 m, with two circles each of diameter 8 m, removed. First we find the area of the large circle:

$$A = \pi \cdot r \cdot r \approx 3.14 \cdot 8 \text{ m} \cdot 8 \text{ m} = 200.96 \text{ m}^2$$

Then we find the area of one of the small circles: The radius is $\frac{8 \text{ m}}{2} = 4$ m.

$$A = \pi \cdot r \cdot r \approx 3.14 \cdot 4 \text{ m} \cdot 4 \text{ m} = 50.24 \text{ m}^2$$

We multiply this area by 2 to find the area of the two small circles:

$$2 \cdot 50.24 \text{ m}^2 = 100.48 \text{ m}^2$$

Finally we subtract to find the area of the shaded region:

$$200.96 \text{ m}^2 - 100.48 \text{ m}^2 = 100.48 \text{ m}^2$$

The area of the shaded region is 100.48 m^2.

27. The shaded region consists of one-half of a circle with diameter 2.8 cm and a triangle with base 2.8 cm and height 2.8 cm. First we find the area of the semicircle. The radius is $\frac{2.8 \text{ cm}}{2} = 1.4$ cm.

$$A = \frac{1}{2} \cdot \pi \cdot r \cdot r \approx \frac{1}{2} \cdot 3.14 \cdot 1.4 \text{ cm} \cdot 1.4 \text{ cm} = 3.0772 \text{ cm}^2$$

Then we find the area of the triangle:

$$A = \frac{1}{2} \cdot b \cdot h = \frac{1}{2} \cdot 2.8 \text{ cm} \cdot 2.8 \text{ cm} = 3.92 \text{ cm}^2$$

Finally we add to find the area of the shaded region:

$$3.0772 \text{ cm}^2 + 3.92 \text{ cm}^2 = 6.9972 \text{ cm}^2$$

The area of the shaded region is 6.9972 cm^2.

29. The shaded area consists of a rectangle of dimensions 11.4 in. by 14.6 in., with the area of two semicircles, each of diameter 11.4 in., removed. This is equivalent to removing one circle with diameter 11.4 in. from the rectangle. First we find the area of the rectangle:

$$l \cdot w = (11.4 \text{ in.}) \cdot (14.6 \text{ in.}) = 166.44 \text{ in}^2$$

Then we find the area of the circle. The radius is $\frac{11.4 \text{ in.}}{2} = 5.7$ in.

$$\pi \cdot r \cdot r \approx 3.14 \cdot 5.7 \text{ in.} \cdot 5.7 \text{ in.} = 102.0186 \text{ in}^2$$

Finally we subtract to find the area of the shaded region:

$$166.44 \text{ in}^2 - 102.0186 \text{ in}^2 = 64.4214 \text{ in}^2$$

31. Discussion and Writing Exercise

33. 0.875

  a)   Move the decimal point   0.87.5
        2 places to the right.   ⌊＿↑

  b)   Add a percent symbol   87.5%

0.875 = 87.5%

35. $0.\overline{6}$

　a)　Move the decimal point　　　　$0.66.\overline{6}$
　　　2 places to the right.　　　　　$\boxed{}\!\uparrow$

　b)　Add a percent symbol　　　　$66.\overline{6}\%$

　$0.\overline{6} = 66.\overline{6}\%$

37. a) Find decimal notation using long division.

$$
\begin{array}{r}
0.375 \\
8\,\overline{)\,3.000} \\
\underline{2\,4} \\
6\,0 \\
\underline{5\,6} \\
4\,0 \\
\underline{4\,0} \\
0
\end{array}
$$

$\dfrac{3}{8} = 0.375$

b) Convert the decimal notation to percent notation. Move the decimal point two places to the right, and write a % symbol.

　$0.37.5$
　$\boxed{}\!\uparrow$

$\dfrac{3}{8} = 37.5\%$

39. a) Find decimal notation using long division.

$$
\begin{array}{r}
0.66 \\
3\,\overline{)\,2.00} \\
\underline{1\,8} \\
2\,0 \\
\underline{1\,8} \\
2
\end{array}
$$

$\dfrac{2}{3} = 0.\overline{6}$

b) Convert the decimal notation to percent notation. Move the decimal point two places to the right, and write a % symbol.

　$0.66.\overline{6}$
　$\boxed{}\!\uparrow$

$\dfrac{2}{3} = 66.\overline{6}\%$

41. $3\dfrac{7}{8} = 3 + \dfrac{7}{8} \approx 3 + 1 = 4$

43. $13\dfrac{1}{6} = 13 + \dfrac{1}{6} \approx 13 + 0 = 13$

45. $\dfrac{4}{5} + 3\dfrac{7}{8} = \dfrac{4}{5} + 3 + \dfrac{7}{8} \approx 1 + 3 + 1 = 5$

47. $\dfrac{2}{3} + \dfrac{7}{15} + \dfrac{8}{9} \approx \dfrac{1}{2} + \dfrac{1}{2} + 1 = 2$

49. $\dfrac{57}{100} - \dfrac{1}{10} + \dfrac{9}{1000} \approx \dfrac{1}{2} - 0 + 0 = \dfrac{1}{2}$

51. $11\dfrac{29}{80} + 10\dfrac{14}{15} \cdot 24\dfrac{2}{17} \approx 11\dfrac{1}{2} + 11 \cdot 24 =$
　　　$11\dfrac{1}{2} + 264 = 275\dfrac{1}{2}$

53. Find $3927 \div 1250$ using a calculator.
　$\dfrac{3927}{1250} = 3.1416$

55. The height of the stack of tennis balls is three tines the diameter of one ball, or $3 \cdot d$.

The circumference of one ball is given by $\pi \cdot d$.

The circumference of one ball is greater than the height of the stack of balls, because $\pi > 3$.

Exercise Set 5.5

1. $V = l \cdot w \cdot h$
　$V = 12$ cm $\cdot 8$ cm $\cdot 8$ cm
　$V = 12 \cdot 64$ cm^3
　$V = 768$ cm^3

　$SA = 2lw + 2lh + 2wh$
　　$= 2 \cdot 12$ cm $\cdot 8$ cm $+ 2 \cdot 12$ cm $\cdot 8$ cm $+$
　　　$2 \cdot 8$ cm $\cdot 8$ cm
　　$= 192$ cm^2 $+ 192$ cm^2 $+ 128$ cm^2
　　$= 512$ cm^2

3. $V = l \cdot w \cdot h$
　$V = 7.5$ in. $\cdot 2$ in. $\cdot 3$ in.
　$V = 7.5 \cdot 6$ in^3
　$V = 45$ in^3

　$SA = 2lw + 2lh + 2wh$
　　$= 2 \cdot 7.5$ in. $\cdot 2$ in. $+ 2 \cdot 7.5$ in. $\cdot 3$ in. $+$
　　　$2 \cdot 2$ in. $\cdot 3$ in.
　　$= 30$ in^2 $+ 45$ in^2 $+ 12$ in^2
　　$= 87$ in^2

5. $V = l \cdot w \cdot h$
　$V = 10$ m $\cdot 5$ m $\cdot 1.5$ m
　$V = 10 \cdot 7.5$ m^3
　$V = 75$ m^3

　$SA = 2lw + 2lh + 2wh$
　　$= 2 \cdot 10$ m $\cdot 5$ m $+ 2 \cdot 10$ m $\cdot 1.5$ m $+$
　　　$2 \cdot 5$ m $\cdot 1.5$ m
　　$= 100$ m^2 $+ 30$ m^2 $+ 15$ m^2
　　$= 145$ m^2

7. $V = l \cdot w \cdot h$
　$V = 6\dfrac{1}{2}$ yd $\cdot 5\dfrac{1}{2}$ yd $\cdot 10$ yd
　$V = \dfrac{13}{2} \cdot \dfrac{11}{2} \cdot 10$ yd^3
　$V = \dfrac{715}{2}$ yd^3
　$V = 357\dfrac{1}{2}$ yd^3

$$SA = 2lw + 2lh + 2wh$$
$$= 2 \cdot 6\frac{1}{2} \text{ yd} \cdot 5\frac{1}{2} \text{ yd} + 2 \cdot 6\frac{1}{2} \text{ yd} \cdot 10 \text{ yd} +$$
$$2 \cdot 5\frac{1}{2} \text{ yd} \cdot 10 \text{ yd}$$
$$= 2 \cdot \frac{13}{2} \cdot \frac{11}{2} \text{ yd}^2 + 2 \cdot \frac{13}{2} \cdot 10 \text{ yd}^2 +$$
$$2 \cdot \frac{11}{2} \cdot 10 \text{ yd}^2$$
$$= \frac{143}{2} \text{ yd}^2 + 130 \text{ yd}^2 + 110 \text{ yd}^2$$
$$= 311\frac{1}{2} \text{ yd}^2$$

9. $V = Bh = \pi \cdot r^2 \cdot h$
$\approx 3.14 \times 8 \text{ in.} \times 8 \text{ in.} \times 4 \text{ in.}$
$\approx 803.84 \text{ in}^3$

11. $V = Bh = \pi \cdot r^2 \cdot h$
$\approx 3.14 \times 5 \text{ cm} \times 5 \text{ cm} \times 4.5 \text{ cm}$
$\approx 353.25 \text{ cm}^3$

13. $V = Bh = \pi \cdot r^2 \cdot h$
$\approx \frac{22}{7} \times 210 \text{ yd} \times 210 \text{ yd} \times 300 \text{ yd}$
$\approx 41,580,000 \text{ yd}^3$

15. $V = \frac{4}{3} \cdot \pi \cdot r^3$
$\approx \frac{4}{3} \times 3.14 \times (100 \text{ in.})^3$
$\approx \frac{4 \times 3.14 \times 1,000,000 \text{ in}^3}{3}$
$\approx 4,186,666\frac{2}{3} \text{ in}^3$

17. $V = \frac{4}{3} \cdot \pi \cdot r^3$
$\approx \frac{4}{3} \times 3.14 \times (3.1 \text{ m})^3$
$\approx \frac{4 \times 3.14 \times 29.791 \text{ m}^3}{3}$
$\approx 124.72 \text{ m}^3$

19. $V = \frac{4}{3} \cdot \pi \cdot r^3$
$\approx \frac{4}{3} \times \frac{22}{7} \times \left(7\frac{3}{4} \text{ ft}\right)^3$
$= \frac{4}{3} \times \frac{22}{7} \times \left(\frac{31}{4} \text{ ft}\right)^3$
$\approx \frac{4 \times 22 \times 29,791 \text{ ft}^3}{3 \times 7 \times 64}$
$\approx 1950\frac{101}{168} \text{ ft}^3$

21. $V = \frac{1}{3} \cdot \pi \cdot r^2 \cdot h$
$\approx \frac{1}{3} \times 3.14 \times 33 \text{ ft} \times 33 \text{ ft} \times 100 \text{ ft}$
$\approx 113,982 \text{ ft}^3$

23. $V = \frac{1}{3} \cdot \pi \cdot r^2 \cdot h$
$\approx \frac{1}{3} \times \frac{22}{7} \times 1.4 \text{ cm} \times 1.4 \text{ cm} \times 12 \text{ cm}$
$\approx 24.64 \text{ cm}^3$

25. We must find the radius of the base in order to use the formula for the volume of a circular cylinder.
$$r = \frac{d}{2} = \frac{0.7 \text{ yd}}{2} = 0.35 \text{ yd}$$
$$V = Bh = \pi \cdot r^2 \cdot h$$
$$\approx 3.14 \times 0.35 \text{ yd} \times 0.35 \text{ yd} \times 1.1 \text{yd}$$
$$\approx 0.423115 \text{ yd}^3$$

27. We must find the radius of the silo in order to use the formula for the volume of a circular cylinder.
$$r = \frac{d}{2} = \frac{6 \text{ m}}{2} = 3 \text{ m}$$
$$V = Bh = \pi \cdot r^2 \cdot h$$
$$\approx 3.14 \times 3 \text{ m} \times 3 \text{ m} \times 13 \text{ m}$$
$$\approx 367.38 \text{ m}^3$$

29. First we find the radius of the ball:
$$r = \frac{d}{2} = \frac{6.5 \text{ cm}}{2} = 3.25 \text{ cm}$$
Then we find the volume, using the formula for the volume of a sphere.
$$V = \frac{4}{3} \cdot \pi \cdot r^3$$
$$\approx \frac{4}{3} \cdot 3.14 \cdot (3.25 \text{ cm})^3$$
$$\approx 143.72 \text{ cm}^3$$

31. First we find the radius of the earth:
$$\frac{3980 \text{ mi}}{2} = 1990 \text{ mi}$$
Then we find the volume, using the formula for the volume of a sphere.
$$V = \frac{4}{3} \cdot \pi \cdot r^3$$
$$\approx \frac{4}{3} \cdot 3.14 \cdot (1990 \text{ mi})^3$$
$$\approx 32,993,440,000 \text{ mi}^3$$

33. First we find the radius of the can.
$$r = \frac{d}{2} = \frac{6.5 \text{ cm}}{2} = 3.25 \text{ cm}$$
The height of the can is the length of the diameters of 3 tennis balls.
$$h = 3(6.5 \text{ cm}) = 19.5 \text{ cm}$$
Now we find the volume.
$$V = Bh = \pi \cdot r^2 \cdot h$$
$$\approx 3.14 \times 3.25 \text{ cm} \times 3.25 \text{ cm} \times 19.5 \text{ cm}$$
$$\approx 646.74 \text{ cm}^3$$

35. $V = Bh = \pi \cdot r^2 \cdot h$

$\approx \dfrac{22}{7} \cdot 14 \text{ cm} \cdot 14 \text{ cm} \cdot 100 \text{ cm}$

$\approx 61,600 \text{ cm}^3$

37. A cube is a rectangular solid.

$V = l \cdot w \cdot h$

$= 18 \text{ yd} \cdot 18 \text{ yd} \cdot 18 \text{ yd}$

$= 5832 \text{ yd}^3$

39. Discussion and Writing Exercise

41. Interest $= P \cdot r \cdot t$

$= \$600 \times 6.4\% \times \dfrac{1}{2}$

$= \dfrac{\$600 \times 0.064}{2}$

$= \$19.20$

The interest is \$19.20.

43. $10^3 = 10 \cdot 10 \cdot 10 = 1000$

45. $7^2 = 7 \cdot 7 = 49$

47. *Rephrase:*

$$\underbrace{\begin{matrix}\text{Sales}\\\text{tax}\end{matrix}}\quad \text{is} \quad \underbrace{\begin{matrix}\text{what}\\\text{percent}\end{matrix}}\quad\text{of}\quad\underbrace{\begin{matrix}\text{purchase}\\\text{price?}\end{matrix}}$$

$$\downarrow \qquad \downarrow \qquad \downarrow \qquad \downarrow \qquad \downarrow$$

Translate: $\quad 878 \quad = \quad r \quad \times \quad 17,560$

To solve the equation we divide on both sides by 17,560.

$\dfrac{878}{17,560} = \dfrac{r \times 17,560}{17,560}$

$0.05 = r$

$5\% = r$

The sales tax rate is 5%.

49. First find the volume of one one-dollar bill in cubic inches:

$V = l \cdot w \cdot h$

$V = 6.0625 \text{ in.} \times 2.3125 \text{ in.} \times 0.0041 \text{ in.}$

$V = 0.05748 \text{ in}^3 \qquad$ Rounding

Then multiply to find the volume of one million one-dollar bills in cubic inches:

$1,000,000 \times 0.05748 \text{ in}^3 = 57,480 \text{ in}^3$

Thus the volume of one million one-dollar bills is about $57,480 \text{ in}^3$.

51. Radius of water stream: $\dfrac{2 \text{ cm}}{2} = 1 \text{ cm}$

To convert 30 m to centimeters, think: 1 meter is 100 times as large as 1 centimeter. Thus, we move the decimal point 2 places to the right:

$30 \text{ m} = 3000 \text{ cm}$

$V = Bh = \pi \cdot r^2 \cdot h$

$\approx 3.141593 \cdot 1 \text{ cm} \cdot 1 \text{ cm} \cdot 3000 \text{ cm}$

$\approx 9425 \text{ cm}^3$

Now we convert 9425 cm^3 to liters:

$9425 \text{ cm}^3 = 9425 \text{ cm}^3 \cdot \dfrac{1 \text{ L}}{1000 \text{ cm}^3}$

$= 9.425 \text{ L}$

There is about 9.425 L of water in the hose.

53. Find the diameter of the earth at the equator:

$$C = \pi \cdot d$$

$$24,901.55 \text{ mi} \approx 3.14 \cdot d$$

$$7930 \text{ mi} \approx d$$

Find the diameter of the earth through the north and south poles:

$$C = \pi \cdot d$$

$$24,859.82 \text{ mi} \approx 3.14 \cdot d$$

$$7917 \text{ mi} \approx d$$

Find the average of these two diameters:

$$\dfrac{7930 \text{ mi} + 7917 \text{ mi}}{2} = 7923.5 \text{ mi}$$

Use this average to estimate the volume of the earth:

$$r = \dfrac{d}{2} = \dfrac{7923.5 \text{ mi}}{2} = 3961.75 \text{ mi}$$

$$V = \dfrac{4}{3} \cdot \pi \cdot r^3$$

$$\approx \dfrac{4 \times 3.14 \times (3961.75 \text{ mi})^3}{3}$$

$$\approx 260,000,000,000 \text{ mi}^3$$

55. The length of a diagonal of the cube is the length of the diameter of the sphere, 1 m. Visualize a triangle whose hypotenuse is a diagonal of the cube and with one leg a side s of the cube and the other leg a diagonal c of a side of the cube.

We want to find the length of a side s of the cube in order to find the volume. We begin by using the Pythagorean theorem to find c.

$s^2 + s^2 = c^2$

$2s^2 = c^2$

$\sqrt{2s^2} = c$

Now use the Pythagorean theorem again to find s.

$c^2 + s^2 = 1^2$

$(\sqrt{2s^2})^2 + s^2 = 1 \qquad$ Subtracting $\sqrt{2s^2}$ for c

$2s^2 + s^2 = 1$

$3s^2 = 1$

$s^2 = \dfrac{1}{3}$

$s = \sqrt{\dfrac{1}{3}} \approx 0.577$

Next we find the volume of the cube.

$V = l \cdot w \cdot h$

$= 0.577 \text{ m} \cdot 0.577 \text{ m} \cdot 0.577 \text{ m}$

$= 0.192 \text{ m}^3$

Find the volume of the sphere. The radius is $\dfrac{1 \text{ m}}{2}$, or 0.5 m.

$$V = \frac{4}{3} \cdot \pi \cdot r^3$$

$$\approx \frac{4}{3} \times 3.14 \times (0.5 \text{ m})^3$$

$$\approx \frac{4 \times 3.14 \times 0.125 \text{ m}^3}{3}$$

$$\approx 0.523 \text{ m}^3$$

Finally we subtract to find how much more volume is in the sphere.

$$0.523 \text{ m}^3 - 0.192 \text{ m}^3 = 0.331 \text{ m}^3$$

There is 0.331 m^3 more volume in the sphere.

Exercise Set 5.6

1. Two angles are complementary if the sum of their measures is 90°.

$$90° - 11° = 79°.$$

The measure of a complement is 79°.

3. Two angles are complementary if the sum of their measures is 90°.

$$90° - 67° = 23°.$$

The measure of a complement is 23°.

5. Two angles are complementary if the sum of their measures is 90°.

$$90° - 58° = 32°.$$

The measure of a complement is 32°.

7. Two angles are complementary if the sum of their measures is 90°.

$$90° - 29° = 61°.$$

The measure of a complement is 61°.

9. Two angles are supplementary if the sum of their measures is 180°.

$$180° - 3° = 177°.$$

The measure of a supplement is 177°.

11. Two angles are supplementary if the sum of their measures is 180°.

$$180° - 139° = 41°.$$

The measure of a supplement is 41°.

13. Two angles are supplementary if the sum of their measures is 180°.

$$180° - 85° = 95°.$$

The measure of a supplement is 95°.

15. Two angles are supplementary if the sum of their measures is 180°.

$$180° - 102° = 78°.$$

The measure of a supplement is 78°.

17. The segments have different lengths. They are not congruent.

19. $m\angle G = m\angle R$, so $\angle G \cong \angle R$.

21. Since $\angle 2$ and $\angle 5$ are vertical angles, $m\angle 2 = 67°$. Likewise, $\angle 1$ and $\angle 4$ are vertical angles, so $m\angle 4 = 80°$.

$$\begin{aligned} m\angle 1 + m\angle 2 + m\angle 3 &= 180° \\ 80° + 67° + m\angle 3 &= 180° \quad \text{Substituting} \\ 147° + m\angle 3 &= 180° \\ m\angle 3 &= 180° - 147° \\ m\angle 3 &= 33° \end{aligned}$$

Since $\angle 3$ and $\angle 6$ are vertical angles, $m\angle 6 = 33°$.

23. a) The pairs of corresponding angles are

$\angle 1$ and $\angle 3$,

$\angle 2$ and $\angle 4$,

$\angle 8$ and $\angle 6$,

$\angle 7$ and $\angle 5$.

b) The interior angles are $\angle 2$, $\angle 3$, $\angle 6$, and $\angle 7$.

c) The pairs of alternate interior angles are

$\angle 2$ and $\angle 6$,

$\angle 3$ and $\angle 7$.

25. $\angle 4$ and $\angle 6$ are vertical angles, so $m\angle 6 = 125°$.

$\angle 4$ and $\angle 2$ are corresponding angles. By Property 1, $m\angle 2 = 125°$.

$\angle 6$ and $\angle 8$ are corresponding angles. By Property 1, $m\angle 8 = 125°$.

$\angle 2$ and $\angle 3$ are interior angles on the same side of the transversal. Using Property 4 and $m\angle 2 = 125°$, $m\angle 3 = 55°$.

$\angle 6$ and $\angle 7$ are interior angles on the same side of the transversal. Using Property 4 and $m\angle 6 = 125°$, $m\angle 7 = 55°$.

$\angle 3$ and $\angle 5$ are vertical angles, so $m\angle 5 = 55°$.

$\angle 7$ and $\angle 1$ are vertical angles, so $m\angle 1 = 55°$.

27. Considering the transversal $\overleftrightarrow{BC}$, $\angle ABE$ and $\angle DCE$ are alternate interior angles. By Property 2, $\angle ABE \cong \angle DCE$. Then $m\angle ABE = m\angle DCE = 95°$.

Considering the transversal $\overleftrightarrow{AD}$, $\angle BAE$ and $\angle CDE$ are alternate interior angles. By Property 2, $\angle BAE \cong \angle CDE$. We cannot determine the measure of these angles.

$\angle AEB$ and $\angle DEC$ are vertical angles, so $\angle AEB \cong \angle DEC$. We cannot determine the measure of these angles.

$\angle BED$ and $\angle AEC$ are also vertical angles, so $\angle BED \cong \angle AEC$. We cannot determine their measures.

29. Considering the transversal $\overleftrightarrow{CE}$, $\angle AEC$ and $\angle DCE$ are alternate interior angles. By Property 2, $\angle AEC \cong \angle DCE$. Then $m\angle AEC = m\angle DCE = 50°$.

Considering the transversal $\overleftrightarrow{DE}$, $\angle BED$ and $\angle EDC$ are alternate interior angles. By Property 2, $\angle BED \cong \angle EDC$. Then $m\angle BED = m\angle EDC = 41°$.

31. $6 \times 1\frac{7}{8} = 6 \times \frac{15}{8}$

$$= \frac{6 \times 15}{8}$$

$$= \frac{2 \times 3 \times 15}{2 \times 4}$$

$$= \frac{2}{2} \times \frac{3 \times 15}{4}$$

$$= \frac{45}{4}$$

$$= 11\frac{1}{4}$$

33. $8\frac{3}{7} \times 14 = \frac{59}{7} \times 14$

$$= \frac{59 \times 14}{7}$$

$$= \frac{59 \times 2 \times 7}{7 \times 1}$$

$$= \frac{7}{7} \times \frac{59 \times 2}{1}$$

$$= 118$$

Exercise Set 5.7

1. The notation tells us the way in which the vertices of the two triangles are matched.

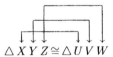

$\triangle ABC \cong \triangle RST$ means

$\angle A \cong \angle R$	and	$\overline{AB} \cong \overline{RS}$
$\angle B \cong \angle S$		$\overline{AC} \cong \overline{RT}$
$\angle C \cong \angle T$		$\overline{BC} \cong \overline{ST}$

3. The notation tells us the way in which the vertices of the two triangles are matched.

$\triangle DEF \cong \triangle GHK$

$\triangle DEF \cong \triangle GHK$ means

$\angle D \cong \angle G$	and	$\overline{DE} \cong \overline{GH}$
$\angle E \cong \angle H$		$\overline{DF} \cong \overline{GK}$
$\angle F \cong \angle K$		$\overline{EF} \cong \overline{HK}$

5. The notation tells us the way in which the two triangles are matched.

$\triangle XYZ \cong \triangle UVW$

$\triangle XYZ \cong \triangle UVW$ means

$\angle X \cong \angle U$	and	$\overline{XY} \cong \overline{UV}$
$\angle Y \cong \angle V$		$\overline{XZ} \cong \overline{UW}$
$\angle Z \cong \angle W$		$\overline{YZ} \cong \overline{VW}$

7. The notation tells us the way in which the vertices of the two triangles are matched.

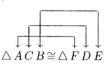

$\triangle ACB \cong \triangle FDE$

$\triangle ACB \cong \triangle FDE$ means

$\angle A \cong \angle F$	and	$\overline{AC} \cong \overline{FD}$
$\angle C \cong \angle D$		$\overline{AB} \cong \overline{FE}$
$\angle B \cong \angle E$		$\overline{CB} \cong \overline{DE}$

9. The notation tells us the way in which the vertices of the two triangles are matched.

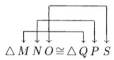

$\triangle MNO \cong \triangle QPS$

$\triangle MNO \cong \triangle QPS$ means

$\angle M \cong \angle Q$	and	$\overline{MN} \cong \overline{QP}$
$\angle N \cong \angle P$		$\overline{MO} \cong \overline{QS}$
$\angle O \cong \angle S$		$\overline{NO} \cong \overline{PS}$

11. We cannot determine from the information given that two sides of one triangle and the included angle are congruent to two sides and the included angle of the other triangle. Therefore, we cannot use the SAS Property.

13. Two sides of one triangle and the included angle are congruent to two sides and the included angle of the other triangle. They are congruent by the SAS Property.

15. Two sides of one triangle and the included angle are congruent to two sides and the included angle of the other triangle. They are congruent by the SAS Property.

17. We cannot determine from the information given that three sides of one triangle are congruent to three sides of the other triangle. Therefore, we cannot use the SSS Property.

19. Three sides of one triangle are congruent to three sides of the other triangle. They are congruent by the SSS Property.

21. Three sides of one triangle are congruent to three sides of the other triangle. They are congruent by the SSS Property.

23. Two angles and the included side are of one triangle are congruent to two angles and the included side of the other triangle. They are congruent by the ASA Property.

25. Two angles and the included side are of one triangle are congruent to two angles and the included side of the other triangle. They are congruent by the ASA Property.

27. The vertical angles are congruent so two angles and the included side are of one triangle are congruent to two angles and the included side of the other triangle. They are congruent by the ASA Property.

29. Two angles and the included side are of one triangle are congruent to two angles and the included side of the other triangle. They are congruent by the ASA Property.

31. Two sides of one triangle and the included angle are congruent to two sides and the included angle of the other triangle. They are congruent by the SAS Property.

33. Three sides of one triangle are congruent to three sides of the other triangle. In addition, two sides of one triangle and the included angle are congruent to two sides and the included angle of the other triangle. Therefore, we can use either the SSS Property or the SAS Property to show that they are congruent.

35. Since R is the midpoint of $\overline{PT}$, $\overline{PR} \cong \overline{TR}$.

Since R is the midpoint of $\overline{QS}$, $\overline{RQ} \cong \overline{RS}$.

$\angle PRQ$ and $\angle TRS$ are vertical angles, so $\angle PRQ \cong \angle TRS$.

Two sides and the included angle of $\triangle PRQ$ are congruent to two sides and the included angle of $\triangle TRS$, so $\triangle PRQ \cong \triangle TRS$ by the SAS Property.

37. Since $GL \perp KM$, $m\angle GLK = m\angle GLM = 90°$. Then $\angle GLK \cong \angle GLM$.

Since L is the midpoint of $\overline{KM}$, $\overline{KL} \cong \overline{LM}$.

$\overline{GL} \cong \overline{GL}$.

Two sides and the included angle of $\triangle KLG$ are congruent to two sides and the included angle of $\triangle MLG$, so $\triangle KLG \cong \triangle MLG$ by the SAS Property.

39. The information given tells us that $\overline{AE} \cong \overline{CB}$ and $\overline{AB} \cong \overline{CD}$.

Since B is the midpoint of $\overline{ED}$, $\overline{EB} \cong \overline{BD}$.

Three sides of $\triangle AEB$ are congruent to three sides of $\triangle CDB$, so $\triangle AEB \cong \triangle CDB$ by the SSS Property.

41. The information given tells us that $\overline{HK} \cong \overline{KJ}$ and $\overline{GK} \cong \overline{LK}$.

Since $\overline{GK} \perp \overline{LJ}$, $m\angle HKL = m\angle GKJ = 90°$.

Then $\angle HKL \cong \angle GKJ$.

Two sides and the included angle of $\triangle LKH$ are congruent to two sides and the included angle of $\triangle GKJ$, so $\triangle LKH \cong \triangle GKJ$ by the SAS Property. This means that the remaining corresponding parts of the two triangles are congruent. That is, $\angle HLK \cong \angle JGK$, $\angle LHK \cong \angle GJK$, and $\overline{LH} \cong \overline{GJ}$.

43. Two angles and the included side of $\triangle PED$ are congruent to two angles and the included side of $\triangle PFG$, so $\triangle PED \cong \triangle PFG$ by the ASA Property. Then corresponding parts of the two triangles are congruent, so $\overline{EP} \cong \overline{FP}$. Therefore, P is the midpoint of $\overline{EF}$.

45. $\angle A$ and $\angle C$ are opposite angles, so $m\angle A = 70°$ by Property 2.

$\angle C$ and $\angle B$ are consecutive angles, so the are supplementary by Property 4. Then

$$m\angle B = 180° - m\angle C$$
$$m\angle B = 180° - 70°$$
$$m\angle B = 110°.$$

$\angle B$ and $\angle D$ are opposite angles, so $m\angle D = 110°$ by Property 2.

47. $\angle M$ and $\angle K$ are opposite angles, so $m\angle M = 71°$ by Property 2.

$\angle K$ and $\angle L$ are consecutive angles, so the are supplementary by Property 4. Then

$$m\angle L = 180° - m\angle K$$
$$m\angle L = 180° - 71°$$
$$m\angle L = 109°.$$

$\angle J$ and $\angle L$ are opposite angles, so $m\angle J = 109°$ by Property 2.

49. $\overline{ON}$ and $\overline{TU}$ are opposite sides of the parallelogram. So are $\overline{OT}$ and $\overline{NU}$. The opposite sides of a parallelogram are congruent (Property 3), so $TU = 9$ and $NU = 15$.

51. $\overline{JM}$ and $\overline{KL}$ are opposite sides of the parallelogram. So are $\overline{JK}$ and $\overline{ML}$. The opposite sides of a parallelogram are congruent (Property 3). Then $KL = 3\frac{1}{2}$ and $JK + LM = 22 - 3\frac{1}{2} - 3\frac{1}{2} = 15$. Thus, $JK = LM = \frac{1}{2} \cdot 15 = 7\frac{1}{2}$.

53. The diagonals of a parallelogram bisect each other (Property 5). Then

$$AC = 2 \cdot AB = 2 \cdot 14 = 28$$
$$ED = 2 \cdot BD = 2 \cdot 19 = 38$$

55. 0.452   0.45.2   Move the decimal point 2 places to the right.

Write a % symbol: 45.2%

$0.452 = 45.2\%$

57. We multiply by 1 to get 100 in the denominator.

$$\frac{11}{20} = \frac{11}{20} \cdot \frac{5}{5} = \frac{55}{100} = 55\%$$

59. The ratio of the amount spent in Florida to the total amount spent is $\frac{2.7}{13.1}$. This can also be expressed as follows:

$$\frac{2.7}{13.1} = \frac{2.7}{13.1} \cdot \frac{10}{10} = \frac{27}{131}$$

The ratio of the total amount spent to the amount spent in Florida is $\frac{13.1}{2.7}$, or $\frac{131}{27}$.

61.

$$
\begin{array}{r}
1.7\,5 \\
12\,\overline{)\,2\,1.0\,0} \\
1\,2\,0\,0 \\
\hline
9\,0\,0 \\
8\,4\,0 \\
\hline
6\,0 \\
6\,0 \\
\hline
0
\end{array}
$$

The answer is 1.75.

63. To divide by 100, move the decimal point 2 places to the left.

23.4   .23.4

$23.4 \div 100 = 0.234$

65.
$$
\begin{array}{r}
3.1\,4 \quad \text{(2 decimal places)} \\
\times\ 4.\,4\,1 \quad \text{(2 decimal places)} \\
\hline
3\,1\,4 \\
1\ 2\ 5\ 6\ 0 \\
1\ 2\ 5\ 6\ 0\ 0 \\
\hline
1\,3.\,8\,4\,7\,4 \quad \text{(4 decimal places)}
\end{array}
$$

Round

13. 8 4 $\boxed{7}$ 4   to the nearest hundredth.

┘└———— Thousandths digit is 5 or higher.

↓

13. 8 5   Round up.

Exercise Set 5.8

1. Vertex R is matched with vertex A, vertex S is matched with vertex B, and vertex T is matched with vertex C. Then

$$
\begin{array}{lll}
\overline{RS} \longleftrightarrow \overline{AB} & \text{and} & \angle R \longleftrightarrow \angle A \\
\overline{ST} \longleftrightarrow \overline{BC} & & \angle S \longleftrightarrow \angle B \\
\overline{TR} \longleftrightarrow \overline{CA} & & \angle T \longleftrightarrow \angle C
\end{array}
$$

3. Vertex C is matched with vertex W, vertex B is matched with vertex J, and vertex S is matched with vertex Z. Then

$$
\begin{array}{lll}
\overline{CB} \longleftrightarrow \overline{WJ} & \text{and} & \angle C \longleftrightarrow \angle W \\
\overline{BS} \longleftrightarrow \overline{JZ} & & \angle B \longleftrightarrow \angle J \\
\overline{SC} \longleftrightarrow \overline{ZW} & & \angle S \longleftrightarrow \angle Z
\end{array}
$$

5. The notation tells us the way in which the vertices are matched.

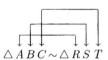

$\triangle ABC \sim \triangle RST$

$\triangle ABC \sim \triangle RST$ means

$$
\begin{array}{ll}
\angle A \cong \angle R & \\
\angle B \cong \angle S & \text{and} \quad \dfrac{AB}{RS} = \dfrac{AC}{RT} = \dfrac{BC}{ST}. \\
\angle C \cong \angle T &
\end{array}
$$

7. The notation tells us the way in which the vertices are matched.

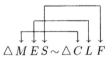

$\triangle MES \sim \triangle CLF$

$\triangle MES \sim \triangle CLF$ means

$$
\begin{array}{ll}
\angle M \cong \angle C & \\
\angle E \cong \angle L & \text{and} \quad \dfrac{ME}{CL} = \dfrac{MS}{CF} = \dfrac{ES}{LF}. \\
\angle S \cong \angle F &
\end{array}
$$

9. If we match P with N, S with D, and Q with M, the corresponding angles will be congruent. That is, $\triangle PSQ \sim \triangle NDM$. Then

$$
\frac{PS}{ND} = \frac{PQ}{NM} = \frac{SQ}{DM}.
$$

11. If we match T with G, A with F, and W with C, the corresponding angles will be congruent. That is, $\triangle TAW \sim \triangle GFC$. Then

$$
\frac{TA}{GF} = \frac{TW}{GC} = \frac{AW}{FC}.
$$

13. Since $\triangle ABC \sim \triangle PQR$, the corresponding sides are proportional. Then

$$
\frac{3}{6} = \frac{4}{PR} \quad \text{and} \quad \frac{3}{6} = \frac{5}{QR}
$$

$$
\begin{array}{ll}
3(PR) = 6 \cdot 4 & \quad 3(QR) = 6 \cdot 5 \\
3(PR) = 24 & \quad 3(QR) = 30 \\
PR = 8 & \quad QR = 10
\end{array}
$$

15. Recall that if a transversal intersects two parallel lines, then the alternate interior angles are congruent. Thus,

$$\angle A \cong \angle B \text{ and } \angle D \cong \angle C.$$

Since $\angle AED$ and $\angle CEB$ are vertical angles, they are congruent. Thus,

$$\angle AED \cong \angle CEB.$$

Then $\triangle AED \sim \triangle CEB$, and the lengths of the corresponding sides are proportional.

$$
\frac{AD}{CB} = \frac{ED}{EC}
$$

$$
\frac{7}{21} = \frac{6}{EC}
$$

$$
7 \cdot EC = 126
$$

$$
EC = 18
$$

17. If we use the sun's rays to represent the third side of a triangle in a drawing of the situation, we see that we have similar triangles. We let $h =$ the height of the tree.

The ratio of h to 4 is the same as the ratio of 27 to 3. We have the proportion

$$
\frac{h}{4} = \frac{27}{3}.
$$

Solve: $3 \cdot h = 4 \cdot 27$

$$
h = \frac{4 \cdot 27}{3}
$$

$$
h = 36
$$

The tree is 36 ft tall.

19. Since the ratio of d to 25 ft is the same as the ratio of 40 ft to 10 ft, we have the proportion

$$
\frac{d}{25} = \frac{40}{10}.
$$

Solve: $10 \cdot d = 25 \cdot 40$

$$
d = \frac{25 \cdot 40}{10}
$$

$$
d = 100
$$

The distance across the river is 100 ft.

21. Discussion and Writing Exercise

23. $2\frac{4}{5} \times 10\frac{1}{2} = \frac{14}{5} \times \frac{21}{2} = \frac{14 \times 21}{5 \times 2} =$

$\frac{2 \times 7 \times 21}{5 \times 2} = \frac{\cancel{2} \times 7 \times 21}{5 \times \cancel{2}} = \frac{147}{5} = 29\frac{2}{5}$

25. $8 \times 9\frac{3}{4} = \frac{8}{1} \times \frac{39}{4} = \frac{8 \times 39}{1 \times 4} = \frac{2 \times 4 \times 39}{1 \times 4} =$

$\frac{2 \times \cancel{4} \times 39}{1 \times \cancel{4}} = \frac{78}{1} = 78$

Chapter 6

Introduction to Real Numbers and Algebraic Expressions

Exercise Set 6.1

1. Substitute 34 for n: $600(34) = 20,400$, so \$20,400 is collected if 34 students enroll.

Substitute 78 for n: $600(78) = 46,800$, so \$46,800 is collected if 78 students enroll.

Substitute 250 for n: $600(250) = 150,000$, so \$150,000 is collected if 250 students enroll.

3. Substitute 45 m for b and 86 m for h, and carry out the multiplication:

$$A = \frac{1}{2}bh = \frac{1}{2}(45 \text{ m})(86 \text{ m})$$
$$= \frac{1}{2}(45)(86)(\text{m})(\text{m})$$
$$= 1935 \text{ m}^2$$

5. Substitute 65 for r and 4 for t, and carry out the multiplication:
$$d = rt = 65 \cdot 4 = 260 \text{ mi}$$

7. We substitute 6 ft for l and 4 ft for w in the formula for the area of a rectangle.
$$A = lw = (6 \text{ ft})(4 \text{ ft})$$
$$= (6)(4)(\text{ft})(\text{ft})$$
$$= 24 \text{ ft}^2$$

9. $8x = 8 \cdot 7 = 56$

11. $\dfrac{a}{b} = \dfrac{24}{3} = 8$

13. $\dfrac{3p}{q} = \dfrac{3 \cdot 2}{6} = \dfrac{6}{6} = 1$

15. $\dfrac{x+y}{5} = \dfrac{10+20}{5} = \dfrac{30}{5} = 6$

17. $\dfrac{x-y}{8} = \dfrac{20-4}{8} = \dfrac{16}{8} = 2$

19. $b + 7$, or $7 + b$

21. $c - 12$

23. $4 + q$, or $q + 4$

25. $a + b$, or $b + a$

27. $x \div y$, or $\dfrac{x}{y}$, or x/y, or $x \cdot \dfrac{1}{y}$

29. $x + w$, or $w + x$

31. $n - m$

33. $x + y$, or $y + x$

35. $2z$

37. $3m$

39. Let s represent your salary. Then we have $89\% s$, or $0.89s$.

41. The distance traveled is the product of the speed and the time. Thus, Danielle traveled $65t$ miles.

43. $\$50 - x$

45. Discussion and Writing Exercise

47. $A = l \cdot w = (28.6 \text{ ft}) \cdot (12.5 \text{ ft})$
$= (28.6) \cdot (12.5) \cdot \text{ft} \cdot \text{ft}$
$= 357.5 \text{ ft}^2$

49. $A = s \cdot s = 234 \text{ mi} \cdot 234 \text{ mi}$
$= 234 \cdot 234 \cdot \text{mi} \cdot \text{mi}$
$= 54,756 \text{ mi}^2$

51. $A = \dfrac{1}{2} \cdot h \cdot (a + b)$
$= \dfrac{1}{2} \cdot 7.4 \text{ ft} \cdot (5.8 \text{ ft} + 12.3 \text{ ft})$
$= \dfrac{7.4 \cdot 18.1}{2} \cdot \text{ft}^2$
$= 66.97 \text{ ft}^2$

53. $x + 3y$

55. $2x - 3$

Exercise Set 6.2

1. The integer -1286 corresponds to 1286 ft below sea level; the integer 14,410 corresponds to 14,410 ft above sea level.

3. The integer 24 corresponds to $24°$ above zero; the integer -2 corresponds to $2°$ below zero.

5. The integer $-5,600,000,000,000$ corresponds to the total public debt of \$5,600,000,000,000.

7. The integer -34 describes the situation from the Alley Cats' point of view. The integer 34 describes the situation from the Strikers' point of view.

9. The number $\dfrac{10}{3}$ can be named $3\dfrac{1}{3}$, or $3.3\overline{3}$. The graph is $\dfrac{1}{3}$ of the way from 3 to 4.

11. The graph of -5.2 is $\dfrac{2}{10}$ of the way from -5 to -6.

13. We first find decimal notation for $\dfrac{7}{8}$. Since $\dfrac{7}{8}$ means $7 \div 8$, we divide.

$$
\begin{array}{r}
0.8\,7\,5 \\
8\overline{\smash{)}7.0\,0\,0} \\
\underline{6\,4} \\
6\,0 \\
\underline{5\,6} \\
4\,0 \\
\underline{4\,0} \\
0
\end{array}
$$

Thus $\dfrac{7}{8} = 0.875$, so $-\dfrac{7}{8} = -0.875$.

15. $\dfrac{5}{6}$ means $5 \div 6$, so we divide.

$$
\begin{array}{r}
0.8\,3\,3\,\ldots \\
6\overline{\smash{)}5.0\,0\,0} \\
\underline{4\,8} \\
2\,0 \\
\underline{1\,8} \\
2\,0 \\
\underline{1\,8} \\
0
\end{array}
$$

We have $\dfrac{5}{6} = 0.8\overline{3}$.

17. $\dfrac{7}{6}$ means $7 \div 6$, so we divide.

$$
\begin{array}{r}
1.1\,6\,6\,\ldots \\
6\overline{\smash{)}7.0\,0\,0} \\
\underline{6} \\
1\,0 \\
\underline{6} \\
4\,0 \\
\underline{3\,6} \\
4\,0 \\
\underline{3\,6} \\
4
\end{array}
$$

We have $\dfrac{7}{6} = 1.1\overline{6}$.

19. $\dfrac{2}{3}$ means $2 \div 3$, so we divide.

$$
\begin{array}{r}
0.6\,6\,6\,\ldots \\
3\overline{\smash{)}2.0\,0\,0} \\
\underline{1\,8} \\
2\,0 \\
\underline{1\,8} \\
2\,0 \\
\underline{1\,8} \\
2
\end{array}
$$

We have $\dfrac{2}{3} = 0.\overline{6}$.

21. We first find decimal notation for $\dfrac{1}{2}$. Since $\dfrac{1}{2}$ means $1 \div 2$, we divide.

$$
\begin{array}{r}
0.5 \\
2\overline{\smash{)}1.0} \\
\underline{1\,0} \\
0
\end{array}
$$

Thus $\dfrac{1}{2} = 0.5$, so $-\dfrac{1}{2} = -0.5$

23. $\dfrac{1}{10}$ means $1 \div 10$, so we divide.

$$
\begin{array}{r}
0.1 \\
1\,0\overline{\smash{)}1.0} \\
\underline{1\,0} \\
0
\end{array}
$$

We have $\dfrac{1}{10} = 0.1$

25. Since 8 is to the right of 0, we have $8 > 0$.

27. Since -8 is to the left of 3, we have $-8 < 3$.

29. Since -8 is to the left of 8, we have $-8 < 8$.

31. Since -8 is to the left of -5, we have $-8 < -5$.

33. Since -5 is to the right of -11, we have $-5 > -11$.

35. Since -6 is to the left of -5, we have $-6 < -5$.

37. Since 2.14 is to the right of 1.24, we have $2.14 > 1.24$.

39. Since -14.5 is to the left of 0.011, we have $-14.5 < 0.011$.

41. Since -12.88 is to the left of -6.45, we have $-12.88 < -6.45$.

43. Convert to decimal notation $\dfrac{5}{12} = 0.4166\ldots$ and $\dfrac{11}{25} = 0.44$. Since $0.4166\ldots$ is to the left of 0.44, $\dfrac{5}{12} < \dfrac{11}{25}$.

45. $-3 \geq -11$ is true since $-3 > -11$ is true.

47. $0 \geq 8$ is false since neither $0 > 8$ nor $0 = 8$ is true.

49. $x < -6$ has the same meaning as $-6 > x$.

51. $y \geq -10$ has the same meaning as $-10 \leq y$.

53. The distance of -3 from 0 is 3, so $|-3| = 3$.

55. The distance of 10 from 0 is 10, so $|10| = 10$.

57. The distance of 0 from 0 is 0, so $|0| = 0$.

59. The distance of -24 from 0 is 24, so $|-24| = 24$.

61. The distance of $-\frac{2}{3}$ from 0 is $\frac{2}{3}$, so $\left|-\frac{2}{3}\right| = \frac{2}{3}$.

63. The distance of $\frac{0}{4}$ from 0 is $\frac{0}{4}$, or 0, so $\left|\frac{0}{4}\right| = 0$.

65. The distance of $-3\frac{5}{8}$ from 0 is $3\frac{5}{8}$, so $\left|-3\frac{5}{8}\right| = 3\frac{5}{8}$.

67. Discussion and Writing Exercise

69. $10 = 2 \cdot 5$

$35 = 5 \cdot 7$

$\text{LCM} = 2 \cdot 5 \cdot 7 = 70$

71. $4 = 2 \cdot 2$

$10 = 2 \cdot 5$

$18 = 2 \cdot 3 \cdot 3$

$\text{LCM} = 2 \cdot 2 \cdot 3 \cdot 3 \cdot 5 = 180$

73. $\dfrac{5}{8} + \dfrac{7}{12}$ LCM is 24.

$= \dfrac{5}{8} \cdot \dfrac{3}{3} + \dfrac{7}{12} \cdot \dfrac{2}{2}$

$= \dfrac{15}{24} + \dfrac{14}{24}$

$= \dfrac{29}{24}$

75. $-\dfrac{2}{3}, \dfrac{1}{2}, -\dfrac{3}{4}, -\dfrac{5}{6}, \dfrac{3}{8}, \dfrac{1}{6}$ can be written in decimal notation as $-0.\overline{6}, 0.5, -0.75, -0.8\overline{3}, 0.375, 0.1\overline{6}$, respectively. Listing from least to greatest, we have

$-\dfrac{5}{6}, -\dfrac{3}{4}, -\dfrac{2}{3}, \dfrac{1}{6}, \dfrac{3}{8}, \dfrac{1}{2}$.

77. $0.\overline{1} = \dfrac{0.\overline{3}}{3} = \dfrac{\frac{1}{3}}{3} = \dfrac{1}{3} \cdot \dfrac{1}{3} = \dfrac{1}{9}$

79. First consider $0.\overline{5}$.

$0.\overline{5} = 0.\overline{3} \cdot \dfrac{5}{3} = \dfrac{1}{3} \cdot \dfrac{5}{3} = \dfrac{5}{9}$

Then, $5.\overline{5} = 5 + 0.\overline{5} = 5 + \dfrac{5}{9} = 5\dfrac{5}{9}$, or $\dfrac{50}{9}$.

Exercise Set 6.3

1. $2 + (-9)$ The absolute values are 2 and 9. The difference is $9 - 2$, or 7. The negative number has the larger absolute value, so the answer is negative. $2 + (-9) = -7$

3. $-11 + 5$ The absolute values are 11 and 5. The difference is $11 - 5$, or 6. The negative number has the larger absolute value, so the answer is negative. $-11 + 5 = -6$

5. $-8 + 8$ A negative and a positive number. The numbers have the same absolute value. The sum is 0. $-8 + 8 = 0$

7. $-3 + (-5)$ Two negatives. Add the absolute values, getting 8. Make the answer negative. $-3 + (-5) = -8$

9. $-7 + 0$ One number is 0. The answer is the other number. $-7 + 0 = -7$

11. $0 + (-27)$ One number is 0. The answer is the other number. $0 + (-27) = -27$

13. $17 + (-17)$ A negative and a positive number. The numbers have the same absolute value. The sum is 0. $17 + (-17) = 0$

15. $-17 + (-25)$ Two negatives. Add the absolute values, getting 42. Make the answer negative. $-17 + (-25) = -42$

17. $18 + (-18)$ A positive and a negative number. The numbers have the same absolute value. The sum is 0. $18 + (-18) = 0$

19. $-28 + 28$ A negative and a positive number. The numbers have the same absolute value. The sum is 0. $-28 + 28 = 0$

21. $8 + (-5)$ The absolute values are 8 and 5. The difference is $8 - 5$, or 3. The positive number has the larger absolute value, so the answer is positive. $8 + (-5) = 3$

23. $-4 + (-5)$ Two negatives. Add the absolute values, getting 9. Make the answer negative. $-4 + (-5) = -9$

25. $13 + (-6)$ The absolute values are 13 and 6. The difference is $13 - 6$, or 7. The positive number has the larger absolute value, so the answer is positive. $13 + (-6) = 7$

27. $-25 + 25$ A negative and a positive number. The numbers have the same absolute value. The sum is 0. $-25 + 25 = 0$

29. $53 + (-18)$ The absolute values are 53 and 18. The difference is $53 - 18$, or 35. The positive number has the larger absolute value, so the answer is positive. $53 + (-18) = 35$

31. $-8.5 + 4.7$ The absolute values are 8.5 and 4.7. The difference is $8.5 - 4.7$, or 3.8. The negative number has the larger absolute value, so the answer is negative. $-8.5 + 4.7 = -3.8$

33. $-2.8 + (-5.3)$ Two negatives. Add the absolute values, getting 8.1. Make the answer negative. $-2.8 + (-5.3) = -8.1$

35. $-\dfrac{3}{5} + \dfrac{2}{5}$ The absolute values are $\dfrac{3}{5}$ and $\dfrac{2}{5}$. The difference is $\dfrac{3}{5} - \dfrac{2}{5}$, or $\dfrac{1}{5}$. The negative number has the larger absolute value, so the answer is negative. $-\dfrac{3}{5} + \dfrac{2}{5} = -\dfrac{1}{5}$

37. $-\dfrac{2}{9} + \left(-\dfrac{5}{9}\right)$ Two negatives. Add the absolute values, getting $\dfrac{7}{9}$. Make the answer negative. $-\dfrac{2}{9} + \left(-\dfrac{5}{9}\right) = -\dfrac{7}{9}$

39. $-\dfrac{5}{8} + \dfrac{1}{4}$ The absolute values are $\dfrac{5}{8}$ and $\dfrac{1}{4}$. The difference is $\dfrac{5}{8} - \dfrac{2}{8}$, or $\dfrac{3}{8}$. The negative number has the larger absolute value, so the answer is negative. $-\dfrac{5}{8} + \dfrac{1}{4} = -\dfrac{3}{8}$

41. $-\dfrac{5}{8} + \left(-\dfrac{1}{6} \right)$ Two negatives. Add the absolute values,

getting $\dfrac{15}{24} + \dfrac{4}{24}$, or $\dfrac{19}{24}$. Make the answer negative.

$$-\dfrac{5}{8} + \left(-\dfrac{1}{6} \right) = -\dfrac{19}{24}$$

43. $-\dfrac{3}{8} + \dfrac{5}{12}$ The absolute values are $\dfrac{3}{8}$ and $\dfrac{5}{12}$. The difference

is $\dfrac{10}{24} - \dfrac{9}{24}$, or $\dfrac{1}{24}$. The positive number has the larger

absolute value, so the answer is positive. $-\dfrac{3}{8} + \dfrac{5}{12} = \dfrac{1}{24}$

45. $76 + (-15) + (-18) + (-6)$

     a) Add the negative numbers: $-15 + (-18) + (-6) = -39$

     b) Add the results: $76 + (-39) = 37$

47. $-44 + \left(-\dfrac{3}{8} \right) + 95 + \left(-\dfrac{5}{8} \right)$

     a) Add the negative numbers: $-44 + \left(-\dfrac{3}{8} \right) + \left(-\dfrac{5}{8} \right) = -45$

     b) Add the results: $-45 + 95 = 50$

49. We add from left to right.

$$
\begin{aligned}
& 98 + (-54) + 113 + (-998) + 44 + (-612) \\
=\ & \quad\ 44 \quad + 113 + (-998) + 44 + (-612) \\
=\ & \qquad\qquad 157 + (-998) + 44 + (-612) \\
=\ & \qquad\qquad\qquad -841 \quad + 44 + (-612) \\
=\ & \qquad\qquad\qquad\qquad\quad -797 + (-612) \\
=\ & \qquad\qquad\qquad\qquad\qquad\qquad -1409
\end{aligned}
$$

51. The additive inverse of 24 is -24 because $24 + (-24) = 0$.

53. The additive inverse of -26.9 is 26.9 because $-26.9 + 26.9 = 0$.

55. If $x = 8$, then $-x = -8$. (The opposite of 8 is -8.)

57. If $x = -\dfrac{13}{8}$ then $-x = -\left(-\dfrac{13}{8} \right) = \dfrac{13}{8}$. (The opposite of $-\dfrac{13}{8}$ is $\dfrac{13}{8}$.)

59. If $x = -43$ then $-(-x) = -(-(-43)) = -43$. (The opposite of the opposite of -43 is -43.)

61. If $x = \dfrac{4}{3}$ then $-(-x) = -\left(-\dfrac{4}{3} \right) = \dfrac{4}{3}$. (The opposite of the opposite of $\dfrac{4}{3}$ is $\dfrac{4}{3}$.)

63. $-(-24) = 24$ (The opposite of -24 is 24.)

65. $-\left(-\dfrac{3}{8} \right) = \dfrac{3}{8}$ (The opposite of $-\dfrac{3}{8}$ is $\dfrac{3}{8}$.)

67. Let $E =$ the elevation of Mauna Kea above sea level.

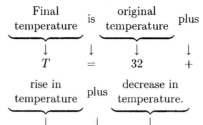

We carry out the addition.

$$E = 33,480 + (-19,684) = 13,796$$

The elevation of Mauna Kea is 13,796 ft above sea level.

69. Let $T =$ the final temperature. We will express the rise in temperature as a positive number and a decrease in the temperature as a negative number.

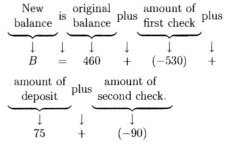

We add from left to right.

$$
\begin{aligned}
T &= 32 + 15 + (-50) \\
&= 47 + (-50) \\
&= -3
\end{aligned}
$$

The final temperature was $-3°$F.

71. Let $S =$ the sum of the profits and losses. We add the five numbers in the bar graph to find S.

$S = \$10,500 + (-\$16,600) + (-\$12,800) + (-\$9600) + \$8200 = -\$20,300$

The sum of the profits and losses is $-\$20,300$.

73. Let $B =$ the new balance in the account. We will express the deposit as a positive number and the amounts of the checks as negative numbers.

New balance	is	original balance	plus	amount of first check	plus
↓	↓	↓	↓	↓	↓
B	$=$	460	$+$	(-530)	$+$

amount of deposit	plus	amount of second check.
↓	↓	↓
75	$+$	(-90)

We add from left to right.

$$
\begin{aligned}
B &= 460 + (-530) + 75 + (-90) \\
&= -70 + 75 + (-90) \\
&= 5 + (-90) \\
&= -85
\end{aligned}
$$

The balance in the account is $-\$85$.

75. Discussion and Writing Exercise

77. Round 87,452 to the nearest ten.

$$8\ 7,\ 4\ 5\ \boxed{2}$$
$$\uparrow$$

The digit 5 is in the tens place. Consider the next digit to the right. Since the digit, 2, is 4 or lower, round down, meaning that 5 tens stays as 5 tens. Then change the digit to the right of the tens digit to zero.

The answer is 87,450.

79. Round 87,452 to the nearest thousand.

8 7, $\boxed{4}$ 5 2

↑

The digit 7 is in the thousands place. Consider the next digit to the right. Since the digit, 4, is 4 or lower, round down, meaning that 7 thousands stays as 7 thousands. Then change all the digits to the right of the thousands digit to zeros.

The answer is 87,000.

81. $\dfrac{9}{4} - \dfrac{5}{6}$ LCM is 12.

$= \dfrac{9}{4} \cdot \dfrac{3}{3} - \dfrac{5}{6} \cdot \dfrac{2}{2}$

$= \dfrac{27}{12} - \dfrac{10}{12}$

$= \dfrac{17}{12}$

83. When x is positive, the opposite of x, $-x$, is negative, so $-x$ is negative for all positive numbers x.

85. If a is positive, $-a$ is negative. Thus $-a + b$, the sum of two negative numbers, is negative. The correct answer is (b).

Exercise Set 6.4

1. $2 - 9 = 2 + (-9) = -7$

3. $0 - 4 = 0 + (-4) = -4$

5. $-8 - (-2) = -8 + 2 = -6$

7. $-11 - (-11) = -11 + 11 = 0$

9. $12 - 16 = 12 + (-16) = -4$

11. $20 - 27 = 20 + (-27) = -7$

13. $-9 - (-3) = -9 + 3 = -6$

15. $-40 - (-40) = -40 + 40 = 0$

17. $7 - 7 = 7 + (-7) = 0$

19. $7 - (-7) = 7 + 7 = 14$

21. $8 - (-3) = 8 + 3 = 11$

23. $-6 - 8 = -6 + (-8) = -14$

25. $-4 - (-9) = -4 + 9 = 5$

27. $1 - 8 = 1 + (-8) = -7$

29. $-6 - (-5) = -6 + 5 = -1$

31. $8 - (-10) = 8 + 10 = 18$

33. $0 - 10 = 0 + (-10) = -10$

35. $-5 - (-2) = -5 + 2 = -3$

37. $-7 - 14 = -7 + (-14) = -21$

39. $0 - (-5) = 0 + 5 = 5$

41. $-8 - 0 = -8 + 0 = -8$

43. $7 - (-5) = 7 + 5 = 12$

45. $2 - 25 = 2 + (-25) = -23$

47. $-42 - 26 = -42 + (-26) = -68$

49. $-71 - 2 = -71 + (-2) = -73$

51. $24 - (-92) = 24 + 92 = 116$

53. $-50 - (-50) = -50 + 50 = 0$

55. $-\dfrac{3}{8} - \dfrac{5}{8} = -\dfrac{3}{8} + \left(-\dfrac{5}{8}\right) = -\dfrac{8}{8} = -1$

57. $\dfrac{3}{4} - \dfrac{2}{3} = \dfrac{3}{4} + \left(-\dfrac{2}{3}\right) = \dfrac{9}{12} + \left(-\dfrac{8}{12}\right) = \dfrac{1}{12}$

59. $-\dfrac{3}{4} - \dfrac{2}{3} = -\dfrac{3}{4} + \left(-\dfrac{2}{3}\right) = -\dfrac{9}{12} + \left(-\dfrac{8}{12}\right) = -\dfrac{17}{12}$

61. $-\dfrac{5}{8} - \left(-\dfrac{3}{4}\right) = -\dfrac{5}{8} + \dfrac{3}{4} = -\dfrac{5}{8} + \dfrac{6}{8} = \dfrac{1}{8}$

63. $6.1 - (-13.8) = 6.1 + 13.8 = 19.9$

65. $-2.7 - 5.9 = -2.7 + (-5.9) = -8.6$

67. $0.99 - 1 = 0.99 + (-1) = -0.01$

69. $-79 - 114 = -79 + (-114) = -193$

71. $0 - (-500) = 0 + 500 = 500$

73. $-2.8 - 0 = -2.8 + 0 = -2.8$

75. $7 - 10.53 = 7 + (-10.53) = -3.53$

77. $\dfrac{1}{6} - \dfrac{2}{3} = \dfrac{1}{6} + \left(-\dfrac{2}{3}\right) = \dfrac{1}{6} + \left(-\dfrac{4}{6}\right) = -\dfrac{3}{6}$, or $-\dfrac{1}{2}$

79. $-\dfrac{4}{7} - \left(-\dfrac{10}{7}\right) = -\dfrac{4}{7} + \dfrac{10}{7} = \dfrac{6}{7}$

81. $-\dfrac{7}{10} - \dfrac{10}{15} = -\dfrac{7}{10} + \left(-\dfrac{10}{15}\right) = -\dfrac{21}{30} + \left(-\dfrac{20}{30}\right) = -\dfrac{41}{30}$

83. $\dfrac{1}{5} - \dfrac{1}{3} = \dfrac{1}{5} + \left(-\dfrac{1}{3}\right) = \dfrac{3}{15} + \left(-\dfrac{5}{15}\right) = -\dfrac{2}{15}$

85. $18 - (-15) - 3 - (-5) + 2 = 18 + 15 + (-3) + 5 + 2 = 37$

87. $-31 + (-28) - (-14) - 17 = (-31) + (-28) + 14 + (-17) = -62$

89. $-34 - 28 + (-33) - 44 = (-34) + (-28) + (-33) + (-44) = -139$

91. $-93 - (-84) - 41 - (-56) = (-93) + 84 + (-41) + 56 = 6$

93. $-5 - (-30) + 30 + 40 - (-12) = (-5) + 30 + 30 + 40 + 12 = 107$

95. $132 - (-21) + 45 - (-21) = 132 + 21 + 45 + 21 = 219$

97. Let D = the difference in elevation.

Difference in elevation | is | larger depth | minus | smaller depth.

$$D = 11,033 - 8648$$

We carry out the subtraction.

$$D = 11,033 - 8648 = 2385$$

The difference in elevation is 2385 m.

99. Let A = the amount owed.

Amount owed | is | amount of charge | minus | amount of return.

$$A = 476.89 - 128.95$$

We subtract.

$$A = 476.89 - 128.95 = 347.94$$

Laura owes \$347.94.

101. a) We subtract the number of home runs allowed from the number of home runs hit.

Home run differential $= 197 - 120 = 77$

b) We subtract the number of home runs allowed from the number of home runs hit.

Home run differential $= 153 - 194 = -41$

103. Let D = the difference in elevation.

Difference in elevation | is | higher elevation | minus | lower elevation.

$$D = -132 - (-515)$$

We carry out the subtraction.

$$D = -132 - (-515) = -132 + 515 = 383$$

Lake Assal is 383 ft lower than the Valdes Peninsula.

105. Discussion and Writing Exercise

107. $\dfrac{3}{4} \cdot \dfrac{8}{9} = \dfrac{3 \cdot 8}{4 \cdot 9}$

$\phantom{\dfrac{3}{4} \cdot \dfrac{8}{9}} = \dfrac{3 \cdot 2 \cdot 4}{4 \cdot 3 \cdot 3}$

$\phantom{\dfrac{3}{4} \cdot \dfrac{8}{9}} = \dfrac{3 \cdot 4}{3 \cdot 4} \cdot \dfrac{2}{3}$

$\phantom{\dfrac{3}{4} \cdot \dfrac{8}{9}} = \dfrac{2}{3}$

109. $\dfrac{15}{16} \cdot \dfrac{6}{25} = \dfrac{15 \cdot 6}{16 \cdot 25}$

$\phantom{\dfrac{15}{16} \cdot \dfrac{6}{25}} = \dfrac{3 \cdot 5 \cdot 2 \cdot 3}{2 \cdot 8 \cdot 5 \cdot 5}$

$\phantom{\dfrac{15}{16} \cdot \dfrac{6}{25}} = \dfrac{2 \cdot 5}{2 \cdot 5} \cdot \dfrac{3 \cdot 3}{8 \cdot 5}$

$\phantom{\dfrac{15}{16} \cdot \dfrac{6}{25}} = \dfrac{9}{40}$

111. $\dfrac{18}{25} = \dfrac{18}{25} \cdot \dfrac{4}{4} = \dfrac{72}{100} = 72\%$

113. Use a calculator.

$$123,907 - 433,789 = -309,882$$

115. False. $3 - 0 = 3, 0 - 3 = -3, 3 - 0 \neq 0 - 3$

117. True

119. True by definition of opposites.

121. a) We add the values assigned to the cards.

$-1 + (-1) + 1 + 1 + 1 + (-1) + (-1) +$
$0 + (-1) + (-1) + 1 = -2$

b) Yes, because the final count is negative.

Exercise Set 6.5

1. -8

3. -48

5. -24

7. -72

9. 16

11. 42

13. -120

15. -238

17. 1200

19. 98

21. -72

23. -12.4

25. 30

27. 21.7

29. $\dfrac{2}{3} \cdot \left(-\dfrac{3}{5}\right) = -\left(\dfrac{2 \cdot 3}{3 \cdot 5}\right) = -\left(\dfrac{2}{5} \cdot \dfrac{3}{3}\right) = -\dfrac{2}{5}$

31. $-\dfrac{3}{8} \cdot \left(-\dfrac{2}{9}\right) = \dfrac{3 \cdot 2}{8 \cdot 9} = \dfrac{3 \cdot 2 \cdot 1}{4 \cdot 2 \cdot 3 \cdot 3} = \dfrac{3 \cdot 2}{3 \cdot 2} \cdot \dfrac{1}{4 \cdot 3} = \dfrac{1}{12}$

33. -17.01

35. $-\dfrac{5}{9} \cdot \dfrac{3}{4} = -\left(\dfrac{5 \cdot 3}{9 \cdot 4}\right) = -\dfrac{5 \cdot 3}{3 \cdot 3 \cdot 4} = -\dfrac{5}{3 \cdot 4} \cdot \dfrac{3}{3} = -\dfrac{5}{12}$

37. $7 \cdot (-4) \cdot (-3) \cdot 5 = 7 \cdot 12 \cdot 5 = 7 \cdot 60 = 420$

39. $-\dfrac{2}{3} \cdot \dfrac{1}{2} \cdot \left(-\dfrac{6}{7}\right) = -\dfrac{2}{6} \cdot \left(-\dfrac{6}{7}\right) = \dfrac{2 \cdot 6}{7 \cdot 6} = \dfrac{2}{7} \cdot \dfrac{6}{6} = \dfrac{2}{7}$

41. $-3 \cdot (-4) \cdot (-5) = 12 \cdot (-5) = -60$

43. $-2 \cdot (-5) \cdot (-3) \cdot (-5) = 10 \cdot 15 = 150$

45. $-\dfrac{2}{45}$

47. $-7 \cdot (-21) \cdot 13 = 147 \cdot 13 = 1911$

49. $-4 \cdot (-1.8) \cdot 7 = (7.2) \cdot 7 = 50.4$

51. $-\dfrac{1}{9} \cdot \left(-\dfrac{2}{3}\right) \cdot \left(\dfrac{5}{7}\right) = \dfrac{2}{27} \cdot \dfrac{5}{7} = \dfrac{10}{189}$

53. $4 \cdot (-4) \cdot (-5) \cdot (-12) = -16 \cdot (60) = -960$

55. $0.07 \cdot (-7) \cdot 6 \cdot (-6) = 0.07 \cdot 6 \cdot (-7) \cdot (-6) = 0.42 \cdot (42) = 17.64$

57. $\left(-\dfrac{5}{6}\right)\left(\dfrac{1}{8}\right)\left(-\dfrac{3}{7}\right)\left(-\dfrac{1}{7}\right) = \left(-\dfrac{5}{48}\right)\left(\dfrac{3}{49}\right) = -\dfrac{5 \cdot 3}{16 \cdot 3 \cdot 49} =$
$-\dfrac{5}{16 \cdot 49} \cdot \dfrac{3}{3} = -\dfrac{5}{784}$

59. 0, The product of 0 and any real number is 0.

61. $(-8)(-9)(-10) = 72(-10) = -720$

63. $(-6)(-7)(-8)(-9)(-10) = 42 \cdot 72 \cdot (-10) = 3024 \cdot (-10) = -30,240$

65. $\begin{aligned}(-3x)^2 &= (-3 \cdot 7)^2 &&\text{Substituting}\\ &= (-21)^2 &&\text{Multiplying inside the}\\ & &&\text{parentheses}\\ &= (-21)(-21) &&\text{Evaluating the power}\\ &= 441\end{aligned}$

$\begin{aligned}-3x^2 &= -3(7)^2 &&\text{Substituting}\\ &= -3 \cdot 49 &&\text{Evaluating the power}\\ &= -147\end{aligned}$

67. When $x = 2$: $\begin{aligned}5x^2 &= 5(2)^2 &&\text{Substituting}\\ &= 5 \cdot 4 &&\text{Evaluating the power}\\ &= 20\end{aligned}$

When $x = -2$: $\begin{aligned}5x^2 &= 5(-2)^2 &&\text{Substituting}\\ &= 5 \cdot 4 &&\text{Evaluating the power}\\ &= 20\end{aligned}$

69. Let $w =$ the total weight change. Since Dave's weight decreases 2 lb each week for 10 weeks we have
$$w = 10 \cdot (-2) = -20.$$
Thus, the total weight change is -20 lb.

71. This is a multistep problem. First we find the number of degrees the temperature dropped. Since it dropped 3°C each minute for 18 minutes we have a drop d given by
$$d = 18 \cdot (-3) = -54.$$
Now let $T =$ the temperature at 10:18 AM.
$$T = 0 + (-54) = -54$$
The temperature was -54°C at 10:18 AM.

73. This is a multistep problem. First we find the total decrease in price. Since it decreased \$1.38 each hour for 8 hours we have a decrease in price d given by
$$d = 8(-\$1.38) = -\$11.04.$$
Now let $P =$ the price of the stock after 8 hours.
$$P = \$23.75 + (-\$11.04) = \$12.71$$
After 8 hours the price of the stock was \$12.71.

75. This is a multistep problem. First we find the total distance the diver rises. Since the diver rises 7 meters each minute for 9 minutes, the total distance d the diver rises is given by
$$d = 9 \cdot 7 = 63.$$

Now let $E =$ the diver's elevation after 9 minutes.
$$E = -95 + 63 = -32$$
The diver's elevation is -32 m, or 32 m below the surface.

77. Discussion and Writing Exercise

79. $\begin{aligned}12 \div \dfrac{3}{4} &= 12 \cdot \dfrac{4}{3}\\ &= \dfrac{12 \cdot 4}{3}\\ &= \dfrac{3 \cdot 4 \cdot 4}{3 \cdot 1}\\ &= \dfrac{3}{3} \cdot \dfrac{4 \cdot 4}{1}\\ &= 16\end{aligned}$

81. $\begin{aligned}\dfrac{5}{6} \div 30 &= \dfrac{5}{6} \cdot \dfrac{1}{30}\\ &= \dfrac{5 \cdot 1}{6 \cdot 30}\\ &= \dfrac{5 \cdot 1}{6 \cdot 5 \cdot 6}\\ &= \dfrac{5}{5} \cdot \dfrac{1}{6 \cdot 6}\\ &= \dfrac{1}{36}\end{aligned}$

83. Move the decimal point 2 places to the right and write a percent symbol.
$$1.4 = 140\%$$

85. If a is positive and b is negative, then ab is negative and thus $-ab$ is positive. The correct answer is (a).

87. To locate $2x$, start at 0 and measure off two adjacent lengths of x to the right of 0.

To locate $3x$, start at 0 and measure off three adjacent lengths of x to the right of 0.

To locate $2y$, start at 0 and measure off two adjacent lengths of y to the right of 0.

To locate $-x$, start at 0 and measure off the length x to the left of 0.

To locate $-y$, start at 0 and measure off the length y to the left of 0.

To locate $x + y$, start at 0 and measure off the length x to the right of 0 followed by the length y immediately to the right of x. (We could also measure off y followed by x.)

To locate $x - y$, start at 0 and measure off the length x to the right of 0. Then, from that point, measure off the length y going to the left.

To locate $x - 2y$, first locate $x - y$ as described above. Then, from that point, measure off another length y going to the left.

Exercise Set 6.6

1. $48 \div (-6) = -8$ Check: $-8(-6) = 48$

3. $\dfrac{28}{-2} = -14$ Check: $-14(-2) = 28$

5. $\dfrac{-24}{8} = -3$ Check: $-3 \cdot 8 = -24$

7. $\dfrac{-36}{-12} = 3$ Check: $3(-12) = -36$

9. $\dfrac{-72}{9} = -8$ Check: $-8 \cdot 9 = -72$

11. $-100 \div (-50) = 2$ Check: $2(-50) = -100$

13. $-108 \div 9 = -12$ Check: $9(-12) = -108$

15. $\dfrac{200}{-25} = -8$ Check: $-8(-25) = 200$

17. Not defined

19. $\dfrac{-23}{-2} = \dfrac{23}{2}$ Check: $\dfrac{23}{2}(-2) = -23$

21. The reciprocal of $\dfrac{15}{7}$ is $\dfrac{7}{15}$ because $\dfrac{15}{7} \cdot \dfrac{7}{15} = 1$.

23. The reciprocal of $-\dfrac{47}{13}$ is $-\dfrac{13}{47}$ because $\left(-\dfrac{47}{13}\right)\left(-\dfrac{13}{47}\right) = 1$.

25. The reciprocal of 13 is $\dfrac{1}{13}$ because $13 \cdot \dfrac{1}{13} = 1$.

27. The reciprocal of 4.3 is $\dfrac{1}{4.3}$ because $4.3 \cdot \dfrac{1}{4.3} = 1$.

29. The reciprocal of $-\dfrac{1}{7.1}$ is -7.1 because $\left(-\dfrac{1}{7.1}\right)(-7.1) = 1$.

31. The reciprocal of $\dfrac{p}{q}$ is $\dfrac{q}{p}$ because $\dfrac{p}{q} \cdot \dfrac{q}{p} = 1$.

33. The reciprocal of $\dfrac{1}{4y}$ is $4y$ because $\dfrac{1}{4y} \cdot 4y = 1$.

35. The reciprocal of $\dfrac{2a}{3b}$ is $\dfrac{3b}{2a}$ because $\dfrac{2a}{3b} \cdot \dfrac{3b}{2a} = 1$.

37. $4 \cdot \dfrac{1}{17}$

39. $8 \cdot \left(-\dfrac{1}{13}\right)$

41. $13.9 \cdot \left(-\dfrac{1}{1.5}\right)$

43. $x \cdot y$

45. $(3x + 4)\left(\dfrac{1}{5}\right)$

47. $(5a - b)\left(\dfrac{1}{5a + b}\right)$

49. $\dfrac{3}{4} \div \left(-\dfrac{2}{3}\right) = \dfrac{3}{4} \cdot \left(-\dfrac{3}{2}\right) = -\dfrac{9}{8}$

51. $-\dfrac{5}{4} \div \left(-\dfrac{3}{4}\right) = -\dfrac{5}{4} \cdot \left(-\dfrac{4}{3}\right) = \dfrac{20}{12} = \dfrac{5 \cdot 4}{3 \cdot 4} = \dfrac{5}{3}$

53. $-\dfrac{2}{7} \div \left(-\dfrac{4}{9}\right) = -\dfrac{2}{7} \cdot \left(-\dfrac{9}{4}\right) = \dfrac{18}{28} = \dfrac{9 \cdot 2}{14 \cdot 2} = \dfrac{9}{14}$

55. $-\dfrac{3}{8} \div \left(-\dfrac{8}{3}\right) = -\dfrac{3}{8} \cdot \left(-\dfrac{3}{8}\right) = \dfrac{9}{64}$

57. $-6.6 \div 3.3 = -2$ Do the long division. Make the answer negative.

59. $\dfrac{-11}{-13} = \dfrac{11}{13}$ The opposite of a number divided by the opposite of another number is the quotient of the two numbers.

61. $\dfrac{48.6}{-3} = -16.2$ Do the long division. Make the answer negative.

63. $\dfrac{-9}{17 - 17} = \dfrac{-9}{0}$ Division by 0 is not defined.

65. $\dfrac{-4}{54} \approx -0.074 \approx -7.4\%$

67. $\dfrac{97}{229} \approx 0.424 \approx 42.4\%$

69. Discussion and Writing Exercise

71.
```
   0.0 9 0 9 ...
11 )1.0 0 0 0
   9 9
   ‾‾‾‾
     1 0 0
       9 9
       ‾‾‾
         1
```
$\dfrac{1}{11} = 0.\overline{09}$

73. We substitute \$3000 for P, 3%, or 0.03, for r, 1 for n, and 3 for t in the compound interest formula.

$$A = P \cdot \left(1 + \dfrac{r}{n}\right)^{n \cdot t}$$
$$= \$3000 \cdot \left(1 + \dfrac{0.03}{1}\right)^{1 \cdot 3}$$
$$= \$3000(1.03)^3$$
$$\approx \$3278.18$$

75. We find $\dfrac{1}{-10.5}$ using a calculator. If a reciprocal key is available, enter -10.5 and then use the reciprocal key. If a reciprocal key is not available, find $\dfrac{1}{-10.5}$, or $1 \div (-10.5)$. In either case, the result is $-0.\overline{095238}$.

Use the reciprocal key again or enter $\dfrac{1}{\frac{1}{-10.5}}$ to find the reciprocal of the first result. The calculator returns -10.5, the original number.

77. $-a$ is positive and b is negative, so $\dfrac{-a}{b}$ is the quotient of a positive and a negative number and, thus, is negative.

79. a is negative and $-b$ is positive, so $\dfrac{a}{-b}$ is the quotient of a negative number and a positive number and, thus, is negative. Then $-\left(\dfrac{a}{-b}\right)$ is the opposite of a negative number and, thus, is positive.

81. $-a$ and $-b$ are both positive, so $\dfrac{-a}{-b}$ is the quotient of two positive numbers and, thus, is positive. Then $-\left(\dfrac{-a}{-b}\right)$ is the opposite of a positive number and, thus, is negative.

Exercise Set 6.7

1. Note that $5y = 5 \cdot y$. We multiply by 1, using y/y as an equivalent expression for 1:

$$\frac{3}{5} = \frac{3}{5} \cdot 1 = \frac{3}{5} \cdot \frac{y}{y} = \frac{3y}{5y}$$

3. Note that $15x = 3 \cdot 5x$. We multiply by 1, using $5x/5x$ as an equivalent expression for 1:

$$\frac{2}{3} = \frac{2}{3} \cdot 1 = \frac{2}{3} \cdot \frac{5x}{5x} = \frac{10x}{15x}$$

5.
$$-\frac{24a}{16a} = -\frac{3 \cdot 8a}{2 \cdot 8a}$$
$$= -\frac{3}{2} \cdot \frac{8a}{8a}$$
$$= -\frac{3}{2} \cdot 1 \qquad \left(\frac{8a}{8a} = 1\right)$$
$$= -\frac{3}{2} \qquad \text{Identity property of 1}$$

7.
$$-\frac{42ab}{36ab} = -\frac{7 \cdot 6ab}{6 \cdot 6ab}$$
$$= -\frac{7}{6} \cdot \frac{6ab}{6ab}$$
$$= -\frac{7}{6} \cdot 1 \qquad \left(\frac{6ab}{6ab} = 1\right)$$
$$= -\frac{7}{6} \qquad \text{Identity property of 1}$$

9. $8 + y$, commutative law of addition

11. nm, commutative law of multiplication

13. $xy + 9$, commutative law of addition
$9 + yx$, commutative law of multiplication

15. $c + ab$, commutative law of addition
$ba + c$, commutative law of multiplication

17. $(a + b) + 2$, associative law of addition

19. $8(xy)$, associative law of multiplication

21. $a + (b + 3)$, associative law of addition

23. $(3a)b$, associative law of multiplication

25. a) $(a + b) + 2 = a + (b + 2)$, associative law of addition
b) $(a + b) + 2 = (b + a) + 2$, commutative law of addition
c) $(a + b) + 2 = (b + a) + 2$ Using the commutative law first,
$= b + (a + 2)$ then the associative law

There are other correct answers.

27. a) $5 + (v + w) = (5 + v) + w$, associative law of addition
b) $5 + (v + w) = 5 + (w + v)$, commutative law of addition
c) $5 + (v + w) = 5 + (w + v)$ Using the commutative law first,
$= (5 + w) + v$ then the associative law

There are other correct answers.

29. a) $(xy)3 = x(y3)$, associative law of multiplication
b) $(xy)3 = (yx)3$, commutative law of multiplication
c) $(xy)3 = (yx)3$ Using the commutative law first,
$= y(x3)$ then the associative law

There are other correct answers.

31. a) $7(ab) = (7a)b$
b) $7(ab) = (7a)b = b(7a)$
c) $7(ab) = 7(ba) = (7b)a$

There are other correct answers.

33. $2(b + 5) = 2 \cdot b + 2 \cdot 5 = 2b + 10$

35. $7(1 + t) = 7 \cdot 1 + 7 \cdot t = 7 + 7t$

37. $6(5x + 2) = 6 \cdot 5x + 6 \cdot 2 = 30x + 12$

39. $7(x + 4 + 6y) = 7 \cdot x + 7 \cdot 4 + 7 \cdot 6y = 7x + 28 + 42y$

41. $7(x - 3) = 7 \cdot x - 7 \cdot 3 = 7x - 21$

43. $-3(x - 7) = -3 \cdot x - (-3) \cdot 7 = -3x - (-21) = -3x + 21$

45. $\dfrac{2}{3}(b - 6) = \dfrac{2}{3} \cdot b - \dfrac{2}{3} \cdot 6 = \dfrac{2}{3}b - 4$

47. $7.3(x - 2) = 7.3 \cdot x - 7.3 \cdot 2 = 7.3x - 14.6$

49. $-\dfrac{3}{5}(x - y + 10) = -\dfrac{3}{5} \cdot x - \left(-\dfrac{3}{5}\right) \cdot y + \left(-\dfrac{3}{5}\right) \cdot 10 =$
$-\dfrac{3}{5}x - \left(-\dfrac{3}{5}y\right) + (-6) = -\dfrac{3}{5}x + \dfrac{3}{5}y - 6$

51. $-9(-5x - 6y + 8) = -9(-5x) - (-9)6y + (-9)8$
$= 45x - (-54y) + (-72) = 45x + 54y - 72$

53. $-4(x - 3y - 2z) = -4 \cdot x - (-4)3y - (-4)2z$
$= -4x - (-12y) - (-8z) = -4x + 12y + 8z$

55. $3.1(-1.2x + 3.2y - 1.1) = 3.1(-1.2x) + (3.1)3.2y - 3.1(1.1)$
$= -3.72x + 9.92y - 3.41$

57. $4x + 3z$ Parts are separated by plus signs. The terms are $4x$ and $3z$.

59. $7x + 8y - 9z = 7x + 8y + (-9z)$ Separating parts with plus signs

The terms are $7x$, $8y$, and $-9z$.

61. $2x + 4 = 2 \cdot x + 2 \cdot 2 = 2(x + 2)$

63. $30 + 5y = 5 \cdot 6 + 5 \cdot y = 5(6 + y)$

65. $14x + 21y = 7 \cdot 2x + 7 \cdot 3y = 7(2x + 3y)$

67. $5x + 10 + 15y = 5 \cdot x + 5 \cdot 2 + 5 \cdot 3y = 5(x + 2 + 3y)$

69. $8x - 24 = 8 \cdot x - 8 \cdot 3 = 8(x - 3)$

71. $32 - 4y = 4 \cdot 8 - 4 \cdot y = 4(8 - y)$

73. $8x + 10y - 22 = 2 \cdot 4x + 2 \cdot 5y - 2 \cdot 11 = 2(4x + 5y - 11)$

75. $ax - a = a \cdot x - a \cdot 1 = a(x - 1)$

77. $ax - ay - az = a \cdot x - a \cdot y - a \cdot z = a(x - y - z)$

79. $18x - 12y + 6 = 6 \cdot 3x - 6 \cdot 2y + 6 \cdot 1 = 6(3x - 2y + 1)$

81. $\dfrac{2}{3}x - \dfrac{5}{3}y + \dfrac{1}{3} = \dfrac{1}{3} \cdot 2x - \dfrac{1}{3} \cdot 5y + \dfrac{1}{3} \cdot 1 =$

$\dfrac{1}{3}(2x - 5y + 1)$

83. $9a + 10a = (9 + 10)a = 19a$

85. $10a - a = 10a - 1 \cdot a = (10 - 1)a = 9a$

87. $2x + 9z + 6x = 2x + 6x + 9z = (2 + 6)x + 9z = 8x + 9z$

89. $7x + 6y^2 + 9y^2 = 7x + (6 + 9)y^2 = 7x + 15y^2$

91. $41a + 90 - 60a - 2 = 41a - 60a + 90 - 2$
$= (41 - 60)a + (90 - 2)$
$= -19a + 88$

93. $23 + 5t + 7y - t - y - 27$
$= 23 - 27 + 5t - 1 \cdot t + 7y - 1 \cdot y$
$= (23 - 27) + (5 - 1)t + (7 - 1)y$
$= -4 + 4t + 6y$, or $4t + 6y - 4$

95. $\dfrac{1}{2}b + \dfrac{1}{2}b = \left(\dfrac{1}{2} + \dfrac{1}{2}\right)b = 1b = b$

97. $2y + \dfrac{1}{4}y + y = 2y + \dfrac{1}{4}y + 1 \cdot y = \left(2 + \dfrac{1}{4} + 1\right)y = 3\dfrac{1}{4}y$, or
$\dfrac{13}{4}y$

99. $11x - 3x = (11 - 3)x = 8x$

101. $6n - n = (6 - 1)n = 5n$

103. $y - 17y = (1 - 17)y = -16y$

105. $-8 + 11a - 5b + 6a - 7b + 7$
$= 11a + 6a - 5b - 7b - 8 + 7$
$= (11 + 6)a + (-5 - 7)b + (-8 + 7)$
$= 17a - 12b - 1$

107. $9x + 2y - 5x = (9 - 5)x + 2y = 4x + 2y$

109. $11x + 2y - 4x - y = (11 - 4)x + (2 - 1)y = 7x + y$

111. $2.7x + 2.3y - 1.9x - 1.8y = (2.7 - 1.9)x + (2.3 - 1.8)y = 0.8x + 0.5y$

113. $\dfrac{13}{2}a + \dfrac{9}{5}b - \dfrac{2}{3}a - \dfrac{3}{10}b - 42$

$= \left(\dfrac{13}{2} - \dfrac{2}{3}\right)a + \left(\dfrac{9}{5} - \dfrac{3}{10}\right)b - 42$

$= \left(\dfrac{39}{6} - \dfrac{4}{6}\right)a + \left(\dfrac{18}{10} - \dfrac{3}{10}\right)b - 42$

$= \dfrac{35}{6}a + \dfrac{15}{10}b - 42$

$= \dfrac{35}{6}a + \dfrac{3}{2}b - 42$

115. Discussion and Writing Exercise

117. $x + 27 = 36$
$x + 27 - 27 = 36 - 27$ Subtracting 27 on both sides
$x + 0 = 9$
$x = 9$

The solution is 9.

119. $4^3 = 4 \cdot 4 \cdot 4 = 64$

121. $15^2 = 15 \cdot 15 = 225$

123. No; for any replacement other than 5 the two expressions do not have the same value. For example, let $t = 2$. Then $3 \cdot 2 + 5 = 6 + 5 = 11$, but $3 \cdot 5 + 2 = 15 + 2 = 17$.

125. Yes; commutative law of addition

127. $q + qr + qrs + qrst$ There are no like terms.
$= q \cdot 1 + q \cdot r + q \cdot rs + q \cdot rst$
$= q(1 + r + rs + rst)$ Factoring

Exercise Set 6.8

1. $-(2x + 7) = -2x - 7$ Changing the sign of each term

3. $-(8 - x) = -8 + x$ Changing the sign of each term

5. $-4a + 3b - 7c$

7. $-6x + 8y - 5$

9. $-3x + 5y + 6$

11. $8x + 6y + 43$

13. $9x - (4x + 3) = 9x - 4x - 3$ Removing parentheses by changing the sign of every term
$= 5x - 3$ Collecting like terms

15. $2a - (5a - 9) = 2a - 5a + 9 = -3a + 9$

17. $2x + 7x - (4x + 6) = 2x + 7x - 4x - 6 = 5x - 6$

19. $2x - 4y - 3(7x - 2y) = 2x - 4y - 21x + 6y = -19x + 2y$

21. $15x - y - 5(3x - 2y + 5z)$
$= 15x - y - 15x + 10y - 25z$ Multiplying each term in parentheses by -5
$= 9y - 25z$

23. $(3x + 2y) - 2(5x - 4y) = 3x + 2y - 10x + 8y = -7x + 10y$

25. $\quad (12a - 3b + 5c) - 5(-5a + 4b - 6c)$
$= 12a - 3b + 5c + 25a - 20b + 30c$
$= 37a - 23b + 35c$

27. $[9 - 2(5 - 4)] = [9 - 2 \cdot 1] \qquad$ Computing $5 - 4$
$\qquad\qquad\qquad = [9 - 2] \qquad$ Computing $2 \cdot 1$
$\qquad\qquad\qquad = 7$

29. $8[7 - 6(4 - 2)] = 8[7 - 6(2)] = 8[7 - 12] = 8[-5] = -40$

31. $\quad [4(9 - 6) + 11] - [14 - (6 + 4)]$
$= [4(3) + 11] - [14 - 10]$
$= [12 + 11] - [14 - 10]$
$= 23 - 4$
$= 19$

33. $\quad [10(x + 3) - 4] + [2(x - 1) + 6]$
$= [10x + 30 - 4] + [2x - 2 + 6]$
$= [10x + 26] + [2x + 4]$
$= 10x + 26 + 2x + 4$
$= 12x + 30$

35. $\quad [7(x + 5) - 19] - [4(x - 6) + 10]$
$= [7x + 35 - 19] - [4x - 24 + 10]$
$= [7x + 16] - [4x - 14]$
$= 7x + 16 - 4x + 14$
$= 3x + 30$

37. $\quad 3\{[7(x - 2) + 4] - [2(2x - 5) + 6]\}$
$= 3\{[7x - 14 + 4] - [4x - 10 + 6]\}$
$= 3\{[7x - 10] - [4x - 4]\}$
$= 3\{7x - 10 - 4x + 4\}$
$= 3\{3x - 6\}$
$= 9x - 18$

39. $\quad 4\{[5(x - 3) + 2] - 3[2(x + 5) - 9]\}$
$= 4\{[5x - 15 + 2] - 3[2x + 10 - 9]\}$
$= 4\{[5x - 13] - 3[2x + 1]\}$
$= 4\{5x - 13 - 6x - 3\}$
$= 4\{-x - 16\}$
$= -4x - 64$

41. $8 - 2 \cdot 3 - 9 = 8 - 6 - 9$ Multiplying
$\qquad\qquad\qquad = 2 - 9 \qquad$ Doing all additions and subtractions in order from
$\qquad\qquad\qquad = -7 \qquad$ left to right

43. $(8 - 2 \cdot 3) - 9 = (8 - 6) - 9$ Multiplying inside the parentheses
$\qquad\qquad\qquad = 2 - 9 \qquad$ Subtracting inside the parentheses
$\qquad\qquad\qquad = -7$

45. $[(-24) \div (-3)] \div \left(-\dfrac{1}{2}\right) = 8 \div \left(-\dfrac{1}{2}\right) = 8 \cdot (-2) = -16$

47. $16 \cdot (-24) + 50 = -384 + 50 = -334$

49. $2^4 + 2^3 - 10 = 16 + 8 - 10 = 24 - 10 = 14$

51. $5^3 + 26 \cdot 71 - (16 + 25 \cdot 3) = 5^3 + 26 \cdot 71 - (16 + 75) =$
$5^3 + 26 \cdot 71 - 91 = 125 + 26 \cdot 71 - 91 = 125 + 1846 - 91 =$
$1971 - 91 = 1880$

53. $4 \cdot 5 - 2 \cdot 6 + 4 = 20 - 12 + 4 = 8 + 4 = 12$

55. $4^3/8 = 64/8 = 8$

57. $8(-7) + 6(-5) = -56 - 30 = -86$

59. $19 - 5(-3) + 3 = 19 + 15 + 3 = 34 + 3 = 37$

61. $9 \div (-3) + 16 \div 8 = -3 + 2 = -1$

63. $-4^2 + 6 = -16 + 6 = -10$

65. $-8^2 - 3 = -64 - 3 = -67$

67. $12 - 20^3 = 12 - 8000 = -7988$

69. $2 \cdot 10^3 - 5000 = 2 \cdot 1000 - 5000 = 2000 - 5000 = -3000$

71. $6[9 - (3 - 4)] = 6[9 - (-1)] = 6[9 + 1] = 6[10] = 60$

73. $-1000 \div (-100) \div 10 = 10 \div 10 = 1$

75. $8 - (7 - 9) = 8 - (-2) = 8 + 2 = 10$

77. $\dfrac{10 - 6^2}{9^2 + 3^2} = \dfrac{10 - 36}{81 + 9} = \dfrac{-26}{90} = -\dfrac{13}{45}$

79. $\dfrac{3(6 - 7) - 5 \cdot 4}{6 \cdot 7 - 8(4 - 1)} = \dfrac{3(-1) - 5 \cdot 4}{42 - 8 \cdot 3} = \dfrac{-3 - 20}{42 - 24} = -\dfrac{23}{18}$

81. $\dfrac{2^3 - 3^2 + 12 \cdot 5}{-32 \div (-16) \div (-4)} = \dfrac{8 - 9 + 12 \cdot 5}{-32 \div (-16) \div (-4)} =$
$\dfrac{8 - 9 + 60}{2 \div (-4)} = \dfrac{8 - 9 + 60}{-\frac{1}{2}} = \dfrac{-1 + 60}{-\frac{1}{2}} = \dfrac{59}{-\frac{1}{2}} =$
$59(-2) = -118$

83. Discussion and Writing Exercise

85. $\quad 65\% = \dfrac{65}{100}$
$\qquad\qquad = \dfrac{5 \cdot 13}{5 \cdot 20} = \dfrac{5}{5} \cdot \dfrac{13}{20}$
$\qquad\qquad = \dfrac{13}{20}$

87. $\quad 37.5\% = \dfrac{37.5}{100}$
$\qquad\qquad = \dfrac{37.5}{100} \cdot \dfrac{10}{10}$
$\qquad\qquad = \dfrac{375}{1000}$
$\qquad\qquad = \dfrac{3 \cdot 125}{8 \cdot 125}$
$\qquad\qquad = \dfrac{3}{8} \cdot \dfrac{125}{125}$
$\qquad\qquad = \dfrac{3}{8}$

89. $d = 2 \cdot r = 2 \cdot 15 \text{ yd} = 30 \text{ yd}$

$\quad C = 2 \cdot \pi \cdot r$
$\qquad \approx 2 \times 3.14 \times 15 \text{ yd}$
$\qquad = 94.2 \text{ yd}$

$\quad A = \pi \cdot r \cdot r$
$\qquad \approx 3.14 \times 15 \text{ yd} \times 15 \text{ yd}$
$\qquad = 706.5 \text{ yd}^2$

91. $d = 2 \cdot r = 2 \cdot 9\frac{1}{2}$ mi $= 2 \cdot \frac{19}{2}$ mi $= 19$ mi

$C = 2 \cdot \pi \cdot r$

$\approx 2 \times 3.14 \times 9\frac{1}{2}$ mi

$= 2 \times 3.14 \times \frac{19}{2}$ mi

$= 59.66$ mi

$A = \pi \cdot r \cdot r$

$\approx 3.14 \times 9\frac{1}{2}$ mi $\times 9\frac{1}{2}$ mi

$= 3.14 \times \frac{19}{2}$ mi $\times \frac{19}{2}$ mi

$= 283.385$ mi^2

93. $6y + 2x - 3a + c = 6y - (-2x) - 3a - (-c) = 6y - (-2x + 3a - c)$

95. $6m + 3n - 5m + 4b = 6m - (-3n) - 5m - (-4b) =$
$6m - (-3n + 5m - 4b)$

97. $\{x - [f - (f - x)] + [x - f]\} - 3x$
$= \{x - [f - f + x] + [x - f]\} - 3x$
$= \{x - [x] + [x - f]\} - 3x$
$= \{x - x + x - f\} - 3x = x - f - 3x = -2x - f$

99. a) $x^2 + 3 = 7^2 + 3 = 49 + 3 = 52;$

$x^2 + 3 = (-7)^2 + 3 = 49 + 3 = 52;$

$x^2 + 3 = (-5.013)^2 + 3 = 25.130169 + 3 = 28.130169$

b) $1 - x^2 = 1 - 5^2 = 1 - 25 = -24;$

$1 - x^2 = 1 - (-5)^2 = 1 - 25 = -24;$

$1 - x^2 = 1 - (-10.455)^2 = 1 - 109.307025 =$
-108.307025

101. $\dfrac{-15 + 20 + 50 + (-82) + (-7) + (-2)}{6} = \dfrac{-36}{6} = -6$

Chapter 7

Solving Equations and Inequalities

Exercise Set 7.1

1. $\underline{x + 17 = 32}$ Writing the equation

   $15 + 17 \ ? \ 32$ Substituting 15 for x
   $32 \ |$ TRUE

   Since the left-hand and right-hand sides are the same, 15 is a solution of the equation.

3. $\underline{x - 7 = 12}$ Writing the equation

   $21 - 7 \ ? \ 12$ Substituting 21 for x
   $14 \ |$ FALSE

   Since the left-hand and right-hand sides are not the same, 21 is not a solution of the equation.

5. $\underline{6x = 54}$ Writing the equation

   $6(-7) \ ? \ 54$ Substituting
   $-42 \ |$ FALSE

   -7 is not a solution of the equation.

7. $\underline{\dfrac{x}{6} = 5}$ Writing the equation

   $\dfrac{30}{6} \ ? \ 5$ Substituting
   $5 \ |$ TRUE

   5 is a solution of the equation.

9. $\underline{5x + 7 = 107}$

   $5 \cdot 19 + 7 \ ? \ 107$ Substituting
   $95 + 7 \ |$
   $102 \ |$ FALSE

   19 is not a solution of the equation.

11. $\underline{7(y - 1) = 63}$

   $7(-11 - 1) \ ? \ 63$ Substituting
   $7(-12) \ |$
   $-84 \ |$ FALSE

   -11 is not a solution of the equation.

13. $x + 2 = 6$
   $x + 2 - 2 = 6 - 2$ Subtracting 2 on both sides
   $x = 4$ Simplifying

   Check: $\underline{x + 2 = 6}$

   $4 + 2 \ ? \ 6$
   $6 \ |$ TRUE

   The solution is 4.

15. $x + 15 = -5$
   $x + 15 - 15 = -5 - 15$ Subtracting 15 on both sides
   $x = -20$

   Check: $\underline{x + 15 = -5}$

   $-20 + 15 \ ? \ -5$
   $-5 \ |$ TRUE

   The solution is -20.

17. $x + 6 = -8$
   $x + 6 - 6 = -8 - 6$
   $x = -14$

   Check: $\underline{x + 6 = -8}$

   $-14 + 6 \ ? \ -8$
   $-8 \ |$ TRUE

   The solution is -14.

19. $x + 16 = -2$
   $x + 16 - 16 = -2 - 16$
   $x = -18$

   Check: $\underline{x + 16 = -2}$

   $-18 + 16 \ ? \ -2$
   $-2 \ |$ TRUE

   The solution is -18.

21. $x - 9 = 6$
   $x - 9 + 9 = 6 + 9$
   $x = 15$

   Check: $\underline{x - 9 = 6}$

   $15 - 9 \ ? \ 6$
   $6 \ |$ TRUE

   The solution is 15.

23. $x - 7 = -21$
   $x - 7 + 7 = -21 + 7$
   $x = -14$

   Check: $\underline{x - 7 = -21}$

   $-14 - 7 \ ? \ -21$
   $-21 \ |$ TRUE

   The solution is -14.

25. $5 + t = 7$
   $-5 + 5 + t = -5 + 7$
   $t = 2$

   Check: $\underline{5 + t = 7}$

   $5 + 2 \ ? \ 7$
   $7 \ |$ TRUE

   The solution is 2.

27.
$$-7 + y = 13$$
$$7 + (-7) + y = 7 + 13$$
$$y = 20$$

Check:
$$\begin{array}{c|c} -7 + y = 13 \\ \hline -7 + 20 \ ? \ 13 \\ 13 \ \Big| \end{array} \quad \text{TRUE}$$

The solution is 20.

29.
$$-3 + t = -9$$
$$3 + (-3) + t = 3 + (-9)$$
$$t = -6$$

Check:
$$\begin{array}{c|c} -3 + t = -9 \\ \hline -3 + (-6) \ ? \ -9 \\ -9 \ \Big| \end{array} \quad \text{TRUE}$$

The solution is -6.

31.
$$x + \frac{1}{2} = 7$$
$$x + \frac{1}{2} - \frac{1}{2} = 7 - \frac{1}{2}$$
$$x = 6\frac{1}{2}$$

Check:
$$\begin{array}{c|c} x + \frac{1}{2} = 7 \\ \hline 6\frac{1}{2} + \frac{1}{2} \ ? \ 7 \\ 7 \ \Big| \end{array} \quad \text{TRUE}$$

The solution is $6\frac{1}{2}$.

33.
$$12 = a - 7.9$$
$$12 + 7.9 = a - 7.9 + 7.9$$
$$19.9 = a$$

Check:
$$\begin{array}{c|c} 12 = a - 7.9 \\ \hline 12 \ ? \ 19.9 - 7.9 \\ \Big| \ 12 \end{array} \quad \text{TRUE}$$

The solution is 19.9.

35.
$$r + \frac{1}{3} = \frac{8}{3}$$
$$r + \frac{1}{3} - \frac{1}{3} = \frac{8}{3} - \frac{1}{3}$$
$$r = \frac{7}{3}$$

Check:
$$\begin{array}{c|c} r + \frac{1}{3} = \frac{8}{3} \\ \hline \frac{7}{3} + \frac{1}{3} \ ? \ \frac{8}{3} \\ \frac{8}{3} \ \Big| \end{array} \quad \text{TRUE}$$

The solution is $\frac{7}{3}$.

37.
$$m + \frac{5}{6} = -\frac{11}{12}$$
$$m + \frac{5}{6} - \frac{5}{6} = -\frac{11}{12} - \frac{5}{6}$$
$$m = -\frac{11}{12} - \frac{5}{6} \cdot \frac{2}{2}$$
$$m = -\frac{11}{12} - \frac{10}{12}$$
$$m = -\frac{21}{12} = -\frac{\cancel{3} \cdot 7}{\cancel{3} \cdot 4}$$
$$m = -\frac{7}{4}$$

Check:
$$\begin{array}{c|c} m + \frac{5}{6} = -\frac{11}{12} \\ \hline -\frac{7}{4} + \frac{5}{6} \ ? \ -\frac{11}{12} \\ -\frac{21}{12} + \frac{10}{12} \ \Big| \\ -\frac{11}{12} \ \Big| \end{array} \quad \text{TRUE}$$

The solution is $-\frac{7}{4}$.

39.
$$x - \frac{5}{6} = \frac{7}{8}$$
$$x - \frac{5}{6} + \frac{5}{6} = \frac{7}{8} + \frac{5}{6}$$
$$x = \frac{7}{8} \cdot \frac{3}{3} + \frac{5}{6} \cdot \frac{4}{4}$$
$$x = \frac{21}{24} + \frac{20}{24}$$
$$x = \frac{41}{24}$$

Check:
$$\begin{array}{c|c} x - \frac{5}{6} = \frac{7}{8} \\ \hline \frac{41}{24} - \frac{5}{6} \ ? \ \frac{7}{8} \\ \frac{41}{24} - \frac{20}{24} \ \Big| \ \frac{21}{24} \\ \frac{21}{24} \ \Big| \end{array} \quad \text{TRUE}$$

The solution is $\frac{41}{24}$.

41.
$$-\frac{1}{5} + z = -\frac{1}{4}$$
$$\frac{1}{5} - \frac{1}{5} + z = \frac{1}{5} - \frac{1}{4}$$
$$z = \frac{1}{5} \cdot \frac{4}{4} - \frac{1}{4} \cdot \frac{5}{5}$$
$$z = \frac{4}{20} - \frac{5}{20}$$
$$z = -\frac{1}{20}$$

Check: $$-\frac{1}{5} + z = -\frac{1}{4}$$

$$-\frac{1}{5} + \left(-\frac{1}{20}\right) \;?\; -\frac{1}{4}$$

$$-\frac{4}{20} + \left(-\frac{1}{20}\right) \;\Big|\; -\frac{5}{20}$$

$$-\frac{5}{20} \qquad \text{TRUE}$$

The solution is $-\dfrac{1}{20}$.

43. $$x + 2.3 = 7.4$$
$$x + 2.3 - 2.3 = 7.4 - 2.3$$
$$x = 5.1$$

Check: $$x + 2.3 = 7.4$$
$$5.1 + 2.3 \;?\; 7.4$$
$$7.4 \;\Big|\; \qquad \text{TRUE}$$

The solution is 5.1.

45. $$7.6 = x - 4.8$$
$$7.6 + 4.8 = x - 4.8 + 4.8$$
$$12.4 = x$$

Check: $$7.6 = x - 4.8$$
$$7.6 \;?\; 12.4 - 4.8$$
$$\Big|\; 7.6 \qquad \text{TRUE}$$

The solution is 12.4.

47. $$-9.7 = -4.7 + y$$
$$4.7 + (-9.7) = 4.7 + (-4.7) + y$$
$$-5 = y$$

Check: $$-9.7 = -4.7 + y$$
$$-9.7 \;?\; -4.7 + (-5)$$
$$\Big|\; -9.7 \qquad \text{TRUE}$$

The solution is -5.

49. $$5\frac{1}{6} + x = 7$$

$$-5\frac{1}{6} + 5\frac{1}{6} + x = -5\frac{1}{6} + 7$$

$$x = -\frac{31}{6} + \frac{42}{6}$$

$$x = \frac{11}{6}, \text{ or } 1\frac{5}{6}$$

Check: $$5\frac{1}{6} + x = 7$$

$$5\frac{1}{6} + 1\frac{5}{6} \;?\; 7$$

$$7 \;\Big|\; \qquad \text{TRUE}$$

The solution is $\dfrac{11}{6}$, or $1\dfrac{5}{6}$.

51. $$q + \frac{1}{3} = -\frac{1}{7}$$

$$q + \frac{1}{3} - \frac{1}{3} = -\frac{1}{7} - \frac{1}{3}$$

$$q = -\frac{1}{7} \cdot \frac{3}{3} - \frac{1}{3} \cdot \frac{7}{7}$$

$$q = -\frac{3}{21} - \frac{7}{21}$$

$$q = -\frac{10}{21}$$

Check: $$q + \frac{1}{3} = -\frac{1}{7}$$

$$-\frac{10}{21} + \frac{1}{3} \;?\; -\frac{1}{7}$$

$$-\frac{10}{21} + \frac{7}{21} \;\Big|\; -\frac{3}{21}$$

$$-\frac{3}{21} \qquad \text{TRUE}$$

The solution is $-\dfrac{10}{21}$.

53. Discussion and Writing Exercise

55. $-3 + (-8)$ Two negative numbers. We add the absolute values, getting 11, and make the answer negative.

$$-3 + (-8) = -11$$

57. $-\dfrac{2}{3} \cdot \dfrac{5}{8} = -\dfrac{2 \cdot 5}{3 \cdot 8} = -\dfrac{\cancel{2} \cdot 5}{3 \cdot \cancel{2} \cdot 4} = -\dfrac{5}{12}$

59. $\dfrac{2}{3} \div \left(-\dfrac{4}{9}\right) = \dfrac{2}{3} \cdot \left(-\dfrac{9}{4}\right) = -\dfrac{2 \cdot 9}{3 \cdot 4} = -\dfrac{\cancel{2} \cdot \cancel{3} \cdot 3}{\cancel{3} \cdot \cancel{2} \cdot 2} = -\dfrac{3}{2}$

61. $-\dfrac{2}{3} - \left(-\dfrac{5}{8}\right) = -\dfrac{2}{3} + \dfrac{5}{8}$

$$= -\frac{2}{3} \cdot \frac{8}{8} + \frac{5}{8} \cdot \frac{3}{3}$$

$$= -\frac{16}{24} + \frac{15}{24}$$

$$= -\frac{1}{24}$$

63. The translation is $\$83 - x$.

65. $$-356.788 = -699.034 + t$$
$$699.034 + (-356.788) = 699.034 + (-699.034) + t$$
$$342.246 = t$$

The solution is 342.246.

67.
$$x + \frac{4}{5} = -\frac{2}{3} - \frac{4}{15}$$
$$x + \frac{4}{5} = -\frac{2}{3} \cdot \frac{5}{5} - \frac{4}{15} \qquad \text{Adding on the right side}$$
$$x + \frac{4}{5} = -\frac{10}{15} - \frac{4}{15}$$
$$x + \frac{4}{5} = -\frac{14}{15}$$
$$x + \frac{4}{5} - \frac{4}{5} = -\frac{14}{15} - \frac{4}{5}$$
$$x = -\frac{14}{15} - \frac{4}{5} \cdot \frac{3}{3}$$
$$x = -\frac{14}{15} - \frac{12}{15}$$
$$x = -\frac{26}{15}$$

The solution is $-\dfrac{26}{15}$.

69.
$$16 + x - 22 = -16$$
$$x - 6 = -16 \qquad \text{Adding on the left side}$$
$$x - 6 + 6 = -16 + 6$$
$$x = -10$$

The solution is -10.

71.
$$x + 3 = 3 + x$$
$$x + 3 - 3 = 3 + x - 3$$
$$x = x$$

$x = x$ is true for all real numbers. Thus the solution is all real numbers.

73.
$$-\frac{3}{2} + x = -\frac{5}{17} - \frac{3}{2}$$
$$\frac{3}{2} - \frac{3}{2} + x = \frac{3}{2} - \frac{5}{17} - \frac{3}{2}$$
$$x = \left(\frac{3}{2} - \frac{3}{2}\right) - \frac{5}{17}$$
$$x = -\frac{5}{17}$$

The solution is $-\dfrac{5}{17}$.

75.
$$|x| + 6 = 19$$
$$|x| + 6 - 6 = 19 - 6$$
$$|x| = 13$$

x represents a number whose distance from 0 is 13. Thus $x = -13$ or $x = 13$.

The solutions are -13 and 13.

Exercise Set 7.2

1.
$$6x = 36$$
$$\frac{6x}{6} = \frac{36}{6} \qquad \text{Dividing by 6 on both sides}$$
$$1 \cdot x = 6 \qquad \text{Simplifying}$$
$$x = 6 \qquad \text{Identity property of 1}$$

Check: $\dfrac{6x = 36}{}$
$$6 \cdot 6 \ ? \ 36$$
$$36 \ | \qquad \text{TRUE}$$

The solution is 6.

3.
$$5x = 45$$
$$\frac{5x}{5} = \frac{45}{5} \qquad \text{Dividing by 5 on both sides}$$
$$1 \cdot x = 9 \qquad \text{Simplifying}$$
$$x = 9 \qquad \text{Identity property of 1}$$

Check: $\dfrac{5x = 45}{}$
$$5 \cdot 9 \ ? \ 45$$
$$45 \ | \qquad \text{TRUE}$$

The solution is 9.

5.
$$84 = 7x$$
$$\frac{84}{7} = \frac{7x}{7} \qquad \text{Dividing by 7 on both sides}$$
$$12 = 1 \cdot x$$
$$12 = x$$

Check: $\dfrac{84 = 7x}{}$
$$84 \ ? \ 7 \cdot 12$$
$$| \ 84 \qquad \text{TRUE}$$

The solution is 12.

7.
$$-x = 40$$
$$-1 \cdot x = 40$$
$$\frac{-1 \cdot x}{-1} = \frac{40}{-1}$$
$$1 \cdot x = -40$$
$$x = -40$$

Check: $\dfrac{-x = 40}{}$
$$-(-40) \ ? \ 40$$
$$40 \ | \qquad \text{TRUE}$$

The solution is -40.

9.
$$-x = -1$$
$$-1 \cdot x = -1$$
$$\frac{-1 \cdot x}{-1} = \frac{-1}{-1}$$
$$1 \cdot x = 1$$
$$x = 1$$

Check: $\dfrac{-x = -1}{}$
$$-(1) \ ? \ -1$$
$$-1 \ | \qquad \text{TRUE}$$

The solution is 1.

11.
$$7x = -49$$
$$\frac{7x}{7} = \frac{-49}{7}$$
$$1 \cdot x = -7$$
$$x = -7$$

Check: $\dfrac{7x = -49}{}$
$$7(-7) \ ? \ -49$$
$$-49 \ | \qquad \text{TRUE}$$

The solution is -7.

13. $-12x = 72$

$$\frac{-12x}{-12} = \frac{72}{-12}$$

$$1 \cdot x = -6$$

$$x = -6$$

Check: $\dfrac{-12x = 72}{-12(-6) \ ? \ 72}$

$72 \ \Big|$ TRUE

The solution is -6.

15. $-21x = -126$

$$\frac{-21x}{-21} = \frac{-126}{-21}$$

$$1 \cdot x = 6$$

$$x = 6$$

Check: $\dfrac{-21x = -126}{-21 \cdot 6 \ ? \ -126}$

$-126 \ \Big|$ TRUE

The solution is 6.

17. $\dfrac{t}{7} = -9$

$$7 \cdot \frac{1}{7}t = 7 \cdot (-9)$$

$$1 \cdot t = -63$$

$$t = -63$$

Check: $\dfrac{\dfrac{t}{7} = -9}{\dfrac{-63}{7} \ ? \ -9}$

$\phantom{\dfrac{-63}{7} \ ? \ }-9 \ \Big|$ TRUE

The solution is -63.

19. $\dfrac{3}{4}x = 27$

$$\frac{4}{3} \cdot \frac{3}{4}x = \frac{4}{3} \cdot 27$$

$$1 \cdot x = \frac{4 \cdot \cancel{3} \cdot 3 \cdot 3}{\cancel{3} \cdot 1}$$

$$x = 36$$

Check: $\dfrac{\dfrac{3}{4}x = 27}{\dfrac{3}{4} \cdot 36 \ ? \ 27}$

$\phantom{\dfrac{3}{4} \cdot 36 \ ? \ }27 \ \Big|$ TRUE

The solution is 36.

21. $\dfrac{-t}{3} = 7$

$$3 \cdot \frac{1}{3} \cdot (-t) = 3 \cdot 7$$

$$-t = 21$$

$$-1 \cdot (-1 \cdot t) = -1 \cdot 21$$

$$1 \cdot t = -21$$

$$t = -21$$

Check: $\dfrac{\dfrac{-t}{3} = 7}{\dfrac{-(-21)}{3} \ ? \ 7}$

$\phantom{\dfrac{-(-21)}{3} \ ? \ }\dfrac{21}{3}$

$\phantom{\dfrac{-(-21)}{3} \ ? \ }7 \ \Big|$ TRUE

The solution is -21.

23. $-\dfrac{m}{3} = \dfrac{1}{5}$

$$-\frac{1}{3} \cdot m = \frac{1}{5}$$

$$-3 \cdot \left(-\frac{1}{3} \cdot m\right) = -3 \cdot \frac{1}{5}$$

$$m = -\frac{3}{5}$$

Check: $\dfrac{-\dfrac{m}{3} = \dfrac{1}{5}}{}$

$-\dfrac{-\dfrac{3}{5}}{3} \ ? \ \dfrac{1}{5}$

$-\left(-\dfrac{3}{5} \div 3\right)$

$-\left(-\dfrac{3}{5} \cdot \dfrac{1}{3}\right)$

$-\left(-\dfrac{1}{5}\right)$

$\dfrac{1}{5} \ \Big|$ TRUE

The solution is $-\dfrac{3}{5}$.

25. $-\dfrac{3}{5}r = \dfrac{9}{10}$

$$-\frac{5}{3} \cdot \left(-\frac{3}{5}r\right) = -\frac{5}{3} \cdot \frac{9}{10}$$

$$1 \cdot r = -\frac{\cancel{5} \cdot \cancel{3} \cdot 3}{\cancel{3} \cdot \cancel{5} \cdot 2}$$

$$r = -\frac{3}{2}$$

Check: $\dfrac{-\dfrac{3}{5}r = \dfrac{9}{10}}{-\dfrac{3}{5} \cdot \left(-\dfrac{3}{2}\right) \ ? \ \dfrac{9}{10}}$

$\phantom{-\dfrac{3}{5} \cdot \left(-\dfrac{3}{2}\right) \ ? \ }\dfrac{9}{10} \ \Big|$ TRUE

The solution is $-\dfrac{3}{2}$.

27.
$$-\frac{3}{2}r = -\frac{27}{4}$$
$$-\frac{2}{3} \cdot \left(-\frac{3}{2}r\right) = -\frac{2}{3} \cdot \left(-\frac{27}{4}\right)$$
$$1 \cdot r = \frac{\cancel{2} \cdot \cancel{3} \cdot 3 \cdot 3}{\cancel{3} \cdot \cancel{2} \cdot 2}$$
$$r = \frac{9}{2}$$

Check:
$$\begin{array}{c|c} -\dfrac{3}{2}r = -\dfrac{27}{4} \\ \hline -\dfrac{3}{2} \cdot \dfrac{9}{2} \ ? \ -\dfrac{27}{4} \\ -\dfrac{27}{4} \ \Big| \end{array} \quad \text{TRUE}$$

The solution is $\frac{9}{2}$.

29. $6.3x = 44.1$
$$\frac{6.3x}{6.3} = \frac{44.1}{6.3}$$
$$1 \cdot x = 7$$
$$x = 7$$
Check:
$$\begin{array}{c|c} 6.3x = 44.1 \\ \hline 6.3 \cdot 7 \ ? \ 44.1 \\ 44.1 \ \Big| \end{array} \quad \text{TRUE}$$
The solution is 7.

31. $-3.1y = 21.7$
$$\frac{-3.1y}{-3.1} = \frac{21.7}{-3.1}$$
$$1 \cdot y = -7$$
$$y = -7$$
Check:
$$\begin{array}{c|c} 3.1y = 21.7 \\ \hline -3.1(-7) \ ? \ 21.7 \\ 21.7 \ \Big| \end{array} \quad \text{TRUE}$$
The solution is -7.

33. $38.7m = 309.6$
$$\frac{38.7m}{38.7} = \frac{309.6}{38.7}$$
$$1 \cdot m = 8$$
$$m = 8$$
Check:
$$\begin{array}{c|c} 38.7m = 309.6 \\ \hline 38.7 \cdot 8 \ ? \ 309.6 \\ 309.6 \ \Big| \end{array} \quad \text{TRUE}$$
The solution is 8.

35.
$$-\frac{2}{3}y = -10.6$$
$$-\frac{3}{2} \cdot \left(-\frac{2}{3}y\right) = -\frac{3}{2} \cdot (-10.6)$$
$$1 \cdot y = \frac{31.8}{2}$$
$$y = 15.9$$

Check:
$$\begin{array}{c|c} -\dfrac{2}{3}y = -10.6 \\ \hline -\dfrac{2}{3} \cdot (15.9) \ ? \ -10.6 \\ -\dfrac{31.8}{3} \\ -10.6 \ \Big| \end{array} \quad \text{TRUE}$$

The solution is 15.9.

37.
$$\frac{-x}{5} = 10$$
$$5 \cdot \frac{-x}{5} = 5 \cdot 10$$
$$-x = 50$$
$$-1 \cdot (-x) = -1 \cdot 50$$
$$x = -50$$

Check:
$$\begin{array}{c|c} \dfrac{-x}{5} = 10 \\ \hline \dfrac{-(-50)}{5} \ ? \ 10 \\ \dfrac{50}{5} \\ 10 \ \Big| \end{array} \quad \text{TRUE}$$

The solution is -50.

39.
$$-\frac{t}{2} = 7$$
$$2 \cdot \left(-\frac{t}{2}\right) = 2 \cdot 7$$
$$-t = 14$$
$$-1 \cdot (-t) = -1 \cdot 14$$
$$t = -14$$

Check:
$$\begin{array}{c|c} -\dfrac{t}{2} = 7 \\ \hline -\dfrac{-14}{2} \ ? \ 7 \\ -(-7) \ \Big| \\ 7 \ \Big| \end{array} \quad \text{TRUE}$$

The solution is -14.

41. Discussion and Writing Exercise

43. $3x + 4x = (3 + 4)x = 7x$

45. $-4x + 11 - 6x + 18x = (-4 - 6 + 18)x + 11 = 8x + 11$

47. $3x - (4 + 2x) = 3x - 4 - 2x = x - 4$

49. $8y - 6(3y + 7) = 8y - 18y - 42 = -10y - 42$

51. The translation is $8r$ miles.

53. $-0.2344m = 2028.732$
$$\frac{-0.2344m}{-0.2344} = \frac{2028.732}{-0.2344}$$
$$1 \cdot m = -8655$$
$$m = -8655$$
The solution is -8655.

55. For all x, $0 \cdot x = 0$. There is no solution to $0 \cdot x = 9$.

57.
$$2|x| = -12$$
$$\frac{2|x|}{2} = \frac{-12}{2}$$
$$1 \cdot |x| = -6$$
$$|x| = -6$$

Absolute value cannot be negative. The equation has no solution.

59.
$$3x = \frac{b}{a}$$
$$\frac{1}{3} \cdot 3x = \frac{1}{3} \cdot \frac{b}{a}$$
$$x = \frac{b}{3a}$$

The solution is $\dfrac{b}{3a}$.

61.
$$\frac{a}{b}x = 4$$
$$\frac{b}{a} \cdot \frac{a}{b}x = \frac{b}{a} \cdot 4$$
$$x = \frac{4b}{a}$$

The solution is $\dfrac{4b}{a}$.

Exercise Set 7.3

1.
$$5x + 6 = 31$$

$5x + 6 - 6 = 31 - 6$	Subtracting 6 on both sides
$5x = 25$	Simplifying
$\dfrac{5x}{5} = \dfrac{25}{5}$	Dividing by 5 on both sides
$x = 5$	Simplifying

Check:
$$\frac{5x + 6 = 31}{}$$
$$5 \cdot 5 + 6 \ ? \ 31$$
$$25 + 6 \ \big|$$
$$31 \ \big| \qquad \text{TRUE}$$

The solution is 5.

3.
$$8x + 4 = 68$$

$8x + 4 - 4 = 68 - 4$	Subtracting 4 on both sides
$8x = 64$	Simplifying
$\dfrac{8x}{8} = \dfrac{64}{8}$	Dividing by 8 on both sides
$x = 8$	Simplifying

Check:
$$\frac{8x + 4 = 68}{}$$
$$8 \cdot 8 + 4 \ ? \ 68$$
$$64 + 4 \ \big|$$
$$68 \ \big| \qquad \text{TRUE}$$

The solution is 8.

5.
$$4x - 6 = 34$$

$4x - 6 + 6 = 34 + 6$	Adding 6 on both sides
$4x = 40$	
$\dfrac{4x}{4} = \dfrac{40}{4}$	Dividing by 4 on both sides
$x = 10$	

Check:
$$\frac{4x - 6 = 34}{}$$
$$4 \cdot 10 - 6 \ ? \ 34$$
$$40 - 6 \ \big|$$
$$34 \ \big| \qquad \text{TRUE}$$

The solution is 10.

7.
$$3x - 9 = 33$$
$$3x - 9 + 9 = 33 + 9$$
$$3x = 42$$
$$\frac{3x}{3} = \frac{42}{3}$$
$$x = 14$$

Check:
$$\frac{3x - 9 = 33}{}$$
$$3 \cdot 14 - 9 \ ? \ 33$$
$$42 - 9 \ \big|$$
$$33 \ \big| \qquad \text{TRUE}$$

The solution is 14.

9.
$$7x + 2 = -54$$
$$7x + 2 - 2 = -54 - 2$$
$$7x = -56$$
$$\frac{7x}{7} = \frac{-56}{7}$$
$$x = -8$$

Check:
$$\frac{7x + 2 = -54}{}$$
$$7(-8) + 2 \ ? \ -54$$
$$-56 + 2 \ \big|$$
$$-54 \ \big| \qquad \text{TRUE}$$

The solution is -8.

11.
$$-45 = 6y + 3$$
$$-45 - 3 = 6y + 3 - 3$$
$$-48 = 6y$$
$$\frac{-48}{6} = \frac{6y}{6}$$
$$-8 = y$$

Check:
$$\frac{-45 = 6y + 3}{}$$
$$-45 \ ? \ 6(-8) + 3$$
$$\big| \ -48 + 3$$
$$\big| \ -45 \qquad \text{TRUE}$$

The solution is -8.

13.
$$-4x + 7 = 35$$
$$-4x + 7 - 7 = 35 - 7$$
$$-4x = 28$$
$$\frac{-4x}{-4} = \frac{28}{-4}$$
$$x = -7$$

Check: $$-4x + 7 = 35$$
$$-4(-7) + 7 \ ? \ 35$$
$$28 + 7$$
$$35 \quad \text{TRUE}$$

The solution is -7.

15. $$-7x - 24 = -129$$
$$-7x - 24 + 24 = -129 + 24$$
$$-7x = -105$$
$$\frac{-7x}{-7} = \frac{-105}{-7}$$
$$x = 15$$

Check: $$-7x - 24 = -129$$
$$-7 \cdot 15 - 24 \ ? \ -129$$
$$-105 - 24$$
$$-129 \quad \text{TRUE}$$

The solution is 15.

17. $5x + 7x = 72$
$$12x = 72 \qquad \text{Collecting like terms}$$
$$\frac{12x}{12} = \frac{72}{12} \qquad \text{Dividing by 12 on both sides}$$
$$x = 6$$

Check: $$5x + 7x = 72$$
$$5 \cdot 6 + 7 \cdot 6 \ ? \ 72$$
$$30 + 42$$
$$72 \quad \text{TRUE}$$

The solution is 6.

19. $8x + 7x = 60$
$$15x = 60 \qquad \text{Collecting like terms}$$
$$\frac{15x}{15} = \frac{60}{15} \qquad \text{Dividing by 15 on both sides}$$
$$x = 4$$

Check: $$8x + 7x = 60$$
$$8 \cdot 4 + 7 \cdot 4 \ ? \ 60$$
$$32 + 28$$
$$60 \quad \text{TRUE}$$

The solution is 4.

21. $4x + 3x = 42$
$$7x = 42$$
$$\frac{7x}{7} = \frac{42}{7}$$
$$x = 6$$

Check: $$4x + 3x = 42$$
$$4 \cdot 6 + 3 \cdot 6 \ ? \ 42$$
$$24 + 18$$
$$42 \quad \text{TRUE}$$

The solution is 6.

23. $$-6y - 3y = 27$$
$$-9y = 27$$
$$\frac{-9y}{-9} = \frac{27}{-9}$$
$$y = -3$$

Check: $$-6y - 3y = 27$$
$$-6(-3) - 3(-3) \ ? \ 27$$
$$18 + 9$$
$$27 \quad \text{TRUE}$$

The solution is -3.

25. $$-7y - 8y = -15$$
$$-15y = -15$$
$$\frac{-15y}{-15} = \frac{-15}{-15}$$
$$y = 1$$

Check: $$-7y - 8y = -15$$
$$-7 \cdot 1 - 8 \cdot 1 \ ? \ -15$$
$$-7 - 8$$
$$-15 \quad \text{TRUE}$$

The solution is 1.

27. $$x + \frac{1}{3}x = 8$$
$$\left(1 + \frac{1}{3}\right)x = 8$$
$$\frac{4}{3}x = 8$$
$$\frac{3}{4} \cdot \frac{4}{3}x = \frac{3}{4} \cdot 8$$
$$x = 6$$

Check: $$x + \frac{1}{3}x = 8$$
$$6 + \frac{1}{3} \cdot 6 \ ? \ 8$$
$$6 + 2$$
$$8 \quad \text{TRUE}$$

The solution is 6.

29. $10.2y - 7.3y = -58$
$$2.9y = -58$$
$$\frac{2.9y}{2.9} = \frac{-58}{2.9}$$
$$y = -20$$

Check: $$10.2y - 7.3y = -58$$
$$10.2(-20) - 7.3(-20) \ ? \ -58$$
$$-204 + 146$$
$$-58 \quad \text{TRUE}$$

The solution is -20.

31. $8y - 35 = 3y$

$\qquad 8y = 3y + 35 \qquad$ Adding 35 and simplifying

$8y - 3y = 35 \qquad$ Subtracting $3y$ and simplifying

$\qquad 5y = 35 \qquad$ Collecting like terms

$\dfrac{5y}{5} = \dfrac{35}{5} \qquad$ Dividing by 5

$\qquad y = 7$

Check: $\quad \dfrac{8y - 35 = 3y}{}$

$\qquad 8 \cdot 7 - 35 \ ? \ 3 \cdot 7$

$\qquad 56 - 35 \ \big| \ 21$

$\qquad \qquad 21 \ \big| \qquad \qquad$ TRUE

The solution is 7.

33. $8x - 1 = 23 - 4x$

$8x + 4x = 23 + 1 \qquad$ Adding 1 and $4x$ and simplifying

$\qquad 12x = 24 \qquad$ Collecting like terms

$\dfrac{12x}{12} = \dfrac{24}{12} \qquad$ Dividing by 12

$\qquad x = 2$

Check: $\quad \dfrac{8x - 1 = 23 - 4x}{}$

$\qquad 8 \cdot 2 - 1 \ ? \ 23 - 4 \cdot 2$

$\qquad 16 - 1 \ \big| \ 23 - 8$

$\qquad \quad 15 \ \big| \ 15 \qquad \qquad$ TRUE

The solution is 2.

35. $2x - 1 = 4 + x$

$2x - x = 4 + 1 \qquad$ Adding 1 and $-x$

$\qquad x = 5 \qquad$ Collecting like terms

Check: $\quad \dfrac{2x - 1 = 4 + x}{}$

$\qquad 2 \cdot 5 - 1 \ ? \ 4 + 5$

$\qquad 10 - 1 \ \big| \ 9$

$\qquad \quad 9 \ \big| \qquad \qquad$ TRUE

The solution is 5.

37. $6x + 3 = 2x + 11$

$6x - 2x = 11 - 3$

$\qquad 4x = 8$

$\dfrac{4x}{4} = \dfrac{8}{4}$

$\qquad x = 2$

Check: $\quad \dfrac{6x + 3 = 2x + 11}{}$

$\qquad 6 \cdot 2 + 3 \ ? \ 2 \cdot 2 + 11$

$\qquad 12 + 3 \ \big| \ 4 + 11$

$\qquad \quad 15 \ \big| \ 15 \qquad \qquad$ TRUE

The solution is 2.

39. $5 - 2x = 3x - 7x + 25$

$5 - 2x = -4x + 25$

$4x - 2x = 25 - 5$

$\qquad 2x = 20$

$\dfrac{2x}{2} = \dfrac{20}{2}$

$\qquad x = 10$

Check: $\quad \dfrac{5 - 2x = 3x - 7x + 25}{}$

$\qquad 5 - 2 \cdot 10 \ ? \ 3 \cdot 10 - 7 \cdot 10 + 25$

$\qquad 5 - 20 \ \big| \ 30 - 70 + 25$

$\qquad \ -15 \ \big| \ -40 + 25$

$\qquad \qquad \big| \ -15 \qquad \qquad$ TRUE

The solution is 10.

41. $4 + 3x - 6 = 3x + 2 - x$

$3x - 2 = 2x + 2 \qquad$ Collecting like terms on each side

$3x - 2x = 2 + 2$

$\qquad x = 4$

Check: $\quad \dfrac{4 + 3x - 6 = 3x + 2 - x}{}$

$\qquad 4 + 3 \cdot 4 - 6 \ ? \ 3 \cdot 4 + 2 - 4$

$\qquad 4 + 12 - 6 \ \big| \ 12 + 2 - 4$

$\qquad \quad 16 - 6 \ \big| \ 14 - 4$

$\qquad \qquad 10 \ \big| \ 10 \qquad \qquad$ TRUE

The solution is 4.

43. $4y - 4 + y + 24 = 6y + 20 - 4y$

$5y + 20 = 2y + 20$

$5y - 2y = 20 - 20$

$\qquad 3y = 0$

$\qquad y = 0$

Check: $\quad \dfrac{4y - 4 + y + 24 = 6y + 20 - 4y}{}$

$\qquad 4 \cdot 0 - 4 + 0 + 24 \ ? \ 6 \cdot 0 + 20 - 4 \cdot 0$

$\qquad 0 - 4 + 0 + 24 \ \big| \ 0 + 20 - 0$

$\qquad \qquad \quad 20 \ \big| \ 20 \qquad \qquad$ TRUE

The solution is 0.

45. $\dfrac{7}{2}x + \dfrac{1}{2}x = 3x + \dfrac{3}{2} + \dfrac{5}{2}x$

The least common multiple of all the denominators is 2. We multiply by 2 on both sides.

$2\left(\dfrac{7}{2}x + \dfrac{1}{2}x\right) = 2\left(3x + \dfrac{3}{2} + \dfrac{5}{2}x\right)$

$2 \cdot \dfrac{7}{2}x + 2 \cdot \dfrac{1}{2}x = 2 \cdot 3x + 2 \cdot \dfrac{3}{2} + 2 \cdot \dfrac{5}{2}x$

$7x + x = 6x + 3 + 5x$

$8x = 11x + 3$

$8x - 11x = 3$

$-3x = 3$

$\dfrac{-3x}{-3} = \dfrac{3}{-3}$

$x = -1$

Check:

$$\frac{7}{2}x + \frac{1}{2}x = 3x + \frac{3}{2} + \frac{5}{2}x$$

$$\frac{7}{2}(-1) + \frac{1}{2}(-1) \ ? \ 3(-1) + \frac{3}{2} + \frac{5}{2}(-1)$$

$$-\frac{7}{2} - \frac{1}{2} \ \Big| \ -3 + \frac{3}{2} - \frac{5}{2}$$

$$-4 \ \Big| \ -\frac{8}{2}$$

$$\Big| \ -4 \quad \text{TRUE}$$

The solution is -1.

47. $\frac{2}{3} + \frac{1}{4}t = \frac{1}{3}$

The least common multiple of all the denominators is 12. We multiply by 12 on both sides.

$$12\Big(\frac{2}{3} + \frac{1}{4}t\Big) = 12 \cdot \frac{1}{3}$$

$$12 \cdot \frac{2}{3} + 12 \cdot \frac{1}{4}t = 12 \cdot \frac{1}{3}$$

$$8 + 3t = 4$$

$$3t = 4 - 8$$

$$3t = -4$$

$$\frac{3t}{3} = \frac{-4}{3}$$

$$t = -\frac{4}{3}$$

Check:

$$\frac{2}{3} + \frac{1}{4}t = \frac{1}{3}$$

$$\frac{2}{3} + \frac{1}{4}\Big(-\frac{4}{3}\Big) \ ? \ \frac{1}{3}$$

$$\frac{2}{3} - \frac{1}{3} \ \Big|$$

$$\frac{1}{3} \ \Big| \quad \text{TRUE}$$

The solution is $-\frac{4}{3}$.

49. $\frac{2}{3} + 3y = 5y - \frac{2}{15}, \quad \text{LCM is 15}$

$$15\Big(\frac{2}{3} + 3y\Big) = 15\Big(5y - \frac{2}{15}\Big)$$

$$15 \cdot \frac{2}{3} + 15 \cdot 3y = 15 \cdot 5y - 15 \cdot \frac{2}{15}$$

$$10 + 45y = 75y - 2$$

$$10 + 2 = 75y - 45y$$

$$12 = 30y$$

$$\frac{12}{30} = \frac{30y}{30}$$

$$\frac{2}{5} = y$$

Check:

$$\frac{2}{3} + 3y = 5y - \frac{2}{15}$$

$$\frac{2}{3} + 3 \cdot \frac{2}{5} \ ? \ 5 \cdot \frac{2}{5} - \frac{2}{15}$$

$$\frac{2}{3} + \frac{6}{5} \ \Big| \ 2 - \frac{2}{15}$$

$$\frac{10}{15} + \frac{18}{15} \ \Big| \ \frac{30}{15} - \frac{2}{15}$$

$$\frac{28}{15} \ \Big| \ \frac{28}{15} \quad \text{TRUE}$$

The solution is $\frac{2}{5}$.

51. $\frac{5}{3} + \frac{2}{3}x = \frac{25}{12} + \frac{5}{4}x + \frac{3}{4}, \quad \text{LCM is 12}$

$$12\Big(\frac{5}{3} + \frac{2}{3}x\Big) = 12\Big(\frac{25}{12} + \frac{5}{4}x + \frac{3}{4}\Big)$$

$$12 \cdot \frac{5}{3} + 12 \cdot \frac{2}{3}x = 12 \cdot \frac{25}{12} + 12 \cdot \frac{5}{4}x + 12 \cdot \frac{3}{4}$$

$$20 + 8x = 25 + 15x + 9$$

$$20 + 8x = 15x + 34$$

$$20 - 34 = 15x - 8x$$

$$-14x = 7x$$

$$\frac{-14}{7} = \frac{7x}{7}$$

$$-2 = x$$

Check:

$$\frac{5}{3} + \frac{2}{3}x = \frac{25}{12} + \frac{5}{4}x + \frac{3}{4}$$

$$\frac{5}{3} + \frac{2}{3}(-2) \ ? \ \frac{25}{12} + \frac{5}{4}(-2) + \frac{3}{4}$$

$$\frac{5}{3} - \frac{4}{3} \ \Big| \ \frac{25}{12} - \frac{5}{2} + \frac{3}{4}$$

$$\frac{1}{3} \ \Big| \ \frac{25}{12} - \frac{30}{12} + \frac{9}{12}$$

$$\Big| \ \frac{4}{12}$$

$$\Big| \ \frac{1}{3} \quad \text{TRUE}$$

The solution is -2.

53. $2.1x + 45.2 = 3.2 - 8.4x$

Greatest number of decimal places is 1

$$10(2.1x + 45.2) = 10(3.2 - 8.4x)$$

Multiplying by 10 to clear decimals

$$10(2.1x) + 10(45.2) = 10(3.2) - 10(8.4x)$$

$$21x + 452 = 32 - 84x$$

$$21x + 84x = 32 - 452$$

$$105x = -420$$

$$\frac{105x}{105} = \frac{-420}{105}$$

$$x = -4$$

Check: $\dfrac{2.1x + 45.2 = 3.2 - 8.4x}{}$

$$2.1(-4) + 45.2 \ ? \ 3.2 - 8.4(-4)$$
$$-8.4 + 45.2 \ | \ 3.2 + 33.6$$
$$36.8 \ | \ 36.8 \qquad \text{TRUE}$$

The solution is -4.

55. $\qquad 1.03 - 0.62x = 0.71 - 0.22x$

Greatest number of decimal places is 2

$$100(1.03 - 0.62x) = 100(0.71 - 0.22x)$$

Multiplying by 100 to clear decimals

$$100(1.03) - 100(0.62x) = 100(0.71) - 100(0.22x)$$
$$103 - 62x = 71 - 22x$$
$$32 = 40x$$
$$\frac{32}{40} = \frac{40x}{40}$$
$$\frac{4}{5} = x, \text{ or}$$
$$0.8 = x$$

Check: $\dfrac{1.03 - 0.62x = 0.71 - 0.22x}{}$

$$1.03 - 0.62(0.8) \ ? \ 0.71 - 0.22(0.8)$$
$$1.03 - 0.496 \ | \ 0.71 - 0.176$$
$$0.534 \ | \ 0.534 \qquad \text{TRUE}$$

The solution is $\dfrac{4}{5}$, or 0.8.

57. $\qquad \dfrac{2}{7}x - \dfrac{1}{2}x = \dfrac{3}{4}x + 1$, LCM is 28

$$28\left(\frac{2}{7}x - \frac{1}{2}x\right) = 28\left(\frac{3}{4}x + 1\right)$$
$$28 \cdot \frac{2}{7}x - 28 \cdot \frac{1}{2}x = 28 \cdot \frac{3}{4}x + 28 \cdot 1$$
$$8x - 14x = 21x + 28$$
$$-6x = 21x + 28$$
$$-6x - 21x = 28$$
$$-27x = 28$$
$$x = -\frac{28}{27}$$

Check: $\qquad \dfrac{2}{7}x - \dfrac{1}{2}x = \dfrac{3}{4}x + 1$

$$\dfrac{\frac{2}{7}\left(-\frac{28}{27}\right) - \frac{1}{2}\left(-\frac{28}{27}\right) \ ? \ \frac{3}{4}\left(-\frac{28}{27}\right) + 1}{}$$
$$-\frac{8}{27} + \frac{14}{27} \ \bigg| \ -\frac{21}{27} + 1$$
$$\frac{6}{27} \ \bigg| \ \frac{6}{27} \qquad \text{TRUE}$$

The solution is $-\dfrac{28}{27}$.

59. $\quad 3(2y - 3) = 27$

$$6y - 9 = 27 \qquad \text{Using a distributive law}$$
$$6y = 27 + 9 \qquad \text{Adding 9}$$
$$6y = 36$$
$$y = 6 \qquad \text{Dividing by 6}$$

Check: $\dfrac{3(2y - 3) = 27}{}$

$$3(2 \cdot 6 - 3) \ ? \ 27$$
$$3(12 - 3) \ \bigg|$$
$$3 \cdot 9 \ \bigg|$$
$$27 \ \bigg| \qquad \text{TRUE}$$

The solution is 6.

61. $\qquad 40 = 5(3x + 2)$

$$40 = 15x + 10 \qquad \text{Using a distributive law}$$
$$40 - 10 = 15x$$
$$30 = 15x$$
$$2 = x$$

Check: $\dfrac{40 = 5(3x + 2)}{}$

$$40 \ ? \ 5(3 \cdot 2 + 2)$$
$$\bigg| \ 5(6 + 2)$$
$$\bigg| \ 5 \cdot 8$$
$$\bigg| \ 40 \qquad \text{TRUE}$$

The solution is 2.

63. $\qquad -23 + y = y + 25$

$$-y - 23 + y = -y + y + 25$$
$$-23 = 25 \qquad \text{FALSE}$$

The equation has no solution.

65. $\qquad -23 + x = x - 23$

$$-x - 23 + x = -x + x - 23$$
$$-23 = -23 \qquad \text{TRUE}$$

All real numbers are solutions.

67. $\quad 2(3 + 4m) - 9 = 45$

$$6 + 8m - 9 = 45 \qquad \text{Collecting like terms}$$
$$8m - 3 = 45$$
$$8m = 45 + 3$$
$$8m = 48$$
$$m = 6$$

Check: $\dfrac{2(3 + 4m) - 9 = 45}{}$

$$2(3 + 4 \cdot 6) - 9 \ ? \ 45$$
$$2(3 + 24) - 9 \ \bigg|$$
$$2 \cdot 27 - 9 \ \bigg|$$
$$54 - 9 \ \bigg|$$
$$45 \ \bigg| \qquad \text{TRUE}$$

The solution is 6.

69. $5r - (2r + 8) = 16$

$5r - 2r - 8 = 16$

$3r - 8 = 16$ $\qquad$ Collecting like terms

$3r = 16 + 8$

$3r = 24$

$r = 8$

Check: $\quad \dfrac{5r - (2r + 8) = 16}{}$

$\begin{array}{c|c} 5 \cdot 8 - (2 \cdot 8 + 8) \ ? \ 16 \\ 40 - (16 + 8) \quad \Big| \\ 40 - 24 \quad \Big| \\ 16 \quad \Big| \end{array}$ $\quad$ TRUE

The solution is 8.

71. $6 - 2(3x - 1) = 2$

$6 - 6x + 2 = 2$

$8 - 6x = 2$

$8 - 2 = 6x$

$6 = 6x$

$1 = x$

Check: $\quad \dfrac{6 - 2(3x - 1) = 2}{}$

$\begin{array}{c|c} 6 - 2(3 \cdot 1 - 1) \ ? \ 2 \\ 6 - 2(3 - 1) \quad \Big| \\ 6 - 2 \cdot 2 \quad \Big| \\ 6 - 4 \quad \Big| \\ 2 \quad \Big| \end{array}$ $\quad$ TRUE

The solution is 1.

73. $5x + 5 - 7x = 15 - 12x + 10x - 10$

$-2x + 5 = 5 - 2x$ $\quad$ Collecting like terms

$2x - 2x + 5 = 2x + 5 - 2x$ $\quad$ Adding $2x$

$5 = 5$ $\quad$ TRUE

All real numbers are solutions.

75. $22x - 5 - 15x + 3 = 10x - 4 - 3x + 11$

$7x - 2 = 7x + 7$ $\quad$ Collecting like terms

$-7x + 7x - 2 = -7x + 7x + 7$

$-2 = 7$ $\quad$ FALSE

The equation has no solution.

77. $5(d + 4) = 7(d - 2)$

$5d + 20 = 7d - 14$

$20 + 14 = 7d - 5d$

$34 = 2d$

$17 = d$

Check: $\quad \dfrac{5(d + 4) = 7(d - 2)}{}$

$\begin{array}{c|c} 5(17 + 4) \ ? \ 7(17 - 2) \\ 5 \cdot 21 \quad \Big| \quad 7 \cdot 15 \\ 105 \quad \Big| \quad 105 \end{array}$ $\quad$ TRUE

The solution is 17.

79. $8(2t + 1) = 4(7t + 7)$

$16t + 8 = 28t + 28$

$16t - 28t = 28 - 8$

$-12t = 20$

$t = -\dfrac{20}{12}$

$t = -\dfrac{5}{3}$

Check: $\quad \dfrac{8(2t + 1) = 4(7t + 7)}{}$

$\begin{array}{c|c} 8\left(2\left(-\dfrac{5}{3}\right) + 1\right) \ ? \ 4\left(7\left(-\dfrac{5}{3}\right) + 7\right) \\ 8\left(-\dfrac{10}{3} + 1\right) \quad \Big| \quad 4\left(-\dfrac{35}{3} + 7\right) \\ 8\left(-\dfrac{7}{3}\right) \quad \Big| \quad 4\left(-\dfrac{14}{3}\right) \\ -\dfrac{56}{3} \quad \Big| \quad -\dfrac{56}{3} \end{array}$ $\quad$ TRUE

The solution is $-\dfrac{5}{3}$.

81. $3(r - 6) + 2 = 4(r + 2) - 21$

$3r - 18 + 2 = 4r + 8 - 21$

$3r - 16 = 4r - 13$

$13 - 16 = 4r - 3r$

$-3 = r$

Check: $\quad \dfrac{3(r - 6) + 2 = 4(r + 2) - 21}{}$

$\begin{array}{c|c} 3(-3 - 6) + 2 \ ? \ 4(-3 + 2) - 21 \\ 3(-9) + 2 \quad \Big| \quad 4(-1) - 21 \\ -27 + 2 \quad \Big| \quad -4 - 21 \\ -25 \quad \Big| \quad -25 \end{array}$ $\quad$ TRUE

The solution is -3.

83. $19 - (2x + 3) = 2(x + 3) + x$

$19 - 2x - 3 = 2x + 6 + x$

$16 - 2x = 3x + 6$

$16 - 6 = 3x + 2x$

$10 = 5x$

$2 = x$

Check: $\quad \dfrac{19 - (2x + 3) = 2(x + 3) + x}{}$

$\begin{array}{c|c} 19 - (2 \cdot 2 + 3) \ ? \ 2(2 + 3) + 2 \\ 19 - (4 + 3) \quad \Big| \quad 2 \cdot 5 + 2 \\ 19 - 7 \quad \Big| \quad 10 + 2 \\ 12 \quad \Big| \quad 12 \end{array}$ $\quad$ TRUE

The solution is 2.

85. $2[4 - 2(3 - x)] - 1 = 4[2(4x - 3) + 7] - 25$

$2[4 - 6 + 2x] - 1 = 4[8x - 6 + 7] - 25$

$2[-2 + 2x] - 1 = 4[8x + 1] - 25$

$-4 + 4x - 1 = 32x + 4 - 25$

$4x - 5 = 32x - 21$

$-5 + 21 = 32x - 4x$

$16 = 28x$

$\dfrac{16}{28} = x$

$\dfrac{4}{7} = x$

The check is left to the student.

The solution is $\dfrac{4}{7}$.

87. $11 - 4(x + 1) - 3 = 11 + 2(4 - 2x) - 16$

$11 - 4x - 4 - 3 = 11 + 8 - 4x - 16$

$4 - 4x = 3 - 4x$

$4x + 4 - 4x = 4x + 3 - 4x$

$4 = 3 \qquad \text{FALSE}$

The equation has no solution.

89. $22x - 1 - 12x = 5(2x - 1) + 4$

$22x - 1 - 12x = 10x - 5 + 4$

$10x - 1 = 10x - 1$

$-10x + 10x - 1 = -10x + 10x - 1$

$-1 = -1 \qquad \text{TRUE}$

All real numbers are solutions.

91. $0.7(3x + 6) = 1.1 - (x + 2)$

$2.1x + 4.2 = 1.1 - x - 2$

$10(2.1x + 4.2) = 10(1.1 - x - 2) \quad \text{Clearing decimals}$

$21x + 42 = 11 - 10x - 20$

$21x + 42 = -10x - 9$

$21x + 10x = -9 - 42$

$31x = -51$

$x = -\dfrac{51}{31}$

The check is left to the student.

The solution is $-\dfrac{51}{31}$.

93. Discussion and Writing Exercise

95. Do the long division. The answer is negative.

```
        6 . 5
3. 4⌄‾2 2.1⌄0‾
       2 0 4
       ‾‾‾‾‾
       1 7 0
       1 7 0
       ‾‾‾‾‾
           0
```

$-22.1 \div 3.4 = -6.5$

97. $7x - 21 - 14y = 7 \cdot x - 7 \cdot 3 - 7 \cdot 2y = 7(x - 3 - 2y)$

99. Since we are using a calculator we will not clear the decimals.

$0.008 + 9.62x - 42.8 = 0.944x + 0.0083 - x$

$9.62x - 42.792 = -0.056x + 0.0083$

$9.62x + 0.056x = 0.0083 + 42.792$

$9.676x = 42.8003$

$x = \dfrac{42.8003}{9.676}$

$x \approx 4.4233464$

The solution is approximately 4.4233464.

101. First we multiply to remove the parentheses.

$\dfrac{2}{3}\left(\dfrac{7}{8} - 4x\right) - \dfrac{5}{8} = \dfrac{3}{8}$

$\dfrac{7}{12} - \dfrac{8}{3}x - \dfrac{5}{8} = \dfrac{3}{8}, \text{ LCM is } 24$

$24\left(\dfrac{7}{12} - \dfrac{8}{3}x - \dfrac{5}{8}\right) = 24 \cdot \dfrac{3}{8}$

$24 \cdot \dfrac{7}{12} - 24 \cdot \dfrac{8}{3}x - 24 \cdot \dfrac{5}{8} = 9$

$14 - 64x - 15 = 9$

$-1 - 64x = 9$

$-64x = 10$

$x = -\dfrac{10}{64}$

$x = -\dfrac{5}{32}$

The solution is $-\dfrac{5}{32}$.

Exercise Set 7.4

1. a) We substitute 1900 for a and calculate B.

$B = 30a = 30 \cdot 1900 = 57,000$

The minimum furnace output is 57,000 Btu's.

b) $B = 30a$

$\dfrac{B}{30} = \dfrac{30a}{30} \quad \text{Dividing by 30}$

$\dfrac{B}{30} = a$

3. a) We substitute 8 for t and calculate M.

$M = \dfrac{1}{5} \cdot 8 = \dfrac{8}{5}, \text{ or } 1\dfrac{3}{5}$

The storm is $1\dfrac{3}{5}$ miles away.

b) $M = \dfrac{1}{5}t$

$5 \cdot M = 5 \cdot \dfrac{1}{5}t$

$5M = t$

5. a) We substitute 21,345 for n and calculate f.

$f = \dfrac{21,345}{15} = 1423$

There are 1423 full-time equivalent students.

b) $$f = \frac{n}{15}$$
$$15 \cdot f = 15 \cdot \frac{n}{15}$$
$$15f = n$$

7. We substitute 84 for c and 8 for w and calculate D.
$$D = \frac{c}{w} = \frac{84}{8} = 10.5$$
The calorie density is 10.5 calories per oz.

9. We substitute 7 for n and calculate N.
$$N = n^2 - n = 7^2 - 7 = 49 - 7 = 42$$
42 games are played.

11. $y = 5x$
$$\frac{y}{5} = \frac{5x}{5}$$
$$\frac{y}{5} = x$$

13. $a = bc$
$$\frac{a}{b} = \frac{bc}{b}$$
$$\frac{a}{b} = c$$

15. $y = 13 + x$
$$y - 13 = 13 + x - 13$$
$$y - 13 = x$$

17. $y = x + b$
$$y - b = x + b - b$$
$$y - b = x$$

19. $y = 5 - x$
$$y - 5 = 5 - x - 5$$
$$y - 5 = -x$$
$$-1 \cdot (y - 5) = -1 \cdot (-x)$$
$$-y + 5 = x, \text{ or}$$
$$5 - y = x$$

21. $y = a - x$
$$y - a = a - x - a$$
$$y - a = -x$$
$$-1 \cdot (y - a) = -1 \cdot (-x)$$
$$-y + a = x, \text{ or}$$
$$a - y = x$$

23. $8y = 5x$
$$\frac{8y}{8} = \frac{5x}{8}$$
$$y = \frac{5x}{8}, \text{ or } \frac{5}{8}x$$

25. $By = Ax$
$$\frac{By}{A} = \frac{Ax}{A}$$
$$\frac{By}{A} = x$$

27. $W = mt + b$
$$W - b = mt + b - b$$
$$W - b = mt$$
$$\frac{W - b}{m} = \frac{mt}{m}$$
$$\frac{W - b}{m} = t$$

29. $y = bx + c$
$$y - c = bx + c - c$$
$$y - c = bx$$
$$\frac{y - c}{b} = \frac{bx}{b}$$
$$\frac{y - c}{b} = x$$

31. $A = \frac{a + b + c}{3}$
$$3A = a + b + c \quad \text{Multiplying by 3}$$
$$3A - a - c = b \qquad \text{Subtracting } a \text{ and } c$$

33. $A = at + b$
$$A - b = at \qquad \text{Subtracting } b$$
$$\frac{A - b}{a} = t \qquad \text{Dividing by } a$$

35. $A = bh$
$$\frac{A}{b} = \frac{bh}{b} \qquad \text{Dividing by } b$$
$$\frac{A}{b} = h$$

37. $P = 2l + 2w$
$$P - 2l = 2l + 2w - 2l \qquad \text{Subtracting } 2l$$
$$P - 2l = 2w$$
$$\frac{P - 2l}{2} = \frac{2w}{2} \qquad \text{Dividing by 2}$$
$$\frac{P - 2l}{2} = w, \text{ or}$$
$$\frac{1}{2}P - l = w$$

39. $A = \frac{a + b}{2}$
$$2A = a + b \quad \text{Multiplying by 2}$$
$$2A - b = a \qquad \text{Subtracting } b$$

41. $F = ma$
$$\frac{F}{m} = \frac{ma}{m} \qquad \text{Dividing by } m$$
$$\frac{F}{m} = a$$

43. $E = mc^2$
$$\frac{E}{m} = \frac{mc^2}{m} \qquad \text{Dividing by } m$$
$$\frac{E}{m} = c^2$$

45. $Ax + By = c$

$$Ax = c - By \quad \text{Subtracting } By$$

$$\frac{Ax}{A} = \frac{c - By}{A} \quad \text{Dividing by } A$$

$$x = \frac{c - By}{A}$$

47. $v = \dfrac{3k}{t}$

$$tv = t \cdot \frac{3k}{t} \quad \text{Multiplying by } t$$

$$tv = 3k$$

$$\frac{tv}{v} = \frac{3k}{v} \quad \text{Dividing by } v$$

$$t = \frac{3k}{v}$$

49. Discussion and Writing Exercise

51. $2a - b = 2 \cdot 2 - 3 = 4 - 3 = 1$

53. $0.082 + (-9.407) = -9.325$

55. $-45.8 - (-32.6) = -45.8 + 32.6 = -13.2$

57. $-\dfrac{2}{3} + \dfrac{5}{6} = -\dfrac{2}{3} \cdot \dfrac{2}{2} + \dfrac{5}{6}$

$$= -\frac{4}{6} + \frac{5}{6}$$

$$= \frac{1}{6}$$

59. a) We substitute 120 for w, 67 for h, and 23 for a and calculate K.

$$K = 917 + 6(w + h - a)$$
$$K = 917 + 6(120 + 67 - 23)$$
$$K = 917 + 6(164)$$
$$K = 917 + 984$$
$$K = 1901 \text{ calories}$$

b) Solve for a:

$$K = 917 + 6(w + h - a)$$
$$K = 917 + 6w + 6h - 6a$$
$$K + 6a = 917 + 6w + 6h$$
$$6a = 917 + 6w + 6h - K$$
$$a = \frac{917 + 6w + 6h - K}{6}$$

Solve for h:

$$K = 917 + 6(w + h - a)$$
$$K = 917 + 6w + 6h - 6a$$
$$K - 917 - 6w + 6a = 6h$$
$$\frac{K - 917 - 6w + 6a}{6} = h$$

Solve for w:

$$K = 917 + 6(w + h - a)$$
$$K = 917 + 6w + 6h - 6a$$
$$K - 917 - 6h + 6a = 6w$$
$$\frac{K - 917 - 6h + 6a}{6} = w$$

61.

$$A = \frac{1}{2}ah + \frac{1}{2}bh$$

$$2A = 2\left(\frac{1}{2}ah + \frac{1}{2}bh\right) \quad \text{Clearing the fractions}$$

$$2A = ah + bh$$

$$2A - ah = bh \quad \text{Subtracting } ah$$

$$\frac{2A - ah}{h} = b \quad \text{Dividing by } h$$

$$A = \frac{1}{2}ah + \frac{1}{2}bh$$

$$2A = ah + bh \quad \text{Clearing fractions as above}$$

$$2A = h(a + b) \quad \text{Factoring}$$

$$\frac{2A}{a + b} = h \quad \text{Dividing by } a + b$$

63. $A = lw$

When l and w both double, we have

$$2l \cdot 2w = 4lw = 4A,$$

so A quadruples.

65. $A = \dfrac{1}{2}bh$

When b increases by 4 units we have

$$\frac{1}{2}(b + 4)h = \frac{1}{2}bh + 2h = A + 2h,$$

so A increases by $2h$ units.

Exercise Set 7.5

1. Translate.

$$\underbrace{\text{What percent}}_{p} \text{ of } \underbrace{180}_{180} \text{ is } \underbrace{36?}_{36}$$

$$p \qquad \cdot \quad 180 \ = \ 36$$

Solve. We divide by 36 on both sides and convert the answer to percent notation.

$$p \cdot 180 = 36$$
$$\frac{p \cdot 180}{180} = \frac{36}{180}$$
$$p = 0.2$$
$$p = 20\%$$

Thus, 36 is 20% of 180. The answer is 20%.

3. Translate.

$$45 \text{ is } 30\% \text{ of what?}$$

$$45 \ = \ 30\% \quad \cdot \qquad b$$

Solve. We solve the equation.

$$45 = 30\% \cdot b$$
$$45 = 0.3b \quad \text{Converting to decimal notation}$$
$$\frac{45}{0.3} = \frac{b}{0.3}$$
$$150 = b$$

Thus, 45 is 30% of 150. The answer is 150.

5. Translate.

$$\begin{array}{ccccc} \text{What} & \text{is} & 65\% & \text{of} & 840? \\ \downarrow & \downarrow & \downarrow & \downarrow & \downarrow \\ a & = & 65\% & \cdot & 840 \end{array}$$

Solve. We convert 65% to decimal notation and multiply.

$$a = 65\% \cdot 840$$
$$a = 0.65 \times 840$$
$$a = 546$$

Thus, 546 is 65% of 840. The answer is 546.

7. Translate.

$$\begin{array}{ccccc} 30 & \text{is} & \text{what percent} & \text{of} & 125? \\ \downarrow & \downarrow & \downarrow & \downarrow & \downarrow \\ 30 & = & p & \cdot & 125 \end{array}$$

Solve. We solve the equation.

$$30 = p \cdot 125$$
$$\frac{30}{125} = \frac{p \cdot 125}{125}$$
$$0.24 = p$$
$$24\% = p$$

Thus, 30 is 24% of 125. The answer is 24%.

9. Translate.

$$\begin{array}{ccccc} 12\% & \text{of} & \text{what number} & \text{is} & 0.3? \\ \downarrow & \downarrow & \downarrow & \downarrow & \downarrow \\ 12\% & \cdot & b & = & 0.3 \end{array}$$

Solve. We solve the equation.

$$12\% \cdot b = 0.3$$
$$0.12b = 0.3 \qquad \text{Converting to decimal notation}$$
$$\frac{b}{0.12} = \frac{0.3}{0.12}$$
$$b = 2.5$$

Thus, 12% of 2.5 is 0.3. The answer is 2.5.

11. Translate.

$$\begin{array}{ccccc} 2 & \text{is} & \text{what percent} & \text{of} & 40? \\ \downarrow & \downarrow & \downarrow & \downarrow & \downarrow \\ 2 & = & p & \cdot & 40 \end{array}$$

Solve. We divide by 40 on both sides and convert the answer to percent notation.

$$2 = p \cdot 40$$
$$\frac{2}{40} = \frac{p \cdot 40}{40}$$
$$0.05 = p$$
$$5\% = p$$

Thus, 2 is 5% of 40. The answer is 5%.

13. Translate.

$$\begin{array}{ccccc} \text{What percent} & \text{of} & 68 & \text{is} & 17? \\ \downarrow & & \downarrow & \downarrow & \downarrow \\ p & \cdot & 68 & = & 17 \end{array}$$

Solve. We divide by 68 on both sides and then convert to percent notation.

$$p \cdot 68 = 17$$
$$p = \frac{17}{68}$$
$$p = 0.25 = 25\%$$

The answer is 25%.

15. Translate.

$$\begin{array}{ccccc} \text{What} & \text{is} & 35\% & \text{of} & 240? \\ \downarrow & \downarrow & \downarrow & \downarrow & \downarrow \\ a & = & 35\% & \cdot & 240 \end{array}$$

Solve. We convert 35% to decimal notation and multiply.

$$a = 35\% \cdot 240$$
$$a = 0.35 \cdot 240$$
$$a = 84$$

The answer is 84.

17. Translate.

$$\begin{array}{ccccc} \text{What percent} & \text{of} & 125 & \text{is} & 30? \\ \downarrow & & \downarrow & \downarrow & \downarrow \\ p & \cdot & 125 & = & 30 \end{array}$$

Solve. We divide by 125 on both sides and then convert to percent notation.

$$p \cdot 125 = 30$$
$$p = \frac{30}{125}$$
$$p = 0.24 = 24\%$$

The answer is 24%.

19. Translate.

$$\begin{array}{ccccc} \text{What percent} & \text{of} & 300 & \text{is} & 48? \\ \downarrow & & \downarrow & \downarrow & \downarrow \\ p & \cdot & 300 & = & 48 \end{array}$$

Solve. We divide by 300 on both sides and then convert to percent notation.

$$p \cdot 300 = 48$$
$$p = \frac{48}{300}$$
$$p = 0.16 = 16\%$$

The answer is 16%.

21. Translate.

$$\begin{array}{ccccc} 14 & \text{is} & 30\% & \text{of} & \text{what number?} \\ \downarrow & \downarrow & \downarrow & \downarrow & \downarrow \\ 14 & = & 30\% & \cdot & b \end{array}$$

Solve. We solve the equation.

$$14 = 0.3b \qquad (30\% = 0.3)$$
$$\frac{14}{0.3} = b$$
$$46.\overline{6} = b$$

The answer is $46.\overline{6}$, or $46\frac{2}{3}$, or $\frac{140}{3}$.

23. *Translate*.

What is 2% of 40?

$$\downarrow \quad \downarrow \quad \downarrow \quad \downarrow \quad \downarrow$$
$$a \quad = \quad 2\% \quad \cdot \quad 40$$

***Solve*.** We convert 2% to decimal notation and multiply.

$$a = 2\% \cdot 40$$
$$a = 0.02 \cdot 40$$
$$a = 0.8$$

The answer is 0.8.

25. *Translate*.

0.8 is 16% of what number?

$$\downarrow \quad \downarrow \quad \downarrow \quad \downarrow \qquad \downarrow$$
$$0.8 \quad = \quad 16\% \quad \cdot \qquad b$$

***Solve*.** We solve the equation.

$$0.8 = 0.16b \quad (16\% = 0.16)$$
$$\frac{0.8}{0.16} = b$$
$$5 = b$$

The answer is 5.

27. *Translate*.

54 is 135% of what number?

$$\downarrow \quad \downarrow \quad \downarrow \quad \downarrow \qquad \downarrow$$
$$54 \quad = \quad 135\% \quad \cdot \qquad b$$

***Solve*.** We solve the equation.

$$54 = 1.35b \quad (135\% = 1.35)$$
$$\frac{54}{1.35} = b$$
$$40 = b$$

The answer is 40.

29. First we reword and translate.

What is 3% of $6600?

$$\downarrow \quad \downarrow \quad \downarrow \quad \downarrow \quad \downarrow$$
$$a \quad = \quad 3\% \quad \cdot \quad 6600$$

***Solve*.** We convert 3% to decimal notation and multiply.

$$a = 3\% \cdot 6600 = 0.03 \cdot 6600 = 198$$

The price of the dog is $198.

31. First we reword and translate.

What is 24% of $6600?

$$\downarrow \quad \downarrow \quad \downarrow \quad \downarrow \quad \downarrow$$
$$a \quad = \quad 24\% \quad \cdot \quad 6600$$

***Solve*.** We convert 24% to decimal notation and multiply.

$$a = 24\% \cdot 6600 = 0.24 \cdot 6600 = 1584$$

Veterinarian expenses are $1584.

33. First we reword and translate.

What is 8% of $6600?

$$\downarrow \quad \downarrow \quad \downarrow \quad \downarrow \quad \downarrow$$
$$a \quad = \quad 8\% \quad \cdot \quad 6600$$

***Solve*.** We convert 8% to decimal notation and multiply.

$$a = 8\% \cdot 6600 = 0.08 \cdot 6600 = 528$$

The cost of supplies is $528.

35. To find the percent of cars manufactured in the U.S., we first reword and translate.

11.9 million is what percent of 17.4 million?

$$\downarrow \qquad \downarrow \qquad \downarrow \qquad \downarrow \qquad \downarrow$$
$$11.9 \quad = \quad p \qquad \cdot \qquad 17.4$$

***Solve*.** We divide by 17.4 on both sides and convert to percent notation.

$$11.9 = p \cdot 17.4$$
$$\frac{11.9}{17.4} = p$$
$$0.684 \approx p$$
$$68.4\% \approx p$$

About 68.4% of the cars were manufactured in the U.S.

To find the percent of cars manufactured in Asia, we first reword and translate.

4.5 million is what percent of 17.4 million?

$$\downarrow \qquad \downarrow \qquad \downarrow \qquad \downarrow \qquad \downarrow$$
$$4.5 \quad = \quad p \qquad \cdot \qquad 17.4$$

***Solve*.** We divide by 17.4 on both sides and convert to percent notation.

$$4.5 = p \cdot 17.4$$
$$\frac{4.5}{17.4} = p$$
$$0.259 \approx p$$
$$25.9\% \approx p$$

About 25.9% of the cars were manufactured in Asia.

To find the percent of cars manufactured in Europe, we subtract the two percents found above from 100%.

$$100\% - 68.4\% - 25.9\% = 5.7\%$$

About 5.7% of the cars were manufactured in Europe.

37. First we reword and translate.

193 is 32% of what number?

$$\downarrow \quad \downarrow \quad \downarrow \quad \downarrow \qquad \downarrow$$
$$193 \quad = \quad 32\% \quad \cdot \qquad b$$

***Solve*.** We solve the equation.

$$193 = 0.32 \cdot b \quad (32\% = 0.32)$$
$$\frac{193}{0.32} = b$$
$$603 \approx b$$

Sammy Sosa had 603 at-bats.

39. First we reword and translate.

What is 8% of $3500?

$$\downarrow \quad \downarrow \quad \downarrow \quad \downarrow \quad \downarrow$$
$$a \quad = \quad 8\% \quad \cdot \quad 3500$$

***Solve*.** We convert 8% to decimal notation and multiply.

$$a = 8\% \cdot 3500 = 0.08 \cdot 3500 = 280$$

Sarah will pay $280 in interest.

41. a) First we reword and translate.

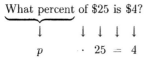

$$p \quad \cdot \ 25 \ = \ 4$$

Solve. We divide by 25 on both sides and convert to percent notation.

$$p \cdot 25 = 4$$
$$\frac{p \cdot 25}{25} = \frac{4}{25}$$
$$p = 0.16$$
$$p = 16\%$$

The tip was 16% of the cost of the meal.

b) We add to find the total cost of the meal, including tip:

$$\$25 + \$4 = \$29$$

43. a) First we reword and translate.

What is 15% of \$25?
↓   ↓   ↓   ↓   ↓
$a \quad = 15\% \ \cdot \quad 25$

Solve. We convert 15% to decimal notation and multiply.

$$a = 15\% \cdot 25$$
$$a = 0.15 \times 25$$
$$a = 3.75$$

The tip was \$3.75.

b) We add to find the total cost of the meal, including tip:

$$\$25 + \$3.75 = \$28.75$$

45. a) First we reword and translate.

15% of what is \$4.32?
↓   ↓   ↓   ↓   ↓
$15\% \ \cdot \quad b \quad = \ 4.32$

Solve. We solve the equation.

$$15\% \cdot b = 4.32$$
$$0.15 \cdot b = 4.32$$
$$\frac{0.15 \cdot b}{0.15} = \frac{4.32}{0.15}$$
$$b = 28.8$$

The cost of the meal before the tip was \$28.80.

b) We add to find the total cost of the meal, including tip:

$$\$28.80 + \$4.32 = \$33.12$$

47. First we reword and translate.

8% of what is 16?
↓   ↓   ↓   ↓   ↓
$8\% \ \cdot \quad b \quad = \ 16$

Solve. We solve the equation.

$$8\% \cdot b = 16$$
$$0.08 \cdot b = 16$$
$$\frac{0.08 \cdot b}{0.08} = \frac{16}{0.08}$$
$$b = 200$$

There were 200 women in the original study.

49. First we reword and translate.

What is 16.5% of 191?
↓   ↓   ↓   ↓   ↓
$a \quad = 16.5\% \ \cdot \ 191$

Solve. We convert 16.5% to decimal notation and multiply.

$$a = 16.5\% \cdot 191$$
$$a = 0.165 \cdot 191$$
$$a = 31.515 \approx 31.5$$

About 31.5 lb of the author's body weight is fat.

51. We subtract to find the increase.

$$\$735 - \$430 = \$305$$

The increase is \$305.

Now we find the percent increase.

\$305 is what percent of \$430?
↓   ↓   ↓   ↓   ↓
$305 \ = \qquad p \qquad \cdot \ 430$

We divide by 430 on both sides and then convert to percent notation.

$$305 = p \cdot 430$$
$$\frac{305}{430} = p$$
$$0.71 \approx p$$
$$71\% \approx p$$

The percent increase is about 71%.

53. First we find the increase in the rate for smokers.

Rate increase is 100% of \$780.
↓   ↓   ↓   ↓   ↓
$a \qquad = 100\% \ \cdot \ \ 780$

We convert 100% to decimal notation and multiply.

$$a = 100\% \cdot \$780 = 1 \cdot 780 = 780$$

The rate increase is \$780.

Now we add the rate increase to the rate for nonsmokers to find the rate for smokers.

$$\$780 + \$780 = \$1560$$

55. We subtract to find the increase.

$$\$2955 - \$1645 = \$1310$$

The increase is \$1310.

Now we find the percent increase.

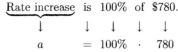

↓   ↓   ↓   ↓   ↓
$1310 \ = \qquad p \qquad \cdot \ 1645$

We divide by 1645 on both sides and then convert to percent notation.

$$1310 = p \cdot 1645$$
$$\frac{1310}{1645} = p$$
$$0.80 \approx p$$
$$80\% \approx p$$

The percent increase is about 80%.

57. Discussion and Writing Exercise

59. $-3 - 8 = -3 + (-8) = -11$

61. $-\frac{3}{5} + \frac{1}{5}$ The absolute values are $\frac{3}{5}$ and $\frac{1}{5}$. The difference is $\frac{3}{5} - \frac{1}{5}$, or $\frac{2}{5}$. The negative number has the larger absolute value, so the answer is negative. $-\frac{3}{5} + \frac{1}{5} = -\frac{2}{5}$

63.
$$-5a + 3c - 2(c - 3a)$$
$$= -5a + 3c - 2 \cdot c - 2(-3a)$$
$$= -5a + 3c - 2c + 6a$$
$$= (-5 + 6)a + (3 - 2)c$$
$$= 1 \cdot a + 1 \cdot c$$
$$= a + c$$

65. $-6.5 + 2.6 = -3.9$ The absolute values are 6.5 and 2.6. The difference is 3.9. The negative number has the larger absolute value, so the answer is negative, -3.9.

67. Since 6 ft = 6 × 1 ft = 6 × 12 in. = 72 in., we can express 6 ft, 4 in. as 72 in. + 4 in., or 76 in.

Translate. We reword the problem.

96.1% of what is 76 in.?
$$\downarrow \quad \downarrow \quad \downarrow \quad \downarrow \quad \downarrow$$
$$96.1\% \quad \cdot \quad b \quad = \quad 76$$

Solve. We solve the equation.
$$96.1\% \cdot b = 76$$
$$0.961 \cdot b = 76$$
$$\frac{0.961 \cdot b}{0.961} = \frac{76}{0.961}$$
$$b \approx 79$$

Note that 79 in. = 72 in. + 7 in. = 6 ft, 7 in.

Jaraan's final adult height will be about 6 ft, 7 in.

Exercise Set 7.6

1. *Familiarize*. Using the labels on the drawing in the text, we let $x = $ the length of the shorter piece, in inches, and $3x = $ the length of the longer piece, in inches.

Translate. We reword the problem.

The length of the shorter piece plus the length of the longer piece is 240 ft.
$$\downarrow \qquad\qquad \downarrow \qquad \downarrow \qquad\qquad \downarrow \quad \downarrow$$
$$x \qquad\qquad + \qquad 3x \qquad\qquad = \quad 240$$

Solve. We solve the equation.
$$x + 3x = 240$$
$$4x = 240 \quad \text{Collecting like terms}$$
$$\frac{4x}{4} = \frac{240}{4}$$
$$x = 60$$

If x is 60, then $3x = 3 \cdot 60$, or 180.

Check. 180 is three times 60, and $60 + 180 = 240$. The answer checks.

State. The lengths of the pieces are 60 in. and 180 in.

3. *Familiarize*. Let $c = $ the cost of one box of Wheaties.

Translate.

Total cost is Number of boxes times Price of one box
$$\downarrow \qquad \downarrow \qquad \downarrow \qquad \downarrow \qquad \downarrow$$
$$14.68 \quad = \quad 4 \quad \cdot \quad c$$

Solve. We solve the equation.
$$14.68 = 4 \cdot c$$
$$\frac{14.68}{4} = c \quad \text{Dividing by 4}$$
$$3.67 = c$$

Check. If one box of Wheaties costs \$3.67, then 4 boxes cost 4(\$3.67), or \$14.68. The answer checks.

State. One box of Wheaties costs \$3.67.

5. *Familiarize*. Let $d = $ the amount spent on women's dresses, in billions of dollars.

Translate.

Amount spent on blouses was \$0.2 billion more than amount spent on dresses
$$\downarrow \qquad \downarrow \qquad \downarrow \qquad \downarrow \qquad \downarrow$$
$$6.5 \quad = \quad 0.2 \quad + \quad d$$

Solve. We solve the equation.
$$6.5 = 0.2 + d$$
$$6.5 - 0.2 = 0.2 + d - 0.2 \quad \text{Subtracting 0.2}$$
$$6.3 = d$$

Check. If we add \$0.2 billion to \$6.3 billion, we get \$6.5 billion. The answer checks.

State. \$6.3 billion was spent on dresses.

7. *Familiarize*. Let $d = $ the musher's distance from Nome, in miles. Then $2d = $ the distance from Anchorage, in miles. This is the number of miles the musher has completed. The sum of the two distances is the length of the race, 1049 miles.

Translate.

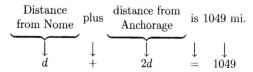

Distance from Nome plus distance from Anchorage is 1049 mi.
$$\downarrow \qquad \downarrow \qquad \downarrow \qquad \downarrow \quad \downarrow$$
$$d \qquad + \qquad 2d \qquad = \quad 1049$$

Carry out. We solve the equation.

$$d + 2d = 1049$$
$$3d = 1049 \qquad \text{Collecting like terms}$$
$$\frac{3d}{3} = \frac{1049}{3}$$
$$d = \frac{1049}{3}$$

If $d = \dfrac{1049}{3}$, then $2d = 2 \cdot \dfrac{1049}{3} = \dfrac{2098}{3} = 699\dfrac{1}{3}$.

Check. $\dfrac{2098}{3}$ is twice $\dfrac{1049}{3}$, and $\dfrac{1049}{3} + \dfrac{2098}{3} = \dfrac{3147}{3} = 1049$. The result checks.

State. The musher has traveled $699\dfrac{1}{3}$ miles.

9. ***Familiarize***. Using the labels on the drawing in the text, we let $x =$ the smaller number and $x + 1 =$ the larger number.

Translate. We reword the problem.

$$\underbrace{\text{First number}}_{\downarrow} + \underbrace{\text{second number}}_{\downarrow} \text{ is } \overbrace{547}^{\downarrow\ \downarrow}$$
$$x \qquad + \qquad (x + 1) \qquad = 547$$

Solve. We solve the equation.

$$x + (x + 1) = 547$$
$$2x + 1 = 547 \qquad \text{Collecting like terms}$$
$$2x + 1 - 1 = 547 - 1 \qquad \text{Subtracting 1}$$
$$2x = 546$$
$$\frac{2x}{2} = \frac{546}{2} \qquad \text{Dividing by 2}$$
$$x = 273$$

If x is 273, then $x + 1$ is 274.

Check. 273 and 274 are consecutive integers, and their sum is 547. The answer checks.

State. The page numbers are 273 and 274.

11. ***Familiarize***. Let $a =$ the first number. Then $a + 1 =$ the second number, and $a + 2 =$ the third number.

Translate. We reword the problem.

$$\underbrace{\text{First number}}_{\downarrow} + \underbrace{\text{second number}}_{\downarrow} + \underbrace{\text{third number}}_{\downarrow} \text{ is } 126$$
$$a \qquad + \qquad (a + 1) \qquad + \qquad (a + 2) \qquad = 114$$

Solve. We solve the equation.

$$a + (a + 1) + (a + 2) = 126$$
$$3a + 3 = 126 \qquad \text{Collecting like terms}$$
$$3a + 3 - 3 = 126 - 3$$
$$3a = 123$$
$$\frac{3a}{3} = \frac{123}{3}$$
$$a = 41$$

If a is 41, then $a + 1$ is 42 and $a + 2$ is 43.

Check. 41, 42, and 43 are consecutive integers, and their sum is 126. The answer checks.

State. The numbers are 41, 42, and 43.

13. ***Familiarize***. Let $x =$ the first odd integer. Then $x + 2 =$ the next odd integer and $(x + 2) + 2$, or $x + 4 =$ the third odd integer.

Translate. We reword the problem.

$$\underbrace{\text{First odd integer}}_{\downarrow} + \underbrace{\text{second odd integer}}_{\downarrow} + \underbrace{\text{third odd integer}}_{\downarrow} \text{ is } \overbrace{189}^{\downarrow\ \downarrow}$$
$$x \qquad + \qquad (x + 2) \qquad + \qquad (x + 4) \qquad = 189$$

Solve. We solve the equation.

$$x + (x + 2) + (x + 4) = 189$$
$$3x + 6 = 189 \qquad \text{Collecting like terms}$$
$$3x + 6 - 6 = 189 - 6$$
$$3x = 183$$
$$\frac{3x}{3} = \frac{183}{3}$$
$$x = 61$$

If x is 61, then $x + 2$ is 63 and $x + 4$ is 65.

Check. 61, 63, and 65 are consecutive odd integers, and their sum is 189. The answer checks.

State. The integers are 61, 63, and 65.

15. ***Familiarize***. Using the labels on the drawing in the text, we let $w =$ the width and $3w + 6 =$ the length. The perimeter P of a rectangle is given by the formula $2l + 2w = P$, where $l =$ the length and $w =$ the width.

Translate. Substitute $3w + 6$ for l and 124 for P:

$$2l + 2w = P$$
$$2(3w + 6) + 2w = 124$$

Solve. We solve the equation.

$$2(3w + 6) + 2w = 124$$
$$6w + 12 + 2w = 124$$
$$8w + 12 = 124$$
$$8w + 12 - 12 = 124 - 12$$
$$8w = 112$$
$$\frac{8w}{8} = \frac{112}{8}$$
$$w = 14$$

The possible dimensions are $w = 14$ ft and $l = 3w + 6 = 3(14) + 6$, or 48 ft.

Check. The length, 48 ft, is 6 ft more than three times the width, 14 ft. The perimeter is $2(48 \text{ ft}) + 2(14 \text{ ft}) = 96 \text{ ft} + 28 \text{ ft} = 124$ ft. The answer checks.

State. The width is 14 ft, and the length is 48 ft.

17. ***Familiarize***. Let $p =$ the regular price of the shoes. At 15% off, Amy paid 85% of the regular price.

Translate.

$$\$63.75 \text{ is } 85\% \text{ of } \underbrace{\text{the regular price.}}_{\downarrow}$$
$$63.75 \ = 0.85 \ \cdot \qquad p$$

Solve. We solve the equation.

$$63.75 = 0.85p$$

$$\frac{63.75}{0.08} = p \qquad \text{Dividing both sides by 0.85}$$

$$75 = p$$

Check. 85% of $75, or 0.85($75), is $63.75. The answer checks.

State. The regular price was $75.

19. *Familiarize*. Let $b =$ the price of the book itself. When the sales tax rate is 5%, the tax paid on the book is 5% of b, or $0.05b$.

Translate.

$$\underbrace{\text{Price of book}}_{b} \; \text{plus} \; \underbrace{\text{sales tax}}_{0.05b} \; \text{is} \; \underbrace{\$89.25.}_{}$$

$$b + 0.05b = 89.25$$

Solve. We solve the equation.

$$b + 0.05b = 89.25$$

$$1.05b = 89.25$$

$$b = \frac{89.25}{1.05}$$

$$b = 85$$

Check. 5% of $85, or 0.05($85), is $4.25 and $85 + $4.25 is $89.25, the total cost. The answer checks.

State. The book itself cost $85.

21. *Familiarize*. Let $n =$ the number of visits required for a total parking cost of $27.00. The parking cost for each $1\frac{1}{2}$ hour visit is $1.50 for the first hour plus $1.00 for part of a second hour, or $2.50. Then the total parking cost for n visits is $2.50n$ dollars.

Translate. We reword the problem.

$$\underbrace{\text{Total parking cost}}_{2.50n} \; \text{is} \; \underbrace{\$27.00.}_{= \; 27.00}$$

Solve. We solve the equation.

$$2.5n = 27$$

$$10(2.5n) = 10(27) \qquad \text{Clearing the decimal}$$

$$25n = 270$$

$$\frac{25n}{25} = \frac{270}{25}$$

$$n = 10.8$$

If the total parking cost is $27.00 for 10.8 visits, then the cost will be more than $27.00 for 11 or more visits.

Check. The parking cost for 10 visits is $2.50(10), or $25, and the parking cost for 11 visits is $2.50(11), or $27.50. Since 11 is the smallest number for which the parking cost exceeds $27.00, the answer checks.

State. The minimum number of weekly visits for which it is worthwhile to buy a parking pass is 11.

23. *Familiarize*. Let $x =$ the measure of the first angle. Then $3x =$ the measure of the second angle, and $x + 40 =$ the

measure of the third angle. Recall that the sum of measures of the angles of a triangle is 180°.

Translate.

$$\underbrace{\text{Measure of first angle}}_{x} + \underbrace{\text{measure of second angle}}_{3x} + \underbrace{\text{measure of third angle}}_{(x + 40)} \; \text{is} \; \underbrace{180.}_{= \; 180}$$

Solve. We solve the equation.

$$x + 3x + (x + 40) = 180$$

$$5x + 40 = 180$$

$$5x + 40 - 40 = 180 - 40$$

$$5x = 140$$

$$\frac{5x}{5} = \frac{140}{5}$$

$$x = 28$$

Possible answers for the angle measures are as follows:

First angle: $x = 28°$

Second angle: $3x = 3(28) = 84°$

Third angle: $x + 40 = 28 + 40 = 68°$

Check. Consider 28°, 84°, and 68°. The second angle is three times the first, and the third is 40° more than the first. The sum, 28° + 84° + 68°, is 180°. These numbers check.

State. The measures of the angles are 28°, 84°, and 68°.

25. *Familiarize*. Using the labels on the drawing in the text, we let $x =$ the measure of the first angle, $x + 5 =$ the measure of the second angle, and $3x + 10 =$ the measure of the third angle. Recall that the sum of measures of the angles of a triangle is 180°.

Translate.

$$\underbrace{\text{Measure of first angle}}_{x} + \underbrace{\text{measure of second angle}}_{(x + 5)} + \underbrace{\text{measure of third angle}}_{(3x + 10)} \; \text{is} \; \underbrace{180.}_{= \; 180}$$

Solve. We solve the equation.

$$x + (x + 5) + (3x + 10) = 180$$

$$5x + 15 = 180$$

$$5x + 15 - 15 = 180 - 15$$

$$5x = 165$$

$$\frac{5x}{5} = \frac{165}{5}$$

$$x = 33$$

Possible answers for the angle measures are as follows:

First angle: $x = 33°$

Second angle: $x + 5 = 33 + 5 = 38°$

Third angle: $3x + 10 = 3(33) + 10 = 109°$

Check. The second angle is 5° more than the first, and the third is 10° more than 3 times the first. The sum, 33° + 38° + 109°, is 180°. The numbers check.

State. The measures of the angles are 33°, 38°, and 109°.

27. Familiarize. Let a = the amount Sarah invested. The investment grew by 28% of a, or $0.28a$.

Translate.

$$\underbrace{\text{Amount invested}}_{a} \underbrace{\text{plus}}_{+} \underbrace{\text{amount of growth}}_{0.28a} \underbrace{\text{is}}_{=} \underbrace{\$448.}_{448}$$

Solve. We solve the equation.

$$a + 0.28a = 448$$
$$1.28a = 448$$
$$a = 350$$

Check. 28% of \$350 is 0.28(\$350), or \$98, and \$350 + \$98 = \$448. The answer checks.

State. Sarah invested \$350.

29. Familiarize. Let b = the balance in the account at the beginning of the month. The balance grew by 2% of b, or $0.02b$.

Translate.

$$\underbrace{\text{Original balance}}_{b} \underbrace{\text{plus}}_{+} \underbrace{\text{amount of growth}}_{0.02b} \underbrace{\text{is}}_{=} \underbrace{\$870.}_{870}$$

Solve. We solve the equation.

$$b + 0.02b = 870$$
$$1.02b = 870$$
$$b \approx \$852.94$$

Check. 2% of \$852.94 is 0.02(\$852.94), or \$17.06, and \$852.94 + \$17.06 = \$870. The answer checks.

State. The balance at the beginning of the month was \$852.94.

31. Familiarize. The total cost is the initial charge plus the mileage charge. Let d = the distance, in miles, that Courtney can travel for \$12. The mileage charge is the cost per mile times the number of miles traveled or $0.75d$.

Translate.

$$\underbrace{\text{Initial charge}}_{3} \underbrace{\text{plus}}_{+} \underbrace{\text{mileage charge}}_{0.75d} \underbrace{\text{is}}_{=} \underbrace{\$12.}_{12}$$

Solve. We solve the equation.

$$3 + 0.75d = 12$$
$$0.75d = 9$$
$$d = 12$$

Check. A 12-mi taxi ride from the airport would cost \$3 + 12(\$0.75), or \$3 + \$9, or \$12. The answer checks.

State. Courtney can travel 12 mi from the airport for \$12.

33. Familiarize. Let c = the cost of the meal before the tip. We know that the cost of the meal before the tip plus the tip, 15% of the cost, is the total cost, \$41.40.

Translate.

$$\underbrace{\text{Cost of meal}}_{c} \underbrace{\text{plus}}_{+} \underbrace{\text{tip}}_{15\%c} \underbrace{\text{is}}_{=} \underbrace{\$41.40}_{41.40}$$

Solve. We solve the equation.

$$c + 15\%c = 41.40$$
$$c + 0.15c = 41.40$$
$$1c + 0.15c = 41.40$$
$$1.15c = 41.40$$
$$\frac{1.15c}{1.15} = \frac{41.40}{1.15}$$
$$c = 36$$

Check. We find 15% of \$36 and add it to \$36:

$15\% \times \$36 = 0.15 \times \$36 = \$5.40$ and $\$36 + \$5.40 = \$41.40$. The answer checks.

State. The cost of the meal before the tip was added was \$36.

35. Discussion and Writing Exercise

37.
$$-\frac{4}{5} - \frac{3}{8} = -\frac{4}{5} + \left(-\frac{3}{8}\right)$$
$$= -\frac{32}{40} + \left(-\frac{15}{40}\right)$$
$$= -\frac{47}{40}$$

39.
$$-\frac{4}{5} \cdot \frac{3}{8} = -\frac{4 \cdot 3}{5 \cdot 8}$$
$$= -\frac{4 \cdot 3}{5 \cdot 2 \cdot 4}$$
$$= -\frac{\cancel{4} \cdot 3}{5 \cdot 2 \cdot \cancel{4}}$$
$$= -\frac{3}{10}$$

41. $\frac{1}{10} \div \left(-\frac{1}{100}\right) = \frac{1}{10} \cdot \left(-\frac{100}{1}\right) = -\frac{1 \cdot 100}{10 \cdot 1} =$
$$-\frac{\cancel{1} \cdot \cancel{10} \cdot 10}{\cancel{10} \cdot \cancel{1} \cdot 1} = -\frac{10}{1} = -10$$

43. $-25.6(-16) = 409.6$

45. $-25.6 + (-16) = -41.6$

47. Familiarize. Let a = the original number of apples. Then $\frac{1}{3}a$, $\frac{1}{4}a$, $\frac{1}{8}a$, and $\frac{1}{5}a$ are given to four people, respectively. The fifth and sixth people get 10 apples and 1 apple, respectively.

Translate. We reword the problem.

$$\underbrace{\text{The total number of apples}}_{\frac{1}{3}a + \frac{1}{4}a + \frac{1}{8}a + \frac{1}{5}a + 10 + 1} \underbrace{\text{is}}_{=} \underbrace{a}_{a}$$

Solve. We solve the equation.

$$\frac{1}{3}a + \frac{1}{4}a + \frac{1}{8}a + \frac{1}{5}a + 10 + 1 = a, \quad \text{LCD is } 120$$

$$120\left(\frac{1}{3}a + \frac{1}{4}a + \frac{1}{8}a + \frac{1}{5}a + 11\right) = 120 \cdot a$$

$$40a + 30a + 15a + 24a + 1320 = 120a$$

$$109a + 1320 = 120a$$

$$1320 = 11a$$

$$120 = a$$

Check. If the original number of apples was 120, then the first four people got $\frac{1}{3} \cdot 120, \frac{1}{4} \cdot 120, \frac{1}{8} \cdot 120$, and $\frac{1}{5} \cdot 120$, or 40, 30, 15, and 24 apples, respectively. Adding all the apples we get $40 + 30 + 15 + 24 + 10 + 1$, or 120. The result checks.

State. There were originally 120 apples in the basket.

49. Divide the largest triangle into three triangles, each with a vertex at the center of the circle and with height x as shown.

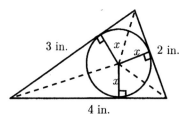

3 in. x 2 in.

4 in.

Then the sum of the areas of the three smaller triangles is the area of the original triangle. We have:

$$\frac{1}{2} \cdot 3x + \frac{1}{2} \cdot 2x + \frac{1}{2} \cdot 4x = 2.9047$$

$$2\left(\frac{1}{2} \cdot 3x + \frac{1}{2} \cdot 2x + \frac{1}{2} \cdot 4x\right) = 2(2.9047)$$

$$3x + 2x + 4x = 5.8094$$

$$9x = 5.8094$$

$$x \approx 0.65$$

Thus, x is about 0.65 in.

51. *Familiarize.* Let p = the price of the gasoline as registered on the pump. Then the sales tax will be $9\%p$.

Translate. We reword the problem.

Price on pump plus sales tax is $10

$$p \quad + \quad 9\%p \quad = \quad 10$$

Solve. We solve the equation.

$$p + 9\%p = 10$$

$$1p + 0.09p = 10$$

$$1.09p = 10$$

$$\frac{1.09p}{1.09} = \frac{10}{1.09}$$

$$p \approx 9.17$$

Check. We find 9% of $9.17 and add it to $9.17:

$$9\% \times \$9.17 = 0.09 \times \$9.17 \approx \$0.83$$

Then $9.17 + $0.83 = $10, so $9.17 checks.

State. The attendant should have filled the tank until the pump read $9.17, not $9.10.

Exercise Set 7.7

1. $x > -4$

   a) Since $4 > -4$ is true, 4 is a solution.

   b) Since $0 > -4$ is true, 0 is a solution.

   c) Since $-4 > -4$ is false, -4 is not a solution.

   d) Since $6 > -4$ is true, 6 is a solution.

   e) Since $5.6 > -4$ is true, 5.6 is a solution.

3. $x \geq 6.8$

   a) Since $-6 \geq 6.8$ is false, -6 is not a solution.

   b) Since $0 \geq 6.8$ is false, 0 is not a solution.

   c) Since $6 \geq 6.8$ is false, 6 is not a solution.

   d) Since $8 \geq 6.8$ is true, 8 is a solution.

   e) Since $-3\frac{1}{2} \geq 6.8$ is false, $-3\frac{1}{2}$ is not a solution.

5. The solutions of $x > 4$ are those numbers greater than 4. They are shown on the graph by shading all points to the right of 4. The open circle at 4 indicates that 4 is not part of the graph.

$x > 4$

   −5 −4 −3 −2 −1 0 1 2 3 4 5

7. The solutions of $t < -3$ are those numbers less than -3. They are shown on the graph by shading all points to the left of -3. The open circle at -3 indicates that -3 is not part of the graph.

$t < -3$

   −5 −4 −3 −2 −1 0 1 2 3 4 5

9. The solutions of $m \geq -1$ are are shown by shading the point for -1 and all points to the right of -1. The closed circle at -1 indicates that -1 is part of the graph.

$m \geq -1$

   −5 −4 −3 −2 −1 0 1 2 3 4 5

11. In order to be a solution of the inequality $-3 < x \leq 4$, a number must be a solution of both $-3 < x$ and $x \leq 4$. The solution set is graphed as follows:

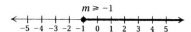

$-3 < x \leq 4$

   −5 −4 −3 −2 −1 0 1 2 3 4 5

The open circle at -3 means that -3 is not part of the graph. The closed circle at 4 means that 4 is part of the graph.

13. In order to be a solution of the inequality $0 < x < 3$, a number must be a solution of both $0 < x$ and $x < 3$. The solution set is graphed as follows:

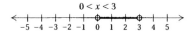

$$0 < x < 3$$

The open circles at 0 and at 3 mean that 0 and 3 are not part of the graph.

15.
$$x + 7 > 2$$
$$x + 7 - 7 > 2 - 7 \qquad \text{Subtracting 7}$$
$$x > -5 \qquad \text{Simplifying}$$

The solution set is $\{x | x > -5\}$.

The graph is as follows:

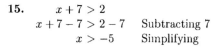

17.
$$x + 8 \le -10$$
$$x + 8 - 8 \le -10 - 8 \qquad \text{Subtracting 8}$$
$$x \le -18 \qquad \text{Simplifying}$$

The solution set is $\{x | x \le -18\}$.

The graph is as follows:

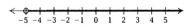

$$-18$$

19.
$$y - 7 > -12$$
$$y - 7 + 7 > -12 + 7 \qquad \text{Adding 7}$$
$$y > -5 \qquad \text{Simplifying}$$

The solution set is $\{y | y > -5\}$.

21.
$$2x + 3 > x + 5$$
$$2x + 3 - 3 > x + 5 - 3 \qquad \text{Subtracting 3}$$
$$2x > x + 2 \qquad \text{Simplifying}$$
$$2x - x > x + 2 - x \qquad \text{Subtracting } x$$
$$x > 2 \qquad \text{Simplifying}$$

The solution set is $\{x | x > 2\}$.

23.
$$3x + 9 \le 2x + 6$$
$$3x + 9 - 9 \le 2x + 6 - 9 \qquad \text{Subtracting 9}$$
$$3x \le 2x - 3 \qquad \text{Simplifying}$$
$$3x - 2x \le 2x - 3 - 2x \qquad \text{Subtracting } 2x$$
$$x \le -3 \qquad \text{Simplifying}$$

The solution set is $\{x | x \le -3\}$.

25.
$$5x - 6 < 4x - 2$$
$$5x - 6 + 6 < 4x - 2 + 6$$
$$5x < 4x + 4$$
$$5x - 4x < 4x + 4 - 4x$$
$$x < 4$$

The solution set is $\{x | x < 4\}$.

27.
$$-9 + t > 5$$
$$-9 + t + 9 > 5 + 9$$
$$t > 14$$

The solution set is $\{t | t > 14\}$.

29.
$$y + \frac{1}{4} \le \frac{1}{2}$$
$$y + \frac{1}{4} - \frac{1}{4} \le \frac{1}{2} - \frac{1}{4}$$
$$y \le \frac{2}{4} - \frac{1}{4} \qquad \text{Obtaining a common de-}$$
$$\qquad \qquad \qquad \text{nominator}$$
$$y \le \frac{1}{4}$$

The solution set is $\left\{ y \middle| y \le \frac{1}{4} \right\}$.

31.
$$x - \frac{1}{3} > \frac{1}{4}$$
$$x - \frac{1}{3} + \frac{1}{3} > \frac{1}{4} + \frac{1}{3}$$
$$x > \frac{3}{12} + \frac{4}{12} \qquad \text{Obtaining a common de-}$$
$$\qquad \qquad \qquad \text{nominator}$$
$$x > \frac{7}{12}$$

The solution set is $\left\{ x \middle| x > \frac{7}{12} \right\}$.

33.
$$5x < 35$$
$$\frac{5x}{5} < \frac{35}{5} \qquad \text{Dividing by 5}$$
$$x < 7$$

The solution set is $\{x | x < 7\}$. The graph is as follows:

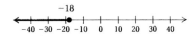

35.
$$-12x > -36$$
$$\frac{-12x}{-12} < \frac{-36}{-12} \qquad \text{Dividing by } -12$$
$$\qquad \qquad \quad \text{The symbol has to be reversed.}$$
$$x < 3 \qquad \qquad \text{Simplifying}$$

The solution set is $\{x | x < 3\}$. The graph is as follows:

37.
$$5y \ge -2$$
$$\frac{5y}{5} \ge \frac{-2}{5} \qquad \text{Dividing by 5}$$
$$y \ge -\frac{2}{5}$$

The solution set is $\left\{ y \middle| y \ge -\frac{2}{5} \right\}$.

39.
$$-2x \le 12$$
$$\frac{-2x}{-2} \ge \frac{12}{-2} \qquad \text{Dividing by } -2$$
$$\qquad \qquad \quad \text{The symbol has to be reversed.}$$
$$x \ge -6 \qquad \qquad \text{Simplifying}$$

The solution set is $\{x | x \ge -6\}$.

41.
$$-4y \ge -16$$
$$\frac{-4y}{-4} \le \frac{-16}{-4} \qquad \text{Dividing by } -4$$
$$\qquad \qquad \quad \text{The symbol has to be reversed.}$$
$$y \le 4 \qquad \qquad \text{Simplifying}$$

The solution set is $\{y | y \le 4\}$.

43. $-3x < -17$

$\dfrac{-3x}{-3} > \dfrac{-17}{-3}$ Dividing by -3

⌐— The symbol has to be reversed.

$x > \dfrac{17}{3}$ Simplifying

The solution set is $\left\{x\middle|x > \dfrac{17}{3}\right\}$.

45. $-2y > \dfrac{1}{7}$

$-\dfrac{1}{2}\cdot(-2y) < -\dfrac{1}{2}\cdot\dfrac{1}{7}$

⌐— The symbol has to be reversed.

$y < -\dfrac{1}{14}$

The solution set is $\left\{y\middle|y < -\dfrac{1}{14}\right\}$.

47. $-\dfrac{6}{5} \le -4x$

$-\dfrac{1}{4}\cdot\left(-\dfrac{6}{5}\right) \ge -\dfrac{1}{4}\cdot(-4x)$

$\dfrac{6}{20} \ge x$

$\dfrac{3}{10} \ge x$, or $x \le \dfrac{3}{10}$

The solution set is $\left\{x\middle|\dfrac{3}{10} \ge x\right\}$, or $\left\{x\middle|x \le \dfrac{3}{10}\right\}$.

49. $4 + 3x < 28$

$-4 + 4 + 3x < -4 + 28$ Adding -4

$3x < 24$ Simplifying

$\dfrac{3x}{3} < \dfrac{24}{3}$ Dividing by 3

$x < 8$

The solution set is $\{x|x < 8\}$.

51. $3x - 5 \le 13$

$3x - 5 + 5 \le 13 + 5$ Adding 5

$3x \le 18$

$\dfrac{3x}{3} \le \dfrac{18}{3}$ Dividing by 3

$x \le 6$

The solution set is $\{x|x \le 6\}$.

53. $13x - 7 < -46$

$13x - 7 + 7 < -46 + 7$

$13x < -39$

$\dfrac{13x}{13} < \dfrac{-39}{13}$

$x < -3$

The solution set is $\{x|x < -3\}$.

55. $30 > 3 - 9x$

$30 - 3 > 3 - 9x - 3$ Subtracting 3

$27 > -9x$

$\dfrac{27}{-9} < \dfrac{-9x}{-9}$ Dividing by -9

⌐— The symbol has to be reversed.

$-3 < x$

The solution set is $\{x| - 3 < x\}$, or $\{x|x > -3\}$.

57. $4x + 2 - 3x \le 9$

$x + 2 \le 9$ Collecting like terms

$x + 2 - 2 \le 9 - 2$

$x \le 7$

The solution set is $\{x|x \le 7\}$.

59. $-3 < 8x + 7 - 7x$

$-3 < x + 7$ Collecting like terms

$-3 - 7 < x + 7 - 7$

$-10 < x$

The solution set is $\{x| - 10 < x\}$, or $\{x|x > -10\}$.

61. $6 - 4y > 4 - 3y$

$6 - 4y + 4y > 4 - 3y + 4y$ Adding $4y$

$6 > 4 + y$

$-4 + 6 > -4 + 4 + y$ Adding -4

$2 > y$, or $y < 2$

The solution set is $\{y|2 > y\}$, or $\{y|y < 2\}$.

63. $5 - 9y \le 2 - 8y$

$5 - 9y + 9y \le 2 - 8y + 9y$

$5 \le 2 + y$

$-2 + 5 \le -2 + 2 + y$

$3 \le y$, or $y \ge 3$

The solution set is $\{y|3 \le y\}$, or $\{y|y \ge 3\}$.

65. $19 - 7y - 3y < 39$

$19 - 10y < 39$ Collecting like terms

$-19 + 19 - 10y < -19 + 39$

$-10y < 20$

$\dfrac{-10y}{-10} > \dfrac{20}{-10}$

⌐— The symbol has to be reversed.

$y > -2$

The solution set is $\{y|y > -2\}$.

67. $2.1x + 45.2 > 3.2 - 8.4x$

$10(2.1x + 45.2) > 10(3.2 - 8.4x)$ Multiplying by 10 to clear decimals

$21x + 452 > 32 - 84x$

$21x + 84x > 32 - 452$ Adding $84x$ and subtracting 452

$105x > -420$

$x > -4$ Dividing by 105

The solution set is $\{x|x > -4\}$.

69. $\dfrac{x}{3} - 2 \le 1$

$3\left(\dfrac{x}{3} - 2\right) \le 3\cdot 1$ Multiplying by 3 to to clear the fraction

$x - 6 \le 3$ Simplifying

$x \le 9$ Adding 6

The solution set is $\{x|x \le 9\}$.

71. $\dfrac{y}{5} + 1 \le \dfrac{2}{5}$

$5\left(\dfrac{y}{5} + 1\right) \le 5\cdot\dfrac{2}{5}$ Clearing fractions

$y + 5 \le 2$

$y \le -3$ Subtracting 5

The solution set is $\{y|y \le -3\}$.

73. $3(2y - 3) < 27$
$$6y - 9 < 27 \qquad \text{Removing parentheses}$$
$$6y < 36 \qquad \text{Adding 9}$$
$$y < 6 \qquad \text{Dividing by 6}$$
The solution set is $\{y | y < 6\}$.

75. $2(3 + 4m) - 9 \geq 45$
$$6 + 8m - 9 \geq 45 \qquad \text{Removing parentheses}$$
$$8m - 3 \geq 45 \qquad \text{Collecting like terms}$$
$$8m \geq 48 \qquad \text{Adding 3}$$
$$m \geq 6 \qquad \text{Dividing by 8}$$
The solution set is $\{m | m \geq 6\}$.

77. $8(2t + 1) > 4(7t + 7)$
$$16t + 8 > 28t + 28$$
$$16t - 28t > 28 - 8$$
$$-12t > 20$$
$$t < -\frac{20}{12} \qquad \begin{array}{l}\text{Dividing by } -12 \text{ and}\\\text{reversing the symbol}\end{array}$$
$$t < -\frac{5}{3}$$
The solution set is $\left\{ t \left| t < -\frac{5}{3} \right. \right\}$.

79. $3(r - 6) + 2 < 4(r + 2) - 21$
$$3r - 18 + 2 < 4r + 8 - 21$$
$$3r - 16 < 4r - 13$$
$$-16 + 13 < 4r - 3r$$
$$-3 < r, \text{ or } r > -3$$
The solution set is $\{r | r > -3\}$.

81. $0.8(3x + 6) \geq 1.1 - (x + 2)$
$$2.4x + 4.8 \geq 1.1 - x - 2$$
$$10(2.4x + 4.8) \geq 10(1.1 - x - 2) \qquad \text{Clearing decimals}$$
$$24x + 48 \geq 11 - 10x - 20$$
$$24x + 48 \geq -10x - 9 \qquad \text{Collecting like terms}$$
$$24x + 10x \geq -9 - 48$$
$$34x \geq -57$$
$$x \geq -\frac{57}{34}$$
The solution set is $\left\{ x \left| x \geq -\frac{57}{34} \right. \right\}$.

83. $\frac{5}{3} + \frac{2}{3}x < \frac{25}{12} + \frac{5}{4}x + \frac{3}{4}$

The number 12 is the least common multiple of all the denominators. We multiply by 12 on both sides.
$$12\left(\frac{5}{3} + \frac{2}{3}x \right) < 12\left(\frac{25}{12} + \frac{5}{4}x + \frac{3}{4} \right)$$
$$12 \cdot \frac{5}{3} + 12 \cdot \frac{2}{3}x < 12 \cdot \frac{25}{12} + 12 \cdot \frac{5}{4}x + 12 \cdot \frac{3}{4}$$
$$20 + 8x < 25 + 15x + 9$$
$$20 + 8x < 34 + 15x$$
$$20 - 34 < 15x - 8x$$
$$-14 < 7x$$
$$-2 < x, \text{ or } x > -2$$
The solution set is $\{x | x > -2\}$.

85. Discussion and Writing Exercise

87. $-56 + (-18)$ Two negative numbers. Add the absolute values and make the answer negative.
$$-56 + (-18) = -74$$

89. $-\frac{3}{4} + \frac{1}{8}$ One negative and one positive number. Find the difference of the absolute values. Then make the answer negative, since the negative number has the larger absolute value.
$$-\frac{3}{4} + \frac{1}{8} = -\frac{6}{8} + \frac{1}{8} = -\frac{5}{8}$$

91. $-56 - (-18) = -56 + 18 = -38$

93. $-2.3 - 7.1 = -2.3 + (-7.1) = -9.4$

95. $5 - 3^2 + (8 - 2)^2 \cdot 4 = 5 - 3^2 + 6^2 \cdot 4$
$$= 5 - 9 + 36 \cdot 4$$
$$= 5 - 9 + 144$$
$$= -4 + 144$$
$$= 140$$

97. $5(2x - 4) - 3(4x + 1) = 10x - 20 - 12x - 3 = -2x - 23$

99. $|x| < 3$

a) Since $|0| = 0$ and $0 < 3$ is true, 0 is a solution.

b) Since $|-2| = 2$ and $2 < 3$ is true, -2 is a solution.

c) Since $|-3| = 3$ and $3 < 3$ is false, -3 is not a solution.

d) Since $|4| = 4$ and $4 < 3$ is false, 4 is not a solution.

e) Since $|3| = 3$ and $3 < 3$ is false, 3 is not a solution.

f) Since $|1.7| = 1.7$ and $1.7 < 3$ is true, 1.7 is a solution.

g) Since $|-2.8| = 2.8$ and $2.8 < 3$ is true, -2.8 is a solution.

101. $x + 3 \leq 3 + x$
$$x - x \leq 3 - 3 \qquad \text{Subtracting } x \text{ and 3}$$
$$0 \leq 0$$
We get an inequality that is true for all values of x, so the inequality is true for all real numbers.

Exercise Set 7.8

1. $n \geq 7$

3. $w > 2$ kg

5. 90 mph $< s < 110$ mph

7. $a \leq 1,200,000$

9. $c \leq \$1.50$

11. $x > 8$

13. $y \leq -4$

15. $n \geq 1300$

17. $a \leq 500$ L

19. $3x + 2 < 13$, or $2 + 3x < 13$

21. *Familiarize*. Let s represent the score on the fourth test.

***Translate*.**

$\underbrace{\text{The average score}}$ $\underbrace{\text{is at least}}$ $80.$

$$\frac{82 + 76 + 78 + s}{4} \qquad \geq \qquad 80$$

***Solve*.**

$$\frac{82 + 76 + 78 + s}{4} \geq 80$$

$$4\left(\frac{82 + 76 + 78 + s}{4}\right) \geq 4 \cdot 80$$

$$82 + 76 + 78 + s \geq 320$$

$$236 + s \geq 320$$

$$s \geq 84$$

***Check*.** As a partial check we show that the average is at least 80 when the fourth test score is 84.

$$\frac{82 + 76 + 78 + 84}{4} = \frac{320}{4} = 80$$

***State*.** The student will get at least a B if the score on the fourth test is at least 84. The solution set is $\{s | x \geq 84\}$.

23. *Familiarize*. We use the formula for converting Celsius temperatures to Fahrenheit temperatures, $F = \frac{9}{5}C + 32$.

***Translate*.**

$\underbrace{\text{Fahrenheit temperature}}$ $\underbrace{\text{is less than}}$ $1945.4.$

$$\frac{9}{5}C + 32 \qquad \leq \qquad 1945.4$$

***Solve*.**

$$\frac{9}{5}C + 32 < 1945.4$$

$$\frac{9}{5}C < 1913.4$$

$$\frac{5}{9} \cdot \frac{9}{5}C < \frac{5}{9}(1913.4)$$

$$C < 1063$$

***Check*.** As a partial check we can show that the Fahrenheit temperature is less than 1945.4° for a Celsius temperature less than 1063° and is greater than 1945.4° for a Celsius temperature greater than 1063°.

$$F = \frac{9}{5} \cdot 1062 + 32 = 1943.6 < 1945.4$$

$$F = \frac{9}{5} \cdot 1064 + 32 = 1947.2 > 1945.4$$

***State*.** Gold stays solid for temperatures less than 1063°C. The solution set is $\{C | C < 1063°\}$.

25. *Familiarize*. $R = -0.075t + 3.85$

In the formula R represents the world record and t represents the years since 1930. When $t = 0$ (1930), the record was $-0.075 \cdot 0 + 3.85$, or 3.85 minutes. When $t = 2$ (1932), the record was $-0.075(2) + 3.85$, or 3.7 minutes. For what values of t will $-0.075t + 3.85$ be less than 3.5?

***Translate*.** The record is to be less than 3.5. We have the inequality

$$R < 3.5.$$

To find the t values which satisfy this condition we substitute $-0.075t + 3.85$ for R.

$$-0.075t + 3.85 < 3.5$$

***Solve*.**

$$-0.075t + 3.85 < 3.5$$

$$-0.075t < 3.5 - 3.85$$

$$-0.075t < -0.35$$

$$t > \frac{-0.35}{-0.075}$$

$$t > 4\frac{2}{3}$$

***Check*.** With inequalities it is impossible to check each solution. But we can check to see if the solution set we obtained seems reasonable.

When $t = 4\frac{1}{2}$, $R = -0.075(4.5) + 3.85$, or 3.5125.

When $t = 4\frac{2}{3}$, $R = -0.075\left(\frac{14}{3}\right) + 3.85$, or 3.5.

When $t = 4\frac{3}{4}$, $R = -0.075(4.75) + 3.85$, or 3.49375.

Since $r = 3.5$ when $t = 4\frac{2}{3}$ and R decreases as t increases, R will be less than 3.5 when t is greater than $4\frac{2}{3}$.

***State*.** The world record will be less than 3.5 minutes more than $4\frac{2}{3}$ years after 1930. If we let $Y = $ the year, then the solution set is $\{Y | Y \geq 1935\}$.

27. *Familiarize*. As in the drawing in the text, we let $L = $ the length of the envelope. Recall that the area of a rectangle is the product of the length and the width.

***Translate*.**

$\underbrace{\text{Length}}$ times $\underbrace{\text{width}}$ $\underbrace{\text{is at least}}$ $\underbrace{17\frac{1}{2} \text{ in}^2}$

$$L \qquad \cdot \qquad 3\frac{1}{2} \qquad \geq \qquad 17\frac{1}{2}$$

***Solve*.**

$$L \cdot 3\frac{1}{2} \geq 17\frac{1}{2}$$

$$L \cdot \frac{7}{2} \geq \frac{35}{2}$$

$$L \cdot \frac{7}{2} \cdot \frac{2}{7} \geq \frac{35}{2} \cdot \frac{2}{7}$$

$$L \geq 5$$

The solution set is $\{L | L \geq 5\}$.

***Check*.** We can obtain a partial check by substituting a number greater than or equal to 5 in the inequality. For example, when $L = 6$:

$$L \cdot 3\frac{1}{2} = 6 \cdot 3\frac{1}{2} = 6 \cdot \frac{7}{2} = 21 \geq 17\frac{1}{2}$$

The result appears to be correct.

***State*.** Lengths of 5 in. or more will satisfy the constraints. The solution set is $\{L | L \geq 5 \text{ in.}\}$.

29. *Familiarize*. Let $c = $ the number of copies Myra has made. The total cost of the copies is the setup fee of $5 plus $4 times the number of copies, or $4 \cdot c$.

Translate.

Setup fee	plus	copying cost	cannot exceed	$65.
↓	↓	↓	↓	↓
5	+	4c	≤	65

Solve. We solve the inequality.

$$5 + 4c \leq 65$$
$$4c \leq 60$$
$$c \leq 15$$

Check. As a partial check, we show that Myra can have 15 copies made and not exceed her $65 budget.

$$\$5 + \$4 \cdot 15 = 5 + 60 = \$65$$

State. Myra can have 15 or fewer copies made and stay within her budget.

31. *Familiarize.* Let m represent the length of a telephone call, in minutes.

Translate.

$0.75 charge	plus	charge for time used	is at least	$3.00.
↓	↓	↓	↓	↓
0.75	+	0.45m	≥	3

Solve. We solve the inequality.

$$0.75 + 0.45m \geq 3$$
$$0.45m \geq 2.25$$
$$m \geq 5$$

Check. As a partial check, we can show that if a call lasts 5 minutes it costs at least $3.00:

$$\$0.75 + \$0.45(5) = \$0.75 + \$2.25 = \$3.00.$$

State. Simon's calls last at least 5 minutes each.

33. *Familiarize.* Let c = the number of courses for which Angelica registers. Her total tuition is the $35 registration fee plus $375 times the number of courses for which she registers, or $375 · c$.

Translate.

Registration fee	plus	fee for courses	cannot exceed	$1000.
↓	↓	↓	↓	↓
35	+	375 · c	≤	1000

Solve. We solve the inequality.

$$35 + 375c \leq 1000$$
$$375c \leq 965$$
$$c \leq 2.57\overline{3}$$

Check. Although the solution set of the inequality is all numbers less than or equal to $2.57\overline{3}$, since c represents the number of courses for which Angelica registers, we round down to 2. If she registers for 2 courses, her tuition is $35 + $375 \cdot 2$, or $785 which does not exceed $1000. If she registers for 3 courses, her tuition is $35 + $375 \cdot 3$, or $1160 which exceeds $1000.

State. Angelica can register for at most 2 courses.

35. *Familiarize.* Let s = the number of servings of fruits or vegetables Dale eats on Saturday.

Translate.

Average number of fruit or vegetable servings	is at least	5.
↓	↓	↓
$\dfrac{4+6+7+4+6+4+s}{7}$	≥	5

Solve. We first multiply by 7 to clear the fraction.

$$7\left(\frac{4+6+7+4+6+4+s}{7}\right) \geq 7 \cdot 5$$
$$4+6+7+4+6+4+s \geq 35$$
$$31 + s \geq 35$$
$$s \geq 4$$

Check. As a partial check, we show that Dale can eat 4 servings of fruits or vegetables on Saturday and average at least 5 servings per day for the week:

$$\frac{4+6+7+4+6+4+4}{7} = \frac{35}{7} = 5$$

State. Dale should eat at least 4 servings of fruits or vegetables on Saturday.

37. *Familiarize.* We first make a drawing. We let l represent the length, in feet.

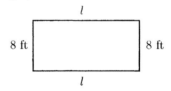

The perimeter is $P = 2l + 2w$, or $2l + 2 \cdot 8$, or $2l + 16$.

Translate. We translate to 2 inequalities.

The perimeter	is at least	200 ft.
↓	↓	↓
2l + 16	≥	200

The perimeter	is at most	200 ft.
↓	↓	↓
2l + 16	≤	200

Solve. We solve each inequality.

$$2l + 16 \geq 200 \qquad 2l + 16 \leq 200$$
$$2l \geq 184 \qquad\qquad 2l \leq 184$$
$$l \geq 92 \qquad\qquad\quad l \leq 92$$

Check. We check to see if the solutions seem reasonable.

When $l = 91$ ft, $P = 2 \cdot 91 + 16$, or 198 ft.

When $l = 92$ ft, $P = 2 \cdot 92 + 16$, or 200 ft.

When $l = 93$ ft, $P = 2 \cdot 93 + 16$, or 202 ft.

From these calculations, it appears that the solutions are correct.

State. Lengths greater than or equal to 92 ft will make the perimeter at least 200 ft. Lengths less than or equal to 92 ft will make the perimeter at most 200 ft.

39. *Familiarize*. Using the label on the drawing in the text, we let L represent the length.

The area is the length times the width, or $4L$.

***Translate*.**

$$\underbrace{\text{Area}}_{\downarrow \atop 4L} \quad \underbrace{\text{is less than}}_{\downarrow \atop <} \quad \underbrace{86 \text{ cm}^2.}_{\downarrow \atop 86}$$

***Solve*.**

$$4L < 86$$
$$L < 21.5$$

***Check*.** We check to see if the solution seems reasonable.

When $L = 22$, the area is $22 \cdot 4$, or 88 cm^2.

When $L = 21.5$, the area is $21.5(4)$, or 86 cm^2.

When $L = 21$, the area is $21 \cdot 4$, or 84 cm^2.

From these calculations, it would appear that the solution is correct.

***State*.** The area will be less than 86 cm^2 for lengths less than 21.5 cm.

41. *Familiarize*. Let $v =$ the blue book value of the car. Since the car was repaired, we know that $8500 does not exceed $0.8v$ or, in other words, $0.8v$ is at least $8500.

***Translate*.**

$$\underbrace{\substack{80\% \text{ of the} \\ \text{blue book value}}}_{\downarrow \atop 0.8v} \quad \underbrace{\text{is at least}}_{\downarrow \atop \geq} \quad \underbrace{\$8500.}_{\downarrow \atop 8500}$$

***Solve*.**

$$0.8v \geq 8500$$
$$v \geq \frac{8500}{0.8}$$
$$v \geq 10,625$$

***Check*.** As a partial check, we show that 80% of $10,625 is at least $8500:

$$0.8(\$10,625) = \$8500$$

***State*.** The blue book value of the car was at least $10,625.

43. *Familiarize*. Let $r =$ the amount of fat in a serving of the regular peanut butter, in grams. If reduced fat peanut butter has at least 25% less fat than regular peanut butter, then it has at most 75% as much fat as the regular peanut butter.

***Translate*.**

$$\underbrace{12 \text{ g of fat}}_{\downarrow \atop 12} \quad \underbrace{\text{is at most}}_{\downarrow \atop \leq} \quad \underbrace{75\%}_{\downarrow \atop 0.75} \quad \underbrace{\text{of}}_{\downarrow \atop \cdot} \quad \underbrace{\substack{\text{the amount of} \\ \text{fat in regular} \\ \text{peanut butter.}}}_{\downarrow \atop r}$$

***Solve*.**

$$12 \leq 0.75r$$
$$16 \leq r$$

***Check*.** As a partial check, we show that 12 g of fat does not exceed 75% of 16 g of fat:

$$0.75(16) = 12$$

***State*.** Regular peanut butter contains at least 16 g of fat per serving.

45. *Familiarize*. Let $w =$ the number of weeks after July 1. After w weeks the water level has dropped $\frac{2}{3}w$ ft.

***Translate*.**

$$\underbrace{\substack{\text{Original} \\ \text{depth}}}_{\downarrow \atop 25} \; \underbrace{\text{minus}}_{\downarrow \atop -} \; \underbrace{\substack{\text{drop in} \\ \text{water level}}}_{\downarrow \atop \frac{2}{3}w} \; \underbrace{\substack{\text{does not} \\ \text{exceed}}}_{\downarrow \atop \leq} \; \underbrace{21 \text{ ft.}}_{\downarrow \atop 21}$$

***Solve*.** We solve the inequality.

$$25 - \frac{2}{3}w \leq 21$$
$$-\frac{2}{3}w \leq -4$$
$$w \geq -\frac{3}{2}(-4)$$
$$w \geq 6$$

***Check*.** As a partial check we show that the water level is 21 ft 6 weeks after July 1.

$$25 - \frac{2}{3} \cdot 6 = 25 - 4 = 21 \text{ ft}$$

Since the water level continues to drop during the weeks after July 1, the answer seems reasonable.

***State*.** The water level will not exceed 21 ft for dates at least 6 weeks after July 1.

47. *Familiarize*. Let $h =$ the height of the triangle, in ft. Recall that the formula for the area of a triangle with base b and height h is $A = \frac{1}{2}bh$.

***Translate*.**

$$\underbrace{\text{Area}}_{\downarrow \atop \frac{1}{2}\left(1\frac{1}{2}\right)h} \quad \underbrace{\text{is at least}}_{\downarrow \atop \geq} \quad \underbrace{3 \text{ ft}^2.}_{\downarrow \atop 3}$$

***Solve*.** We solve the inequality.

$$\frac{1}{2}\left(1\frac{1}{2}\right)h \geq 3$$
$$\frac{1}{2} \cdot \frac{3}{2} \cdot h \geq 3$$
$$\frac{3}{4}h \geq 3$$
$$h \geq \frac{4}{3} \cdot 3$$
$$h \geq 3$$

***Check*.** As a partial check, we show that the area of the triangle is 3 ft^2 when the height is 4 ft.

$$\frac{1}{2}\left(1\frac{1}{2}\right)(4) = \frac{1}{2} \cdot \frac{3}{2} \cdot \frac{4}{1} = 3$$

***State*.** The height should be at least 4 ft.

49. *Familiarize.* The average number of calls per week is the sum of the calls for the three weeks divided by the number of weeks, 3. We let c represent the number of calls made during the third week.

Translate. The average of the three weeks is given by

$$\frac{17 + 22 + c}{3}.$$

Since the average must be at least 20, this means that it must be greater than or equal to 20. Thus, we can translate the problem to the inequality

$$\frac{17 + 22 + c}{3} \geq 20.$$

Solve. We first multiply by 3 to clear the fraction.

$$3\left(\frac{17 + 22 + c}{3}\right) \geq 3 \cdot 20$$
$$17 + 22 + c \geq 60$$
$$39 + c \geq 60$$
$$c \geq 21$$

Check. Suppose c is a number greater than or equal to 21. Then by adding 17 and 22 on both sides of the inequality we get

$$17 + 22 + c \geq 17 + 22 + 21$$
$$17 + 22 + c \geq 60$$

so

$$\frac{17 + 22 + c}{3} \geq \frac{60}{3}, \text{ or } 20.$$

State. 21 calls or more will maintain an average of at least 20 for the three-week period.

51. Discussion and Writing Exercise.

53.
$$-3 + 2(-5)^2(-3) - 7 = -3 + 2(25)(-3) - 7$$
$$= -3 + 50(-3) - 7$$
$$= -3 - 150 - 7$$
$$= -153 - 7$$
$$= -160$$

55. $23(2x - 4) - 15(10 - 3x) = 46x - 92 - 150 + 45x = 91x - 242$

57. *Familiarize.* We use the formula $F = \frac{9}{5}C + 32$.

Translate. We are interested in temperatures such that $5° < F < 15°$. Substituting for F, we have:

$$5 < \frac{9}{5}C + 32 < 15$$

Solve.

$$5 < \frac{9}{5}C + 32 < 15$$
$$5 \cdot 5 < 5\left(\frac{9}{5}C + 32\right) < 5 \cdot 15$$
$$25 < 9C + 160 < 75$$
$$-135 < 9C < -85$$
$$-15 < C < -9\frac{4}{9}$$

Check. The check is left to the student.

State. Green ski wax works best for temperatures between $-15°C$ and $-9\frac{4}{9}°C$.

59. *Familiarize.* Let $f =$ the fat content of a serving of regular tortilla chips, in grams. A product that contains 60% less fat than another product has 40% of the fat content of that product. If Reduced Fat Tortilla Pops cannot be labeled lowfat, then they contain at least 3 g of fat.

Translate.

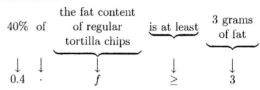

40% of	the fat content of regular tortilla chips	is at least	3 grams of fat
↓ ↓	↓	↓	↓
0.4 ·	f	$\geq$	3

Solve.

$$0.4f \geq 3$$
$$f \geq 7.5$$

Check. As a partial check, we show that 40% of 7.5 g is not less than 3 g.

$$0.4(7.5) = 3$$

State. A serving of regular tortilla chips contains at least 7.5 g of fat.

Chapter 8

Graphs of Linear Equations

Exercise Set 8.1

1. The section of the circle graph for women representing heart disease shows that 47.6% of women 65 and older die of heart disease.

3. The section of the circle graph for men representing heart disease shows that 44.6% of men 65 and older die of heart disease.

5. *Familiarize.* The circle graph shows that 33.0% of men 65 and older die of cancer. Let c = the number of men who would be expected to die of cancer.

Translate. We reword and translate.

$$\begin{array}{ccccc} \text{What} & \text{is} & 33.0\% & \text{of} & 150,000? \\ \downarrow & \downarrow & \downarrow & \downarrow & \downarrow \\ c & = & 33.0\% & \cdot & 150,000 \end{array}$$

Solve. We carry out the computation.

$$c = 33.0\% \cdot 150,000 = 0.330 \cdot 150,000 = 49,500$$

Check. We repeat the calculation. The answer checks.

State. In a group of 150,000 men age 65 and older, 49,500 of them would be expected to die of cancer.

7. *Familiarize.* The circle graph shows that 12.6% of women 65 and older die of a stroke. Let s = the number of women who would be expected to die of a stroke.

Translate. We reword and translate.

$$\begin{array}{ccccc} \text{What} & \text{is} & 12.6\% & \text{of} & 150,000? \\ \downarrow & \downarrow & \downarrow & \downarrow & \downarrow \\ s & = & 12.6\% & \cdot & 150,000 \end{array}$$

Solve. We carry out the computation.

$$s = 12.6\% \cdot 150,000 = 0.126 \cdot 150,000 = 18,900$$

Check. We repeat the calculation. The answer checks.

State. In a group of 150,000 women age 65 and older, 18,900 of them would be expected to die of a stroke.

9. We go to the top of the bar that is above the body weight 200 lb. Then we move horizontally from the top of the bar to the vertical scale listing numbers of drinks. It appears approximately 6 drinks will give a 200-lb person a blood-alcohol level of 0.10%.

11. We see that the bars for weights above 200 lb extend beyond the 6 drink level. Thus, the weight of someone who can consume 6 drinks without reaching a blood-alcohol level of 0.10% is greater than 200 lb.

13. From $3\frac{1}{2}$ on the vertical scale we move horizontally until we reach a bar whose top is above the horizontal line on which we are moving. The first such bar corresponds to a body weight of 120 lb. Thus, we can conclude an individual weighs more than 120 lb if $3\frac{1}{2}$ drinks are consumed without reaching a blood-alcohol level of 0.10%.

15. First locate 1995 on the horizontal axis and then move up to the line. Now move across to the vertical scale and read that there were approximately 17,000 alcohol-related deaths in 1995.

17. The lowest point on the graph occurs above 1998. Thus, the lowest number of deaths occurred in 1998.

19. In Exercise 15 we found that there were approximately 17,000 alcohol-related deaths in 1995. To find the number of alcohol-related deaths in 1998, first locate 1998 on the horizontal scale and then move up to the line. Now move across to the vertical scale and read that there were approximately 16,000 alcohol-related deaths in 1998. We subtract to find the decrease:

$$17,000 - 16,000 = 1000$$

Thus, alcohol-related deaths decreased by about 1000 from 1995 to 1998.

21. $(2,5)$ is 2 units right and 5 units up.

$(-1,3)$ is 1 unit left and 3 units up.

$(3,-2)$ is 3 units right and 2 units down.

$(-2,-4)$ is 2 units left and 4 units down.

$(0,4)$ is 0 units left or right and 4 units up.

$(0,-5)$ is 0 units left or right and 5 units down.

$(5,0)$ is 5 units right and 0 units up or down.

$(-5,0)$ is 5 units left and 0 units up or down.

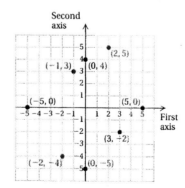

23. Since the first coordinate is negative and the second coordinate positive, the point $(-5,3)$ is located in quadrant II.

25. Since the first coordinate is positive and the second coordinate negative, the point $(100,-1)$ is in quadrant IV.

27. Since both coordinates are negative, the point $(-6,-29)$ is in quadrant III.

29. Since both coordinates are positive, the point $(3.8, 9.2)$ is in quadrant I.

31. Since the first coordinate is negative and the second coordinate is positive, the point $\left(-\dfrac{1}{3}, \dfrac{15}{7}\right)$ is in quadrant II.

33. Since the first coordinate is positive and the second coordinate is negative, the point $\left(12\dfrac{7}{8}, -1\dfrac{1}{2}\right)$ is in quadrant IV.

35. A point with both coordinates positive is in quadrant I.

37. In quadrant II, the first coordinate is negative and the second coordinate is positive.

39.

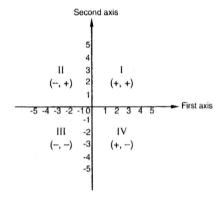

If the first coordinate is positive, then the point must be in either quadrant I or quadrant IV.

41. If the first and second coordinates are equal, they must either be both positive or both negative. The point must be in either quadrant I (both positive) or quadrant III (both negative).

43.

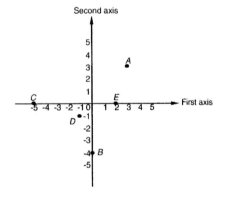

Point A is 3 units right and 3 units up. The coordinates of A are $(3, 3)$.

Point B is 0 units left or right and 4 units down. The coordinates of B are $(0, -4)$.

Point C is 5 units left and 0 units up or down. The coordinates of C are $(-5, 0)$.

Point D is 1 unit left and 1 unit down. The coordinates of D are $(-1, -1)$.

Point E is 2 units right and 0 units up or down. The coordinates of E are $(2, 0)$.

45. Discussion and Writing Exercise

47. The distance of -12 from 0 is 12, so $|-12| = 12$.

49. The distance of 0 from 0 is 0, so $|0| = 0$.

51. The distance of -3.4 from 0 is 3.4, so $|-3.4| = 3.4$.

53. The distance of $\dfrac{2}{3}$ from 0 is $\dfrac{2}{3}$, so $\left|\dfrac{2}{3}\right| = \dfrac{2}{3}$.

55. ***Familiarize***. Let $p =$ the average price of a ticket to a Boston Red Sox game in 2000. Then the price in 2001 was $p + 27.4\%p$, or $p + 0.274p$, or $1.274p$.

Translate.

$$\underbrace{\text{The price in 2001}}_{\displaystyle 1.274p} \;\; \underset{=}{\text{was}} \;\; \underset{36.08}{\$36.08.}$$

Solve. We solve the equation.

$$1.274p = 36.08$$
$$p = \dfrac{36.08}{1.274}$$
$$p \approx 28.32$$

Check. 27.4% of \$28.32 is about \$7.76 and \$28.32 + \$7.76 = \$36.08. The answer checks.

State. The average price of a ticket to a Boston Red Sox game was about \$28.32 in 2000.

57.

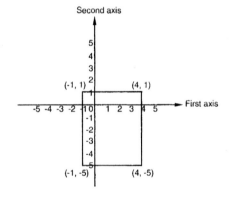

The coordinates of the fourth vertex are $(-1, -5)$.

59. Answers may vary.

We select eight points such that the sum of the coordinates for each point is 6.

$$
\begin{array}{ll}
(-1, 7) & -1 + 7 = 6 \\
(0, 6) & 0 + 6 = 6 \\
(1, 5) & 1 + 5 = 6 \\
(2, 4) & 2 + 4 = 6 \\
(3, 3) & 3 + 3 = 6 \\
(4, 2) & 4 + 2 = 6 \\
(5, 1) & 5 + 1 = 6 \\
(6, 0) & 6 + 0 = 6
\end{array}
$$

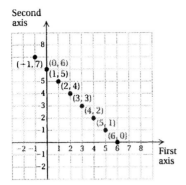

61.

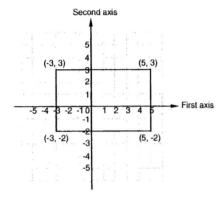

The length is 8, and the width is 5.

$P = 2l + 2w$

$P = 2 \cdot 8 + 2 \cdot 5 = 16 + 10 = 26$

Exercise Set 8.2

1. We substitute 2 for x and 9 for y (alphabetical order of variables).

$$\begin{array}{c|c} y = 3x - 1 \\ \hline 9 \ ? \ 3 \cdot 2 - 1 \\ \quad \mid \ 6 - 1 \\ \quad \mid \ 5 & \text{FALSE} \end{array}$$

Since $9 = 5$ is false, the pair $(2, 9)$ is not a solution.

3. We substitute 4 for x and 2 for y.

$$\begin{array}{c|c} 2x + 3y = 12 \\ \hline 2 \cdot 4 + 3 \cdot 2 \ ? \ 12 \\ 8 + 6 \ \mid \\ \quad 14 \ \mid & \text{FALSE} \end{array}$$

Since $14 = 12$ is false, the pair $(4, 2)$ is not a solution.

5. We substitute 3 for a and -1 for b.

$$\begin{array}{c|c} 3a - 4b = 13 \\ \hline 3 \cdot 3 - 4(-1) \ ? \ 13 \\ 9 + 4 \ \mid \\ \quad 13 \ \mid & \text{TRUE} \end{array}$$

Since $13 = 13$ is true, the pair $(3, -1)$ is a solution.

7. To show that a pair is a solution, we substitute, replacing x with the first coordinate and y with the second coordinate in each pair.

$$\begin{array}{c|c} y = x - 5 \\ \hline -1 \ ? \ 4 - 5 \\ \quad \mid \ -1 \quad \text{TRUE} \end{array} \qquad \begin{array}{c|c} y = x - 5 \\ \hline -4 \ ? \ 1 - 5 \\ \quad \mid \ -4 \quad \text{TRUE} \end{array}$$

In each case the substitution results in a true equation. Thus, $(4, -1)$ and $(1, -4)$ are both solutions of $y = x - 5$. We graph these points and sketch the line passing through them.

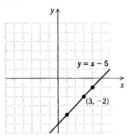

The line appears to pass through $(3, -2)$ also. We check to determine if $(3, -2)$ is a solution of $y = x - 5$.

$$\begin{array}{c|c} y = x - 5 \\ \hline -2 \ ? \ 3 - 5 \\ \quad \mid \ -2 \quad \text{TRUE} \end{array}$$

Thus, $(3, -2)$ is another solution. There are other correct answers, including $(-1, -6)$, $(2, -3)$, $(0, -5)$, $(5, 0)$, and $(6, 1)$.

9. To show that a pair is a solution, we substitute, replacing x with the first coordinate and y with the second coordinate in each pair.

$$\begin{array}{c|c} y = \frac{1}{2}x + 3 \\ \hline 5 \ ? \ \frac{1}{2} \cdot 4 + 3 \\ \quad \mid \ 2 + 3 \\ \quad \mid \ 5 \quad \text{TRUE} \end{array} \qquad \begin{array}{c|c} y = \frac{1}{2}x + 3 \\ \hline 2 \ ? \ \frac{1}{2}(-2) + 3 \\ \quad \mid \ -1 + 3 \\ \quad \mid \ 2 \quad \text{TRUE} \end{array}$$

In each case the substitution results in a true equation. Thus, $(4, 5)$ and $(-2, 2)$ are both solutions of $y = \frac{1}{2}x + 3$. We graph these points and sketch the line passing through them.

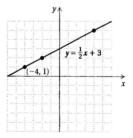

The line appears to pass through $(-4, 1)$ also. We check to determine if $(-4, 1)$ is a solution of $y = \frac{1}{2}x + 3$.

$$y = \frac{1}{2}x + 3$$

$$
\begin{array}{c|l}
\hline
1 \; ? \; \dfrac{1}{2}(-4) + 3 & \\
 -2 + 3 & \\
 1 & \text{TRUE}
\end{array}
$$

Thus, $(-4, 1)$ is another solution. There are other correct answers, including $(-6, 0)$, $(0, 3)$, $(2, 4)$, and $(6, 6)$.

11. To show that a pair is a solution, we substitute, replacing x with the first coordinate and y with the second coordinate in each pair.

$$
\begin{array}{c|l}
\hline
\multicolumn{2}{c}{4x - 2y = 10} \\
\hline
4 \cdot 0 - 2(-5) \; ? \; 10 & \\
 10 & \text{TRUE}
\end{array}
$$

$$
\begin{array}{c|l}
\hline
\multicolumn{2}{c}{4x - 2y = 10} \\
\hline
4 \cdot 4 - 2 \cdot 3 \; ? \; 10 & \\
 16 - 6 & \\
 10 & \text{TRUE}
\end{array}
$$

In each case the substitution results in a true equation. Thus, $(0, -5)$ and $(4, 3)$ are both solutions of $4x - 2y = 10$. We graph these points and sketch the line passing through them.

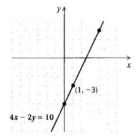

The line appears to pass through $(1, -3)$ also. We check to determine if $(1, -3)$ is a solution of $4x - 2y = 10$.

$$
\begin{array}{c|l}
\hline
\multicolumn{2}{c}{4x - 2y = 10} \\
\hline
4 \cdot 1 - 2(-3) \; ? \; 10 & \\
 4 + 6 & \\
 10 & \text{TRUE}
\end{array}
$$

Thus, $(1, -3)$ is another solution. There are other correct answers, including $(2, -1)$, $(3, 1)$, and $(5, 5)$.

13. $y = x + 1$

The equation is in the form $y = mx + b$. The y-intercept is $(0, 1)$. We find five other pairs.

When $x = -2$, $y = -2 + 1 = -1$.

When $x = -1$, $y = -1 + 1 = 0$.

When $x = 1$, $y = 1 + 1 = 2$.

When $x = 2$, $y = 2 + 1 = 3$.

When $x = 3$, $y = 3 + 1 = 4$.

x	y
-2	-1
-1	0
0	1
1	2
2	3
3	4

Plot these points, draw the line they determine, and label the graph $y = x + 1$.

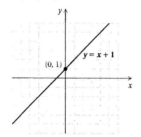

15. $y = x$

The equation is equivalent to $y = x + 0$. The y-intercept is $(0, 0)$. We find five other points.

When $x = -2$, $y = -2$.

When $x = -1$, $y = -1$.

When $x = 1$, $y = 1$.

When $x = 2$, $y = 2$.

When $x = 3$, $y = 3$.

x	y
-2	-2
-1	-1
0	0
1	1
2	2
3	3

Plot these points, draw the line they determine, and label the graph $y = x$.

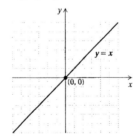

17. $y = \frac{1}{2}x$

The equation is equivalent to $y = \frac{1}{2}x + 0$. The y-intercept is $(0, 0)$. We find two other points.

When $x = -2$, $y = \frac{1}{2}(-2) = -1$.

When $x = 4$, $y = \frac{1}{2} \cdot 4 = 2$.

x	y
-2	-1
0	0
4	2

Plot these points, draw the line they determine, and label the graph $y = \frac{1}{2}x$.

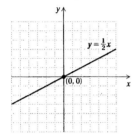

19. $y = x - 3$

The equation is equivalent to $y = x + (-3)$. The y-intercept is $(0, -3)$. We find two other points.

When $x = -2$, $y = -2 - 3 = -5$.

When $x = 4$, $y = 4 - 3 = 1$.

x	y
-2	-5
0	-3
4	1

Plot these points, draw the line they determine, and label the graph $y = x - 3$.

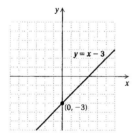

21. $y = 3x - 2 = 3x + (-2)$

The y-intercept is $(0, -2)$. We find two other points.

When $x = -2$, $y = 3(-2) + 2 = -6 + 2 = -4$.

When $x = 1$, $y = 3 \cdot 1 + 2 = 3 + 2 = 5$.

x	y
-2	-4
0	-2
1	5

Plot these points, draw the line they determine, and label the graph $y = 3x + 2$.

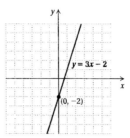

23. $y = \frac{1}{2}x + 1$

The y-intercept is $(0, 1)$. We find two other points using multiples of 2 for x to avoid fractions.

When $x = -4$, $y = \frac{1}{2}(-4) + 1 = -2 + 1 = -1$.

When $x = 4$, $y = \frac{1}{2} \cdot 4 + 1 = 2 + 1 = 3$.

x	y
-4	-1
0	1
4	3

Plot these points, draw the line they determine, and label the graph $y = \frac{1}{2}x + 1$.

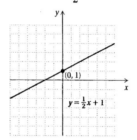

25. $x + y = -5$
$$y = -x - 5$$
$$y = -x + (-5)$$

The y-intercept is $(0, -5)$. We find two other points.

When $x = -4$, $y = -(-4) - 5 = 4 - 5 = -1$.

When $x = -1$, $y = -(-1) - 5 = 1 - 5 = -4$.

x	y
-4	-1
0	-5
-1	-4

Plot these points, draw the line they determine, and label the graph $x + y = -5$.

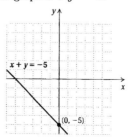

27. $y = \dfrac{5}{3}x - 2 = \dfrac{5}{3}x + (-2)$

The y-intercept is $(0, -2)$. We find two other points using multiples of 3 for x to avoid fractions.

When $x = -3$, $y = \dfrac{5}{3}(-3) - 2 = -5 - 2 = -7$.

When $x = 3$, $y = \dfrac{5}{3} \cdot 3 - 2 = 5 - 2 = 3$.

x	y
-3	-7
0	-2
3	3

Plot these points, draw the line they determine, and label the graph $y = \dfrac{5}{3}x - 2$.

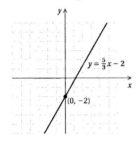

29. $x + 2y = 8$

$2y = -x + 8$

$y = -\dfrac{1}{2}x + 4$

The y-intercept is $(0, 4)$. We find two other points using multiples of 2 for x to avoid fractions.

When $x = -2$, $y = -\dfrac{1}{2}(-2) + 4 = 1 + 4 = 5$.

When $x = 4$, $y = -\dfrac{1}{2} \cdot 4 + 4 = -2 + 4 = 2$.

x	y
-2	5
0	4
4	2

Plot these points, draw the line they determine, and label the graph $x + 2y = 8$.

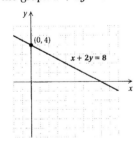

31. $y = \dfrac{3}{2}x + 1$

The y-intercept is $(0, 1)$. We find two other points using multiples of 2 for x to avoid fractions.

When $x = -4$, $y = \dfrac{3}{2}(-4) + 1 = -6 + 1 = -5$.

When $x = 2$, $y = \dfrac{3}{2} \cdot 2 + 1 = 3 + 1 = 4$.

x	y
-4	-5
0	1
2	4

Plot these points, draw the line they determine, and label the graph $y = \dfrac{3}{2}x + 1$.

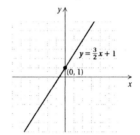

33. $8x - 2y = -10$

$-2y = -8x - 10$

$y = 4x + 5$

The y-intercept is $(0, 5)$. We find two other points.

When $x = -2$, $y = 4(-2) + 5 = -8 + 5 = -3$.

When $x = -1$, $y = 4(-1) + 5 = -4 + 5 = 1$.

x	y
-2	-3
-1	1
0	5

Plot these points, draw the line they determine, and label the graph $8x - 2y = -10$.

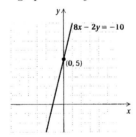

35. $8y + 2x = -4$

$8y = -2x - 4$

$y = -\dfrac{1}{4}x - \dfrac{1}{2}$

$y = -\dfrac{1}{4}x + \left(-\dfrac{1}{2}\right)$

The y-intercept is $\left(0, -\dfrac{1}{2}\right)$. We find two other points.

When $x = -2$, $y = -\dfrac{1}{4}(-2) - \dfrac{1}{2} = \dfrac{1}{2} - \dfrac{1}{2} = 0$.

When $x = 2$, $y = -\dfrac{1}{4} \cdot 2 - \dfrac{1}{2} = -\dfrac{1}{2} - \dfrac{1}{2} = -1$.

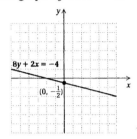

x	y
-2	0
0	$-\dfrac{1}{2}$
2	-1

Plot these points, draw the line they determine, and label the graph $8y + 2x = -4$.

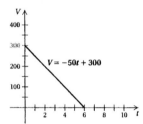

37. a) We substitute 0, 4, and 6 for t and then calculate V.

If $t = 0$, then $V = -50 \cdot 0 + 300 = \$300$.

If $t = 4$, then $V = -50 \cdot 4 + 300 = -200 + 300 = \$100$.

If $t = 6$, then $V = -50 \cdot 6 + 300 = -300 + 300 = \$0$.

b) We plot the three ordered pairs we found in part (a). Note the negative t- and V-values have no meaning in this problem.

To use the graph to estimate the value of the software after 5 years we must determine which V-value is paired with $t = 5$. We locate 5 on the t-axis, go up to the graph, and then find the value on the V-axis that corresponds to that point. It appears that after 5 years the value of the software is \$50.

c) Substitute 150 for V and then solve for t.
$$V = -50t + 300$$
$$150 = -50t + 300$$
$$-150 = -50t$$
$$3 = t$$

The value of the software is \$150 after 3 years.

39. a) When $d = 1$, $N = 0.1(1) + 7 = 0.1 + 7 = 7.1$ gal.

In 1996, $d = 1996 - 1991 = 5$. When $d = 5$, $N = 0.1(5) + 7 = 0.5 + 7 = 7.5$ gal.

In 2001, $d = 2001 - 1991 = 10$. When $d = 10$, $N = 0.1(10) + 7 = 1 + 7 = 8$ gal.

In 2011, $d = 2011 - 1991 = 20$. When $d = 20$, $N = 0.1(20) + 7 = 2 + 7 = 9$ gal.

b) Plot the four ordered pairs we found in part (a). Note that negative d- and N-values have no meaning in this problem.

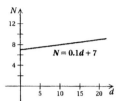

To use the graph to estimate what tea consumption was in 1997 we must determine which N-value is paired with 1997, or with $d = 6$. We locate 6 on the d-axis, go up to the graph, and then find the value on the N-axis that corresponds to that point. It appears that tea consumption was about 7.6 gallons in 1997.

c) Substitute 8.5 for N and then solve for d.
$$N = 0.1d + 7$$
$$8.5 = 0.1d + 7$$
$$1.5 = 0.1d$$
$$15 = d$$

Tea consumption will be about 8.5 gallons 15 years after 1991, or in 2006.

41. Discussion and Writing Exercise

43.
$$63 = 9x$$
$$\frac{63}{9} = \frac{9x}{9} \quad \text{Dividing by 9 on both sides}$$
$$7 = 1 \cdot x \quad \text{Simplifying}$$
$$7 = x \quad \text{Identity property of 1}$$
The solution is 7.

45.
$$13x = -52$$
$$\frac{13x}{13} = \frac{-52}{13} \quad \text{Dividing by 13 on both sides}$$
$$1 \cdot x = -4 \quad \text{Simplifying}$$
$$x = -4 \quad \text{Identity property of 1}$$
The solution is -4.

47.
$$\frac{1}{10}x = \frac{2}{5}$$
$$10 \cdot \frac{1}{10}x = 10 \cdot \frac{2}{5} \quad \begin{array}{l}\text{Multiplying by 10}\\\text{to clear fractions}\end{array}$$
$$1 \cdot x = 4 \quad \text{Simplifying}$$
$$x = 4 \quad \text{Identity property of 1}$$
The solution is 4.

49. First we find decimal notation for $\frac{7}{8}$.

$$
\begin{array}{r}
0.875 \\
8\overline{\smash{)}7.000} \\
\underline{64} \\
60 \\
\underline{56} \\
40 \\
\underline{40} \\
0
\end{array}
$$

Since $\frac{7}{8} = 0.875$, then $-\frac{7}{8} = -0.875$.

51.

$$
\begin{array}{r}
1.828125 \\
64\overline{\smash{)}117.000000} \\
\underline{64} \\
530 \\
\underline{512} \\
180 \\
\underline{128} \\
520 \\
\underline{512} \\
80 \\
\underline{64} \\
160 \\
\underline{128} \\
320 \\
\underline{320} \\
0
\end{array}
$$

$\frac{117}{64} = 1.828125$

53. Note that the sum of the coordinates of each point on the graph is 5. Thus, we have $x + y = 5$, or $y = -x + 5$.

55. Note that each y-coordinate is 2 more than the corresponding x-coordinate. Thus, we have $y = x + 2$.

Exercise Set 8.3

1. (a) The graph crosses the y-axis at $(0, 5)$, so the y-intercept is $(0, 5)$.

    (b) The graph crosses the x- axis at $(2, 0)$, so the x-intercept is $(2, 0)$.

3. (a) The graph crosses the y-axis at $(0, -4)$, so the y-intercept is $(0, -4)$.

    (b) The graph crosses the x-axis at $(3, 0)$, so the x-intercept is $(3, 0)$.

5. $3x + 5y = 15$

    (a) To find the y-intercept, let $x = 0$. This is the same as covering up the x-term and then solving.

$$5y = 15$$
$$y = 3$$

The y-intercept is $(0, 3)$.

    (b) To find the x-intercept, let $y = 0$. This is the same as covering up the y-term and then solving.

$$3x = 15$$
$$x = 5$$

The x-intercept is $(5, 0)$.

7. $7x - 2y = 28$

    (a) To find the y-intercept, let $x = 0$. This is the same as covering up the x-term and then solving.

$$-2y = 28$$
$$y = -14$$

The y-intercept is $(0, -14)$.

    (b) To find the x-intercept, let $y = 0$. This is the same as covering up the y-term and then solving.

$$7x = 28$$
$$x = 4$$

The x-intercept is $(4, 0)$.

9. $-4x + 3y = 10$

    (a) To find the y-intercept, let $x = 0$. This is the same as covering up the x-term and then solving.

$$3y = 10$$
$$y = \frac{10}{3}$$

The y-intercept is $\left(0, \frac{10}{3}\right)$.

    (b) To find the x-intercept, let $y = 0$. This is the same as covering up the y-term and then solving.

$$-4x = 10$$
$$x = -\frac{5}{2}$$

The x-intercept is $\left(-\frac{5}{2}, 0\right)$.

11. $6x - 3 = 9y$

$6x - 9y = 3$ Writing the equation in the form $Ax + By = C$

    (a) To find the y-intercept, let $x = 0$. This is the same as covering up the x-term and then solving.

$$-9y = 3$$
$$y = -\frac{1}{3}$$

The y-intercept is $\left(0, -\frac{1}{3}\right)$.

(b) To find the x-intercept, let $y = 0$. This is the same as covering up the y-term and then solving.

$$6x = 3$$
$$x = \frac{1}{2}$$

The x-intercept is $\left(\frac{1}{2}, 0\right)$.

13. $x + 3y = 6$

To find the x-intercept, let $y = 0$. Then solve for x.

$$x + 3y = 6$$
$$x + 3 \cdot 0 = 6$$
$$x = 6$$

Thus, $(6, 0)$ is the x-intercept.

To find the y-intercept, let $x = 0$. Then solve for y.

$$x + 3y = 6$$
$$0 + 3y = 6$$
$$3y = 6$$
$$y = 2$$

Thus, $(0, 2)$ is the y-intercept.

Plot these points and draw the line.

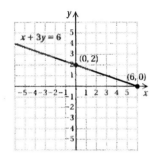

A third point should be used as a check. We substitute any value for x and solve for y.

We let $x = 3$. Then

$$x + 3y = 6$$
$$3 + 3y = 6$$
$$3y = 3$$
$$y = 1$$

The point $(3, 1)$ is on the graph, so the graph is probably correct.

15. $-x + 2y = 4$

To find the x-intercept, let $y = 0$. Then solve for x.

$$-x + 2y = 4$$
$$-x + 2 \cdot 0 = 4$$
$$-x = 4$$
$$x = -4$$

Thus, $(-4, 0)$ is the x-intercept.

To find the y-intercept, let $x = 0$. Then solve for y.

$$-x + 2y = 4$$
$$-0 + 2y = 4$$
$$2y = 4$$
$$y = 2$$

Thus, $(0, 2)$ is the y-intercept.

Plot these points and draw the line.

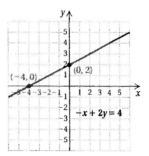

A third point should be used as a check. We substitute any value for x and solve for y.

We let $x = 4$. Then

$$-x + 2y = 4$$
$$-4 + 2y = 4$$
$$2y = 8$$
$$y = 4$$

The point $(4, 4)$ is on the graph, so the graph is probably correct.

17. $3x + y = 6$

To find the x-intercept, let $y = 0$. Then solve for x.

$$3x + y = 6$$
$$3x + 0 = 6$$
$$3x = 6$$
$$x = 2$$

Thus, $(2, 0)$ is the x-intercept.

To find the y-intercept, let $x = 0$. Then solve for y.

$$3x + y = 6$$
$$3 \cdot 0 + y = 6$$
$$y = 6$$

Thus, $(0, 6)$ is the y-intercept.

Plot these points and draw the line.

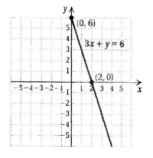

A third point should be used as a check. We substitute any value for x and solve for y.

We let $x = 1$. Then

$$3x + y = 6$$
$$3 \cdot 1 + y = 6$$
$$3 + y = 6$$
$$y = 3$$

The point $(1, 3)$ is on the graph, so the graph is probably correct.

19. $2y - 2 = 6x$

To find the x-intercept, let $y = 0$. Then solve for x.

$$2y - 2 = 6x$$
$$2 \cdot 0 - 2 = 6x$$
$$-2 = 6x$$
$$-\frac{1}{3} = x$$

Thus, $\left(-\frac{1}{3}, 0 \right)$ is the x-intercept.

To find the y-intercept, let $x = 0$. Then solve for y.

$$2y - 2 = 6x$$
$$2y - 2 = 6 \cdot 0$$
$$2y - 2 = 0$$
$$2y = 2$$
$$y = 1$$

Thus, $(0, 1)$ is the y-intercept.

It is helpful to plot another point since the intercepts are so close together. This point can also serve as a check.

We let $x = 1$. Then

$$2y - 2 = 6x$$
$$2y - 2 = 6 \cdot 1$$
$$2y - 2 = 6$$
$$2y = 8$$
$$y = 4$$

Plot the point $(1, 4)$ and the intercepts and draw the line.

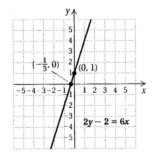

21. $3x - 9 = 3y$

To find the x-intercept, let $y = 0$. Then solve for x.

$$3x - 9 = 3y$$
$$3x - 9 = 3 \cdot 0$$
$$3x - 9 = 0$$
$$3x = 9$$
$$x = 3$$

Thus, $(3, 0)$ is the x-intercept.

To find the y-intercept, let $x = 0$. Then solve for y.

$$3x - 9 = 3y$$
$$3 \cdot 0 - 9 = 3y$$
$$-9 = 3y$$
$$-3 = y$$

Thus, $(0, -3)$ is the y-intercept.

Plot these points and draw the line.

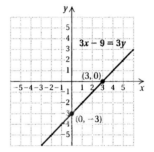

A third point should be used as a check. We substitute any value for x and solve for y.

We let $x = 1$. Then

$$3x - 9 = 3y$$
$$3 \cdot 1 - 9 = 3y$$
$$3 - 9 = 3y$$
$$-6 = 3y$$
$$-2 = y$$

The point $(1, -2)$ is on the graph, so the graph is probably correct.

23. $2x - 3y = 6$

To find the x-intercept, let $y = 0$. Then solve for x.

$$2x - 3y = 6$$
$$2x - 3 \cdot 0 = 6$$
$$2x = 6$$
$$x = 3$$

Thus, $(3, 0)$ is the x-intercept.

To find the y-intercept, let $x = 0$. Then solve for y.

$$2x - 3y = 6$$
$$2 \cdot 0 - 3y = 6$$
$$-3y = 6$$
$$y = -2$$

Thus, $(0, -2)$ is the y-intercept.

Plot these points and draw the line.

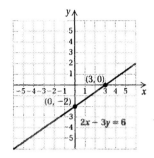

A third point should be used as a check. We substitute any value for x and solve for y.

We let $x = -3$.

$$2x - 3y = 6$$
$$2(-3) - 3y = 6$$
$$-6 - 3y = 6$$
$$-3y = 12$$
$$y = -4$$

The point $(-3, -4)$ is on the graph, so the graph is probably correct.

25. $4x + 5y = 20$

To find the x-intercept, let $y = 0$. Then solve for x.

$$4x + 5y = 20$$
$$4x + 5 \cdot 0 = 20$$
$$4x = 20$$
$$x = 5$$

Thus, $(5, 0)$ is the x-intercept.

To find the y-intercept, let $x = 0$. Then solve for y.

$$4x + 5y = 20$$
$$4 \cdot 0 + 5y = 20$$
$$5y = 20$$
$$y = 4$$

Thus, $(0, 4)$ is the y-intercept.

Plot these points and draw the graph.

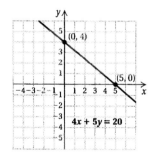

A third point should be used as a check. We substitute any value for x and solve for y.

We let $x = 4$. Then

$$4x + 5y = 20$$
$$4 \cdot 4 + 5y = 20$$
$$16 + 5y = 20$$
$$5y = 4$$
$$y = \frac{4}{5}$$

The point $\left(4, \frac{4}{5}\right)$ is on the graph, so the graph is probably correct.

27. $2x + 3y = 8$

To find the x-intercept, let $y = 0$. Then solve for x.

$$2x + 3y = 8$$
$$2x + 3 \cdot 0 = 8$$
$$2x = 8$$
$$x = 4$$

Thus, $(4, 0)$ is the x-intercept.

To find the y-intercept, let $x = 0$. Then solve for y.

$$2x + 3y = 8$$
$$2 \cdot 0 + 3y = 8$$
$$3y = 8$$
$$y = \frac{8}{3}$$

Thus, $\left(0, \frac{8}{3}\right)$ is the y-intercept.

Plot these points and draw the graph.

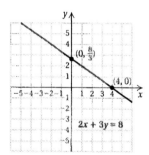

A third point should be used as a check.

We let $x = 1$. Then

$$2x + 3y = 8$$
$$2 \cdot 1 + 3y = 8$$
$$2 + 3y = 8$$
$$3y = 6$$
$$y = 2$$

The point $(1, 2)$ is on the graph, so the graph is probably correct.

29. $x - 3 = y$

To find the x-intercept, let $y = 0$. Then solve for x.

$$x - 3 = y$$
$$x - 3 = 0$$
$$x = 3$$

Thus, $(3, 0)$ is the x-intercept.

To find the y-intercept, let $x = 0$. Then solve for y.

$$x - 3 = y$$
$$0 - 3 = y$$
$$-3 = y$$

Thus, $(0, -3)$ is the y-intercept.

Plot these points and draw the line.

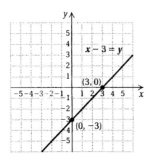

A third point should be used as a check.

We let $x = -2$. Then

$$x - 3 = y$$
$$-2 - 3 = y$$
$$-5 = y$$

The point $(-2, -5)$ is on the graph, so the graph is probably correct.

31. $3x - 2 = y$

To find the x-intercept, let $y = 0$. Then solve for x.

$$3x - 2 = y$$
$$3x - 2 = 0$$
$$3x = 2$$
$$x = \frac{2}{3}$$

Thus, $\left(\frac{2}{3}, 0\right)$ is the x-intercept.

To find the y-intercept, let $x = 0$. Then solve for y.

$$3x - 2 = y$$
$$3 \cdot 0 - 2 = y$$
$$-2 = y$$

Thus, $(0, -2)$ is the y-intercept.

Plot these points and draw the line.

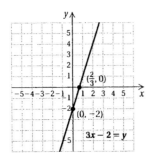

A third point should be used as a check.

We let $x = 2$. Then

$$3x - 2 = y$$
$$3 \cdot 2 - 2 = y$$
$$6 - 2 = y$$
$$4 = y$$

The point $(2, 4)$ is on the graph, so the graph is probably correct.

33. $6x - 2y = 12$

To find the x-intercept, let $y = 0$. Then solve for x.

$$6x - 2y = 12$$
$$6x - 2 \cdot 0 = 12$$
$$6x = 12$$
$$x = 2$$

Thus, $(2, 0)$ is the x-intercept.

To find the y-intercept, let $x = 0$. Then solve for y.

$$6x - 2y = 12$$
$$6 \cdot 0 - 2y = 12$$
$$-2y = 12$$
$$y = -6$$

Thus, $(0, -6)$ is the y-intercept.

Plot these points and draw the line.

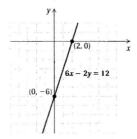

We use a third point as a check.

We let $x = 1$. Then

$$6x - 2y = 12$$
$$6 \cdot 1 - 2y = 12$$
$$6 - 2y = 12$$
$$-2y = 6$$
$$y = -3$$

The point $(1, -3)$ is on the graph, so the graph is probably correct.

35. $3x + 4y = 5$

To find the x-intercept, let $y = 0$. Then solve for x.

$$3x + 4y = 5$$
$$3x + 4 \cdot 0 = 5$$
$$3x = 5$$
$$x = \frac{5}{3}$$

Thus, $\left(\frac{5}{3}, 0\right)$ is the x-intercept.

To find the y-intercept, let $x = 0$. Then solve for y.

$$3x + 4y = 5$$
$$3 \cdot 0 + 4y = 5$$
$$4y = 5$$
$$y = \frac{5}{4}$$

Thus, $\left(0, \frac{5}{4}\right)$ is the y-intercept.

It is helpful to plot another point since the intercepts are so close together. This point can also serve as a check.

We let $x = 3$. Then

$$3x + 4y = 5$$
$$3 \cdot 3 + 4y = 5$$
$$9 + 4y = 5$$
$$4y = -4$$
$$y = -1$$

Plot the point $(3, -1)$ and the intercepts and draw the line.

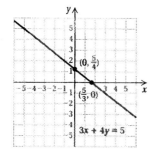

37. $y = -3 - 3x$

To find the x-intercept, let $y = 0$. Then solve for x.

$$y = -3 - 3x$$
$$0 = -3 - 3x$$
$$3x = -3$$
$$x = -1$$

Thus, $(-1, 0)$ is the x-intercept.

To find the y-intercept, let $x = 0$. Then solve for y.

$$y = -3 - 3x$$
$$y = -3 - 3 \cdot 0$$
$$y = -3$$

Thus, $(0, -3)$ is the y-intercept.

Plot these points and draw the graph.

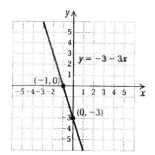

We use a third point as a check.

We let $x = -2$. Then

$$y = -3 - 3x$$
$$y = -3 - 3 \cdot (-2)$$
$$y = -3 + 6$$
$$y = 3$$

The point $(-2, 3)$ is on the graph, so the graph is probably correct.

39. $y - 3x = 0$

To find the x-intercept, let $y = 0$. Then solve for x.

$$0 - 3x = 0$$
$$-3x = 0$$
$$x = 0$$

Thus, $(0, 0)$ is the x-intercept. Note that this is also the y-intercept.

In order to graph the line, we will find a second point.

When $x = 1$, $y - 3 \cdot 1 = 0$
$$y - 3 = 0$$
$$y = 3$$

Plot the points and draw the graph.

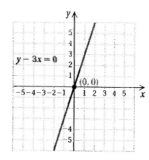

We use a third point as a check.

We let $x = -1$. Then

$$y - 3(-1) = 0$$
$$y + 3 = 0$$
$$y = -3$$

The point $(-1, -3)$ is on the graph, so the graph is probably correct.

41. $x = -2$

Any ordered pair $(-2, y)$ is a solution. The variable x must be -2, but y can be any number we choose. A few solutions are listed below. Plot these points and draw the line.

x	y
-2	-2
-2	0
-2	4

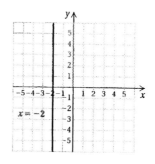

43. $y = 2$

Any ordered pair $(x, 2)$ is a solution. The variable y must be 2, but x can be any number we choose. A few solutions are listed below. Plot these points and draw the line.

x	y
-3	2
0	2
2	2

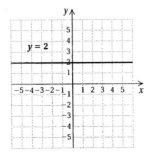

45. $x = 2$

Any ordered pair $(2, y)$ is a solution. The variable x must be 2, but y can be any number we choose. A few solutions are listed below. Plot these points and draw the line.

x	y
2	-1
2	4
2	5

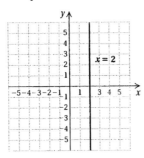

47. $y = 0$

Any ordered pair $(x, 0)$ is a solution. The variable y must be 0, but x can be any number we choose. A few solutions are listed below. Plot these points and draw the line.

x	y
-5	0
-1	0
3	0

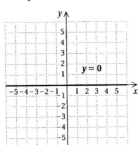

49. $x = \dfrac{3}{2}$

Any ordered pair $\left(\dfrac{3}{2}, y\right)$ is a solution. The variable x must be $\dfrac{3}{2}$, but y can be any number we choose. A few solutions are listed below. Plot these points and draw the line.

x	y
$\dfrac{3}{2}$	-2
$\dfrac{3}{2}$	0
$\dfrac{3}{2}$	4

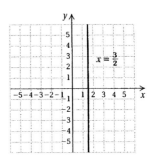

51. $3y = -5$

$y = -\dfrac{5}{3}$ Solving for y

Any ordered pair $\left(x, -\dfrac{5}{3}\right)$ is a solution. A few solutions are listed below. Plot these points and draw the line.

x	y
-3	$-\dfrac{5}{3}$
0	$-\dfrac{5}{3}$
2	$-\dfrac{5}{3}$

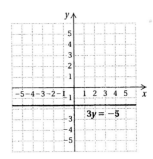

53. $4x + 3 = 0$

$4x = -3$

$x = -\dfrac{3}{4}$ Solving for x

Any ordered pair $\left(-\dfrac{3}{4}, y\right)$ is a solution. A few solutions are listed below. Plot these points and draw the line.

x	y
$-\dfrac{3}{4}$	-2
$-\dfrac{3}{4}$	0
$-\dfrac{3}{4}$	3

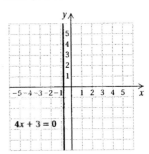

55. $48 - 3y = 0$

$-3y = -48$

$y = 16$ Solving for y

Any ordered pair $(x, 16)$ is a solution. A few solutions are listed below. Plot these points and draw the line.

x	y
-4	16
0	16
2	16

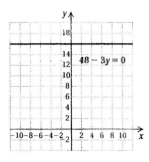

57. Note that every point on the horizontal line passing through $(0, -1)$ has -1 as the y-coordinate. Thus, the equation of the line is $y = -1$.

59. Note that every point on the vertical line passing through $(4, 0)$ has 4 as the x-coordinate. Thus, the equation of the line is $x = 4$.

61. Discussion and Writing Exercise

63. *Familiarize.* Let p = the percent of desserts sold that will be pie.

Translate. We reword the problem.

40 is what percent of 250?

$$40 = p \cdot 250$$

Solve. We solve the equation.

$$40 = p \cdot 250$$
$$\frac{40}{250} = \frac{p \cdot 250}{250}$$
$$0.16 = p$$
$$16\% = p$$

Check. We can find 16% of 250:

$$16\% \cdot 250 = 0.16 \cdot 250 = 40$$

The answer checks.

State. 16% of the desserts sold will be pie.

65. $-1.6x < 64$

$$\frac{-1.6x}{-1.6} > \frac{64}{-1.6} \qquad \begin{array}{l}\text{Dividing by } -1.6 \text{ and reversing} \\ \text{the inequality symbol}\end{array}$$

$$x > -40$$

The solution set is $\{x | x > -40\}$.

67. $x + (x - 1) < (x + 2) - (x + 1)$
$$2x - 1 < x + 2 - x - 1$$
$$2x - 1 < 1$$
$$2x < 2$$
$$x < 1$$

The solution set is $\{x | x < 1\}$.

69. A line parallel to the x-axis has an equation of the form $y = b$. Since the y-coordinate of one point on the line is -4, then $b = -4$ and the equation is $y = -4$.

71. Substitute -4 for x and 0 for y.

$$3(-4) + k = 5 \cdot 0$$
$$-12 + k = 0$$
$$k = 12$$

Exercise Set 8.4

1. We consider (x_1, y_1) to be $(-3, 5)$ and (x_2, y_2) to be $(4, 2)$.

$$m = \frac{y_2 - y_1}{x_2 - x_1} = \frac{2 - 5}{4 - (-3)} = \frac{-3}{7} = -\frac{3}{7}$$

3. We can choose any two points. We consider (x_1, y_1) to be $(-3, -1)$ and (x_2, y_2) to be $(0, 1)$.

$$m = \frac{y_2 - y_1}{x_2 - x_1} = \frac{1 - (-1)}{0 - (-3)} = \frac{2}{3}$$

5. We can choose any two points. We consider (x_1, y_1) to be $(-4, -2)$ and (x_2, y_2) to be $(4, 4)$.

$$m = \frac{y_2 - y_1}{x_2 - x_1} = \frac{4 - (-2)}{4 - (-4)} = \frac{6}{8} = \frac{3}{4}$$

7. We consider (x_1, y_1) to be $(-4, -2)$ and (x_2, y_2) to be $(3, -2)$.

$$m = \frac{y_2 - y_1}{x_2 - x_1} = \frac{-2 - (-2)}{3 - (-4)} = \frac{0}{7} = 0$$

9. We plot $(-2, 4)$ and $(3, 0)$ and draw the line containing these points.

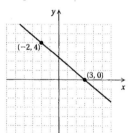

To find the slope, consider (x_1, y_1) to be $(-2, 4)$ and (x_2, y_2) to be $(3, 0)$.

$$m = \frac{y_2 - y_1}{x_2 - x_1} = \frac{0 - 4}{3 - (-2)} = \frac{-4}{5} = -\frac{4}{5}$$

11. We plot $(-4, 0)$ and $(-5, -3)$ and draw the line containing these points.

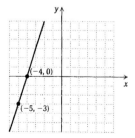

To find the slope, consider (x_1, y_1) to be $(-4, 0)$ and (x_2, y_2) to be $(-5, -3)$.

$$m = \frac{y_2 - y_1}{x_2 - x_1} = \frac{-3 - 0}{-5 - (-4)} = \frac{-3}{-1} = 3$$

13. We plot $(-4, 2)$ and $(2, -3)$ and draw the line containing these points.

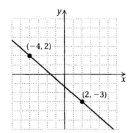

To find the slope, consider (x_1, y_1) to be $(-4, 2)$ and (x_2, y_2) to be $(2, -3)$.

$$m = \frac{y_2 - y_1}{x_2 - x_1} = \frac{-3 - 2}{2 - (-4)} = \frac{-5}{6} = -\frac{5}{6}$$

15. We plot $(5, 3)$ and $(-3, -4)$ and draw the line containing these points.

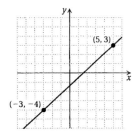

To find the slope, consider (x_1, y_1) to be $(5, 3)$ and (x_2, y_2) to be $(-3, -4)$.

$$m = \frac{y_2 - y_1}{x_2 - x_1} = \frac{-4 - 3}{-3 - 5} = \frac{-7}{-8} = \frac{7}{8}$$

17. $m = \dfrac{-\dfrac{1}{2} - \dfrac{3}{2}}{2 - 5} = \dfrac{-2}{-3} = \dfrac{2}{3}$

19. $m = \dfrac{-2 - 3}{4 - 4} = \dfrac{-5}{0}$

Since division by 0 is not defined, the slope is not defined.

21. $m = \dfrac{\text{rise}}{\text{run}} = \dfrac{2.4}{8.2} = \dfrac{2.4}{8.2} \cdot \dfrac{10}{10} = \dfrac{24}{82}$

$\qquad = \dfrac{\cancel{2} \cdot 12}{\cancel{2} \cdot 41} = \dfrac{12}{41}$

23. $m = \dfrac{\text{rise}}{\text{run}} = \dfrac{56}{258} = \dfrac{\cancel{2} \cdot 28}{\cancel{2} \cdot 129} = \dfrac{28}{129}$

25. Long's Peak rises $14{,}255 \text{ ft} - 9600 \text{ ft} = 4655 \text{ ft}$.

Grade $= \dfrac{4655}{15{,}840} \approx 0.294 \approx 29.4\%$

27. The rate of change is the slope of the line. We can use any two ordered pairs to find the slope. We choose $(2, 50)$ and $(8, 200)$.

Rate of change $= \dfrac{200 \text{ mi} - 50 \text{ mi}}{8 \text{ gal} - 2 \text{ gal}} = \dfrac{150 \text{ mi}}{6 \text{ gal}} =$

25 miles per gallon

29. The rate of change is the slope of the line. We can use any two ordered pairs to find the slope. We choose $(2, 2000)$ and $(4, 1000)$. (Note that units on the vertical axis are given in thousands.)

Rate of change $= \dfrac{\$1000 - \$2000}{4 \text{ yr} - 2 \text{ yr}} = \dfrac{-\$1000}{2 \text{ yr}} = -\$500$ per year

31. The rate of change is the slope of the line. We can use any two ordered pairs to find the slope. We choose $(1990, 550, 043)$ and $(2000, 626, 932)$.

Rate of change $= \dfrac{626{,}932 - 550{,}043}{2000 - 1990} = \dfrac{76{,}889}{10} \approx$

7689 people per year

33. $y = -10x + 7$

The equation is in the form $y = mx + b$, where $m = -10$. Thus, the slope is -10.

35. $y = 3.78x - 4$

The equation is in the form $y = mx + b$, where $m = 3.78$. Thus, the slope is 3.78.

37. We solve for y, obtaining an equation of the form $y = mx + b$.

$$3x - y = 4$$
$$-y = -3x + 4$$
$$-1(-y) = -1(-3x + 4)$$
$$y = 3x - 4$$

The slope is 3.

39. We solve for y, obtaining an equation of the form $y = mx + b$.

$$x + 5y = 10$$
$$5y = -x + 10$$
$$y = \frac{1}{5}(-x + 10)$$
$$y = -\frac{1}{5}x + 2$$

The slope is $-\dfrac{1}{5}$.

41. We solve for y, obtaining an equation of the form $y = mx + b$.

$$3x + 2y = 6$$
$$2y = -3x + 6$$
$$y = \frac{1}{2}(-3x + 6)$$
$$y = -\frac{3}{2}x + 3$$

The slope is $-\dfrac{3}{2}$.

43. We solve for y, obtaining an equation of the form $y = mx + b$.

$$5x - 7y = 14$$
$$-7y = -5x + 14$$
$$y = -\frac{1}{7}(-5x + 14)$$
$$y = \frac{5}{7}x - 2$$

The slope is $\frac{5}{7}$.

45. $y = -2.74x$

The equation is in the form $y = mx + b$, where $m = -2.74$. Thus, the slope is -2.74.

47. We solve for y, obtaining an equation of the form $y = mx + b$.

$$9x = 3y + 5$$
$$9x - 5 = 3y$$
$$\frac{1}{3}(9x - 5) = y$$
$$3x - \frac{5}{3} = y$$

The slope is 3.

49. We solve for y, obtaining an equation of the form $y = mx + b$.

$$5x - 4y + 12 = 0$$
$$5x + 12 = 4y$$
$$\frac{1}{4}(5x + 12) = y$$
$$\frac{5}{4}x + 3 = y$$

The slope is $\frac{5}{4}$.

51. $y = 4$

The equation can be thought of as $y = 0 \cdot x + 4$, so the slope is 0.

53. Discussion and Writing Exercise

55.
$$15x = -60$$
$$\frac{15x}{15} = \frac{-60}{15} \quad \text{Dividing by 15 on both sides}$$
$$1 \cdot x = -4 \quad \text{Simplifying}$$
$$x = -4 \quad \text{Identity property of 1}$$

The solution is -4.

57.
$$-x = 37$$
$$-1 \cdot x = 37$$
$$\frac{-1 \cdot x}{-1} = \frac{37}{-1} \quad \text{Dividing by } -1 \text{ on both sides}$$
$$1 \cdot x = -37 \quad \text{Simplifying}$$
$$x = -37 \quad \text{Identity property of 1}$$

The solution is -37.

59. *Translate*.

What is 15% of $23.80?
$$\downarrow \quad \downarrow \quad \downarrow \quad \downarrow \quad \downarrow$$
$$a \quad = \quad 15\% \quad \cdot \quad 23.80$$

Solve. We convert to decimal notation and multiply.

$$a = 15\% \cdot 23.80 = 0.15 \cdot 23.80 = 3.57$$

The answer is $3.57.

61. *Familiarize*. Let $p =$ the percent of the cost of the meal represented by the tip.

Translate. We reword the problem.

$8.50 is what percent of $42.50?
$$\downarrow \quad \downarrow \qquad \downarrow \qquad \downarrow \quad \downarrow$$
$$8.50 \quad = \qquad p \qquad \cdot \quad 42.50$$

Solve. We solve the equation.

$$8.50 = p \cdot 42.50$$
$$0.2 = p$$
$$20\% = p$$

Check. We can find 20% of 42.50.

$$20\% \cdot 42.50 = 0.2 \cdot 42.50 = 8.50$$

The answer checks.

State. The tip was 20% of the cost of the meal.

63. *Familiarize*. Let $c =$ the cost of the meal before the tip was added. Then the tip is $15\% \cdot c$.

Translate. We reword the problem.

Cost of meal plus tip is total cost
$$\downarrow \qquad \downarrow \quad \downarrow \quad \downarrow \qquad \downarrow$$
$$c \qquad + \quad 15\% \cdot c = \quad 51.92$$

Solve. We solve the equation.

$$c + 15\% \cdot c = 51.92$$
$$1 \cdot c + 0.15c = 51.92$$
$$1.15c = 51.92$$
$$c \approx 45.15$$

Check. We can find 15% of 45.15 and then add this to 45.15.

$$15\% \cdot 45.15 = 0.15 \cdot 45.15 \approx 6.77 \text{ and } 45.15 + 6.77 = 51.92$$

The answer checks.

State. Before the tip the meal cost $45.15.

65. $y = 0.35x - 7$

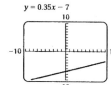

67. $y = x^3 - 5$

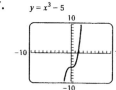

Chapter 9

Polynomials: Operations

Exercise Set 9.1

---

1. 3^4 means $3 \cdot 3 \cdot 3 \cdot 3$.

3. $(-1.1)^5$ means $(-1.1)(-1.1)(-1.1)(-1.1)(-1.1)$.

5. $\left(\dfrac{2}{3}\right)^4$ means $\left(\dfrac{2}{3}\right)\left(\dfrac{2}{3}\right)\left(\dfrac{2}{3}\right)\left(\dfrac{2}{3}\right)$.

7. $(7p)^2$ means $(7p)(7p)$.

9. $8k^3$ means $8 \cdot k \cdot k \cdot k$.

11. $a^0 = 1$, $a \neq 0$

13. $b^1 = b$

15. $\left(\dfrac{2}{3}\right)^0 = 1$

17. $8.38^0 = 1$

19. $(ab)^1 = ab$

21. $ab^1 = a \cdot b^1 = ab$

23. $m^3 = 3^3 = 3 \cdot 3 \cdot 3 = 27$

25. $p^1 = 19^1 = 19$

27. $x^4 = 4^4 = 4 \cdot 4 \cdot 4 \cdot 4 = 256$

29. $y^2 - 7 = 10^2 - 7$
$\qquad = 100 - 7$ Evaluating the power
$\qquad = 93$ Subtracting

31. $x^1 + 3 = 7^1 + 3$
$\qquad = 7 + 3$ $(7^1 = 7)$
$\qquad = 10$

$\qquad x^0 + 3 = 7^0 + 3$
$\qquad\qquad = 1 + 3$ $(7^0 = 1)$
$\qquad\qquad = 4$

33. $A = \pi r^2 \approx 3.14 \times (34 \text{ ft})^2$
$\qquad \approx 3.14 \times 1156 \text{ ft}^2$ Evaluating the power
$\qquad \approx 3629.84 \text{ ft}^2$

35. $3^{-2} = \dfrac{1}{3^2} = \dfrac{1}{9}$

37. $10^{-3} = \dfrac{1}{10^3} = \dfrac{1}{1000}$

39. $7^{-3} = \dfrac{1}{7^3} = \dfrac{1}{343}$

41. $a^{-3} = \dfrac{1}{a^3}$

43. $\dfrac{1}{8^{-2}} = 8^2 = 64$

45. $\dfrac{1}{y^{-4}} = y^4$

47. $\dfrac{1}{z^{-n}} = z^n$

49. $\dfrac{1}{4^3} = 4^{-3}$

51. $\dfrac{1}{x^3} = x^{-3}$

53. $\dfrac{1}{a^5} = a^{-5}$

55. $2^4 \cdot 2^3 = 2^{4+3} = 2^7$

57. $8^5 \cdot 8^9 = 8^{5+9} = 8^{14}$

59. $x^4 \cdot x^3 = x^{4+3} = x^7$

61. $9^{17} \cdot 9^{21} = 9^{17+21} = 9^{38}$

63. $(3y)^4(3y)^8 = (3y)^{4+8} = (3y)^{12}$

65. $(7y)^1(7y)^{16} = (7y)^{1+16} = (7y)^{17}$

67. $3^{-5} \cdot 3^8 = 3^{-5+8} = 3^3$

69. $x^{-2} \cdot x = x^{-2+1} = x^{-1} = \dfrac{1}{x}$

71. $x^{14} \cdot x^3 = x^{14+3} = x^{17}$

73. $x^{-7} \cdot x^{-6} = x^{-7+(-6)} = x^{-13} = \dfrac{1}{x^{13}}$

75. $a^{11} \cdot a^{-3} \cdot a^{-18} = a^{11+(-3)+(-18)} = a^{-10} = \dfrac{1}{a^{10}}$

77. $t^8 \cdot t^{-8} = t^{8+(-8)} = t^0 = 1$

79. $\dfrac{7^5}{7^2} = 7^{5-2} = 7^3$

81. $\dfrac{8^{12}}{8^6} = 8^{12-6} = 8^6$

83. $\dfrac{y^9}{y^5} = y^{9-5} = y^4$

85. $\dfrac{16^2}{16^8} = 16^{2-8} = 16^{-6} = \dfrac{1}{16^6}$

87. $\dfrac{m^6}{m^{12}} = m^{6-12} = m^{-6} = \dfrac{1}{m^6}$

89. $\dfrac{(8x)^6}{(8x)^{10}} = (8x)^{6-10} = (8x)^{-4} = \dfrac{1}{(8x)^4}$

91. $\frac{(2y)^9}{(2y)^9} = (2y)^{9-9} = (2y)^0 = 1$

93. $\frac{x}{x^{-1}} = x^{1-(-1)} = x^2$

95. $\frac{x^7}{x^{-2}} = x^{7-(-2)} = x^9$

97. $\frac{z^{-6}}{z^{-2}} = z^{-6-(-2)} = z^{-4} = \frac{1}{z^4}$

99. $\frac{x^{-5}}{x^{-8}} = x^{-5-(-8)} = x^3$

101. $\frac{m^{-9}}{m^{-9}} = m^{-9-(-9)} = m^0 = 1$

103. $5^2 = 5 \cdot 5 = 25$

$5^{-2} = \frac{1}{5^2} = \frac{1}{25}$

$\left(\frac{1}{5}\right)^2 = \frac{1}{5} \cdot \frac{1}{5} = \frac{1}{25}$

$\left(\frac{1}{5}\right)^{-2} = \frac{1}{\left(\frac{1}{5}\right)^2} = \frac{1}{\frac{1}{25}} = 1 \cdot \frac{25}{1} = 25$

$-5^2 = -(5)(5) = -25$

$(-5)^2 = (-5)(-5) = 25$

$-\left(-\frac{1}{5}\right)^2 = -\left(-\frac{1}{5}\right)\left(-\frac{1}{5}\right) = -\frac{1}{25}$

$\left(-\frac{1}{5}\right)^{-2} = \frac{1}{\left(-\frac{1}{5}\right)^2} = \frac{1}{\frac{1}{25}} = 1 \cdot \frac{25}{1} = 25$

105. Discussion and Writing Exercise

107. $64\%t$, or $0.64t$

109.

```
              6 4 .
2 4.3∧⟌ 1 5 5 5.2 ∧
        1 4 5 8
          9 7 2
          9 7 2
              0
```

The answer is 64.

111.
$$3x - 4 + 5x - 10x = x - 8$$
$$-2x - 4 = x - 8 \qquad \text{Collecting like terms}$$
$$-2x - 4 + 4 = x - 8 + 4 \quad \text{Adding 4}$$
$$-2x = x - 4$$
$$-2x - x = x - 4 - x \quad \text{Subtracting } x$$
$$-3x = -4$$
$$\frac{-3x}{-3} = \frac{-4}{-3} \qquad \text{Dividing by } -3$$
$$x = \frac{4}{3}$$

The solution is $\frac{4}{3}$.

113. *Familiarize*. Let x = the length of the shorter piece. Then $2x$ = the length of the longer piece.

***Translate*.**

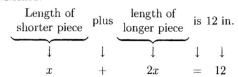

Length of shorter piece	plus	length of longer piece	is	12 in.
↓	↓	↓	↓	↓
x	$+$	$2x$	$=$	12

***Solve*.**
$$x + 2x = 12$$
$$3x = 12$$
$$\frac{3x}{3} = \frac{12}{3}$$
$$x = 4$$

If $x = 4$, $2x = 2 \cdot 4 = 8$.

***Check*.** The longer piece, 8 in., is twice as long as the shorter piece, 4 in. Also, 4 in. + 8 in. = 12 in., the total length of the sandwich. The answer checks.

***State*.** The lengths of the pieces are 4 in. and 8 in.

115. Let $y_1 = (x+1)^2$ and $y_2 = x^2 + 1$. A graph of the equations or a table of values shows that $(x + 1)^2 = x^2 + 1$ is not correct.

117. Let $y_1 = (5x)^0$ and $y_2 = 5x^0$. A graph of the equations or a table of values shows that $(5x)^0 = 5x^0$ is not correct.

119. $(y^{2x})(y^{3x}) = y^{2x+3x} = y^{5x}$

121. $\frac{a^{6t}(a^{7t})}{a^{9t}} = \frac{a^{6t+7t}}{a^{9t}} = \frac{a^{13t}}{a^{9t}} = a^{13t-9t} = a^{4t}$

123. $\frac{(0.8)^5}{(0.8)^3(0.8)^2} = \frac{(0.8)^5}{(0.8)^{3+2}} = \frac{(0.8)^5}{(0.8)^5} = 1$

125. Since the bases are the same, the expression with the larger exponent is larger. Thus, $3^5 > 3^4$.

127. Since the exponents are the same, the expression with the larger base is larger. Thus, $4^3 < 5^3$.

129. Choose any number except 0. For example, let $x = 1$.
$$3x^2 = 3 \cdot 1^2 = 3 \cdot 1 = 3, \text{ but}$$
$$(3x)^2 = (3 \cdot 1)^2 = 3^2 = 9.$$

Exercise Set 9.2

1. $(2^3)^2 = 2^{3 \cdot 2} = 2^6$

3. $(5^2)^{-3} = 5^{2(-3)} = 5^{-6} = \frac{1}{5^6}$

5. $(x^{-3})^{-4} = x^{(-3)(-4)} = x^{12}$

7. $(a^{-2})^9 = a^{-2 \cdot 9} = a^{-18} = \frac{1}{a^{18}}$

9. $(t^{-3})^{-6} = t^{(-3)(-6)} = t^{18}$

11. $(t^4)^{-3} = t^{4(-3)} = t^{-12} = \frac{1}{t^{12}}$

13. $(x^{-2})^{-4} = x^{-2)(-4)} = x^8$

15. $(ab)^3 = a^3b^3$ Raising each factor to
the third power

17. $(ab)^{-3} = a^{-3}b^{-3} = \dfrac{1}{a^3b^3}$

19. $(mn^2)^{-3} = m^{-3}(n^2)^{-3} = m^{-3}n^{2(-3)} =$
$m^{-3}n^{-6} = \dfrac{1}{m^3n^6}$

21. $(4x^3)^2 = 4^2(x^3)^2$ Raising each factor to the
second power
$= 16x^6$

23. $(3x^{-4})^2 = 3^2(x^{-4})^2 = 3^2x^{-4\cdot2} = 9x^{-8} = \dfrac{9}{x^8}$

25. $(x^4y^5)^{-3} = (x^4)^{-3}(y^5)^{-3} = x^{4(-3)}y^{5(-3)} =$
$x^{-12}y^{-15} = \dfrac{1}{x^{12}y^{15}}$

27. $(x^{-6}y^{-2})^{-4} = (x^{-6})^{-4}(y^{-2})^{-4} = x^{(-6)(-4)}y^{(-2)(-4)} =$
$x^{24}y^8$

29. $(a^{-2}b^7)^{-5} = (a^{-2})^{-5}(b^7)^{-5} = a^{10}b^{-35} = \dfrac{a^{10}}{b^{35}}$

31. $(5r^{-4}t^3)^2 = 5^2(r^{-4})^2(t^3)^2 = 25r^{-4\cdot2}t^{3\cdot2} =$
$25r^{-8}t^6 = \dfrac{25t^6}{r^8}$

33. $(a^{-5}b^7c^{-2})^3 = (a^{-5})^3(b^7)^3(c^{-2})^3 =$
$a^{-5\cdot3}b^{7\cdot3}c^{-2\cdot3} = a^{-15}b^{21}c^{-6} = \dfrac{b^{21}}{a^{15}c^6}$

35. $(3x^3y^{-8}z^{-3})^2 = 3^2(x^3)^2(y^{-8})^2(z^{-3})^2 =$
$9x^6y^{-16}z^{-6} = \dfrac{9x^6}{y^{16}z^6}$

37. $\left(\dfrac{y^3}{2}\right)^2 = \dfrac{(y^3)^2}{2^2} = \dfrac{y^6}{4}$

39. $\left(\dfrac{a^2}{b^3}\right)^4 = \dfrac{(a^2)^4}{(b^3)^4} = \dfrac{a^8}{b^{12}}$

41. $\left(\dfrac{y^2}{2}\right)^{-3} = \dfrac{(y^2)^{-3}}{2^{-3}} = \dfrac{y^{-6}}{2^{-3}} = \dfrac{\frac{1}{y^6}}{\frac{1}{2^3}} = \dfrac{1}{y^6}\cdot\dfrac{2^3}{1} = \dfrac{8}{y^6}$

43. $\left(\dfrac{7}{x^{-3}}\right)^2 = \dfrac{7^2}{(x^{-3})^2} = \dfrac{49}{x^{-6}} = 49x^6$

45. $\left(\dfrac{x^2y}{z}\right)^3 = \dfrac{(x^2)^3y^3}{z^3} = \dfrac{x^6y^3}{z^3}$

47. $\left(\dfrac{a^2b}{cd^3}\right)^{-2} = \dfrac{(a^2)^{-2}b^{-2}}{c^{-2}(d^3)^{-2}} = \dfrac{a^{-4}b^{-2}}{c^{-2}d^{-6}} = \dfrac{\frac{1}{a^4}\cdot\frac{1}{b^2}}{\frac{1}{c^2}\cdot\frac{1}{d^6}} = \dfrac{\frac{1}{a^4b^2}}{\frac{1}{c^2d^6}} =$
$\dfrac{1}{a^4b^2}\cdot\dfrac{c^2d^6}{1} = \dfrac{c^2d^6}{a^4b^2}$

49. $2.8,000,000,000.$
⌞_____⌟ 10 places
Large number, so the exponent is positive.
$28,000,000,000 = 2.8 \times 10^{10}$

51. $9.07,000,000,000,000,000.$
⌞_____⌟ 17 places
Large number, so the exponent is positive.
$907,000,000,000,000,000 = 9.07 \times 10^{17}$

53. $0.000003.04$
⌞_____⌝ 6 places
Small number, so the exponent is negative.
$0.00000304 = 3.04 \times 10^{-6}$

55. $0.00000001.8$
⌞_____⌝ 8 places
Small number, so the exponent is negative.
$0.000000018 = 1.8 \times 10^{-8}$

57. $1.00,000,000,000.$
⌞_____⌟ 11 places
Large number, so the exponent is positive.
$100,000,000,000 = 1.0 \times 10^{11} = 10^{11}$

59. 281 million $= 281,000,000$
$2.81,000,000.$
⌞_____⌟ 8 places
Large number, so the exponent is positive.
281 million $= 2.81 \times 10^8$

61. $\dfrac{1}{10,000,000} = 0.0000001$
$0.0000001.$
⌞_____⌝ 7 places
Small number, so the exponent is negative.
$\dfrac{1}{10,000,000} = 1 \times 10^{-7}$, or 10^{-7}

63. 8.74×10^7
Positive exponent, so the answer is a large number.
$8.7400000.$
⌞_____⌝ 7 places
$8.74 \times 10^7 = 87,400,000$

65. 5.704×10^{-8}
Negative exponent, so the answer is a small number.
$0.00000005.704$
⌞_____⌟ 8 places
$5.704 \times 10^{-8} = 0.00000005704$

67. $10^7 = 1 \times 10^7$

Positive exponent, so the answer is a large number.

1.0000000.

$\rfloor\underline{\qquad}\uparrow$ 7 places

$10^7 = 10,000,000$

69. $10^{-5} = 1 \times 10^{-5}$

Negative exponent, so the answer is a small number.

0.00001.

$\uparrow\underline{\qquad}\rfloor$ 5 places

$10^{-5} = 0.00001$

71. $(3 \times 10^4)(2 \times 10^5) = (3 \cdot 2) \times (10^4 \cdot 10^5)$
$$= 6 \times 10^9$$

73. $(5.2 \times 10^5)(6.5 \times 10^{-2}) = (5.2 \cdot 6.5) \times (10^5 \cdot 10^{-2})$
$$= 33.8 \times 10^3$$

The answer at this stage is 33.8×10^3 but this is not scientific notation since 33.8 is not a number between 1 and 10. We convert 33.8 to scientific notation and simplify.

$33.8 \times 10^3 = (3.38 \times 10^1) \times 10^3 = 3.38 \times (10^1 \times 10^3) = 3.38 \times 10^4$

The answer is 3.38×10^4.

75. $(9.9 \times 10^{-6})(8.23 \times 10^{-8}) = (9.9 \cdot 8.23) \times (10^{-6} \cdot 10^{-8})$
$$= 81.477 \times 10^{-14}$$

The answer at this stage is 81.477×10^{-14}. We convert 81.477 to scientific notation and simplify.

$81.477 \times 10^{-14} = (8.1477 \times 10^1) \times 10^{-14} = 8.1477 \times (10^1 \times 10^{-14}) = 8.1477 \times 10^{-13}$.

The answer is 8.1477×10^{-13}.

77. $\dfrac{8.5 \times 10^8}{3.4 \times 10^{-5}} = \dfrac{8.5}{3.4} \times \dfrac{10^8}{10^{-5}}$
$$= 2.5 \times 10^{8-(-5)}$$
$$= 2.5 \times 10^{13}$$

79. $(3.0 \times 10^6) \div (6.0 \times 10^9) = \dfrac{3.0 \times 10^6}{6.0 \times 10^9}$
$$= \dfrac{3.0}{6.0} \times \dfrac{10^6}{10^9}$$
$$= 0.5 \times 10^{6-9}$$
$$= 0.5 \times 10^{-3}$$

The answer at this stage is 0.5×10^{-3}. We convert 0.5 to scientific notation and simplify.

$0.5 \times 10^{-3} = (5.0 \times 10^{-1}) \times 10^{-3} = 5.0 \times (10^{-1} \times 10^{-3}) = 5.0 \times 10^{-4}$

81. $\dfrac{7.5 \times 10^{-9}}{2.5 \times 10^{12}} = \dfrac{7.5}{2.5} \times \dfrac{10^{-9}}{10^{12}}$
$$= 3.0 \times 10^{-9-12}$$
$$= 3.0 \times 10^{-21}$$

83. There are 60 seconds in one minute and 60 minutes in one hour, so there are 60(60), or 3600 seconds in one hour. There are 24 hours in one day and 365 days in one year, so there are 3600(24)(365), or 31,536,000 seconds in one year.

$$4,200,000 \times 31,536,000$$
$$= (4.2 \times 10^6) \times (3.1536 \times 10^7)$$
$$= (4.2 \times 3.1536) \times (10^6 \times 10^7)$$
$$\approx 13.25 \times 10^{13}$$
$$\approx (1.325 \times 10) \times 10^{13}$$
$$\approx 1.325 \times (10 \times 10^{13})$$
$$\approx 1.325 \times 10^{14}$$

About 1.325×10^{14} cubic feet of water is discharged from the Amazon River in 1 yr.

85. $\dfrac{1.908 \times 10^{24}}{6 \times 10^{21}} = \dfrac{1.908}{6} \times \dfrac{10^{24}}{10^{21}}$
$$= 0.318 \times 10^3$$
$$= (3.18 \times 10^{-1}) \times 10^3$$
$$= 3.18 \times (10^{-1} \times 10^3)$$
$$= 3.18 \times 10^2$$

The mass of Jupiter is 3.18×10^2 times the mass of Earth.

87. 10 billion trillion $= 1 \times 10 \times 10^9 \times 10^{12}$
$$= 1 \times 10^{22}$$

There are 1×10^{22} stars in the known universe.

89. We divide the mass of the sun by the mass of earth.
$$\dfrac{1.998 \times 10^{27}}{6 \times 10^{21}} = 0.333 \times 10^6$$
$$= (3.33 \times 10^{-1}) \times 10^6$$
$$= 3.33 \times 10^5$$

The mass of the sun is 3.33×10^5 times the mass of Earth.

91. First we divide the distance from the earth to the moon by 3 days to find the number of miles per day the space vehicle travels. Note that $240,000 = 2.4 \times 10^5$.
$$\dfrac{2.4 \times 10^5}{3} = 0.8 \times 10^5 = 8 \times 10^4$$

The space vehicle travels 8×10^4 miles per day. Now divide the distance from the earth to Mars by 8×10^4 to find how long it will take the space vehicle to reach Mars. Note that $35,000,000 = 3.5 \times 10^7$.
$$\dfrac{3.5 \times 10^7}{8 \times 10^4} = 0.4375 \times 10^3 = 4.375 \times 10^2$$

It takes 4.375×10^2 days for the space vehicle to travel from the earth to Mars.

93. Discussion and Writing Exercise

95. $9x - 36 = 9 \cdot x - 9 \cdot 4 = 9(x - 4)$

97. $3s + 3t + 24 = 3 \cdot s + 3 \cdot t + 3 \cdot 8 = 3(s + t + 8)$

99. $2x - 4 - 5x + 8 = x - 3$

$\qquad -3x + 4 = x - 3 \qquad$ Collecting like terms

$\qquad -3x + 4 - 4 = x - 3 - 4 \qquad$ Subtracting 4

$\qquad -3x = x - 7$

$\qquad -3x - x = x - 7 - x \qquad$ Subtracting x

$\qquad -4x = -7$

$\qquad \dfrac{-4x}{-4} = \dfrac{-7}{-4} \qquad$ Dividing by -4

$\qquad x = \dfrac{7}{4}$

The solution is $\dfrac{7}{4}$.

101. $8(2x + 3) - 2(x - 5) = 10$

$\qquad 16x + 24 - 2x + 10 = 10 \qquad$ Removing parentheses

$\qquad 14x + 34 = 10 \qquad$ Collecting like terms

$\qquad 14x + 34 - 34 = 10 - 34 \qquad$ Subtracting 34

$\qquad 14x = -24$

$\qquad \dfrac{14x}{14} = \dfrac{-24}{14} \qquad$ Dividing by 14

$\qquad x = -\dfrac{12}{7} \qquad$ Simplifying

The solution is $-\dfrac{12}{7}$.

103. $y = x - 5$

The equation is equivalent to $y = x + (-5)$. The y-intercept is $(0, -5)$. We find two other points.

When $x = 2$, $y = 2 - 5 = -3$.

When $x = 4$, $y = 4 - 5 = -1$.

x	y
0	-5
2	-3
4	-1

Plot these points, draw the line they determine, and label the graph $y = x - 5$.

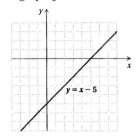

105. $\dfrac{(5.2 \times 10^6)(6.1 \times 10^{-11})}{1.28 \times 10^{-3}} = \dfrac{(5.2 \cdot 6.1)}{1.28} \times \dfrac{(10^6 \cdot 10^{-11})}{10^{-3}}$

$\qquad = 24.78125 \times 10^{-2}$

$\qquad = (2.478125 \times 10^1) \times 10^{-2}$

$\qquad = 2.478125 \times 10^{-1}$

107. $\dfrac{(5^{12})^2}{5^{25}} = \dfrac{5^{24}}{5^{25}} = 5^{24-25} = 5^{-1} = \dfrac{1}{5}$

109. $\dfrac{(3^5)^4}{3^5 \cdot 3^4} = \dfrac{3^{5 \cdot 4}}{3^{5+4}} = \dfrac{3^{20}}{3^9} = 3^{20-9} = 3^{11}$

111. $\dfrac{49^{18}}{7^{35}} = \dfrac{(7^2)^{18}}{7^{35}} = \dfrac{7^{36}}{7^{35}} = 7$

113. $\dfrac{(0.4)^5}{\left((0.4)^3\right)^2} = \dfrac{(0.4)^5}{(0.4)^6} = (0.4)^{-1} = \dfrac{1}{0.4}$, or 2.5

115. False; let $x = 2$, $y = 3$, $m = 4$, and $n = 2$:

$\qquad 2^4 \cdot 3^2 = 16 \cdot 9 = 144$, but

$\qquad (2 \cdot 3)^{4 \cdot 2} = 6^8 = 1,679,616$

117. False; let $x = 5$, $y = 3$, and $m = 2$:

$\qquad (5 - 3)^2 = 2^2 = 4$, but

$\qquad 5^2 - 3^2 = 25 - 9 = 16$

Exercise Set 9.3

1. $-5x + 2 = -5 \cdot 4 + 2 = -20 + 2 = -18$;

$\qquad -5x + 2 = -5(-1) + 2 = 5 + 2 = 7$

3. $2x^2 - 5x + 7 = 2 \cdot 4^2 - 5 \cdot 4 + 7 = 2 \cdot 16 - 20 + 7 = 32 - 20 + 7 = 19$;

$\qquad 2x^2 - 5x + 7 = 2(-1)^2 - 5(-1) + 7 = 2 \cdot 1 + 5 + 7 = 2 + 5 + 7 = 14$

5. $x^3 - 5x^2 + x = 4^3 - 5 \cdot 4^2 + 4 = 64 - 5 \cdot 16 + 4 = 64 - 80 + 4 = -12$;

$\qquad x^3 - 5x^2 + x = (-1)^3 - 5(-1)^2 + (-1) = -1 - 5 \cdot 1 - 1 = -1 - 5 - 1 = -7$

7. $3x + 5 = 3(-2) + 5 = -6 + 5 = -1$;

$\qquad 3x + 5 = 3 \cdot 0 + 5 = 0 + 5 = 5$

9. $x^2 - 2x + 1 = (-2)^2 - 2(-2) + 1 = 4 + 4 + 1 = 9$;

$\qquad x^2 - 2x + 1 = 0^2 - 2 \cdot 0 + 1 = 0 - 0 + 1 = 1$

11. $-3x^3 + 7x^2 - 3x - 2 = -3(-2)^3 + 7(-2)^2 - 3(-2) - 2 = -3(-8) + 7(4) - 3(-2) - 2 = 24 + 28 + 6 - 2 = 56$;

$\qquad -3x^3 + 7x^2 - 3x - 2 = -3 \cdot 0^3 + 7 \cdot 0^2 - 3 \cdot 0 - 2 = -3 \cdot 0 + 7 \cdot 0 - 0 - 2 = 0 + 0 - 0 - 2 = -2$

13. We evaluate the polynomial for $t = 10$:

$\qquad S = 11.12t^2 = 11.12(10)^2 = 11.12(100) = 1112$

The skydiver has fallen approximately 1112 ft.

15. a) In 2002, $t = 0$.

$\qquad E = 0.19(0) + 3.93 = 0 + 3.93 = 3.93$

The consumption in 2000 was 3.93 million gigawatt hours.

In 2001, $t = 2001 - 2000 = 1$.

$\qquad E = 0.19(1) + 3.93 = 0.19 + 3.93 = 4.12$

The consumption in 2001 was 4.12 million gigawatt hours.

In 2003, $t = 2003 - 2000 = 3$.

$\qquad E = 0.19(3) + 3.93 = 0.57 + 3.93 = 4.5$

The consumption in 2003 will be 4.5 million gigawatt hours.

In 2005, $t = 2005 - 2000 = 5$.

$E = 0.19(5) + 3.93 = 0.95 + 3.93 = 4.88$

The consumption in 2005 will be 4.88 million gigawatt hours.

In 2008, $t = 2008 - 2000 = 8$.

$E = 0.19(8) + 3.93 = 1.52 + 3.93 = 5.45$

The consumption in 2008 will be 5.45 million gigawatt hours.

In 2010, $t = 2010 - 2000 = 10$.

$E = 0.19(10) + 3.93 = 1.9 + 3.93 = 5.83$

The consumption in 2010 will be 5.83 million gigawatt hours.

b) It appears that the points $(0, 3.93)$, $(1, 4.12)$, $(3, 4.5)$, $(5, 4.88)$, $(8, 5.45)$, and $(10, 5.83)$ are on the graph, so the results check.

17. We evaluate the polynomial for $x = 75$:

$$\begin{aligned} R = 280x - 0.4x^2 &= 280(75) - 0.4(75)^2 \\ &= 280(75) - 0.4(5625) \\ &= 21,000 - 2250 \\ &= 18,750 \end{aligned}$$

The total revenue from the sale of 75 TVs is \$18,750.

We evaluate the polynomial for $x = 100$:

$$\begin{aligned} R = 280x - 0.4x^2 &= 280(100) - 0.4(100)^2 \\ &= 280(100) - 0.4(10,000) \\ &= 28,000 - 4000 \\ &= 24,000 \end{aligned}$$

The total revenue from the sale of 100 TVs is \$24,000.

19. Locate -3 on the x-axis. Then move vertically to the graph and horizontally to the y-axis. It appears that the y-value that is paired with -3 is -4. Thus, the value of $y = 5 - x^2$ is -4 when $x = -3$.

Locate -1 on the x-axis. Then move vertically to the graph and horizontally to the y-axis. It appears that the y-value that is paired with -1 is 4. Thus, the value of $y = 5 - x^2$ is 4 when $x = -1$.

Locate 0 on the x-axis. Then move vertically to the graph. We arrive at a point on the y-axis with the y-value 5. Thus, the value of $5 - x^2$ is 5 when $x = 0$.

Locate 1.5 on the x-axis. Then move vertically to the graph and horizontally to the y-axis. It appears that the y-value that is paired with 1.5 is 2.75. Thus, the value of $y = 5 - x^2$ is 2.75 when $x = 1.5$.

Locate 2 on the x-axis. Then move vertically to the graph and horizontally to the y-axis. It appears that the y-value that is paired with 2 is 1. Thus, the value of $y = 5 - x^2$ is 1 when $x = 2$.

21. We evaluate the polynomial for $x = 20$:

$$\begin{aligned} N &= -0.00006(20)^3 + 0.006(20)^2 - 0.1(20) + 1.9 \\ &= -0.00006(8000) + 0.006(400) - 0.1(20) + 1.9 \\ &= -0.48 + 2.4 - 2.0 + 1.9 \\ &= 1.82 \end{aligned}$$

There are about 1.82 million or 1,820,000 hearing-impaired Americans of age 20.

We evaluate the polynomial for $x = 40$:

$$\begin{aligned} N &= -0.00006(40)^3 + 0.006(40)^2 - 0.1(40) + 1.9 \\ &= -0.00006(64,000) + 0.006(1600) - 0.1(40) + 1.9 \\ &= -3.84 + 9.6 - 4.0 + 1.9 \\ &= 3.66 \end{aligned}$$

There are about 3.66 million, or 3,660,000, hearing-impaired Americans of age 40.

23. Locate 10 on the horizontal axis. From there move vertically to the graph and then horizontally to the M-axis. This locates an M-value of about 9. Thus, about 9 words were memorized in 10 minutes.

25. Locate 8 on the horizontal axis. From there move vertically to the graph and then horizontally to the M-axis. This locates an M-value of about 6. Thus, the value of $-0.001t^3 + 0.1t^2$ for $t = 8$ is approximately 6.

27. Locate 13 on the horizontal axis. It is halfway between 12 and 14. From there move vertically to the graph and then horizontally to the M-axis. This locates an M-value of about 15. Thus, the value of $-0.001t^3 + 0.1t^2$ when t is 13 is approximately 15.

29. $2 - 3x + x^2 = 2 + (-3x) + x^2$

The terms are 2, $-3x$, and x^2.

31. $5x^3 + 6x^2 - 3x^2$

Like terms: $6x^2$ and $-3x^2$ Same variable and exponent

33. $2x^4 + 5x - 7x - 3x^4$

Like terms: $2x^4$ and $-3x^4$ Same variable and
Like terms: $5x$ and $-7x$ exponent

35. $3x^5 - 7x + 8 + 14x^5 - 2x - 9$

Like terms: $3x^5$ and $14x^5$
Like terms: $-7x$ and $-2x$
Like terms: 8 and -9 Constant terms are like terms.

37. $-3x + 6$

The coefficient of $-3x$, the first term, is -3.

The coefficient of 6, the second term, is 6.

39. $5x^2 + 3x + 3$

The coefficient of $5x^2$, the first term, is 5.

The coefficient of $3x$, the second term, is 3.

The coefficient of 3, the third term, is 3.

41. $-5x^4 + 6x^3 - 3x^2 + 8x - 2$

The coefficient of $-5x^4$, the first term, is -5.

The coefficient of $6x^3$, the second term, is 6.

The coefficient of $-3x^2$, the third term, is -3.

The coefficient of $8x$, the fourth term, is 8.

The coefficient of -2, the fifth term, is -2.

43. $2x - 5x = (2 - 5)x = -3x$

45. $x - 9x = 1x - 9x = (1 - 9)x = -8x$

47. $5x^3 + 6x^3 + 4 = (5 + 6)x^3 + 4 = 11x^3 + 4$

49. $5x^3 + 6x - 4x^3 - 7x = (5 - 4)x^3 + (6 - 7)x =$
$1x^3 + (-1)x = x^3 - x$

51. $6b^5 + 3b^2 - 2b^5 - 3b^2 = (6 - 2)b^5 + (3 - 3)b^2 =$
$4b^5 + 0b^2 = 4b^5$

53. $\dfrac{1}{4}x^5 - 5 + \dfrac{1}{2}x^5 - 2x - 37 =$
$\left(\dfrac{1}{4} + \dfrac{1}{2}\right)x^5 - 2x + (-5 - 37) = \dfrac{3}{4}x^5 - 2x - 42$

55. $6x^2 + 2x^4 - 2x^2 - x^4 - 4x^2 =$
$6x^2 + 2x^4 - 2x^2 - 1x^4 - 4x^2 =$
$(6 - 2 - 4)x^2 + (2 - 1)x^4 = 0x^2 + 1x^4 =$
$0 + x^4 = x^4$

57. $\dfrac{1}{4}x^3 - x^2 - \dfrac{1}{6}x^2 + \dfrac{3}{8}x^3 + \dfrac{5}{16}x^3 =$
$\dfrac{1}{4}x^3 - 1x^2 - \dfrac{1}{6}x^2 + \dfrac{3}{8}x^3 + \dfrac{5}{16}x^3 =$
$\left(\dfrac{1}{4} + \dfrac{3}{8} + \dfrac{5}{16}\right)x^3 + \left(-1 - \dfrac{1}{6}\right)x^2 =$
$\left(\dfrac{4}{16} + \dfrac{6}{16} + \dfrac{5}{16}\right)x^3 + \left(-\dfrac{6}{6} - \dfrac{1}{6}\right)x^2 = \dfrac{15}{16}x^3 - \dfrac{7}{6}x^2$

59. $x^5 + x + 6x^3 + 1 + 2x^2 = x^5 + 6x^3 + 2x^2 + x + 1$

61. $5y^3 + 15y^9 + y - y^2 + 7y^8 =$
$15y^9 + 7y^8 + 5y^3 - y^2 + y$

63. $3x^4 - 5x^6 - 2x^4 + 6x^6 = x^4 + x^6 = x^6 + x^4$

65. $-2x + 4x^3 - 7x + 9x^3 + 8 = -9x + 13x^3 + 8 =$
$13x^3 - 9x + 8$

67. $3x + 3x + 3x - x^2 - 4x^2 = 9x - 5x^2 = -5x^2 + 9x$

69. $-x + \dfrac{3}{4} + 15x^4 - x - \dfrac{1}{2} - 3x^4 = -2x + \dfrac{1}{4} + 12x^4 =$
$12x^4 - 2x + \dfrac{1}{4}$

71. $2x - 4 = 2x^1 - 4x^0$
The degree of $2x$ is 1.
The degree of -4 is 0.
The degree of the polynomial is 1, the largest exponent.

73. $3x^2 - 5x + 2 = 3x^2 - 5x^1 + 2x^0$
The degree of $3x^2$ is 2.
The degree of $-5x$ is 1.
The degree of 2 is 0.
The degree of the polynomial is 2, the largest exponent.

75. $-7x^3 + 6x^2 + 3x + 7 = -7x^3 + 6x^2 + 3x^1 + 7x^0$
The degree of $-7x^3$ is 3.
The degree of $6x^2$ is 2.
The degree of $3x$ is 1.
The degree of 7 is 0.
The degree of the polynomial is 3, the largest exponent.

77. $x^2 - 3x + x^6 - 9x^4 = x^2 - 3x^1 + x^6 - 9x^4$
The degree of x^2 is 2.
The degree of $-3x$ is 1.
The degree of x^6 is 6.
The degree of $-9x^4$ is 4.
The degree of the polynomial is 6, the largest exponent.

79. See the answer section in the text.

81. In the polynomial $x^3 - 27$, there are no x^2 or x terms. The x^2 term (or second-degree term) and the x term (or first-degree term) are missing.

83. In the polynomial $x^4 - x$, there are no x^3, x^2, or x^0 terms. The x^3 term (or third-degree term), the x^2 term (or second-degree term), and the x^0 term (or zero-degree term) are missing.

85. No terms are missing in the polynomial
$2x^3 - 5x^2 + x - 3$.

87. $x^3 - 27 = x^3 + 0x^2 + 0x - 27$
$x^3 - 27 = x^3 \hspace{2.5cm} - 27$

89. $x^4 - x = x^4 + 0x^3 + 0x^2 - x + 0x^0$
$x^4 - x = x^4 \hspace{3cm} - x$

91. There are no missing terms.

93. The polynomial $x^2 - 10x + 25$ is a *trinomial* because it has just three terms.

95. The polynomial $x^3 - 7x^2 + 2x - 4$ is *none of these* because it has more than three terms.

97. The polynomial $4x^2 - 25$ is a *binomial* because it has just two terms.

99. The polynomial $40x$ is a *monomial* because it has just one term.

101. Discussion and Writing Exercise

103. *Familiarize.* Let a = the number of apples the campers had to begin with. Then the first camper ate $\dfrac{1}{3}a$ apples and $a - \dfrac{1}{3}a$, or $\dfrac{2}{3}a$, apples were left. The second camper ate $\dfrac{1}{3}\left(\dfrac{2}{3}a\right)$, or $\dfrac{2}{9}a$, apples, and $\dfrac{2}{3}a - \dfrac{2}{9}a$, or $\dfrac{4}{9}a$, apples were left. The third camper ate $\dfrac{1}{3}\left(\dfrac{4}{9}a\right)$, or $\dfrac{4}{27}a$, apples, and $\dfrac{4}{9}a - \dfrac{4}{27}a$, or $\dfrac{8}{27}a$, apples were left.

Translate. We write an equation for the number of apples left after the third camper eats.

Number of apples left is 8.

$$\frac{8}{27}a = 8$$

Solve. We solve the equation.

$$\frac{8}{27}a = 8$$
$$a = \frac{27}{8} \cdot 8$$
$$a = 27$$

Check. If the campers begin with 27 apples, then the first camper eats $\frac{1}{3} \cdot 27$, or 9, and $27 - 9$, or 18, are left. The second camper then eats $\frac{1}{3} \cdot 18$, or 6 apples and $18 - 6$, or 12, are left. Finally, the third camper eats $\frac{1}{3} \cdot 12$, or 4 apples and $12 - 4$, or 8, are left. The answer checks.

State. The campers had 27 apples to begin with.

105. $\frac{1}{8} - \frac{5}{6} = \frac{1}{8} + \left(-\frac{5}{6}\right)$, LCM is 24

$$= \frac{1}{8} \cdot \frac{3}{3} + \left(-\frac{5}{6}\right)\left(\frac{4}{4}\right)$$
$$= \frac{3}{24} + \left(-\frac{20}{24}\right)$$
$$= -\frac{17}{24}$$

107. $5.6 - 8.2 = 5.6 + (-8.2) = -2.6$

109. $C = ab - r$

$C + r = ab$ Adding r

$\dfrac{C + r}{a} = \dfrac{ab}{a}$ Dividing by a

$\dfrac{C + r}{a} = b$ Simplifying

111. $3x - 15y + 63 = 3 \cdot x - 3 \cdot 5y + 3 \cdot 21 = 3(x - 5y + 21)$

113. $(3x^2)^3 + 4x^2 \cdot 4x^4 - x^4(2x)^2 + [(2x)^2]^3 - 100x^2(x^2)^2$
$= 27x^6 + 4x^2 \cdot 4x^4 - x^4 \cdot 4x^2 + (2x)^6 - 100x^2 \cdot x^4$
$= 27x^6 + 16x^6 - 4x^6 + 64x^6 - 100x^6$
$= 3x^6$

115. $(5m^5)^2 = 5^2 m^{5 \cdot 2} = 25m^{10}$

The degree is 10.

117. Graph $y = 5 - x^2$. Then use VALUE from the CALC menu to find the y-values that correspond to $x = -3$, $x = -1$, x=0, $x = 1.5$, and $x = 2$. As before, we find that these values are -4, 4, 5, 2.75, and 1, respectively.

119. Graph $y = -0.00006x^3 + 0.006x^2 - 0.1x + 1.9$. Then use VALUE from the CALC menu to find the y-values that correspond to $x = 20$ and $x = 40$. As before, we find that these values are 1.82 and 3.66, respectively. These results represent 1,820,000 and 3,660,000 hearing-impaired Americans.

Exercise Set 9.4

1. $(3x + 2) + (-4x + 3) = (3 - 4)x + (2 + 3) = -x + 5$

3. $(-6x + 2) + (x^2 + x - 3) =$
$x^2 + (-6 + 1)x + (2 - 3) = x^2 - 5x - 1$

5. $(x^2 - 9) + (x^2 + 9) = (1 + 1)x^2 + (-9 + 9) = 2x^2$

7. $(3x^2 - 5x + 10) + (2x^2 + 8x - 40) =$
$(3 + 2)x^2 + (-5 + 8)x + (10 - 40) = 5x^2 + 3x - 30$

9. $(1.2x^3 + 4.5x^2 - 3.8x) + (-3.4x^3 - 4.7x^2 + 23) =$
$(1.2 - 3.4)x^3 + (4.5 - 4.7)x^2 - 3.8x + 23 =$
$-2.2x^3 - 0.2x^2 - 3.8x + 23$

11. $(1 + 4x + 6x^2 + 7x^3) + (5 - 4x + 6x^2 - 7x^3) =$
$(1 + 5) + (4 - 4)x + (6 + 6)x^2 + (7 - 7)x^3 =$
$6 + 0x + 12x^2 + 0x^3 = 6 + 12x^2$, or $12x^2 + 6$

13. $\left(\frac{1}{4}x^4 + \frac{2}{3}x^3 + \frac{5}{8}x^2 + 7\right) + \left(-\frac{3}{4}x^4 + \frac{3}{8}x^2 - 7\right) =$
$\left(\frac{1}{4} - \frac{3}{4}\right)x^4 + \frac{2}{3}x^3 + \left(\frac{5}{8} + \frac{3}{8}\right)x^2 + (7 - 7) =$
$-\frac{2}{4}x^4 + \frac{2}{3}x^3 + \frac{8}{8}x^2 + 0 =$
$-\frac{1}{2}x^4 + \frac{2}{3}x^3 + x^2$

15. $(0.02x^5 - 0.2x^3 + x + 0.08) + (-0.01x^5 + x^4 - 0.8x - 0.02) =$
$(0.02 - 0.01)x^5 + x^4 - 0.2x^3 + (1 - 0.8)x + (0.08 - 0.02) =$
$0.01x^5 + x^4 - 0.2x^3 + 0.2x + 0.06$

17. $9x^8 - 7x^4 + 2x^2 + 5) + (8x^7 + 4x^4 - 2x) +$
$(-3x^4 + 6x^2 + 2x - 1) = 9x^8 + 8x^7 + (-7 + 4 - 3)x^4 +$
$(2 + 6)x^2 + (-2 + 2)x + (5 - 1) =$
$9x^8 + 8x^7 - 6x^4 + 8x^2 + 4$

19. Rewrite the problem so the coefficients of like terms have the same number of decimal places.

$$\begin{array}{r}
0.15x^4 + 0.10x^3 - 0.90x^2 \\
- 0.01x^3 + 0.01x^2 + x \\
1.25x^4 + 0.11x^2 + 0.01 \\
0.27x^3 + 0.99 \\
-0.35x^4 + 15.00x^2 - 0.03 \\
\hline
1.05x^4 + 0.36x^3 + 14.22x^2 + x + 0.97
\end{array}$$

21. We change the sign of the term inside the parentheses.
$-(-5x) = 5x$

23. We change the sign of every term inside the parentheses.
$-(-x^2 + 10x - 2) = x^2 - 10x + 2$

25. We change the sign of every term inside the parentheses.

$-(12x^4 - 3x^3 + 3) = -12x^4 + 3x^3 - 3$

27. We change the sign of every term inside parentheses.

$-(3x - 7) = -3x + 7$

29. We change the sign of every term inside parentheses.

$-(4x^2 - 3x + 2) = -4x^2 + 3x - 2$

31. We change the sign of every term inside parentheses.

$-\left(-4x^4 + 6x^2 + \dfrac{3}{4}x - 8\right) = 4x^4 - 6x^2 - \dfrac{3}{4}x + 8$

33. $(3x + 2) - (-4x + 3) = 3x + 2 + 4x - 3$

Changing the sign of every term inside parentheses

$= 7x - 1$

35. $(-6x + 2) - (x^2 + x - 3) = -6x + 2 - x^2 - x + 3$

$\qquad\qquad\qquad\qquad\qquad\quad = -x^2 - 7x + 5$

37. $(x^2 - 9) - (x^2 + 9) = x^2 - 9 - x^2 - 9 = -18$

39. $(6x^4 + 3x^3 - 1) - (4x^2 - 3x + 3)$

$= 6x^4 + 3x^3 - 1 - 4x^2 + 3x - 3$

$= 6x^4 + 3x^3 - 4x^2 + 3x - 4$

41. $(1.2x^3 + 4.5x^2 - 3.8x) - (-3.4x^3 - 4.7x^2 + 23)$

$= 1.2x^3 + 4.5x^2 - 3.8x + 3.4x^3 + 4.7x^2 - 23$

$= 4.6x^3 + 9.2x^2 - 3.8x - 23$

43. $\dfrac{5}{8}x^3 - \dfrac{1}{4}x - \dfrac{1}{3} - \left(-\dfrac{1}{8}x^3 + \dfrac{1}{4}x - \dfrac{1}{3}\right)$

$= \dfrac{5}{8}x^3 - \dfrac{1}{4}x - \dfrac{1}{3} + \dfrac{1}{8}x^3 - \dfrac{1}{4}x + \dfrac{1}{3}$

$= \dfrac{6}{8}x^3 - \dfrac{2}{4}x$

$= \dfrac{3}{4}x^3 - \dfrac{1}{2}x$

45. $(0.08x^3 - 0.02x^2 + 0.01x) - (0.02x^3 + 0.03x^2 - 1)$

$= 0.08x^3 - 0.02x^2 + 0.01x - 0.02x^3 - 0.03x^2 + 1$

$= 0.06x^3 - 0.05x^2 + 0.01x + 1$

47.

$x^2 + 5x + 6$
$\underline{x^2 + 2x}$

$x^2 + 5x + 6$
$\underline{-x^2 - 2x}\qquad$ Changing signs
$\quad\ \ 3x + 6\qquad$ Adding

49.

$5x^4 + 6x^3 - 9x^2$
$\underline{-6x^4 - 6x^3 \qquad\quad + 8x + 9}$

$5x^4 + 6x^3 - 9x^2$
$\underline{6x^4 + 6x^3 \qquad\ - 8x - 9}\quad$ Changing signs
$11x^4 + 12x^3 - 9x^2 - 8x - 9\quad$ Adding

51.

$x^5 \qquad\qquad\qquad\ - 1$
$\underline{x^5 - x^4 + x^3 - x^2 + x - 1}$

$x^5 \qquad\qquad\qquad\ - 1$
$\underline{-x^5 + x^4 - x^3 + x^2 - x + 1}\quad$ Changing signs
$\qquad x^4 - x^3 + x^2 - x\qquad$ Adding

53.

The area of a rectangle is the product of the length and width. The sum of the areas is found as follows:

$$\begin{array}{ccccccc} \text{Area} & + & \text{Area} & + & \text{Area} & + & \text{Area} \\ \text{of } A & & \text{of } B & & \text{of } C & & \text{of } D \end{array}$$

$= \ 3x \cdot x \ + \ x \cdot x \ + \ x \cdot x \ + \ 4 \cdot x$

$= \ 3x^2 \ + \ x^2 \ + \ x^2 \ + \ 4x$

$= \ 5x^2 \ + \ 4x$

A polynomial for the sum of the areas is $5x^2 + 4x$.

55. We add the lengths of the sides:

$4a + 7 + a + \dfrac{1}{2}a + 3 + a + 2a + 3a$

$= \left(4 + 1 + \dfrac{1}{2} + 1 + 2 + 3\right)a + (7 + 3)$

$= 11\dfrac{1}{2}a + 10,\ \text{or}\ \dfrac{23}{2}a + 10$

57.

The length and width of the figure can be expressed as $r + 11$ and $r + 9$, respectively. The area of this figure (a rectangle) is the product of the length and width. An algebraic expression for the area is $(r + 11) \cdot (r + 9)$.

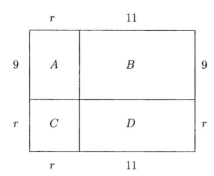

The area of the figure can also be found by adding the areas of the four rectangles A, B, C, and D. The area of a rectangle is the product of the length and the width.

$$
\begin{array}{ccccccc}
\begin{array}{c}\text{Area}\\\text{of }A\end{array} & + & \begin{array}{c}\text{Area}\\\text{of }B\end{array} & + & \begin{array}{c}\text{Area}\\\text{of }C\end{array} & + & \begin{array}{c}\text{Area}\\\text{of }D\end{array} \\
= \quad 9 \cdot r & + & 11 \cdot 9 & + & r \cdot r & + & 11 \cdot r \\
= \quad 9r & + & 99 & + & r^2 & + & 11r
\end{array}
$$

A second algebraic expression for the area of the figure is $9r + 99 + r^2 + 11r$, or $r^2 + 20r + 99$.

59.

The length and width of the figure can each be expressed as $x + 3$. The area can be expressed as $(x + 3) \cdot (x + 3)$, or $(x + 3)^2$.

Another way to express the area is to find an expression for the sum of the areas of the four rectangles A, B, C, and D. The area of each rectangle is the product of its length and width.

$$
\begin{array}{ccccccc}
\begin{array}{c}\text{Area}\\\text{of }A\end{array} & + & \begin{array}{c}\text{Area}\\\text{of }B\end{array} & + & \begin{array}{c}\text{Area}\\\text{of }C\end{array} & + & \begin{array}{c}\text{Area}\\\text{of }D\end{array} \\
= \quad x \cdot x & + & 3 \cdot x & + & 3 \cdot x & + & 3 \cdot 3 \\
= \quad x^2 & + & 3x & + & 3x & + & 9
\end{array}
$$

Then a second algebraic expression for the area of the figure is $x^2 + 3x + 3x + 9$, or $x^2 + 6x + 9$.

61.

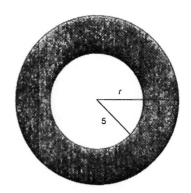

Familiarize. Recall that the area of a circle is the product of π and the square of the radius, r^2.

$$A = \pi r^2$$

Translate.

$$
\begin{array}{ccc}
\begin{array}{c}\text{Area of circle}\\\text{with radius }r\end{array} & - & \begin{array}{c}\text{Area of circle}\\\text{with radius }5\end{array} = \begin{array}{c}\text{Shaded}\\\text{area}\end{array} \\
\pi \cdot r^2 & - & \pi \cdot 5^2 \quad = \text{Shaded area}
\end{array}
$$

Carry out. We simplify the expression.

$$\pi \cdot r^2 - \pi \cdot 5^2 = \pi r^2 - 25\pi$$

Check. We can go over our calculations. We can also assign some value to r, say 7, and carry out the computation in two ways.

Difference of areas: $\pi \cdot 7^2 - \pi \cdot 5^2 = 49\pi - 25\pi = 24\pi$

Substituting in the polynomial: $\pi \cdot 7^2 - 25\pi = 49\pi - 25\pi = 24\pi$

Since the results are the same, our solution is probably correct.

State. A polynomial for the shaded area is $\pi r^2 - 25\pi$.

63. Familiarize. We label the figure with additional information.

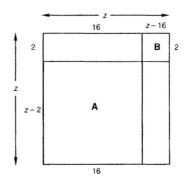

Translate.

Area of shaded sections = Area of A + Area of B

Area of shaded sections = $16(z - 2) + 2(z - 16)$

Carry out. We simplify the expression.

$16(z - 2) + 2(z - 16) = 16z - 32 + 2z - 32 = 18z - 64$

Check. We can go over the calculations. We can also assign some value to z, say 30, and carry out the computation in two ways.

Sum of areas:

$$16 \cdot 28 + 2 \cdot 14 = 448 + 28 = 476$$

Substituting in the polynomial:

$$18 \cdot 30 - 64 = 540 - 64 = 476$$

Since the results are the same, our solution is probably correct.

State. A polynomial for the shaded area is $18z - 64$.

65. Discussion and Writing Exercise

67. $8x + 3x = 66$

$\quad\quad 11x = 66$ Collecting like terms

$\quad\quad \dfrac{11x}{11} = \dfrac{66}{11}$ Dividing by 11

$\quad\quad\quad x = 6$

The solution is 6.

69. $\dfrac{3}{8}x + \dfrac{1}{4} - \dfrac{3}{4}x = \dfrac{11}{16} + x$, LCM is 16

$16\left(\dfrac{3}{8}x + \dfrac{1}{4} - \dfrac{3}{4}x\right) = 16\left(\dfrac{11}{16} + x\right)$ Clearing fractions

$\quad\quad 6x + 4 - 12x = 11 + 16x$

$\quad\quad\quad -6x + 4 = 11 + 16x$ Collecting like terms

$-6x + 4 - 4 = 11 + 16x - 4$ Subtracting 4

$\quad\quad\quad -6x = 7 + 16x$

$-6x - 16x = 7 + 16x - 16x$ Subtracting $16x$

$\quad\quad -22x = 7$

$\quad\quad \dfrac{-22x}{-22} = \dfrac{7}{-22}$ Dividing by -22

$\quad\quad\quad x = -\dfrac{7}{22}$

The solution is $-\dfrac{7}{22}$.

71. $1.5x - 2.7x = 22 - 5.6x$

$10(1.5x - 2.7x) = 10(22 - 5.6x)$ Clearing decimals

$\quad 15x - 27x = 220 - 56x$

$\quad\quad -12x = 220 - 56x$ Collecting like terms

$\quad\quad 44x = 220$ Adding $56x$

$\quad\quad x = \dfrac{220}{44}$ Dividing by 44

$\quad\quad x = 5$ Simplifying

The solution is 5.

73. $6(y - 3) - 8 = 4(y + 2) + 5$

$6y - 18 - 8 = 4y + 8 + 5$ Removing parentheses

$\quad 6y - 26 = 4y + 13$ Collecting like terms

$6y - 26 + 26 = 4y + 13 + 26$ Adding 26

$\quad\quad 6y = 4y + 39$

$6y - 4y = 4y + 39 - 4y$ Subtracting $4y$

$\quad\quad 2y = 39$

$\quad\quad \dfrac{2y}{2} = \dfrac{39}{2}$ Dividing by 2

$\quad\quad y = \dfrac{39}{2}$

The solution is $\dfrac{39}{2}$.

75. $3x - 7 \leq 5x + 13$

$\quad -2x - 7 \leq 13$ Subtracting $5x$

$\quad\quad -2x \leq 20$ Adding 7

$\quad\quad x \geq -10$ Dividing by -2 and reversing the inequality symbol

The solution set is $\{x | x \geq -10\}$.

77. Familiarize. The surface area is $2lw + 2lh + 2wh$, where $l = $ length, $w = $ width, and $h = $ height of the rectangular solid. Here we have $l = 3$, $w = w$, and $h = 7$.

Translate. We substitute in the formula above.

$$2 \cdot 3 \cdot w + 2 \cdot 3 \cdot 7 + 2 \cdot w \cdot 7$$

Carry out. We simplify the expression.

$$2 \cdot 3 \cdot w + 2 \cdot 3 \cdot 7 + 2 \cdot w \cdot 7$$
$$= 6w + 42 + 14w$$
$$= 20w + 42$$

Check. We can go over the calculations. We can also assign some value to w, say 6, and carry out the computation in two ways.

Using the formula: $2 \cdot 3 \cdot 6 + 2 \cdot 3 \cdot 7 + 2 \cdot 6 \cdot 7 = 36 + 42 + 84 = 162$

Substituting in the polynomial: $20 \cdot 6 + 42 = 120 + 42 = 162$

Since the results are the same, our solution is probably correct.

State. A polynomial for the surface area is $20w + 42$.

79. Familiarize. The surface area is $2lw + 2lh + 2wh$, where $l = $ length, $w = $ width, and $h = $ height of the rectangular solid. Here we have $l = x$, $w = x$, and $h = 5$.

Translate. We substitute in the formula above.

$$2 \cdot x \cdot x + 2 \cdot x \cdot 5 + 2 \cdot x \cdot 5$$

Carry out. We simplify the expression.

$$2 \cdot x \cdot x + 2 \cdot x \cdot 5 + 2 \cdot x \cdot 5$$
$$= 2x^2 + 10x + 10x$$
$$= 2x^2 + 20x$$

Check. We can go over the calculations. We can also assign some value to x, say 3, and carry out the computation in two ways.

Using the formula: $2 \cdot 3 \cdot 3 + 2 \cdot 3 \cdot 5 + 2 \cdot 3 \cdot 5 = 18 + 30 + 30 = 78$

Substituting in the polynomial: $2 \cdot 3^2 + 20 \cdot 3 = 2 \cdot 9 + 60 = 18 + 60 = 78$

Since the results are the same, our solution is probably correct.

State. A polynomial for the surface area is $2x^2 + 20x$.

81.

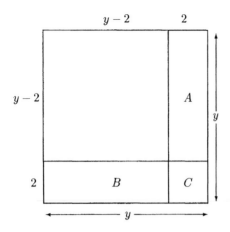

The shaded area is $(y-2)^2$. We find it as follows:

$$\begin{array}{ccccccc} \text{Shaded} & = & \text{Area of} & - & \text{Area} & - & \text{Area} & - & \text{Area} \\ \text{area} & & \text{square} & & \text{of } A & & \text{of } B & & \text{of } C \end{array}$$

$(y-2)^2 = \quad y^2 \quad -2(y-2)-2(y-2)- \quad 2 \cdot 2$

$(y-2)^2 = y^2 - 2y + 4 - 2y + 4 - 4$

$(y-2)^2 = y^2 - 4y + 4$

83. $(7y^2 - 5y + 6) - (3y^2 + 8y - 12) + (8y^2 - 10y + 3)$
$= 7y^2 - 5y + 6 - 3y^2 - 8y + 12 + 8y^2 - 10y + 3$
$= 12y^2 - 23y + 21$

85. $(-y^4 - 7y^3 + y^2) + (-2y^4 + 5y - 2) - (-6y^3 + y^2)$
$= -y^4 - 7y^3 + y^2 - 2y^4 + 5y - 2 + 6y^3 - y^2$
$= -3y^4 - y^3 + 5y - 2$

Exercise Set 9.5

1. $(8x^2)(5) = (8 \cdot 5)x^2 = 40x^2$

3. $(-x^2)(-x) = (-1x^2)(-1x) = (-1)(-1)(x^2 \cdot x) = x^3$

5. $(8x^5)(4x^3) = (8 \cdot 4)(x^5 \cdot x^3) = 32x^8$

7. $(0.1x^6)(0.3x^5) = (0.1)(0.3)(x^6 \cdot x^5) = 0.03x^{11}$

9. $\left(-\frac{1}{5}x^3\right)\left(-\frac{1}{3}x\right) = \left(-\frac{1}{5}\right)\left(-\frac{1}{3}\right)(x^3 \cdot x) = \frac{1}{15}x^4$

11. $(-4x^2)(0) = 0$ Any number multiplied by 0 is 0.

13. $(3x^2)(-4x^3)(2x^6) = (3)(-4)(2)(x^2 \cdot x^3 \cdot x^6) = -24x^{11}$

15. $2x(-x + 5) = 2x(-x) + 2x(5)$
$= -2x^2 + 10x$

17. $-5x(x - 1) = -5x(x) - 5x(-1)$
$= -5x^2 + 5x$

19. $x^2(x^3 + 1) = x^2(x^3) + x^2(1)$
$= x^5 + x^2$

21. $3x(2x^2 - 6x + 1) = 3x(2x^2) + 3x(-6x) + 3x(1)$
$= 6x^3 - 18x^2 + 3x$

23. $-6x^2(x^2 + x) = -6x^2(x^2) - 6x^2(x)$
$= -6x^4 - 6x^3$

25. $3y^2(6y^4 + 8y^3) = 3y^2(6y^4) + 3y^2(8y^3)$
$= 18y^6 + 24y^5$

27. $(x + 6)(x + 3) = (x + 6)x + (x + 6)3$
$= x \cdot x + 6 \cdot x + x \cdot 3 + 6 \cdot 3$
$= x^2 + 6x + 3x + 18$
$= x^2 + 9x + 18$

29. $(x + 5)(x - 2) = (x + 5)x + (x + 5)(-2)$
$= x \cdot x + 5 \cdot x + x(-2) + 5(-2)$
$= x^2 + 5x - 2x - 10$
$= x^2 + 3x - 10$

31. $(x - 4)(x - 3) = (x - 4)x + (x - 4)(-3)$
$= x \cdot x - 4 \cdot x + x(-3) - 4(-3)$
$= x^2 - 4x - 3x + 12$
$= x^2 - 7x + 12$

33. $(x + 3)(x - 3) = (x + 3)x + (x + 3)(-3)$
$= x \cdot x + 3 \cdot x + x(-3) + 3(-3)$
$= x^2 + 3x - 3x - 9$
$= x^2 - 9$

35. $(5 - x)(5 - 2x) = (5 - x)5 + (5 - x)(-2x)$
$= 5 \cdot 5 - x \cdot 5 + 5(-2x) - x(-2x)$
$= 25 - 5x - 10x + 2x^2$
$= 25 - 15x + 2x^2$

37. $(2x + 5)(2x + 5) = (2x + 5)2x + (2x + 5)5$
$= 2x \cdot 2x + 5 \cdot 2x + 2x \cdot 5 + 5 \cdot 5$
$= 4x^2 + 10x + 10x + 25$
$= 4x^2 + 20x + 25$

39. $\left(x - \frac{5}{2}\right)\left(x + \frac{2}{5}\right) = \left(x - \frac{5}{2}\right)x + \left(x - \frac{5}{2}\right)\frac{2}{5}$
$= x \cdot x - \frac{5}{2} \cdot x + x \cdot \frac{2}{5} - \frac{5}{2} \cdot \frac{2}{5}$
$= x^2 - \frac{5}{2}x + \frac{2}{5}x - 1$
$= x^2 - \frac{25}{10}x + \frac{4}{10}x - 1$
$= x^2 - \frac{21}{10}x - 1$

41. $(x - 2.3)(x + 4.7) = (x - 2.3)x + (x - 2.3)4.7$
$= x \cdot x - 2.3 \cdot x + x \cdot 4.7 - 2.3(4.7)$
$= x^2 - 2.3x + 4.7x - 10.81$
$= x^2 + 2.4x - 10.81$

43. Illustrate $x(x + 5)$ as the area of a rectangle with width x and length $x + 5$.

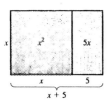

45. Illustrate $(x + 1)(x + 2)$ as the area of a rectangle with width $x + 1$ and length $x + 2$.

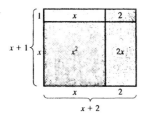

47. Illustrate $(x + 5)(x + 3)$ as the area of a rectangle with length $x + 5$ and width $x + 3$.

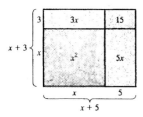

49. Illustrate $(3x+2)(3x+2)$ as the area of a square with sides of length $3x + 2$.

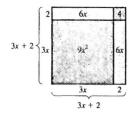

51.
$$(x^2 + x + 1)(x - 1)$$
$$= (x^2 + x + 1)x + (x^2 + x + 1)(-1)$$
$$= x^2 \cdot x + x \cdot x + 1 \cdot x + x^2(-1) + x(-1) + 1(-1)$$
$$= x^3 + x^2 + x - x^2 - x - 1$$
$$= x^3 - 1$$

53.
$$(2x + 1)(2x^2 + 6x + 1)$$
$$= 2x(2x^2 + 6x + 1) + 1(2x^2 + 6x + 1)$$
$$= 2x \cdot 2x^2 + 2x \cdot 6x + 2x \cdot 1 + 1 \cdot 2x^2 + 1 \cdot 6x + 1 \cdot 1$$
$$= 4x^3 + 12x^2 + 2x + 2x^2 + 6x + 1$$
$$= 4x^3 + 14x^2 + 8x + 1$$

55.
$$(y^2 - 3)(3y^2 - 6y + 2)$$
$$= y^2(3y^2 - 6y + 2) - 3(3y^2 - 6y + 2)$$
$$= y^2 \cdot 3y^2 + y^2(-6y) + y^2 \cdot 2 - 3 \cdot 3y^2 - 3(-6y) - 3 \cdot 2$$
$$= 3y^4 - 6y^3 + 2y^2 - 9y^2 + 18y - 6$$
$$= 3y^4 - 6y^3 - 7y^2 + 18y - 6$$

57.
$$(x^3 + x^2)(x^3 + x^2 - x)$$
$$= x^3(x^3 + x^2 - x) + x^2(x^3 + x^2 - x)$$
$$= x^3 \cdot x^3 + x^3 \cdot x^2 + x^3(-x) + x^2 \cdot x^3 + x^2 \cdot x^2 + x^2(-x)$$
$$= x^6 + x^5 - x^4 + x^5 + x^4 - x^3$$
$$= x^6 + 2x^5 - x^3$$

59.
$$(-5x^3 - 7x^2 + 1)(2x^2 - x)$$
$$= (-5x^3 - 7x^2 + 1)2x^2 + (-5x^3 - 7x^2 + 1)(-x)$$
$$= -5x^3 \cdot 2x^2 - 7x^2 \cdot 2x^2 + 1 \cdot 2x^2 - 5x^3(-x) - 7x^2(-x) +$$
$$1(-x)$$
$$= -10x^5 - 14x^4 + 2x^2 + 5x^4 + 7x^3 - x$$
$$= -10x^5 - 9x^4 + 7x^3 + 2x^2 - x$$

61.

$$
\begin{array}{ll}
\quad 1 + x + x^2 & \text{Line up like terms} \\
\underline{-1 - x + x^2} & \text{in columns} \\
\quad x^2 + x^3 + x^4 & \text{Multiplying the top row by } x^2 \\
-\quad x - x^2 - x^3 & \text{Multiplying by } -x \\
\underline{-1 - x - x^2 \qquad} & \text{Multiplying by } -1 \\
-1 - 2x - x^2 \quad + x^4 &
\end{array}
$$

63.

$$
\begin{array}{l}
\quad 2t^2 - t - 4 \\
\underline{\quad 3t^2 + 2t - 1} \\
-\ 2t^2 + t + 4 \qquad \text{Multiplying by } -1 \\
4t^3 - 2t^2 - 8t \qquad \text{Multiplying by } 2t \\
\underline{6t^4 - 3t^3 - 12t^2 \qquad} \text{Multiplying by } 3t^2 \\
6t^4 + t^3 - 16t^2 - 7t + 4
\end{array}
$$

65.

$$
\begin{array}{l}
\ x \quad -x^3 \quad +x^5 \\
\underline{-1 +x^2 \quad +x^4} \qquad \text{Rewriting in ascending order} \\
\quad x^5 - x^7 + x^9 \qquad \text{Multiplying by } x^4 \\
\quad x^3 - x^5 + x^7 \qquad \text{Multiplying by } x^2 \\
\underline{-x + x^3 - x^5 \qquad} \text{Multiplying by } -1 \\
-x + 2x^3 - x^5 \quad + x^9
\end{array}
$$

67.

$$
\begin{array}{l}
\ x^3 + x^2 + x + 1 \\
\underline{\qquad\qquad x - 1} \\
-x^3 - x^2 - x - 1 \\
\underline{x^4 + x^3 + x^2 + x} \\
x^4 \qquad\qquad -1
\end{array}
$$

69. We will multiply horizontally while still aligning like terms.

$$(x + 1)(x^3 + 7x^2 + 5x + 4)$$

$$
\begin{array}{ll}
= x^4 + 7x^3 + 5x^2 + 4x & \text{Multiplying by } x \\
\underline{\ + x^3 + 7x^2 + 5x + 4} & \text{Multiplying by } 1 \\
= x^4 + 8x^3 + 12x^2 + 9x + 4 &
\end{array}
$$

71. We will multiply horizontally while still aligning like terms.

$$\left(x - \frac{1}{2}\right)\left(2x^3 - 4x^2 + 3x - \frac{2}{5}\right)$$

$$= 2x^4 - 4x^3 + 3x^2 - \frac{2}{5}x$$

$$\qquad - x^3 + 2x^2 - \frac{3}{2}x + \frac{1}{5}$$

$$\overline{\quad 2x^4 - 5x^3 + 5x^2 - \frac{19}{10}x + \frac{1}{5}}$$

73. Discussion and Writing Exercise

75. $-\dfrac{1}{4} - \dfrac{1}{2} = -\dfrac{1}{4} - \dfrac{1}{2} \cdot \dfrac{2}{2} = -\dfrac{1}{4} - \dfrac{2}{4} = -\dfrac{3}{4}$

77. $(10 - 2)(10 + 2) = 8 \cdot 12 = 96$

79. $15x - 18y + 12 = 3 \cdot 5x - 3 \cdot 6y + 3 \cdot 4 =$
$3(5x - 6y + 4)$

81. $-9x - 45y + 15 = -3 \cdot 3x - 3 \cdot 15y - 3(-5) =$
$-3(3x + 15y - 5)$

83. $y = \dfrac{1}{2}x - 3$

The equation is equivalent to $y = \dfrac{1}{2}x + (-3)$. The y-intercept is $(0, -3)$. We find two other points, using multiples of 2 for x to avoid fractions.

When $x = -2$, $y = \dfrac{1}{2}(-2) - 3 = -1 - 3 = -4$.

When $x = 4$, $y = \dfrac{1}{2} \cdot 4 - 3 = 2 - 3 = -1$.

x	y
0	-3
-2	-4
4	-1

Plot these points, draw the line they determine, and label the graph $y = \dfrac{1}{2}x - 3$.

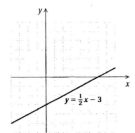

85. The shaded area is the area of the large rectangle, $6y(14y - 5)$ less the area of the unshaded rectangle, $3y(3y + 5)$. We have:

$$6y(14y - 5) - 3y(3y + 5)$$
$$= 84y^2 - 30y - 9y^2 - 15y$$
$$= 75y^2 - 45y$$

87.

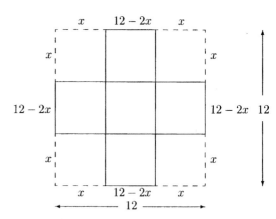

The dimensions, in inches, of the box are $12 - 2x$ by $12 - 2x$ by x. The volume is the product of the dimensions (volume = length × width × height):

$$\text{Volume} = (12 - 2x)(12 - 2x)x$$
$$= (144 - 48x + 4x^2)x$$
$$= 144x - 48x^2 + 4x^3 \text{ in}^3, \text{ or}$$
$$4x^3 - 48x^2 + 144x \text{ in}^3$$

The outside surface area is the sum of the area of the bottom and the areas of the four sides. The dimensions, in inches, of the bottom are $12 - 2x$ by $12 - 2x$, and the dimensions, in inches, of each side are x by $12 - 2x$.

$$\frac{\text{Surface}}{\text{area}} = \frac{\text{Area of bottom} +}{4 \cdot \text{Area of each side}}$$
$$= (12 - 2x)(12 - 2x) + 4 \cdot x(12 - 2x)$$
$$= 144 - 24x - 24x + 4x^2 + 48x - 8x^2$$
$$= 144 - 48x + 4x^2 + 48x - 8x^2$$
$$= 144 - 4x^2 \text{ in}^2, \text{ or } -4x^2 + 144 \text{ in}^2$$

89. Let $n =$ the missing number.

	x	3
n	nx	$3n$
x	x^2	$3x$

The area of the figure is $x^2 + 3x + nx + 3n$. This is equivalent to $x^2 + 8x + 15$, so we have $3x + nx = 8x$ and $3n = 15$. Solving either equation for n, we find that the missing number is 5.

91. We have a rectangular solid with dimensions x m by x m by $x + 2$ m with a rectangular solid piece with dimensions 6 m by 5 m by 7 m cut out of it.

$$\text{Volume} = \frac{\text{Volume of}}{\text{large solid}} - \frac{\text{Volume of}}{\text{small solid}}$$
$$= (x \text{ m})(x \text{ m})(x + 2 \text{ m}) - (6 \text{ m})(5 \text{ m})(7 \text{ m})$$
$$= x^2(x + 2) \text{ m}^3 - 210 \text{ m}^3$$
$$= x^3 + 2x^2 - 210 \text{ m}^3$$

93. $(x-2)(x-7)-(x-7)(x-2)$

First observe that, by the commutative law of multiplication, $(x-2)(x-7)$ and $(x-7)(x-2)$ are equivalent expressions. Then when we subtract $(x-7)(x-2)$ from $(x-2)(x-7)$, the result is 0.

95. $(x-a)(x-b)\cdots(x-x)(x-y)(x-z)$
$= (x-a)(x-b)\cdots 0 \cdot (x-y)(x-z)$
$= 0$

Exercise Set 9.6

---

1. $(x+1)(x^2+3)$

 F $\quad$ O $\quad$ I $\quad$ L
$= x \cdot x^2 + x \cdot 3 + 1 \cdot x^2 + 1 \cdot 3$
$= x^3 + 3x + x^2 + 3$

3. $(x^3+2)(x+1)$

 F $\quad$ O $\quad$ I $\quad$ L
$= x^3 \cdot x + x^3 \cdot 1 + 2 \cdot x + 2 \cdot 1$
$= x^4 + x^3 + 2x + 2$

5. $(y+2)(y-3)$

 F $\quad$ O $\quad$ I $\quad$ L
$= y \cdot y + y \cdot (-3) + 2 \cdot y + 2 \cdot (-3)$
$= y^2 - 3y + 2y - 6$
$= y^2 - y - 6$

7. $(3x+2)(3x+2)$

 F $\quad$ O $\quad$ I $\quad$ L
$= 3x \cdot 3x + 3x \cdot 2 + 2 \cdot 3x + 2 \cdot 2$
$= 9x^2 + 6x + 6x + 4$
$= 9x^2 + 12x + 4$

9. $(5x-6)(x+2)$

 F $\quad$ O $\quad$ I $\quad$ L
$= 5x \cdot x + 5x \cdot 2 + (-6) \cdot x + (-6) \cdot 2$
$= 5x^2 + 10x - 6x - 12$
$= 5x^2 + 4x - 12$

11. $(3t-1)(3t+1)$

 F $\quad$ O $\quad$ I $\quad$ L
$= 3t \cdot 3t + 3t \cdot 1 + (-1) \cdot 3t + (-1) \cdot 1$
$= 9t^2 + 3t - 3t - 1$
$= 9t^2 - 1$

13. $(4x-2)(x-1)$

 F $\quad$ O $\quad$ I $\quad$ L
$= 4x \cdot x + 4x \cdot (-1) + (-2) \cdot x + (-2) \cdot (-1)$
$= 4x^2 - 4x - 2x + 2$
$= 4x^2 - 6x + 2$

15. $\left(p - \dfrac{1}{4}\right)\left(p + \dfrac{1}{4}\right)$

 F $\quad$ O $\quad$ I $\quad$ L
$= p \cdot p + p \cdot \dfrac{1}{4} + \left(-\dfrac{1}{4}\right) \cdot p + \left(-\dfrac{1}{4}\right) \cdot \dfrac{1}{4}$
$= p^2 + \dfrac{1}{4}p - \dfrac{1}{4}p - \dfrac{1}{16}$
$= p^2 - \dfrac{1}{16}$

17. $(x-0.1)(x+0.1)$

 F $\quad$ O $\quad$ I $\quad$ L
$= x \cdot x + x \cdot (0.1) + (-0.1) \cdot x + (-0.1)(0.1)$
$= x^2 + 0.1x - 0.1x - 0.01$
$= x^2 - 0.01$

19. $(2x^2+6)(x+1)$

 F $\quad$ O $\quad$ I $\quad$ L
$= 2x^3 + 2x^2 + 6x + 6$

21. $(-2x+1)(x+6)$

 F $\quad$ O $\quad$ I $\quad$ L
$= -2x^2 - 12x + x + 6$
$= -2x^2 - 11x + 6$

23. $(a+7)(a+7)$

 F $\quad$ O $\quad$ I $\quad$ L
$= a^2 + 7a + 7a + 49$
$= a^2 + 14a + 49$

25. $(1+2x)(1-3x)$

 F $\quad$ O $\quad$ I $\quad$ L
$= 1 - 3x + 2x - 6x^2$
$= 1 - x - 6x^2$

27. $(x^2+3)(x^3-1)$

 F $\quad$ O $\quad$ I $\quad$ L
$= x^5 - x^2 + 3x^3 - 3$

29. $(3x^2-2)(x^4-2)$

 F $\quad$ O $\quad$ I $\quad$ L
$= 3x^6 - 6x^2 - 2x^4 + 4$

31. $(2.8x-1.5)(4.7x+9.3)$

 F $\qquad$ O $\qquad$ I $\qquad$ L
$= 2.8x(4.7x) + 2.8x(9.3) - 1.5(4.7x) - 1.5(9.3)$
$= 13.16x^2 + 26.04x - 7.05x - 13.95$
$= 13.16x^2 + 18.99x - 13.95$

33. $(3x^5+2)(2x^2+6)$

 F $\quad$ O $\quad$ I $\quad$ L
$= 6x^7 + 18x^5 + 4x^2 + 12$

35. $(8x^3+1)(x^3+8)$

 F $\quad$ O $\quad$ I $\quad$ L
$= 8x^6 + 64x^3 + x^3 + 8$
$= 8x^6 + 65x^3 + 8$

37. $(4x^2 + 3)(x - 3)$

$\qquad$ F $\qquad$ O $\qquad$ I $\qquad$ L

$= 4x^3 - 12x^2 + 3x - 9$

39. $(4y^4 + y^2)(y^2 + y)$

$\qquad$ F $\qquad$ O $\qquad$ I $\qquad$ L

$= 4y^6 + 4y^5 + y^4 + y^3$

41. $\quad (x + 4)(x - 4)$ $\quad$ Product of sum and
$\qquad\qquad\qquad\qquad$ difference of two terms

$= x^2 - 4^2$

$= x^2 - 16$

43. $\quad (2x + 1)(2x - 1)$ $\quad$ Product of sum and
$\qquad\qquad\qquad\qquad$ difference of two terms

$= (2x)^2 - 1^2$

$= 4x^2 - 1$

45. $\quad (5m - 2)(5m + 2)$ $\quad$ Product of sum and
$\qquad\qquad\qquad\qquad$ difference of two terms

$= (5m)^2 - 2^2$

$= 25m^2 - 4$

47. $\quad (2x^2 + 3)(2x^2 - 3)$ $\quad$ Product of sum and
$\qquad\qquad\qquad\qquad$ difference of two terms

$= (2x^2)^2 - 3^2$

$= 4x^4 - 9$

49. $\quad (3x^4 - 4)(3x^4 + 4)$

$= (3x^4)^2 - 4^2$

$= 9x^8 - 16$

51. $\quad (x^6 - x^2)(x^6 + x^2)$

$= (x^6)^2 - (x^2)^2$

$= x^{12} - x^4$

53. $\quad (x^4 + 3x)(x^4 - 3x)$

$= (x^4)^2 - (3x)^2$

$= x^8 - 9x^2$

55. $\quad (x^{12} - 3)(x^{12} + 3)$

$= (x^{12})^2 - 3^2$

$= x^{24} - 9$

57. $\quad (2y^8 + 3)(2y^8 - 3)$

$= (2y^8)^2 - 3^2$

$= 4y^{16} - 9$

59. $\quad \left(\dfrac{5}{8}x - 4.3\right)\left(\dfrac{5}{8}x + 4.3\right)$

$= \left(\dfrac{5}{8}x\right)^2 - (4.3)^2$

$= \dfrac{25}{64}x^2 - 18.49$

61. $\quad (x + 2)^2 = x^2 + 2 \cdot x \cdot 2 + 2^2$ $\quad$ Square of a binomial
$\qquad\qquad\qquad\qquad\qquad\qquad\qquad$ sum

$\qquad\qquad = x^2 + 4x + 4$

63. $\quad (3x^2 + 1)$ $\quad$ Square of a binomial sum

$= (3x^2)^2 + 2 \cdot 3x^2 \cdot 1 + 1^2$

$= 9x^4 + 6x^2 + 1$

65. $\quad \left(a - \dfrac{1}{2}\right)^2$ $\quad$ Square of a binomial sum

$= a^2 - 2 \cdot a \cdot \dfrac{1}{2} + \left(\dfrac{1}{2}\right)^2$

$= a^2 - a + \dfrac{1}{4}$

67. $\quad (3 + x)^2 = 3^2 + 2 \cdot 3 \cdot x + x^2$

$\qquad\qquad\quad = 9 + 6x + x^2$

69. $\quad (x^2 + 1)^2 = (x^2)^2 + 2 \cdot x^2 \cdot 1 + 1^2$

$\qquad\qquad\qquad = x^4 + 2x^2 + 1$

71. $\quad (2 - 3x^4)^2 = 2^2 - 2 \cdot 2 \cdot 3x^4 + (3x^4)^2$

$\qquad\qquad\qquad = 4 - 12x^4 + 9x^8$

73. $\quad (5 + 6t^2)^2 = 5^2 + 2 \cdot 5 \cdot 6t^2 + (6t^2)^2$

$\qquad\qquad\qquad = 25 + 60t^2 + 36t^4$

75. $\quad \left(x - \dfrac{5}{8}\right)^2 = x^2 - 2 \cdot x \cdot \dfrac{5}{8} + \left(\dfrac{5}{8}\right)^2$

$\qquad\qquad\qquad = x^2 - \dfrac{5}{4}x + \dfrac{25}{64}$

77. $\quad (3 - 2x^3)^2 = 3^2 - 2 \cdot 3 \cdot 2x^3 + (2x^3)^2$

$\qquad\qquad\qquad = 9 - 12x^3 + 4x^6$

79. $\quad 4x(x^2 + 6x - 3)$ $\quad$ Product of a monomial and
$\qquad\qquad\qquad\qquad$ a trinomial

$= 4x \cdot x^2 + 4x \cdot 6x + 4x(-3)$

$= 4x^3 + 24x^2 - 12x$

81. $\quad \left(2x^2 - \dfrac{1}{2}\right)\left(2x^2 - \dfrac{1}{2}\right)$ $\quad$ Square of a binomial
$\qquad\qquad\qquad\qquad\qquad\qquad$ difference

$= (2x^2)^2 - 2 \cdot 2x^2 \cdot \dfrac{1}{2} + \left(\dfrac{1}{2}\right)^2$

$= 4x^4 - 2x^2 + \dfrac{1}{4}$

83. $\quad (-1 + 3p)(1 + 3p)$

$= (3p - 1)(3p + 1)$ $\quad$ Product of the sum and
$\qquad\qquad\qquad\qquad$ difference of two terms

$= (3p)^2 - 1^2$

$= 9p^2 - 1$

85. $\quad 3t^2(5t^3 - t^2 + t)$ $\quad$ Product of a monomial and
$\qquad\qquad\qquad\qquad$ a trinomial

$= 3t^2 \cdot 5t^3 + 3t^2(-t^2) + 3t^2 \cdot t$

$= 15t^5 - 3t^4 + 3t^3$

87. $(6x^4 + 4)^2$ Square of a binomial sum
$= (6x^4)^2 + 2 \cdot 6x^4 \cdot 4 + 4^2$
$= 36x^8 + 48x^4 + 16$

89. $(3x + 2)(4x^2 + 5)$ Product of two binomials;
                             use FOIL
$= 3x \cdot 4x^2 + 3x \cdot 5 + 2 \cdot 4x^2 + 2 \cdot 5$
$= 12x^3 + 15x + 8x^2 + 10$

91. $(8 - 6x^4)^2$ Square of a binomial difference
$= 8^2 - 2 \cdot 8 \cdot 6x^4 + (6x^4)^2$
$= 64 - 96x^4 + 36x^8$

93.
$$\begin{array}{r} t^2+t+1 \\ t-1 \\ \hline -t^2-t-1 \\ t^3+t^2+t \\ \hline t^3 \qquad -1 \end{array}$$

95. $3^2 + 4^2 = 9 + 16 = 25$
$(3 + 4)^2 = 7^2 = 49$

97. $9^2 - 5^2 = 81 - 25 = 56$
$(9 - 5)^2 = 4^2 = 16$

99.

We can find the shaded area in two ways.

Method 1: The figure is a square with side $a + 1$, so the area is $(a + 1)^2 = a^2 + 2a + 1$.

Method 2: We add the areas of A, B, C, and D.

$1 \cdot a + 1 \cdot 1 + 1 \cdot a + a \cdot a = a + 1 + a + a^2 = a^2 + 2a + 1$.

Either way we find that the total shaded area is $a^2 + 2a + 1$.

101.

We can find the shaded area in two ways.

Method 1: The figure is a rectangle with dimensions $t + 6$ by $t + 4$, so the area is $(t + 6)(t + 4) =$
$t^2 + 4t + 6t + 24 = t^2 + 10t + 24$.

Method 2: We add the areas of A, B, C, and D.

$t \cdot t + t \cdot 6 + 6 \cdot 4 + 4 \cdot t = t^2 + 6t + 24 + 4t = t^2 + 10t + 24$.

Either way, we find that the total shaded area is
$t^2 + 10t + 24$.

103. Discussion and Writing Exercise

105. *Familiarize.* Let t = the number of watts used by the television set. Then $10t$ = the number of watts used by the lamps, and $40t$ = the number of watts used by the air conditioner.

Translate.

Lamp watts	+	Air conditioner watts	+	Television watts	=	Total watts
$10t$	+	$40t$	+	t	=	2550

Solve. We solve the equation.

$10t + 40t + t = 2550$
$51t = 2550$
$t = 50$

The possible solution is:

Television, t: 50 watts

Lamps, $10t$: $10 \cdot 50$, or 500 watts

Air conditioner, $40t$: $40 \cdot 50$, or 2000 watts

Check. The number of watts used by the lamps, 500, is 10 times 50, the number used by the television. The number of watts used by the air conditioner, 2000, is 40 times 50, the number used by the television. Also, $50 + 500 + 2000 = 2550$, the total wattage used.

State. The television uses 50 watts, the lamps use 500 watts, and the air conditioner uses 2000 watts.

107.
$3(x - 2) = 5(2x + 7)$
$3x - 6 = 10x + 35$ Removing parentheses
$3x - 6 + 6 = 10x + 35 + 6$ Adding 6
$3x = 10x + 41$
$3x - 10x = 10x + 41 - 10x$ Subtracting $10x$
$-7x = 41$
$\dfrac{-7x}{-7} = \dfrac{41}{-7}$ Dividing by -7
$x = -\dfrac{41}{7}$

The solution is $-\dfrac{41}{7}$.

109. $3x - 2y = 12$

$$-2y = -3x + 12 \quad \text{Subtracting } 3x$$

$$\frac{-2y}{-2} = \frac{-3x + 12}{-2} \quad \text{Dividing by } -2$$

$$y = \frac{3x - 12}{2}, \text{ or}$$

$$y = \frac{3}{2}x - 6$$

111. $5x(3x - 1)(2x + 3)$

$$= 5x(6x^2 + 7x - 3) \quad \text{Using FOIL}$$

$$= 30x^3 + 35x^2 - 15x$$

113. $[(a - 5)(a + 5)]^2$

$$= (a^2 - 25)^2 \quad \text{Finding the product of a sum}$$
$$\text{and difference of same two terms}$$

$$= a^4 - 50a^2 + 625 \quad \text{Squaring a binomial}$$

115. $(3t^4 - 2)^2 1(3t^4 + 2)^2$

$$= [(3t^4 - 2)(3t^4 + 2)]^2$$

$$= (9t^8 - 4)^2$$

$$= 81t^{16} - 72t^8 + 16$$

117. $(x + 2)(x - 5) = (x + 1)(x - 3)$

$$x^2 - 5x + 2x - 10 = x^2 - 3x + x - 3$$

$$x^2 - 3x - 10 = x^2 - 2x - 3$$

$$-3x - 10 = -2x - 3 \quad \text{Adding } -x^2$$

$$-3x + 2x = 10 - 3 \quad \text{Adding } 2x \text{ and } 10$$

$$-x = 7$$

$$x = -7$$

The solution is -7.

119. See the answer section in the text.

121. Enter $y_1 = (x - 1)^2$ and $y_2 = x^2 - 2x + 1$. Then compare the graphs or the y_1-and y_2-values in a table. It appears that the graphs are the same and that the y_1-and y_2-values are the same, so $(x - 1)^2 = x^2 - 2x + 1$ is correct.

123. Enter $y_1 = (x - 3)(x + 3)$ and $y_2 = x^2 - 6$. Then compare the graphs or the y_1-and y_2-values in a table. The graphs are not the same nor are the y_1-and y_2-values, so $(x - 3)(x + 3) = x^2 - 6$ is not correct.

Exercise Set 9.7

1. We replace x by 3 and y by -2.

$$x^2 - y^2 + xy = 3^2 - (-2)^2 + 3(-2) = 9 - 4 - 6 = -1$$

3. We replace x by 3 and y by -2.

$$x^2 - 3y^2 + 2xy = 3^2 - 3(-2)^2 + 2 \cdot 3(-2) =$$
$$9 - 3 \cdot 4 + 2 \cdot 3(-2) = 9 - 12 - 12 = -15$$

5. We replace x by 3, y by -2, and z by -5.

$$8xyz = 8 \cdot 3 \cdot (-2) \cdot (-5) = 240$$

7. We replace x by 3, y by -2, and z by -5.

$$xyz^2 - z = 3(-2)(-5)^2 - (-5) = 3(-2)(25) - (-5) =$$
$$-150 + 5 = -145$$

9. We replace h by 165 and A by 20.

$$C = 0.041h - 0.018A - 2.69$$

$$= 0.041(165) - 0.018(20) - 2.69$$

$$= 6.765 - 0.36 - 2.69$$

$$= 6.405 - 2.69$$

$$= 3.715$$

The lung capacity of a 20-year-old woman who is 165 cm tall is 3.715 liters.

11. Evaluate the polynomial for $h = 50$, $v = 40$, and $t = 2$.

$$h = h_0 + vt - 4.9t^2$$

$$= 50 + 40 \cdot 2 - 4.9(2)^2$$

$$= 50 + 80 - 19.6$$

$$= 110.4$$

The rocket will be 110.4 m above the ground 2 seconds after blast off.

13. Replace h by 4.7, r by 1.2, and π by 3.14.

$$S = 2\pi rh + 2\pi r^2$$

$$\approx 2(3.14)(1.2)(4.7) + 2(3.14)(1.2)^2$$

$$\approx 2(3.14)(1.2)(4.7) + 2(3.14)(1.44)$$

$$\approx 35.4192 + 9.0432$$

$$\approx 44.46$$

The surface area of the can is about 44.46 in^2.

15. Evaluate the polynomial for $h = 7\frac{1}{2}$, or $\frac{15}{2}$, $r = 1\frac{1}{4}$, or $\frac{5}{4}$, and $\pi \approx 3.14$.

$$S = 2\pi rh + \pi r^2$$

$$\approx 2(3.14)\left(\frac{5}{4}\right)\left(\frac{15}{2}\right) + (3.14)\left(\frac{5}{4}\right)^2$$

$$\approx 2(3.14)\left(\frac{5}{4}\right)\left(\frac{15}{2}\right) + (3.14)\left(\frac{25}{16}\right)$$

$$\approx 58.875 + 4.90625$$

$$\approx 63.78125$$

The surface area is about 63.78125 in^2.

17. $x^3y - 2xy + 3x^2 - 5$

Term	Coefficient	Degree	
x^3y	1	4	(Think: $x^3y = x^3y^1$)
$-2xy$	-2	2	(Think: $-2xy = -2x^1y^1$)
$3x^2$	3	2	
-5	-5	0	(Think: $-5 = -5x^0$)

The degree of the polynomial is the degree of the term of highest degree. The term of highest degree is x^3y. Its degree is 4. The degree of the polynomial is 4.

19. $17x^2y^3 - 3x^3yz - 7$

Term	Coefficient	Degree	
$17x^2y^3$	17	5	
$-3x^3yz$	-3	5	(Think: $-3x^3yz = -3x^3y^1z^1$)
-7	-7	0	(Think: $-7 = -7x^0$)

The terms of highest degree are $17x^2y^3$ and $-3x^3yz$. Each has degree 5. The degree of the polynomial is 5.

21. $a + b - 2a - 3b = (1 - 2)a + (1 - 3)b = -a - 2b$

23. $3x^2y - 2xy^2 + x^2$

There are *no* like terms, so none of the terms can be collected.

25. $\quad 6au + 3av + 14au + 7av$
$= (6 + 14)au + (3 + 7)av$
$= 20au + 10av$

27. $\quad 2u^2v - 3uv^2 + 6u^2v - 2uv^2$
$= (2 + 6)u^2v + (-3 - 2)uv^2$
$= 8u^2v - 5uv^2$

29. $\quad (2x^2 - xy + y^2) + (-x^2 - 3xy + 2y^2)$
$= (2 - 1)x^2 + (-1 - 3)xy + (1 + 2)y^2$
$= x^2 - 4xy + 3y^2$

31. $\quad (r - 2s + 3) + (2r + s) + (s + 4)$
$= (1 + 2)r + (-2 + 1 + 1)s + (3 + 4)$
$= 3r + 0s + 7$
$= 3r + 7$

33. $\quad (b^3a^2 - 2b^2a^3 + 3ba + 4) + (b^2a^3 - 4b^3a^2 + 2ba - 1)$
$= (1 - 4)b^3a^2 + (-2 + 1)b^2a^3 + (3 + 2)ba + (4 - 1)$
$= -3b^3a^2 - b^2a^3 + 5ba + 3$, or
$\quad -a^3b^2 - 3a^2b^3 + 5ab + 3$

35. $\quad (a^3 + b^3) - (a^2b - ab^2 + b^3 + a^3)$
$= a^3 + b^3 - a^2b + ab^2 - b^3 - a^3$
$= (1 - 1)a^3 - a^2b + ab^2 + (1 - 1)b^3$
$= -a^2b + ab^2$

37. $\quad (xy - ab - 8) - (xy - 3ab - 6)$
$= xy - ab - 8 - xy + 3ab + 6$
$= (1 - 1)xy + (-1 + 3)ab + (-8 + 6)$
$= 2ab - 2$

39. $\quad (-2a + 7b - c) - (-3b + 4c - 8d)$
$= -2a + 7b - c + 3b - 4c + 8d$
$= -2a + (7 + 3)b + (-1 - 4)c + 8d$
$= -2a + 10b - 5c + 8d$

41. $\quad \overset{\text{F} \quad\quad \text{O} \quad\quad \text{I} \quad\quad \text{L}}{(3z - u)(2z + 3u) = 6z^2 + 9zu - 2uz - 3u^2}$
$= 6z^2 + 7zu - 3u^2$

43. $\quad \overset{\text{F} \quad\quad \text{O} \quad\quad \text{I} \quad\quad \text{L}}{(a^2b - 2)(a^2b - 5) = a^4b^2 - 5a^2b - 2a^2b + 10}$
$= a^4b^2 - 7a^2b + 10$

45. $\quad (a^3 + bc)(a^3 - bc) = (a^3)^2 - (bc)^2$
$\quad\quad\quad\quad [(A + B)(A - B) = A^2 - B^2]$
$= a^6 - b^2c^2$

47.
$$\begin{array}{r} y^4x + y^2 + 1 \\ y^2 + 1 \\ \hline y^4x + y^2 + 1 \\ y^6x + y^4 \quad\quad + y^2 \\ \hline y^6x + y^4 + y^4x + 2y^2 + 1 \end{array}$$

49. $\quad (3xy - 1)(4xy + 2)$
$\overset{\text{F} \quad\quad \text{O} \quad\quad \text{I} \quad\quad \text{L}}{}$
$= 12x^2y^2 + 6xy - 4xy - 2$
$= 12x^2y^2 + 2xy - 2$

51. $\quad (3 - c^2d^2)(4 + c^2d^2)$
$\overset{\text{F} \quad\quad\quad \text{O} \quad\quad \text{I} \quad\quad\quad \text{L}}{}$
$= 12 + 3c^2d^2 - 4c^2d^2 - c^4d^4$
$= 12 - c^2d^2 - c^4d^4$

53. $\quad (m^2 - n^2)(m + n)$
$\overset{\text{F} \quad\quad \text{O} \quad\quad \text{I} \quad\quad \text{L}}{}$
$= m^3 + m^2n - mn^2 - n^3$

55. $\quad (xy + x^5y^5)(x^4y^4 - xy)$
$\overset{\text{F} \quad\quad\quad \text{O} \quad\quad \text{I} \quad\quad\quad \text{L}}{}$
$= x^5y^5 - x^2y^2 + x^9y^9 - x^6y^6$
$= x^9y^9 - x^6y^6 + x^5y^5 - x^2y^2$

57. $\quad (x + h)^2$
$= x^2 + 2xh + h^2 \quad [(A + B)^2 = A^2 + 2AB + B^2]$

59. $\quad (r^3t^2 - 4)^2$
$= (r^3t^2)^2 - 2 \cdot r^3t^2 \cdot 4 + 4^2$
$\quad\quad\quad\quad [(A - B)^2 = A^2 - 2AB + B^2]$
$= r^6t^4 - 8r^3t^2 + 16$

61. $\quad (p^4 + m^2n^2)^2$
$= (p^4)^2 + 2 \cdot p^4 \cdot m^2n^2 + (m^2n^2)^2$
$\quad\quad\quad\quad [(A + B)^2 = A^2 + 2AB + B^2]$
$= p^8 + 2p^4m^2n^2 + m^4n^4$

63. $\quad \left(2a^3 - \dfrac{1}{2}b^3\right)^2$
$= (2a^3)^2 - 2 \cdot 2a^3 \cdot \dfrac{1}{2}b^3 + \left(\dfrac{1}{2}b^3\right)^2$
$\quad\quad\quad\quad [(A - B)^2 = A^2 - 2AB + B^2]$
$= 4a^6 - 2a^3b^3 + \dfrac{1}{4}b^6$

65. $3a(a - 2b)^2 = 3a(a^2 - 4ab + 4b^2)$
$= 3a^3 - 12a^2b + 12ab^2$

67. $(2a - b)(2a + b) = (2a)^2 - b^2 = 4a^2 - b^2$

69. $(c^2 - d)(c^2 + d) = (c^2)^2 - d^2$
$= c^4 - d^2$

71. $(ab + cd^2)(ab - cd^2) = (ab)^2 - (cd^2)^2$
$= a^2b^2 - c^2d^4$

73. $\quad (x + y - 3)(x + y + 3)$
$= [(x + y) - 3][(x + y) + 3]$
$= (x + y)^2 - 3^2$
$= x^2 + 2xy + y^2 - 9$

75. $\quad [x + y + z][x - (y + z)]$
$= [x + (y + z)][x - (y + z)]$
$= x^2 - (y + z)^2$
$= x^2 - (y^2 + 2yz + z^2)$
$= x^2 - y^2 - 2yz - z^2$

77. $\quad (a + b + c)(a - b - c)$
$= [a + (b + c)][a - (b + c)]$
$= a^2 - (b + c)^2$
$= a^2 - (b^2 + 2bc + c^2)$
$= a^2 - b^2 - 2bc - c^2$

79. Discussion and Writing Exercise

81. The first coordinate is positive and the second coordinate is negative, so $(2, -5)$ is in quadrant IV.

83. Both coordinates are positive, so $(16, 23)$ is in quadrant I.

85. $2x = -10$

$x = -5$

Any ordered pair $(-5, y)$ is a solution. The variable x must be -5, but y can be any number we choose. A few solutions are listed below. Plot these points and draw the line.

x	y
-5	-3
-5	0
-5	4

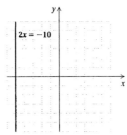

87. $8y - 16 = 0$

$8y = 16$

$y = 2$

Any ordered pair $(x, 2)$ is a solution. The variable y must be 2, but x can be any number we choose. A few solutions are listed below. Plot these points and draw the line.

x	y
-4	2
0	2
3	2

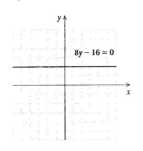

89. It is helpful to add additional labels to the figure.

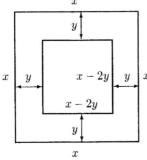

The area of the large square is $x \cdot x$, or x^2. The area of the small square is $(x - 2y)(x - 2y)$, or $(x - 2y)^2$.

$$\begin{array}{ccc} \text{Area of shaded} & = & \text{Area of large} & - & \text{Area of small} \\ \text{region} & & \text{square} & & \text{square} \end{array}$$

$$\begin{array}{ccc} \text{Area of shaded} & = & x^2 & - & (x - 2y)^2 \\ \text{region} & & & & \end{array}$$

$$= x^2 - (x^2 - 4xy + 4y^2)$$
$$= x^2 - x^2 + 4xy - 4y^2$$
$$= 4xy - 4y^2$$

91. It is helpful to add additional labels to the figure.

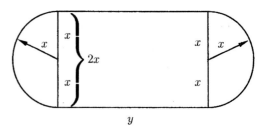

The two semicircles make a circle with radius x. The area of that circle is πx^2. The area of the rectangle is $2x \cdot y$. The sum of the two regions, $\pi x^2 + 2xy$, is the area of the shaded region.

93. The lateral surface area of the outer portion of the solid is the lateral surface area of a right circular cylinder with radius n and height h. The lateral surface area of the inner portion is the lateral surface area of a right circular cylinder with radius m and height h. Recall that the formula for the lateral surface area of a right circular cylinder with radius r and height h is $2\pi rh$.

The surface area of the top is the area of a circle with radius n less the area of a circle with radius m. The surface area of the bottom is the same as the surface area of the top.

Thus, the surface area of the solid is

$$2\pi nh + 2\pi mh + 2\pi n^2 - 2\pi m^2.$$

95. The height of the observatory is 40 ft and its radius is $30/2$, or 15 ft, so the surface area is $2\pi rh + \pi r^2 \approx 2(3.14)(15)(40) + (3.14)(15)^2 \approx 4474.5 \text{ ft}^2$. Since $4474.5 \text{ ft}^2 / 250 \text{ ft}^2 = 17.898$, 18 gallons of paint should be purchased.

97. Substitute $10,400 for P, 8.5% or 0.085 for r, and 5 for t.

$P(1 + r)^t$

$= \$10,400(1 + 0.085)^5$

$= \$10,400(1.085)^5$

$\approx \$15,638.03$

Exercise Set 9.8

1. $\dfrac{24x^4}{8} = \dfrac{24}{8} \cdot x^4 = 3x^4$

Check: We multiply.

$3x^4 \cdot 8 = 24x^4$

3. $\dfrac{25x^3}{5x^2} = \dfrac{25}{5} \cdot \dfrac{x^3}{x^2} = 5x^{3-2} = 5x$

Check: We multiply.

$5x \cdot 5x^2 = 25x^3$

5. $\dfrac{-54x^{11}}{-3x^8} = \dfrac{-54}{-3} \cdot \dfrac{x^{11}}{x^8} = 18x^{11-8} = 18x^3$

Check: We multiply.

$18x^3(-3x^8) = -54x^{11}$

7. $\dfrac{64a^5b^4}{16a^2b^3} = \dfrac{64}{16} \cdot \dfrac{a^5}{a^2} \cdot \dfrac{b^4}{b^3} = 4a^{5-2}b^{4-3} = 4a^3b$

Check: We multiply.
$$(4a^3b)(16a^2b^3) = 64a^5b^4$$

9. $\dfrac{24x^4 - 4x^3 + x^2 - 16}{8}$

$= \dfrac{24x^4}{8} - \dfrac{4x^3}{8} + \dfrac{x^2}{8} - \dfrac{16}{8}$

$= 3x^4 - \dfrac{1}{2}x^3 + \dfrac{1}{8}x^2 - 2$

Check: We multiply.

$$
\begin{array}{r}
3x^4 - \dfrac{1}{2}x^3 + \dfrac{1}{8}x^2 - 2 \\
\underline{\qquad\qquad\qquad\qquad 8} \\
24x^4 - 4x^3 + x^2 - 16
\end{array}
$$

11. $\dfrac{u - 2u^2 - u^5}{u}$

$= \dfrac{u}{u} - \dfrac{2u^2}{u} - \dfrac{u^5}{u}$

$= 1 - 2u - u^4$

Check: We multiply.

$$
\begin{array}{r}
1 - 2u - u^4 \\
\underline{\qquad\qquad\quad u} \\
u - 2u^2 - u^5
\end{array}
$$

13. $(15t^3 + 24t^2 - 6t) \div (3t)$

$= \dfrac{15t^3 + 24t^2 - 6t}{3t}$

$= \dfrac{15t^3}{3t} + \dfrac{24t^2}{3t} - \dfrac{6t}{3t}$

$= 5t^2 + 8t - 2$

Check: We multiply.

$$
\begin{array}{r}
5t^2 + 8t - 2 \\
\underline{\qquad\qquad 3t} \\
15t^3 + 24t^2 - 6t
\end{array}
$$

15. $(20x^6 - 20x^4 - 5x^2) \div (-5x^2)$

$= \dfrac{20x^6 - 20x^4 - 5x^2}{-5x^2}$

$= \dfrac{20x^6}{-5x^2} - \dfrac{20x^4}{-5x^2} - \dfrac{5x^2}{-5x^2}$

$= -4x^4 - (-4x^2) - (-1)$

$= -4x^4 + 4x^2 + 1$

Check: We multiply.

$$
\begin{array}{r}
-4x^4 + 4x^2 + 1 \\
\underline{\qquad\qquad\qquad -5x^2} \\
20x^6 - 20x^4 - 5x^2
\end{array}
$$

17. $(24x^5 - 40x^4 + 6x^3) \div (4x^3)$

$= \dfrac{24x^5 - 40x^4 + 6x^3}{4x^3}$

$= \dfrac{24x^5}{4x^3} - \dfrac{40x^4}{4x^3} + \dfrac{6x^3}{4x^3}$

$= 6x^2 - 10x + \dfrac{3}{2}$

Check: We multiply.

$$
\begin{array}{r}
6x^2 - 10x + \dfrac{3}{2} \\
\underline{\qquad\qquad\qquad 4x^3} \\
24x^5 - 40x^4 + 6x^3
\end{array}
$$

19. $\dfrac{18x^2 - 5x + 2}{2}$

$= \dfrac{18x^2}{2} - \dfrac{5x}{2} + \dfrac{2}{2}$

$= 9x^2 - \dfrac{5}{2}x + 1$

Check: We multiply.

$$
\begin{array}{r}
9x^2 - \dfrac{5}{2}x + 1 \\
\underline{\qquad\qquad\qquad 2} \\
18x^2 - 5x + 2
\end{array}
$$

21. $\dfrac{12x^3 + 26x^2 + 8x}{2x}$

$= \dfrac{12x^3}{2x} + \dfrac{26x^2}{2x} + \dfrac{8x}{2x}$

$= 6x^2 + 13x + 4$

Check: We multiply.

$$
\begin{array}{r}
6x^2 + 13x + 4 \\
\underline{\qquad\qquad\qquad 2x} \\
12x^3 + 26x^2 + 8x
\end{array}
$$

23. $\dfrac{9r^2s^2 + 3r^2s - 6rs^2}{3rs}$

$= \dfrac{9r^2s^2}{3rs} + \dfrac{3r^2s}{3rs} - \dfrac{6rs^2}{3rs}$

$= 3rs + r - 2s$

Check: We multiply.

$$
\begin{array}{r}
3rs + r - 2s \\
\underline{\qquad\qquad\qquad 3rs} \\
9r^2s^2 + 3r^2s - 6rs^2
\end{array}
$$

25.

$$
\begin{array}{r}
x + 2 \\
x + 2 \overline{\smash{\big)}\, x^2 + 4x + 4} \\
\underline{x^2 + 2x} \\
2x + 4 \quad \leftarrow (x^2 + 4x) - (x^2 + 2x) \\
\underline{2x + 4} \\
0 \quad \leftarrow (2x + 4) - (2x + 4)
\end{array}
$$

The answer is $x + 2$.

27.
$$
\begin{array}{r}
x - 5 \\
x-5\,\overline{\big)\,x^2-10x-25} \\
\underline{x^2-5x} \\
-5x-25 \leftarrow (x^2-10x)-(x^2-5x) \\
\underline{-5x+25} \\
-50 \leftarrow (-5x-25)-(-5x+25)
\end{array}
$$

The answer is $x - 5 + \dfrac{-50}{x-5}$.

29.
$$
\begin{array}{r}
x - 2 \\
x+6\,\overline{\big)\,x^2+4x-14} \\
\underline{x^2+6x} \\
-2x-14 \leftarrow (x^2+4x)-(x^2+6x) \\
\underline{-2x-12} \\
-2 \leftarrow (-2x-14)-(-2x-12)
\end{array}
$$

The answer is $x - 2 + \dfrac{-2}{x+6}$.

31.
$$
\begin{array}{r}
x - 3 \\
x+3\,\overline{\big)\,x^2+0x-9} \leftarrow \text{Filling in the missing term} \\
\underline{x^2+3x} \\
-3x-9 \leftarrow x^2-(x^2+3x) \\
\underline{-3x-9} \\
0 \leftarrow (-3x-9)-(-3x-9)
\end{array}
$$

The answer is $x - 3$.

33.
$$
\begin{array}{r}
x^4-x^3+x^2-x+1 \\
x+1\,\overline{\big)\,x^5+0x^4+0x^3+0x^2+0x+1} \leftarrow \text{Filling in missing} \\
\text{terms} \\
\underline{x^5+x^4} \\
-x^4 \leftarrow x^5-(x^5+x^4) \\
\underline{-x^4-x^3} \\
x^3 \leftarrow -x^4-(-x^4-x^3) \\
\underline{x^3+x^2} \\
-x^2 \leftarrow x^3-(x^3+x^2) \\
\underline{-x^2-x} \\
x+1 \leftarrow -x^2-(-x^2-x) \\
\underline{x+1} \\
0 \leftarrow (x+1)-(x+1)
\end{array}
$$

The answer is $x^4 - x^3 + x^2 - x + 1$.

35.
$$
\begin{array}{r}
2x^2-7x+4 \\
4x+3\,\overline{\big)\,8x^3-22x^2-5x+12} \\
\underline{8x^3+6x^2} \\
-28x^2-5x \leftarrow (8x^3-22x^2)-(8x^3+6x^2) \\
\underline{-28x^2-21x} \\
16x+12 \leftarrow (-28x^2-5x)- \\
(-28x^2-21x) \\
\underline{16x+12} \\
0 \leftarrow (16x+12)-(16x+12)
\end{array}
$$

The answer is $2x^2 - 7x + 4$.

37.
$$
\begin{array}{r}
x^3-6 \\
x^3-7\,\overline{\big)\,x^6-13x^3+42} \\
\underline{x^6-7x^3} \\
-6x^3+42 \leftarrow (x^6-13x^3)-(x^6-7x^3) \\
\underline{-6x^3+42} \\
0 \leftarrow (-6x^3+42)-(-6x^3+42)
\end{array}
$$

The answer is $x^3 - 6$.

39.
$$
\begin{array}{r}
x^3+2x^2+4x+8 \\
x-2\,\overline{\big)\,x^4+0x^3+0x^2+0x-16} \\
\underline{x^4-2x^3} \\
2x^3 \leftarrow x^4-(x^4-2x^3) \\
\underline{2x^3-4x^2} \\
4x^2 \leftarrow 2x^3-(2x^3-4x^2) \\
\underline{4x^2-8x} \\
8x-16 \leftarrow 4x^2-(4x^2-8x) \\
\underline{8x-16} \\
0 \leftarrow (8x-16)-(8x-16)
\end{array}
$$

The answer is $x^3 + 2x^2 + 4x + 8$.

41.
$$
\begin{array}{r}
t^2+1 \\
t-1\,\overline{\big)\,t^3-t^2+t-1} \\
\underline{t^3-t^2} \leftarrow (t^3-t^2)-(t^3-t^2) \\
0+t-1 \\
\underline{t-1} \leftarrow (t-1)-(t-1) \\
0
\end{array}
$$

The answer is $t^2 + 1$.

43. $(x^3 - 2x^2 + 2x - 5) \div (x - 1)$

$$
\begin{array}{r|rrrr}
1 & 1 & -2 & 2 & -5 \\
& & 1 & -1 & 1 \\
\hline
& 1 & -1 & 1 & -4
\end{array}
$$

The answer is $x^2 - x + 1$, R -4, or $x^2 - x + 1 + \dfrac{-4}{x-1}$.

45. $(a^2 + 11a - 19) \div (a + 4) =$
$(a^2 + 11a - 19) \div [a - (-4)]$

$$
\begin{array}{r|rrr}
-4 & 1 & 11 & -19 \\
& & -4 & -28 \\
\hline
& 1 & 7 & -47
\end{array}
$$

The answer is $a + 7$, R -47, or $a + 7 + \dfrac{-47}{a+4}$.

47. $(x^3 - 7x^2 - 13x + 3) \div (x - 2)$

$$
\begin{array}{r|rrrr}
2 & 1 & -7 & -13 & 3 \\
& & 2 & -10 & -46 \\
\hline
& 1 & -5 & -23 & -43
\end{array}
$$

The answer is $x^2 - 5x - 23$, R -43, or $x^2 - 5x - 23 + \dfrac{-43}{x-2}$.

49. $(3x^3 + 7x^2 - 4x + 3) \div (x + 3) =$
$(3x^3 + 7x^2 - 4x + 3) \div [x - (-3)]$

$$
\begin{array}{r|rrrr}
-3 & 3 & 7 & -4 & 3 \\
& & -9 & 6 & -6 \\
\hline
& 3 & -2 & 2 & -3
\end{array}
$$

The answer is $3x^2 - 2x + 2$, R -3, or $3x^2 - 2x + 2 + \dfrac{-3}{x+3}$.

51. $(y^3 - 3y + 10) \div (y - 2) =$
$(y^3 + 0y^2 - 3y + 10) \div (y - 2)$

$$
\begin{array}{r|rrrr}
2 & 1 & 0 & -3 & 10 \\
& & 2 & 4 & 2 \\
\hline
& 1 & 2 & 1 & 12
\end{array}
$$

The answer is $y^2 + 2y + 1$, R 12, or $y^2 + 2y + 1 + \dfrac{12}{y-2}$.

53. $(3x^4 - 25x^2 - 18) \div (x - 3) =$

$(3x^4 + 0x^3 - 25x^2 + 0x - 18) \div (x - 3)$

$$
\begin{array}{r|rrrrr}
3 & 3 & 0 & -25 & 0 & -18 \\
  &   & 9 & 27 & 6 & 18 \\
\hline
  & 3 & 9 & 2 & 6 & \,|\quad 0
\end{array}
$$

The answer is $3x^3 + 9x^2 + 2x + 6$.

55. $(x^3 - 8) \div (x - 2) = (x^3 + 0x^2 + 0x - 8) \div (x - 2)$

$$
\begin{array}{r|rrrr}
2 & 1 & 0 & 0 & -8 \\
  &   & 2 & 4 & 8 \\
\hline
  & 1 & 2 & 4 & \,|\quad 0
\end{array}
$$

The answer is $x^2 + 2x + 4$.

57. $(y^4 - 16) \div (y - 2) =$

$(y^4 + 0y^3 + 0y^2 + 0y - 16) \div (y - 2)$

$$
\begin{array}{r|rrrrr}
2 & 1 & 0 & 0 & 0 & -16 \\
  &   & 2 & 4 & 8 & 16 \\
\hline
  & 1 & 2 & 4 & 8 & \,|\quad 0
\end{array}
$$

The answer is $y^3 + 2y^2 + 4y + 8$.

59. Discussion and Writing Exercise

61. $17 - 45 = 17 + (-45) = -28$

63. $-2.3 - (-9.1) = -2.3 + 9.1 = 6.8$

65. *Familiarize.* Let $w =$ the width. Then $w + 15 =$ the length. We draw a picture.

We will use the fact that the perimeter is 640 ft to find w (the width). Then we can find $w + 15$ (the length) and multiply the length and the width to find the area.

Translate.

Width+Width+ Length + Length =Perimeter
$w \;+\; w \;+(w + 15)+(w + 15)=\;\;\; 640$

Solve.

$$
\begin{aligned}
w + w + (w + 15) + (w + 15) &= 640 \\
4w + 30 &= 640 \\
4w &= 610 \\
w &= 152.5
\end{aligned}
$$

If the width is 152.5, then the length is $152.5+15$, or 167.5. The area is $(167.5)(152.5)$, or $25,543.75$ ft^2.

Check. The length, 167.5 ft, is 15 ft greater than the width, 152.5 ft. The perimeter is $152.5 + 152.5 + 167.5 + 167.5$, or 640 ft. We should also recheck the computation we used to find the area. The answer checks.

State. The area is $25,543.75$ ft^2.

67.

$$
\begin{aligned}
-6(2 - x) + 10(5x - 7) &= 10 \\
-12 + 6x + 50x - 70 &= 10 \\
56x - 82 &= 10 \quad \text{Collecting like terms} \\
56x - 82 + 82 &= 10 + 82 \quad \text{Adding 82} \\
56x &= 92 \\
\frac{56x}{56} &= \frac{92}{56} \quad \text{Dividing by 56} \\
x &= \frac{23}{14}
\end{aligned}
$$

The solution is $\frac{23}{14}$.

69. $4x - 12 + 24y = 4 \cdot x - 4 \cdot 3 + 4 \cdot 6y = 4(x - 3 + 6y)$

71.

$$
\require{enclose}
\begin{array}{r}
x^2 + 5 \\
x^2 + 4 \enclose{longdiv}{x^4 + 9x^2 + 20} \\
\underline{x^4 + 4x^2 } \\
5x^2 + 20 \\
\underline{5x^2 + 20} \\
0
\end{array}
$$

The answer is $x^2 + 5$.

73.

$$
\begin{array}{r}
a + 3 \\
5a^2 - 7a - 2 \enclose{longdiv}{5a^3 + 8a^2 - 23a - 1} \\
\underline{5a^3 - 7a^2 - 2a } \\
15a^2 - 21a - 1 \\
\underline{15a^2 - 21a - 6} \\
5
\end{array}
$$

The answer is $a + 3 + \dfrac{5}{5a^2 - 7a - 2}$.

75. We rewrite the dividend in descending order.

$$
\begin{array}{r}
2x^2 + x - 3 \\
3x^3 - 2x - 1 \enclose{longdiv}{6x^5 + 3x^4 - 13x^3 - 4x^2 + 5x + 3} \\
\underline{6x^5 - 4x^3 - 2x^2 } \\
3x^4 - 9x^3 - 2x^2 + 5x \\
\underline{3x^4 - 2x^2 - x } \\
-9x^3 + 6x + 3 \\
\underline{-9x^3 + 6x + 3} \\
0
\end{array}
$$

The answer is $2x^2 + x - 3$.

77.

$$
\begin{array}{r}
a^5 + a^4 b + a^3 b^2 + a^2 b^3 + ab^4 + b^5 \\
a - b \enclose{longdiv}{a^6 + 0a^5 b + 0a^4 b^2 + 0a^3 b^3 + 0a^2 b^4 + 0ab^5 - b^6} \\
\underline{a^6 - a^5 b } \\
a^5 b \\
\underline{a^5 b - a^4 b^2 } \\
a^4 b^2 \\
\underline{a^4 b^2 - a^3 b^3 } \\
a^3 b^3 \\
\underline{a^3 b^3 - a^2 b^4 } \\
a^2 b^4 \\
\underline{a^2 b^4 - ab^5 } \\
ab^5 - b^6 \\
\underline{ab^5 - b^6} \\
0
\end{array}
$$

The answer is $a^5 + a^4 b + a^3 b^2 + a^2 b^3 + ab^4 + b^5$.

79.

$$\begin{array}{r} x + 5 \\ x - 1 \overline{\smash{\big)}\; x^2 + 4x + c} \\ \underline{x^2 - x} \\ 5x + c \\ \underline{5x - 5} \\ c + 5 \end{array}$$

We set the remainder equal to 0.

$c + 5 = 0$

$c = -5$

Thus, c must be -5.

81.

$$\begin{array}{r} c^2 x + (-2c + c^2) \\ x - 1 \overline{\smash{\big)}\; c^2 x^2 - 2cx + 1} \\ \underline{c^2 x^2 - c^2 x} \\ (-2c + c^2)x + 1 \\ \underline{(-2c + c^2)x - (-2c + c^2)} \\ 1 + (-2c + c^2) \end{array}$$

We set the remainder equal to 0.

$c^2 - 2c + 1 = 0$

$(c - 1)^2 = 0$

$c = 1$

Thus, c must be 1.

Chapter 10

Polynomials: Factoring

1. Answers may vary. $8x^3 = (4x^2)(2x) = (-8)(-x^3) = (2x^2)(4x)$

3. Answers may vary. $-10a^6 = (-5a^5)(2a) = (10a^3)(-a^3) = (-2a^2)(5a^4)$

5. Answers may vary. $24x^4 = (6x)(4x^3) = (-3x^2)(-8x^2) = (2x^3)(12x)$

7. $x^2 - 6x = x \cdot x - x \cdot 6$ Factoring each term
   $= x(x - 6)$ Factoring out the common factor x

9. $2x^2 + 6x = 2x \cdot x + 2x \cdot 3$ Factoring each term
   $= 2x(x + 3)$ Factoring out the common factor $2x$

11. $x^3 + 6x^2 = x^2 \cdot x + x^2 \cdot 6$ Factoring each term
    $= x^2(x + 6)$ Factoring out x^2

13. $8x^4 - 24x^2 = 8x^2 \cdot x^2 - 8x^2 \cdot 3$
    $= 8x^2(x^2 - 3)$ Factoring out $8x^2$

15. $2x^2 + 2x - 8 = 2 \cdot x^2 + 2 \cdot x - 2 \cdot 4$
    $= 2(x^2 + x - 4)$ Factoring out 2

17. $17x^5y^3 + 34x^3y^2 + 51xy$
    $= 17xy \cdot x^4y^2 + 17xy \cdot 2x^2y + 17xy \cdot 3$
    $= 17xy(x^4y^2 + 2x^2y + 3)$

19. $6x^4 - 10x^3 + 3x^2 = x^2 \cdot 6x^2 - x^2 \cdot 10x + x^2 \cdot 3$
    $= x^2(6x^2 - 10x + 3)$

21. $x^5y^5 + x^4y^3 + x^3y^3 - x^2y^2$
    $= x^2y^2 \cdot x^3y^3 + x^2y^2 \cdot x^2y + x^2y^2 \cdot xy + x^2y^2(-1)$
    $= x^2y^2(x^3y^3 + x^2y + xy - 1)$

23. $2x^7 - 2x^6 - 64x^5 + 4x^3$
    $= 2x^3 \cdot x^4 - 2x^3 \cdot x^3 - 2x^3 \cdot 32x^2 + 2x^3 \cdot 2$
    $= 2x^3(x^4 - x^3 - 32x^2 + 2)$

25. $1.6x^4 - 2.4x^3 + 3.2x^2 + 6.4x$
    $= 0.8x(2x^3) - 0.8x(3x^2) + 0.8x(4x) + 0.8x(8)$
    $= 0.8x(2x^3 - 3x^2 + 4x + 8)$

27. $\frac{5}{3}x^6 + \frac{4}{3}x^5 + \frac{1}{3}x^4 + \frac{1}{3}x^3$
    $= \frac{1}{3}x^3(5x^3) + \frac{1}{3}x^3(4x^2) + \frac{1}{3}x^3(x) + \frac{1}{3}x^3(1)$
    $= \frac{1}{3}x^3(5x^3 + 4x^2 + x + 1)$

29. Factor: $x^2(x + 3) + 2(x + 3)$
    The binomial $x + 3$ is common to both terms:
    $x^2(x + 3) + 2(x + 3) = (x^2 + 2)(x + 3)$

31. $\quad 5a^3(2a - 7) - (2a - 7)$
    $= 5a^3(2a - 7) - 1(2a - 7)$
    $= (5a^3 - 1)(2a - 7)$

33. $\quad x^3 + 3x^2 + 2x + 6$
    $= (x^3 + 3x^2) + (2x + 6)$
    $= x^2(x + 3) + 2(x + 3)$ Factoring each binomial
    $= (x^2 + 2)(x + 3)$ Factoring out the common factor $x + 3$

35. $\quad 2x^3 + 6x^2 + x + 3$
    $= (2x^3 + 6x^2) + (x + 3)$
    $= 2x^2(x + 3) + 1(x + 3)$ Factoring each binomial
    $= (2x^2 + 1)(x + 3)$

37. $8x^3 - 12x^2 + 6x - 9 = 4x^2(2x - 3) + 3(2x - 3)$
    $= (4x^2 + 3)(2x - 3)$

39. $\quad 12x^3 - 16x^2 + 3x - 4$
    $= 4x^2(3x - 4) + 1(3x - 4)$ Factoring 1 out of the second binomial
    $= (4x^2 + 1)(3x - 4)$

41. $\quad 5x^3 - 5x^2 - x + 1$
    $= (5x^3 - 5x^2) + (-x + 1)$
    $= 5x^2(x - 1) - 1(x - 1)$ Check: $-1(x-1)=-x+1$
    $= (5x^2 - 1)(x - 1)$

43. $x^3 + 8x^2 - 3x - 24 = x^2(x + 8) - 3(x + 8)$
    $= (x^2 - 3)(x + 8)$

45. $2x^3 - 8x^2 - 9x + 36 = 2x^2(x - 4) - 9(x - 4)$
    $= (2x^2 - 9)(x - 4)$

47. Discussion and Writing Exercise

49. $\quad -2x < 48$
    $x > -24$ Dividing by -2 and reversing the inequality symbol
    The solution set is $\{x|x > -24\}$.

51. $\dfrac{-108}{-4} = 27$ (The quotient of two negative numbers is positive.)

53. $(y + 5)(y + 7) = y^2 + 7y + 5y + 35$ Using FOIL
    $= y^2 + 12y + 35$

55. $(y+7)(y-7) = y^2 - 7^2 = y^2 - 49$
$$[(A+B))(A-B) = A^2 - B^2]$$

57. $x + y = 4$

To find the x-intercept, let $y = 0$. Then solve for x.
$$x + y = 4$$
$$x + 0 = 4$$
$$x = 4$$
The x-intercept is $(4, 0)$.

To find the y-intercept, let $x = 0$. Then solve for y.
$$x + y = 4$$
$$0 + y = 4$$
$$y = 4$$
The y-intercept is $(0, 4)$.

Plot these points and draw the line.

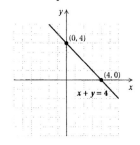

A third point should be used as a check. We substitute any value for x and solve for y. We let $x = 2$. Then
$$x + y = 4$$
$$2 + y = 4$$
$$y = 2$$
The point $(2, 2)$ is on the graph, so the graph is probably correct.

59. $5x - 3y = 15$

To find the x-intercept, let $y = 0$. Then solve for x.
$$5x - 3y = 15$$
$$5x - 3 \cdot 0 = 15$$
$$5x = 15$$
$$x = 3$$
The x-intercept is $(3, 0)$.

To find the y-intercept, let $x = 0$. Then solve for y.
$$5x - 3y = 15$$
$$5 \cdot 0 - 3y = 15$$
$$-3y = 15$$
$$y = -5$$
The y-intercept is $(0, -5)$.

Plot these points and draw the line.

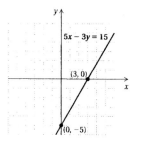

A third point should be used as a check. We substitute any value for x and solve for y. We let $x = 6$. Then
$$5x - 3y = 15$$
$$5 \cdot 6 - 3y = 15$$
$$30 - 3y = 15$$
$$-3y = -15$$
$$y = 5$$
The point $(6, 5)$ is on the graph, so the graph is probably correct.

61. $4x^5 + 6x^3 + 6x^2 + 9 = 2x^3(2x^2 + 3) + 3(2x^2 + 3)$
$$= (2x^3 + 3)(2x^2 + 3)$$

63. $x^{12} + x^7 + x^5 + 1 = x^7(x^5 + 1) + (x^5 + 1)$
$$= (x^7 + 1)(x^5 + 1)$$

65. $p^3 + p^2 - 3p + 10 = p^2(p + 1) - (3p - 10)$

This polynomial is not factorable using factoring by grouping.

Exercise Set 10.2

1. $x^2 + 8x + 15$

Since the constant term and coefficient of the middle term are both positive, we look for a factorization of 15 in which both factors are positive. Their sum must be 8.

Pairs of factors	Sums of factors
1, 15	16
3, 5	8

The numbers we want are 3 and 5.

$x^2 + 8x + 15 = (x + 3)(x + 5)$.

3. $x^2 + 7x + 12$

Since the constant term is positive and the coefficient of the middle term is positive, we look for a factorization of 12 in which both factors are positive. Their sum must be 7.

Pairs of factors	Sums of factors
1, 12	13
2, 6	8
3, 4	7

The numbers we want are 3 and 4.

$x^2 + 7x + 12 = (x + 3)(x + 4)$.

5. $x^2 - 6x + 9$

Since the constant term is positive and the coefficient of the middle term is negative, we look for a factorization of 9 in which both factors are negative. Their sum must be -6.

Pairs of factors	Sums of factors
$-1, -9$	-10
$-3, -3$	-6

The numbers we want are -3 and -3.
$x^2 - 6x + 9 = (x-3)(x-3)$, or $(x-3)^2$.

7. $x^2 - 5x - 14$

Since the constant term is negative, we look for a factorization of -14 in which one factor is positive and one factor is negative. Their sum must be -5, the coefficient of the middle term.

Pairs of factors	Sums of factors
$-1, \ 14$	13
$1, -14$	-13
$-2, \ 7$	5
$2, \ -7$	-5

The numbers we want are 2 and -7.
$x^2 - 5x - 14 = (x+2)(x-7)$.

9. $b^2 + 5b + 4$

Since the constant term is positive and the coefficient of the middle term is positive, we look for a factorization of 4 in which both factors are positive. Their sum must be 5.

Pairs of factors	Sums of factors
$1, \ 4$	5
$2, \ 2$	4

The numbers we want are 1 and 4.
$b^2 + 5b + 4 = (b+1)(b+4)$.

11. $x^2 + \frac{2}{3}x + \frac{1}{9}$

Since the constant term is positive and the coefficient of the middle term is positive, we look for a factorization of $\frac{1}{9}$ in which both factors are positive. Their sum must be $\frac{2}{3}$.

Pairs of factors	Sums of factors
$1, \dfrac{1}{9}$	$\dfrac{10}{9}$
$\dfrac{1}{3}, \dfrac{1}{3}$	$\dfrac{2}{3}$

The numbers we want are $\frac{1}{3}$ and $\frac{1}{3}$.
$x^2 + \frac{2}{3}x + \frac{1}{9} = \left(x + \frac{1}{3}\right)\left(x + \frac{1}{3}\right)$, or $\left(x + \frac{1}{3}\right)^2$.

13. $d^2 - 7d + 10$

Since the constant term is positive and the coefficient of the middle term is negative, we look for a factorization of 10 in which both factors are negative. Their sum must be -7.

Pairs of factors	Sums of factors
$-1, -10$	-11
$-2, \ -5$	-7

The numbers we want are -2 and -5.
$d^2 - 7d + 10 = (d-2)(d-5)$.

15. $y^2 - 11y + 10$

Since the constant term is positive and the coefficient of the middle term is negative, we look for a factorization of 10 in which both factors are negative. Their sum must be -11.

Pairs of factors	Sums of factors
$-1, -10$	-11
$-2, \ -5$	-7

The numbers we want are -1 and -10.
$y^2 - 11y + 10 = (y-1)(y-10)$.

17. $x^2 + x + 1$

Since the constant term and the coefficient of the middle term are both positive, we look for a factorization of 1 in which both factors are positive. The sum must be 1. The only possible pair of factors is 1 and 1, but their sum is not 1. Thus, this polynomial is not factorable into binomials. It is prime.

19. $x^2 - 7x - 18$

Since the constant term is negative, we look for a factorization of -18 in which one factor is positive and one factor is negative. Their sum must be -7, the coefficient of the middle term.

Pairs of factors	Sums of factors
$-1, \ 18$	17
$1, \ -18$	-17
$-2, \ 9$	7
$2, \ -9$	-7
$-3, \ 6$	3
$3, \ -6$	-3

The numbers we want are 2 and -9.
$x^2 - 7x - 18 = (x+2)(x-9)$.

21. $x^3 - 6x^2 - 16x = x(x^2 - 6x - 16)$

After factoring out the common factor, x, we consider $x^2 - 6x - 16$. Since the constant term is negative, we look for a factorization of -16 in which one factor is positive and one factor is negative. Their sum must be -6, the coefficient of the middle term.

Pairs of factors	Sums of factors
−1, 16	15
1, −16	−15
−2, 8	6
2, −8	−6
−4, 4	0

The numbers we want are 2 and −8.

Then $x^2 - 6x - 16 = (x + 2)(x - 8)$, so $x^3 - 6x^2 - 16x = x(x + 2)(x - 8)$.

23. $y^3 - 4y^2 - 45y = y(y^2 - 4y - 45)$

After factoring out the common factor, y, we consider $y^2 - 4y - 45$. Since the constant term is negative, we look for a factorization of −45 in which one factor is positive and one factor is negative. Their sum must be −4, the coefficient of the middle term.

Pairs of factors	Sums of factors
−1, 45	44
1, −45	−44
−3, 15	12
3, −15	−12
−5, 9	4
5, −9	−4

The numbers we want are 5 and −9.

Then $y^2 - 4y - 45 = (y + 5)(y - 9)$, so $y^3 - 4y^2 - 45y = y(y + 5)(y - 9)$.

25. $-2x - 99 + x^2 = x^2 - 2x - 99$

Since the constant term is negative, we look for a factorization of −99 in which one factor is positive and one factor is negative. Their sum must be −2, the coefficient of the middle term.

Pairs of factors	Sums of factors
−1, 99	98
1, −99	−98
−3, 33	30
3, −33	−30
−9, 11	2
9, −11	−2

The numbers we want are 9 and −11.

$-2x - 99 + x^2 = (x + 9)(x - 11)$.

27. $c^4 + c^2 - 56$

Consider this trinomial as $(c^2)^2 + c^2 - 56$. We look for numbers p and q such that $c^4 + c^2 - 56 = (c^2 + p)(c^2 + q)$. Since the constant term is negative, we look for a factorization of −56 in which one factor is positive and one factor is negative. Their sum must be 1.

Pairs of factors	Sums of factors
−1, 56	55
1, −56	−55
−2, 28	26
2, −28	−26
−4, 14	12
4, −14	−12
−7, 8	1
7, −8	−1

The numbers we want are −7 and 8.

$c^4 + c^2 - 56 = (c^2 - 7)(c^2 + 8)$.

29. $a^4 + 2a^2 - 35$

Consider this trinomial as $(a^2)^2 + 2a^2 - 35$. We look for numbers p and q such that $a^4 + 2a^2 - 35 = (a^2 + p)(a^2 + q)$. Since the constant term is negative, we look for a factorization of −35 in which one factor is positive and one factor is negative. Their sum must be 2.

Pairs of factors	Sums of factors
−1, 35	34
1, −35	−34
−5, 7	2
5, −7	−2

The numbers we want are −5 and 7.

$a^4 + 2a^2 - 35 = (a^2 - 5)(a^2 + 7)$.

31. $x^2 + x - 42$

Since the constant term is negative, we look for a factorization of −42 in which one factor is positive and one factor is negative. Their sum must be 1, the coefficient of the middle term.

Pairs of factors	Sums of factors
−1, 42	41
1, −42	−41
−2, 21	19
2, −21	−19
−3, 14	11
3, −14	−11
−6, 7	1
6, −7	−1

The numbers we want are −6 and 7.

$x^2 + x - 42 = (x - 6)(x + 7)$.

33. $7 - 2p + p^2 = p^2 - 2p + 7$

Since the constant term is positive and the coefficient of the middle term is negative, we look for a factorization of 7 in which both factors are negative. The sum must be −2. The only possible pair of factors is −1 and −7, but their sum is not −2. Thus, this polynomial is not factorable into binomials. It is prime.

35. $x^2 + 20x + 100$

We look for two factors, both positive, whose product is 100 and whose sum is 20.

They are 10 and 10. $10 \cdot 10 = 100$ and $10 + 10 = 20$.

$x^2 + 20x + 100 = (x + 10)(x + 10)$, or $(x + 10)^2$.

37. $30 + 7x - x^2 = -x^2 + 7x + 30 = -1(x^2 - 7x - 30)$

Now we factor $x^2 - 7x - 30$. Since the constant term is negative, we look for a factorization of -30 in which one factor is positive and one factor is negative. Their sum must be -7, the coefficient of the middle term.

Pairs of factors	Sums of factors
$-1,\ \ 30$	29
$1,\ -30$	-29
$-2,\ \ 15$	13
$2,\ -15$	-13
$-3,\ \ 10$	7
$3,\ -10$	-7
$-5,\ \ \ 6$	1
$5,\ \ -6$	-1

The numbers we want are 3 and -10. Then

$x^2 - 7x - 30 = (x + 3)(x - 10)$, so we have:

$-x^2 + 7x + 30$

$= -1(x + 3)(x - 10)$

$= (-x - 3)(x - 10)$ Multiplying $x + 3$ by -1

$= (x + 3)(-x + 10)$ Multiplying $x - 10$ by -1

39. $24 - a^2 - 10a = -a^2 - 10a + 24 = -1(a^2 + 10a - 24)$

Now we factor $a^2 + 10a - 24$. Since the constant term is negative, we look for a factorization of -24 in which one factor is positive and one factor is negative. Their sum must be 10, the coefficient of the middle term.

Pairs of factors	Sums of factors
$-1,\ \ 24$	23
$1,\ -24$	-23
$-2,\ \ 12$	10
$2,\ -12$	-10
$-3,\ \ \ 8$	5
$3,\ \ -8$	-5
$-4,\ \ \ 6$	2
$4,\ \ -6$	-2

The numbers we want are -2 and 12. Then

$a^2 + 10a - 24 = (a - 2)(a + 12)$, so we have:

$-a^2 - 10a + 24$

$= -1(a - 2)(a + 12)$

$= (-a + 2)(a + 12)$ Multiplying $a - 2$ by -1

$= (a - 2)(-a - 12)$ Multiplying $a + 12$ by -1

41. $x^4 - 21x^3 - 100x^2 = x^2(x^2 - 21x - 100)$

After factoring out the common factor, x^2, we consider $x^2 - 21x - 100$. We look for two factors, one positive and one negative, whose product is -100 and whose sum is

-21. They are 4 and -25. $4 \cdot (-25) = -100$ and $4 + (-25) = -21$.

Then $x^2 - 21x - 100 = (x + 4)(x - 25)$, so $x^4 - 21x^3 - 100x^2 = x^2(x + 4)(x - 25)$.

43. $x^2 - 21x - 72$

We look for two factors, one positive and one negative, whose product is -72 and whose sum is -21. They are 3 and -24.

$x^2 - 21x - 72 = (x + 3)(x - 24)$.

45. $x^2 - 25x + 144$

We look for two factors, both negative, whose product is 144 and whose sum is -25. They are -9 and -16.

$x^2 - 25x + 144 = (x - 9)(x - 16)$.

47. $a^2 + a - 132$

We look for two factors, one positive and one negative, whose product is -132 and whose sum is 1. They are -11 and 12.

$a^2 + a - 132 = (a - 11)(a + 12)$.

49. $120 - 23x + x^2 = x^2 - 23x + 120$

We look for two factors, both negative, whose product is 120 and whose sum is -23. They are -8 and -15.

$x^2 - 23x + 120 = (x - 8)(x - 15)$.

51. First write the polynomial in descending order and factor out -1.

$108 - 3x - x^2 = -x^2 - 3x + 108 = -1(x^2 + 3x - 108)$

Now we factor the polynomial $x^2 + 3x - 108$. We look for two factors, one positive and one negative, whose product is -108 and whose sum is 3. They are -9 and 12.

$x^2 + 3x - 108 = (x - 9)(x + 12)$

The final answer must include -1 which was factored out above.

$-x^2 - 3x + 108$

$= -1(x - 9)(x + 12)$

$= (-x + 9)(x + 12)$ Multiplying $x - 9$ by -1

$= (x - 9)(-x - 12)$ Multiplying $x + 12$ by -1

53. $y^2 - 0.2y - 0.08$

We look for two factors, one positive and one negative, whose product is -0.08 and whose sum is -0.2. They are -0.4 and 0.2.

$y^2 - 0.2y - 0.08 = (y - 0.4)(y + 0.2)$.

55. $p^2 + 3pq - 10q^2 = p^2 + 3pq - 10q^2$

Think of $3q$ as a "coefficient" of p. Then we look for factors of $-10q^2$ whose sum is $3q$. They are $5q$ and $-2q$.

$p^2 + 3pq - 10q^2 = (p + 5q)(p - 2q)$.

57. $84 - 8t - t^2 = -t^2 - 8t + 84 = -1(t^2 + 8t - 84)$

Now we factor $t^2 + 8t - 84$. We look for two factors, one positive and one negative, whose product is -84 and whose sum is 8. They are 14 and -6.

Then $t^2 + 8t - 84 = (t + 14)(t - 6)$, so we have:

$$-t^2 - 8t + 84$$
$$= -1(t + 14)(t - 6)$$
$$= (-t - 14)(t - 6) \qquad \text{Multiplying } t + 14 \text{ by } -1$$
$$= (t + 14)(-t + 6) \qquad \text{Multiplying } t - 6 \text{ by } -1$$

59. $m^2 + 5mn + 4n^2 = m^2 + 5nm + 4n^2$

We look for factors of $4n^2$ whose sum is $5n$. They are $4n$ and n.

$$m^2 + 5mn + 4n^2 = (m + 4n)(m + n)$$

61. $s^2 - 2st - 15t^2 = s^2 - 2ts - 15t^2$

We look for factors of $-15t^2$ whose sum is $-2t$. They are $-5t$ and $3t$.

$$s^2 - 2st - 15t^2 = (s - 5t)(s + 3t)$$

63. $6a^{10} - 30a^9 - 84a^8 = 6a^8(a^2 - 5a - 14)$

After factoring out the common factor, $6a^8$, we consider $a^2 - 5a - 14$. We look for two factors, one positive and one negative, whose product is -14 and whose sum is -5. They are 2 and -7.

$a^2 - 5a - 14 = (a + 2)(a - 7)$, so $6a^{10} - 30a^9 - 84a^8 = 6a^8(a + 2)(a - 7)$.

65. Discussion and Writing Exercise

67. Discussion and Writing Exercise

69. $8x(2x^2 - 6x + 1) = 8x \cdot 2x^2 - 8x \cdot 6x + 8x \cdot 1 = 16x^3 - 48x^2 + 8x$

71. $(7w + 6)^2 = (7w)^2 + 2 \cdot 7w \cdot 6 + 6^2 = 49w^2 + 84w + 36$

73. $(4w - 11)(4w + 11) = (4w)^2 - (11)^2 = 16w^2 - 121$

75. $3x - 8 = 0$

$$3x = 8 \qquad \text{Adding 8 on both sides}$$
$$x = \frac{8}{3} \qquad \text{Dividing by 3 on both sides}$$

The solution is $\frac{8}{3}$.

77. Familiarize. Let $n =$ the number of people arrested the year before.

Translate. We reword the problem.

$$\underbrace{\begin{array}{c} \text{Number} \\ \text{arrested the} \\ \text{year before} \end{array}}_{n} \; \underset{-}{\text{less}} \; \underset{1.2\%}{1.2\%} \; \underset{\cdot}{\text{of}} \; \underbrace{\begin{array}{c} \text{that} \\ \text{number} \end{array}}_{n} \; \underset{=}{\text{is}} \; \underset{29,200}{29,200}.$$

Carry out. We solve the equation.

$$n - 1.2\% \cdot n = 29,200$$
$$1 \cdot n - 0.012n = 29,200$$
$$0.988n = 29,200$$
$$n \approx 29,555 \quad \text{Rounding}$$

Check. 1.2% of $29,555$ is $0.012(29,555) \approx 355$ and $29,555 - 355 = 29,200$. The answer checks.

State. Approximately 29,555 people were arrested the year before.

79. $y^2 + my + 50$

We look for pairs of factors whose product is 50. The sum of each pair is represented by m.

Pairs of factors whose product is -50	Sums of factors
$1, \quad 50$	51
$-1, -50$	-51
$2, \quad 25$	27
$-2, -25$	-27
$5, \quad 10$	15
$-5, -10$	-15

The polynomial $y^2 + my + 50$ can be factored if m is 51, -51, 27, -27, 15, or -15.

81. $x^2 - \frac{1}{2}x - \frac{3}{16}$

We look for two factors, one positive and one negative, whose product is $-\frac{3}{16}$ and whose sum is $-\frac{1}{2}$.

They are $-\frac{3}{4}$ and $\frac{1}{4}$.

$$-\frac{3}{4} \cdot \frac{1}{4} = -\frac{3}{16} \text{ and } -\frac{3}{4} + \frac{1}{4} = -\frac{2}{4} = -\frac{1}{2}.$$
$$x^2 - \frac{1}{2}x - \frac{3}{16} = \left(x - \frac{3}{4}\right)\left(x + \frac{1}{4}\right)$$

83. $x^2 + \frac{30}{7}x - \frac{25}{7}$

We look for two factors, one positive and one negative, whose product is $-\frac{25}{7}$ and whose sum is $\frac{30}{7}$.

They are 5 and $-\frac{5}{7}$.

$$5 \cdot \left(-\frac{5}{7}\right) = -\frac{25}{7} \text{ and } 5 + \left(-\frac{5}{7}\right) = \frac{35}{7} + \left(-\frac{5}{7}\right) = \frac{30}{7}.$$
$$x^2 + \frac{30}{7}x - \frac{25}{7} = (x + 5)\left(x - \frac{5}{7}\right)$$

85. $b^{2n} + 7b^n + 10$

Consider this trinomial as $(b^n)^2 + 7b^n + 10$. We look for numbers p and q such that $b^{2n} + 7b^n + 10 = (b^n + p)(b^n + q)$. We find two factors, both positive, whose product is 10 and whose sum is 7. They are 5 and 2.

$$b^{2n} + 7b^n + 10 = (b^n + 5)(b^n + 2)$$

87. We first label the drawing with additional information.

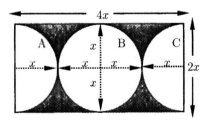

$4x$ represents the length of the rectangle and $2x$ the width. The area of the rectangle is $4x \cdot 2x$, or $8x^2$.

The area of semicircle A is $\frac{1}{2}\pi x^2$.

The area of circle B is πx^2.

The area of semicircle C is $\frac{1}{2}\pi x^2$.

$$\begin{array}{lll} \text{Area of} \\ \text{shaded region} \end{array} = \begin{array}{l} \text{Area of} \\ \text{rectangle} \end{array} - \begin{array}{c} \text{Area} \\ \text{of} \\ A \end{array} - \begin{array}{c} \text{Area} \\ \text{of} \\ B \end{array} - \begin{array}{c} \text{Area} \\ \text{of} \\ C \end{array}$$

$$\begin{array}{lll} \text{Area of} \\ \text{shaded region} \end{array} = \quad 8x^2 \quad - \frac{1}{2}\pi x^2 - \pi x^2 - \frac{1}{2}\pi x^2$$

$$= 8x^2 - 2\pi x^2$$

$$= 2x^2(4 - \pi)$$

The shaded area can be represented by $2x^2(4 - \pi)$.

89. First consider all the factorizations of 36 that contain three factors. We also find the sum of the factors in each factorization.

Factorization	Sum of Factors
$1 \cdot 1 \cdot 36$	38
$1 \cdot 2 \cdot 18$	21
$1 \cdot 3 \cdot 12$	16
$1 \cdot 4 \cdot 9$	14
$1 \cdot 6 \cdot 6$	13
$2 \cdot 2 \cdot 9$	13
$2 \cdot 3 \cdot 6$	11
$3 \cdot 3 \cdot 4$	10

We can conclude that the number on the house next door is 13, because two sums are 13. This is what causes the census taker to be puzzled. She cannot determine which trio of factors gives the children's ages. When the mother supplies the additional information that there is an oldest child, the census taker knows that the ages of the children cannot be 1, 6, and 6 because there is not an oldest child in this group. Therefore, the children's ages must be 2, 2, and 9.

Exercise Set 10.3

1. $2x^2 - 7x - 4$

(1) Look for a common factor. There is none (other than 1 or -1).

(2) Factor the first term, $2x^2$. The only possibility is $2x, x$. The desired factorization is of the form:

$$(2x + \quad)(x + \quad)$$

(3) Factor the last term, -4, which is negative. The possibilities are -4, 1 and 4, -1 and 2, -2.

(4) Look for combinations of factors from steps (2) and (3) such that the sum of their products is the middle term, $-7x$. We try some possibilities:

$$(2x - 4)(x + 1) = 2x^2 - 2x - 4$$
$$(2x + 4)(x - 1) = 2x^2 + 2x - 4$$
$$(2x + 2)(x - 2) = 2x^2 - 2x - 4$$
$$(2x + 1)(x - 4) = 2x^2 - 7x - 4$$

The factorization is $(2x + 1)(x - 4)$.

3. $5x^2 - x - 18$

(1) There is no common factor (other than 1 or -1).

(2) Factor the first term, $5x^2$. The only possibility is $5x, x$. The desired factorization is of the form:

$$(5x + \quad)(x + \quad)$$

(3) Factor the last term, -18. The possibilities are -18, 1 and 18, -1 and -9, 2 and 9, -2 and -6, 3 and 6, -3.

(4) Look for combinations of factors from steps (2) and (3) such that the sum of their products is the middle term, x. We try some possibilities:

$$(5x - 18)(x + 1) = 5x^2 - 13x - 18$$
$$(5x + 18)(x - 1) = 5x^2 + 13x - 18$$
$$(5x + 9)(x - 2) = 5x^2 - x - 18$$

The factorization is $(5x + 9)(x - 2)$.

5. $6x^2 + 23x + 7$

(1) There is no common factor (other than 1 or -1).

(2) Factor the first term, $6x^2$. The possibilities are $6x$, x and $3x$, $2x$. We have these as possibilities for factorizations:

$$(6x + \quad)(x + \quad) \text{ and } (3x + \quad)(2x + \quad)$$

(3) Factor the last term, 7. The possibilities are 7, 1 and -7, -1.

(4) Look for combinations of factors from steps (2) and (3) such that the sum of their products is the middle term, $23x$. Since all signs are positive, we need consider only plus signs. We try some possibilities:

$$(6x + 7)(x + 1) = 6x^2 + 13x + 7$$
$$(3x + 7)(2x + 1) = 6x^2 + 17x + 7$$
$$(6x + 1)(x + 7) = 6x^2 + 43x + 7$$
$$(3x + 1)(2x + 7) = 6x^2 + 23x + 7$$

The factorization is $(3x + 1)(2x + 7)$.

7. $3x^2 + 4x + 1$

(1) There is no common factor (other than 1 or -1).

(2) Factor the first term, $3x^2$. The only possibility is $3x, x$. The desired factorization is of the form:

$$(3x + \quad)(x + \quad)$$

(3) Factor the last term, 1. The possibilities are 1, 1 and -1, -1.

(4) Look for combinations of factors from steps (2) and (3) such that the sum of their products is the middle term, $4x$. Since all signs are positive, we need consider only plus signs. There is only one such possibility:

$$(3x + 1)(x + 1) = 3x^2 + 4x + 1$$

The factorization is $(3x + 1)(x + 1)$.

9. $4x^2 + 4x - 15$

(1) There is no common factor (other than 1 or −1).

(2) Factor the first term, $4x^2$. The possibilities are $4x$, x and $2x$, $2x$. We have these as possibilities for factorizations:
$$(4x+ \quad)(x+ \quad) \text{ and } (2x+ \quad)(2x+ \quad)$$

(3) Factor the last term, −15. The possibilities are 15, −1 and −15, 1 and 5, −3 and −5, 3.

(4) We try some possibilities:
$$(4x + 15)(x - 1) = 4x^2 + 11x - 15$$
$$(2x + 15)(2x - 1) = 4x^2 + 28x - 15$$
$$(4x - 15)(x + 1) = 4x^2 - 11x - 15$$
$$(2x - 15)(2x + 1) = 4x^2 - 28x - 15$$
$$(4x + 5)(x - 3) = 4x^2 - 7x - 15$$
$$(2x + 5)(2x - 3) = 4x^2 + 4x - 15$$

The factorization is $(2x + 5)(2x - 3)$.

11. $2x^2 - x - 1$

(1) There is no common factor (other than 1 or −1).

(2) Factor the first term, $2x^2$. The only possibility is $2x$, x. The desired factorization is of the form:
$$(2x+ \quad)(x+ \quad)$$

(3) Factor the last term, −1. The only possibility is −1, 1.

(4) We try the possibilities:
$$(2x - 1)(x + 1) = 2x^2 + x - 1$$
$$(2x + 1)(x - 1) = 2x^2 - x - 1$$

The factorization is $(2x + 1)(x - 1)$.

13. $9x^2 + 18x - 16$

(1) There is no common factor (other than 1 or −1).

(2) Factor the first term, $9x^2$. The possibilities are $9x$, x and $3x$, $3x$. We have these as possibilities for factorizations:
$$(9x+ \quad)(x+ \quad) \text{ and } (3x+ \quad)(3x+ \quad)$$

(3) Factor the last term, −16. The possibilities are 16, −1 and −16, 1 and 8, −2 and −8, 2 and 4, −4.

(4) We try some possibilities:
$$(9x + 16)(x - 1) = 9x^2 + 7x - 16$$
$$(3x + 16)(3x - 1) = 9x^2 + 45x - 16$$
$$(9x - 16)(x + 1) = 9x^2 - 7x - 16$$
$$(3x - 16)(3x + 1) = 9x^2 - 45x - 16$$
$$(9x + 8)(x - 2) = 9x^2 - 10x - 16$$
$$(3x + 8)(3x - 2) = 9x^2 + 18x - 16$$

The factorization is $(3x + 8)(3x - 2)$.

15. $3x^2 - 5x - 2$

(1) There is no common factor (other than 1 or −1).

(2) Factor the first term, $3x^2$. The only possibility is $3x$, x. The desired factorization is of the form:
$$(3x+ \quad)(x+ \quad)$$

(3) Factor the last term, −2. The possibilities are 2, −1 and −2 and 1.

(4) We try some possibilities:
$$(3x + 2)(x - 1) = 3x^2 - x - 2$$
$$(3x - 2)(x + 1) = 3x^2 + x - 2$$
$$(3x - 1)(x + 2) = 3x^2 + 5x - 2$$
$$(3x + 1)(x - 2) = 3x^2 - 5x - 2$$

The factorization is $(3x + 1)(x - 2)$.

17. $12x^2 + 31x + 20$

(1) There is no common factor (other than 1 or −1).

(2) Factor the first term, $12x^2$. The possibilities are $12x$, x and $6x$, $2x$ and $4x$, $3x$. We have these as possibilities for factorizations:
$$(12x+ \quad)(x+ \quad) \text{ and } (6x+ \quad)(2x+ \quad) \text{ and } (4x+ \quad)(3x+ \quad)$$

(3) Factor the last term, 20. Since all signs are positive, we need consider only positive pairs of factors. Those factor pairs are 20, 1 and 10, 2 and 5, 4.

(4) We can immediately reject all possibilities in which either factor has a common factor, such as $(12x+20)$ or $(6x+4)$, because we determined at the outset that there are no common factors. We try some of the remaining possibilities:
$$(12x + 1)(x + 20) = 12x^2 + 241x + 20$$
$$(12x + 5)(x + 4) = 12x^2 + 53x + 20$$
$$(6x + 1)(2x + 20) = 12x^2 + 122x + 20$$
$$(4x + 5)(3x + 4) = 12x^2 + 31x + 20$$

The factorization is $(4x + 5)(3x + 4)$.

19. $14x^2 + 19x - 3$

(1) There is no common factor (other than 1 or −1).

(2) Factor the first term, $14x^2$. The possibilities are $14x$, x and $7x$, $2x$. We have these as possibilities for factorizations:
$$(14x+ \quad)(x+ \quad) \text{ and } (7x+ \quad)(2x+ \quad)$$

(3) Factor the last term, −3. The possibilities are −1, 3 and 3, −1.

(4) We try some possibilities:
$$(14x - 1)(x + 3) = 14x^2 + 41x - 3$$
$$(7x - 1)(2x + 3) = 7x^2 + 19x - 3$$

The factorization is $(7x - 1)(2x + 3)$.

21. $9x^2 + 18x + 8$

(1) There is no common factor (other than 1 or -1).

(2) Factor the first term, $9x^2$. The possibilities are $9x$, x and $3x$, $3x$. We have these as possibilities for factorizations:

$$(9x+\quad)(x+\quad) \text{ and } (3x+\quad)(3x+\quad)$$

(3) Factor the last term, 8. Since all signs are positive, we need consider only positive pairs of factors. Those factor pairs are 8, 1 and 4, 2.

(4) We try some possibilities:

$$(9x+8)(x+1) = 9x^2 + 17x + 8$$
$$(3x+8)(3x+1) = 9x^2 + 27x + 8$$
$$(9x+4)(x+2) = 9x^2 + 22x + 8$$
$$(3x+4)(3x+2) = 9x^2 + 18x + 8$$

The factorization is $(3x+4)(3x+2)$.

23. $49 - 42x + 9x^2 = 9x^2 - 42x + 49$

(1) There is no common factor (other than 1 or -1).

(2) Factor the first term, $9x^2$. The possibilities are $9x$, x and $3x$, $3x$. We have these as possibilities for factorizations:

$$(9x+\quad)(x+\quad) \text{ and } (3x+\quad)(3x+\quad)$$

(3) Factor 49. Since 49 is positive and the middle term is negative, we need consider only negative pairs of factors. Those factor pairs are -49, -1 and -7, -7.

(4) We try some possibilities:

$$(9x-49)(x-1) = 9x^2 - 58x + 49$$
$$(3x-49)(3x-1) = 9x^2 - 150x + 49$$
$$(9x-7)(x-7) = 9x^2 - 70x + 49$$
$$(3x-7)(3x-7) = 9x^2 - 42x + 49$$

The factorization is $(3x-7)(3x-7)$, or $(3x-7)^2$. This can also be expressed as follows:
$$(3x-7)^2 = (-1)^2(3x-7)^2 = [-1 \cdot (3x-7)]^2 =$$
$$(-3x+7)^2, \text{ or } (7-3x)^2$$

25. $24x^2 + 47x - 2$

(1) There is no common factor (other than 1 or -1).

(2) Factor the first term, $24x^2$. The possibilities are $24x$, x and $12x$, $2x$ and $6x$, $4x$ and $3x$, $8x$. We have these as possibilities for factorizations:

$$(24x+\quad)(x+\quad) \text{ and } (12x+\quad)(2x+\quad) \text{ and }$$
$$(6x+\quad)(4x+\quad) \text{ and } (3x+\quad)(8x+\quad)$$

(3) Factor the last term, -2. The possibilities are 2, -1 and -2, 1.

(4) We can immediately reject all possibilities in which either factor has a common factor, such as $(24x+2)$ or $(12x-2)$, because we determined at the outset that there are no common factors. We try some of the remaining possibilities:

$$(24x-1)(x+2) = 24x^2 + 47x - 2$$

The factorization is $(24x-1)(x+2)$.

27. $35x^2 - 57x - 44$

(1) There is no common factor (other than 1 or -1).

(2) Factor the first term, $35x^2$. The possibilities are $35x$, x and $7x$, $5x$. We have these as possibilities for factorizations:

$$(35x+\quad)(x+\quad) \text{ and } (7x+\quad)(5x+\quad)$$

(3) Factor the last term, -44. The possibilities are 1, -44 and -1, 44 and 2, -22 and -2, 22 and 4, -11, and -4, 11.

(4) We try some possibilities:

$$(35x+1)(x-44) = 35x^2 - 1539x - 44$$
$$(7x+1)(5x-44) = 35x^2 - 303x - 44$$
$$(35x+2)(x-22) = 35x^2 - 768x - 44$$
$$(7x+2)(5x-22) = 35x^2 - 144x - 44$$
$$(35x+4)(x-11) = 35x^2 - 381x - 44$$
$$(7x+4)(5x-11) = 35x^2 - 57x - 44$$

The factorization is $(7x+4)(5x-11)$.

29. $20 + 6x - 2x^2 = -2x^2 + 6x + 20$

We factor out the common factor, -2. Factoring out -2 rather than 2 gives us a positive leading coefficient.

$$-2(x^2 - 3x - 10)$$

Then we factor the trinomial $x^2 - 3x - 10$. We look for a pair of factors whose product is -10 and whose sum is -3. The numbers are -5 and 2. The factorization of $x^2 - 3x - 10$ is $(x-5)(x+2)$. Then $20 + 6x - 2x^2 = -2(x-5)(x+2)$. If we think of -2 and $-1 \cdot 2$ then we can write other correct factorizations:

$$20 + 6x - 2x^2$$
$$= 2(-x+5)(x+2) \qquad \text{Multiplying } x-5 \text{ by } -1$$
$$= 2(x-5)(-x-2) \qquad \text{Multiplying } x+2 \text{ by } -1$$

Note that we can also express $2(-x+5)(x+2)$ as $2(5-x)(x+2)$ since $-x+5 = 5-x$ by the commutative law of addition.

31. $12x^2 + 28x - 24$

(1) We factor out the common factor, 4:
$$4(3x^2 + 7x - 6)$$

Then we factor the trinomial $3x^2 + 7x - 6$.

(2) Factor $3x^2$. The only possibility is $3x$, x. The desired factorization is of the form:

$$(3x+\quad)(x+\quad)$$

(3) Factor -6. The possibilities are 6, -1 and -6, 1 and 3, -2 and -3, 2.

(4) We can immediately reject all possibilities in which either factor has a common factor, such as $(3x + 6)$ or $(3x - 3)$, because we factored out the largest common factor at the outset. We try some of the remaining possibilities:

$$(3x - 1)(x + 6) = 3x^2 + 17x - 6$$
$$(3x - 2)(x + 3) = 3x^2 + 7x - 6$$

The factorization of $3x^2 + 7x - 6$ is $(3x - 2)(x + 3)$. We must include the common factor in order to get a factorization of the original trinomial.

$$12x^2 + 28x - 24 = 4(3x - 2)(x + 3)$$

33. $30x^2 - 24x - 54$

(1) We factor out the common factor, 6:
$6(5x^2 - 4x - 9)$

Then we factor the trinomial $5x^2 - 4x - 9$.

(2) Factor $5x^2$. The only possibility is $5x$, x. The desired factorization is of the form:

$$(5x+ \quad)(x+ \quad)$$

(3) Factor -9. The possibilities are 9, -1 and -9, 1 and 3, -3.

(4) We try some possibilities:

$$(5x + 9)(x - 1) = 5x^2 + 4x - 9$$
$$(5x - 9)(x + 1) = 5x^2 - 4x - 9$$

The factorization of $5x^2 - 4x - 9$ is $(5x - 9)(x + 1)$. We must include the common factor in order to get a factorization of the original trinomial.

$$30x^2 - 24x - 54 = 6(5x - 9)(x + 1)$$

35. $4y + 6y^2 - 10 = 6y^2 + 4y - 10$

(1) We factor out the common factor, 2:
$2(3y^2 + 2y - 5)$

Then we factor the trinomial $3y^2 + 2y - 5$.

(2) Factor $3y^2$. The only possibility is $3y$, y. The desired factorization is of the form:

$$(3y+ \quad)(y+ \quad)$$

(3) Factor -5. The possibilities are 5, -1 and -5, 1.

(4) We try some possibilities:

$$(3y + 5)(y - 1) = 3y^2 + 2y - 5$$

Then $3y^2 + 2y - 5 = (3y + 5)(y - 1)$, so $6y^2 + 4y - 10 = 2(3y + 5)(y - 1)$.

37. $3x^2 - 4x + 1$

(1) There is no common factor (other than 1 or -1).

(2) Factor the first term, $3x^2$. The only possibility is $3x$, x. The desired factorization is of the form:

$$(3x+ \quad)(x+ \quad)$$

(3) Factor the last term, 1. Since 1 is positive and the middle term is negative, we need consider only negative factor pairs. The only such pair is -1, -1.

(4) There is only one possibility:

$$(3x - 1)(x - 1) = 3x^2 - 4x + 1$$

The factorization is $(3x - 1)(x - 1)$.

39. $12x^2 - 28x - 24$

(1) We factor out the common factor, 4:
$4(3x^2 - 7x - 6)$

Then we factor the trinomial $3x^2 - 7x - 6$.

(2) Factor $3x^2$. The only possibility is $3x$, x. The desired factorization is of the form:

$$(3x+ \quad)(x+ \quad)$$

(3) Factor -6. The possibilities are 6, -1 and -6, 1 and 3, -2 and -3, 2.

(4) We can immediately reject all possibilities in which either factor has a common factor, such as $(3x - 6)$ or $(3x + 3)$, because we factored out the largest common factor at the outset. We try some of the remaining possibilities:

$$(3x - 1)(x + 6) = 3x^2 + 17x - 6$$
$$(3x - 2)(x + 3) = 3x^2 + 7x - 6$$
$$(3x + 2)(x - 3) = 3x^2 - 7x - 6$$

Then $3x^2 - 7x - 6 = (3x + 2)(x - 3)$, so $12x^2 - 28x - 24 = 4(3x + 2)(x - 3)$.

41. $-1 + 2x^2 - x = 2x^2 - x - 1$

(1) There is no common factor (other than 1 or -1).

(2) Factor the first term, $2x^2$. The only possibility is $2x$, x. The desired factorization is of the form:

$$(2x+ \quad)(x+ \quad)$$

(3) Factor -1. The only possibility is 1, -1.

(4) We try some possibilities:

$$(2x + 1)(x - 1) = 2x^2 - x - 1$$

The factorization is $(2x + 1)(x - 1)$.

43. $9x^2 - 18x - 16$

(1) There is no common factor (other than 1 or -1).

(2) Factor the first term, $9x^2$. The possibilities are $9x$, x and $3x$, $3x$. We have these as possibilities for factorizations:

$$(9x+ \quad)(x+ \quad) \text{ and } (3x+ \quad)(3x+ \quad)$$

(3) Factor the last term, -16. The possibilities are 16, -1 and -16, 1 and 8, -2 and -8, 2 and 4, -4.

(4) We try some possibilities:

$$(9x + 16)(x - 1) = 9x^2 + 7x - 16$$
$$(3x + 16)(3x - 1) = 9x^2 + 45x - 16$$

$$(9x + 8)(x - 2) = 9x^2 - 10x - 16$$
$$(3x + 8)(3x - 2) = 9x^2 + 18x - 16$$
$$(3x - 8)(3x + 2) = 9x^2 - 18x - 16$$

The factorization is $(3x - 8)(3x + 2)$.

45. $15x^2 - 25x - 10$

(1) Factor out the common factor, 5:

$5(3x^2 - 5x - 2)$

Then we factor the trinomial $3x^2 - 5x - 2$. This was done in Exercise 15. We know that $3x^2 - 5x - 2 = (3x + 1)(x - 2)$, so $15x^2 - 25x - 10 = 5(3x + 1)(x - 2)$.

47. $12p^3 + 31p^2 + 20p$

(1) We factor out the common factor, p:

$p(12p^2 + 31p + 20)$

Then we factor the trinomial $12p^2 + 31p + 20$. This was done in Exercise 17 although the variable is x in that exercise. We know that $12p^2 + 31p + 20 = (3p + 4)(4p + 5)$, so $12p^3 + 31p^2 + 20p = p(3p + 4)(4p + 5)$.

49.
$$16 + 18x - 9x^2 = -9x^2 + 18x + 16$$
$$= -1(9x^2 - 18x - 16)$$
$$= -1(3x - 8)(3x + 2) \quad \text{Using the result from Exercise 43}$$

Other correct factorizations are:

$16 + 18x - 9x^2$

$= (-3x + 8)(3x + 2) \qquad$ Multiplying $3x - 8$ by -1

$= (3x - 8)(-3x - 2) \qquad$ Multiplying $3x + 2$ by -1

We can also express $(-3x + 8)(3x + 2)$ as $(8 - 3x)(3x + 2)$ since $-3x + 8 = 8 - 3x$ by the commutative law of addition.

51. $-15x^2 + 19x - 6 = -1(15x^2 - 19x + 6)$

Now we factor $15x^2 - 19x + 6$.

(1) There is no common factor (other than 1 or -1).

(2) Factor the first term, $15x^2$. The possibilities are $15x$, x and $5x$, $3x$. We have these as possibilities for factorizations:

$(15x+ \quad)(x+ \quad)$ and $(5x+ \quad)(3x+ \quad)$

(3) Factor the last term, 6. The possibilities are 6, 1 and -6, -1 and 3, 2 and -3, -2.

(4) We try some possibilities:

$$(15x + 1)(x + 6) = 15x^2 + 91x + 6$$
$$(5x + 3)(3x + 2) = 15x^2 - 19x + 6$$
$$(5x - 3)(3x - 2) = 15x^2 - 19x + 6$$

The factorization of $15x^2 - 19x + 6$ is $(5x - 3)(3x - 2)$. Then $-15x^2 + 19x - 6 = -1(5x - 3)(3x - 2)$. Other correct factorizations are:

$-15x^2 + 19x - 6$

$= (-5x + 3)(3x - 2) \qquad$ Multiplying $5x - 3$ by -1

$= (5x - 3)(-3x + 2) \qquad$ Multiplying $3x - 2$ by -1

Note that we can also express $(-5x + 3)(3x - 2)$ as $(3 - 5x)(3x - 2)$ since $-5x + 3 = 3 - 5x$ by the commutative law of addition. Similarly, we can express $(5x - 3)(-3x + 2)$ as $(5x - 3)(2 - 3x)$.

53. $14x^4 + 19x^3 - 3x^2$

(1) Factor out the common factor, x^2: $x^2(14x^2 + 19x - 3)$

Then we factor the trinomial $14x^2 + 19x - 3$. This was done in Exercise 19. We know that $14x^2 + 19x - 3 = (7x - 1)(2x + 3)$, so $14x^4 + 19x^3 - 3x^2 = x^2(7x - 1)(2x + 3)$.

55. $168x^3 - 45x^2 + 3x$

(1) Factor out the common factor, $3x$:

$3x(56x^2 - 15x + 1)$

Then we factor the trinomial $56x^2 - 15x + 1$.

(2) Factor $56x^2$. The possibilities are $56x$, x and $28x$, $2x$ and $14x$, $4x$ and $7x$, $8x$. We have these as possibilities for factorizations:

$(56x+ \quad)(x+ \quad)$ and $(28x+ \quad)(2x+ \quad)$ and $(14x+ \quad)(4x+ \quad)$ and $(7x+ \quad)(8x+ \quad)$

(3) Factor 1. Since 1 is positive and the middle term is negative we need consider only the negative factor pair -1, -1.

(4) We try some possibilities:

$$(56x - 1)(x - 1) = 56x^2 - 57x + 1$$
$$(28x - 1)(2x - 1) = 56x^2 - 30x + 1$$
$$(14x - 1)(4x - 1) = 56x^2 - 18x + 1$$
$$(7x - 1)(8x - 1) = 56x^2 - 15x + 1$$

Then $56x^2 - 15x + 1 = (7x - 1)(8x - 1)$, so $168x^3 - 45x^2 + 3x = 3x(7x - 1)(8x - 1)$.

57. $15x^4 - 19x^2 + 6 = 15(x^2)^2 - 19x^2 + 6$

(1) There is no common factor (other than 1 or -1).

(2) Factor the first term, $15x^4$. The possibilities are $15x^2$, x^2 and $5x^2$, $3x^2$. We have these as possibilities for factorizations:

$(15x^2+ \quad)(x^2+ \quad)$ and $(5x^2+ \quad)(3x^2+ \quad)$

(3) Factor 6. Since 6 is positive and the middle term is negative, we need consider only negative factor pairs. Those pairs are -6, -1 and -3, -2.

(4) We can immediately reject all possibilities in which either factor has a common factor, such as $(15x^2 - 6)$ or $(3x^2 - 3)$, because we determined at the outset that there is no common factor. We try some of the remaining possibilities:

$$(15x^2 - 1)(x^2 - 6) = 15x^4 - 91x^2 + 6$$
$$(15x^2 - 2)(x^2 - 3) = 15x^4 - 47x^2 + 6$$
$$(5x^2 - 6)(3x^2 - 1) = 15x^4 - 23x^2 + 6$$
$$(5x^2 - 3)(3x^2 - 2) = 15x^4 - 19x^2 + 6$$

The factorization is $(5x^2 - 3)(3x^2 - 2)$.

59. $25t^2 + 80t + 64$

(1) There is no common factor (other than 1 or -1).

(2) Factor the first term, $25t^2$. The possibilities are $25t$, t and $5t$, $5t$. We have these as possibilities for factorizations:

$$(25t+\quad)(t+\quad) \text{ and } (5t+\quad)(5t+\quad)$$

(3) Factor the last term, 64. Since all signs are positive, we need consider only positive pairs of factors. Those factor pairs are 64, 1 and 32, 2 and 16, 4 and 8, 8.

(4) We try some possibilities:

$$(25t+64)(t+1) = 25t^2 + 89t + 64$$
$$(5t+32)(5t+2) = 25t^2 + 170t + 64$$
$$(25t+16)(t+4) = 25t^2 + 116t + 64$$
$$(5t+8)(5t+8) = 25t^2 + 80t + 64$$

The factorization is $(5t+8)(5t+8)$ or $(5t+8)^2$.

61. $6x^3 + 4x^2 - 10x$

(1) Factor out the common factor, $2x$: $2x(3x^2 + 2x - 5)$

Then we factor the trinomial $3x^2 + 2x - 5$. We did this in Exercise 35 (after we factored 2 out of the original trinomial). We know that $3x^2 + 2x - 5 = (3x+5)(x-1)$, so $6x^3 + 4x^2 - 10x = 2x(3x+5)(x-1)$.

63. $25x^2 + 79x + 64$

We follow the same procedure as in Exercise 59. None of the possibilities works. Thus, $25x^2 + 79x + 64$ is not factorable. It is prime.

65. $6x^2 - 19x - 5$

(1) There is no common factor (other than 1 or -1).

(2) Factor the first term, $6x^2$. The possibilities are $6x$, x and $3x$, $2x$. We have these as possibilities for factorizations:

$$(6x+\quad)(x+\quad) \text{ and } (3x+\quad)(2x+\quad)$$

(3) Factor the last term, -5. The possibilities are -5, 1 and 5, -1.

(4) We try some possibilities:

$$(6x-5)(x+1) = 6x^2 + x - 5$$
$$(6x+5)(x-1) = 6x^2 - x - 5$$
$$(6x+1)(x-5) = 6x^2 - 29x - 5$$
$$(6x-1)(x+5) = 6x^2 + 29x - 5$$
$$(3x-5)(2x+1) = 6x^2 - 7x - 5$$
$$(3x+5)(2x-1) = 6x^2 + 7x - 5$$
$$(3x+1)(2x-5) = 6x^2 - 13x - 5$$
$$(3x-1)(2x+5) = 6x^2 + 13x - 5$$

None of the possibilities works. Thus, $6x^2 - 19x - 5$ is not factorable. It is prime.

67. $12m^2 - mn - 20n^2$

(1) There is no common factor (other than 1 or -1).

(2) Factor the first term, $12m^2$. The possibilities are $12m$, m and $6m$, $2m$ and $3m$, $4m$. We have these as possibilities for factorizations:

$$(12m+\quad)(m+\quad) \text{ and } (6m+\quad)(2m+\quad)$$
$$\text{and } (3m+\quad)(4m+\quad)$$

(3) Factor the last term, $-20n^2$. The possibilities are $20n$, $-n$ and $-20n$, n and $10n$, $-2n$ and $-10n$, $2n$ and $5n$, $-4n$ and $-5n$, $4n$.

(4) We can immediately reject all possibilities in which either factor has a common factor, such as $(12m + 20n)$ or $(4m - 2n)$, because we determined at the outset that there is no common factor. We try some of the remaining possibilities:

$$(12m-n)(m+20n) = 12m^2 + 239mn - 20n^2$$
$$(12m+5n)(m-4n) = 12m^2 - 43mn - 20n^2$$
$$(3m-20n)(4m+n) = 12m^2 - 77mn - 20n^2$$
$$(3m-4n)(4m+5n) = 12m^2 - mn - 20n^2$$

The factorization is $(3m-4n)(4m+5n)$.

69. $6a^2 - ab - 15b^2$

(1) There is no common factor (other than 1 or -1).

(2) Factor the first term, $6a^2$. The possibilities are $6a$, a and $3a$, $2a$. We have these as possibilities for factorizations:

$$(6a+\quad)(a+\quad) \text{ and } (3a+\quad)(2a+\quad)$$

(3) Factor the last term, $-15b^2$. The possibilities are $15b$, $-b$ and $-15b$, b and $5b$, $-3b$ and $-5b$, $3b$.

(4) We can immediately reject all possibilities in which either factor has a common factor, such as $(6a+15b)$ or $(3a - 3b)$, because we determined at the outset that there is no common factor. We try some of the remaining possibilities:

$$(6a-b)(a+15b) = 6a^2 + 89ab - 15b^2$$
$$(3a-b)(2a+15b) = 6a^2 + 43ab - 15b^2$$
$$(6a+5b)(a-3b) = 6a^2 - 13ab - 15b^2$$
$$(3a+5b)(2a-3b) = 6a^2 + ab - 15b^2$$
$$(3a-5b)(2a+3b) = 6a^2 - ab - 15b^2$$

The factorization is $(3a-5b)(2a+3b)$.

71. $9a^2 + 18ab + 8b^2$

(1) There is no common factor (other than 1 or -1).

(2) Factor the first term, $9a^2$. The possibilities are $9a$, a and $3a$, $3a$. We have these as possibilities for factorizations:

$$(9a+\quad)(a+\quad) \text{ and } (3a+\quad)(3a+\quad)$$

(3) Factor $8b^2$. Since all signs are positive, we need consider only pairs of factors with positive coefficients. Those factor pairs are $8b$, b and $4b$, $2b$.

(4) We try some possibilities:

$$(9a + 8b)(a + b) = 9a^2 + 17ab + 8b^2$$
$$(3a + 8b)(3a + b) = 9a^2 + 27ab + 8b^2$$
$$(9a + 4b)(a + 2b) = 9a^2 + 22ab + 8b^2$$
$$(3a + 4b)(3a + 2b) = 9a^2 + 18ab + 8b^2$$

The factorization is $(3a + 4b)(3a + 2b)$.

73. $35p^2 + 34pq + 8q^2$

(1) There is no common factor (other than 1 or -1).

(2) Factor the first term, $35p^2$. The possibilities are $35p$, p and $7p$, $5p$. We have these as possibilities for factorizations:

$$(35p+)(p+) \text{ and } (7p+)(5p+)$$

(3) Factor $8q^2$. Since all signs are positive, we need consider only pairs of factors with positive coefficients. Those factor pairs are $8q$, q and $4q$, $2q$.

(4) We try some possibilities:

$$(35p + 8q)(p + q) = 35p^2 + 43pq + 8q^2$$
$$(7p + 8q)(5p + q) = 35p^2 + 47pq + 8q^2$$
$$(35p + 4q)(p + 2q) = 35p^2 + 74pq + 8q^2$$
$$(7p + 4q)(5p + 2q) = 35p^2 + 34pq + 8p^2$$

The factorization is $(7p + 4q)(5p + 2q)$.

75. $18x^2 - 6xy - 24y^2$

(1) Factor out the common factor, 6:

$$6(3x^2 - xy - 4y^2)$$

Then we factor the trinomial $3x^2 - xy - 4y^2$.

(2) Factor $3x^2$. The only possibility is $3x$, x. The desired factorization is of the form:

$$(3x+)(x+)$$

(3) Factor $-4y^2$. The possibilities are $4y$, $-y$ and $-4y$, y and $2y$, $-2y$.

(4) We try some possibilities:

$$(3x + 4y)(x - y) = 3x^2 + xy - 4y^2$$
$$(3x - 4y)(x + y) = 3x^2 - xy - 4y^2$$

Then $3x^2 - xy - 4y^2 = (3x - 4y)(x + y)$, so $18x^2 - 6xy - 24y^2 = 6(3x - 4y)(x + y)$.

77. Discussion and Writing Exercise

79.
$$A = pq - 7$$
$$A + 7 = pq \qquad \text{Adding 7}$$
$$\frac{A + 7}{p} = q \qquad \text{Dividing by } p$$

81. $3x + 2y = 6$
$$2y = 6 - 3x \quad \text{Subtracting } 3x$$
$$y = \frac{6 - 3x}{2} \quad \text{Dividing by 2}$$

83. $5 - 4x < -11$
$$-4x < -16 \quad \text{Subtracting 5}$$
$$x > 4 \qquad \text{Dividing by } -4 \text{ and reversing the inequality symbol}$$

The solution set is $\{x | x > 4\}$.

85. Graph: $y = \dfrac{2}{5}x - 1$

Because the equation is in the form $y = mx + b$, we know the y-intercept is $(0, -1)$. We find two other points on the line, substituting multiples of 5 for x to avoid fractions.

When $x = -5$, $y = \dfrac{2}{5}(-5) - 1 = -2 - 1 = -3$.

When $x = 5$, $y = \dfrac{2}{5}(5) - 1 = 2 - 1 = 1$.

x	y
0	-1
-5	-3
5	1

87. $(3x - 5)(3x + 5) = (3x)^2 - 5^2 = 9x^2 - 25$

89. $20x^{2n} + 16x^n + 3 = 20(x^n)^2 + 16x^n + 3$

(1) There is no common factor (other than 1 and -1).

(2) Factor the first term, $20x^{2n}$. The possibilities are $20x^n$, x^n and $10x^n$, $2x^n$ and $5x^n$, $4x^n$. We have these as possibilities for factorizations:

$$(20x^n+)(x^n+) \text{ and } (10x^n+)(2x^n+)$$
$$\text{and } (5x^n+)(4x^n+)$$

(3) Factor the last term, 3. Since all signs are positive, we need consider only the positive factor pair 3, 1.

(4) We try some possibilities:

$$(20x^n + 3)(x^n + 1) = 20x^{2n} + 23x^n + 3$$
$$(10x^n + 3)(2x^n + 1) = 20x^{2n} + 16x^n + 3$$

The factorization is $(10x^n + 3)(2x^n + 1)$.

91. $3x^{6a} - 2x^{3a} - 1 = 3(x^{3a})^2 - 2x^{3a} - 1$

(1) There is no common factor (other than 1 or -1).

(2) Factor the first term, $3x^{6a}$. The only possibility is $3x^{3a}$, x^{3a}. The desired factorization is of the form:

$$(3x^{3a}+)(x^{3a}+)$$

(3) Factor the last term, -1. The only possibility is -1, 1.

(4) We try the possibilities:

$$(3x^{3a} - 1)(x^{3a} + 1) = 3x^{6a} + 2x^{3a} - 1$$
$$(3x^{3a} + 1)(x^{3a} - 1) = 3x^{6a} - 2x^{3a} - 1$$

The factorization is $(3x^{3a} + 1)(x^{3a} - 1)$.

93.-101. Left to the student

Exercise Set 10.4

1. $x^2 + 2x + 7x + 14 = x(x+2) + 7(x+2)$
$$= (x+7)(x+2)$$

3. $x^2 - 4x - x + 4 = x(x-4) - 1(x-4)$
$$= (x-1)(x-4)$$

5. $6x^2 + 4x + 9x + 6 = 2x(3x+2) + 3(3x+2)$
$$= (2x+3)(3x+2)$$

7. $3x^2 - 4x - 12x + 16 = x(3x-4) - 4(3x-4)$
$$= (x-4)(3x-4)$$

9. $35x^2 - 40x + 21x - 24 = 5x(7x-8) + 3(7x-8)$
$$= (5x+3)(7x-8)$$

11. $4x^2 + 6x - 6x - 9 = 2x(2x+3) - 3(2x+3)$
$$= (2x-3)(2x+3)$$

13. $2x^4 + 6x^2 + 5x^2 + 15 = 2x^2(x^2+3) + 5(x^2+3)$
$$= (2x^2+5)(x^2+3)$$

15. $2x^2 + 7x - 4$

(1) First factor out a common factor, if any. There is none (other than 1 or -1).

(2) Multiply the leading coefficient, 2 and the constant, -4: $2(-4) = -8$.

(3) Look for a factorization of -8 in which the sum of the factors is the coefficient of the middle term, 7.

Pairs of factors	Sums of factors
$-1, \ \ 8$	7
$1, \ -8$	-7
$-2, \ \ 4$	2
$2, \ -4$	-2

(4) Split the middle term: $7x = -1x + 8x$

(5) Factor by grouping:
$$2x^2 + 7x - 4 = 2x^2 - x + 8x - 4$$
$$= x(2x-1) + 4(2x-1)$$
$$= (x+4)(2x-1)$$

17. $3x^2 - 4x - 15$

(1) First factor out a common factor, if any. There is none (other than 1 or -1).

(2) Multiply the leading coefficient, 3, and the constant, -15: $3(-15) = -45$.

(3) Look for a factorization of -45 in which the sum of the factors is the coefficient of the middle term, -4.

Pairs of factors	Sums of factors
$-1, \ \ 45$	44
$1, \ -45$	-44
$-3, \ \ 15$	12
$3, \ -15$	-12
$-5, \ \ 9$	4
$5, \ -9$	-4

(4) Split the middle term: $-4x = 5x - 9x$

(5) Factor by grouping:
$$3x^2 - 4x - 15 = 3x^2 + 5x - 9x - 15$$
$$= x(3x+5) - 3(3x+5)$$
$$= (x-3)(3x+5)$$

19. $6x^2 + 23x + 7$

(1) First factor out a common factor, if any. There is none (other than 1 or -1).

(2) Multiply the leading coefficient, 6, and the constant, 7: $6 \cdot 7 = 42$.

(3) Look for a factorization of 42 in which the sum of the factors is the coefficient of the middle term, 23. We only need to consider positive factors.

Pairs of factors	Sums of factors
$1, \ \ 42$	43
$2, \ \ 21$	23
$3, \ \ 14$	17
$6, \ \ 7$	13

(4) Split the middle term: $23x = 2x + 21x$

(5) Factor by grouping:
$$6x^2 + 23x + 7 = 6x^2 + 2x + 21x + 7$$
$$= 2x(3x+1) + 7(3x+1)$$
$$= (2x+7)(3x+1)$$

21. $3x^2 - 4x + 1$

(1) First factor out a common factor, if any. There is none (other than 1 or -1).

(2) Multiply the leading coefficient, 3, and the constant, 1: $3 \cdot 1 = 3$.

(3) Look for a factorization of 3 in which the sum of the factors is the coefficient of the middle term, -4. The numbers we want are -1 and -3: $-1 \cdot (-3) = 3$ and $-1 + (-3) = -4$.

(4) Split the middle term: $-4x = -1x - 3x$

(5) Factor by grouping:
$$3x^2 - 4x + 1 = 3x^2 - x - 3x + 1$$
$$= x(3x-1) - 1(3x-1)$$
$$= (x-1)(3x-1)$$

23. $4x^2 - 4x - 15$

(1) First factor out a common factor, if any. There is none (other than 1 or -1).

(2) Multiply the leading coefficient, 4, and the constant, -15: $4(-15) = -60$.

(3) Look for a factorization of -60 in which the sum of the factors is the coefficient of the middle term, -4.

Pairs of factors	Sums of factors
−1, 60	59
1, −60	−59
−2, 30	28
2, −30	−28
−3, 20	17
3, −20	−17
−4, 15	11
4, −15	−11
−5, 12	7
5, −12	−7
−6, 10	4
6, −10	−4

(4) Split the middle term: $-4x = 6x - 10x$

(5) Factor by grouping:
$$4x^2 - 4x - 15 = 4x^2 + 6x - 10x - 15$$
$$= 2x(2x + 3) - 5(2x + 3)$$
$$= (2x - 5)(2x + 3)$$

25. $2x^2 + x - 1$

(1) First factor out a common factor, if any. There is none (other than 1 or −1).

(2) Multiply the leading coefficient, 2, and the constant, −1: $2(-1) = -2$.

(3) Look for a factorization of −2 in which the sum of the factors is the coefficient of the middle term, 1. The numbers we want are 2 and −1: $2(-1) = -2$ and $2 - 1 = 1$.

(4) Split the middle term: $x = 2x - 1x$

(5) Factor by grouping:
$$2x^2 + x - 1 = 2x^2 + 2x - x - 1$$
$$= 2x(x + 1) - 1(x + 1)$$
$$= (2x - 1)(x + 1)$$

27. $9x^2 - 18x - 16$

(1) First factor out a common factor, if any. There is none (other than 1 or −1).

(2) Multiply the leading coefficient, 9, and the constant, −16: $9(-16) = -144$.

(3) Look for a factorization of −144, so the sum of the factors is the coefficient of the middle term, −18.

Pairs of factors	Sums of factors
−1, 144	143
1, −144	−143
−2, 72	70
2, −72	−70
−3, 48	45
3, −48	−45
−4, 36	32
4, −36	−32
−6, 24	18
6, −24	−18
−8, 18	10
8, −18	−10
−9, 16	7
9, −16	−7
−12, 12	0

(4) Split the middle term: $-18x = 6x - 24x$

(5) Factor by grouping:
$$9x^2 - 18x - 16 = 9x^2 + 6x - 24x - 16$$
$$= 3x(3x + 2) - 8(3x + 2)$$
$$= (3x - 8)(3x + 2)$$

29. $3x^2 + 5x - 2$

(1) First factor out a common factor, if any. There is none (other than 1 or −1).

(2) Multiply the leading coefficient, 3, and the constant, −2: $3(-2) = -6$.

(3) Look for a factorization of −6 in which the sum of the factors is the coefficient of the middle term, 5. The numbers we want are −1 and 6: $-1(6) = -6$ and $-1 + 6 = 5$.

(4) Split the middle term: $5x = -1x + 6x$

(5) Factor by grouping:
$$3x^2 + 5x - 2 = 3x^2 - x + 6x - 2$$
$$= x(3x - 1) + 2(3x - 1)$$
$$= (x + 2)(3x - 1)$$

31. $12x^2 - 31x + 20$

(1) First factor out a common factor, if any. There is none (other than 1 or −1).

(2) Multiply the leading coefficient, 12, and the constant, 20: $12 \cdot 20 = 240$.

(3) Look for a factorization of 240 in which the sum of the factors is the coefficient of the middle term, −31. We only need to consider negative factors.

Pairs of factors	Sums of factors
$-1, -240$	-241
$-2, -120$	-122
$-3, -8$	-83
$-4, -60$	-64
$-5, -48$	-53
$-6, -40$	-46
$-8, -30$	-38
$-10, -24$	-34
$-12, -20$	-32
$-15, -16$	-31

(4) Split the middle term: $-31x = -15x - 16x$

(5) Factor by grouping:
$$12x^2 - 31x + 20 = 12x^2 - 15x - 16x + 20$$
$$= 3x(4x - 5) - 4(4x - 5)$$
$$= (3x - 4)(4x - 5)$$

33. $14x^2 - 19x - 3$

(1) First factor out a common factor, if any. There is none (other than 1 or -1).

(2) Multiply the leading coefficient, 14, and the constant, -3: $14(-3) = -42$.

(3) Look for a factorization of -42 so that the sum of the factors is the coefficient of the middle term, -19.

Pairs of factors	Sums of factors
$-1, 42$	41
$1, -42$	-41
$-2, 21$	19
$2, -21$	-19
$-3, 14$	11
$3, -14$	-11
$-6, 7$	1
$6, -7$	-1

(4) Split the middle term: $-19x = 2x - 21x$

(5) Factor by grouping:
$$14x^2 - 19x - 3 = 14x^2 + 2x - 21x - 3$$
$$= 2x(7x + 1) - 3(7x + 1)$$
$$= (2x - 3)(7x + 1)$$

35. $9x^2 + 18x + 8$

(1) First factor out a common factor, if any. There is none (other than 1 or -1).

(2) Multiply the leading coefficient, 9, and the constant, 8: $9 \cdot 8 = 72$.

(3) Look for a factorization of 72 in which the sum of the factors is the coefficient of the middle term, 18. We only need to consider positive factors.

Pairs of factors	Sums of factors
$1, 72$	73
$2, 36$	38
$3, 24$	27
$4, 18$	22
$6, 12$	18
$8, 9$	17

(4) Split the middle term: $18x = 6x + 12x$

(5) Factor by grouping:
$$9x^2 + 18x + 8 = 9x^2 + 6x + 12x + 8$$
$$= 3x(3x + 2) + 4(3x + 2)$$
$$= (3x + 4)(3x + 2)$$

37. $49 - 42x + 9x^2 = 9x^2 - 42x + 49$

(1) First factor out a common factor, if any. There is none (other than 1 or -1).

(2) Multiply the leading coefficient, 9, and the constant, 49: $9 \cdot 49 = 441$.

(3) Look for a factorization of 441 in which the sum of the factors is the coefficient of the middle term, -42. We only need to consider negative factors.

Pairs of factors	Sums of factors
$-1, -441$	-442
$-3, -147$	-150
$-7, -63$	-70
$-9, -49$	-58
$-21, -21$	-42

(4) Split the middle term: $-42x = -21x - 21x$

(5) Factor by grouping:
$$9x^2 - 42x + 49 = 9x^2 - 21x - 21x + 49$$
$$= 3x(3x - 7) - 7(3x - 7)$$
$$= (3x - 7)(3x - 7), \text{ or}$$
$$(3x - 7)^2$$

39. $24x^2 - 47x - 2$

(1) First factor out a common factor, if any. There is none (other than 1 or -1).

(2) Multiply the leading coefficient, 24, and the constant, -2: $24(-2) = -48$.

(3) Look for a factorization of -48 in which the sum of the factors is the coefficient of the middle term, -47. The numbers we want are -48 and 1: $-48 \cdot 1 = -48$ and $-48 + 1 = -47$.

(4) Split the middle term: $-47x = -48x + 1x$

(5) Factor by grouping:
$$24x^2 - 47x - 2 = 24x^2 - 48x + x - 2$$
$$= 24x(x - 2) + 1(x - 2)$$
$$= (24x + 1)(x - 2)$$

41. $5 - 9a^2 - 12a = -9a^2 - 12a + 5 = -1(9a^2 + 12a - 5)$

Now we factor $9a^2 + 12a - 5$.

(1) We have already factored out the common factor, -1, to make the leading coefficient positive.

(2) Multiply the leading coefficient, 9, and the constant, -5: $9(-5) = -45$.

(3) Look for a factorization of -45 in which the sum of the factors is the coefficient of the middle term, 12. The numbers we want are 15 and -3: $15(-3) = -45$ and $15 + (-3) = 12$.

(4) Split the middle term: $12a = 15a - 3a$

(5) Factor by grouping:
$$9a^2 + 12a - 5 = 9a^2 + 15a - 3a - 5$$
$$= 3a(3a + 5) - (3a + 5)$$
$$= (3a - 1)(3a + 5)$$

Then we have
$$5 - 9a^2 - 12a$$
$$= -1(3a - 1)(3a + 5)$$
$$= (-3a + 1)(3a + 5) \qquad \text{Multiplying } 3a-1 \text{ by } -1$$
$$= (3a - 1)(-3a - 5) \qquad \text{Multiplying } 3a+5 \text{ by } -1$$

Note that we can also express $(-3a + 1)(3a + 5)$ as $(1 - 3a)(3a + 5)$ since $-3a + 1 = 1 - 3a$ by the commutative law of addition.

43. $20 + 6x - 2x^2 = -2x^2 + 6x + 20$

(1) Factor out the common factor -2. We factor out -2 rather than 2 in order to make the leading coefficient of the trinomial factor positive.
$$-2x^2 + 6x + 20 = -2(x^2 - 3x - 10)$$

To factor $x^2 - 3x - 10$, we look for two factors of -10 whose sum is -3. The numbers we want are -5 and 2. Then $x^2 - 3x - 10 = (x - 5)(x + 2)$, so we have:
$$20 + 6x - 2x^2$$
$$= -2(x - 5)(x + 2)$$
$$= 2(-x + 5)(x + 2) \qquad \text{Multiplying } x - 5 \text{ by } -1$$
$$= 2(x - 5)(-x - 2) \qquad \text{Multiplying } x + 2 \text{ by } -1$$

Note that we can also express $2(-x + 5)(x + 2)$ as $2(5 - x)(x + 2)$ since $-x + 5 = 5 - x$ by the commutative law of addition.

45. $12x^2 + 28x - 24$

(1) Factor out the common factor, 4:
$$12x^2 + 28x - 24 = 4(3x^2 + 7x - 6)$$

(2) Now we factor the trinomial $3x^2 + 7x - 6$. Multiply the leading coefficient, 3, and the constant, -6: $3(-6) = -18$.

(3) Look for a factorization of -18 in which the sum of the factors is the coefficient of the middle term, 7. The numbers we want are 9 and -2: $9(-2) = -18$ and $9 + (-2) = 7$.

(4) Split the middle term: $7x = 9x - 2x$

(5) Factor by grouping:
$$3x^2 + 7x - 6 = 3x^2 + 9x - 2x - 6$$
$$= 3x(x + 3) - 2(x + 3)$$
$$= (3x - 2)(x + 3)$$

We must include the common factor to get a factorization of the original trinomial.
$$12x^2 + 28x - 24 = 4(3x - 2)(x + 3)$$

47. $30x^2 - 24x - 54$

(1) Factor out the common factor, 6.
$$30x^2 - 24x - 54 = 6(5x^2 - 4x - 9)$$

(2) Now we factor the trinomial $5x^2 - 4x - 9$. Multiply the leading coefficient, 5, and the constant, -9: $5(-9) = -45$.

(3) Look for a factorization of -45 in which the sum of the factors is the coefficient of the middle term, -4. The numbers we want are -9 and 5: $-9 \cdot 5 = -45$ and $-9 + 5 = -4$.

(4) Split the middle term: $-4x = -9x + 5x$

(5) Factor by grouping:
$$5x^2 - 4x - 9 = 5x^2 - 9x + 5x - 9$$
$$= x(5x - 9) + (5x - 9)$$
$$= (x + 1)(5x - 9)$$

We must include the common factor to get a factorization of the original trinomial.
$$30x^2 - 24x - 54 = 6(x + 1)(5x - 9)$$

49. $4y + 6y^2 - 10 = 6y^2 + 4y - 10$

(1) Factor out the common factor, 2.
$$6y^2 + 4y - 10 = 2(3y^2 + 2y - 5)$$

(2) Now we factor the trinomial $3y^2 + 2y - 5$. Multiply the leading coefficient, 3, and the constant, -5: $3(-5) = -15$.

(3) Look for a factorization of -15 in which the sum of the factors is the coefficient of the middle term, 2. The numbers we want are 5 and -3: $5(-3) = -15$ and $5 + (-3) = 2$.

(4) Split the middle term: $2y = 5y - 3y$

(5) Factor by grouping:
$$3y^2 + 2y - 5 = 3y^2 + 5y - 3y - 5$$
$$= y(3y + 5) - (3y + 5)$$
$$= (y - 1)(3y + 5)$$

We must include the common factor to get a factorization of the original trinomial.
$$4y + 6y^2 - 10 = 2(y - 1)(3y + 5)$$

51. $3x^2 - 4x + 1$

(1) There is no common factor (other than 1 or -1).

(2) Multiply the leading coefficient, 3, and the constant, 1: $3 \cdot 1 = 3$.

(3) Look for a factorization of 3 in which the sum of the factors is the coefficient of the middle term, -4. The numbers we want are -1 and -3: $-1(-3) = 3$ and $-1 + (-3) = -4$.

(4) Split the middle term: $-4x = -1x - 3x$

(5) Factor by grouping:
$$3x^2 - 4x + 1 = 3x^2 - x - 3x + 1$$
$$= x(3x - 1) - (3x - 1)$$
$$= (x - 1)(3x - 1)$$

53. $12x^2 - 28x - 24$

(1) Factor out the common factor, 4:
$$12x^2 - 28x - 24 = 4(3x^2 - 7x - 6)$$

(2) Now we factor the trinomial $3x^2 - 7x - 6$. Multiply the leading coefficient, 3, and the constant, -6: $3(-6) = -18$.

(3) Look for a factorization of -18 in which the sum of the factors is the coefficient of the middle term, -7. The numbers we want are -9 and 2: $-9 \cdot 2 = -18$ and $-9 + 2 = -7$.

(4) Split the middle term: $-7x = -9x + 2x$

(5) Factor by grouping:
$$3x^2 - 7x - 6 = 3x^2 - 9x + 2x - 6$$
$$= 3x(x - 3) + 2(x - 3)$$
$$= (3x + 2)(x - 3)$$

We must include the common factor to get a factorization of the original trinomial.
$$12x^2 - 28x - 24 = 4(3x + 2)(x - 3)$$

55. $-1 + 2x^2 - x = 2x^2 - x - 1$

(1) There is no common factor (other than 1 or -1).

(2) Multiply the leading coefficient, 2, and the constant, -1: $2(-1) = -2$.

(3) Look for a factorization of -2 in which the sum of the factors is the coefficient of the middle term, -1. The numbers we want are -2 and 1: $-2 \cdot 1 = -2$ and $-2 + 1 = -1$.

(4) Split the middle term: $-x = -2x + 1x$

(5) Factor by grouping:
$$2x^2 - x - 1 = 2x^2 - 2x + x - 1$$
$$= 2x(x - 1) + (x - 1)$$
$$= (2x + 1)(x - 1)$$

57. $9x^2 + 18x - 16$

(1) There is no common factor (other than 1 or -1).

(2) Multiply the leading coefficient, 9, and the constant, -16: $9(-16) = -144$.

(3) Look for a factorization of -144 in which the sum of the factors is the coefficient of the middle term, 18. The numbers we want are 24 and -6: $24(-6) = -144$ and $24 + (-6) = 18$.

(4) Split the middle term: $18x = 24x - 6x$

(5) Factor by grouping:
$$9x^2 + 18x - 16 = 9x^2 + 24x - 6x - 16$$
$$= 3x(3x + 8) - 2(3x + 8)$$
$$= (3x - 2)(3x + 8)$$

59. $15x^2 - 25x - 10$

(1) Factor out the common factor, 5:
$$15x^2 - 25x - 10 = 5(3x^2 - 5x - 2)$$

(2) Now we factor the trinomial $3x^2 - 5x - 2$. Multiply the leading coefficient, 3, and the constant, -2: $3(-2) = -6$.

(3) Look for a factorization of -6 in which the sum of the factors is the coefficient of the middle term, -5. The numbers we want are -6 and 1: $-6 \cdot 1 = -6$ and $-6 + 1 = -5$.

(4) Split the middle term: $-5x = -6x + 1x$

(5) Factor by grouping:
$$3x^2 - 5x - 2 = 3x^2 - 6x + x - 2$$
$$= 3x(x - 2) + (x - 2)$$
$$= (3x + 1)(x - 2)$$

We must include the common factor to get a factorization of the original trinomial.
$$15x^2 - 25x - 10 = 5(3x + 1)(x - 2)$$

61. $12p^3 + 31p^2 + 20p$

(1) Factor out the common factor, p:
$$12p^3 + 31p^2 + 20p = p(12p^2 + 31p + 20)$$

(2) Now we factor the trinomial $12p^2 + 31p + 20$. Multiply the leading coefficient, 12, and the constant, 20: $12 \cdot 20 = 240$.

(3) Look for a factorization of 240 in which the sum of the factors is the coefficient of the middle term, 31. The numbers we want are 15 and 16: $15 \cdot 16 = 240$ and $15 + 16 = 31$.

(4) Split the middle term: $31p = 15p + 16p$

(5) Factor by grouping:
$$12p^2 + 31p + 20 = 12p^2 + 15p + 16p + 20$$
$$= 3p(4p + 5) + 4(4p + 5)$$
$$= (3p + 4)(4p + 5)$$

We must include the common factor to get a factorization of the original trinomial.
$$12p^3 + 31p^2 + 20p = p(3p + 4)(4p + 5)$$

63. $4 - x - 5x^2 = -5x^2 - x + 4$

(1) Factor out -1 to make the leading coefficient positive:
$$-5x^2 - x + 4 = -1(5x^2 + x - 4)$$

(2) Now we factor the trinomial $5x^2 + x - 4$. Multiply the leading coefficient, 5, and the constant, -4: $5(-4) = -20$.

(3) Look for a factorization of -20 in which the sum of the factors is the coefficient of the middle term, 1. The numbers we want are 5 and -4: $5(-4) = -20$ and $5 + (-4) = 1$.

(4) Split the middle term: $x = 5x - 4x$

(5) Factor by grouping:
$$5x^2 + x - 4 = 5x^2 + 5x - 4x - 4$$
$$= 5x(x+1) - 4(x+1)$$
$$= (5x - 4)(x + 1)$$

We must include the common factor to get a factorization of the original trinomial.
$$4 - x - 5x^2$$
$$= -1(5x - 4)(x + 1)$$
$$= (-5x + 4)(x + 1) \qquad \text{Multiplying } 5x - 4 \text{ by } -1$$
$$= (5x - 4)(-x - 1) \qquad \text{Multiplying } x + 1 \text{ by } -1$$

Note that we can also express $(-5x + 4)(x + 1)$ as $(4 - 5x)(x + 1)$ since $-5x + 4 = 4 - 5x$ by the commutative law of addition.

65. $33t - 15 - 6t^2 = -6t^2 + 33t - 15$

(1) Factor out the common factor, -3. We factor out -3 rather than 3 in order to make the leading coefficient of the trinomial factor positive.
$$-6t^2 + 33t - 15 = -3(2t^2 - 11t + 5)$$

(2) Now we factor the trinomial $2t^2 - 11t + 5$. Multiply the leading coefficient, 2, and the constant, 5: $2 \cdot 5 = 10$.

(3) Look for a factorization of 10 in which the sum of the factors is the coefficient of the middle term, -11. The numbers we want are -1 and -10: $-1(-10) = 10$ and $-1 + (-10) = -11$.

(4) Split the middle term: $-11t = -1t - 10t$

(5) Factor by grouping:
$$2t^2 - 11t + 5 = 2t^2 - t - 10t + 5$$
$$= t(2t - 1) - 5(2t - 1)$$
$$= (t - 5)(2t - 1)$$

We must include the common factor to get a factorization of the original trinomial.
$$33t - 15 - 6t^2$$
$$= -3(t - 5)(2t - 1)$$
$$= 3(-t + 5)(2t - 1) \qquad \text{Multiplying } t - 5 \text{ by } -1$$
$$= 3(t - 5)(-2t + 1) \qquad \text{Multiplying } 2t - 1 \text{ by } -1$$

Note that we can also express $3(-t + 5)(2t - 1)$ as $3(5 - t)(2t - 1)$ since $-t + 5 = 5 - t$ by the commutative law of addition. Similarly, we can express $3(t - 5)(-2t + 1)$ as $3(t - 5)(1 - 2t)$.

67. $14x^4 + 19x^3 - 3x^2$

(1) Factor out the common factor, x^2:
$$14x^4 + 19x^3 - 3x^2 = x^2(14x^2 + 19x - 3)$$

(2) Now we factor the trinomial $14x^2 + 19x - 3$. Multiply the leading coefficient, 14, and the constant, -3: $14(-3) = -42$.

(3) Look for a factorization of -42 in which the sum of the factors is the coefficient of the middle term, 19. The numbers we want are 21 and -2: $21(-2) = -42$ and $21 + (-2) = 19$.

(4) Split the middle term: $19x = 21x - 2x$

(5) Factor by grouping:
$$14x^2 + 19x - 3 = 14x^2 + 21x - 2x - 3$$
$$= 7x(2x + 3) - (2x + 3)$$
$$= (7x - 1)(2x + 3)$$

We must include the common factor to get a factorization of the original trinomial.
$$14x^4 + 19x^3 - 3x^2 = x^2(7x - 1)(2x + 3)$$

69. $168x^3 - 45x^2 + 3x$

(1) Factor out the common factor, $3x$:
$$168x^3 - 45x^2 + 3x = 3x(56x^2 - 15x + 1)$$

(2) Now we factor the trinomial $56x^2 - 15x + 1$. Multiply the leading coefficient, 56, and the constant, 1: $56 \cdot 1 = 56$.

(3) Look for a factorization of 56 in which the sum of the factors is the coefficient of the middle term, -15. The numbers we want are -7 and -8: $-7(-8) = 56$ and $-7 + (-8) = -15$.

(4) Split the middle term: $-15x = -7x - 8x$

(5) Factor by grouping:
$$56x^2 - 15x + 1 = 56x^2 - 7x - 8x + 1$$
$$= 7x(8x - 1) - (8x - 1)$$
$$= (7x - 1)(8x - 1)$$

We must include the common factor to get a factorization of the original trinomial.
$$168x^3 - 45x^2 + 3x = 3x(7x - 1)(8x - 1)$$

71. $15x^4 - 19x^2 + 6$

(1) There are no common factors (other than 1 or -1).

(2) Multiply the leading coefficient, 15, and the constant, 6: $15 \cdot 6 = 90$.

(3) Look for a factorization of 90 in which the sum of the factors is the coefficient of the middle term, -19. The numbers we want are -9 and -10: $-9(-10) = 90$ and $-9 + (-10) = -19$.

(4) Split the middle term: $-19x^2 = -9x^2 - 10x^2$

(5) Factor by grouping:
$$15x^4 - 19x^2 + 6 = 15x^4 - 9x^2 - 10x^2 + 6$$
$$= 3x^2(5x^2 - 3) - 2(5x^2 - 3)$$
$$= (3x^2 - 2)(5x^2 - 3)$$

73. $25t^2 + 80t + 64$

(1) There are no common factors (other than 1 or -1).

(2) Multiply the leading coefficient, 25, and the constant, 64: $25 \cdot 64 = 1600$.

(3) Look for a factorization of 1600 in which the sum of the factors is the coefficient of the middle term, 80. The numbers we want are 40 and 40: $40 \cdot 40 = 1600$ and $40 + 40 = 80$.

(4) Split the middle term: $80t = 40t + 40t$

(5) Factor by grouping:
$$25t^2 + 80t + 64 = 25t^2 + 40t + 40t + 64$$
$$= 5t(5t + 8) + 8(5t + 8)$$
$$= (5t + 8)(5t + 8), \text{ or}$$
$$(5t + 8)^2$$

75. $6x^3 + 4x^2 - 10x$

(1) Factor out the common factor, $2x$:
$$6x^3 + 4x^2 - 10x = 2x(3x^2 + 2x - 5)$$

(2) - (5) Now we factor the trinomial $3x^2 + 2x - 5$. We did this in Exercise 49, using the variable y rather than x. We found that $3x^2 + 2x - 5 = (x-1)(3x+5)$. We must include the common factor to get a factorization of the original trinomial.
$$6x^3 + 4x^2 - 10x = 2x(x - 1)(3x + 5)$$

77. $25x^2 + 79x + 64$

(1) There are no common factors (other than 1 or -1).

(2) Multiply the leading coefficient, 25, and the constant, 64: $25 \cdot 64 = 1600$.

(3) Look for a factorization of 1600 in which the sum of the factors is the coefficient of the middle term, 79. It is not possible to find such a pair of numbers. Thus, $25x^2 + 79x + 64$ cannot be factored into a product of binomial factors. It is prime.

79. $6x^2 - 19x - 5$

(1) There are no common factors (other than 1 or -1).

(2) Multiply the leading coefficient, 6, and the constant, -5: $6(-5) = -30$.

(3) Look for a factorization of -30 in which the sum of the factors is the coefficient of the middle term, -19. There is no such pair of numbers. Thus, $6x^2 - 19x - 5$ cannot be factored into a product of binomial factors. It is prime.

81. $12m^2 - mn - 20n^2$

(1) There are no common factors (other than 1 or -1).

(2) Multiply the leading coefficient, 12, and the constant, -20: $12(-20) = -240$.

(3) Look for a factorization of -240 in which the sum of the factors is the coefficient of the middle term, -1. The numbers we want are 15 and -16: $15(-16) = -240$ and $15 + (-16) = -1$.

(4) Split the middle term: $-mn = 15mn - 16mn$

(5) Factor by grouping:
$$12m^2 - mn - 20n^2$$
$$= 12m^2 + 15mn - 16mn - 20n^2$$
$$= 3m(4m + 5n) - 4n(4m + 5n)$$
$$= (3m - 4n)(4m + 5n)$$

83. $6a^2 - ab - 15b^2$

(1) There are no common factors (other than 1 or -1).

(2) Multiply the leading coefficient, 6, and the constant, -15: $6(-15) = -90$.

(3) Look for a factorization of -90 in which the sum of the factors is the coefficient of the middle term, -1. The numbers we want are -10 and 9: $-10 \cdot 9 = -90$ and $-10 + 9 = -1$.

(4) Split the middle term: $-ab = -10ab + 9ab$

(5) Factor by grouping:
$$6a^2 - ab - 15b^2 = 6a^2 - 10ab + 9ab - 15b^2$$
$$= 2a(3a - 5b) + 3b(3a - 5b)$$
$$= (2a + 3b)(3a - 5b)$$

85. $9a^2 - 18ab + 8b^2$

(1) There are no common factors (other than 1 or -1).

(2) Multiply the leading coefficient, 9, and the constant, 8: $9 \cdot 8 = 72$.

(3) Look for a factorization of 72 in which the sum of the factors is the coefficient of the middle term, -18. The numbers we want are -6 and -12: $-6(-12) = 72$ and $-6 + (-12) = -18$.

(4) Split the middle term: $-18ab = -6ab - 12ab$

(5) Factor by grouping:
$$9a^2 - 18ab + 8b^2 = 9a^2 - 6ab - 12ab + 8b^2$$
$$= 3a(3a - 2b) - 4b(3a - 2b)$$
$$= (3a - 4b)(3a - 2b)$$

87. $35p^2 + 34pq + 8q^2$

(1) There are no common factors (other than 1 or -1).

(2) Multiply the leading coefficient, 35, and the constant, 8: $35 \cdot 8 = 280$.

(3) Look for a factorization of 280 in which the sum of the factors is the coefficient of the middle term, 34. The numbers we want are 14 and 20: $14 \cdot 20 = 280$ and $14 + 20 = 34$.

(4) Split the middle term: $34pq = 14pq + 20pq$

(5) Factor by grouping:
$$35p^2 + 34pq + 8q^2 = 35p^2 + 14pq + 20pq + 8q^2$$
$$= 7p(5p + 2q) + 4q(5p + 2q)$$
$$= (7p + 4q)(5p + 2q)$$

89. $18x^2 - 6xy - 24y^2$

(1) Factor out the common factor, 6.
$$18x^2 - 6xy - 24y^2 = 6(3x^2 - xy - 4y^2)$$

(2) Now we factor the trinomial $3x^2 - xy - 4y^2$. Multiply the leading coefficient, 3, and the constant, -4: $3(-4) = -12$.

(3) Look for a factorization of -12 in which the sum of the factors is the coefficient of the middle term, -1. The numbers we want are -4 and 3: $-4 \cdot 3 = -12$ and $-4 + 3 = -1$.

(4) Split the middle term: $-xy = -4xy + 3xy$

(5) Factor by grouping:
$$3x^2 - xy - 4y^2 = 3x^2 - 4xy + 3xy - 4y^2$$
$$= x(3x - 4y) + y(3x - 4y)$$
$$= (x + y)(3x - 4y)$$

We must include the common factor to get a factorization of the original trinomial.
$$18x^2 - 6xy - 24y^2 = 6(x + y)(3x - 4y)$$

91. $60x + 18x^2 - 6x^3 = -6x^3 + 18x^2 + 60x$

(1) Factor out the common factor, $-6x$. We factor out $-6x$ rather than $6x$ in order to have a positive leading coefficient in the trinomial factor.
$$-6x^3 + 18x^2 + 60x = -6x(x^2 - 3x - 10)$$

(2) - (5) We factor $x^2 - 3x - 10$ as we did in Exercise 43, getting the $(x - 5)(x + 2)$. Then we have:
$$60x + 18x^2 - 6x^3$$
$$= -6x(x - 5)(x + 2)$$
$$= 6x(-x + 5)(x + 2)$$
$$\text{Multiplying } x - 5 \text{ by } -1$$
$$= 6x(x - 5)(-x - 2)$$

Note that we can express $6x(-x + 5)(x + 2)$ as $6x(5 - x)(x + 2)$ since $-x + 5 = 5 - x$ by the commutative law of addition.

93. $35x^5 - 57x^4 - 44x^3$

(1) We first factor out the common factor, x^3.
$$x^3(35x^2 - 57x - 44)$$

(2) Now we factor the trinomial $35x^2 - 57x - 44$. Multiply the leading coefficient, 35, and the constant, -44: $35(-44) = -1540$.

(3) Look for a factorization of -1540 in which the sum of the factors is the coefficient of the middle term, -57.

Pairs of factors	Sums of factors
$7, -220$	-213
$10, -154$	-144
$11, -140$	-129
$14, -110$	-96
$20,\ -77$	-57

(4) Split the middle term: $-57x = 20x - 77x$

(5) Factor by grouping: ·
$$35x^2 - 57x - 44 = 35x^2 + 20x - 77x - 44$$
$$= 5x(7x + 4) - 11(7x + 4)$$
$$= (5x - 11)(7x + 4)$$

We must include the common factor to get a factorization of the original trinomial.
$$35x^5 - 57x^4 - 44x^3 = x^3(5x - 11)(7x + 4)$$

95. Discussion and Writing Exercise

97. $-10x > 1000$
$$\frac{-10x}{-10} < \frac{1000}{-10} \quad \text{Dividing by } -10 \text{ and reversing the inequality symbol}$$
$$x < -100$$
The solution set is $\{x | x < -100\}$.

99. $6 - 3x \geq -18$
$$-3x \geq -24 \quad \text{Subtracting 6}$$
$$x \leq 8 \quad \text{Dividing by } -3 \text{ and reversing the inequality symbol}$$
The solution set is $\{x | x \leq 8\}$.

101.
$$\frac{1}{2}x - 6x + 10 \leq x - 5x$$
$$2\left(\frac{1}{2}x - 6x + 10\right) \leq 2(x - 5x) \quad \text{Multiplying by 2 to clear the fraction}$$
$$x - 12x + 20 \leq 2x - 10x$$
$$-11x + 20 \leq -8x \quad \text{Collecting like terms}$$
$$20 \leq 3x \quad \text{Adding } 11x$$
$$\frac{20}{3} \leq x \quad \text{Dividing by 3}$$
The solution set is $\left\{x | x \geq \frac{20}{3}\right\}$.

103. $3x - 6x + 2(x - 4) > 2(9 - 4x)$
$$3x - 6x + 2x - 8 > 18 - 8x \quad \text{Removing parentheses}$$
$$-x - 8 > 18 - 8x \quad \text{Collecting like terms}$$
$$7x > 26 \quad \text{Adding } 8x \text{ and } 8$$
$$x > \frac{26}{7} \quad \text{Dividing by 7}$$
The solution set is $\left\{x | x > \frac{26}{7}\right\}$.

105. *Familiarize*. We will use the formula $C = 2\pi r$, where C is circumference and r is radius, to find the radius in kilometers. Then we will multiply that number by 0.62 to find the radius in miles.

***Translate*.**
$$\underbrace{\text{Circumference}}_{40,000} = \underbrace{2 \cdot \pi \cdot \text{radius}}_{2(3.14)r}$$
$$40,000 \approx 2(3.14)r$$

***Solve*.** First we solve the equation.
$$40,000 \approx 2(3.14)r$$
$$40,000 \approx 6.28r$$
$$6369 \approx r$$

Then we multiply to find the radius in miles:
$$6369(0.62) \approx 3949$$

***Check*.** If $r = 6369$, then $2\pi r = 2(3.14)(6369) \approx 40,000$. We should also recheck the multiplication we did to find the radius in miles. Both values check.

***State*.** The radius of the earth is about 6369 km or 3949 mi. (These values may differ slightly if a different approximation is used for π.)

107. $9x^{10} - 12x^5 + 4$

(a) First factor out a common factor, if any. There is none (other than 1 or −1).

(b) Multiply the leading coefficient, 9, and the constant, 4: $9 \cdot 4 = 36$.

(c) Look for a factorization of 36 in which the sum of the factors is the coefficient of the middle term, −12. The factors we want are −6 and −6.

(d) Split the middle term: $-12x^5 = -6x^5 - 6x^5$

(e) Factor by grouping:
$$9x^{10} - 12x^5 + 4 = 9x^{10} - 6x^5 - 6x^5 + 4$$
$$= 3x^5(3x^5 - 2) - 2(3x^5 - 2)$$
$$= (3x^5 - 2)(3x^5 - 2), \text{ or}$$
$$= (3x^5 - 2)^2$$

109. $16x^{10} + 8x^5 + 1$

(a) First factor out a common factor, if any. There is none (other than 1 or −1).

(b) Multiply the leading coefficient, 16, and the constant, 1: $16 \cdot 1 = 16$.

(c) Look for a factorization of 16 in which the sum of the factors is the coefficient of the middle term, 8. The factors we want are 4 and 4.

(d) Split the middle term: $8x^5 = 4x^5 + 4x^5$

(e) Factor by grouping:
$$16x^{10} + 8x^5 + 1 = 16x^{10} + 4x^5 + 4x^5 + 1$$
$$= 4x^5(4x^5 + 1) + 1(4x^5 + 1)$$
$$= (4x^5 + 1)(4x^5 + 1), \text{ or}$$
$$= (4x^5 + 1)^2$$

111.-119. Left to the student

Exercise Set 10.5

1. $x^2 - 14x + 49$

(a) We know that x^2 and 49 are squares.

(b) There is no minus sign before either x^2 or 49.

(c) If we multiply the square roots, x and 7, and double the product, we get $2 \cdot x \cdot 7 = 14x$. This is the opposite of the remaining term, $-14x$.

Thus, $x^2 - 14x + 49$ is a trinomial square.

3. $x^2 + 16x - 64$

Both x^2 and 64 are squares, but there is a minus sign before 64. Thus, $x^2 + 16x - 64$ is not a trinomial square.

5. $x^2 - 2x + 4$

(a) Both x^2 and 4 are squares.

(b) There is no minus sign before either x^2 or 4.

(c) If we multiply the square roots, x and 2, and double the product, we get $2 \cdot x \cdot 2 = 4x$. This is neither the remaining term nor its opposite.

Thus, $x^2 - 2x + 4$ is not a trinomial square.

7. $9x^2 - 36x + 24$

Only one term is a square. Thus, $9x^2 - 36x + 24$ is not a trinomial square.

9. $x^2 - 14x + 49 = x^2 - 2 \cdot x \cdot 7 + 7^2 = (x - 7)^2$
$$= A^2 - 2 \; A \; B + B^2 = (A - B)^2$$

11. $x^2 + 16x + 64 = x^2 + 2 \cdot x \cdot 8 + 8^2 = (x + 8)^2$
$$= A^2 + 2 \; A \; B + B^2 = (A + B)^2$$

13. $x^2 - 2x + 1 = x^2 - 2 \cdot x \cdot 1 + 1^2 = (x - 1)^2$

15. $4 + 4x + x^2 = x^2 + 4x + 4$ (Changing the order)
$$= x^2 + 2 \cdot x \cdot 2 + 2^2$$
$$= (x + 2)^2$$

17. $q^4 - 6q^2 + 9 = (q^2)^2 - 2 \cdot q^2 \cdot 3 + 3^2 = (q^2 - 3)^2$

19. $49 + 56y + 16y^2 = 16y^2 + 56y + 49$
$$= (4y)^2 + 2 \cdot 4y \cdot 7 + 7^2$$
$$= (4y + 7)^2$$

21. $2x^2 - 4x + 2 = 2(x^2 - 2x + 1)$
$$= 2(x^2 - 2 \cdot x \cdot 1 + 1^2)$$
$$= 2(x - 1)^2$$

23. $x^3 - 18x^2 + 81x = x(x^2 - 18x + 81)$
$$= x(x^2 - 2 \cdot x \cdot 9 + 9^2)$$
$$= x(x - 9)^2$$

25. $12q^2 - 36q + 27 = 3(4q^2 - 12q + 9)$
$$= 3[(2q)^2 - 2 \cdot 2q \cdot 3 + 3^2]$$
$$= 3(2q - 3)^2$$

27. $49 - 42x + 9x^2 = 7^2 - 2 \cdot 7 \cdot 3x + (3x)^2$
$$= (7 - 3x)^2$$

29. $5y^4 + 10y^2 + 5 = 5(y^4 + 2y^2 + 1)$
$$= 5[(y^2)^2 + 2 \cdot y^2 \cdot 1 + 1^2]$$
$$= 5(y^2 + 1)^2$$

31. $1 + 4x^4 + 4x^2 = 1^2 + 2 \cdot 1 \cdot 2x^2 + (2x^2)^2$
$$= (1 + 2x^2)^2$$

33. $4p^2 + 12pq + 9q^2 = (2p)^2 + 2 \cdot 2p \cdot 3q + (3q)^2$
$$= (2p + 3q)^2$$

35. $a^2 - 6ab + 9b^2 = a^2 - 2 \cdot a \cdot 3b + (3b)^2$
$$= (a - 3b)^2$$

37. $81a^2 - 18ab + b^2 = (9a)^2 - 2 \cdot 9a \cdot b + b^2$
$$= (9a - b)^2$$

39. $36a^2 + 96ab + 64b^2 = 4(9a^2 + 24ab + 16b^2)$
$$= 4[(3a)^2 + 2 \cdot 3a \cdot 4b + (4b)^2]$$
$$= 4(3a + 4b)^2$$

41. $x^2 - 4$

  (a) The first expression is a square: x^2

     The second expression is a square: $4 = 2^2$

  (b) The terms have different signs.

     $x^2 - 4$ is a difference of squares.

43. $x^2 + 25$

  The terms do not have different signs.

  $x^2 + 25$ is not a difference of squares.

45. $x^2 - 45$

  The number 45 is not a square.

  $x^2 - 45$ is not a difference of squares.

47. $16x^2 - 25y^2$

  (a) The first expression is a square: $16x^2 = (4x)^2$

     The second expression is a square: $25y^2 = (5y)^2$

  (b) The terms have different signs.

     $16x^2 - 25y^2$ is a difference of squares.

49. $y^2 - 4 = y^2 - 2^2 = (y + 2)(y - 2)$

51. $p^2 - 9 = p^2 - 3^2 = (p + 3)(p - 3)$

53. $-49 + t^2 = t^2 - 49 = t^2 - 7^2 = (t + 7)(t - 7)$

55. $a^2 - b^2 = (a + b)(a - b)$

57. $25t^2 - m^2 = (5t)^2 - m^2 = (5t + m)(5t - m)$

59. $100 - k^2 = 10^2 - k^2 = (10 + k)(10 - k)$

61. $16a^2 - 9 = (4a)^2 - 3^2 = (4a + 3)(4a - 3)$

63. $4x^2 - 25y^2 = (2x)^2 - (5y)^2 = (2x + 5y)(2x - 5y)$

65. $8x^2 - 98 = 2(4x^2 - 49) = 2[(2x)^2 - 7^2] =$
    $2(2x + 7)(2x - 7)$

67. $36x - 49x^3 = x(36 - 49x^2) = x[6^2 - (7x)^2] =$
    $x(6 + 7x)(6 - 7x)$

69. $49a^4 - 81 = (7a^2)^2 - 9^2 = (7a^2 + 9)(7a^2 - 9)$

71. $a^4 - 16$

  $= (a^2)^2 - 4^2$

  $= (a^2 + 4)(a^2 - 4)$ Factoring a difference of squares

  $= (a^2 + 4)(a + 2)(a - 2)$ Factoring further: $a^2 - 4$ is a difference of squares.

73. $5x^4 - 405$

    $5(x^4 - 81)$

  $= 5[(x^2)^2 - 9^2]$

  $= 5(x^2 + 9)(x^2 - 9)$

  $= 5(x^2 + 9)(x + 3)(x - 3)$ Factoring $x^2 - 9$

75. $1 - y^8$

  $= 1^2 - (y^4)^2$

  $= (1 + y^4)(1 - y^4)$

  $= (1 + y^4)(1 + y^2)(1 - y^2)$ Factoring $1 - y^4$

  $= (1 + y^4)(1 + y^2)(1 + y)(1 - y)$ Factoring $1 - y^2$

77. $x^{12} - 16$

  $= (x^6)^2 - 4^2$

  $= (x^6 + 4)(x^6 - 4)$

  $= (x^6 + 4)(x^3 + 2)(x^3 - 2)$ Factoring $x^6 - 4$

79. $y^2 - \dfrac{1}{16} = y^2 - \left(\dfrac{1}{4}\right)^2$

     $= \left(y + \dfrac{1}{4}\right)\left(y - \dfrac{1}{4}\right)$

81. $25 - \dfrac{1}{49}x^2 = 5^2 - \left(\dfrac{1}{7}x\right)^2$

     $= \left(5 + \dfrac{1}{7}x\right)\left(5 - \dfrac{1}{7}x\right)$

83. $16m^4 - t^4$

  $= (4m^2)^2 - (t^2)^2$

  $= (4m^2 + t^2)(4m^2 - t^2)$

  $= (4m^2 + t^2)(2m + t)(2m - t)$ Factoring $4m^2 - t^2$

85. Discussion and Writing exercise

87. $-110 \div 10$ The quotient of a negative number and a positive number is negative.

  $-110 \div 10 = -11$

89. $-\dfrac{2}{3} \div \dfrac{4}{5} = -\dfrac{2}{3} \cdot \dfrac{5}{4} = -\dfrac{10}{12} = -\dfrac{2 \cdot 5}{2 \cdot 6} = -\dfrac{\cancel{2} \cdot 5}{\cancel{2} \cdot 6} = -\dfrac{5}{6}$

91. $-64 \div (-32)$ The quotient of two negative numbers is a positive number.

  $-64 \div (-32) = 2$

93. The shaded region is a square with sides of length $x - y - y$, or $x - 2y$. Its area is $(x - 2y)(x - 2y)$, or $(x - 2y)^2$. Multiplying, we get the polynomial $x^2 - 4xy + 4y^2$.

95. $y^5 \cdot y^7 = y^{5+7} = y^{12}$

97. $y - 6x = 6$

  To find the x-intercept, let $y = 0$. Then solve for x.

    $y - 6x = 6$

    $0 - 6x = 6$

      $-6x = 6$

        $x = -1$

  The x-intercept is $(-1, 0)$.

  To find the y-intercept, let $x = 0$. Then solve for y.

    $y - 6x = 6$

    $y - 6 \cdot 0 = 6$

        $y = 6$

  The y-intercept is $(0, 6)$.

Plot these points and draw the line.

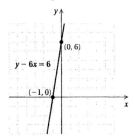

A third point should be used as a check. We substitute any value for x and solve for y. We let $x = -2$. Then

$$y - 6x = 6$$
$$y - 6(-2) = 6$$
$$y + 12 = 6$$
$$y = -6$$

The point $(-2, -6)$ is on the graph, so the graph is probably correct.

99. $49x^2 - 216$

There is no common factor. Also, $49x^2$ is a square, but 216 is not so this expression is not a difference of squares. It is not factorable. It is prime.

101. $x^2 + 22x + 121 = x^2 + 2 \cdot x \cdot 11 + 11^2$
$$= (x + 11)^2$$

103. $18x^3 + 12x^2 + 2x = 2x(9x^2 + 6x + 1)$
$$= 2x[(3x)^2 + 2 \cdot 3x \cdot 1 + 1^2]$$
$$= 2x(3x + 1)^2$$

105. $x^8 - 2^8$
$$= (x^4 + 2^4)(x^4 - 2^4)$$
$$= (x^4 + 2^4)(x^2 + 2^2)(x^2 - 2^2)$$
$$= (x^4 + 2^4)(x^2 + 2^2)(x + 2)(x - 2), \text{ or}$$
$$= (x^4 + 16)(x^2 + 4)(x + 2)(x - 2)$$

107. $3x^5 - 12x^3 = 3x^3(x^2 - 4) = 3x^3(x + 2)(x - 2)$

109. $18x^3 - \dfrac{8}{25}x = 2x\left(9x^2 - \dfrac{4}{25}\right) = 2x\left(3x + \dfrac{2}{5}\right)\left(3x - \dfrac{2}{5}\right)$

111. $0.49p - p^3 = p(0.49 - p^2) = p(0.7 + p)(0.7 - p)$

113. $0.64x^2 - 1.21 = (0.8x)^2 - (1.1)^2 = (0.8x + 1.1)(0.8x - 1.1)$

115. $(x+3)^2 - 9 = [(x+3)+3][(x+3)-3] = (x+6)x, \text{ or } x(x+6)$

117. $x^2 - \left(\dfrac{1}{x}\right)^2 = \left(x + \dfrac{1}{x}\right)\left(x - \dfrac{1}{x}\right)$

119. $81 - b^{4k} = 9^2 - (b^{2k})^2$
$$= (9 + b^{2k})(9 - b^{2k})$$
$$= (9 + b^{2k})[3^2 - (b^k)^2]$$
$$= (9 + b^{2k})(3 + b^k)(3 - b^k)$$

121. $9b^{2n} + 12b^n + 4 = (3b^n)^2 + 2 \cdot 3b^n \cdot 2 + 2^2 =$
$(3b^n + 2)^2$

123. $(y + 3)^2 + 2(y + 3) + 1$
$$= (y + 3)^2 + 2 \cdot (y + 3) \cdot 1 + 1^2$$
$$= [(y + 3) + 1]^2$$
$$= (y + 4)^2$$

125. If $cy^2 + 6y + 1$ is the square of a binomial, then $2 \cdot a \cdot 1 = 6$ where $a^2 = c$. Then $a = 3$, so $c = a^2 = 3^2 = 9$. (The polynomial is $9y^2 + 6y + 1$.)

127. Enter $y_1 = x^2 + 9$ and $y_2 = (x + 3)(x + 3)$ and look at a table of values. The y_1-and y_2-values are not the same, so the factorization is not correct.

129. Enter $y_1 = x^2 + 9$ and $y_2 = (x + 3)^2$ and look at a table of values. The y_1-and y_2-values are not the same, so the factorization is not correct.

Exercise Set 10.6

1. $z^3 + 27 = z^3 + 3^3$
$$= (z + 3)(z^2 - 3z + 9)$$
$$A^3 + B^3 = (A + B)(A^2 - AB + B^2)$$

3. $x^3 - 1 = x^3 - 1^3$
$$= (x - 1)(x^2 + x + 1)$$
$$A^3 - B^3 = (A - B)(A^2 + AB + B^2)$$

5. $y^3 + 125 = y^3 + 5^3$
$$= (y + 5)(y^2 - 5y + 25)$$
$$A^3 + B^3 = (A + B)(A^2 - AB + B^2)$$

7. $8a^3 + 1 = (2a)^3 + 1^3$
$$= (2a + 1)(4a^2 - 2a + 1)$$
$$A^3 + B^3 = (A + B)(A^2 - AB + B^2)$$

9. $y^3 - 8 = y^3 - 2^3$
$$= (y - 2)(y^2 + 2y + 4)$$
$$A^3 - B^3 = (A - B)(A^2 + AB + B^2)$$

11. $8 - 27b^3 = 2^3 - (3b)^3$
$$= (2 - 3b)(4 + 6b + 9b^2)$$

13. $64y^3 + 1 = (4y)^3 + 1^3$
$$= (4y + 1)(16y^2 - 4y + 1)$$

15. $8x^3 + 27 = (2x)^3 + 3^3$
$$= (2x + 3)(4x^2 - 6x + 9)$$

17. $a^3 - b^3 = (a - b)(a^2 + ab + b^2)$

19. $a^3 + \dfrac{1}{8} = a^3 + \left(\dfrac{1}{2}\right)^3$
$$= \left(a + \dfrac{1}{2}\right)\left(a^2 - \dfrac{1}{2}a + \dfrac{1}{4}\right)$$

21. $2y^3 - 128 = 2(y^3 - 64)$
$$= 2(y^3 - 4^3)$$
$$= 2(y - 4)(y^2 + 4y + 16)$$

23. $24a^3 + 3 = 3(8a^3 + 1)$
$= 3[(2a)^3 + 1^3]$
$= 3(2a + 1)(4a^2 - 2a + 1)$

25. $rs^3 + 64r = r(s^3 + 64)$
$= r(s^3 + 4^3)$
$= r(s + 4)(s^2 - 4s + 16)$

27. $5x^3 - 40z^3 = 5(x^3 - 8z^3)$
$= 5[x^3 - (2z)^3]$
$= 5(x - 2z)(x^2 + 2xz + 4z^2)$

29. $x^3 + 0.001 = x^3 + (0.1)^3$
$= (x + 0.1)(x^2 - 0.1x + 0.01)$

31. $64x^6 - 8t^6 = 8(8x^6 - t^6)$
$= 8[(2x^2)^3 - (t^2)^3]$
$= 8(2x^2 - t^2)(4x^4 + 2x^2t^2 + t^4)$

33. $2y^4 - 128y = 2y(y^3 - 64)$
$= 2y(y^3 - 4^3)$
$= 2y(y - 4)(y^2 + 4y + 16)$

35. $z^6 - 1$
$= (z^3)^2 - 1^2$ Writing as a difference of squares
$= (z^3 + 1)(z^3 - 1)$ Factoring a difference of squares
$= (z + 1)(z^2 - z + 1)(z - 1)(z^2 + z + 1)$

Factoring a sum and a difference of cubes

37. $t^6 + 64y^6 = (t^2)^3 + (4y^2)^3$
$= (t^2 + 4y^2)(t^4 - 4t^2y^2 + 16y^4)$

39. $(7y^{-5})^3 = 7^3(y^{-5})^3 = 343y^{-5 \cdot 3} = 343y^{-15} = \dfrac{343}{y^{15}}$

41. $\left(\dfrac{x^3}{4}\right)^{-2} = \left(\dfrac{4}{x^3}\right)^2 = \dfrac{4^2}{(x^3)^2} = \dfrac{16}{x^{3 \cdot 2}} = \dfrac{16}{x^6}$

43. $\left(w - \dfrac{1}{3}\right)^2 = w^2 - 2 \cdot w \cdot \dfrac{1}{3} + \left(\dfrac{1}{3}\right)^2 = w^2 - \dfrac{2}{3}w + \dfrac{1}{9}$

45. $(a + b)^3 = (-2 + 3)^3 = 1^3 = 1$
$a^3 + b^3 = (-2)^3 + 3^3 = -8 + 27 = 19$
$(a + b)(a^2 - ab + b^2) =$
$(-2 + 3)[(-2)^2 - (-2)(3) + 3^2] =$
$1(4 + 6 + 9) = 1 \cdot 19 = 19$
$(a + b)(a^2 + ab + b^2) =$
$(-2 + 3)[(-2)^2 + (-2)(3) + 3^2] =$
$1(4 - 6 + 9) = 1 \cdot 7 = 7$
$(a + b)(a + b)(a + b) =$
$(-2 + 3)(-2 + 3)(-2 + 3) = 1 \cdot 1 \cdot 1 = 1$

47. $x^{6a} + y^{3b} = (x^{2a})^3 + (y^b)^3$
$= (x^{2a} + y^b)(x^{4a} - x^{2a}y^b + y^{2b})$

49. $3x^{3a} + 24y^{3b} = 3(x^{3a} + 8y^{3b})$
$= 3[(x^a)^3 + (2y^b)^3]$
$= 3(x^a + 2y^b)(x^{2a} - 2x^ay^b + 4y^{2b})$

51. $\dfrac{1}{24}x^3y^3 + \dfrac{1}{3}z^3 = \dfrac{1}{3}\left(\dfrac{1}{8}x^3y^3 + z^3\right)$
$= \dfrac{1}{3}\left[\left(\dfrac{1}{2}xy\right)^3 + z^3\right]$
$= \dfrac{1}{3}\left(\dfrac{1}{2}xy + z\right)\left(\dfrac{1}{4}x^2y^2 - \dfrac{1}{2}xyz + z^2\right)$

53. $(x + y)^3 - x^3$
$= [(x + y) - x][(x + y)^2 + x(x + y) + x^2]$
$= (x + y - x)(x^2 + 2xy + y^2 + x^2 + xy + x^2)$
$= y(3x^2 + 3xy + y^2)$

55. $(a + 2)^3 - (a - 2)^3$
$= [(a+2) - (a-2)][(a+2)^2 + (a+2)(a-2) + (a-2)^2]$
$= (a+2-a+2)(a^2+4a+4+a^2-4+a^2-4a+4)$
$= 4(3a^2 + 4)$

Exercise Set 10.7

1. $3x^2 - 192 = 3(x^2 - 64)$ 3 is a common factor
$= 3(x^2 - 8^2)$ Difference of squares
$= 3(x + 8)(x - 8)$

3. $a^2 + 25 - 10a = a^2 - 10a + 25$
$= a^2 - 2 \cdot a \cdot 5 + 5^2$ Trinomial square
$= (a - 5)^2$

5. $2x^2 - 11x + 12$

There is no common factor (other than 1). This polynomial has three terms, but it is not a trinomial square. Multiply the leading coefficient and the constant, 2 and 12: $2 \cdot 12 = 24$. Try to factor 24 so that the sum of the factors is -11. The numbers we want are -3 and -8: $-3(-8) = 24$ and $-3 + (-8) = -11$. Split the middle term and factor by grouping.

$2x^2 - 11x + 12 = 2x^2 - 3x - 8x + 12$
$= x(2x - 3) - 4(2x - 3)$
$= (x - 4)(2x - 3)$

7. $x^3 + 24x^2 + 144x$
$= x(x^2 + 24x + 144)$ x is a common factor
$= x(x^2 + 2 \cdot x \cdot 12 + 12^2)$ Trinomial square
$= x(x + 12)^2$

9. $x^3 + 3x^2 - 4x - 12$
$= x^2(x + 3) - 4(x + 3)$ Factoring by grouping
$= (x^2 - 4)(x + 3)$
$= (x + 2)(x - 2)(x + 3)$ Factoring the difference of squares

11. $48x^2 - 3 = 3(16x^2 - 1)$ 3 is a common factor

$\quad\quad\quad\quad = 3[(4x)^2 - 1^2]$ Difference of squares

$\quad\quad\quad\quad = 3(4x + 1)(4x - 1)$

13. $\quad\; 9x^3 + 12x^2 - 45x$

$\quad = 3x(3x^2 + 4x - 15)$ $3x$ is a common factor

$\quad = 3x(3x - 5)(x + 3)$ Factoring the trinomial

15. $x^2 + 4$ is a *sum* of squares with no common factor. It cannot be factored. It is prime.

17. $x^4 + 7x^2 - 3x^3 - 21x = x(x^3 + 7x - 3x^2 - 21)$

$\quad\quad\quad\quad\quad\quad\quad\quad\quad = x[x(x^2 + 7) - 3(x^2 + 7)]$

$\quad\quad\quad\quad\quad\quad\quad\quad\quad = x[(x - 3)(x^2 + 7)]$

$\quad\quad\quad\quad\quad\quad\quad\quad\quad = x(x - 3)(x^2 + 7)$

19. $\quad\; x^5 - 14x^4 + 49x^3$

$\quad = x^3(x^2 - 14x + 49)$ x^3 is a common factor

$\quad = x^3(x^2 - 2 \cdot x \cdot 7 + 7^2)$ Trinomial square

$\quad = x^3(x - 7)^2$

21. $\quad\; 20 - 6x - 2x^2$

$\quad = -2(-10 + 3x + x^2)$ -2 is a common factor

$\quad = -2(x^2 + 3x - 10)$ Writing in descending order

$\quad = -2(x + 5)(x - 2),$ Using trial and error

$\quad$ or $2(-x - 5)(x - 2),$

$\quad$ or $2(x + 5)(-x + 2)$

23. $x^2 - 6x + 1$

There is no common factor (other than 1 or -1). This is not a trinomial square, because $-6x \neq 2 \cdot x \cdot 1$ and $-6x \neq -2 \cdot x \cdot 1$. We try factoring using the refined trial and error procedure. We look for two factors of 1 whose sum is -6. There are none. The polynomial cannot be factored. It is prime.

25. $\quad\; 4x^4 - 64$

$\quad = 4(x^4 - 16)$ 4 is a common factor

$\quad = 4[(x^2)^2 - 4^2]$ Difference of squares

$\quad = 4(x^2 + 4)(x^2 - 4)$ Difference of squares

$\quad = 4(x^2 + 4)(x + 2)(x - 2)$

27. $\quad\; 1 - y^8$ Difference of squares

$\quad = (1 + y^4)(1 - y^4)$ Difference of squares

$\quad = (1 + y^4)(1 + y^2)(1 - y^2)$ Difference of squares

$\quad = (1 + y^4)(1 + y^2)(1 + y)(1 - y)$

29. $\quad\; x^5 - 4x^4 + 3x^3$

$\quad = x^3(x^2 - 4x + 3)$ x^3 is a common factor

$\quad = x^3(x - 3)(x - 1)$ Factoring the trinomial using trial and error

31. $\quad\; \dfrac{1}{81}x^6 - \dfrac{8}{27}x^3 + \dfrac{16}{9}$

$\quad = \dfrac{1}{9}\left(\dfrac{1}{9}x^6 - \dfrac{8}{3}x^3 + 16\right)$ $\dfrac{1}{9}$ is a common factor

$\quad = \dfrac{1}{9}\left[\left(\dfrac{1}{3}x^3\right)^2 - 2 \cdot \dfrac{1}{3}x^3 \cdot 4 + 4^2\right]$ Trinomial square

$\quad \doteq \dfrac{1}{9}\left(\dfrac{1}{3}x^3 - 4\right)^2$

33. $\quad\; mx^2 + my^2$

$\quad = m(x^2 + y^2)$ m is a common factor

The factor with more than one term cannot be factored further, so we have factored completely.

35. $9x^2y^2 - 36xy = 9xy(xy - 4)$

37. $2\pi rh + 2\pi r^2 = 2\pi r(h + r)$

39. $\quad\; (a + b)(x - 3) + (a + b)(x + 4)$

$\quad = (a + b)[(x - 3) + (x + 4)]$ $(a + b)$ is a common factor

$\quad = (a + b)(2x + 1)$

41. $(x - 1)(x + 1) - y(x + 1) = (x + 1)(x - 1 - y)$

$\quad\quad\quad\quad\quad\quad\quad\quad\quad\quad\quad$ $(x + 1)$ is a common factor

43. $\quad\; n^2 + 2n + np + 2p$

$\quad = n(n + 2) + p(n + 2)$ Factoring by grouping

$\quad = (n + p)(n + 2)$

45. $\quad\; 6q^2 - 3q + 2pq - p$

$\quad = 3q(2q - 1) + p(2q - 1)$ Factoring by grouping

$\quad = (3q + p)(2q - 1)$

47. $\quad\; 4b^2 + a^2 - 4ab$

$\quad = a^2 - 4ab + 4b^2$ Rearranging

$\quad = a^2 - 2 \cdot a \cdot 2b + (2b)^2$ Trinomial square

$\quad \doteq (a - 2b)^2$

(Note that if we had rewritten the polynomial as $4b^2 - 4ab + a^2$, we might have written the result as $(2b - a)^2$. The two factorizations are equivalent.)

49. $\quad\; 16x^2 + 24xy + 9y^2$

$\quad = (4x)^2 + 2 \cdot 4x \cdot 3y + (3y)^2$ Trinomial square

$\quad = (4x + 3y)^2$

51. $\quad\; 49m^4 - 112m^2n + 64n^2$

$\quad = (7m^2)^2 - 2 \cdot 7m^2 \cdot 8n + (8n)^2$ Trinomial square

$\quad = (7m^2 - 8n)^2$

53. $\quad\; y^4 + 10y^2z^2 + 25z^4$

$\quad = (y^2)^2 + 2 \cdot y^2 \cdot 5z^2 + (5z^2)^2$ Trinomial square

$\quad = (y^2 + 5z^2)^2$

55. $\quad\; \dfrac{1}{4}a^2 + \dfrac{1}{3}ab + \dfrac{1}{9}b^2$

$\quad = \left(\dfrac{1}{2}a\right)^2 + 2 \cdot \dfrac{1}{2}a \cdot \dfrac{1}{3}b + \left(\dfrac{1}{3}b\right)^2$

$\quad = \left(\dfrac{1}{2}a + \dfrac{1}{3}b\right)^2$

57. $a^2 - ab - 2b^2 = (a - 2b)(a + b)$ Using trial and error

59. $\quad 2mn - 360n^2 + m^2$
$= m^2 + 2mn - 360n^2 \quad$ Rewriting
$= (m + 20n)(m - 18n) \quad$ Using trial and error

61. $m^2n^2 - 4mn - 32 = (mn - 8)(mn + 4)$ Using trial and error

63. $\quad r^5s^2 - 10r^4s + 16r^3$
$= r^3(r^2s^2 - 10rs + 16) \quad r^3$ is a common factor
$= r^3(rs - 2)(rs - 8) \quad$ Using trial and error

65. $\quad a^5 + 4a^4b - 5a^3b^2$
$= a^3(a^2 + 4ab - 5b^2) \quad a^3$ is a common factor
$= a^3(a + 5b)(a - b) \quad$ Factoring the trinomial

67. $\quad a^2 - \dfrac{1}{25}b^2$
$= a^2 - \left(\dfrac{1}{5}b\right)^2 \quad$ Difference of squares
$= \left(a + \dfrac{1}{5}b\right)\left(a - \dfrac{1}{5}b\right)$

69. $\quad 7x^6 - 7y^6$
$= 7(x^6 - y^6) \quad 7$ is a common factor
$= 7[(x^3)^2 - (y^3)^2] \quad$ Difference of squares
$= 7(x^3 + y^3)(x^3 - y^3)$
$= 7(x + y)(x^2 - xy + y^2)(x - y)(x^2 + xy + y^2)$
Factoring the sum of cubes and the difference of cubes

71. $\quad 16 - p^4q^4$
$= 4^2 - (p^2q^2)^2 \quad$ Difference of squares
$= (4 + p^2q^2)(4 - p^2q^2) \quad 4 - p^2q^2$ is a difference of squares
$= (4 + p^2q^2)(2 + pq)(2 - pq)$

73. $\quad 1 - 16x^{12}y^{12}$
$= 1^2 - (4x^6y^6)^2 \quad$ Difference of squares
$= (1 + 4x^6y^6)(1 - 4x^6y^6) \quad 1 - 4x^6y^6$ is a difference of squares
$= (1 + 4x^6y^6)(1 + 2x^3y^3)(1 - 2x^3y^3)$

75. $\quad q^3 + 8q^2 - q - 8$
$= q^2(q + 8) - (q + 8) \quad$ Factoring by grouping
$= (q^2 - 1)(q + 8)$
$= (q + 1)(q - 1)(q + 8) \quad$ Factoring the difference of squares

77. $\quad 112xy + 49x^2 + 64y^2$
$= 49x^2 + 112xy + 64y^2 \quad$ Rearranging
$= (7x)^2 + 2 \cdot 7x \cdot 8y + (8y)^2 \quad$ Trinomial square
$= (7x + 8y)^2$

79. Discussion and Writing Exercise

81. The highest point on the graph lies above 1999, so CD sales were highest in 1999.

83. The point on the graph corresponding to 779 million lies above 1996, so CD sales were 779 million in 1996.

85. From the graph we see that 847 million CDs were sold in 1998 and 939 million were sold in 1999. First we subtract to find the amount of the increase, in millions.

$$\begin{array}{r} 9\ 3\ 9 \\ -\ 8\ 4\ 7 \\ \hline 9\ 2 \end{array}$$

Now we find the percent increase.

$\underset{92}{92}$ is $\underbrace{\text{what percent}}$ of 847?
$\quad \downarrow \quad \downarrow \qquad \downarrow \qquad \downarrow \quad \downarrow$
$\quad 92 \quad = \qquad p \qquad \cdot \quad 847$

We solve the equation.
$$92 = p \cdot 847$$
$$\frac{92}{847} = p$$
$$0.109 \approx p$$
$$10.9\% \approx p$$

Sales of CDs increased about 10.9% from 1998 to 1999.

87. $\quad \dfrac{7}{5} \div \left(-\dfrac{11}{10}\right)$
$= \dfrac{7}{5} \cdot \left(-\dfrac{10}{11}\right) \quad$ Multiplying by the reciprocal of the divisor
$= -\dfrac{7 \cdot 10}{5 \cdot 11}$
$= -\dfrac{7 \cdot 5 \cdot 2}{5 \cdot 11} = -\dfrac{7 \cdot 2}{11} \cdot \dfrac{5}{5}$
$= -\dfrac{14}{11}$

89. $\quad A = aX + bX - 7$
$A + 7 = aX + bX$
$A + 7 = X(a + b)$
$\dfrac{A + 7}{a + b} = X$

91. $a^4 - 2a^2 + 1 = (a^2)^2 - 2 \cdot a^2 \cdot 1 + 1^2$
$= (a^2 - 1)^2$
$= [(a + 1)(a - 1)]^2$
$= (a + 1)^2(a - 1)^2$

93. $12.25x^2 - 7x + 1 = (3.5x)^2 - 2 \cdot (3.5x) \cdot 1 + 1^2$
$= (3.5x - 1)^2$

95. $5x^2 + 13x + 7.2$

Multiply the leading coefficient and the constant, 5 and 7.2: $5(7.2) = 36$. Try to factor 36 so that the sum of the factors is 13. The numbers we want are 9 and 4. Split the middle term and factor by grouping:

$$5x^2 + 13x + 7.2 = 5x^2 + 9x + 4x + 7.2$$
$$= 5x(x + 1.8) + 4(x + 1.8)$$
$$= (5x + 4)(x + 1.8)$$

97. $18 + y^3 - 9y - 2y^2$
$$= y^3 - 2y^2 - 9y + 18$$
$$= y^2(y - 2) - 9(y - 2)$$
$$= (y^2 - 9)(y - 2)$$
$$= (y + 3)(y - 3)(y - 2)$$

99. $a^3 + 4a^2 + a + 4 = a^2(a + 4) + 1(a + 4)$
$$= (a^2 + 1)(a + 4)$$

101. $x^3 - x^2 - 4x + 4 = x^2(x - 1) - 4(x - 1)$
$$= (x^2 - 4)(x - 1)$$
$$= (x + 2)(x - 2)(x - 1)$$

103. $y^2(y - 1) - 2y(y - 1) + (y - 1)$
$$= (y - 1)(y^2 - 2y + 1)$$
$$= (y - 1)(y - 1)^2$$
$$= (y - 1)^3$$

105. $(y + 4)^2 + 2x(y + 4) + x^2$
$$= (y + 4)^2 + 2 \cdot (y + 4) \cdot x + x^2 \quad \text{Trinomial square}$$
$$= (y + 4 + x)^2$$

Exercise Set 10.8

1. $(x + 4)(x + 9) = 0$

$x + 4 = 0 \quad$ or $\quad x + 9 = 0 \quad$ Using the principle of zero products

$x = -4 \quad$ or $\qquad x = -9 \quad$ Solving the two equations separately

Check:

For -4
$$\frac{(x + 4)(x + 9) = 0}{(-4 + 4)(-4 + 9) \ ? \ 0}$$
$$\begin{array}{c|c} 0 \cdot 5 & \\ 0 & \text{TRUE} \end{array}$$

For -9
$$\frac{(x + 4)(x + 9) = 0}{(9 + 4)(-9 + 9) \ ? \ 0}$$
$$\begin{array}{c|c} 13 \cdot 0 & \\ 0 & \text{TRUE} \end{array}$$

The solutions are -4 and -9.

3. $(x + 3)(x - 8) = 0$

$x + 3 = 0 \quad$ or $\quad x - 8 = 0 \quad$ Using the principle of zero products

$x = -3 \quad$ or $\qquad x = 8$

Check:

For -3
$$\frac{(x + 3)(x - 8) = 0}{(-3 + 3)(-3 - 8) \ ? \ 0}$$
$$\begin{array}{c|c} 0(-11) & \\ 0 & \text{TRUE} \end{array}$$

For 8
$$\frac{(x + 3)(x - 8) = 0}{(8 + 3)(8 - 8) \ ? \ 0}$$
$$\begin{array}{c|c} 11 \cdot 0 & \\ 0 & \text{TRUE} \end{array}$$

The solutions are -3 and 8.

5. $(x + 12)(x - 11) = 0$

$x + 12 = 0 \quad$ or $\quad x - 11 = 0$

$x = -12 \quad$ or $\qquad x = 11$

The solutions are -12 and 11.

7. $x(x + 3) = 0$

$x = 0 \quad$ or $\quad x + 3 = 0$

$x = 0 \quad$ or $\qquad x = -3$

The solutions are 0 and -3.

9. $0 = y(y + 18)$

$y = 0 \quad$ or $\quad y + 18 = 0$

$y = 0 \quad$ or $\qquad y = -18$

The solutions are 0 and -18.

11. $(2x + 5)(x + 4) = 0$

$2x + 5 = 0 \quad$ or $\quad x + 4 = 0$

$2x = -5 \quad$ or $\qquad x = -4$

$x = -\dfrac{5}{2} \quad$ or $\qquad x = -4$

The solutions are $-\dfrac{5}{2}$ and -4.

13. $(5x + 1)(4x - 12) = 0$

$5x + 1 = 0 \quad$ or $\quad 4x - 12 = 0$

$5x = -1 \quad$ or $\qquad 4x = 12$

$x = -\dfrac{1}{5} \quad$ or $\qquad x = 3$

The solutions are $-\dfrac{1}{5}$ and 3.

15. $(7x - 28)(28x - 7) = 0$

$7x - 28 = 0 \quad$ or $\quad 28x - 7 = 0$

$7x = 28 \quad$ or $\qquad 28x = 7$

$x = 4 \quad$ or $\qquad x = \dfrac{7}{28} = \dfrac{1}{4}$

The solutions are 4 and $\dfrac{1}{4}$.

17. $2x(3x - 2) = 0$

$2x = 0 \quad$ or $\quad 3x - 2 = 0$

$x = 0 \quad$ or $\qquad 3x = 2$

$x = 0 \quad$ or $\qquad x = \dfrac{2}{3}$

The solutions are 0 and $\dfrac{2}{3}$.

19. $\left(\dfrac{1}{5} + 2x\right)\left(\dfrac{1}{9} - 3x\right) = 0$

$\dfrac{1}{5} + 2x = 0 \qquad$ or $\quad \dfrac{1}{9} - 3x = 0$

$\qquad 2x = -\dfrac{1}{5} \quad$ or $\qquad -3x = -\dfrac{1}{9}$

$\qquad x = -\dfrac{1}{10} \quad$ or $\qquad x = \dfrac{1}{27}$

The solutions are $-\dfrac{1}{10}$ and $\dfrac{1}{27}$.

21. $(0.3x - 0.1)(0.05x + 1) = 0$

$0.3x - 0.1 = 0 \qquad$ or $\quad 0.05x + 1 = 0$

$\qquad 0.3x = 0.1 \quad$ or $\qquad 0.05x = -1$

$\qquad x = \dfrac{0.1}{0.3} \quad$ or $\qquad x = -\dfrac{1}{0.05}$

$\qquad x = \dfrac{1}{3} \quad$ or $\qquad x = -20$

The solutions are $\dfrac{1}{3}$ and -20.

23. $9x(3x - 2)(2x - 1) = 0$

$9x = 0 \quad$ or $\quad 3x - 2 = 0 \quad$ or $\quad 2x - 1 = 0$

$x = 0 \quad$ or $\qquad 3x = 2 \quad$ or $\qquad 2x = 1$

$x = 0 \quad$ or $\qquad x = \dfrac{2}{3} \quad$ or $\qquad x = \dfrac{1}{2}$

The solutions are 0, $\dfrac{2}{3}$, and $\dfrac{1}{2}$.

25. $x^2 + 6x + 5 = 0$

$(x + 5)(x + 1) = 0 \qquad$ Factoring

$x + 5 = 0 \quad$ or $\quad x + 1 = 0 \qquad$ Using the principle of zero products

$\qquad x = -5 \quad$ or $\qquad x = -1$

The solutions are -5 and -1.

27. $x^2 + 7x - 18 = 0$

$(x + 9)(x - 2) = 0 \qquad$ Factoring

$x + 9 = 0 \quad$ or $\quad x - 2 = 0 \quad$ Using the principle of zero products

$\qquad x = -9 \quad$ or $\qquad x = 2$

The solutions are -9 and 2.

29. $x^2 - 8x + 15 = 0$

$(x - 5)(x - 3) = 0$

$x - 5 = 0 \quad$ or $\quad x - 3 = 0$

$\qquad x = 5 \quad$ or $\qquad x = 3$

The solutions are 5 and 3.

31. $x^2 - 8x = 0$

$x(x - 8) = 0$

$x = 0 \quad$ or $\quad x - 8 = 0$

$x = 0 \quad$ or $\qquad x = 8$

The solutions are 0 and 8.

33. $x^2 + 18x = 0$

$x(x + 18) = 0$

$x = 0 \quad$ or $\quad x + 18 = 0$

$x = 0 \quad$ or $\qquad x = -18$

The solutions are 0 and -18.

35. $\qquad x^2 = 16$

$\qquad x^2 - 16 = 0 \qquad$ Subtracting 16

$(x - 4)(x + 4) = 0$

$x - 4 = 0 \quad$ or $\quad x + 4 = 0$

$\qquad x = 4 \quad$ or $\qquad x = -4$

The solutions are 4 and -4.

37. $\qquad 9x^2 - 4 = 0$

$(3x - 2)(3x + 2) = 0$

$3x - 2 = 0 \quad$ or $\quad 3x + 2 = 0$

$\qquad 3x = 2 \quad$ or $\qquad 3x = -2$

$\qquad x = \dfrac{2}{3} \quad$ or $\qquad x = -\dfrac{2}{3}$

The solutions are $\dfrac{2}{3}$ and $-\dfrac{2}{3}$.

39. $0 = 6x + x^2 + 9$

$0 = x^2 + 6x + 9 \qquad$ Writing in descending order

$0 = (x + 3)(x + 3)$

$x + 3 = 0 \quad$ or $\quad x + 3 = 0$

$\qquad x = -3 \quad$ or $\qquad x = -3$

There is only one solution, -3.

41. $\qquad x^2 + 16 = 8x$

$\qquad x^2 - 8x + 16 = 0 \qquad$ Subtracting $8x$

$(x - 4)(x - 4) = 0$

$x - 4 = 0 \quad$ or $\quad x - 4 = 0$

$\qquad x = 4 \quad$ or $\qquad x = 4$

There is only one solution, 4.

43. $\qquad 5x^2 = 6x$

$\qquad 5x^2 - 6x = 0$

$x(5x - 6) = 0$

$x = 0 \quad$ or $\quad 5x - 6 = 0$

$x = 0 \quad$ or $\qquad 5x = 6$

$x = 0 \quad$ or $\qquad x = \dfrac{6}{5}$

The solutions are 0 and $\dfrac{6}{5}$.

45.
$$6x^2 - 4x = 10$$
$$6x^2 - 4x - 10 = 0$$
$$2(3x^2 - 2x - 5) = 0$$
$$2(3x - 5)(x + 1) = 0$$
$$3x - 5 = 0 \quad \text{or} \quad x + 1 = 0$$
$$3x = 5 \quad \text{or} \quad x = -1$$
$$x = \frac{5}{3} \quad \text{or} \quad x = -1$$
The solutions are $\frac{5}{3}$ and -1.

47.
$$12y^2 - 5y = 2$$
$$12y^2 - 5y - 2 = 0$$
$$(4y + 1)(3y - 2) = 0$$
$$4y + 1 = 0 \quad \text{or} \quad 3y - 2 = 0$$
$$4y = -1 \quad \text{or} \quad 3y = 2$$
$$y = -\frac{1}{4} \quad \text{or} \quad y = \frac{2}{3}$$
The solutions are $-\frac{1}{4}$ and $\frac{2}{3}$.

49.
$$t(3t + 1) = 2$$
$$3t^2 + t = 2 \qquad \text{Multiplying on the left}$$
$$3t^2 + t - 2 = 0 \qquad \text{Subtracting 2}$$
$$(3t - 2)(t + 1) = 0$$
$$3t - 2 = 0 \quad \text{or} \quad t + 1 = 0$$
$$3t = 2 \quad \text{or} \quad t = -1$$
$$t = \frac{2}{3} \quad \text{or} \quad t = -1$$
The solutions are $\frac{2}{3}$ and -1.

51.
$$100y^2 = 49$$
$$100y^2 - 49 = 0$$
$$(10y + 7)(10y - 7) = 0$$
$$10y + 7 = 0 \quad \text{or} \quad 10y - 7 = 0$$
$$10y = -7 \quad \text{or} \quad 10y = 7$$
$$y = -\frac{7}{10} \quad \text{or} \quad y = \frac{7}{10}$$
The solutions are $-\frac{7}{10}$ and $\frac{7}{10}$.

53.
$$x^2 - 5x = 18 + 2x$$
$$x^2 - 5x - 18 - 2x = 0 \qquad \text{Subtracting 18 and } 2x$$
$$x^2 - 7x - 18 = 0$$
$$(x - 9)(x + 2) = 0$$
$$x - 9 = 0 \quad \text{or} \quad x + 2 = 0$$
$$x = 9 \quad \text{or} \quad x = -2$$
The solutions are 9 and -2.

55.
$$10x^2 - 23x + 12 = 0$$
$$(5x - 4)(2x - 3) = 0$$
$$5x - 4 = 0 \quad \text{or} \quad 2x - 3 = 0$$
$$5x = 4 \quad \text{or} \quad 2x = 3$$
$$x = \frac{4}{5} \quad \text{or} \quad x = \frac{3}{2}$$
The solutions are $\frac{4}{5}$ and $\frac{3}{2}$.

57. We let $y = 0$ and solve for x.
$$0 = x^2 + 3x - 4$$
$$0 = (x + 4)(x - 1)$$
$$x + 4 = 0 \quad \text{or} \quad x - 1 = 0$$
$$x = -4 \quad \text{or} \quad x = 1$$
The x-intercepts are $(-4, 0)$ and $(1, 0)$.

59. We let $y = 0$ and solve for x
$$0 = 2x^2 + x - 10$$
$$0 = (2x + 5)(x - 2)$$
$$2x + 5 = 0 \quad \text{or} \quad x - 2 = 0$$
$$2x = -5 \quad \text{or} \quad x = 2$$
$$x = -\frac{5}{2} \quad \text{or} \quad x = 2$$
The x-intercepts are $\left(-\frac{5}{2}, 0\right)$ and $(2, 0)$.

61. We let $y = 0$ and solve for x.
$$0 = x^2 - 2x - 15$$
$$0 = (x - 5)(x + 3)$$
$$x - 5 = 0 \quad or \quad x + 3 = 0$$
$$x = 5 \quad or \quad x = -3$$
The x-intercepts are $(5, 0)$ and $(-3, 0)$.

63. The solutions of the equation are the first coordinates of the x-intercepts of the graph. From the graph we see that the x-intercepts are $(-1, 0)$ and $(4, 0)$, so the solutions of the equation are -1 and 4.

65. The solutions of the equation are the first coordinates of the x-intercepts of the graph. From the graph we see that the x-intercepts are $(-1, 0)$ and $(3, 0)$, so the solutions of the equation are -1 and 3.

67. Discussion and Writing Exercise

69. $(a + b)^2$

71. The two numbers have different signs, so their quotient is negative.
$$144 \div -9 = -16$$

73.
$$-\frac{5}{8} \div \frac{3}{16} = -\frac{5}{8} \cdot \frac{16}{3}$$
$$= -\frac{5 \cdot 16}{8 \cdot 3}$$
$$= -\frac{5 \cdot 8 \cdot 2}{8 \cdot 3}$$
$$= -\frac{10}{3}$$

75.
$$b(b+9) = 4(5+2b)$$
$$b^2 + 9b = 20 + 8b$$
$$b^2 + 9b - 8b - 20 = 0$$
$$b^2 + b - 20 = 0$$
$$(b+5)(b-4) = 0$$
$$b+5 = 0 \quad \text{or} \quad b-4 = 0$$
$$b = -5 \quad \text{or} \quad b = 4$$

The solutions are -5 and 4.

77.
$$(t-3)^2 = 36$$
$$t^2 - 6t + 9 = 36$$
$$t^2 - 6t - 27 = 0$$
$$(t-9)(t+3) = 0$$
$$t-9 = 0 \quad \text{or} \quad t+3 = 0$$
$$t = 9 \quad \text{or} \quad t = -3$$

The solutions are 9 and -3.

79.
$$x^2 - \frac{1}{64} = 0$$
$$\left(x - \frac{1}{8}\right)\left(x + \frac{1}{8}\right) = 0$$
$$x - \frac{1}{8} = 0 \quad \text{or} \quad x + \frac{1}{8} = 0$$
$$x = \frac{1}{8} \quad \text{or} \quad x = -\frac{1}{8}$$

The solutions are $\frac{1}{8}$ and $-\frac{1}{8}$.

81.
$$\frac{5}{16}x^2 = 5$$
$$\frac{5}{16}x^2 - 5 = 0$$
$$5\left(\frac{1}{16}x^2 - 1\right) = 0$$
$$5\left(\frac{1}{4}x - 1\right)\left(\frac{1}{4}x + 1\right) = 0$$
$$\frac{1}{4}x - 1 = 0 \quad \text{or} \quad \frac{1}{4}x + 1 = 0$$
$$\frac{1}{4}x = 1 \quad \text{or} \quad \frac{1}{4}x = -1$$
$$x = 4 \quad \text{or} \quad x = -4$$

The solutions are 4 and -4.

83. (a)
$$x = -3 \quad \text{or} \quad x = 4$$
$$x+3 = 0 \quad \text{or} \quad x-4 = 0$$
$$(x+3)(x-4) = 0 \quad \text{Principle of zero}$$
$$\qquad\qquad\qquad\qquad\text{products}$$
$$x^2 - x - 12 = 0 \quad \text{Multiplying}$$

(b)
$$x = -3 \quad \text{or} \quad x = -4$$
$$x+3 = 0 \quad \text{or} \quad x+4 = 0$$
$$(x+3)(x+4) = 0$$
$$x^2 + 7x + 12 = 0$$

(c)
$$x = \frac{1}{2} \quad \text{or} \quad x = \frac{1}{2}$$
$$x - \frac{1}{2} = 0 \quad \text{or} \quad x - \frac{1}{2} = 0$$
$$\left(x - \frac{1}{2}\right)\left(x - \frac{1}{2}\right) = 0$$
$$x^2 - x + \frac{1}{4} = 0, \quad \text{or}$$
$$4x^2 - 4x + 1 = 0 \quad \text{Multiplying by 4}$$

(d)
$$(x-5)(x+5) = 0$$
$$x^2 - 25 = 0$$

(e)
$$(x-0)(x-0.1)\left(x - \frac{1}{4}\right) = 0$$
$$x\left(x - \frac{1}{10}\right)\left(x - \frac{1}{4}\right) = 0$$
$$x\left(x^2 - \frac{7}{20}x + \frac{1}{40}\right) = 0$$
$$x^3 - \frac{7}{20}x^2 + \frac{1}{40}x = 0, \quad \text{or}$$
$$40x^3 - 14x^2 + x = 0 \quad \text{Multiplying by 40}$$

85. 2.33, 6.77

87. -9.15, -4.59

89. 0, 2.74

Exercise Set 10.9

1. _Familiarize_. Let w = the width of the table, in feet. Then $6w$ = the length, in feet. Recall that the area of a rectangle is Length · Width.

Translate.

$$\underbrace{\text{The area of the table}}_{6w \cdot w} \;\; \underset{=}{\text{is}} \;\; \underbrace{24 \text{ ft}^2.}_{24}$$

Solve. We solve the equation.
$$6w \cdot w = 24$$
$$6w^2 = 24$$
$$6w^2 - 24 = 0$$
$$6(w^2 - 4) = 0$$
$$6(w+2)(w-2) = 0$$
$$w+2 = 0 \quad \text{or} \quad w-2 = 0$$
$$w = -2 \quad \text{or} \quad w = 2$$

Check. Since the width must be positive, -2 cannot be a solution. If the width is 2 ft, then the length is $6 \cdot 2$ ft, or 12 ft, and the area is 12 ft · 2 ft = 24 ft^2. These numbers check.

State. The table is 12 ft long and 2 ft wide.

3. _Familiarize_. We make a drawing. Let w = the width, in cm. Then $2w + 2$ = the length, in cm.

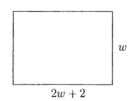

$2w + 2$

Recall that the area of a rectangle is length times width.

Translate. We reword the problem.

Length   times   width   is $\underbrace{144 \text{ cm}^2}$.

$\downarrow$ $\downarrow$ $\downarrow$ $\downarrow$ $\downarrow$

$(2w + 2)$ $\cdot$ w $=$ 144

Solve. We solve the equation.

$$(2w + 2)w = 144$$
$$2w^2 + 2w = 144$$
$$2w^2 + 2w - 144 = 0$$
$$2(w^2 + w - 72) = 0$$
$$2(w + 9)(w - 8) = 0$$
$$w + 9 = 0 \quad or \quad w - 8 = 0$$
$$w = -9 \ or \qquad w = 8$$

Check. Since the width must be positive, -9 cannot be a solution. If the width is 8 cm, then the length is $2 \cdot 8 + 2$, or 18 cm, and the area is $8 \cdot 18$, or 144 cm^2. Thus, 8 checks.

State. The width is 8 cm, and the length is 18 cm.

5. Familiarize. Using the labels shown on the drawing in the text, we let h = the height, in cm, and $h + 10$ = the base, in cm. Recall that the formula for the area of a triangle is $\frac{1}{2} \cdot$ (base) $\cdot$ (height).

Translate.

$\frac{1}{2}$ times   base   times height   is $\underbrace{28 \text{ cm}^2}$.

$\downarrow$ $\downarrow$ $\downarrow$ $\downarrow$ $\downarrow$ $\downarrow$ $\downarrow$

$\frac{1}{2}$ $\cdot$ $(h + 10)$ $\cdot$ h $=$ 28

Solve. We solve the equation.

$$\frac{1}{2}(h + 10)h = 28$$
$$(h + 10)h = 56 \quad \text{Multiplying by 2}$$
$$h^2 + 10h = 56$$
$$h^2 + 10h - 56 = 0$$
$$(h + 14)(h - 4) = 0$$
$$h + 14 = 0 \quad or \quad h - 4 = 0$$
$$h = -14 \ or \qquad h = 4$$

Check. Since the height of the triangle must be positive, -14 cannot be a solution. If the height is 4 cm, then the base is $4 + 10$, or 14 cm, and the area is $\frac{1}{2} \cdot 14 \cdot 4$, or 28 cm^2. Thus, 4 checks.

State. The height of the triangle is 4 cm, and the base is 14 cm.

7. Familiarize. Using the labels show on the drawing in the text, we let h = the height of the triangle, in meters, and

$\frac{1}{2}h$ = the length of the base, in meters. Recall that the formula for the area of a triangle is $\frac{1}{2} \cdot$ (base) $\cdot$ (height).

Translate.

$\frac{1}{2}$ times   base   times   height   is $\underbrace{64 \text{ m}^2}$.

$\downarrow$ $\downarrow$ $\downarrow$ $\downarrow$ $\downarrow$ $\downarrow$ $\downarrow$

$\frac{1}{2}$ $\cdot$ $\frac{1}{2}h$ $\cdot$ h $=$ 64

Solve. We solve the equation.

$$\frac{1}{2} \cdot \frac{1}{2}h \cdot h = 64$$
$$\frac{1}{4}h^2 = 64$$
$$h^2 = 256 \quad \text{Multiplying by 4}$$
$$h^2 - 256 = 0$$
$$(h + 16)(h - 16) = 0$$
$$h + 16 = 0 \quad or \quad h - 16 = 0$$
$$h = -16 \ or \qquad h = 16$$

Check. The height of the triangle cannot be negative, so -16 cannot be a solution. If the height is 16 m, then the length of the base is $\frac{1}{2} \cdot 16$ m, or 8 m, and the area is $\frac{1}{2} \cdot 8$ m $\cdot 16$ m $= 64$ m^2. These numbers check.

State. The length of the base is 8 m, and the height is 16 m.

9. Familiarize. Reread Example 4 in Section 4.3.

Translate. Substitute 14 for n.

$$14^2 - 14 = N$$

Solve. We do the computation on the left.

$$14^2 - 14 = N$$
$$196 - 14 = N$$
$$182 = N$$

Check. We can redo the computation, or we can solve the equation $n^2 - n = 182$. The answer checks.

State. 182 games will be played.

11. Familiarize. Reread Example 4 in Section 4.3.

Translate. Substitute 132 for N.

$$n^2 - n = 132$$

Solve.

$$n^2 - n = 132$$
$$n^2 - n - 132 = 0$$
$$(n - 12)(n + 11) = 0$$
$$n - 12 = 0 \quad or \quad n + 11 = 0$$
$$n = 12 \ or \qquad n = -11$$

Check. The solutions of the equation are 12 and -11. Since the number of teams cannot be negative, -11 cannot be a solution. But 12 checks since $12^2 - 12 = 144 - 12 = 132$.

State. There are 12 teams in the league.

13. *Familiarize*. We will use the formula $N = \frac{1}{2}(n^2 - n)$.

***Translate*.** Substitute 100 for n.

$$N = \frac{1}{2}(100^2 - 100)$$

***Solve*.** We do the computation on the right.

$$N = \frac{1}{2}(10,000 - 100)$$

$$N = \frac{1}{2}(9900)$$

$$N = 4950$$

***Check*.** We can redo the computation, or we can solve the equation $4950 = \frac{1}{2}(n^2 - n)$. The answer checks.

***State*.** 4950 handshakes are possible.

15. *Familiarize*. We will use the formula $N = \frac{1}{2}(n^2 - n)$.

***Translate*.** Substitute 300 for N.

$$300 = \frac{1}{2}(n^2 - n)$$

***Solve*.** We solve the equation.

$$2 \cdot 300 = 2 \cdot \frac{1}{2}(n^2 - n) \qquad \text{Multiplying by 2}$$

$$600 = n^2 - n$$

$$0 = n^2 - n - 600$$

$$0 = (n + 24)(n - 25)$$

$$n + 24 = 0 \quad \text{or} \quad n - 25 = 0$$
$$n = -24 \quad \text{or} \quad n = 25$$

***Check*.** The number of people at a meeting cannot be negative, so -24 cannot be a solution. But 25 checks since $\frac{1}{2}(25^2 - 25) = \frac{1}{2}(625 - 25) = \frac{1}{2} \cdot 600 = 300$.

***State*.** There were 25 people at the party.

17. *Familiarize*. We will use the formula $N = \frac{1}{2}(n^2 - n)$, since toasts can be substituted for handshakes.

***Translate*.** Substitute 190 for N.

$$190 = \frac{1}{2}(n^2 - n)$$

***Solve*.**

$$190 = \frac{1}{2}(n^2 - n)$$

$$380 = n^2 - n \qquad \text{Multiplying by 2}$$

$$0 = n^2 - n - 380$$

$$0 = (n - 20)(n + 19)$$

$$n - 20 = 0 \quad or \quad n + 19 = 0$$
$$n = 20 \ or \qquad n = -19$$

***Check*.** The solutions of the equation are 20 and -19. Since the number of people cannot be negative, -19 cannot be a solution. However, 20 checks since $\frac{1}{2}(20^2 - 20) = \frac{1}{2}(400 - 20) = \frac{1}{2}(380) = 190$.

***State*.** 20 people took part in the toast.

19. *Familiarize*. The page numbers on facing pages are consecutive integers. Let $x =$ the smaller integer. Then $x + 1 =$ the larger integer.

***Translate*.** We reword the problem.

$$\underbrace{\text{Smaller integer}}_{x} \underbrace{\text{times}}_{\cdot} \underbrace{\text{larger integer}}_{(x+1)} \underbrace{\text{is}}_{=} \underbrace{210.}_{210}$$

***Solve*.** We solve the equation.

$$x(x + 1) = 210$$

$$x^2 + x = 210$$

$$x^2 + x - 210 = 0$$

$$(x + 15)(x - 14) = 0$$

$$x + 15 = 0 \quad \text{or} \quad x - 14 = 0$$
$$x = -15 \quad \text{or} \qquad x = 14$$

***Check*.** The solutions of the equation are -15 and 14. Since a page number cannot be negative, -15 cannot be a solution of the original problem. We only need to check 14. When $x = 14$, then $x + 1 = 15$, and $14 \cdot 15 = 210$. This checks.

***State*.** The page numbers are 14 and 15.

21. *Familiarize*. Let $x =$ the smaller even integer. Then $x + 2 =$ the larger even integer.

***Translate*.** We reword the problem.

$$\underbrace{\begin{array}{c}\text{Smaller}\\ \text{even integer}\end{array}}_{x} \underbrace{\text{times}}_{\cdot} \underbrace{\begin{array}{c}\text{larger}\\ \text{even integer}\end{array}}_{(x+2)} \underbrace{\text{is}}_{=} \underbrace{168.}_{168}$$

***Solve*.**

$$x(x + 2) = 168$$

$$x^2 + 2x = 168$$

$$x^2 + 2x - 168 = 0$$

$$(x + 14)(x - 12) = 0$$

$$x + 14 = 0 \quad \text{or} \quad x - 12 = 0$$
$$x = -14 \quad \text{or} \qquad x = 12$$

***Check*.** The solutions of the equation are -14 and 12. When x is -14, then $x + 2$ is -12 and $-14(-12) = 168$. The numbers -14 and -12 are consecutive even integers which are solutions of the problem. When x is 12, then $x + 2$ is 14 and $12 \cdot 14 = 168$. The numbers 12 and 14 are also consecutive even integers which are solutions of the problem.

***State*.** We have two solutions, each of which consists of a pair of numbers: -14 and -12, and 12 and 14.

23. *Familiarize*. Let $x =$ the smaller odd integer. Then $x + 2 =$ the larger odd integer.

***Translate*.** We reword the problem.

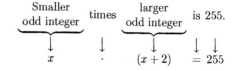

Solve.
$$x(x+2) = 255$$
$$x^2 + 2x = 255$$
$$x^2 + 2x - 255 = 0$$
$$(x-15)(x+17) = 0$$
$$x - 15 = 0 \quad \text{or} \quad x + 17 = 0$$
$$x = 15 \quad \text{or} \quad x = -17$$

Check. The solutions of the equation are 15 and -17. When x is 15, then $x + 2$ is 17 and $15 \cdot 17 = 255$. The numbers 15 and 17 are consecutive odd integers which are solutions to the problem. When x is -17, then $x + 2$ is -15 and $-17(-15) = 255$. The numbers -17 and -15 are also consecutive odd integers which are solutions to the problem.

State. We have two solutions, each of which consists of a pair of numbers: 15 and 17, and -17 and -15.

25. Familiarize. We make a drawing. Let $x =$ the length of the unknown leg. Then $x + 2 =$ the length of the hypotenuse.

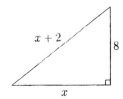

Translate. Use the Pythagorean theorem.
$$a^2 + b^2 = c^2$$
$$8^2 + x^2 = (x+2)^2$$
Solve. We solve the equation.
$$8^2 + x^2 = (x+2)^2$$
$$64 + x^2 = x^2 + 4x + 4$$
$$60 = 4x \qquad \text{Subtracting } x^2 \text{ and } 4$$
$$15 = x$$

Check. When $x = 15$, then $x + 2 = 17$ and $8^2 + 15^2 = 17^2$. Thus, 15 and 17 check.

State. The lengths of the hypotenuse and the other leg are 17 ft and 15 ft, respectively.

27. Familiarize. Consider the drawing in the text. We let $w =$ the width of Main Street, in feet.

Translate. Use the Pythagorean theorem.
$$a^2 + b^2 = c^2$$
$$24^2 + w^2 = 40^2$$
Solve. We solve the equation.
$$24^2 + w^2 = 40^2$$
$$576 + w^2 = 1600$$
$$w^2 - 1024 = 0$$
$$(w+32)(w-32) = 0$$
$$w + 32 = 0 \quad \text{or} \quad w - 32 = 0$$
$$w = -32 \quad \text{or} \quad w = 32$$

Check. The width of the street cannot be negative, so -32 cannot be a solution. If Main Street is 32 ft wide, we have $24^2 + 32^2 = 576 + 1024 = 1600$, which is 40^2. Thus, 32 ft checks.

State. Main Street is 32 ft wide.

29. Familiarize. We make a drawing. Let $l =$ the length of the cable, in ft.

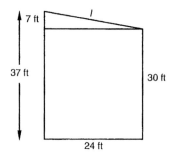

Note that we have a right triangle with hypotenuse l and legs of 24 ft and $37 - 30$, or 7 ft.

Translate. We use the Pythagorean theorem.
$$a^2 + b^2 = c^2$$
$$7^2 + 24^2 = l^2 \quad \text{Substituting}$$
Solve.
$$7^2 + 24^2 = l^2$$
$$49 + 576 = l^2$$
$$625 = l^2$$
$$0 = l^2 - 625$$
$$0 = (l+25)(l-25)$$
$$l + 25 = 0 \quad \text{or} \quad l - 25 = 0$$
$$l = -25 \quad \text{or} \quad l = 25$$

Check. The integer -25 cannot be the length of the cable, because it is negative. When $l = 25$, we have $7^2 + 24^2 = 25^2$. This checks.

State. The cable is 25 ft long.

31. Familiarize. We label the drawing. Let $x =$ the length of a side of the dining room, in ft. Then the dining room has dimensions x by x and the kitchen has dimensions x by 10. The entire rectangular space has dimension x by $x + 10$. Recall that we multiply these dimensions to find the area of the rectangle.

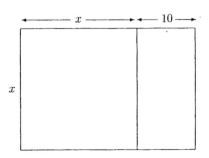

Translate.

$$\underbrace{\text{The area of the rectangular space}} \text{ is } \underbrace{264 \text{ ft}^2.}$$
$$\downarrow\downarrow\downarrow$$
$$x(x+10)=264$$

Solve. We solve the equation.

$$x(x+10) = 264$$
$$x^2 + 10x = 264$$
$$x^2 + 10x - 264 = 0$$
$$(x+22)(x-12) = 0$$
$$x + 22 = 0 \quad or \quad x - 12 = 0$$
$$x = -22 \ or \qquad x = 12$$

Check. Since the length of a side of the dining room must be positive, -22 cannot be a solution. If x is 12 ft, then $x+10$ is 22 ft, and the area of the space is $12 \cdot 22$, or 264 ft^2. The number 12 checks.

State. The dining room is 12 ft by 12 ft, and the kitchen is 12 ft by 10 ft.

33. *Familiarize.* We will use the formula $h = 180t - 16t^2$.

Translate. Substitute 464 for h.

$$464 = 180t - 16t^2$$

Solve. We solve the equation.

$$464 = 180t - 16t^2$$
$$16t^2 - 180t + 464 = 0$$
$$4(4t^2 - 45t + 116) = 0$$
$$4(4t - 29)(t - 4) = 0$$
$$4t - 29 = 0 \quad or \quad t - 4 = 0$$
$$4t = 29 \quad or \qquad t = 4$$
$$t = \frac{29}{4} \quad or \qquad t = 4$$

Check. The solutions of the equation are $\frac{29}{4}$, or $7\frac{1}{4}$, and 4. Since we want to find how many seconds it takes the rocket to *first* reach a height of 464 ft, we check the smaller number, 4. We substitute 4 for t in the formula.

$$h = 180t - 16t^2$$
$$h = 180 \cdot 4 - 16(4)^2$$
$$h = 180 \cdot 4 - 16 \cdot 16$$
$$h = 720 - 256$$
$$h = 464$$

The answer checks.

State. The rocket will first reach a height of 464 ft after 4 seconds.

35. *Familiarize.* Let $x = $ the smaller odd positive integer. Then $x + 2 = $ the larger odd positive integer.

Translate.

$$\underbrace{\text{Square of the smaller odd positive integer}} + \underbrace{\text{Square of the larger odd positive integer}} \text{ is } 74$$
$$\downarrow\downarrow\downarrow\downarrow\downarrow$$
$$x^2+(x+2)^2=74$$

Solve.

$$x^2 + (x+2)^2 = 74$$
$$x^2 + x^2 + 4x + 4 = 74$$
$$2x^2 + 4x - 70 = 0$$
$$2(x^2 + 2x - 35) = 0$$
$$2(x+7)(x-5) = 0$$
$$x + 7 = 0 \quad or \quad x - 5 = 0$$
$$x = -7 \quad or \qquad x = 5$$

Check. The solutions of the equation are -7 and 5. The problem asks for odd positive integers, so -7 cannot be a solution. When x is 5, $x+2$ is 7. The numbers 5 and 7 are consecutive odd positive integers. The sum of their squares, $25 + 49$, is 74. The numbers check.

State. The integers are 5 and 7.

37. Discussion and Writing Exercise

39. $(3x - 5y)(3x + 5y) = (3x)^2 - (5y)^2 = 9x^2 - 25y^2$

41. $(3x + 5y)^2 = (3x)^2 + 2 \cdot 3x \cdot 5y + (5y)^2 = 9x^2 + 30xy + 25y^2$

43. $4x - 16y = 64$

To find the x-intercept, let $y = 0$ and solve for x.

$$4x - 16y = 64$$
$$4x - 16 \cdot 0 = 64$$
$$4x = 64$$
$$x = 16$$

The x-intercept is $(16, 0)$.

To find the y-intercept, let $x = 0$ and solve for y.

$$4x - 16y = 64$$
$$4 \cdot 0 - 16y = 64$$
$$-16y = 64$$
$$y = -4$$

The y-intercept is $(0, -4)$.

45. $x - 1.3y = 6.5$

To find the x-intercept, let $y = 0$ and solve for x.

$$x - 1.3y = 6.5$$
$$x - 1.3(0) = 6.5$$
$$x = 6.5$$

The x-intercept is $(6.5, 0)$.

To find the y-intercept, let $x = 0$ and solve for y.

$$x - 1.3y = 6.5$$
$$0 - 1.3y = 6.5$$
$$-1.3y = 6.5$$
$$y = -5$$

The y-intercept is $(0, -5)$.

47. $y = 4 - 5x$

To find the x-intercept, let $y = 0$ and solve for x.

$$y = 4 - 5x$$
$$0 = 4 - 5x$$
$$5x = 4$$
$$x = \frac{4}{5}$$

The x-intercept is $\left(\frac{4}{5}, 0\right)$.

To find the y-intercept, let $x = 0$ and solve for y.

$$y = 4 - 5x$$
$$y = 4 - 5 \cdot 0$$
$$y = 4$$

The y-intercept is $(0, 4)$.

49. Familiarize. First we can use the Pythagorean theorem to find x, in ft. Then the height of the telephone pole is $x + 5$.

Translate. We use the Pythagorean theorem.
$$a^2 + b^2 = c^2$$
$$\left(\frac{1}{2}x + 1\right)^2 + x^2 = 34^2$$

Solve. We solve the equation.

$$\left(\frac{1}{2}x + 1\right)^2 + x^2 = 34^2$$
$$\frac{1}{4}x^2 + x + 1 + x^2 = 1156$$
$$x^2 + 4x + 4 + 4x^2 = 4624 \qquad \text{Multiplying by 4}$$
$$5x^2 + 4 + 4 = 4624$$
$$5x^2 + 4x - 4620 = 0$$
$$(5x + 154)(x - 30) = 0$$
$$5x + 154 = 0 \qquad or \quad x - 30 = 0$$
$$5x = -154 \quad or \qquad x = 30$$
$$x = -30.8 \quad or \qquad x = 30$$

Check. Since the length x must be positive, -30.8 cannot be a solution. If x is 30 ft, then $\frac{1}{2}x + 1$ is $\frac{1}{2} \cdot 30 + 1$, or 16 ft. Since $16^2 + 30^2 = 1156 = 34^2$, the number 30 checks. When x is 30 ft, then $x + 5$ is 35 ft.

State. The height of the telephone pole is 35 ft.

51. Familiarize. Using the labels shown on the drawing in the text, we let $x =$ the width of the walk. Then the length and width of the rectangle formed by the pool and walk together are $40 + 2x$ and $20 + 2x$, respectively.

Translate.

$$\underbrace{\text{Area}}_{1500} \quad \text{is} \quad \underbrace{\text{length}}_{(40 + 2x)} \quad \text{times} \quad \underbrace{\text{width.}}_{(20 + 2x)}$$

Solve. We solve the equation.

$$1500 = (40 + 2x)(20 + 2x)$$
$$1500 = 2(20 + x) \cdot 2(10 + x) \qquad \text{Factoring 2 out of each factor on the right}$$
$$1500 = 4 \cdot (20 + x)(10 + x)$$
$$375 = (20 + x)(10 + x) \qquad \text{Dividing by 4}$$
$$375 = 200 + 30x + x^2$$
$$0 = x^2 + 30x - 175$$
$$0 = (x + 35)(x - 5)$$
$$x + 35 = 0 \qquad or \quad x - 5 = 0$$
$$x = -35 \quad or \qquad x = 5$$

Check. The solutions of the equation are -35 and 5. Since the width of the walk cannot be negative, -35 is not a solution. When $x = 5$, $40 + 2x = 40 + 2 \cdot 5$, or 50 and $20 + 2x = 20 + 2 \cdot 5$, or 30. The total area of the pool and walk is $50 \cdot 30$, or 1500 ft^2. This checks.

State. The width of the walk is 5 ft.

53. Familiarize. We make a drawing. Let $w =$ the width of the piece of cardboard. Then $2w =$ the length.

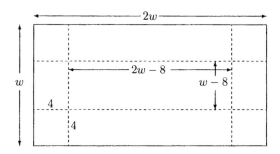

The box will have length $2w - 8$, width $w - 8$, and height 4. Recall that the formula for volume is $V =$ length $\times$ width $\times$ height.

Translate.

$$\underbrace{\text{The volume}}_{(2w - 8)(w - 8)(4)} \quad \underbrace{\text{is}}_{=} \quad \underbrace{616\text{cm}^3.}_{616}$$

Solve. We solve the equation.

$$(2w - 8)(w - 8)(4) = 616$$
$$(2w^2 - 24w + 64)(4) = 616$$
$$8w^2 - 96w + 256 = 616$$
$$8w^2 - 96w - 360 = 0$$
$$8(w^2 - 12w - 45) = 0$$
$$w^2 - 12w - 45 = 0 \qquad \text{Dividing by 8}$$
$$(w - 15)(w + 3) = 0$$
$$w - 15 = 0 \qquad or \quad w + 3 = 0$$
$$w = 15 \quad or \qquad w = -3$$

Check. The width cannot be negative, so we only need to check 15. When $w = 15$, then $2w = 30$ and the dimensions of the box are $30 - 8$ by $15 - 8$ by 4, or 22 by 7 by 4. The volume is $22 \cdot 7 \cdot 4$, or 616.

State. The cardboard is 30 cm by 15 cm.

55. We add labels to the drawing in the text.

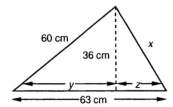

First we will use the Pythagorean theorem to find y.

$$y^2 + 36^2 = 60^2$$
$$y^2 + 1296 = 3600$$
$$y^2 - 2304 = 0$$
$$(y + 48)(y - 48) = 0$$

$$y + 48 = 0 \quad \text{or} \quad y - 48 = 0$$
$$y = -48 \quad \text{or} \quad y = 48$$

Since y cannot be negative, we use $y = 48$. Now we subtract to find z.

$$z = 63 - 48 = 15$$

Now we use the Pythagorean theorem to find x.

$$15^2 + 36^2 = x^2$$
$$225 + 1296 = x^2$$
$$1521 = x^2$$
$$0 = x^2 - 1521$$
$$0 = (x + 39)(x - 39)$$

$$x + 39 = 0 \quad \text{or} \quad x - 39 = 0$$
$$x = -39 \quad \text{or} \quad x = 39$$

Since x cannot be negative, -39 cannot be a solution. Thus, we find that x is 39 cm.

Chapter 11

Rational Expressions and Equations

1. $\dfrac{-3}{2x}$

To determine the numbers for which the rational expression is not defined, we set the denominator equal to 0 and solve:

$$2x = 0$$
$$x = 0$$

The expression is not defined for the replacement number 0.

3. $\dfrac{5}{x-8}$

To determine the numbers for which the rational expression is not defined, we set the denominator equal to 0 and solve:

$$x - 8 = 0$$
$$x = 8$$

The expression is not defined for the replacement number 8.

5. $\dfrac{3}{2y+5}$

Set the denominator equal to 0 and solve:

$$2y + 5 = 0$$
$$2y = -5$$
$$y = -\frac{5}{2}$$

The expression is not defined for the replacement number $-\dfrac{5}{2}$.

7. $\dfrac{x^2 + 11}{x^2 - 3x - 28}$

Set the denominator equal to 0 and solve:

$$x^2 - 3x - 28 = 0$$
$$(x - 7)(x + 4) = 0$$
$$x - 7 = 0 \quad \text{or} \quad x + 4 = 0$$
$$x = 7 \quad \text{or} \qquad x = -4$$

The expression is not defined for the replacement numbers 7 and -4.

9. $\dfrac{m^3 - 2m}{m^2 - 25}$

Set the denominator equal to 0 and solve:

$$m^2 - 25 = 0$$
$$(m + 5)(m - 5) = 0$$
$$m + 5 = 0 \quad \text{or} \quad m - 5 = 0$$
$$m = -5 \quad \text{or} \qquad m = 5$$

The expression is not defined for the replacement numbers -5 and 5.

11. $\dfrac{x-4}{3}$

Since the denominator is the constant 3, there are no replacement numbers for which the expression is not defined.

13. $\dfrac{4x}{4x} \cdot \dfrac{3x^2}{5y} = \dfrac{(4x)(3x^2)}{(4x)(5y)}$ Multiplying the numerators and the denominators

15. $\dfrac{2x}{2x} \cdot \dfrac{x-1}{x+4} = \dfrac{2x(x-1)}{2x(x+4)}$ Multiplying the numerators and the denominators

17. $\dfrac{3-x}{4-x} \cdot \dfrac{-1}{-1} = \dfrac{(3-x)(-1)}{(4-x)(-1)}$, or $\dfrac{-1(3-x)}{-1(4-x)}$

19. $\dfrac{y+6}{y+6} \cdot \dfrac{y-7}{y+2} = \dfrac{(y+6)(y-7)}{(y+6)(y+2)}$

21. $\dfrac{8x^3}{32x} = \dfrac{8 \cdot x \cdot x^2}{8 \cdot 4 \cdot x}$ Factoring numerator and denominator

$\qquad = \dfrac{8x}{8x} \cdot \dfrac{x^2}{4}$ Factoring the rational expression

$\qquad = 1 \cdot \dfrac{x^2}{4} \qquad \left(\dfrac{8x}{8x} = 1\right)$

$\qquad = \dfrac{x^2}{4}$ We removed a factor of 1.

23. $\dfrac{48p^7 q^5}{18p^5 q^4} = \dfrac{8 \cdot 6 \cdot p^5 \cdot p^2 \cdot q^4 \cdot q}{6 \cdot 3 \cdot p^5 \cdot q^4}$ Factoring numerator and denominator

$\qquad = \dfrac{6p^5 q^4}{6p^5 q^4} \cdot \dfrac{8p^2 q}{3}$ Factoring the rational expression

$\qquad = 1 \cdot \dfrac{8p^2 q}{3} \qquad \left(\dfrac{6p^5 q^4}{6p^5 q^4} = 1\right)$

$\qquad = \dfrac{8p^2 q}{3}$ Removing a factor of 1

25. $\dfrac{4x - 12}{4x} = \dfrac{4(x-3)}{4 \cdot x}$

$\qquad = \dfrac{4}{4} \cdot \dfrac{x-3}{x}$

$\qquad = 1 \cdot \dfrac{x-3}{x}$

$\qquad = \dfrac{x-3}{x}$

27. $\dfrac{3m^2 + 3m}{6m^2 + 9m} = \dfrac{3m(m+1)}{3m(2m+3)}$

$\qquad = \dfrac{3m}{3m} \cdot \dfrac{m+1}{2m+3}$

$\qquad = 1 \cdot \dfrac{m+1}{2m+3}$

$\qquad = \dfrac{m+1}{2m+3}$

29. $\dfrac{a^2-9}{a^2+5a+6} = \dfrac{(a-3)(a+3)}{(a+2)(a+3)}$

$\phantom{\dfrac{a^2-9}{a^2+5a+6}} = \dfrac{a-3}{a+2} \cdot \dfrac{a+3}{a+3}$

$\phantom{\dfrac{a^2-9}{a^2+5a+6}} = \dfrac{a-3}{a+2} \cdot 1$

$\phantom{\dfrac{a^2-9}{a^2+5a+6}} = \dfrac{a-3}{a+2}$

31. $\dfrac{a^2-10a+21}{a^2-11a+28} = \dfrac{(a-7)(a-3)}{(a-7)(a-4)}$

$\phantom{\dfrac{a^2-10a+21}{a^2-11a+28}} = \dfrac{a-7}{a-7} \cdot \dfrac{a-3}{a-4}$

$\phantom{\dfrac{a^2-10a+21}{a^2-11a+28}} = 1 \cdot \dfrac{a-3}{a-4}$

$\phantom{\dfrac{a^2-10a+21}{a^2-11a+28}} = \dfrac{a-3}{a-4}$

33. $\dfrac{x^2-25}{x^2-10x+25} = \dfrac{(x-5)(x+5)}{(x-5)(x-5)}$

$\phantom{\dfrac{x^2-25}{x^2-10x+25}} = \dfrac{x-5}{x-5} \cdot \dfrac{x+5}{x-5}$

$\phantom{\dfrac{x^2-25}{x^2-10x+25}} = 1 \cdot \dfrac{x+5}{x-5}$

$\phantom{\dfrac{x^2-25}{x^2-10x+25}} = \dfrac{x+5}{x-5}$

35. $\dfrac{a^2-1}{a-1} = \dfrac{(a-1)(a+1)}{a-1}$

$\phantom{\dfrac{a^2-1}{a-1}} = \dfrac{a-1}{a-1} \cdot \dfrac{a+1}{1}$

$\phantom{\dfrac{a^2-1}{a-1}} = 1 \cdot \dfrac{a+1}{1}$

$\phantom{\dfrac{a^2-1}{a-1}} = a+1$

37. $\dfrac{x^2+1}{x+1}$ cannot be simplified.

Neither the numerator nor the denominator can be factored.

39. $\dfrac{6x^2-54}{4x^2-36} = \dfrac{2\cdot3(x^2-9)}{2\cdot2(x^2-9)}$

$\phantom{\dfrac{6x^2-54}{4x^2-36}} = \dfrac{2(x^2-9)}{2(x^2-9)} \cdot \dfrac{3}{2}$

$\phantom{\dfrac{6x^2-54}{4x^2-36}} = 1 \cdot \dfrac{3}{2}$

$\phantom{\dfrac{6x^2-54}{4x^2-36}} = \dfrac{3}{2}$

41. $\dfrac{6t+12}{t^2-t-6} = \dfrac{6(t+2)}{(t-3)(t+2)}$

$\phantom{\dfrac{6t+12}{t^2-t-6}} = \dfrac{6}{t-3} \cdot \dfrac{t+2}{t+2}$

$\phantom{\dfrac{6t+12}{t^2-t-6}} = \dfrac{6}{t-3} \cdot 1$

$\phantom{\dfrac{6t+12}{t^2-t-6}} = \dfrac{6}{t-3}$

43. $\dfrac{2t^2+6t+4}{4t^2-12t-16} = \dfrac{2(t^2+3t+2)}{4(t^2-3t-4)}$

$\phantom{\dfrac{2t^2+6t+4}{4t^2-12t-16}} = \dfrac{2(t+2)(t+1)}{2\cdot2(t-4)(t+1)}$

$\phantom{\dfrac{2t^2+6t+4}{4t^2-12t-16}} = \dfrac{2(t+1)}{2(t+1)} \cdot \dfrac{t+2}{2(t-4)}$

$\phantom{\dfrac{2t^2+6t+4}{4t^2-12t-16}} = 1 \cdot \dfrac{t+2}{2(t-4)}$

$\phantom{\dfrac{2t^2+6t+4}{4t^2-12t-16}} = \dfrac{t+2}{2(t-4)}$

45. $\dfrac{t^2-4}{(t+2)^2} = \dfrac{(t-2)(t+2)}{(t+2)(t+2)}$

$\phantom{\dfrac{t^2-4}{(t+2)^2}} = \dfrac{t-2}{t+2} \cdot \dfrac{t+2}{t+2}$

$\phantom{\dfrac{t^2-4}{(t+2)^2}} = \dfrac{t-2}{t+2} \cdot 1$

$\phantom{\dfrac{t^2-4}{(t+2)^2}} = \dfrac{t-2}{t+2}$

47. $\dfrac{6-x}{x-6} = \dfrac{-(-6+x)}{x-6}$

$\phantom{\dfrac{6-x}{x-6}} = \dfrac{-1(x-6)}{x-6}$

$\phantom{\dfrac{6-x}{x-6}} = -1 \cdot \dfrac{x-6}{x-6}$

$\phantom{\dfrac{6-x}{x-6}} = -1 \cdot 1$

$\phantom{\dfrac{6-x}{x-6}} = -1$

49. $\dfrac{a-b}{b-a} = \dfrac{-(-a+b)}{b-a}$

$\phantom{\dfrac{a-b}{b-a}} = \dfrac{-1(b-a)}{b-a}$

$\phantom{\dfrac{a-b}{b-a}} = -1 \cdot \dfrac{b-a}{b-a}$

$\phantom{\dfrac{a-b}{b-a}} = -1 \cdot 1$

$\phantom{\dfrac{a-b}{b-a}} = -1$

51. $\dfrac{6t-12}{2-t} = \dfrac{-6(-t+2)}{2-t}$

$\phantom{\dfrac{6t-12}{2-t}} = \dfrac{-6(2-t)}{2-t}$

$\phantom{\dfrac{6t-12}{2-t}} = \dfrac{-6(2-t)}{2-t}$

$\phantom{\dfrac{6t-12}{2-t}} = -6$

53. $\dfrac{x^2-1}{1-x} = \dfrac{(x+1)(x-1)}{-1(-1+x)}$

$\phantom{\dfrac{x^2-1}{1-x}} = \dfrac{(x+1)(x-1)}{-1(x-1)}$

$\phantom{\dfrac{x^2-1}{1-x}} = \dfrac{(x+1)(x-1)}{-1(x-1)}$

$\phantom{\dfrac{x^2-1}{1-x}} = -(x+1)$

$\phantom{\dfrac{x^2-1}{1-x}} = -x-1$

55. $\dfrac{4x^3}{3x} \cdot \dfrac{14}{x} = \dfrac{4x^3 \cdot 14}{3x \cdot x}$ Multiplying the numerators and the denominators

$= \dfrac{4 \cdot x \cdot x \cdot x \cdot 14}{3 \cdot x \cdot x}$ Factoring the numerator and the denominator

$= \dfrac{4 \cdot \cancel{x} \cdot \cancel{x} \cdot x \cdot 14}{3 \cdot \cancel{x} \cdot \cancel{x}}$ Removing a factor of 1

$= \dfrac{56x}{3}$ Simplifying

57. $\dfrac{3c}{d^2} \cdot \dfrac{4d}{6c^3} = \dfrac{3c \cdot 4d}{d^2 \cdot 6c^3}$ Multiplying the numerators and the denominators

$= \dfrac{3 \cdot c \cdot 2 \cdot 2 \cdot d}{d \cdot d \cdot 3 \cdot 2 \cdot c \cdot c \cdot c}$ Factoring the numerator and the denominator

$= \dfrac{\cancel{3} \cdot \cancel{c} \cdot \cancel{2} \cdot 2 \cdot \cancel{d}}{\cancel{d} \cdot d \cdot \cancel{3} \cdot \cancel{2} \cdot \cancel{c} \cdot c \cdot c}$

$= \dfrac{2}{dc^2}$

59. $\dfrac{x^2 - 3x - 10}{(x-2)^2} \cdot \dfrac{x-2}{x-5} = \dfrac{(x^2 - 3x - 10)(x-2)}{(x-2)^2(x-5)}$

$= \dfrac{(x-5)(x+2)(x-2)}{(x-2)(x-2)(x-5)}$

$= \dfrac{(\cancel{x-5})(x+2)(\cancel{x-2})}{(\cancel{x-2})(x-2)(\cancel{x-5})}$

$= \dfrac{x+2}{x-2}$

61. $\dfrac{a^2 - 9}{a^2} \cdot \dfrac{a^2 - 3a}{a^2 + a - 12} = \dfrac{(a-3)(a+3)(a)(a-3)}{a \cdot a(a+4)(a-3)}$

$= \dfrac{(\cancel{a-3})(a+3)(\cancel{a})(a-3)}{\cancel{a} \cdot a(a+4)(\cancel{a-3})}$

$= \dfrac{(a-3)(a+3)}{a(a+4)}$

63. $\dfrac{4a^2}{3a^2 - 12a + 12} \cdot \dfrac{3a - 6}{2a} = \dfrac{4a^2(3a-6)}{(3a^2 - 12a + 12)2a}$

$= \dfrac{2 \cdot 2 \cdot a \cdot a \cdot 3 \cdot (a-2)}{3 \cdot (a-2) \cdot (a-2) \cdot 2 \cdot a}$

$= \dfrac{\cancel{2} \cdot 2 \cdot \cancel{a} \cdot a \cdot \cancel{3} \cdot (\cancel{a-2})}{\cancel{3} \cdot (\cancel{a-2}) \cdot (a-2) \cdot \cancel{2} \cdot \cancel{a}}$

$= \dfrac{2a}{a-2}$

65. $\dfrac{t^4 - 16}{t^4 - 1} \cdot \dfrac{t^2 + 1}{t^2 + 4}$

$= \dfrac{(t^4 - 16)(t^2 + 1)}{(t^4 - 1)(t^2 + 4)}$

$= \dfrac{(t^2 + 4)(t+2)(t-2)(t^2 + 1)}{(t^2 + 1)(t+1)(t-1)(t^2 + 4)}$

$= \dfrac{(\cancel{t^2 + 4})(t+2)(t-2)(\cancel{t^2 + 1})}{(\cancel{t^2 + 1})(t+1)(t-1)(\cancel{t^2 + 4})}$

$= \dfrac{(t+2)(t-2)}{(t+1)(t-1)}$

67. $\dfrac{(x+4)^3}{(x+2)^3} \cdot \dfrac{x^2 + 4x + 4}{x^2 + 8x + 16}$

$= \dfrac{(x+4)^3(x^2 + 4x + 4)}{(x+2)^3(x^2 + 8x + 16)}$

$= \dfrac{(x+4)(x+4)(x+4)(x+2)(x+2)}{(x+2)(x+2)(x+2)(x+4)(x+4)}$

$= \dfrac{(\cancel{x+4})(\cancel{x+4})(x+4)(\cancel{x+2})(\cancel{x+2})}{(\cancel{x+2})(\cancel{x+2})(x+2)(\cancel{x+4})(\cancel{x+4})}$

$= \dfrac{x+4}{x+2}$

69. $\dfrac{5a^2 - 180}{10a^2 - 10} \cdot \dfrac{20a + 20}{2a - 12} = \dfrac{(5a^2 - 180)(20a + 20)}{(10a^2 - 10)(2a - 12)}$

$= \dfrac{5(a+6)(a-6)(2)(10)(a+1)}{10(a+1)(a-1)(2)(a-6)}$

$= \dfrac{5(a+6)(\cancel{a-6})(\cancel{2})(\cancel{10})(\cancel{a+1})}{\cancel{10}(\cancel{a+1})(a-1)(\cancel{2})(\cancel{a-6})}$

$= \dfrac{5(a+6)}{a-1}$

71. Discussion and Writing Exercise

73. *Familiarize.* Let $x =$ the smaller even integer. Then $x + 2 =$ the larger even integer.

Translate. We reword the problem.

Smaller even integer	times	larger even integer	is 360.
↓	↓	↓	↓ ↓
x	$\cdot$	$(x+2)$	$= 360$

Solve.

$$x(x + 2) = 360$$
$$x^2 + 2x = 360$$
$$x^2 + 2x - 360 = 0$$
$$(x + 20)(x - 18) = 0$$
$$x + 20 = 0 \quad \text{or} \quad x - 18 = 0$$
$$x = -20 \quad \text{or} \quad x = 18$$

Check. The solutions of the equation are -20 and 18. When $x = -20$, then $x + 2 = -18$ and $-20(-18) = 360$. The numbers -20 and -18 are consecutive even integers which are solutions to the problem. When $x = 18$, then $x + 2 = 20$ and $18 \cdot 20 = 360$. The numbers 18 and 20 are also consecutive even integers which are solutions to the problem.

State. We have two solutions, each of which consists of a pair of numbers: -20 and -18, and 18 and 20.

75. $x^2 - x - 56$

We look for a pair of numbers whose product is -56 and whose sum is -1. The numbers are -8 and 7.

$$x^2 - x - 56 = (x - 8)(x + 7)$$

77. $x^5 - 2x^4 - 35x^3 = x^3(x^2 - 2x - 35) = x^3(x - 7)(x + 5)$

79. $16 - t^4 = 4^2 - (t^2)^2$ Difference of squares

$= (4 + t^2)(4 - t^2)$

$= (4 + t^2)(2^2 - t^2)$ Difference of squares

$= (4 + t^2)(2 + t)(2 - t)$

81. $x^2 - 9x + 14$

We look for a pair of numbers whose product is 14 and whose sum is -9. The numbers are -2 and -7.

$$x^2 - 9x + 14 = (x - 2)(x - 7)$$

83. $16x^2 - 40xy + 25y^2$

$= (4x)^2 - 2 \cdot 4x \cdot 5y + (5y)^2$ Trinomial square

$= (4x - 5y)^2$

85. $\dfrac{x^4 - 16y^2}{(x^2 + 4y^2)(x - 2y)}$

$= \dfrac{(x^2 + 4y^2)(x + 2y)(x - 2y)}{(x^2 + 4y^2)(x - 2y)}$

$= \dfrac{(x^2 + 4y^2)\,(x + 2y)(x - 2y)}{(x^2 + 4y^2)\,(x - 2y)(1)}$

$= x + 2y$

87. $\dfrac{t^4 - 1}{t^4 - 81} \cdot \dfrac{t^2 - 9}{t^2 + 1} \cdot \dfrac{(t - 9)^2}{(t + 1)^2}$

$= \dfrac{(t^2 + 1)(t + 1)(t - 1)(t + 3)(t - 3)(t - 9)(t - 9)}{(t^2 + 9)(t + 3)(t - 3)(t^2 + 1)(t + 1)(t + 1)}$

$= \dfrac{(t^2 + 1)(t + 1)(t - 1)(t + 3)(t - 3)(t - 9)(t - 9)}{(t^2 + 9)(t + 3)(t - 3)(t^2 + 1)(t + 1)(t + 1)}$

$= \dfrac{(t - 1)(t - 9)(t - 9)}{(t^2 + 9)(t + 1)}$, or $\dfrac{(t - 1)(t - 9)^2}{(t^2 + 9)(t + 1)}$

89. $\dfrac{x^2 - y^2}{(x - y)^2} \cdot \dfrac{x^2 - 2xy + y^2}{x^2 - 4xy - 5y^2}$

$= \dfrac{(x + y)(x - y)(x - y)(x - y)}{(x - y)(x - y)(x - 5y)(x + y)}$

$= \dfrac{(x + y)(x - y)(x - y)(x - y)}{(x - y)(x - y)(x - 5y)(x + y)}$

$= \dfrac{x - y}{x - 5y}$

91. $\dfrac{5(2x + 5) - 25}{10} = \dfrac{10x + 25 - 25}{10}$

$= \dfrac{10x}{10}$

$= x$

You get the same number you selected.

To do a number trick, ask someone to select a number and then perform these operations. The person will probably be surprised that the result is the original number.

Exercise Set 11.2

1. The reciprocal of $\dfrac{4}{x}$ is $\dfrac{x}{4}$ because $\dfrac{4}{x} \cdot \dfrac{x}{4} = 1$.

3. The reciprocal of $x^2 - y^2$ is $\dfrac{1}{x^2 - y^2}$ because

$\dfrac{x^2 - y^2}{1} \cdot \dfrac{1}{x^2 - y^2} = 1$.

5. The reciprocal of $\dfrac{1}{a + b}$ is $a + b$ because $\dfrac{1}{a + b} \cdot (a + b) = 1$.

7. The reciprocal of $\dfrac{x^2 + 2x - 5}{x^2 - 4x + 7}$ is $\dfrac{x^2 - 4x + 7}{x^2 + 2x - 5}$ because

$\dfrac{x^2 + 2x - 5}{x^2 - 4x + 7} \cdot \dfrac{x^2 - 4x + 7}{x^2 + 2x - 5} = 1$.

9. $\dfrac{2}{5} \div \dfrac{4}{3} = \dfrac{2}{3} \cdot \dfrac{3}{4}$ Multiplying by the reciprocal of the divisor

$= \dfrac{2 \cdot 3}{5 \cdot 4}$

$= \dfrac{2 \cdot 3}{5 \cdot 2 \cdot 2}$ Factoring the denominator

$= \dfrac{2 \cdot 3}{5 \cdot 2 \cdot 2}$ Removing a factor of 1

$= \dfrac{3}{10}$ Simplifying

11. $\dfrac{2}{x} \div \dfrac{8}{x} = \dfrac{2}{x} \cdot \dfrac{x}{8}$ Multiplying by the reciprocal of the divisor

$= \dfrac{2 \cdot x}{x \cdot 8}$

$= \dfrac{2 \cdot x \cdot 1}{x \cdot 2 \cdot 4}$ Factoring the numerator and the denominator

$= \dfrac{2 \cdot x \cdot 1}{x \cdot 2 \cdot 4}$ Removing a factor of 1

$= \dfrac{1}{4}$ Simplifying

13. $\dfrac{a}{b^2} \div \dfrac{a^2}{b^3} = \dfrac{a}{b^2} \cdot \dfrac{b^3}{a^2}$ Multiplying by the reciprocal of the divisor

$= \dfrac{a \cdot b^3}{b^2 \cdot a^2}$

$= \dfrac{a \cdot b^2 \cdot b}{b^2 \cdot a \cdot a}$

$= \dfrac{a \cdot b^2 \cdot b}{b^2 \cdot a \cdot a}$

$= \dfrac{b}{a}$

15. $\dfrac{a + 2}{a - 3} \div \dfrac{a - 1}{a + 3} = \dfrac{a + 2}{a - 3} \cdot \dfrac{a + 3}{a - 1}$

$= \dfrac{(a + 2)(a + 3)}{(a - 3)(a - 1)}$

17. $\dfrac{x^2 - 1}{x} \div \dfrac{x + 1}{x - 1} = \dfrac{x^2 - 1}{x} \cdot \dfrac{x - 1}{x + 1}$

$= \dfrac{(x^2 - 1)(x - 1)}{x(x + 1)}$

$= \dfrac{(x - 1)(x + 1)(x - 1)}{x(x + 1)}$

$= \dfrac{(x - 1)(x + 1)(x - 1)}{x(x + 1)}$

$= \dfrac{(x - 1)^2}{x}$

19. $\dfrac{x+1}{6} \div \dfrac{x+1}{3} = \dfrac{x+1}{6} \cdot \dfrac{3}{x+1}$

$= \dfrac{(x+1) \cdot 3}{6(x+1)}$

$= \dfrac{3(x+1)}{2 \cdot 3(x+1)}$

$= \dfrac{1 \cdot \cancel{3}\cancel{(x+1)}}{2 \cdot \cancel{3}\cancel{(x+1)}}$

$= \dfrac{1}{2}$

21. $\dfrac{5x-5}{16} \div \dfrac{x-1}{6} = \dfrac{5x-5}{16} \cdot \dfrac{6}{x-1}$

$= \dfrac{(5x-5) \cdot 6}{16(x-1)}$

$= \dfrac{5(x-1) \cdot 2 \cdot 3}{2 \cdot 8(x-1)}$

$= \dfrac{5\cancel{(x-1)} \cdot \cancel{2} \cdot 3}{\cancel{2} \cdot 8\cancel{(x-1)}}$

$= \dfrac{15}{8}$

23. $\dfrac{-6+3x}{5} \div \dfrac{4x-8}{25} = \dfrac{-6+3x}{5} \cdot \dfrac{25}{4x-8}$

$= \dfrac{(-6+3x) \cdot 25}{5(4x-8)}$

$= \dfrac{3(x-2) \cdot 5 \cdot 5}{5 \cdot 4(x-2)}$

$= \dfrac{3\cancel{(x-2)} \cdot \cancel{5} \cdot 5}{\cancel{5} \cdot 4\cancel{(x-2)}}$

$= \dfrac{15}{4}$

25. $\dfrac{a+2}{a-1} \div \dfrac{3a+6}{a-5} = \dfrac{a+2}{a-1} \cdot \dfrac{a-5}{3a+6}$

$= \dfrac{(a+2)(a-5)}{(a-1)(3a+6)}$

$= \dfrac{(a+2)(a-5)}{(a-1) \cdot 3 \cdot (a+2)}$

$= \dfrac{\cancel{(a+2)}(a-5)}{(a-1) \cdot 3 \cdot \cancel{(a+2)}}$

$= \dfrac{a-5}{3(a-1)}$

27. $\dfrac{x^2-4}{x} \div \dfrac{x-2}{x+2} = \dfrac{x^2-4}{x} \cdot \dfrac{x+2}{x-2}$

$= \dfrac{(x^2-4)(x+2)}{x(x-2)}$

$= \dfrac{(x-2)(x+2)(x+2)}{x(x-2)}$

$= \dfrac{\cancel{(x-2)}(x+2)(x+2)}{x\cancel{(x-2)}}$

$= \dfrac{(x+2)^2}{x}$

29. $\dfrac{x^2-9}{4x+12} \div \dfrac{x-3}{6} = \dfrac{x^2-9}{4x+12} \cdot \dfrac{6}{x-3}$

$= \dfrac{(x^2-9) \cdot 6}{(4x+12)(x-3)}$

$= \dfrac{(x-3)(x+3) \cdot 3 \cdot 2}{2 \cdot 2(x+3)(x-3)}$

$= \dfrac{\cancel{(x-3)}\cancel{(x+3)} \cdot 3 \cdot \cancel{2}}{\cancel{2} \cdot 2\cancel{(x+3)}\cancel{(x-3)}}$

$= \dfrac{3}{2}$

31. $\dfrac{c^2+3c}{c^2+2c-3} \div \dfrac{c}{c+1} = \dfrac{c^2+3c}{c^2+2c-3} \cdot \dfrac{c+1}{c}$

$= \dfrac{(c^2+3c)(c+1)}{(c^2+2c-3)c}$

$= \dfrac{c(c+3)(c+1)}{(c+3)(c-1)c}$

$= \dfrac{\cancel{c}\cancel{(c+3)}(c+1)}{\cancel{(c+3)}(c-1)\cancel{c}}$

$= \dfrac{c+1}{c-1}$

33. $\dfrac{2y^2-7y+3}{2y^2+3y-2} \div \dfrac{6y^2-5y+1}{3y^2+5y-2}$

$= \dfrac{2y^2-7y+3}{2y^2+3y-2} \cdot \dfrac{3y^2+5y-2}{6y^2-5y+1}$

$= \dfrac{(2y^2-7y+3)(3y^2+5y-2)}{(2y^2+3y-2)(6y^2-5y+1)}$

$= \dfrac{(2y-1)(y-3)(3y-1)(y+2)}{(2y-1)(y+2)(3y-1)(2y-1)}$

$= \dfrac{\cancel{(2y-1)}(y-3)\cancel{(3y-1)}\cancel{(y+2)}}{\cancel{(2y-1)}\cancel{(y+2)}\cancel{(3y-1)}(2y-1)}$

$= \dfrac{y-3}{2y-1}$

35. $\dfrac{x^2-1}{4x+4} \div \dfrac{2x^2-4x+2}{8x+8} = \dfrac{x^2-1}{4x+4} \cdot \dfrac{8x+8}{2x^2-4x+2}$

$= \dfrac{(x^2-1)(8x+8)}{(4x+4)(2x^2-4x+2)}$

$= \dfrac{(x+1)(x-1)(2)(4)(x+1)}{4(x+1)(2)(x-1)(x-1)}$

$= \dfrac{(x+1)\cancel{(x-1)}\cancel{(2)}\cancel{(4)}(x+1)}{\cancel{4}\cancel{(x+1)}\cancel{(2)}(x-1)\cancel{(x-1)}}$

$= \dfrac{x+1}{x-1}$

37. Discussion and Writing Exercise

39. **Familiarize.** Let $s =$ Bonnie's score on the last test.

Translate. The average of the four scores must be at least 90. This means it must be greater than or equal to 90. We translate.

$$\dfrac{96+98+89+s}{4} \geq 90$$

Solve. We solve the inequality. First we multiply by 4 to clear the fraction.

$$4\left(\frac{96+98+89+s}{4}\right) \geq 4 \cdot 90$$
$$96+98+89+s \geq 360$$
$$283+s \geq 360$$
$$s \geq 77 \qquad \text{Subtracting 283}$$

Check. We can do a partial check by substituting a value for s less than 77 and a value for s greater than 77.

For $s = 76$: $\dfrac{96+98+89+76}{4} = 89.75 < 90$

For $s = 78$: $\dfrac{96+98+89+78}{4} = 90.25 \leq 90$

Since the average is less than 90 for a value of s less than 77 and greater than or equal to 90 for a value greater than or equal to 77, the answer is probably correct.

State. The scores on the last test that will earn Bonnie an A are $\{s|s \geq 77\}$.

41. $(8x^3 - 3x^2 + 7) - (8x^2 + 3x - 5) =$
$8x^3 - 3x^2 + 7 - 8x^2 - 3x + 5 =$
$8x^3 - 11x^2 - 3x + 12$

43. $(2x^{-3}y^4)^2 = 2^2(x^{-3})^2(y^4)^2$

$\qquad = 2^2 x^{-6} y^8 \qquad$ Multiplying exponents

$\qquad = 4x^{-6}y^8 \qquad (2^2 = 4)$

$\qquad = \dfrac{4y^8}{x^6} \qquad \left(x^{-6} = \dfrac{1}{x^6}\right)$

45. $\left(\dfrac{2x^3}{y^5}\right)^2 = \dfrac{2^2(x^3)^2}{(y^5)^2}$

$\qquad = \dfrac{2^2 x^6}{y^{10}} \qquad$ Multiplying exponents

$\qquad = \dfrac{4x^6}{y^{10}} \qquad (2^2 = 4)$

47. $\dfrac{3a^2 - 5ab - 12b^2}{3ab + 4b^2} \div (3b^2 - ab)$

$= \dfrac{3a^2 - 5ab - 12b^2}{3ab + 4b^2} \cdot \dfrac{1}{3b^2 - ab}$

$= \dfrac{(3a+4b)(a-3b)}{b(3a+4b) \cdot b(3b-a)}$

$= \dfrac{(3a+4b)(-1)(3b-a)}{b(3a+4b) \cdot b(3b-a)}$

$= \dfrac{(3a+4b)(-1)(3b-a)}{b(3a+4b) \cdot b(3b-a)}$

$= -\dfrac{1}{b^2}$

49. The volume V of a rectangular solid is given by the formula $V = l \cdot w \cdot h$, where l = the length, w = the width, and h = the height. We substitute in the formula and solve for h.

$$V = l \cdot w \cdot h$$
$$x - 3 = \frac{x-3}{x-7} \cdot \frac{x+y}{x-7} \cdot h$$
$$\frac{x-7}{x-3} \cdot \frac{x-7}{x+y} \cdot (x-3) = \frac{x-7}{x-3} \cdot \frac{x-7}{x+y} \cdot \frac{x-3}{x-7} \cdot \frac{x+y}{x-7} \cdot h$$
$$\frac{(x-7)^2}{x+y} = h$$

The height is $\dfrac{(x-7)^2}{x+y}$.

Exercise Set 11.3

---

1. $12 = 2 \cdot 2 \cdot 3$

$27 = 3 \cdot 3 \cdot 3$

LCM $= 2 \cdot 2 \cdot 3 \cdot 3 \cdot 3$, or 108

3. $8 = 2 \cdot 2 \cdot 2$

$9 = 3 \cdot 3$

LCM $= 2 \cdot 2 \cdot 2 \cdot 3 \cdot 3$, or 72

5. $6 = 2 \cdot 3$

$9 = 3 \cdot 3$

$21 = 3 \cdot 7$

LCM $= 2 \cdot 3 \cdot 3 \cdot 7$, or 126

7. $24 = 2 \cdot 2 \cdot 2 \cdot 3$

$36 = 2 \cdot 2 \cdot 3 \cdot 3$

$40 = 2 \cdot 2 \cdot 2 \cdot 5$

LCM $= 2 \cdot 2 \cdot 2 \cdot 3 \cdot 3 \cdot 5$, or 360

9. $10 = 2 \cdot 5$

$100 = 2 \cdot 2 \cdot 5 \cdot 5$

$500 = 2 \cdot 2 \cdot 5 \cdot 5 \cdot 5$

LCM $= 2 \cdot 2 \cdot 5 \cdot 5 \cdot 5$, or 500

(We might have observed at the outset that both 10 and 100 are factors of 500, so the LCM is 500.)

11. $24 = 2 \cdot 2 \cdot 2 \cdot 3$

$18 = 2 \cdot 3 \cdot 3$

LCD $= 2 \cdot 2 \cdot 2 \cdot 3 \cdot 3$, or 72

$\dfrac{7}{24} + \dfrac{11}{18} = \dfrac{7}{2 \cdot 2 \cdot 2 \cdot 3} \cdot \dfrac{3}{3} + \dfrac{11}{2 \cdot 3 \cdot 3} \cdot \dfrac{2 \cdot 2}{2 \cdot 2}$

$\qquad = \dfrac{21}{2 \cdot 2 \cdot 2 \cdot 3 \cdot 3} + \dfrac{44}{2 \cdot 2 \cdot 2 \cdot 3 \cdot 3}$

$\qquad = \dfrac{65}{72}$

13. $\dfrac{1}{6} + \dfrac{3}{40}$

$= \dfrac{1}{2 \cdot 3} + \dfrac{3}{2 \cdot 2 \cdot 2 \cdot 5}$

$\qquad$ LCD is $2 \cdot 2 \cdot 2 \cdot 3 \cdot 5$, or 120

$= \dfrac{1}{2 \cdot 3} \cdot \dfrac{2 \cdot 2 \cdot 5}{2 \cdot 2 \cdot 5} + \dfrac{3}{2 \cdot 2 \cdot 2 \cdot 5} \cdot \dfrac{3}{3}$

$= \dfrac{20 + 9}{2 \cdot 2 \cdot 2 \cdot 3 \cdot 5}$

$= \dfrac{29}{120}$

15. $\dfrac{1}{20} + \dfrac{1}{30} + \dfrac{2}{45}$

$= \dfrac{1}{2 \cdot 2 \cdot 5} + \dfrac{1}{2 \cdot 3 \cdot 5} + \dfrac{2}{3 \cdot 3 \cdot 5}$

$\qquad$ LCD is $2 \cdot 2 \cdot 3 \cdot 3 \cdot 5$, or 180

$= \dfrac{1}{2 \cdot 2 \cdot 5} \cdot \dfrac{3 \cdot 3}{3 \cdot 3} + \dfrac{1}{2 \cdot 3 \cdot 5} \cdot \dfrac{2 \cdot 3}{2 \cdot 3} + \dfrac{2}{3 \cdot 3 \cdot 5} \cdot \dfrac{2 \cdot 2}{2 \cdot 2}$

$= \dfrac{9 + 6 + 8}{2 \cdot 2 \cdot 3 \cdot 3 \cdot 5}$

$= \dfrac{23}{180}$

17. $6x^2 = 2 \cdot 3 \cdot x \cdot x$

$12x^3 = 2 \cdot 2 \cdot 3 \cdot x \cdot x \cdot x$

LCM $= 2 \cdot 2 \cdot 3 \cdot x \cdot x \cdot x$, or $12x^3$

19. $2x^2 = 2 \cdot x \cdot x$

$6xy = 2 \cdot 3 \cdot x \cdot y$

$18y^2 = 2 \cdot 3 \cdot 3 \cdot y \cdot y$

LCM $= 2 \cdot 3 \cdot 3 \cdot x \cdot x \cdot y \cdot y$, or $18x^2y^2$

21. $2(y - 3) = 2 \cdot (y - 3)$

$6(y - 3) = 2 \cdot 3 \cdot (y - 3)$

LCM $= 2 \cdot 3 \cdot (y - 3)$, or $6(y - 3)$

23. $t, t + 2, t - 2$

The expressions are not factorable, so the LCM is their product:

LCM $= t(t + 2)(t - 2)$

25. $x^2 - 4 = (x + 2)(x - 2)$

$x^2 + 5x + 6 = (x + 3)(x + 2)$

LCM $= (x + 2)(x - 2)(x + 3)$

27. $t^3 + 4t^2 + 4t = t(t^2 + 4t + 4) = t(t + 2)(t + 2)$

$t^2 - 4t = t(t - 4)$

LCM $= t(t + 2)(t + 2)(t - 4) = t(t + 2)^2(t - 4)$

29. $a + 1 = a + 1$

$(a - 1)^2 = (a - 1)(a - 1)$

$a^2 - 1 = (a + 1)(a - 1)$

LCM $= (a + 1)(a - 1)(a - 1) = (a + 1)(a - 1)^2$

31. $m^2 - 5m + 6 = (m - 3)(m - 2)$

$m^2 - 4m + 4 = (m - 2)(m - 2)$

LCM $= (m - 3)(m - 2)(m - 2) = (m - 3)(m - 2)^2$

33. $2 + 3x = 2 + 3x$

$4 - 9x^2 = (2 + 3x)(2 - 3x)$

$2 - 3x = 2 - 3x$

LCM $= (2 + 3x)(2 - 3x)$

35. $10v^2 + 30v = 10v(v + 3) = 2 \cdot 5 \cdot v(v + 3)$

$5v^2 + 35v + 60 = 5(v^2 + 7v + 12)$

$\qquad = 5(v + 4)(v + 3)$

LCM $= 2 \cdot 5 \cdot v(v + 3)(v + 4) = 10v(v + 3)(v + 4)$

37. $9x^3 - 9x^2 - 18x = 9x(x^2 - x - 2)$

$\qquad = 3 \cdot 3 \cdot x(x - 2)(x + 1)$

$6x^5 - 24x^4 + 24x^3 = 6x^3(x^2 - 4x + 4)$

$\qquad = 2 \cdot 3 \cdot x \cdot x \cdot x(x - 2)(x - 2)$

LCM $= 2 \cdot 3 \cdot 3 \cdot x \cdot x \cdot x(x - 2)(x - 2)(x + 1) =$

$18x^3(x - 2)^2(x + 1)$

39. $x^5 + 4x^4 + 4x^3 = x^3(x^2 + 4x + 4)$

$\qquad = x \cdot x \cdot x(x + 2)(x + 2)$

$3x^2 - 12 = 3(x^2 - 4) = 3(x + 2)(x - 2)$

$2x + 4 = 2(x + 2)$

LCM $= 2 \cdot 3 \cdot x \cdot x \cdot x(x + 2)(x + 2)(x - 2)$

$\qquad = 6x^3(x + 2)^2(x - 2)$

41. Discussion and Writing Exercise

43. $x^2 - 6x + 9 = x^2 - 2 \cdot x \cdot 3 + 3^2 \qquad$ Trinomial square

$\qquad = (x - 3)^2$

45. $x^2 - 9 = x^2 - 3^2 \qquad$ Difference of squares

$\qquad = (x + 3)(x - 3)$

47. $x^2 + 6x + 9 = x^2 + 2 \cdot x \cdot 3 + 3^2 \qquad$ Trinomial square

$\qquad = (x + 3)^2$

49. Locate 1970 on the horizontal axis, go up to the graph, and then go over to the corresponding point on the vertical axis. We read that about 54% of those married in 1970 will divorce.

51. Locate 1990 on the horizontal axis, go up to the graph, and then go over to the corresponding point on the vertical axis. We read that about 74% of those married in 1990 will divorce.

53. Locate 50 on the vertical axis, go across to the graph, and then go down to the corresponding point on the horizontal axis. We read that the divorce percentage was about 50% in 1965.

55. The time it takes Pedro and Maria to meet again at the starting place is the LCM of the times it takes them to complete one round of the course.

$6 = 2 \cdot 3$

$8 = 2 \cdot 2 \cdot 2$

$\text{LCM} = 2 \cdot 2 \cdot 2 \cdot 3, \text{ or } 24$

It takes 24 min.

Exercise Set 11.4

1. $\dfrac{5}{8} + \dfrac{3}{8} + \dfrac{5+3}{8} = \dfrac{8}{8} = 1$

3. $\dfrac{1}{3+x} + \dfrac{5}{3+x} = \dfrac{1+5}{3+x} = \dfrac{6}{3+x}$

5. $\dfrac{x^2 + 7x}{x^2 - 5x} + \dfrac{x^2 - 4x}{x^2 - 5x} = \dfrac{(x^2 + 7x) + (x^2 - 4x)}{x^2 - 5x}$

$= \dfrac{2x^2 + 3x}{x^2 - 5x}$

$= \dfrac{x(2x + 3)}{x(x - 5)}$

$= \dfrac{\cancel{x}(2x + 3)}{\cancel{x}(x - 5)}$

$= \dfrac{2x + 3}{x - 5}$

7. $\dfrac{2}{x} + \dfrac{5}{x^2} = \dfrac{2}{x} + \dfrac{5}{x \cdot x} \qquad \text{LCD} = x \cdot x, \text{ or } x^2$

$= \dfrac{2}{x} \cdot \dfrac{x}{x} + \dfrac{5}{x \cdot x}$

$= \dfrac{2x + 5}{x^2}$

9. $\left.\begin{array}{l} 6r = 2 \cdot 3 \cdot r \\ 8r = 2 \cdot 2 \cdot 2 \cdot r \end{array}\right\} \text{LCD} = 2 \cdot 2 \cdot 2 \cdot 3 \cdot r, \text{ or } 24r$

$\dfrac{5}{6r} + \dfrac{7}{8r} = \dfrac{5}{6r} \cdot \dfrac{4}{4} + \dfrac{7}{8r} \cdot \dfrac{3}{3}$

$= \dfrac{20 + 21}{24r}$

$= \dfrac{41}{24r}$

11. $\left.\begin{array}{l} xy^2 = x \cdot y \cdot y \\ x^2 y = x \cdot x \cdot y \end{array}\right\} \text{LCD} = x \cdot x \cdot y \cdot y, \text{ or } x^2 y^2$

$\dfrac{4}{xy^2} + \dfrac{6}{x^2 y} = \dfrac{4}{xy^2} \cdot \dfrac{x}{x} + \dfrac{6}{x^2 y} \cdot \dfrac{y}{y}$

$= \dfrac{4x + 6y}{x^2 y^2}$

13. $\left.\begin{array}{l} 9t^3 = 3 \cdot 3 \cdot t \cdot t \cdot t \\ 6t^2 = 2 \cdot 3 \cdot t \cdot t \end{array}\right\} \text{LCD} = 2 \cdot 3 \cdot 3 \cdot t \cdot t \cdot t, \text{ or } 18t^3$

$\dfrac{2}{9t^3} + \dfrac{1}{6t^2} = \dfrac{2}{9t^3} \cdot \dfrac{2}{2} + \dfrac{1}{6t^2} \cdot \dfrac{3t}{3t}$

$= \dfrac{4 + 3t}{18t^3}$

15. $\text{LCD} = x^2 y^2$ (See Exercise 11.)

$\dfrac{x + y}{xy^2} + \dfrac{3x + y}{x^2 y} = \dfrac{x + y}{xy^2} \cdot \dfrac{x}{x} + \dfrac{3x + y}{x^2 y} \cdot \dfrac{y}{y}$

$= \dfrac{x(x + y) + y(3x + y)}{x^2 y^2}$

$= \dfrac{x^2 + xy + 3xy + y^2}{x^2 y^2}$

$= \dfrac{x^2 + 4xy + y^2}{x^2 y^2}$

17. The denominators do not factor, so the LCD is their product, $(x - 2)(x + 2)$.

$\dfrac{3}{x - 2} + \dfrac{3}{x + 2} = \dfrac{3}{x - 2} \cdot \dfrac{x + 2}{x + 2} + \dfrac{3}{x + 2} \cdot \dfrac{x - 2}{x - 2}$

$= \dfrac{3(x + 2) + 3(x - 2)}{(x - 2)(x + 2)}$

$= \dfrac{3x + 6 + 3x - 6}{(x - 2)(x + 2)}$

$= \dfrac{6x}{(x - 2)(x + 2)}$

19. $\left.\begin{array}{l} 3x = 3 \cdot x \\ x + 1 = x + 1 \end{array}\right\} \text{LCD} = 3x(x + 1)$

$\dfrac{3}{x + 1} + \dfrac{2}{3x} = \dfrac{3}{x + 1} \cdot \dfrac{3x}{3x} + \dfrac{2}{3x} \cdot \dfrac{x + 1}{x + 1}$

$= \dfrac{9x + 2(x + 1)}{3x(x + 1)}$

$= \dfrac{9x + 2x + 2}{3x(x + 1)}$

$= \dfrac{11x + 2}{3x(x + 1)}$

21. $\left.\begin{array}{l} x^2 - 16 = (x + 4)(x - 4) \\ x - 4 = x - 4 \end{array}\right\} \text{LCD} = (x + 4)(x - 4)$

$\dfrac{2x}{x^2 - 16} + \dfrac{x}{x - 4} = \dfrac{2x}{(x + 4)(x - 4)} + \dfrac{x}{x - 4} \cdot \dfrac{x + 4}{x + 4}$

$= \dfrac{2x + x(x + 4)}{(x + 4)(x - 4)}$

$= \dfrac{2x + x^2 + 4x}{(x + 4)(x - 4)}$

$= \dfrac{x^2 + 6x}{(x + 4)(x - 4)}$

23. $\dfrac{5}{z + 4} + \dfrac{3}{3z + 12} = \dfrac{5}{z + 4} + \dfrac{3}{3(z + 4)} \qquad \text{LCD} = 3(z + 4)$

$= \dfrac{5}{z + 4} \cdot \dfrac{3}{3} + \dfrac{3}{3(z + 4)}$

$= \dfrac{15 + 3}{3(z + 4)} = \dfrac{18}{3(z + 4)}$

$= \dfrac{3 \cdot 6}{3(z + 4)} = \dfrac{\cancel{3} \cdot 6}{\cancel{3}(z + 4)}$

$= \dfrac{6}{z + 4}$

25. $\dfrac{3}{x-1}+\dfrac{2}{(x-1)^2}$ LCD $=(x-1)^2$

$=\dfrac{3}{x-1}\cdot\dfrac{x-1}{x-1}+\dfrac{2}{(x-1)^2}$

$=\dfrac{3(x-1)+2}{(x-1)^2}$

$=\dfrac{3x-3+2}{(x-1)^2}$

$=\dfrac{3x-1}{(x-1)^2}$

27. $\dfrac{4a}{5a-10}+\dfrac{3a}{10a-20}=\dfrac{4a}{5(a-2)}+\dfrac{3a}{2\cdot5(a-2)}$

$\qquad\qquad\qquad\qquad$ LCD $=2\cdot5(a-2)$

$=\dfrac{4a}{5(a-2)}\cdot\dfrac{2}{2}+\dfrac{3a}{2\cdot5(a-2)}$

$=\dfrac{8a+3a}{10(a-2)}$

$=\dfrac{11a}{10(a-2)}$

29. $\dfrac{x+4}{x}+\dfrac{x}{x+4}$ LCD $=x(x+4)$

$=\dfrac{x+4}{x}\cdot\dfrac{x+4}{x+4}+\dfrac{x}{x+4}\cdot\dfrac{x}{x}$

$=\dfrac{(x+4)^2+x^2}{x(x+4)}$

$=\dfrac{x^2+8x+16+x^2}{x(x+4)}$

$=\dfrac{2x^2+8x+16}{x(x+4)}$

31. $\dfrac{4}{a^2-a-2}+\dfrac{3}{a^2+4a+3}$

$=\dfrac{4}{(a-2)(a+1)}+\dfrac{3}{(a+3)(a+1)}$

$\qquad\qquad\qquad$ LCD $=(a-2)(a+1)(a+3)$

$=\dfrac{4}{(a-2)(a+1)}\cdot\dfrac{a+3}{a+3}+\dfrac{3}{(a+3)(a+1)}\cdot\dfrac{a-2}{a-2}$

$=\dfrac{4(a+3)+3(a-2)}{(a-2)(a+1)(a+3)}$

$=\dfrac{4a+12+3a-6}{(a-2)(a+1)(a+3)}$

$=\dfrac{7a+6}{(a-2)(a+1)(a+3)}$

33. $\dfrac{x+3}{x-5}+\dfrac{x-5}{x+3}$ LCD $=(x-5)(x+3)$

$=\dfrac{x+3}{x-5}\cdot\dfrac{x+3}{x+3}+\dfrac{x-5}{x+3}\cdot\dfrac{x-5}{x-5}$

$=\dfrac{(x+3)^2+(x-5)^2}{(x-5)(x+3)}$

$=\dfrac{x^2+6x+9+x^2-10x+25}{(x-5)(x+3)}$

$=\dfrac{2x^2-4x+34}{(x-5)(x+3)}$

35. $\dfrac{a}{a^2-1}+\dfrac{2a}{a^2-a}$

$=\dfrac{a}{(a+1)(a-1)}+\dfrac{2a}{a(a-1)}$

$\qquad\qquad\qquad$ LCD $=a(a+1)(a-1)$

$=\dfrac{a}{(a+1)(a-1)}\cdot\dfrac{a}{a}+\dfrac{2a}{a(a-1)}\cdot\dfrac{a+1}{a+1}$

$=\dfrac{a^2+2a(a+1)}{a(a+1)(a-1)}=\dfrac{a^2+2a^2+2a}{a(a+1)(a-1)}$

$=\dfrac{3a^2+2a}{a(a+1)(a-1)}=\dfrac{a(3a+2)}{a(a+1)(a-1)}$

$=\dfrac{\cancel{a}(3a+2)}{\cancel{a}(a+1)(a-1)}=\dfrac{3a+2}{(a+1)(a-1)}$

37. $\dfrac{7}{8}+\dfrac{5}{-8}=\dfrac{7}{8}+\dfrac{5}{-8}\cdot\dfrac{-1}{-1}$

$=\dfrac{7}{8}+\dfrac{-5}{8}$

$=\dfrac{7+(-5)}{8}$

$=\dfrac{2}{8}=\dfrac{\cancel{2}\cdot1}{4\cdot\cancel{2}}$

$=\dfrac{1}{4}$

39. $\dfrac{3}{t}+\dfrac{4}{-t}=\dfrac{3}{t}+\dfrac{4}{-t}\cdot\dfrac{-1}{-1}$

$=\dfrac{3}{t}+\dfrac{-4}{t}$

$=\dfrac{3+(-4)}{t}$

$=\dfrac{-1}{t}$

$=-\dfrac{1}{t}$

41. $\dfrac{2x+7}{x-6}+\dfrac{3x}{6-x}=\dfrac{2x+7}{x-6}+\dfrac{3x}{6-x}\cdot\dfrac{-1}{-1}$

$=\dfrac{2x+7}{x-6}+\dfrac{-3x}{x-6}$

$=\dfrac{(2x+7)+(-3x)}{x-6}$

$=\dfrac{-x+7}{x-6}$

43. $\dfrac{y^2}{y-3} + \dfrac{9}{3-y} = \dfrac{y^2}{y-3} + \dfrac{9}{3-y} \cdot \dfrac{-1}{-1}$

$\qquad = \dfrac{y^2}{y-3} + \dfrac{-9}{y-3}$

$\qquad = \dfrac{y^2 + (-9)}{y-3}$

$\qquad = \dfrac{y^2 - 9}{y-3}$

$\qquad = \dfrac{(y+3)(y-3)}{y-3}$

$\qquad = \dfrac{(y+3)(y\!\!\!\!\diagup\!\!3)}{1(y\!\!\!\!\diagup\!\!3)}$

$\qquad = y + 3$

45. $\dfrac{b-7}{b^2-16} + \dfrac{7-b}{16-b^2} = \dfrac{b-7}{b^2-16} + \dfrac{7-b}{16-b^2} \cdot \dfrac{-1}{-1}$

$\qquad = \dfrac{b-7}{b^2-16} + \dfrac{b-7}{b^2-16}$

$\qquad = \dfrac{(b-7)+(b-7)}{b^2-16}$

$\qquad = \dfrac{2b-14}{b^2-16}$

47. $\dfrac{a^2}{a-b} + \dfrac{b^2}{b-a} = \dfrac{a^2}{a-b} + \dfrac{b^2}{b-a} \cdot \dfrac{-1}{-1}$

$\qquad = \dfrac{a^2}{a-b} + \dfrac{-b^2}{a-b}$

$\qquad = \dfrac{a^2 + (-b^2)}{a-b}$

$\qquad = \dfrac{a^2 - b^2}{a-b}$

$\qquad = \dfrac{(a+b)(a-b)}{a-b}$

$\qquad = \dfrac{(a+b)(a\!\!\!\!\diagup\!\!b)}{1(a\!\!\!\!\diagup\!\!b)}$

$\qquad = a + b$

49. $\dfrac{x+3}{x-5} + \dfrac{2x-1}{5-x} + \dfrac{2(3x-1)}{x-5}$

$\qquad = \dfrac{x+3}{x-5} + \dfrac{2x-1}{5-x} \cdot \dfrac{-1}{-1} + \dfrac{2(3x-1)}{x-5}$

$\qquad = \dfrac{x+3}{x-5} + \dfrac{1-2x}{x-5} + \dfrac{2(3x-1)}{x-5}$

$\qquad = \dfrac{(x+3)+(1-2x)+(6x-2)}{x-5}$

$\qquad = \dfrac{5x+2}{x-5}$

51. $\dfrac{2(4x+1)}{5x-7} + \dfrac{3(x-2)}{7-5x} + \dfrac{-10x-1}{5x-7}$

$\qquad = \dfrac{2(4x+1)}{5x-7} + \dfrac{3(x-2)}{7-5x} \cdot \dfrac{-1}{-1} + \dfrac{-10x-1}{5x-7}$

$\qquad = \dfrac{2(4x+1)}{5x-7} + \dfrac{-3(x-2)}{5x-7} + \dfrac{-10x-1}{5x-7}$

$\qquad = \dfrac{(8x+2)+(-3x+6)+(-10x-1)}{5x-7}$

$\qquad = \dfrac{-5x+7}{5x-7}$

$\qquad = \dfrac{-1(5x-7)}{5x-7}$

$\qquad = \dfrac{-1(5x\!\!\!\!\diagup\!\!7)}{5x\!\!\!\!\diagup\!\!7}$

$\qquad = -1$

53. $\dfrac{x+1}{(x+3)(x-3)} + \dfrac{4(x-3)}{(x-3)(x+3)} + \dfrac{(x-1)(x-3)}{(3-x)(x+3)}$

$\qquad = \dfrac{x+1}{(x+3)(x-3)} + \dfrac{4(x-3)}{(x-3)(x+3)} + \dfrac{(x-1)(x-3)}{(3-x)(x+3)} \cdot \dfrac{-1}{-1}$

$\qquad = \dfrac{x+1}{(x+3)(x-3)} + \dfrac{4(x-3)}{(x-3)(x+3)} + \dfrac{-1(x^2-4x+3)}{(x-3)(x+3)}$

$\qquad = \dfrac{(x+1)+(4x-12)+(-x^2+4x-3)}{(x+3)(x-3)}$

$\qquad = \dfrac{-x^2+9x-14}{(x+3)(x-3)}$

55. $\dfrac{6}{x-y} + \dfrac{4x}{y^2-x^2}$

$\qquad = \dfrac{6}{x-y} + \dfrac{4x}{(y-x)(y+x)}$

$\qquad = \dfrac{6}{x-y} + \dfrac{4x}{(y-x)(y+x)} \cdot \dfrac{-1}{-1}$

$\qquad = \dfrac{6}{x-y} + \dfrac{-4x}{(x-y)(x+y)}$

$\qquad\qquad [-1(y-x) = x-y; y+x = x+y]$

$\qquad\qquad\qquad \text{LCD} = (x-y)(x+y)$

$\qquad = \dfrac{6}{x-y} \cdot \dfrac{x+y}{x+y} + \dfrac{-4x}{(x-y)(x+y)}$

$\qquad = \dfrac{6(x+y)-4x}{(x-y)(x+y)}$

$\qquad = \dfrac{6x+6y-4x}{(x-y)(x+y)}$

$\qquad = \dfrac{2x+6y}{(x-y)(x+y)}$

57.
$$\frac{4-a}{25-a^2} + \frac{a+1}{a-5}$$

$$= \frac{4-a}{25-a^2} \cdot \frac{-1}{-1} + \frac{a+1}{a-5}$$

$$= \frac{a-4}{a^2-25} + \frac{a+1}{a-5}$$

$$= \frac{a-4}{(a+5)(a-5)} + \frac{a+1}{a-5}$$

$$\text{LCD} = (a+5)(a-5)$$

$$= \frac{a-4}{(a+5)(a-5)} + \frac{a+1}{a-5} \cdot \frac{a+5}{a+5}$$

$$= \frac{a-4}{(a+5)(a-5)} + \frac{(a+1)(a+5)}{(a+5)(a-5)}$$

$$= \frac{(a-4)+(a+1)(a+5)}{(a+5)(a-5)}$$

$$= \frac{a-4+a^2+6a+5}{(a+5)(a-5)}$$

$$= \frac{a^2+7a+1}{(a+5)(a-5)}$$

59.
$$\frac{2}{t^2+t-6} + \frac{3}{t^2-9}$$

$$= \frac{2}{(t+3)(t-2)} + \frac{3}{(t+3)(t-3)}$$

$$\text{LCD} = (t+3)(t-2)(t-3)$$

$$= \frac{2}{(t+3)(t-2)} \cdot \frac{t-3}{t-3} + \frac{3}{(t+3)(t-3)} \cdot \frac{t-2}{t-2}$$

$$= \frac{2(t-3)+3(t-2)}{(t+3)(t-2)(t-3)}$$

$$= \frac{2t-6+3t-6}{(t+3)(t-2)(t-3)}$$

$$= \frac{5t-12}{(t+3)(t-2)(t-3)}$$

61. Discussion and Writing Exercise

63. $(x^2+x)-(x+1) = x^2+x-x-1 = x^2-1$

65. $(2x^4y^3)^{-3} = \dfrac{1}{(2x^4y^3)^3} = \dfrac{1}{2^3(x^4)^3(y^3)^3} = \dfrac{1}{8x^{12}y^9}$

67. $\left(\dfrac{x^{-4}}{y^7}\right)^3 = \dfrac{(x^{-4})^3}{(y^7)^3} = \dfrac{x^{-12}}{y^{21}} = \dfrac{1}{x^{12}y^{21}}$

69. $y = \dfrac{1}{2}x - 5 = \dfrac{1}{2}x + (-5)$

The y-intercept is $(0,-5)$. We find two other pairs.

When $x = 2$, $y = \dfrac{1}{2} \cdot 2 - 5 = 1 - 5 = -4$.

When $x = 4$, $y = \dfrac{1}{2} \cdot 4 - 5 = 2 - 5 = -3$.

x	y
0	-5
2	-4
4	-3

Plot these points, draw the line they determine, and label the graph $y = \dfrac{1}{2}x - 5$.

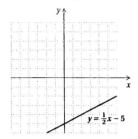

71. $y = 3$

Any ordered pair $(x, 3)$ is a solution. The variable y must be 3, but x can be any number we choose. A few solutions are listed below. Plot these points and draw the line.

x	y
-4	3
0	3
3	3

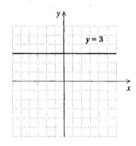

73.
$$3x - 7 = 5x + 9$$
$$-2x - 7 = 9 \qquad \text{Subtracting } 5x$$
$$-2x = 16 \qquad \text{Adding } 7$$
$$x = -8 \qquad \text{Dividing by } -2$$

The solution is -8.

75.
$$x^2 - 8x + 15 = 0$$
$$(x-3)(x-5) = 0$$
$$x - 3 = 0 \text{ or } x - 5 = 0 \quad \text{Principle of zero products}$$
$$x = 3 \text{ or } \qquad x = 5$$

The solutions are 3 and 5.

77. To find the perimeter we add the lengths of the sides:

$$\frac{y+4}{3} + \frac{y+4}{3} + \frac{y-2}{5} + \frac{y-2}{5} \quad \text{LCD} = 3 \cdot 5$$

$$= \frac{y+4}{3} \cdot \frac{5}{5} + \frac{y+4}{3} \cdot \frac{5}{5} + \frac{y-2}{5} \cdot \frac{3}{3} + \frac{y-2}{5} \cdot \frac{3}{3}$$

$$= \frac{5y+20+5y+20+3y-6+3y-6}{3 \cdot 5}$$

$$= \frac{16y+28}{15}$$

To find the area we multiply the length and the width:

$$\left(\frac{y+4}{3}\right)\left(\frac{y-2}{5}\right) = \frac{(y+4)(y-2)}{3 \cdot 5} = \frac{y^2+2y-8}{15}$$

79. $\dfrac{5}{z+2} + \dfrac{4z}{z^2-4} + 2$

$= \dfrac{5}{z+2} + \dfrac{4z}{(z+2)(z-2)} + \dfrac{2}{1}$ LCD $= (z+2)(z-2)$

$= \dfrac{5}{z+2} \cdot \dfrac{z-2}{z-2} + \dfrac{4z}{(z+2)(z-2)} + \dfrac{2}{1} \cdot \dfrac{(z+2)(z-2)}{(z+2)(z-2)}$

$= \dfrac{5z - 10 + 4z + 2(z^2-4)}{(z+2)(z-2)}$

$= \dfrac{5z - 10 + 4z + 2z^2 - 8}{(z+2)(z-2)} = \dfrac{2z^2 + 9z - 18}{(z+2)(z-2)}$

$= \dfrac{(2z-3)(z+6)}{(z+2)(z-2)}$

81. $\dfrac{3z^2}{z^4-4} + \dfrac{5z^2-3}{2z^4+z^2-6}$

$= \dfrac{3z^2}{(z^2+2)(z^2-2)} + \dfrac{5z^2-3}{(2z^2-3)(z^2+2)}$

    LCD $= (z^2+2)(z^2-2)(2z^2-3)$

$= \dfrac{3z^2}{(z^2+2)(z^2-2)} \cdot \dfrac{2z^2-3}{2z^2-3} +$

$\qquad \dfrac{5z^2-3}{(2z^2-3)(z^2+2)} \cdot \dfrac{z^2-2}{z^2-2}$

$= \dfrac{6z^4 - 9z^2 + 5z^4 - 13z^2 + 6}{(z^2+2)(z^2-2)(2z^2-3)}$

$= \dfrac{11z^4 - 22z^2 + 6}{(z^2+2)(z^2-2)(2z^2-3)}$

83.-85. Left to the student

Exercise Set 11.5

1. $\dfrac{7}{x} - \dfrac{3}{x} = \dfrac{7-3}{x} = \dfrac{4}{x}$

3. $\dfrac{y}{y-4} - \dfrac{4}{y-4} = \dfrac{y-4}{y-4} = 1$

5. $\dfrac{2x-3}{x^2+3x-4} - \dfrac{x-7}{x^2+3x-4}$

$= \dfrac{2x - 3 - (x-7)}{x^2+3x-4}$

$= \dfrac{2x - 3 - x + 7}{x^2+3x-4}$

$= \dfrac{x+4}{x^2+3x-4}$

$= \dfrac{x+4}{(x+4)(x-1)}$

$= \dfrac{(x+4) \cdot 1}{(x+4)(x-1)}$

$= \dfrac{1}{x-1}$

7. $\dfrac{a-2}{10} - \dfrac{a+1}{5} = \dfrac{a-2}{10} - \dfrac{a+1}{5} \cdot \dfrac{2}{2}$ LCD $= 10$

$= \dfrac{a-2}{10} - \dfrac{2(a+1)}{10}$

$= \dfrac{(a-2) - 2(a+1)}{10}$

$= \dfrac{a - 2 - 2a - 2}{10}$

$= \dfrac{-a-4}{10}$

9. $\dfrac{4z-9}{3z} - \dfrac{3z-8}{4z} = \dfrac{4z-9}{3z} \cdot \dfrac{4}{4} - \dfrac{3z-8}{4z} \cdot \dfrac{3}{3}$

    LCD $= 3 \cdot 4 \cdot z$, or $12z$

$= \dfrac{16z-36}{12z} - \dfrac{9z-24}{12z}$

$= \dfrac{16z - 36 - (9z-24)}{12z}$

$= \dfrac{16z - 36 - 9z + 24}{12z}$

$= \dfrac{7z-12}{12z}$

11. $\dfrac{4x+2t}{3xt^2} - \dfrac{5x-3t}{x^2t}$ LCD $= 3x^2t^2$

$= \dfrac{4x+2t}{3xt^2} \cdot \dfrac{x}{x} - \dfrac{5x-3t}{x^2t} \cdot \dfrac{3t}{3t}$

$= \dfrac{4x^2 + 2tx}{3x^2t^2} - \dfrac{15xt - 9t^2}{3x^2t^2}$

$= \dfrac{4x^2 + 2tx - (15xt - 9t^2)}{3x^2t^2}$

$= \dfrac{4x^2 + 2tx - 15xt + 9t^2}{3x^2t^2}$

$= \dfrac{4x^2 - 13xt + 9t^2}{3x^2t^2}$

13. $\dfrac{5}{x+5} - \dfrac{3}{x-5}$ LCD $= (x+5)(x-5)$

$= \dfrac{5}{x+5} \cdot \dfrac{x-5}{x-5} - \dfrac{3}{x-5} \cdot \dfrac{x+5}{x+5}$

$= \dfrac{5x-25}{(x+5)(x-5)} - \dfrac{3x+15}{(x+5)(x-5)}$

$= \dfrac{5x - 25 - (3x+15)}{(x+5)(x-5)}$

$= \dfrac{5x - 25 - 3x - 15}{(x+5)(x-5)}$

$= \dfrac{2x-40}{(x+5)(x-5)}$

15. $\dfrac{3}{2t^2 - 2t} - \dfrac{5}{2t - 2}$

$= \dfrac{3}{2t(t-1)} - \dfrac{5}{2(t-1)}$ $\text{LCD} = 2t(t-1)$

$= \dfrac{3}{2t(t-1)} - \dfrac{5}{2(t-1)} \cdot \dfrac{t}{t}$

$= \dfrac{3}{2t(t-1)} - \dfrac{5t}{2t(t-1)}$

$= \dfrac{3 - 5t}{2t(t-1)}$

17. $\dfrac{2s}{t^2 - s^2} - \dfrac{s}{t - s}$ $\text{LCD} = (t-s)(t+s)$

$= \dfrac{2s}{(t-s)(t+s)} - \dfrac{s}{t-s} \cdot \dfrac{t+s}{t+s}$

$= \dfrac{2s}{(t-s)(t+s)} - \dfrac{st + s^2}{(t-s)(t+s)}$

$= \dfrac{2s - (st + s^2)}{(t-s)(t+s)}$

$= \dfrac{2s - st - s^2}{(t-s)(t+s)}$

19. $\dfrac{y-5}{y} - \dfrac{3y-1}{4y} = \dfrac{y-5}{y} \cdot \dfrac{4}{4} - \dfrac{3y-1}{4y}$ $\text{LCD} = 4y$

$= \dfrac{4y - 20}{4y} - \dfrac{3y - 1}{4y}$

$= \dfrac{4y - 20 - (3y - 1)}{4y}$

$= \dfrac{4y - 20 - 3y + 1}{4y}$

$= \dfrac{y - 19}{4y}$

21. $\dfrac{a}{x+a} - \dfrac{a}{x-a}$ $\text{LCD} = (x+a)(x-a)$

$= \dfrac{a}{x+a} \cdot \dfrac{x-a}{x-a} - \dfrac{a}{x-a} \cdot \dfrac{x+a}{x+a}$

$= \dfrac{ax - a^2}{(x+a)(x-a)} - \dfrac{ax + a^2}{(x+a)(x-a)}$

$= \dfrac{ax - a^2 - (ax + a^2)}{(x+a)(x-a)}$

$= \dfrac{ax - a^2 - ax - a^2}{(x+a)(x-a)}$

$= \dfrac{-2a^2}{(x+a)(x-a)}$

23. $\dfrac{11}{6} - \dfrac{5}{-6} = \dfrac{11}{6} - \dfrac{5}{-6} \cdot \dfrac{-1}{-1}$

$= \dfrac{11}{6} - \dfrac{-5}{6}$

$= \dfrac{11 - (-5)}{6}$

$= \dfrac{11 + 5}{6}$

$= \dfrac{16}{6}$

$= \dfrac{8}{3}$

25. $\dfrac{5}{a} - \dfrac{8}{-a} = \dfrac{5}{a} - \dfrac{8}{-a} \cdot \dfrac{-1}{-1}$

$= \dfrac{5}{a} - \dfrac{-8}{a}$

$= \dfrac{5 - (-8)}{a}$

$= \dfrac{5 + 8}{a}$

$= \dfrac{13}{a}$

27. $\dfrac{4}{y-1} - \dfrac{4}{1-y} = \dfrac{4}{y-1} - \dfrac{4}{1-y} \cdot \dfrac{-1}{-1}$

$= \dfrac{4}{y-1} - \dfrac{4(-1)}{(1-y)(-1)}$

$= \dfrac{4}{y-1} - \dfrac{-4}{y-1}$

$= \dfrac{4 - (-4)}{y-1}$

$= \dfrac{4 + 4}{y-1}$

$= \dfrac{8}{y-1}$

29. $\dfrac{3-x}{x-7} - \dfrac{2x-5}{7-x} = \dfrac{3-x}{x-7} - \dfrac{2x-5}{7-x} \cdot \dfrac{-1}{-1}$

$= \dfrac{3-x}{x-7} - \dfrac{(2x-5)(-1)}{(7-x)(-1)}$

$= \dfrac{3-x}{x-7} - \dfrac{5-2x}{x-7}$

$= \dfrac{(3-x) - (5-2x)}{x-7}$

$= \dfrac{3 - x - 5 + 2x}{x-7}$

$= \dfrac{x-2}{x-7}$

31. $\dfrac{a-2}{a^2-25} - \dfrac{6-a}{25-a^2} = \dfrac{a-2}{a^2-25} - \dfrac{6-a}{25-a^2} \cdot \dfrac{-1}{-1}$

$= \dfrac{a-2}{a^2-25} - \dfrac{(6-a)(-1)}{(25-a^2)(-1)}$

$= \dfrac{a-2}{a^2-25} - \dfrac{a-6}{a^2-25}$

$= \dfrac{(a-2)-(a-6)}{a^2-25}$

$= \dfrac{a-2-a+6}{a^2-25}$

$= \dfrac{4}{a^2-25}$

33. $\dfrac{4-x}{x-9} - \dfrac{3x-8}{9-x} = \dfrac{4-x}{x-9} - \dfrac{3x-8}{9-x} \cdot \dfrac{-1}{-1}$

$= \dfrac{4-x}{x-9} - \dfrac{8-3x}{x-9}$

$= \dfrac{(4-x)-(8-3x)}{x-9}$

$= \dfrac{4-x-8+3x}{x-9}$

$= \dfrac{2x-4}{x-9}$

35. $\dfrac{5x}{x^2-9} - \dfrac{4}{3-x}$

$= \dfrac{5x}{(x+3)(x-3)} - \dfrac{4}{3-x}$ $x-3$ and $3-x$ are opposites

$= \dfrac{5x}{(x+3)(x-3)} - \dfrac{4}{3-x} \cdot \dfrac{-1}{-1}$

$= \dfrac{5x}{(x+3)(x-3)} - \dfrac{-4}{x-3}$ LCD $= (x+3)(x-3)$

$= \dfrac{5x}{(x+3)(x-3)} - \dfrac{-4}{x-3} \cdot \dfrac{x+3}{x+3}$

$= \dfrac{5x}{(x+3)(x-3)} - \dfrac{-4x-12}{(x+3)(x-3)}$

$= \dfrac{5x-(-4x-12)}{(x+3)(x-3)}$

$= \dfrac{5x+4x+12}{(x+3)(x-3)}$

$= \dfrac{9x+12}{(x+3)(x-3)}$

37. $\dfrac{t^2}{2t^2-2t} - \dfrac{1}{2t-2}$

$= \dfrac{t^2}{2t(t-1)} - \dfrac{1}{2(t-1)}$ LCD $= 2t(t-1)$

$= \dfrac{t^2}{2t(t-1)} - \dfrac{1}{2(t-1)} \cdot \dfrac{t}{t}$

$= \dfrac{t^2}{2t(t-1)} - \dfrac{t}{2t(t-1)}$

$= \dfrac{t^2-t}{2t(t-1)}$

$= \dfrac{t(t-1)}{2t(t-1)}$

$= \dfrac{\cancel{t}(\cancel{t-1})(1)}{\cancel{2t}(\cancel{t-1})}$

$= \dfrac{1}{2}$

39. $\dfrac{x}{x^2+5x+6} - \dfrac{2}{x^2+3x+2}$

$= \dfrac{x}{(x+3)(x+2)} - \dfrac{2}{(x+2)(x+1)}$

        LCD $= (x+3)(x+2)(x+1)$

$= \dfrac{x}{(x+3)(x+2)} \cdot \dfrac{x+1}{x+1} - \dfrac{2}{(x+2)(x+1)} \cdot \dfrac{x+3}{x+3}$

$= \dfrac{x^2+x}{(x+3)(x+2)(x+1)} - \dfrac{2x+6}{(x+3)(x+2)(x+1)}$

$= \dfrac{x^2+x-(2x+6)}{(x+3)(x+2)(x+1)}$

$= \dfrac{x^2+x-2x-6}{(x+3)(x+2)(x+1)}$

$= \dfrac{x^2-x-6}{(x+3)(x+2)(x+1)}$

$= \dfrac{(x-3)(x+2)}{(x+3)(x+2)(x+1)}$

$= \dfrac{(x-3)\cancel{(x+2)}}{(x+3)\cancel{(x+2)}(x+1)}$

$= \dfrac{x-3}{(x+3)(x+1)}$

41. $\dfrac{3(2x+5)}{x-1} - \dfrac{3(2x-3)}{1-x} + \dfrac{6x+1}{x-1}$

$= \dfrac{3(2x+5)}{x-1} - \dfrac{3(2x-3)}{1-x} \cdot \dfrac{-1}{-1} + \dfrac{6x-1}{x-1}$

$= \dfrac{3(2x+5)}{x-1} - \dfrac{-3(2x-3)}{x-1} + \dfrac{6x-1}{x-1}$

$= \dfrac{(6x+15)-(-6x+9)+(6x-1)}{x-1}$

$= \dfrac{6x+15+6x-9+6x-1}{x-1}$

$= \dfrac{18x+5}{x-1}$

43. $\dfrac{x-y}{x^2-y^2}+\dfrac{x+y}{x^2-y^2}-\dfrac{2x}{x^2-y^2}$

$=\dfrac{x-y+x+y-2x}{x^2-y^2}$

$=\dfrac{0}{x^2-y^2}$

$=0$

45. $\dfrac{2(x-1)}{2x-3}-\dfrac{3(x+2)}{2x-3}-\dfrac{x-1}{3-2x}$

$=\dfrac{2(x-1)}{2x-3}-\dfrac{3(x+2)}{2x-3}-\dfrac{x-1}{3-2x}\cdot\dfrac{-1}{-1}$

$=\dfrac{2(x-1)}{2x-3}-\dfrac{3(x+2)}{2x-3}-\dfrac{1-x}{2x-3}$

$=\dfrac{(2x-2)-(3x+6)-(1-x)}{2x-3}$

$=\dfrac{2x-2-3x-6-1+x}{2x-3}$

$=\dfrac{-9}{2x-3}$

47. $\dfrac{10}{2y-1}-\dfrac{6}{1-2y}+\dfrac{y}{2y-1}+\dfrac{y-4}{1-2y}$

$=\dfrac{10}{2y-1}-\dfrac{6}{1-2y}\cdot\dfrac{-1}{-1}+\dfrac{y}{2y-1}+\dfrac{y-4}{1-2y}\cdot\dfrac{-1}{-1}$

$=\dfrac{10}{2y-1}-\dfrac{-6}{2y-1}+\dfrac{y}{2y-1}+\dfrac{4-y}{2y-1}$

$=\dfrac{10-(-6)+y+4-y}{2y-1}$

$=\dfrac{10+6+y+4-y}{2y-1}$

$=\dfrac{20}{2y-1}$

49. $\dfrac{a+6}{4-a^2}-\dfrac{a+3}{a+2}+\dfrac{a-3}{2-a}$

$=\dfrac{a+6}{(2+a)(2-a)}-\dfrac{a+3}{2+a}+\dfrac{a-3}{2-a}$

$\qquad a+2=2+a;\ \text{LCD}=(2+a)(2-a)$

$=\dfrac{a+6}{(2+a)(2-a)}-\dfrac{a+3}{2+a}\cdot\dfrac{2-a}{2-a}+\dfrac{a-3}{2-a}\cdot\dfrac{2+a}{2+a}$

$=\dfrac{(a+6)-(a+3)(2-a)+(a-3)(2+a)}{(2+a)(2-a)}$

$=\dfrac{a+6-(-a^2-a+6)+(a^2-a-6)}{(2+a)(2-a)}$

$=\dfrac{a+6+a^2+a-6+a^2-a-6}{(2+a)(2-a)}$

$=\dfrac{2a^2+a-6}{(2+a)(2-a)}$

$=\dfrac{(2a-3)(a+2)}{(2+a)(2-a)}$

$=\dfrac{(2a-3)(2+a)}{(2+a)(2-a)}$

$=\dfrac{2a-3}{2-a}$

51. $\dfrac{2z}{1-2z}+\dfrac{3z}{2z+1}-\dfrac{3}{4z^2-1}$

$=\dfrac{2z}{1-2z}\cdot\dfrac{-1}{-1}+\dfrac{3z}{2z+1}-\dfrac{3}{4z^2-1}$

$=\dfrac{-2z}{2z-1}+\dfrac{3z}{2z+1}-\dfrac{3}{(2z-1)(2z+1)}$

$\qquad\qquad \text{LCD}=(2z-1)(2z+1)$

$=\dfrac{-2z}{2z-1}\cdot\dfrac{2z+1}{2z+1}+\dfrac{3z}{2z+1}\cdot\dfrac{2z-1}{2z-1}-$

$\qquad\qquad\qquad\qquad \dfrac{3}{(2z-1)(2z+1)}$

$=\dfrac{(-4z^2-2z)+(6z^2-3z)-3}{(2z-1)(2z+1)}$

$=\dfrac{2z^2-5z-3}{(2z-1)(2z+1)}$

$=\dfrac{(z-3)(2z+1)}{(2z-1)(2z+1)}$

$=\dfrac{(z-3)(2z+1)}{(2z-1)(2z+1)}$

$=\dfrac{z-3}{2z-1}$

53.
$$\frac{1}{x+y} - \frac{1}{x-y} + \frac{2x}{x^2-y^2}$$
$$= \frac{1}{x+y} - \frac{1}{x-y} + \frac{2x}{(x+y)(x-y)}$$
$$\text{LCD} = (x+y)(x-y)$$
$$= \frac{1}{x+y} \cdot \frac{x-y}{x-y} - \frac{1}{x-y} \cdot \frac{x+y}{x+y} \cdot \frac{x+y}{x+y} +$$
$$\frac{2x}{(x+y)(x-y)}$$
$$= \frac{x-y-(x+y)+2x}{(x+y)(x-y)}$$
$$= \frac{x-y-x-y+2x}{(x+y)(x-y)}$$
$$= \frac{2x-2y}{(x+y)(x-y)}$$
$$= \frac{2(x-y)}{(x+y)(x-y)}$$
$$= \frac{2(x-y)}{(x+y)(x-y)}$$
$$= \frac{2}{x+y}$$

55. Discussion and Writing Exercise

57. $\dfrac{x^8}{x^3} = x^{8-3} = x^5$

59. $(a^2 b^{-5})^{-4} = a^{2(-4)} b^{-5(-4)} = a^{-8} b^{20} = \dfrac{b^{20}}{a^8}$

61. $\dfrac{66x^2}{11x^5} = \dfrac{6 \cdot 11 \cdot x^2}{11 \cdot x^2 \cdot x^3} = \dfrac{6}{x^3}$

63. The shaded area has dimensions $x-6$ by $x-3$. Then the area is $(x-6)(x-3)$, or $x^2 - 9x + 18$.

65.
$$\frac{2x+11}{x-3} \cdot \frac{3}{x+4} + \frac{2x+1}{4+x} \cdot \frac{3}{3-x}$$
$$= \frac{6x+33}{(x-3)(x+4)} + \frac{6x+3}{(4+x)(3-x)}$$
$$= \frac{6x+33}{(x-3)(x+4)} + \frac{6x+3}{(4+x)(3-x)} \cdot \frac{-1}{-1}$$
$$= \frac{6x+33}{(x-3)(x+4)} + \frac{-6x-3}{(x+4)(x-3)}$$
$$= \frac{6x+33-6x-3}{(x-3)(x+4)}$$
$$= \frac{30}{(x-3)(x+4)}$$

67.
$$\frac{x}{x^4-y^4} - \left(\frac{1}{x+y}\right)^2$$
$$= \frac{x}{(x^2+y^2)(x+y)(x-y)} - \frac{1}{(x+y)^2}$$
$$\text{LCD} = (x^2+y^2)(x+y)^2(x-y)$$
$$= \frac{x}{(x^2+y^2)(x+y)(x-y)} \cdot \frac{x+y}{x+y} -$$
$$\frac{1}{(x+y)^2} \cdot \frac{(x^2+y^2)(x-y)}{(x^2+y^2)(x-y)}$$
$$= \frac{x(x+y) - (x^2+y^2)(x-y)}{(x^2+y^2)(x+y)^2(x-y)}$$
$$= \frac{x^2+xy - (x^3 - x^2y + xy^2 - y^3)}{(x^2+y^2)(x+y)^2(x-y)}$$
$$= \frac{x^2+xy - x^3 + x^2y - xy^2 + y^3}{(x^2+y^2)(x+y)^2(x-y)}$$

69. Let $l =$ the length of the missing side.
$$\frac{a^2-5a-9}{a-6} + \frac{a^2-6}{a-6} + l = 2a+5$$
$$\frac{2a^2-5a-15}{a-6} + l = 2a+5$$
$$l = 2a+5 - \frac{2a^2-5a-15}{a-6}$$
$$l = \left(2a+5\right) \cdot \frac{a-6}{a-6} - \frac{2a^2-5a-15}{a-6}$$
$$l = \frac{2a^2-7a-30}{a-6} - \frac{2a^2-5a-15}{a-6}$$
$$l = \frac{2a^2-7a-30-(2a^2-5a-15)}{a-6}$$
$$l = \frac{2a^2-7a-30-2a^2+5a+15}{a-6}$$
$$l = \frac{-2a-15}{a-6}$$

The length of the missing side is $\dfrac{-2a-15}{a-6}$.

Now find the area.
$$A = \frac{1}{2} \cdot b \cdot h$$
$$A = \frac{1}{2}\left(\frac{-2a-15}{a-6}\right)\left(\frac{a^2-6}{a-6}\right)$$
$$A = \frac{(-2a-15)(a^2-6)}{2(a-6)^2}, \text{ or}$$
$$A = \frac{-2a^3-15a^2+12a+90}{2a^2-24a+72}$$

71.–73. Left to the student

Exercise Set 11.6

1. $\dfrac{1+\dfrac{9}{16}}{1-\dfrac{3}{4}}$ LCM of the denominators is 16.

$=\dfrac{1+\dfrac{9}{16}}{1-\dfrac{3}{4}}\cdot\dfrac{16}{16}$ Multiplying by 1 using $\dfrac{16}{16}$

$=\dfrac{\left(1+\dfrac{9}{16}\right)16}{\left(1-\dfrac{3}{4}\right)16}$ Multiplying numerator and denominator by 16

$=\dfrac{1(16)+\dfrac{9}{16}(16)}{1(16)-\dfrac{3}{4}(16)}$

$=\dfrac{16+9}{16-12}$

$=\dfrac{25}{4}$

3. $\dfrac{1-\dfrac{3}{5}}{1+\dfrac{1}{5}}$

$=\dfrac{1\cdot\dfrac{5}{5}-\dfrac{3}{5}}{1\cdot\dfrac{5}{5}+\dfrac{1}{5}}$ Getting a common denominator in numerator and in denominator

$=\dfrac{\dfrac{5}{5}-\dfrac{3}{5}}{\dfrac{5}{5}+\dfrac{1}{5}}$

$=\dfrac{\dfrac{2}{5}}{\dfrac{6}{5}}$ Subtracting in numerator; adding in denominator

$=\dfrac{2}{5}\cdot\dfrac{5}{6}$ Multiplying by the reciprocal of the divisor

$=\dfrac{2\cdot5}{5\cdot2\cdot3}$

$=\dfrac{\cancel{2}\cdot\cancel{5}\cdot1}{\cancel{5}\cdot\cancel{2}\cdot3}$

$=\dfrac{1}{3}$

5. $\dfrac{\dfrac{1}{2}+\dfrac{3}{4}}{\dfrac{5}{8}-\dfrac{5}{6}}=\dfrac{\dfrac{1}{2}\cdot\dfrac{2}{2}+\dfrac{3}{4}}{\dfrac{5}{8}\cdot\dfrac{3}{3}-\dfrac{5}{6}\cdot\dfrac{4}{4}}$ Getting a common denominator in numerator and denominator

$=\dfrac{\dfrac{2}{4}+\dfrac{3}{4}}{\dfrac{15}{24}-\dfrac{20}{24}}$

$=\dfrac{\dfrac{5}{4}}{\dfrac{-5}{24}}$ Adding in numerator; subtracting in denominator

$=\dfrac{5}{4}\cdot\dfrac{24}{-5}$ Multiplying by the reciprocal of the divisor

$=\dfrac{5\cdot4\cdot6}{4\cdot(-1)\cdot5}$

$=\dfrac{\cancel{5}\cdot\cancel{4}\cdot6}{\cancel{4}\cdot(-1)\cdot\cancel{5}}$

$=-6$

7. $\dfrac{\dfrac{1}{x}+3}{\dfrac{1}{x}-5}$ LCM of the denominators is x.

$=\dfrac{\dfrac{1}{x}+3}{\dfrac{1}{x}-5}\cdot\dfrac{x}{x}$ Multiplying by 1 using $\dfrac{x}{x}$

$=\dfrac{\left(\dfrac{1}{x}+3\right)x}{\left(\dfrac{1}{x}-5\right)x}$

$=\dfrac{\dfrac{1}{x}\cdot x+3\cdot x}{\dfrac{1}{x}\cdot x-5\cdot x}$

$=\dfrac{1+3x}{1-5x}$

9. $\dfrac{4 - \dfrac{1}{x^2}}{2 - \dfrac{1}{x}}$ LCM of the denominators is x^2.

$= \dfrac{4 - \dfrac{1}{x^2}}{2 - \dfrac{1}{x}} \cdot \dfrac{x^2}{x^2}$

$= \dfrac{\left(4 - \dfrac{1}{x^2}\right)x^2}{\left(2 - \dfrac{1}{x}\right)x^2}$

$= \dfrac{4 \cdot x^2 - \dfrac{1}{x^2} \cdot x^2}{2 \cdot x^2 - \dfrac{1}{x} \cdot x^2}$

$= \dfrac{4x^2 - 1}{2x^2 - x}$

$= \dfrac{(2x + 1)(2x - 1)}{x(2x - 1)}$ Factoring numerator and denominator

$= \dfrac{(2x + 1)(2x\!-\!1)}{x(2x\!-\!1)}$

$= \dfrac{2x + 1}{x}$

11. $\dfrac{8 + \dfrac{8}{d}}{1 + \dfrac{1}{d}} = \dfrac{8 \cdot \dfrac{d}{d} + \dfrac{8}{d}}{1 \cdot \dfrac{d}{d} + \dfrac{1}{d}}$

$= \dfrac{\dfrac{8d + 8}{d}}{\dfrac{d + 1}{d}}$

$= \dfrac{8d + 8}{d} \cdot \dfrac{d}{d + 1}$

$= \dfrac{8(d + 1)(d)}{d(d + 1)}$

$= \dfrac{8(d\!+\!1)(d)}{d(d\!+\!1)(1)}$

$= 8$

13. $\dfrac{\dfrac{x}{8} - \dfrac{8}{x}}{\dfrac{1}{8} + \dfrac{1}{x}}$ LCM of the denominators is $8x$.

$= \dfrac{\dfrac{x}{8} - \dfrac{8}{x}}{\dfrac{1}{8} + \dfrac{1}{x}} \cdot \dfrac{8x}{8x}$

$= \dfrac{\left(\dfrac{x}{8} - \dfrac{8}{x}\right)8x}{\left(\dfrac{1}{8} + \dfrac{1}{x}\right)8x}$

$= \dfrac{\dfrac{x}{8}(8x) - \dfrac{8}{x}(8x)}{\dfrac{1}{8}(8x) + \dfrac{1}{x}(8x)}$

$= \dfrac{x^2 - 64}{x + 8}$

$= \dfrac{(x + 8)(x - 8)}{x + 8}$

$= \dfrac{(x\!+\!8)(x - 8)}{1(x\!+\!8)}$

$= x - 8$

15. $\dfrac{1 + \dfrac{1}{y}}{1 - \dfrac{1}{y^2}} = \dfrac{1 \cdot \dfrac{y}{y} + \dfrac{1}{y}}{1 \cdot \dfrac{y^2}{y^2} - \dfrac{1}{y^2}}$

$= \dfrac{\dfrac{y + 1}{y}}{\dfrac{y^2 - 1}{y^2}}$

$= \dfrac{y + 1}{y} \cdot \dfrac{y^2}{y^2 - 1}$

$= \dfrac{(y + 1)y \cdot y}{y(y + 1)(y - 1)}$

$= \dfrac{(y\!+\!1)y \cdot y}{y(y\!+\!1)(y - 1)}$

$= \dfrac{y}{y - 1}$

17. $\dfrac{\dfrac{1}{5} - \dfrac{1}{a}}{\dfrac{5-a}{5}}$ LCM of the denominators is $5a$.

$= \dfrac{\dfrac{1}{5} - \dfrac{1}{a}}{\dfrac{5-a}{5}} \cdot \dfrac{5a}{5a}$

$= \dfrac{\left(\dfrac{1}{5} - \dfrac{1}{a}\right)5a}{\left(\dfrac{5-a}{5}\right)5a}$

$= \dfrac{\dfrac{1}{5}(5a) - \dfrac{1}{a}(5a)}{a(5-a)}$

$= \dfrac{a-5}{5a-a^2}$

$= \dfrac{a-5}{-a(-5+a)}$

$= \dfrac{1(a-5)}{-a(a-5)}$

$= -\dfrac{1}{a}$

19. $\dfrac{\dfrac{1}{a} + \dfrac{1}{b}}{\dfrac{1}{a^2} - \dfrac{1}{b^2}}$ LCM of the denominators is $a^2 b^2$.

$= \dfrac{\dfrac{1}{a} + \dfrac{1}{b}}{\dfrac{1}{a^2} - \dfrac{1}{b^2}} \cdot \dfrac{a^2 b^2}{a^2 b^2}$

$= \dfrac{\left(\dfrac{1}{a} + \dfrac{1}{b}\right) \cdot a^2 b^2}{\left(\dfrac{1}{a^2} - \dfrac{1}{b^2}\right) \cdot a^2 b^2}$

$= \dfrac{\dfrac{1}{a} \cdot a^2 b^2 + \dfrac{1}{b} \cdot a^2 b^2}{\dfrac{1}{a^2} \cdot a^2 b^2 - \dfrac{1}{b^2} \cdot a^2 b^2}$

$= \dfrac{ab^2 + a^2 b}{b^2 - a^2}$

$= \dfrac{ab(b+a)}{(b+a)(b-a)}$

$= \dfrac{ab(b+a)}{(b+a)(b-a)}$

$= \dfrac{ab}{b-a}$

21. $\dfrac{\dfrac{p}{q} + \dfrac{q}{p}}{\dfrac{1}{p} + \dfrac{1}{q}}$ LCM of the denominators is pq.

$= \dfrac{\left(\dfrac{p}{q} + \dfrac{q}{p}\right) \cdot pq}{\left(\dfrac{1}{p} + \dfrac{1}{q}\right) \cdot pq}$

$= \dfrac{\dfrac{p}{q} \cdot pq + \dfrac{q}{p} \cdot pq}{\dfrac{1}{p} \cdot pq + \dfrac{1}{q} \cdot pq}$

$= \dfrac{p^2 + q^2}{q + p}$

23. $\dfrac{\dfrac{2}{a} + \dfrac{4}{a^2}}{\dfrac{5}{a^3} - \dfrac{3}{a}}$ LCD is a^3

$= \dfrac{\dfrac{2}{a} + \dfrac{4}{a^2}}{\dfrac{5}{a^3} - \dfrac{3}{a}} \cdot \dfrac{a^3}{a^3}$

$= \dfrac{\dfrac{2}{a} \cdot a^3 + \dfrac{4}{a^2} \cdot a^3}{\dfrac{5}{a^3} \cdot a^3 - \dfrac{3}{a} \cdot a^3}$

$= \dfrac{2a^2 + 4a}{5 - 3a^2}$

(Although the numerator can be factored, doing so will not enable us to simplify further.)

25. $\dfrac{\dfrac{2}{7a^4} - \dfrac{1}{14a}}{\dfrac{3}{5a^2} + \dfrac{2}{15a}} = \dfrac{\dfrac{2}{7a^4} \cdot \dfrac{2}{2} - \dfrac{1}{14a} \cdot \dfrac{a^3}{a^3}}{\dfrac{3}{5a^2} \cdot \dfrac{3}{3} + \dfrac{2}{15a} \cdot \dfrac{a}{a}}$

$= \dfrac{\dfrac{4 - a^3}{14a^4}}{\dfrac{9 + 2a}{15a^2}}$

$= \dfrac{4 - a^3}{14a^4} \cdot \dfrac{15a^2}{9 + 2a}$

$= \dfrac{15 \cdot a^2 (4 - a^3)}{14a^2 \cdot a^2 (9 + 2a)}$

$= \dfrac{15(4 - a^3)}{14a^2(9 + 2a)}, \text{ or } \dfrac{60 - 15a^3}{126a^2 + 28a^3}$

27.
$$\frac{\dfrac{a}{b}+\dfrac{c}{d}}{\dfrac{b}{a}+\dfrac{d}{c}}=\frac{\dfrac{a}{b}\cdot\dfrac{d}{d}+\dfrac{c}{d}\cdot\dfrac{b}{b}}{\dfrac{b}{a}\cdot\dfrac{c}{c}+\dfrac{d}{c}\cdot\dfrac{a}{a}}$$

$$=\frac{\dfrac{ad+bc}{bd}}{\dfrac{bc+ad}{ac}}$$

$$=\frac{ad+bc}{bd}\cdot\frac{ac}{bc+ad}$$

$$=\frac{ac(ad+bc)}{bd(bc+ad)}$$

$$=\frac{ac}{bd}\cdot\frac{ad+bc}{bc+ad}$$

$$=\frac{ac}{bd}\cdot1$$

$$=\frac{ac}{bd}$$

29.
$$\frac{\dfrac{x}{5y^3}+\dfrac{3}{10y}}{\dfrac{3}{10y}+\dfrac{x}{5y^3}}$$

Observe that, by the commutative law of addition, the numerator and denominator are equivalent, so the result is 1. We could also simplify this expression as follows:

$$\frac{\dfrac{x}{5y^3}+\dfrac{3}{10y}}{\dfrac{3}{10y}+\dfrac{x}{5y^3}}=\frac{\dfrac{x}{5y^3}+\dfrac{3}{10y}}{\dfrac{3}{10y}+\dfrac{x}{5y^3}}\cdot\frac{10y^3}{10y^3}$$

$$=\frac{\dfrac{x}{5y^3}\cdot10y^3+\dfrac{3}{10y}\cdot10y^3}{\dfrac{3}{10y}\cdot10y^3+\dfrac{x}{5y^3}\cdot10y^3}$$

$$=\frac{2x+3y^2}{3y^2+2x}$$

$$=1$$

31.
$$\frac{\dfrac{3}{x+1}+\dfrac{1}{x}}{\dfrac{2}{x+1}+\dfrac{3}{x}}=\frac{\dfrac{3}{x+1}+\dfrac{1}{x}}{\dfrac{2}{x+1}+\dfrac{3}{x}}\cdot\frac{x(x+1)}{x(x+1)}$$

$$=\frac{\dfrac{3}{x+1}\cdot x(x+1)+\dfrac{1}{x}\cdot x(x+1)}{\dfrac{2}{x+1}\cdot x(x+1)+\dfrac{3}{x}\cdot x(x+1)}$$

$$=\frac{3x+x+1}{2x+3(x+1)}$$

$$=\frac{4x+1}{2x+3x+3}$$

$$=\frac{4x+1}{5x+3}$$

33. Discussion and Writing Exercise

35.
$$(2x^3-4x^2+x-7)+(4x^4+x^3+4x^2+x)$$
$$=4x^4+3x^3+2x-7$$

37. $p^2-10p+25=p^2-2\cdot p\cdot5+5^2$ Trinomial square
$$=(p-5)^2$$

39. $50p^2-100=50(p^2-2)$ Factoring out the common factor

Since p^2-2 cannot be factored, we have factored completely.

41. _Familiarize._ Let $w=$ the width of the rectangle. Then $w+3=$ the length. Recall that the formula for the area of a rectangle is $A=lw$ and the formula for the perimeter of a rectangle is $P=2l+2w$.

Translate. We substitute in the formula for area.
$$10=lw$$
$$10=(w+3)w$$

Solve.
$$10=(w+3)w$$
$$10=w^2+3w$$
$$0=w^2+3w-10$$
$$0=(w+5)(w-2)$$
$$w+5=0\quad\text{or}\quad w-2=0$$
$$w=-5\quad\text{or}\qquad w=2$$

Check. Since the width cannot be negative, we only check 2. If $w=2$, then $w+3=2+3$, or 5. Since $2\cdot5=10$, the given area, the answer checks. Now we find the perimeter:
$$P=2l+2w$$
$$P=2\cdot5+2\cdot2$$
$$P=10+4$$
$$P=14$$

We can check this by repeating the calculation.

State. The perimeter is 14 yd.

43.
$$\frac{\dfrac{1}{\dfrac{2}{x-1}-\dfrac{1}{3x-2}}}{}$$

$$=\frac{\dfrac{1}{\dfrac{2}{x-1}-\dfrac{1}{3x-2}}\cdot\dfrac{(x-1)(3x-2)}{(x-1)(3x-2)}}{}$$

$$=\frac{(x-1)(3x-2)}{\left(\dfrac{2}{x-1}-\dfrac{1}{3x-2}\right)(x-1)(3x-2)}$$

$$=\frac{(x-1)(3x-2)}{\dfrac{2}{x-1}(x-1)(3x-2)-\dfrac{1}{3x-2}(x-1)(3x-2)}$$

$$=\frac{(x-1)(3x-2)}{2(3x-2)-(x-1)}$$

$$=\frac{(x-1)(3x-2)}{6x-4-x+1}$$

$$=\frac{(x-1)(3x-2)}{5x-3}$$

45. $1 + \cfrac{1}{1 + \cfrac{1}{1 + \cfrac{1}{1 + \cfrac{1}{x}}}} = 1 + \cfrac{1}{1 + \cfrac{1}{1 + \cfrac{1}{x+1}{x}}}$

$$= 1 + \cfrac{1}{1 + \cfrac{1}{1 + \cfrac{x}{x+1}}}$$

$$= 1 + \cfrac{1}{1 + \cfrac{1}{\cfrac{x+1+x}{x+1}}}$$

$$= 1 + \cfrac{1}{1 + \cfrac{1}{\cfrac{2x+1}{x+1}}}$$

$$= 1 + \cfrac{1}{1 + \cfrac{x+1}{2x+1}}$$

$$= 1 + \cfrac{1}{\cfrac{2x+1+x+1}{2x+1}}$$

$$= 1 + \cfrac{1}{\cfrac{3x+2}{2x+1}}$$

$$= 1 + \cfrac{2x+1}{3x+2}$$

$$= \cfrac{3x+2+2x+1}{3x+2}$$

$$= \cfrac{5x+3}{3x+2}$$

Exercise Set 11.7

1. $\qquad \dfrac{4}{5} - \dfrac{2}{3} = \dfrac{x}{9}$, LCM = 45

$$45\left(\dfrac{4}{5} - \dfrac{2}{3}\right) = 45 \cdot \dfrac{x}{9}$$

$$45 \cdot \dfrac{4}{5} - 45 \cdot \dfrac{2}{3} = 45 \cdot \dfrac{x}{9}$$

$$36 - 30 = 5x$$

$$6 = 5x$$

$$\dfrac{6}{5} = x$$

Check:

$$\dfrac{\dfrac{4}{5} - \dfrac{2}{3} = \dfrac{x}{9}}{\begin{array}{c|c} \dfrac{4}{5} - \dfrac{2}{3} \ ? & \dfrac{\frac{6}{5}}{9} \\ \dfrac{12}{15} - \dfrac{10}{15} & \dfrac{6}{5} \cdot \dfrac{1}{9} \\ \dfrac{2}{15} & \dfrac{2}{15} \qquad \text{TRUE} \end{array}}$$

This checks, so the solution is $\dfrac{6}{5}$.

3. $\qquad \dfrac{3}{5} + \dfrac{1}{8} = \dfrac{1}{x}$, LCM = $40x$

$$40x\left(\dfrac{3}{5} + \dfrac{1}{8}\right) = 40x \cdot \dfrac{1}{x}$$

$$40x \cdot \dfrac{3}{5} + 40x \cdot \dfrac{1}{8} = 40x \cdot \dfrac{1}{x}$$

$$24x + 5x = 40$$

$$29x = 40$$

$$x = \dfrac{40}{29}$$

Check:

$$\dfrac{\dfrac{3}{5} + \dfrac{1}{8} = \dfrac{1}{x}}{\begin{array}{c|c} \dfrac{3}{5} + \dfrac{1}{8} \ ? & \dfrac{1}{\frac{40}{29}} \\ \dfrac{24}{40} + \dfrac{5}{40} & 1 \cdot \dfrac{29}{40} \\ \dfrac{29}{40} & \dfrac{29}{40} \qquad \text{TRUE} \end{array}}$$

This checks, so the solution is $\dfrac{40}{29}$.

5. $\qquad \dfrac{3}{8} + \dfrac{4}{5} = \dfrac{x}{20}$, LCM = 40

$$40\left(\dfrac{3}{8} + \dfrac{4}{5}\right) = 40 \cdot \dfrac{x}{20}$$

$$40 \cdot \dfrac{3}{8} + 40 \cdot \dfrac{4}{5} = 40 \cdot \dfrac{x}{20}$$

$$15 + 32 = 2x$$

$$47 = 2x$$

$$\dfrac{47}{2} = x$$

Check:

$$\dfrac{\dfrac{3}{8} + \dfrac{4}{5} = \dfrac{x}{20}}{\begin{array}{c|c} \dfrac{3}{8} + \dfrac{4}{5} \ ? & \dfrac{\frac{47}{2}}{20} \\ \dfrac{15}{40} + \dfrac{32}{40} & \dfrac{47}{2} \cdot \dfrac{1}{20} \\ \dfrac{47}{40} & \dfrac{47}{40} \qquad \text{TRUE} \end{array}}$$

This checks, so the solution is $\dfrac{47}{2}$.

7. $\quad \dfrac{1}{x} = \dfrac{2}{3} - \dfrac{5}{6}$, LCM $= 6x$

$$6x \cdot \dfrac{1}{x} = 6x\left(\dfrac{2}{3} - \dfrac{5}{6}\right)$$

$$6x \cdot \dfrac{1}{x} = 6x \cdot \dfrac{2}{3} - 6x \cdot \dfrac{5}{6}$$

$$6 = 4x - 5x$$

$$6 = -x$$

$$-6 = x$$

Check:

$$\dfrac{1}{x} = \dfrac{2}{3} - \dfrac{5}{6}$$

$$\begin{array}{c|c} \dfrac{1}{-6} \;?\; & \dfrac{2}{3} - \dfrac{5}{6} \\ \hline -\dfrac{1}{6} & \dfrac{4}{6} - \dfrac{5}{6} \\ & -\dfrac{1}{6} \quad \text{TRUE} \end{array}$$

This checks, so the solution is -6.

9. $\quad \dfrac{1}{6} + \dfrac{1}{8} = \dfrac{1}{t}$, LCM $= 24t$

$$24t\left(\dfrac{1}{6} + \dfrac{1}{8}\right) = 24t \cdot \dfrac{1}{t}$$

$$24t \cdot \dfrac{1}{6} + 24t \cdot \dfrac{1}{8} = 24t \cdot \dfrac{1}{t}$$

$$4t + 3t = 24$$

$$7t = 24$$

$$t = \dfrac{24}{7}$$

Check:

$$\dfrac{1}{6} + \dfrac{1}{8} = \dfrac{1}{t}$$

$$\begin{array}{c|c} \dfrac{1}{6} + \dfrac{1}{8} \;?\; & \dfrac{1}{24/7} \\ \hline \dfrac{4}{24} + \dfrac{3}{24} & 1 \cdot \dfrac{7}{24} \\ \dfrac{7}{24} & \dfrac{7}{24} \quad \text{TRUE} \end{array}$$

This checks, so the solution is $\dfrac{24}{7}$.

11. $\quad x + \dfrac{4}{x} = -5$, LCM $= x$

$$x\left(x + \dfrac{4}{x}\right) = x(-5)$$

$$x \cdot x + x \cdot \dfrac{4}{x} = x(-5)$$

$$x^2 + 4 = -5x$$

$$x^2 + 5x + 4 = 0$$

$$(x + 4)(x + 1) = 0$$

$$x + 4 = 0 \quad \text{or} \quad x + 1 = 0$$

$$x = -4 \quad \text{or} \quad x = -1$$

Check:

$$\begin{array}{c|c} x + \dfrac{4}{x} = -5 & x + \dfrac{4}{x} = -5 \\ \hline -4 + \dfrac{4}{-4} \;?\; -5 & -1 + \dfrac{4}{-1} \;?\; -5 \\ -4 - 1 & -1 - 4 \\ -5 \quad \text{TRUE} & -5 \quad \text{TRUE} \end{array}$$

Both of these check, so the two solutions are -4 and -1.

13. $\quad \dfrac{x}{4} - \dfrac{4}{x} = 0$, LCM $= 4x$

$$4x\left(\dfrac{x}{4} - \dfrac{4}{x}\right) = 4x \cdot 0$$

$$4x \cdot \dfrac{x}{4} - 4x \cdot \dfrac{4}{x} = 4x \cdot 0$$

$$x^2 - 16 = 0$$

$$(x + 4)(x - 4) = 0$$

$$x + 4 = 0 \quad \text{or} \quad x - 4 = 0$$

$$x = -4 \quad \text{or} \quad x = 4$$

Check:

$$\begin{array}{c|c} \dfrac{x}{4} - \dfrac{4}{x} = 0 & \dfrac{x}{4} - \dfrac{4}{x} = 0 \\ \hline \dfrac{-4}{4} - \dfrac{4}{-4} \;?\; 0 & \dfrac{4}{4} - \dfrac{4}{4} \;?\; 0 \\ -1 - (-1) & 1 - 1 \\ -1 + 1 & 0 \quad \text{TRUE} \\ 0 \quad \text{TRUE} & \end{array}$$

Both of these check, so the two solutions are -4 and 4.

15. $\quad \dfrac{5}{x} = \dfrac{6}{x} - \dfrac{1}{3}$, LCM $= 3x$

$$3x \cdot \dfrac{5}{x} = 3x\left(\dfrac{6}{x} - \dfrac{1}{3}\right)$$

$$3x \cdot \dfrac{5}{x} = 3x \cdot \dfrac{6}{x} - 3x \cdot \dfrac{1}{3}$$

$$15 = 18 - x$$

$$-3 = -x$$

$$3 = x$$

Check:

$$\dfrac{5}{x} = \dfrac{6}{x} - \dfrac{1}{3}$$

$$\begin{array}{c|c} \dfrac{5}{3} \;?\; & \dfrac{6}{3} - \dfrac{1}{3} \\ \hline & \dfrac{5}{3} \quad \text{TRUE} \end{array}$$

This checks, so the solution is 3.

17.
$$\frac{5}{3x} + \frac{3}{x} = 1, \text{ LCM} = 3x$$

$$3x\left(\frac{5}{3x} + \frac{3}{x}\right) = 3x \cdot 1$$

$$3x \cdot \frac{5}{3x} + 3x \cdot \frac{3}{x} = 3x \cdot 1$$

$$5 + 9 = 3x$$

$$14 = 3x$$

$$\frac{14}{3} = x$$

Check:
$$\frac{5}{3x} + \frac{3}{x} = 1$$

$$\begin{array}{c|c} \dfrac{5}{3 \cdot (14/3)} + \dfrac{3}{(14/3)} & ? \ 1 \\[2ex] \dfrac{5}{14} + \dfrac{9}{14} & \\[2ex] \dfrac{14}{14} & \\[2ex] 1 & \text{TRUE} \end{array}$$

This checks, so the solution is $\dfrac{14}{3}$.

19.
$$\frac{t-2}{t+3} = \frac{3}{8}, \text{ LCM} = 8(t+3)$$

$$8(t+3)\left(\frac{t-2}{t+3}\right) = 8(t+3)\left(\frac{3}{8}\right)$$

$$8(t-2) = 3(t+3)$$

$$8t - 16 = 3t + 9$$

$$5t = 25$$

$$t = 5$$

Check:
$$\frac{t-2}{t+3} = \frac{3}{8}$$

$$\begin{array}{c|c} \dfrac{5-2}{5+3} & ? \ \dfrac{3}{8} \\[2ex] \dfrac{3}{8} & \text{TRUE} \end{array}$$

This checks, so the solution is 5.

21.
$$\frac{2}{x+1} = \frac{1}{x-2}, \text{ LCM} = (x+1)(x-2)$$

$$(x+1)(x-2) \cdot \frac{2}{x+1} = (x+1)(x-2) \cdot \frac{1}{x-2}$$

$$2(x-2) = x+1$$

$$2x - 4 = x + 1$$

$$x = 5$$

This checks, so the solution is 5.

23.
$$\frac{x}{6} - \frac{x}{10} = \frac{1}{6}, \text{ LCM} = 30$$

$$30\left(\frac{x}{6} - \frac{x}{10}\right) = 30 \cdot \frac{1}{6}$$

$$30 \cdot \frac{x}{6} - 30 \cdot \frac{x}{10} = 30 \cdot \frac{1}{6}$$

$$5x - 3x = 5$$

$$2x = 5$$

$$x = \frac{5}{2}$$

This checks, so the solution is $\dfrac{5}{2}$.

25.
$$\frac{t+2}{5} - \frac{t-2}{4} = 1, \text{ LCM} = 20$$

$$20\left(\frac{t+2}{5} - \frac{t-2}{4}\right) = 20 \cdot 1$$

$$20\left(\frac{t+2}{5}\right) - 20\left(\frac{t-2}{4}\right) = 20 \cdot 1$$

$$4(t+2) - 5(t-2) = 20$$

$$4t + 8 - 5t + 10 = 20$$

$$-t + 18 = 20$$

$$-t = 2$$

$$t = -2$$

This checks, so the solution is -2.

27.
$$\frac{5}{x-1} = \frac{3}{x+2},$$
$$\text{LCD} = (x-1)(x+2)$$

$$(x-1)(x+2) \cdot \frac{5}{x-1} = (x-1)(x+2) \cdot \frac{3}{x+2}$$

$$5(x+2) = 3(x-1)$$

$$5x + 10 = 3x - 3$$

$$2x = -13$$

$$x = -\frac{13}{2}$$

This checks, so the solution is $-\dfrac{13}{2}$.

29.
$$\frac{a-3}{3a+2} = \frac{1}{5}, \text{ LCM} = 5(3a+2)$$

$$5(3a+2) \cdot \frac{a-3}{3a+2} = 5(3a+2) \cdot \frac{1}{5}$$

$$5(a-3) = 3a+2$$

$$5a - 15 = 3a + 2$$

$$2a = 17$$

$$a = \frac{17}{2}$$

This checks, so the solution is $\dfrac{17}{2}$.

31. $\dfrac{x-1}{x-5} = \dfrac{4}{x-5}$, LCM $= x - 5$

$$(x-5) \cdot \dfrac{x-1}{x-5} = (x-5) \cdot \dfrac{4}{x-5}$$

$$x - 1 = 4$$

$$x = 5$$

The number 5 is not a solution because it makes a denominator zero. Thus, there is no solution.

33. $\dfrac{2}{x+3} = \dfrac{5}{x}$, LCM $= x(x+3)$

$$x(x+3) \cdot \dfrac{2}{x+3} = x(x+3) \cdot \dfrac{5}{x}$$

$$2x = 5(x+3)$$

$$2x = 5x + 15$$

$$-15 = 3x$$

$$-5 = x$$

This checks, so the solution is -5.

35. $\dfrac{x-2}{x-3} = \dfrac{x-1}{x+1}$, LCM $= (x-3)(x+1)$

$$(x-3)(x+1) \cdot \dfrac{x-2}{x-3} = (x-3)(x+1) \cdot \dfrac{x-1}{x+1}$$

$$(x+1)(x-2) = (x-3)(x-1)$$

$$x^2 - x - 2 = x^2 - 4x + 3$$

$$-x - 2 = -4x + 3$$

$$3x = 5$$

$$x = \dfrac{5}{3}$$

This checks, so the solution is $\dfrac{5}{3}$.

37. $\dfrac{1}{x+3} + \dfrac{1}{x-3} = \dfrac{1}{x^2-9}$,

LCM $= (x+3)(x-3)$

$$(x+3)(x-3)\left(\dfrac{1}{x+3} + \dfrac{1}{x-3}\right) = (x+3)(x-3) \cdot \dfrac{1}{(x+3)(x-3)}$$

$$(x-3) + (x+3) = 1$$

$$2x = 1$$

$$x = \dfrac{1}{2}$$

This checks, so the solution is $\dfrac{1}{2}$.

39. $\dfrac{x}{x+4} - \dfrac{4}{x-4} = \dfrac{x^2+16}{x^2-16}$,

LCM $= (x+4)(x-4)$

$$(x+4)(x-4)\left(\dfrac{x}{x+4} - \dfrac{x}{x-4}\right) = (x+4)(x-4) \cdot \dfrac{x^2+16}{(x+4)(x-4)}$$

$$x(x-4) - 4(x+4) = x^2 + 16$$

$$x^2 - 4x - 4x - 16 = x^2 + 16$$

$$x^2 - 8x - 16 = x^2 + 16$$

$$-8x - 16 = 16$$

$$-8x = 32$$

$$x = -4$$

The number -4 is not a solution because it makes a denominator zero. Thus, there is no solution.

41. $\dfrac{4-a}{8-a} = \dfrac{4}{a-8}$ $8-a$ and $a-8$ are opposites

$$\dfrac{4-a}{8-a} \cdot \dfrac{-1}{-1} = \dfrac{4}{a-8}$$

$$\dfrac{a-4}{a-8} = \dfrac{4}{a-8}, \text{ LCM} = a - 8$$

$$(a-8)\left(\dfrac{a-4}{a-8}\right) = (a-8)\left(\dfrac{4}{a-8}\right)$$

$$a - 4 = 4$$

$$a = 8$$

The number 8 is not a solution because it makes a denominator zero. Thus, there is no solution.

43. $2 - \dfrac{a-2}{a+3} = \dfrac{a^2-4}{a+3}$, LCM $= a + 3$

$$(a+3)\left(2 - \dfrac{a-2}{a+3}\right) = (a+3) \cdot \dfrac{a^2-4}{a+3}$$

$$2(a+3) - (a-2) = a^2 - 4$$

$$2a + 6 - a + 2 = a^2 - 4$$

$$0 = a^2 - a - 12$$

$$0 = (a-4)(a+3)$$

$$a - 4 = 0 \text{ or } a + 3 = 0$$

$$a = 4 \text{ or } \quad a = -3$$

Only 4 checks, so the solution is 4.

45. Discussion and Writing Exercise

47. $(a^2b^5)^{-3} = \dfrac{1}{(a^2b^5)^3} = \dfrac{1}{(a^2)^3(b^5)^3} = \dfrac{1}{a^6b^{15}}$

49. $\left(\dfrac{2x}{t^2}\right)^4 = \dfrac{(2x)^4}{(t^2)^4} = \dfrac{2^4x^4}{t^8} = \dfrac{16x^4}{t^8}$

51. $4x^{-5} \cdot 8x^{11} = 4 \cdot 8x^{-5+11} = 32x^6$

53. $5x + 10y = 20$

To find the x-intercept, let $y = 0$. Then solve for x.

$$5x + 10y = 20$$
$$5x + 10 \cdot 0 = 20$$
$$5x = 20$$
$$x = 4$$

The x-intercept is $(4, 0)$.

To find the y-intercept, let $x = 0$. Then solve for y.

$$5x + 10y = 20$$
$$5 \cdot 0 + 10y = 20$$
$$10y = 20$$
$$y = 2$$

The y-intercept is $(0, 2)$.

Plot these points and draw the line.

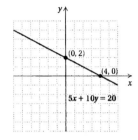

A third point should be used as a check. We substitute any value for x and solve for y. We let $x = -4$. Then

$$5x + 10y = 20$$
$$5(-4) + 10y = 20$$
$$-20 + 10y = 20$$
$$10y = 40$$
$$y = 4.$$

The point $(-4, 4)$ is on the graph, so the graph is probably correct.

55. $10y - 4x = -20$

To find the x-intercept, let $y = 0$. Then solve for x.

$$10y - 4x = -20$$
$$10 \cdot 0 - 4x = -20$$
$$-4x = -20$$
$$x = 5$$

The x-intercept is $(5, 0)$.

To find the y-intercept, let $x = 0$. Then solve for y.

$$10y - 4x = -20$$
$$10y - 4 \cdot 0 = -20$$
$$10y = -20$$
$$y = -2$$

The y-intercept is $(0, -2)$.

Plot these points and draw the line.

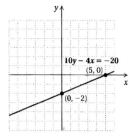

A third point should be used as a check. We substitute any value for x and solve for y. We let $x = -5$. Then

$$10y - 4x = -20$$
$$10y - 4(-5) = -20$$
$$10y + 20 = -20$$
$$10y = -40$$
$$y = -4.$$

The point $(-5, -4)$ is on the graph, so the graph is probably correct.

57.
$$\frac{4}{y-2} - \frac{2y-3}{y^2-4} = \frac{5}{y+2},$$
$$\text{LCM} = (y+2)(y-2)$$

$$(y+2)(y-2)\left(\frac{4}{y-2} - \frac{2y-3}{(y+2)(y-2)}\right) =$$
$$(y+2)(y-2) \cdot \frac{5}{y+2}$$
$$4(y+2) - (2y-3) = 5(y-2)$$
$$4y + 8 - 2y + 3 = 5y - 10$$
$$2y + 11 = 5y - 10$$
$$21 = 3y$$
$$7 = y$$

This checks, so the solution is 7.

59.
$$\frac{x+1}{x+2} = \frac{x+3}{x+4},$$
$$\text{LCM} = (x+2)(x+4)$$
$$(x+2)(x+4)\left(\frac{x+1}{x+2}\right) = (x+2)(x+4)\left(\frac{x+3}{x+4}\right)$$
$$(x+4)(x+1) = (x+2)(x+3)$$
$$x^2 + 5x + 4 = x^2 + 5x + 6$$
$$4 = 6 \quad \text{Subtracting } x^2 \text{ and } 5x$$

We get a false equation, so the original equation has no solution.

61.
$$4a - 3 = \frac{a+13}{a+1}, \text{ LCM} = a+1$$
$$(a+1)(4a-3) = (a+1) \cdot \frac{a+13}{a+1}$$
$$4a^2 + a - 3 = a + 13$$
$$4a^2 - 16 = 0$$
$$4(a+2)(a-2) = 0$$
$$a + 2 = 0 \quad \text{or} \quad a - 2 = 0$$
$$a = -2 \quad \text{or} \quad a = 2$$

Both of these check, so the two solutions are -2 and 2.

63.
$$\frac{y^2-4}{y+3} = 2 - \frac{y-2}{y+3}, \ \text{LCM} = y+3$$

$$(y+3)\cdot\frac{y^2-4}{y+3} = (y+3)\left(2-\frac{y-2}{y+3}\right)$$

$$y^2-4 = 2(y+3)-(y-2)$$

$$y^2-4 = 2y+6-y+2$$

$$y^2-4 = y+8$$

$$y^2-y-12 = 0$$

$$(y-4)(y+3) = 0$$

$$y-4 = 0 \quad \text{or} \quad y+3 = 0$$

$$y = 4 \quad \text{or} \quad y = -3$$

The number 4 is a solution, but -3 is not because it makes a denominator zero.

65. Left to the student

Exercise Set 11.8

---

1. _Familiarize._ The job takes Mandy 4 hours working alone and Omar 5 hours working alone. Then in 1 hour Mandy does $\frac{1}{4}$ of the job and Omar does $\frac{1}{5}$ of the job. Working together, they can do $\frac{1}{4}+\frac{1}{5}$, or $\frac{9}{20}$ of the job in 1 hour. In two hours, Mandy does $2\left(\frac{1}{4}\right)$ of the job and Omar does $2\left(\frac{1}{5}\right)$ of the job. Working together they can do $2\left(\frac{1}{4}\right)+2\left(\frac{1}{5}\right)$, or $\frac{9}{10}$ of the job in 2 hours. In 3 hours they can do $3\left(\frac{1}{4}\right)+3\left(\frac{1}{5}\right)$, or $1\frac{7}{20}$ of the job which is more of the job then needs to be done. The answer is somewhere between 2 hr and 3 hr.

Translate. If they work together t hours, then Mandy does $t\left(\frac{1}{4}\right)$ of the job and Omar does $t\left(\frac{1}{5}\right)$ of the job. We want some number t such that

$$t\left(\frac{1}{4}\right)+t\left(\frac{1}{5}\right) = 1, \ \text{or} \ \frac{t}{4}+\frac{t}{5} = 1.$$

Solve. We solve the equation.
$$\frac{t}{4}+\frac{t}{5} = 1, \ \text{LCM} = 20$$

$$20\left(\frac{t}{4}+\frac{t}{5}\right) = 20\cdot 1$$

$$20\cdot\frac{t}{4}+20\cdot\frac{t}{5} = 20$$

$$5t+4t = 20$$

$$9t = 20$$

$$t = \frac{20}{9}, \ \text{or} \ 2\frac{2}{9}$$

Check. The check can be done by repeating the computations. We also have a partial check in that we expected from our familiarization step that the answer would be between 2 hr and 3 hr.

State. Working together, it takes them $2\frac{2}{9}$ hr to complete the job.

3. _Familiarize._ The job takes Vern 45 min working alone and Nina 60 min working alone. Then in 1 minute Vern does $\frac{1}{45}$ of the job and Nina does $\frac{1}{60}$ of the job. Working together, they can do $\frac{1}{45}+\frac{1}{60}$, or $\frac{7}{180}$ of the job in 1 minute. In 20 minutes, Vern does $\frac{20}{45}$ of the job and Nina does $\frac{20}{60}$ of the job. Working together, they can do $\frac{20}{45}+\frac{20}{60}$, or $\frac{7}{9}$ of the job. In 30 minutes, they can do $\frac{30}{45}+\frac{30}{60}$, or $\frac{7}{6}$ of the job which is more of the job than needs to be done. The answer is somewhere between 20 minutes and 30 minutes.

Translate. If they work together t minutes, then Vern does $t\left(\frac{1}{45}\right)$ of the job and Nina does $t\left(\frac{1}{60}\right)$ of the job. We want some number t such that

$$t\left(\frac{1}{45}\right)+t\left(\frac{1}{60}\right) = 1, \ \text{or} \ \frac{t}{45}+\frac{t}{60} = 1.$$

Solve. We solve the equation.
$$\frac{t}{45}+\frac{t}{60} = 1, \ \text{LCM} = 180$$

$$180\left(\frac{t}{45}+\frac{t}{60}\right) = 180\cdot 1$$

$$180\cdot\frac{t}{45}+180\cdot\frac{t}{60} = 180$$

$$4t+3t = 180$$

$$7t = 180$$

$$t = \frac{180}{7}, \ \text{or} \ 25\frac{5}{7}$$

Check. The check can be done by repeating the computations. We also have a partial check in that we expected from our familiarization step that the answer would be between 20 minutes and 30 minutes.

State. It would take them $25\frac{5}{7}$ minutes to complete the job working together.

5. _Familiarize._ The job takes Kenny Dewitt 9 hours working alone and Betty Wohat 7 hours working alone. Then in 1 hour Kenny does $\frac{1}{9}$ of the job and Betty does $\frac{1}{7}$ of the job. Working together they can do $\frac{1}{9}+\frac{1}{7}$, or $\frac{16}{63}$ of the job in 1 hour. In two hours, Kenny does $2\left(\frac{1}{9}\right)$ of the job and Betty does $2\left(\frac{1}{7}\right)$ of the job. Working together they can do $2\left(\frac{1}{9}\right)+2\left(\frac{1}{7}\right)$, or $\frac{32}{63}$ of the job in two hours. In five hours they can do $5\left(\frac{1}{9}\right)+5\left(\frac{1}{7}\right)$, or $\frac{80}{63}$, or $1\frac{17}{63}$ of the job which is more of the job than needs to be done. The answer is somewhere between 2 hr and 5 hr.

Translate. If they work together t hours, Kenny does $t\left(\dfrac{1}{9}\right)$ of the job and Betty does $t\left(\dfrac{1}{7}\right)$ of the job. We want some number t such that

$$t\left(\frac{1}{9}\right) + t\left(\frac{1}{7}\right) = 1, \text{ or } \frac{t}{9} + \frac{t}{7} = 1.$$

Solve. We solve the equation.

$$\frac{t}{9} + \frac{t}{7} = 1, \text{ LCM} = 63$$

$$63\left(\frac{t}{9} + \frac{t}{7}\right) = 63 \cdot 1$$

$$63 \cdot \frac{t}{9} + 63 \cdot \frac{t}{7} = 63$$

$$7t + 9t = 63$$

$$16t = 63$$

$$t = \frac{63}{16}, \text{ or } 3\frac{15}{16}$$

Check. The check can be done by repeating the computations. We also have a partial check in that we expected from our familiarization step that the answer would be between 2 hr and 5 hr.

State. Working together, it takes them $3\dfrac{15}{16}$ hr to complete the job.

7. Familiarize. Let $t = $ the number of minutes it takes Nicole and Glen to weed the garden, working together.

Translate. We use the work principle.

$$t\left(\frac{1}{50}\right) + t\left(\frac{1}{40}\right) = 1, \text{ or } \frac{t}{50} + \frac{t}{40} = 1$$

Solve. We solve the equation.

$$\frac{t}{50} + \frac{t}{40} = 1, \text{ LCM} = 200$$

$$200\left(\frac{t}{50} + \frac{t}{40}\right) = 200 \cdot 1$$

$$200 \cdot \frac{t}{50} + 200 \cdot \frac{t}{40} = 200$$

$$4t + 5t = 200$$

$$9t = 200$$

$$t = \frac{200}{9}, \text{ or } 22\frac{2}{9}$$

Check. In $\dfrac{200}{9}$ min, the portion of the job done is $\dfrac{1}{50} \cdot \dfrac{200}{9} + \dfrac{1}{40} \cdot \dfrac{200}{9} = \dfrac{4}{9} + \dfrac{5}{9} = 1$. The answer checks.

State. It would take $22\dfrac{2}{9}$ min to weed the garden if Nicole and Glen worked together.

9. Familiarize. Let $t = $ the number of minutes it would take the two machines to copy the dissertation, working together.

Translate. We use the work principle.

$$t\left(\frac{1}{12}\right) + t\left(\frac{1}{20}\right) = 1, \text{ or } \frac{t}{12} + \frac{t}{20} = 1$$

Solve. We solve the equation.

$$\frac{t}{12} + \frac{t}{20} = 1, \text{ LCM} = 60$$

$$60\left(\frac{t}{12} + \frac{t}{20}\right) = 60 \cdot 1$$

$$60 \cdot \frac{t}{12} + 60 \cdot \frac{t}{20} = 60$$

$$5t + 3t = 60$$

$$8t = 60$$

$$t = \frac{15}{2}, \text{ or } 7.5$$

Check. In $\dfrac{15}{2}$ min, the portion of the job done is $\dfrac{1}{12} \cdot \dfrac{15}{2} + \dfrac{1}{20} \cdot \dfrac{15}{2} = \dfrac{5}{8} + \dfrac{3}{8} = 1$. The answer checks.

State. It would take the two machines 7.5 min to copy the dissertation, working together.

11. Familiarize. We complete the table shown in the text.

	d	$=$	r	$\cdot$	t	

	Distance	Speed	Time	
Car	150	r	t	$\rightarrow 150 = r(t)$
Truck	350	$r + 40$	t	$\rightarrow 350 = (r+40)t$

Translate. We apply the formula $d = rt$ along the rows of the table to obtain two equations:

$$150 = rt,$$

$$350 = (r + 40)t$$

Then we solve each equation for t and set the results equal:

Solving $150 = rt$ for t: $t = \dfrac{150}{r}$

Solving $350 = (r + 40)t$ for t: $t = \dfrac{350}{r + 40}$

Thus, we have

$$\frac{150}{r} = \frac{350}{r + 40}.$$

Solve. We multiply by the LCM, $r(r + 40)$.

$$r(r + 40) \cdot \frac{150}{r} = r(r + 40) \cdot \frac{350}{r + 40}$$

$$150(r + 40) = 350r$$

$$150r + 6000 = 350r$$

$$6000 = 200r$$

$$30 = r$$

Check. If r is 30 km/h, then $r + 40$ is 70 km/h. The time for the car is 150/30, or 5 hr. The time for the truck is 350/70, or 5 hr. The times are the same. The values check.

State. The speed of Sarah's car is 30 km/h, and the speed of Rick's truck is 70 km/h.

13. *Familiarize.* We complete the table shown in the text.

$$d = r \cdot t$$

	Distance	Speed	Time
Freight	330	$r - 14$	t
Passenger	400	r	t

Translate. From the rows of the table we have two equations:

$$330 = (r - 14)t,$$
$$400 = rt$$

We solve each equation for t and set the results equal:

Solving $330 = (r - 14)t$ for t: $t = \dfrac{330}{r - 14}$

Solving $400 = rt$ for t: $t = \dfrac{400}{r}$

Thus, we have

$$\frac{330}{r - 14} = \frac{400}{r}.$$

Solve. We multiply by the LCM, $r(r - 14)$.

$$r(r - 14) \cdot \frac{330}{r - 14} = r(r - 14) \cdot \frac{400}{r}$$
$$330r = 400(r - 14)$$
$$330r = 400r - 5600$$
$$-70r = -5600$$
$$r = 80$$

Then substitute 80 for r in either equation to find t:

$$t = \frac{400}{r}$$
$$t = \frac{400}{80} \quad \text{Substituting 80 for } r$$
$$t = 5$$

Check. If $r = 80$, then $r - 14 = 66$. In 5 hr the freight train travels $66 \cdot 5$, or 330 mi, and the passenger train travels $80 \cdot 5$, or 400 mi. The values check.

State. The speed of the passenger train is 80 mph. The speed of the freight train is 66 mph.

15. *Familiarize.* We let r represent the speed going. Then $2r$ is the speed returning. We let t represent the time going. Then $t - 3$ represents the time returning. We organize the information in a table.

$$d = r \cdot t$$

	Distance	Speed	Time
Going	120	r	t
Returning	120	$2r$	$t - 3$

Translate. The rows of the table give us two equations:

$$120 = rt,$$
$$120 = 2r(t - 3)$$

We can solve each equation for r and set the results equal:

Solving $120 = rt$ for r: $r = \dfrac{120}{t}$

Solving $120 = 2r(t - 3)$ for r: $r = \dfrac{120}{2(t - 3)}$, or

$$r = \frac{60}{t - 3}$$

Then $\dfrac{120}{t} = \dfrac{60}{t - 3}$.

Solve. We multiply on both sides by the LCM, $t(t - 3)$.

$$t(t - 3) \cdot \frac{120}{t} = t(t - 3) \cdot \frac{60}{t - 3}$$
$$120(t - 3) = 60t$$
$$120t - 360 = 60t$$
$$-360 = -60t$$
$$6 = t$$

Then substitute 6 for t in either equation to find r, the speed going:

$$r = \frac{120}{t}$$
$$r = \frac{120}{6} \quad \text{Substituting 6 for } t$$
$$r = 20$$

Check. If $r = 20$ and $t = 6$, then $2r = 2 \cdot 20$, or 40 mph and $t - 3 = 6 - 3$, or 3 hr. The distance going is $6 \cdot 20$, or 120 mi. The distance returning is $40 \cdot 3$, or 120 mi. The numbers check.

State. The speed going is 20 mph.

17. *Familiarize.* Let $r =$ Kelly's speed, in km/h, and $t =$ the time the bicyclists travel, in hours. Organize the information in a table.

	Distance	Speed	Time
Hank	42	$r - 5$	t
Kelly	57	r	t

Translate. We can replace the t's in the table above using the formula $r = d/t$.

	Distance	Speed	Time
Hank	42	$r - 5$	$\dfrac{42}{r - 5}$
Kelly	57	r	$\dfrac{57}{r}$

Since the times are the same for both bicyclists, we have the equation

$$\frac{42}{r - 5} = \frac{57}{r}.$$

Solve. We first multiply by the LCD, $r(r - 5)$.

$$r(r - 5) \cdot \frac{42}{r - 5} = r(r - 5) \cdot \frac{57}{r}$$
$$42r = 57(r - 5)$$
$$42r = 57r - 285$$
$$-15r = -285$$
$$r = 19$$

If $r = 19$, then $r - 5 = 14$.

Check. If Hank's speed is 14 km/h and Kelly's speed is 19 km/h, then Hank bicycles 5 km/h slower than Kelly. Hank's time is 42/14, or 3 hr. Kelly's time is 57/19, or 3 hr. Since the times are the same, the answer checks.

State. Hank travels at 14 km/h, and Kelly travels at 19 km/h.

19. Familiarize. Let r = Ralph's speed, in km/h. Then Bonnie's speed is $r + 3$. Also set t = the time, in hours, that Ralph and Bonnie walk. We organize the information in a table.

	Distance	Speed	Time
Ralph	7.5	r	t
Bonnie	12	$r + 3$	t

Translate. We can replace the t's in the table shown above using the formula $r = d/t$.

	Distance	Speed	Time
Ralph	7.5	r	$\dfrac{7.5}{r}$
Bonnie	12	$r + 3$	$\dfrac{12}{r+3}$

Since the times are the same for both walkers, we have the equation

$$\frac{7.5}{r} = \frac{12}{r+3}.$$

Solve. We first multiply by the LCD, $r(r + 3)$.

$$r(r+3) \cdot \frac{7.5}{r} = r(r+3) \cdot \frac{12}{r+3}$$

$$7.5(r+3) = 12r$$

$$7.5r + 22.5 = 12r$$

$$22.5 = 4.5r$$

$$5 = r$$

If $r = 5$, then $r + 3 = 8$.

Check. If Ralph's speed is 5 km/h and Bonnie's speed is 8 km/h, then Bonnie walks 3 km/h faster than Ralph. Ralph's time is 7.5/5, or 1.5 hr. Bonnie's time is 12/8, or 1.5 hr. Since the times are the same, the answer checks.

State. Ralph's speed is 5 km/h, and Bonnie's speed is 8 km/h.

21. Familiarize. Let t = the time it takes Caledonia to drive to town and organize the given information in a table.

	Distance	Speed	Time
Caledonia	15	r	t
Manley	20	r	$t + 1$

Translate. We can replace the r's in the table above using the formula $r = d/t$.

	Distance	Speed	Time
Caledonia	15	$\dfrac{15}{t}$	t
Manley	20	$\dfrac{20}{t+1}$	$t + 1$

Since the speeds are the same for both riders, we have the equation

$$\frac{15}{t} = \frac{20}{t+1}.$$

Solve. We multiply by the LCD, $t(t + 1)$.

$$t(t+1) \cdot \frac{15}{t} = t(t+1) \cdot \frac{20}{t+1}$$

$$15(t+1) = 20t$$

$$15t + 15 = 20t$$

$$15 = 5t$$

$$3 = t$$

If $t = 3$, then $t + 1 = 3 + 1$, or 4.

Check. If Caledonia's time is 3 hr and Manley's time is 4 hr, then Manley's time is 1 hr more than Caledonia's. Caledonia's speed is 15/3, or 5 mph. Manley's speed is 20/4, or 5 mph. Since the speeds are the same, the answer checks.

State. It takes Caledonia 3 hr to drive to town.

23. $\dfrac{10 \text{ divorces}}{18 \text{ marriages}} = \dfrac{10}{18}$ divorce/marriage $=$
$\dfrac{5}{9}$ divorce/marriage

25. $\dfrac{4.6 \text{ km}}{2 \text{ hr}} = 2.3$ km/h

27. Familiarize. A 120-lb person should eat at least 44 g of protein each day, and we wish to find the minimum protein required for a 180-lb person. We can set up ratios. We let p = the minimum number of grams of protein a 180-lb person should eat each day.

Translate. If we assume the rates of protein intake are the same, the ratios are the same and we have an equation.

$$\begin{array}{c} \text{Protein} \to \\ \text{Weight} \to \end{array} \frac{44}{120} = \frac{p}{180} \begin{array}{c} \leftarrow \text{Protein} \\ \leftarrow \text{Weight} \end{array}$$

Solve. We solve the proportion.

$$360 \cdot \frac{44}{120} = 360 \cdot \frac{p}{180} \text{ Multiplying by the LCM, 360}$$

$$3 \cdot 44 = 2 \cdot p$$

$$132 = 2p$$

$$66 = p$$

Check. $\dfrac{44}{120} = \dfrac{4 \cdot 11}{4 \cdot 30} = \dfrac{\cancel{4} \cdot 11}{\cancel{4} \cdot 30} = \dfrac{11}{30}$ and

$\dfrac{66}{180} = \dfrac{6 \cdot 11}{6 \cdot 30} = \dfrac{\cancel{6} \cdot 11}{\cancel{6} \cdot 30} = \dfrac{11}{30}$. The ratios are the same.

State. A 180-lb person should eat a minimum of 66 g of protein each day.

29. *Familiarize.* 10 cc of human blood contains 1.2 grams of hemoglobin, and we wish to find how many grams of hemoglobin are contained in 16 cc of the same blood. We can set up ratios. Let H = the amount of hemoglobin in 16 cc of the same blood.

Translate. Assuming the two ratios are the same, we can translate to a proportion.

$$\text{Grams} \to \frac{H}{16} = \frac{1.2}{10} \leftarrow \text{Grams}$$
$$\text{cm}^3 \to \quad\quad\quad \leftarrow \text{cm}^3$$

Solve. We solve the proportion.

We multiply by 16 to get H alone.

$$16 \cdot \frac{H}{16} = 16 \cdot \frac{1.2}{10}$$
$$H = \frac{19.2}{10}$$
$$H = 1.92$$

Check.
$$\frac{1.92}{16} = 0.12 \qquad \frac{1.2}{10} = 0.12$$
The ratios are the same.

State. 16 cc of the same blood would contain 1.92 grams of hemoglobin.

31. *Familiarize.* Let h = the amount of honey, in pounds, that 35,000 trips to flowers would produce.

Translate. We translate to a proportion.

$$\text{Honey} \to \frac{1}{20,000} = \frac{h}{35,000} \leftarrow \text{Honey}$$
$$\text{Trips} \to \quad\quad\quad\quad \leftarrow \text{Trips}$$

Solve. We solve the proportion.

$$35,000 \cdot \frac{1}{20,000} = 35,000 \cdot \frac{h}{35,000}$$
$$1.75 = h$$

Check. $\frac{1}{20,000} = 0.00005$ and $\frac{1.75}{35,000} = 0.00005$.

The ratios are the same.

State. 35,000 trips to gather nectar will produce 1.75 lb of honey.

33. *Familiarize.* The ratio of the weight of copper to the weight of zinc in a U.S. penny is $\frac{1}{39}$, and we wish to find how much copper is needed if 50 kg of zinc is being turned into pennies. We can set up a second ratio to go with the one we already have. Let C = the amount of copper needed, in kg, if 50 kg of zinc is being turned into pennies.

Translate. We translate to a proportion.

$$\frac{1}{39} = \frac{C}{50}$$

Solve. We solve the proportion.

$$50 \cdot \frac{1}{39} = 50 \cdot \frac{C}{50}$$
$$\frac{50}{39} = C, \text{ or}$$
$$1\frac{11}{39} = C$$

Check. $\frac{50/39}{50} = \frac{1}{39}$, so the ratios are the same.

State. $1\frac{11}{39}$ kg of copper is needed if 50 kg of zinc is turned into pennies.

35. (a) $\frac{96}{266} \approx 0.361$

Suzuki's batting average was 0.361.

(b) Let h = the number of hits Suzuki would get in the 162-game season. We translate to a proportion and solve it.

$$\frac{96}{58} = \frac{h}{162}$$
$$162 \cdot \frac{96}{58} = 162 \cdot \frac{h}{162}$$
$$268 \approx h$$

Suzuki would get 268 hits in the 162-game season.

(c) Let h = the number of hits Suzuki would get if he batted 560 times. We translate to a proportion and solve it.

$$\frac{96}{266} = \frac{h}{560}$$
$$560 \cdot \frac{96}{266} = 560 \cdot \frac{h}{560}$$
$$202 \approx h$$

Suzuki would get 202 hits if he batted 560 times.

37. Let h = the head circumference, in inches. We translate to a proportion and solve it.

$$\frac{6\frac{3}{4}}{21\frac{1}{5}} = \frac{7}{h}$$
$$6\frac{3}{4} \cdot h = 21\frac{1}{5} \cdot 7$$
$$\frac{27}{4} \cdot h = \frac{106}{5} \cdot 7$$
$$h = \frac{4}{27} \cdot \frac{106}{5} \cdot 7$$
$$h \approx 22$$

The head circumference is 22 in.

Now let c = the head circumference, in centimeters. We translate to a proportion and solve it.

$$\frac{6\frac{3}{4}}{53.8} = \frac{7}{c}$$
$$\frac{6.75}{53.8} = \frac{7}{c} \qquad \left(6\frac{3}{4} = 6.75\right)$$
$$6.75 \cdot c = 53.8 \cdot 7$$
$$c = \frac{53.8 \cdot 7}{6.75}$$
$$c \approx 55.8$$

The head circumference is 55.8 cm.

39. Let h = the hat size. We translate to a proportion and solve it.

$$\frac{6\frac{3}{4}}{21\frac{1}{5}} = \frac{h}{22\frac{4}{5}}$$

$$6\frac{3}{4} \cdot 22\frac{4}{5} = 21\frac{1}{5} \cdot h$$

$$\frac{27}{4} \cdot \frac{114}{5} = \frac{106}{5} \cdot h$$

$$\frac{5}{106} \cdot \frac{27}{4} \cdot \frac{114}{5} = h$$

$$7.26 \approx h$$

$$7\frac{1}{4} \approx h$$

The hat size is $7\frac{1}{4}$.

Now let c = the head circumference, in centimeters. We translate to a proportion and solve it. We use the hat size found above in the translation.

$$\frac{6\frac{3}{4}}{53.8} = \frac{7\frac{1}{4}}{c}$$

$$\frac{6.75}{53.8} = \frac{7.25}{c}$$

$$6.75 \cdot c = 53.8 \cdot 7.25$$

$$c = \frac{53.8 \cdot 7.25}{6.75}$$

$$c \approx 57.8$$

The head circumference is 57.8 cm. (Answers may vary slightly.)

41. Let h = the hat size. We translate to a proportion and solve it.

$$\frac{6\frac{3}{4}}{53.8} = \frac{h}{59.8}$$

$$\frac{6.75}{53.8} = \frac{h}{59.8}$$

$$59.8 \cdot \frac{6.75}{53.8} = h$$

$$7.5 \approx h, \text{ or}$$

$$7\frac{1}{2} \approx h$$

The hat size is $7\frac{1}{2}$.

Now let c = the head circumference, in inches. We translate to a proportion and solve it. We use the hat size found above in the translation.

$$\frac{6\frac{3}{4}}{21\frac{1}{5}} = \frac{7\frac{1}{2}}{c}$$

$$6\frac{3}{4} \cdot c = 21\frac{1}{5} \cdot 7\frac{1}{2}$$

$$\frac{27}{4} \cdot c = \frac{106}{5} \cdot \frac{15}{2}$$

$$c = \frac{4}{27} \cdot \frac{106}{5} \cdot \frac{15}{2}$$

$$c \approx 23.6, \text{ or}$$

$$c \approx 23\frac{3}{5}$$

The head circumference is $23\frac{3}{5}$ in.

43. Familiarize. The ratio of blue whales tagged to the total blue whale population, P, is $\frac{500}{P}$. Of the 400 blue whales checked later, 20 were tagged. The ratio of blue whales tagged to blue whales checked is $\frac{20}{400}$.

Translate. Assuming the two ratios are the same, we can translate to a proportion.

Whales tagged
originally $\longrightarrow$ $\dfrac{500}{P} = \dfrac{20}{400}$ $\longleftarrow$ Tagged whales caught later
Whale $\longrightarrow$ $\longleftarrow$ Whales caught
population later

Solve. We solve the equation.

$$400P \cdot \frac{500}{P} = 400P \cdot \frac{20}{400} \quad \text{Multiplying by the LCM, } 400P$$

$$400 \cdot 500 = P \cdot 20$$

$$200,000 = 20P$$

$$10,000 = P$$

Check.

$$\frac{500}{10,000} = \frac{1}{20} \quad \text{and} \quad \frac{20}{400} = \frac{1}{20}.$$

The ratios are the same.

State. The blue whale population is about 10,000.

45. Familiarize. The ratio of the weight of an object on Mars to the weight of an object on earth is 0.4 to 1.

a) We wish to find how much a 12-ton rocket would weigh on Mars.

b) We wish to find how much a 120-lb astronaut would weigh on Mars.

We can set up ratios. We let r = the weight of a 12-ton rocket and a = the weight of a 120-lb astronaut on Mars.

Translate. Assuming the ratios are the same, we can translate to proportions.

a) Weight
on Mars $\rightarrow$ $\dfrac{0.4}{1} = \dfrac{r}{12}$ $\leftarrow$ Weight on Mars
Weight $\rightarrow$ $\leftarrow$ Weight
on earth on earth

b)   Weight   Weight

on Mars $\rightarrow$ $\dfrac{0.4}{1} = \dfrac{a}{120}$ $\leftarrow$ on Mars

Weight $\rightarrow$ $\leftarrow$ Weight

on earth   on earth

Solve. We solve each proportion.

a) $\dfrac{0.4}{1} = \dfrac{r}{12}$ b) $\dfrac{0.4}{1} = \dfrac{1}{120}$

$12(0.4) = r$ $120(0.4) = a$

$4.8 = r$ $48 = a$

Check. $\dfrac{0.4}{1} = 0.4$, $\dfrac{4.8}{12} = 0.4$, and $\dfrac{48}{120} = 0.4$.

The ratios are the same.

State. a) A 12-ton rocket would weigh 4.8 tons on Mars.

b) A 120-lb astronaut would weigh 48 lb on Mars.

47. Familiarize. A sample of 144 firecrackers contained 9 duds, and we wish to find how many duds could be expected in a sample of 3200 firecrackers. We can set up ratios, letting $d =$ the number of duds expected in a sample of 3200 firecrackers.

Translate. Assuming the rates of occurrence of duds are the same, we can translate to a proportion.

Duds $\rightarrow$ $\dfrac{9}{144} = \dfrac{d}{3200}$ $\leftarrow$ Duds

Sample size $\rightarrow$ $\leftarrow$ Sample size

Solve. We solve the equation. We multiply by 3200 to get d alone.

$$3200 \cdot \dfrac{9}{144} = 3200 \cdot \dfrac{d}{3200}$$

$$\dfrac{28{,}800}{144} = d$$

$$200 = d$$

Check.

$\dfrac{9}{144} = 0.0625$ and $\dfrac{200}{3200} = 0.0625$

The ratios are the same.

State. You would expect 200 duds in a sample of 3200 firecrackers.

49. We write a proportion and then solve it.

$$\dfrac{b}{6} = \dfrac{7}{4}$$

$$b = \dfrac{7}{4} \cdot 6 \qquad\qquad \text{Multiplying by 6}$$

$$b = \dfrac{42}{4}$$

$$b = \dfrac{21}{2}, \text{ or } 10.5$$

$\left(\text{Note that the proportions } \dfrac{6}{b} = \dfrac{4}{7}, \dfrac{b}{7} = \dfrac{6}{4}, \text{ or } \dfrac{7}{b} = \dfrac{4}{6} \text{ could also be used.}\right)$

51. We write a proportion and then solve it.

$$\dfrac{4}{f} = \dfrac{6}{4}$$

$$4f \cdot \dfrac{4}{f} = 4f \cdot \dfrac{6}{4}$$

$$16 = 6f$$

$$\dfrac{8}{3} = f \qquad \text{Simplifying}$$

$\left(\text{One of the following proportions could also be used: } \dfrac{f}{4} = \dfrac{4}{6}, \dfrac{4}{f} = \dfrac{9}{6}, \dfrac{f}{4} = \dfrac{6}{9}, \dfrac{4}{9} = \dfrac{f}{6}, \dfrac{9}{4} = \dfrac{6}{f}\right)$

53. We write a proportion and then solve it.

$$\dfrac{h}{7} = \dfrac{10}{6}$$

$$h = \dfrac{10}{6} \cdot 7 \quad \text{Multiplying by 7}$$

$$h = \dfrac{70}{6}$$

$$h = \dfrac{35}{3} \qquad \text{Simplifying}$$

$\left(\text{Note that the proportions } \dfrac{7}{h} = \dfrac{6}{10}, \dfrac{h}{10} = \dfrac{7}{6}, \text{ or } \dfrac{10}{h} = \dfrac{6}{7} \text{ could also be used.}\right)$

55. We write a proportion and then solve it.

$$\dfrac{4}{10} = \dfrac{6}{l}$$

$$10l \cdot \dfrac{4}{10} = 10l \cdot \dfrac{6}{l}$$

$$4l = 60$$

$$l = 15 \text{ ft}$$

$\left(\text{One of the following proportions could also be used: } \dfrac{4}{6} = \dfrac{10}{l}, \dfrac{10}{4} = \dfrac{l}{6}, \text{ or } \dfrac{6}{4} = \dfrac{l}{10}\right)$

57. Discussion and Writing Exercise

59. $x^5 \cdot x^6 = x^{5+6} = x^{11}$

61. $x^{-5} \cdot x^{-6} = x^{-5+(-6)} = x^{-11} = \dfrac{1}{x^{11}}$

63. Graph: $y = 2x - 6$.

We select some x-values and compute y-values.

If $x = 1$, then $y = 2 \cdot 1 - 6 = -4$.

If $x = 3$, then $y = 2 \cdot 3 - 6 = 0$.

If $x = 5$, then $y = 2 \cdot 5 - 6 = 4$.

x	y	(x, y)
1	-4	$(1, -4)$
3	0	$(3, 0)$
5	4	$(5, 4)$

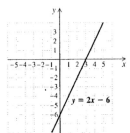

65. Graph: $3x + 2y = 12$.

We can replace either variable with a number and then calculate the other coordinate. We will find the intercepts and one other point.

If $y = 0$, we have:

$$3x + 2 \cdot 0 = 12$$
$$3x = 12$$
$$x = 4$$

The x-intercept is $(4, 0)$.

If $x = 0$, we have:

$$3 \cdot 0 + 2y = 12$$
$$2y = 12$$
$$y = 6$$

The y-intercept is $(0, 6)$.

If $y = -3$, we have:

$$3x + 2(-3) = 12$$
$$3x - 6 = 12$$
$$3x = 18$$
$$x = 6$$

The point $(6, -3)$ is on the graph.

We plot these points and draw a line through them.

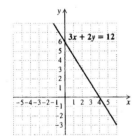

67. Graph: $y = -\dfrac{3}{4}x + 2$

We select some x-values and compute y-values. We use multiples of 4 to avoid fractions.

If $x = -4$, then $y = -\dfrac{3}{4}(-4) + 2 = 5$.

If $x = 0$, then $y = -\dfrac{3}{4} \cdot 0 + 2 = 2$.

If $x = 4$, then $y = -\dfrac{3}{4} \cdot 4 + 2 = -1$.

x	y	(x, y)
-4	5	$(-4, 5)$
0	2	$(0, 2)$
4	-1	$(4, -1)$

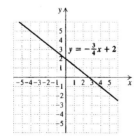

69. Familiarize. Let $t =$ the time it would take for Ann to complete the report working alone. Then $t + 6 =$ the time it would take Betty to complete the report working alone.

In 1 hour they would complete $\dfrac{1}{t} + \dfrac{1}{t+6}$ of the report and in 4 hours they would complete $4\left(\dfrac{1}{t} + \dfrac{1}{t+6}\right)$, or $\dfrac{4}{t} + \dfrac{4}{t+6}$ of the report.

Translate. In 4 hours one entire job is done, so we have

$$\frac{4}{t} + \frac{4}{t+6} = 1.$$

Solve. We solve the equation.

$$\frac{4}{t} + \frac{4}{t+6} = 1, \quad \text{LCM} = t(t+6)$$
$$t(t+6)\left(\frac{4}{t} + \frac{4}{t+6}\right) = t(t+6) \cdot 1$$
$$t(t+6) \cdot \frac{4}{t} + t(t+6) \cdot \frac{4}{t+6} = t^2 + 6t$$
$$4(t+6) + 4t = t^2 + 6t$$
$$4t + 24 + 4t = t^2 + 6t$$
$$0 = t^2 - 2t - 24$$
$$0 = (t-6)(t+4)$$

$$t - 6 = 0 \ or \ t + 4 = 0$$
$$t = 6 \ or \quad t = -4$$

Check. The time cannot be negative, so we check only 6. If it takes Ann 6 hr to complete the report, then it would take Betty $6 + 6$, or 12 hr, to complete the report. In 4 hr Ann does $4 \cdot \dfrac{1}{6}$, or $\dfrac{2}{3}$, of the report, Betty does $4 \cdot \dfrac{1}{12}$, or $\dfrac{1}{3}$, of the report, and together they do $\dfrac{2}{3} + \dfrac{1}{3}$, or 1 entire job. The answer checks.

State. It would take Ann 6 hr and Betty 12 hr to complete the report working alone.

71. Familiarize. Let $t =$ the number of minutes after 5:00 at which the hands of the clock will first be together. While the minute hand moves through t minutes, the hour hand moves through $t/12$ minutes. At 5:00 the hour hand is on the 25-minute mark. We wish to find when a move of the minute hand through t minutes is equal to $25 + t/12$ minutes.

Translate. We use the last sentence of the familiarization step to write an equation.

$$t = 25 + \frac{t}{12}$$

Solve. We solve the equation.

$$t = 25 + \frac{t}{12}$$
$$12 \cdot t = 12\left(25 + \frac{t}{12}\right)$$
$$12t = 300 + t \qquad \text{Multiplying by 12}$$
$$11t = 300$$
$$t = \frac{300}{11} \text{ or } 27\frac{3}{11}$$

Check. At $27\dfrac{3}{11}$ minutes after 5:00, the minute hand is at the $27\dfrac{3}{11}$-minutes mark and the hour hand is at the

$25 + \dfrac{27\frac{3}{11}}{12}$-minute mark. Simplifying $25 + \dfrac{27\frac{3}{11}}{12}$, we get

$$25 + \dfrac{\frac{300}{11}}{12} = 25 + \dfrac{300}{11} \cdot \dfrac{1}{12} = 25 + \dfrac{25}{11} = 25 + 2\dfrac{3}{11} = 27\dfrac{3}{11}.$$

Thus, the hands are together.

State. The hands are first together $27\dfrac{3}{11}$ minutes after 5:00.

73.
$$\dfrac{t}{a} + \dfrac{t}{b} = 1, \text{ LCM} = ab$$
$$ab\left(\dfrac{t}{a} + \dfrac{t}{b}\right) = ab \cdot 1$$
$$ab \cdot \dfrac{t}{a} + ab \cdot \dfrac{t}{b} = ab$$
$$bt + at = ab$$
$$t(b + a) = ab$$
$$t = \dfrac{ab}{b + a}$$

Exercise Set 11.9

1. $y = kx$

$40 = k \cdot 8$ Substituting

$5 = k$ Solving for k

The variation constant is 5.

The equation of variation is $y = 5x$.

3. $y = kx$

$4 = k \cdot 30$ Substituting

$\dfrac{4}{30} = k$, or   Solving for k

$\dfrac{2}{15} = k$ Simplifying

The variation constant is $\dfrac{2}{15}$.

The equation of variation is $y = \dfrac{2}{15}x$.

5. $y = kx$

$0.9 = k \cdot 0.4$ Substituting

$\dfrac{0.9}{0.4} = k$, or

$\dfrac{9}{4} = k$

The variation constant is $\dfrac{9}{4}$.

The equation of variation is $y = \dfrac{9}{4}x$.

7. Let p = the number of people using the cans.

$N = kp$ N varies directly as p.

$60,000 = k \cdot 250$ Substituting

$\dfrac{60,000}{250} = k$ Solving for k

$240 = k$ Variation constant

$N = 240p$ Equation of variation

$N = 240(1,008,000)$ Substituting

$N = 241,920,000$

In Dallas 241,920,000 cans are used each year.

9. $d = kw$ d varies directly as w.

$40 = k \cdot 3$ Substituting

$\dfrac{40}{3} = k$ Variation constant

$d = \dfrac{40}{3}w$ Equation of variation

$d = \dfrac{40}{3} \cdot 5$ Substituting

$d = \dfrac{200}{3}$, or $66\dfrac{2}{3}$

The spring is stretched $66\dfrac{2}{3}$ cm by a 5-kg barbell.

11. Let F = the number of grams of fat and w = the weight.

$F = kw$ F varies directly as w.

$60 = k \cdot 120$ Substituting

$\dfrac{60}{120} = k$, or   Solving for k

$\dfrac{1}{2} = k$ Variation constant

$F = \dfrac{1}{2}w$ Equation of variation

$F = \dfrac{1}{2} \cdot 180$ Substituting

$F = 90$

The maximum daily fat intake for a person weighing 180 lb is 90 g.

13. Let m = the mass of the body.

$W = km$ W varies directly as m.

$64 = k \cdot 96$ Substituting

$\dfrac{64}{96} = k$ Solving for k

$\dfrac{2}{3} = k$ Variation constant

$W = \dfrac{2}{3}m$ Equation of variation

$W = \dfrac{2}{3} \cdot 60$ Substituting

$W = 40$

There are 40 kg of water in a 60-kg person.

15. $y = \dfrac{k}{x}$

$14 = \dfrac{k}{7}$ Substituting

$7 \cdot 14 = k$ Solving for k

$98 = k$

The variation constant is 98.

The equation of variation is $y = \dfrac{98}{x}$.

17. $\quad y = \dfrac{k}{x}$

$\qquad 3 = \dfrac{k}{12} \quad$ Substituting

$\qquad 12 \cdot 3 = k \quad$ Solving for k

$\qquad 36 = k$

The variation constant is 36.

The equation of variation is $y = \dfrac{36}{x}$.

19. $\quad y = \dfrac{k}{x}$

$\qquad 0.1 = \dfrac{k}{0.5} \quad$ Substituting

$\qquad 0.5(0.1) = k \quad$ Solving for k

$\qquad 0.05 = k$

The variation constant is 0.05.

The equation of variation is $y = \dfrac{0.05}{x}$.

21. $\quad T = \dfrac{k}{P} \quad T$ varies inversely as P.

$\qquad 5 = \dfrac{k}{7} \quad$ Substituting

$\qquad 35 = k \quad$ Variation constant

$\qquad T = \dfrac{35}{P} \quad$ Equation of variation

$\qquad T = \dfrac{35}{10} \quad$ Substituting

$\qquad T = 3.5$

It will take 10 bricklayers 3.5 hr to complete the job.

23. $\quad I = \dfrac{k}{R} \quad I$ varies inversely as R.

$\qquad \dfrac{1}{2} = \dfrac{k}{240} \quad$ Substituting

$\qquad 240 \cdot \dfrac{1}{2} = k$

$\qquad 120 = k \quad$ Variation constant

$\qquad I = \dfrac{120}{R} \quad$ Equation of variation

$\qquad I = \dfrac{120}{540} \quad$ Substituting

$\qquad I = \dfrac{2}{9}$

When the resistance is 540 ohms, the current is $\dfrac{2}{9}$ ampere.

25. $\quad P = \dfrac{k}{W} \quad P$ varies inversely as W.

$\qquad 330 = \dfrac{k}{3.2} \quad$ Substituting

$\qquad 1056 = k \quad$ Variation constant

$\qquad P = \dfrac{1056}{W} \quad$ Equation of variation

$\qquad 550 = \dfrac{1056}{W} \quad$ Substituting

$\qquad 550W = 1056 \quad$ Multiplying by W

$\qquad W = \dfrac{1056}{550} \quad$ Dividing by 550

$\qquad W = 1.92 \quad$ Simplifying

A tone with a pitch of 550 vibrations per second has a wavelength of 1.92 ft.

27. $\quad V = \dfrac{k}{P} \quad V$ varies inversely as P.

$\qquad 200 = \dfrac{k}{32} \quad$ Substituting

$\qquad 6400 = k \quad$ Variation constant

$\qquad V = \dfrac{6400}{P} \quad$ Equation of variation

$\qquad V = \dfrac{6400}{40} \quad$ Substituting

$\qquad V = 160$

The volume is 160 cm^3 under a pressure of 40 kg/cm^2.

29. $\quad y = kx^2$

$\qquad 0.15 = k(0.1)^2 \quad$ Substituting

$\qquad 0.15 = 0.01k$

$\qquad \dfrac{0.15}{0.01} = k$

$\qquad 15 = k$

The equation of variation is $y = 15x^2$.

31. $\quad y = \dfrac{k}{x^2}$

$\qquad 0.15 = \dfrac{k}{(0.1)^2} \quad$ Substituting

$\qquad 0.15 = \dfrac{k}{0.01}$

$\qquad 0.15(0.01) = k$

$\qquad 0.0015 = k$

The equation of variation is $y = \dfrac{0.0015}{x^2}$.

33. $\quad y = kxz$

$\qquad 56 = k \cdot 7 \cdot 8 \quad$ Substituting

$\qquad 56 = 56k$

$\qquad 1 = k$

The equation of variation is $y = xz$.

35. $y = kxz^2$

$105 = k \cdot 14 \cdot 5^2$ Substituting

$105 = 350k$

$\dfrac{105}{350} = k$

$\dfrac{3}{10} = k$

The equation of variation is $y = \dfrac{3}{10}xz^2$.

37. $y = k\dfrac{xz}{wp}$

$\dfrac{3}{28} = k\dfrac{3 \cdot 10}{7 \cdot 8}$ Substituting

$\dfrac{3}{28} = k \cdot \dfrac{30}{56}$

$\dfrac{3}{28} \cdot \dfrac{56}{30} = k$

$\dfrac{1}{5} = k$

The equation of variation is $y = \dfrac{xz}{5wp}$.

39. $d = kr^2$

$200 = k \cdot 60^2$ Substituting

$200 = 3600k$

$\dfrac{200}{3600} = k$

$\dfrac{1}{18} = k$

The equation of variation is $d = \dfrac{1}{18}r^2$.

Substitute 72 for d and find r.

$72 = \dfrac{1}{18}r^2$

$1296 = r^2$

$36 = r$

A car can travel 36 mph and still stop in 72 ft.

41. $I = \dfrac{k}{d^2}$

$90 = \dfrac{k}{5^2}$ Substituting

$90 = \dfrac{k}{25}$

$2250 = k$

The equation of variation is $I = \dfrac{2250}{d^2}$.

Substitute 40 for I and find d.

$40 = \dfrac{2250}{d^2}$

$40d^2 = 2250$

$d^2 = 56.25$

$d = 7.5$

The distance from 5 m to 7.5 m is $7.5 - 5$, or 2.5 m, so it is 2.5 m further to a point where the intensity is 40 W/m^2.

43. $E = \dfrac{kR}{I}$

We first find k.

$3.18 = \dfrac{k \cdot 71}{201}$ Substituting

$3.18\left(\dfrac{201}{71}\right) = k$ Multiplying by $\dfrac{201}{71}$

$9 \approx k$

The equation of variation is $E = \dfrac{9R}{I}$.

Substitute 3.18 for E and 300 for I and solve R.

$3.18 = \dfrac{9R}{300}$

$3.18\left(\dfrac{300}{9}\right) = R$ Multiplying by $\dfrac{300}{9}$

$106 = R$

Shawn Estes would have given up 106 earned runs if he had pitched 300 innings.

45. $Q = kd^2$

We first find k.

$225 = k \cdot 5^2$

$225 = 25k$

$9 = k$

The equation of variation is $Q = 9d^2$.

Substitute 9 for d and compute Q.

$Q = 9 \cdot 9^2$

$Q = 9 \cdot 81$

$Q = 729$

729 gallons of water are emptied by a pipe that is 9 in. in diameter.

47. Discussion and Writing Exercise

49. $x^2 - x - 56$

We look for a pair of numbers whose product is -56 and whose sum is -1. The numbers are -8 and 7.

$x^2 - x - 56 = (x - 8)(x + 7)$

51. $x^5 - 2x^4 - 35x^3 = x^3(x^2 - 2x - 35) = x^3(x - 7)(x + 5)$

53. $16 - t^4 = 4^2 - (t^2)^2$ Difference of squares

$= (4 + t^2)(4 - t^2)$

$= (4 + t^2)(2^2 - t^2)$ Difference of squares

$= (4 + t^2)(2 + t)(2 - t)$

55. $x^2 - 9x + 14$

We look for a pair of numbers whose product is 14 and whose sum is -9. The numbers are -2 and -7.

$x^2 - 9x + 14 = (x - 2)(x - 7)$

57. $16x^2 - 40xy + 25y^2$

$= (4x)^2 - 2 \cdot 4x \cdot 5y + (5y)^2$ Trinomial square

$= (4x - 5y)^2$

59. $3x^3 - 3y^3 = 3(x^3 - y^3) = 3(x - y)(x^2 + xy + y^2)$

61. We are told $A = kd^2$, and we know $A = \pi r^2$ so we have:

$$kd^2 = \pi r^2$$

$$kd^2 = \pi \left(\frac{d}{2}\right)^2 \qquad r = \frac{d}{2}$$

$$kd^2 = \frac{\pi d^2}{4}$$

$$k = \frac{\pi}{4} \qquad \text{Variation constant}$$

63. $Q = \dfrac{kp^2}{q^3}$

Q varies directly as the square of p and inversely as the cube of q.

65. Let V represent the volume and p represent the price of a jar of peanut butter. Recall that the formula for the volume of a right circular cylinder with radius r and height h is $V = \pi r^2 h$. The diameter of the smaller jar is 3 in., so its radius is 3 in./2, or $\dfrac{3}{2}$ in. The diameter of the larger jar is 6 in., so its radius is 6 in./2, or 3 in.

$$V = kp$$

$$\pi r^2 h = kp$$

$$\pi \left(\frac{3}{2}\right)^2 (4) = k(1.2) \qquad \text{Substituting}$$

$$7.5\pi = k \qquad \text{Variation constant}$$

$$V = 7.5\pi p \qquad \text{Equation of variation}$$

$$\pi r^2 h = 7.5\pi p$$

$$\pi(3)^2(6) = 7.5\pi p \qquad \text{Substituting}$$

$$54\pi = 7.5\pi p$$

$$7.2 = p$$

The larger jar should cost \$7.20.

Chapter 12

Graphs, Functions, and Applications

1. Yes; each member of the domain is matched to only one member of the range.

3. Yes; each member of the domain is matched to only one member of the range.

5. Yes; each member of the domain is matched to only one member of the range.

7. No; a member of the domain is matched to more than one member of the range. In fact, each member of the domain is matched to 3 members of the range.

9. This correspondence is a function, because each person in the family has only one height.

11. The correspondence is not a function, since it is reasonable to assume that at least one avenue is intersected by more than one road.

13. This correspondence is a function, because each number in the domain, when squared and then increased by 4, corresponds to only one number in the range.

15. $f(x) = x + 5$
   a) $f(4) = 4 + 5 = 9$
   b) $f(7) = 7 + 5 = 12$
   c) $f(-3) = -3 + 5 = 2$
   d) $f(0) = 0 + 5 = 5$
   e) $f(2.4) = 2.4 + 5 = 7.4$
   f) $f\left(\dfrac{2}{3}\right) = \dfrac{2}{3} + 5 = 5\dfrac{2}{3}$

17. $h(p) = 3p$
   a) $h(-7) = 3(-7) = -21$
   b) $h(5) = 3 \cdot 5 = 15$
   c) $h(14) = 3 \cdot 14 = 42$
   d) $h(0) = 3 \cdot 0 = 0$
   e) $h\left(\dfrac{2}{3}\right) = 3 \cdot \dfrac{2}{3} = \dfrac{6}{3} = 2$
   f) $h(a + 1) = 3(a + 1) = 3a + 3$

19. $g(s) = 3s + 4$
   a) $g(1) = 3 \cdot 1 + 4 = 3 + 4 = 7$
   b) $g(-7) = 3(-7) + 4 = -21 + 4 = -17$
   c) $g(6.7) = 3(6.7) + 4 = 20.1 + 4 = 24.1$
   d) $g(0) = 3 \cdot 0 + 4 = 0 + 4 = 4$
   e) $g(-10) = 3(-10) + 4 = -30 + 4 = -26$
   f) $g\left(\dfrac{2}{3}\right) = 3 \cdot \dfrac{2}{3} + 4 = 2 + 4 = 6$

21. $f(x) = 2x^2 - 3x$
   a) $f(0) = 2 \cdot 0^2 - 3 \cdot 0 = 0 - 0 = 0$
   b) $f(-1) = 2(-1)^2 - 3(-1) = 2 + 3 = 5$
   c) $f(2) = 2 \cdot 2^2 - 3 \cdot 2 = 8 - 6 = 2$
   d) $f(10) = 2 \cdot 10^2 - 3 \cdot 10 = 200 - 30 = 170$
   e) $f(-5) = 2(-5)^2 - 3(-5) = 50 + 15 = 65$
   f) $f(4a) = 2(4a)^2 - 3(4a) = 32a^2 - 12a$

23. $f(x) = |x| + 1$
   a) $f(0) = |0| + 1 = 0 + 1 = 1$
   b) $f(-2) = |-2| + 1 = 2 + 1 = 3$
   c) $f(2) = |2| + 1 = 2 + 1 = 3$
   d) $f(-3) = |-3| + 1 = 3 + 1 = 4$
   e) $f(-10) = |-10| + 1 = 10 + 1 = 11$
   f) $f(a - 1) = |a - 1| + 1$

25. $f(x) = x^3$
   a) $f(0) = 0^3 = 0$
   b) $f(-1) = (-1)^3 = -1$
   c) $f(2) = 2^3 = 8$
   d) $f(10) = 10^3 = 1000$
   e) $f(-5) = (-5)^3 = -125$
   f) $f(-10) = (-10)^3 = -1000$

27. $F(x) = 2.75x + 71.48$
   a) $F(32) = 2.75(32) + 71.48$
   $= 88 + 71.48$
   $= 159.48$ cm
   b) $F(35) = 2.75(35) + 71.48$
   $= 96.25 + 71.48$
   $= 167.73$ cm

29. $P(d) = 1 + \dfrac{d}{33}$
   $P(20) = 1 + \dfrac{20}{33} = 1\dfrac{20}{33}$ atm
   $P(30) = 1 + \dfrac{30}{33} = 1\dfrac{10}{11}$ atm
   $P(100) = 1 + \dfrac{100}{33} = 1 + 3\dfrac{1}{33} = 4\dfrac{1}{33}$ atm

31. $W(d) = 0.112d$
   $W(16) = 0.112(16) = 1.792$ cm
   $W(25) = 0.112(25) = 2.8$ cm
   $W(100) = 0.112(100) = 11.2$ cm

33. Graph $f(x) = 3x - 1$

Make a list of function values in a table.

$f(-1) = 3(-1) - 1 = -3 - 1 = -4$

$f(0) = 3 \cdot 0 - 1 = 0 - 1 = -1$

$f(1) = 3 \cdot 1 - 1 = 3 - 1 = 2$

$f(2) = 3 \cdot 2 - 1 = 6 - 1 = 5$

x	$f(x)$
-1	-4
0	-1
1	2
2	5

Plot these points and connect them.

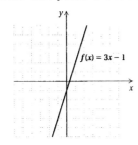

35. Graph $g(x) = -2x + 3$

Make a list of function values in a table.

$g(-1) = -2(-1) + 3 = 2 + 3 = 5$

$g(0) = -2 \cdot 0 + 3 = 0 + 3 = 3$

$g(3) = -2 \cdot 3 + 3 = -6 + 3 = -3$

x	$g(x)$
-1	5
0	3
3	-3

Plot these points and connect them.

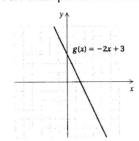

37. Graph $f(x) = \dfrac{1}{2}x + 1$.

Make a list of function values in a table.

$f(-2) = \dfrac{1}{2}(-2) + 1 = -1 + 1 = 0$

$f(0) = \dfrac{1}{2} \cdot 0 + 1 = 0 + 1 = 1$

$f(4) = \dfrac{1}{2} \cdot 4 + 1 = 2 + 1 = 3$

x	$f(x)$
-2	0
0	1
4	3

Plot these points and connect them.

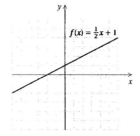

39. Graph $f(x) = 2 - |x|$.

Make a list of function values in a table.

$f(-3) = 2 - |-3| = 2 - 3 = -1$

$f(-2) = 2 - |-2| = 2 - 2 = 0$

$f(-1) = 2 - |-1| = 2 - 1 = 1$

$f(0) = 2 - |0| = 2 - 0 = 2$

$f(1) = 2 - |1| = 2 - 1 = 1$

$f(2) = 2 - |2| = 2 - 2 = 0$

$f(3) = 2 - |3| = 2 - 3 = -1$

x	$f(x)$
-3	-1
-2	0
-1	1
0	2
1	1
2	0
3	-1

Plot these points and connect them.

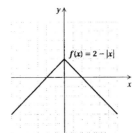

41. Graph $f(x) = x^2$.

Make a list of function values in a table.

$f(-3) = (-3)^2 = 9$

$f(-2) = (-2)^2 = 4$

$f(-1) = (-1)^2 = 1$

$f(0) = 0^2 = 0$

$f(1) = 1^2 = 1$

$f(2) = 2^2 = 4$

$f(3) = 3^2 = 9$

x	$f(x)$
-3	9
-2	4
-1	1
0	0
1	1
2	4
3	9

Plot these points and connect them.

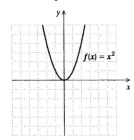

43. Graph $f(x) = x^2 - x - 2$.

Make a list of function values in a table.

$f(-3) = (-3)^2 - (-3) - 2 = 9 + 3 - 2 = 10$

$f(-2) = (-2)^2 - (-2) - 2 = 4 + 2 - 2 = 4$

$f(-1) = (-1)^2 - (-1) - 2 = 1 + 1 - 2 = 0$

$f(0) = 0^2 - 0 - 2 = -2$

$f(1) = 1^2 - 1 - 2 = 1 - 1 - 2 = -2$

$f(2) = 2^2 - 2 - 2 = 4 - 2 - 2 = 0$

$f(3) = 3^2 - 3 - 2 = 9 - 3 - 2 = 4$

x	$f(x)$
-3	10
-2	4
-1	0
0	-2
1	-2
2	0
3	4

Plot these points and connect them.

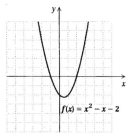

$f(x) = x^2 - x - 2$

45. We can use the vertical line test:

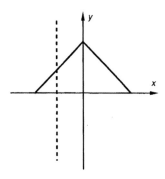

Visualize moving this vertical line across the graph. No vertical line will intersect the graph more than once. Thus, the graph is a graph of a function.

47. We can use the vertical line test:

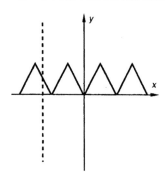

Visualize moving this vertical line across the graph. No vertical line will intersect the graph more than once. Thus, the graph is a graph of a function.

49. We can use the vertical line test.

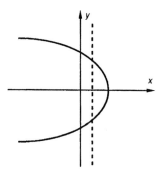

It is possible for a vertical line to intersect the graph more than once. Thus this is not the graph of a function.

51. We can use the vertical line test.

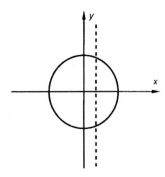

It is possible for a vertical line to intersect the graph more than once. Thus this is not a graph of a function.

53. Locate the point that is directly above 225. Then estimate its second coordinate by moving horizontally from the point to the vertical axis. The rate is about 75 per 10,000 men.

55. Locate 2 on the horizontal axis and move directly up to the graph. Then move across to the vertical axis and read the revenue. The movie revenue for week 2 was about $25 million.

57. Discussion and Writing Exercise

59. Discussion and Writing Exercise

61. *Familiarize*. If x represents the first integer, then $x + 2$ represents the second integer, and $x + 4$ represents the third.

Translate. We write an equation.

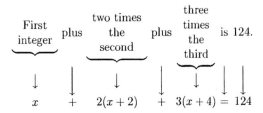

$$x \quad + \quad 2(x+2) \quad + \quad 3(x+4) = 124$$

Carry out. Solve the equation.

$$x + 2(x+2) + 3(x+4) = 124$$
$$x + 2x + 4 + 3x + 12 = 124$$
$$6x + 16 = 124$$
$$6x = 108$$
$$x = 18$$

If $x = 18$, then $x + 2 = 18 + 2$, or 20, and $x + 4 = 18 + 4$, or 22.

Check. 18, 20, and 22 are consecutive even integers. Also, $18 + 2 \cdot 20 + 3 \cdot 22 = 18 + 40 + 66 = 124$. The numbers check.

State. The integers are 18, 20, and 22.

63.
$$S = 2lh + 2lw + 2wh$$
$$S - 2wh = 2lh + 2lw$$
$$S - 2wh = l(2h + 2w)$$
$$\frac{S - 2wh}{2h + 2w} = l$$

65. 9.3×10^{-9}

Negative exponent, so the answer is a small number.

0.000000009.3

9 places

$9.3 \times 10^{-9} = 0.0000000093$

67. 1.075,000,000.

9 places

Large number, so the exponent is positive.

$1,075,000,000 = 1.075 \times 10^{9}$

69. $w^3 + \dfrac{1}{27} = w^3 + \left(\dfrac{1}{3}\right)^3 = \left(w + \dfrac{1}{3}\right)\left(w^2 - \dfrac{1}{3}w + \dfrac{1}{9}\right)$.

71.
$$xy^3 + 64x = x(y^3 + 64)$$
$$= x(y^3 + 4^3)$$
$$= x(y + 4)(y^2 - 4y + 16)$$

73. To find $f(g(-4))$, we first find $g(-4)$:

$g(-4) = 2(-4) + 5 = -8 + 5 = -3$.

Then $f(g(-4)) = f(-3) = 3(-3)^2 - 1 = 3 \cdot 9 - 1 = 27 - 1 = 26$.

To find $g(f(-4))$, we first find $f(-4)$:

$f(-4) = 3(-4)^2 - 1 = 3 \cdot 16 - 1 = 48 - 1 = 47$.

Then $g(f(-4)) = g(47) = 2 \cdot 47 + 5 = 94 + 5 = 99$.

75. We know that $(-1, -7)$ and $(3, 8)$ are both solutions of $g(x) = mx + b$. Substituting, we have

$$-7 = m(-1) + b, \quad \text{or} \quad -7 = -m + b,$$
$$\text{and} \quad 8 = m(3) + b, \quad \text{or} \quad 8 = 3m + b.$$

Solve the first equation for b and substitute that expression into the second equation.

$-7 = -m + b$	First equation
$m - 7 = b$	Solving for b
$8 = 3m + b$	Second equation
$8 = 3m + (m - 7)$	Substituting
$8 = 3m + m - 7$	
$8 = 4m - 7$	
$15 = 4m$	
$\dfrac{15}{4} = m$	

We know that $m - 7 = b$, so $\dfrac{15}{4} - 7 = b$, or $-\dfrac{13}{4} = b$.

We have $m = \dfrac{15}{4}$ and $b = -\dfrac{13}{4}$, so $g(x) = \dfrac{15}{4}x - \dfrac{13}{4}$.

Exercise Set 12.2

1. a) Locate 1 on the horizontal axis and then find the point on the graph for which 1 is the first coordinate. From that point, look to the vertical axis to find the corresponding y-coordinate, 3. Thus, $f(1) = 3$.

b) The domain is the set of all x-values in the graph. It is $\{-4, -3, -2, -1, 0, 1, 2\}$.

c) To determine which member(s) of the domain are paired with 2, locate 2 on the vertical axis. From there look left and right to the graph to find any points for which 2 is the second coordinate. Two such points exist, $(-2, 2)$ and $(0, 2)$. Thus, the x-values for which $f(x) = 2$ are -2 and 0.

d) The range is the set of all y-values in the graph. It is $\{1, 2, 3, 4\}$.

3. a) Locate 1 on the horizontal axis and then find the point on the graph for which 1 is the first coordinate. From that point, look to the vertical axis to find the corresponding y-coordinate, 2. Thus, $f(1) = 2$.

b) The domain is the set of all x-values in the graph. It is $\{-6, -4, -2, 0, 1, 3, 4\}$.

c) To determine which member(s) of the domain are paired with 2, locate 2 on the vertical axis. From there look left and right to the graph to find any points for which 2 is the second coordinate. Two such points exist, $(1, 2)$ and $(3, 2)$. Thus, the x-values for which $f(x) = 2$ are 1 and 3.

d) The range is the set of all y-values in the graph. It is $\{-5, -2, 0, 2, 5\}$.

5. a) Locate 1 on the horizontal axis and then find the point on the graph for which 1 is the first coordinate. From that point, look to the vertical axis to find the corresponding y-coordinate, about 2.5 or $\dfrac{5}{2}$. Thus, $f(1) \approx \dfrac{5}{2}$.

b) The set of all x-values in the graph extends from -3 to 5, so the domain is $\{x| -3 \le x \le 5\}$.

c) To determine which member(s) of the domain are paired with 2, locate 2 on the vertical axis. From there look left and right to the graph to find any points for which 2 is the second coordinate. One such point exists. Its first coordinate appears to be about $2\dfrac{1}{4}$ or $\dfrac{9}{4}$. Thus, the x-value for which $f(x) = 2$ is about $\dfrac{9}{4}$.

d) The set of all y-values in the graph extends from 1 to 4, so the range is $\{y| 1 \le y \le 4\}$.

7. a) Locate 1 on the horizontal axis and the find the point on the graph for which 1 is the first coordinate. From that point, look to the vertical axis to find the corresponding y-coordinate. It appears to be about $2\dfrac{1}{4}$, or $\dfrac{9}{4}$. Thus, $f(1) \approx \dfrac{9}{4}$.

b) The set of all x-values in the graph extends from -4 to 3, so the domain is $\{x| -4 \le x \le 3\}$.

c) To determine which member(s) of the domain are paired with 2, locate 2 on the vertical axis. From there look left and right to the graph to find any points for which 2 is the second coordinate. One such point exists. Its first coordinate is about 0, so the x-value for which $f(x) = 2$ is about 0.

d) The set of all y-values in the graph extends from -5 to 4, so the range is $\{y| -5 \le y \le 4\}$.

9. a) Locate 1 on the horizontal axis and then find the point on the graph for which 1 is the first coordinate. From that point, look to the vertical axis to find the corresponding y-coordinate, 2. Thus, $f(1) = 2$.

b) The set of all x-values in the graph extends from -5 to 4, so the domain is $\{x| -5 \le x \le 4\}$.

c) To determine which member(s) of the domain are paired with 2, locate 2 on the vertical axis. From there look left and right to the graph to find any points for which 2 is the second coordinate. All points in the set $\{x|1 \le x \le 4\}$ satisfy this condition. These are the x-values for which $f(x) = 2$.

d) The set of all y-values in the graph extends from -3 to 2, so the range is $\{y| -3 \le y \le 2\}$.

11. a) Locate 1 on the horizontal axis and then find the point on the graph for which 1 is the first coordinate. From that point, look to the vertical axis to find the corresponding y-coordinate, -1. Thus, $f(1) = -1$.

b) The set of all x-values in the graph extends from -6 to 5, so the domain is $\{x| -6 \le x \le 5\}$.

c) To determine which member(s) of the domain are paired with 2, locate 2 on the vertical axis. From there look left and right to the graph to find any points for which 2 is the second coordinate. Three such points exist, $(-4, 2)$, $(0, 2)$ and $(3, 2)$. Thus, the x-values for which $f(x) = 2$ are -4, 0, and 3.

d) The set of all y-values in the graph extends from -2 to 2, so the range is $\{y| -2 \le y \le 2\}$.

13. $f(x) = \dfrac{2}{x + 3}$

Since $\dfrac{2}{x + 3}$ cannot be calculated when the denominator is 0, we find the x-value that causes $x + 3$ to be 0:

$x + 3 = 0$

$x = -3$ Subtracting 3 on both sides

Thus, -3 is not in the domain of f, while all other real numbers are. The domain of f is

$\{x|x \text{ is a real number } and \ x \ne -3\}$.

15. $f(x) = 2x + 1$

Since we can calculate $2x + 1$ for any real number x, the domain is the set of all real numbers.

17. $f(x) = x^2 + 3$

Since we can calculate $x^2 + 3$ for any real number x, the domain is the set of all real numbers.

19. $f(x) = \dfrac{8}{5x - 14}$

Since $\dfrac{8}{5x - 14}$ cannot be calculated when the denominator is 0, we find the x-value that causes $5x - 14$ to be 0:

$$5x - 14 = 0$$
$$5x = 14$$
$$x = \frac{14}{5}$$

Thus, $\dfrac{14}{5}$ is not in the domain of f, while all other real numbers are. The domain of f is

$$\left\{ x \middle| x \text{ is a real number } and \text{ } x \neq \frac{14}{5} \right\}.$$

21. $f(x) = |x| - 4$

Since we can calculate $|x| - 4$ for any real number x, the domain is the set of all real numbers.

23. $f(x) = \dfrac{4}{|2x - 3|}$

Since $\dfrac{4}{|2x - 3|}$ cannot be calculated when the denominator is 0, we find the x-values that causes $|2x - 3|$ to be 0:

$$|2x - 3| = 0$$
$$2x - 3 = 0$$
$$2x = 3$$
$$x = \frac{3}{2}$$

Thus, $\dfrac{3}{2}$ is not in the domain of f, while all other real numbers are. The domain of f is

$$\left\{ x \middle| x \text{ is a real number } and \text{ } x \neq \frac{3}{2} \right\}.$$

25. $g(x) = \dfrac{1}{x - 1}$

Since $\dfrac{1}{x - 1}$ cannot be calculated when the denominator is 0, we find the x-value that causes $x - 1$ to be 0:

$$x - 1 = 0$$
$$x = 1$$

Thus, 1 is not in the domain of g, while all other real numbers are. The domain of g is

$$\{x | x \text{ is a real number } and \text{ } x \neq 1\}.$$

27. $g(x) = x^2 - 2x + 1$

Since we can calculate $x^2 - 2x + 1$ for any real number x, the domain is the set of all real numbers.

29. $g(x) = x^3 - 1$

Since we can calculate $x^3 - 1$ for any real number x, the domain is the set of all real numbers.

31. $g(x) = \dfrac{7}{20 - 8x}$

Since $\dfrac{7}{20 - 8x}$ cannot be calculated when the denominator is 0, we find the x-values that cause $20 - 8x$ to be 0:

$$20 - 8x = 0$$
$$-8x = -20$$
$$x = \frac{5}{2}$$

Thus, $\dfrac{5}{2}$ is not in the domain of g, while all other real numbers are. The domain of g is

$$\left\{ x \middle| x \text{ is a real number } and \text{ } x \neq \frac{5}{2} \right\}.$$

33. $g(x) = |x + 7|$

Since we can calculate $|x + 7|$ for any real number x, the domain is the set of all real numbers.

35. $g(x) = \dfrac{-2}{|4x + 5|}$

Since $\dfrac{-2}{|4x + 5|}$ cannot be calculated when the denominator is 0, we find the x-value that causes $|4x + 5|$ to be 0:

$$|4x + 5| = 0$$
$$4x + 5 = 0$$
$$4x = -5$$
$$x = -\frac{5}{4}$$

Thus, $-\dfrac{5}{4}$ is not in the domain of g, while all other real numbers are. The domain of g is

$$\left\{ x \middle| x \text{ is a real number } and \text{ } x \neq -\frac{5}{4} \right\}.$$

37. The input -1 has the output -8, so $f(-1) = -8$;

the input 0 has the output 0, so $f(0) = 0$;

the input 1 has the output -2, so $f(1) = -2$.

39. Discussion and Writing Exercise

41.
$$\frac{a^2 - 1}{a + 1} = \frac{(a + 1)(a - 1)}{a + 1}$$
$$= \frac{(a\!\!\!\!\diagup+1)(a - 1)}{(a\!\!\!\!\diagup+1)(1)}$$
$$= \frac{a - 1}{1}$$
$$= a - 1$$

43.
$$\frac{5x - 15}{x^2 - x - 6} = \frac{5(x - 3)}{(x + 2)(x - 3)}$$
$$= \frac{5(x\!\!\!\!\diagup-3)}{(x + 2)(x\!\!\!\!\diagup-3)}$$
$$= \frac{5}{x + 2}$$

45.
$$\begin{array}{r} w + 1 \\ w + 3 \overline{\smash{)}\, w^2 + 4w + 5} \\ \underline{w^2 + 3w} \\ w + 5 \\ \underline{w + 3} \\ 2 \end{array}$$

$(w^2 + 4w) - (w^2 + 3w) = w$

$(w + 5) - (w + 3) = 2$

The answer is $w + 1$, R 2, or $w + 1 + \dfrac{2}{w + 3}$.

47. $(7x - 3)(2x + 9) = 7x \cdot 2x + 7x \cdot 9 - 3 \cdot 2x - 3 \cdot 9$
$$= 14x^2 + 63x - 6x - 27$$
$$= 14x^2 + 57x - 27$$

49. $(9y + 10)^2 = (9y)^2 + 2 \cdot 9y \cdot 10 + 10^2$
$$= 81y^2 + 180y + 100$$

51. We graph each function and determine the range. The range of $f(x) = \dfrac{2}{x+3}$ is $\{y | y \text{ is a real number } and \ y \neq 0\}$; the range of $f(x) = x^2 - 2x + 3$ is $\{y | y \geq 2\}$; the range of $f(x) = |x| - 4$ is $\{y | y \geq -4\}$; the range of $f(x) = |x - 4|$ is $\{y | y \geq 0\}$.

Exercise Set 12.3

---

1. $y = \underset{\uparrow}{4}\, x + \underset{\uparrow}{5}$
$\ \ \ \ y = mx + b$

The slope is 4, and the y-intercept is $(0, 5)$.

3. $f(x) = \underset{\uparrow}{-2}x \underset{\uparrow}{-6}$
$\ \ f(x) = mx + b$

The slope is -2, and the y-intercept is $(0, -6)$.

5. $y = \underset{\uparrow}{-\dfrac{3}{8}}x \underset{\uparrow}{-\dfrac{1}{5}}$
$\ \ \ \ y = mx + b$

The slope is $-\dfrac{3}{8}$, and the y-intercept is $\left(0, -\dfrac{1}{5}\right)$.

7. $g(x) = \underset{\uparrow}{0.5}x \underset{\uparrow}{-9}$
$\ \ g(x) = mx + b$

The slope is 0.5, and the y-intercept is $(0, -9)$.

9. First we find the slope-intercept form of the equation by solving for y. This allows us to determine the slope and y-intercept easily.
$$2x - 3y = 8$$
$$-3y = -2x + 8$$
$$\frac{-3y}{-3} = \frac{-2x + 8}{-3}$$
$$y = \frac{2}{3}x - \frac{8}{3}$$

The slope is $\dfrac{2}{3}$, and the y-intercept is $\left(0, -\dfrac{8}{3}\right)$.

11. First we find the slope-intercept form of the equation by solving for y. This allows us to determine the slope and y-intercept easily.
$$9x = 3y + 6$$
$$9x - 6 = 3y$$
$$\frac{9x - 6}{3} = \frac{3y}{y}$$
$$3x - 2 = y, \text{ or}$$
$$y = 3x - 2$$

The slope is 3, and the y-intercept is $(0, -2)$.

13. First we find the slope-intercept form of the equation by solving for y. This allows us to determine the slope and y-intercept easily.
$$3 - \frac{1}{4}y = 2x$$
$$-\frac{1}{4}y = 2x - 3$$
$$-4\left(-\frac{1}{4}y\right) = -4(2x - 3)$$
$$y = -8x + 12$$

The slope is -8, and the y-intercept is $(0, 12)$.

15. First we find the slope-intercept form of the equation by solving for y. This allows us to determine the slope and y-intercept easily.
$$17y + 4x + 3 = 7 + 4x$$
$$17y + 3 = 7$$
$$17y = 4$$
$$y = \frac{4}{17}, \text{ or}$$
$$y = 0 \cdot x + \frac{4}{17}$$

The slope is 0, and the y-intercept is $\left(0, \dfrac{4}{17}\right)$.

17. We can use any two points on the line, such as $(0, 3)$ and $(4, 1)$.
$$\text{Slope} = \frac{\text{change in } y}{\text{change in } x}$$
$$= \frac{1 - 3}{4 - 0} = \frac{-2}{4} = -\frac{1}{2}$$

19. We can use any two points on the line, such as $(-3, 1)$ and $(3, 3)$.
$$\text{Slope} = \frac{\text{change in } y}{\text{change in } x}$$
$$= \frac{3 - 1}{3 - (-3)} = \frac{2}{6} = \frac{1}{3}$$

21. $\text{Slope} = \dfrac{\text{change in } y}{\text{change in } x} = \dfrac{5 - 9}{4 - 6} = \dfrac{-4}{-2} = 2$

23. $\text{Slope} = \dfrac{\text{change in } y}{\text{change in } x} = \dfrac{-8 - (-4)}{3 - 9} = \dfrac{-4}{-6} = \dfrac{2}{3}$

25. $\text{Slope} = \dfrac{\text{change in } y}{\text{change in } x} = \dfrac{8.7 - 12.4}{-5.2 - (-16.3)} = \dfrac{-3.7}{11.1} = -\dfrac{37}{111} = -\dfrac{1}{3}$

27. $\text{Slope} = \dfrac{0.4}{5} = 0.08 = 8\%$; this can also be expressed as $\dfrac{2}{25}$.

29. $\text{Slope} = \dfrac{43.33}{1238} = 0.035 = 3.5\%$

31. The rate of change can be found using the coordinates of any two points on the line. We use $(2000, 9.7)$ and $(2005, 35)$.

$$\begin{aligned}
\text{Rate} &= \frac{\text{change in volume of e-mail}}{\text{corresponding change in time}} \\
&= \frac{35 - 9.7}{2005 - 2000} \\
&= \frac{25.3}{5} \\
&= 5.06 \text{ billion messages daily per year}
\end{aligned}$$

33. We can use the coordinates of any two points on the line. We'll use $(0, 30)$ and $(3, 3)$.

$$\text{Slope} = \frac{\text{change in } y}{\text{change in } x} = \frac{3 - 30}{3 - 0} = \frac{-27}{3} = -9$$

The rate of change is $-\$900$ per year. That is, the value is decreasing at a rate of $\$900$ per year.

35. We can use the coordinates of any two points on the line. We'll use $(15, 470)$ and $(55, 510)$:

$$\text{Slope} = \frac{\text{change in } y}{\text{change in } x} = \frac{510 - 470}{55 - 15} = \frac{40}{40} = 1$$

The average SAT math score is increasing at a rate of 1 point per thousand dollars of family income.

37. Discussion and Writing Exercise

39.
$$\begin{aligned}
&3^2 - 24 \cdot 56 + 144 \div 12 \\
&= 9 - 24 \cdot 56 + 144 \div 12 \\
&= 9 - 1344 + 144 \div 12 \\
&= 9 - 1344 + 12 \\
&= -1335 + 12 \\
&= -1323
\end{aligned}$$

41.
$$\begin{aligned}
&10\{2x + 3[5x - 2(-3x + y^1 - 2)]\} \\
&= 10\{2x + 3[5x - 2(-3x + y - 2)]\} \\
&= 10\{2x + 3[5x + 6x - 2y + 4]\} \\
&= 10\{2x + 3[11x - 2y + 4]\} \\
&= 10\{2x + 33x - 6y + 12\} \\
&= 10\{35x - 6y + 12\} \\
&= 350x - 60y + 120
\end{aligned}$$

43. *Familiarize.* Let t represent the length of a side of the triangle. Then $t - 5$ represents the length of a side of the square.

Translate.

Perimeter of the square	is the same as	perimeter of the triangle
$\downarrow$	$\downarrow$	$\downarrow$
$4(t-5)$	$=$	$3t$

Solve.
$$\begin{aligned}
4(t - 5) &= 3t \\
4t - 20 &= 3t \\
t - 20 &= 0 \\
t &= 20
\end{aligned}$$

Check. If 20 is the length of a side of the triangle, then the length of a side of the square is $20 - 5$, or 15. The perimeter of the square is $4 \cdot 15$, or 60, and the perimeter of the triangle is $3 \cdot 20$, or 60. The numbers check.

State. The square and triangle have sides of length 15 yd and 20 yd, respectively.

45.
$$\begin{aligned}
c^6 - d^6 &= (c^3)^2 - (d^3)^2 \\
&= (c^3 + d^3)(c^3 - d^3) \\
&= (c + d)(c^2 - cd + d^2)(c - d)(c^2 + cd + d^2)
\end{aligned}$$

47.
$$\require{enclose}
\begin{array}{r}
a - 10 \\
a - 1 \enclose{longdiv}{a^2 - 11a + 6} \\
\underline{a^2 - a} \\
-10a + 6 \quad (a^2 - 11a) - (a^2 - a) = -10a \\
\underline{-10a + 10} \\
-4 \quad (-10a + 6) - (-10a + 10) = -4
\end{array}$$

The answer is $a - 10$, R -4, or $a - 10 + \dfrac{-4}{a - 1}$.

Exercise Set 12.4

1. $x - 2 = y$

To find the x-intercept we let $y = 0$ and solve for x. We have $x - 2 = 0$, or $x = 2$. The x-intercept is $(2, 0)$.

To find the y-intercept we let $x = 0$ and solve for y.
$$\begin{aligned}
x - 2 &= y \\
0 - 2 &= y \\
-2 &= y
\end{aligned}$$

The y-intercept is $(0, -2)$. We plot these points and draw the line.

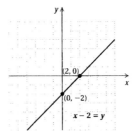

We use a third point as a check. We choose $x = 5$ and solve for y.
$$\begin{aligned}
5 - 2 &= y \\
3 &= y
\end{aligned}$$

We plot $(5, 3)$ and note that it is on the line.

3. $x + 3y = 6$

To find the x-intercept we let $y = 0$ and solve for x.
$$\begin{aligned}
x + 3y &= 6 \\
x + 3 \cdot 0 &= 6 \\
x &= 6
\end{aligned}$$

The x-intercept is $(6, 0)$.

To find the y-intercept we let $x = 0$ and solve for y.

$$x + 3y = 6$$
$$0 + 3y = 6$$
$$3y = 6$$
$$y = 2$$

The y-intercept is $(0, 2)$.

We plot these points and draw the line.

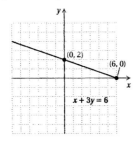

We use a third point as a check. We choose $x = 3$ and solve for y.

$$3 + 3y = 6$$
$$3y = 3$$
$$y = 1$$

We plot $(3, 1)$ and note that it is on the line.

5. $2x + 3y = 6$

To find the x-intercept we let $y = 0$ and solve for x.

$$2x + 3y = 6$$
$$2x + 3 \cdot 0 = 6$$
$$2x = 6$$
$$x = 3$$

The x-intercept is $(3, 0)$.

To find the y-intercept we let $x = 0$ and solve for y.

$$2x + 3y = 6$$
$$2 \cdot 0 + 3y = 6$$
$$3y = 6$$
$$y = 2$$

The y-intercept is $(0, 2)$.

We plot these points and draw the line.

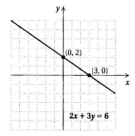

We use a third point as a check. We choose $x = -3$ and solve for y.

$$2(-3) + 3y = 6$$
$$-6 + 3y = 6$$
$$3y = 12$$
$$y = 4$$

We plot $(-3, 4)$ and note that it is on the line.

7. $f(x) = -2 - 2x$

We can think of this equation as $y = -2 - 2x$.

To find the x-intercept we let $f(x) = 0$ and solve for x. We have $0 = -2 - 2x$, or $2x = -2$, or $x = -1$. The x-intercept is $(-1, 0)$.

To find the y-intercept we let $x = 0$ and solve for $f(x)$, or y.

$$y = -2 - 2x$$
$$y = -2 - 2 \cdot 0$$
$$y = -2$$

The y-intercept is $(0, -2)$.

We plot these points and draw the line.

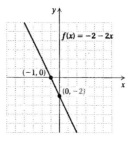

We use a third point as a check. We choose $x = -3$ and calculate y.

$$y = -2 - 2(-3) = -2 + 6 = 4$$

We plot $(-3, 4)$ and note that it is on the line.

9. $5y = -15 + 3x$

To find the x-intercept we let $y = 0$ and solve for x. We have $0 = -15 + 3x$, or $15 = 3x$, or $5 = x$. The x-intercept is $(5, 0)$.

To find the y-intercept we let $x = 0$ and solve for y.

$$5y = -15 + 3x$$
$$5y = -15 + 3 \cdot 0$$
$$5y = -15$$
$$y = -3$$

The y-intercept is $(0, -3)$.

We plot these points and draw the line.

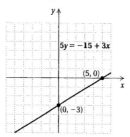

We use a third point as a check. We choose $x = -5$ and solve for y.

$$5y = -15 + 3(-5)$$
$$5y = -15 - 15$$
$$5y = -30$$
$$y = -6$$

We plot $(-5, -6)$ and note that it is on the line.

11. $2x - 3y = 6$

To find the x-intercept we let $y = 0$ and solve for x.

$$2x - 3y = 6$$
$$2x - 3 \cdot 0 = 6$$
$$2x = 6$$
$$x = 3$$

The x-intercept is $(3, 0)$.

To find the y-intercept we let $x = 0$ and solve for y.

$$2x - 3y = 6$$
$$2 \cdot 0 - 3y = 6$$
$$-3y = 6$$
$$y = -2$$

The y-intercept is $(0, -2)$.

We plot these points and draw the line.

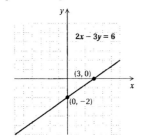

We use a third point as a check. We choose $x = -3$ and solve for y.

$$2(-3) - 3y = 6$$
$$-6 - 3y = 6$$
$$-3y = 12$$
$$y = -4$$

We plot $(-3, -4)$ and note that it is on the line.

13. $2.8y - 3.5x = -9.8$

To find the x-intercept we let $y = 0$ and solve for x.

$$2.8y - 3.5x = -9.8$$
$$2.8(0) - 3.5x = -9.8$$
$$-3.5x = -9.8$$
$$x = 2.8$$

The x-intercept is $(2.8, 0)$.

To find the y-intercept we let $x = 0$ and solve for y.

$$2.8y - 3.5x = -9.8$$
$$2.8y - 3.5(0) = -9.8$$
$$2.8y = -9.8$$
$$y = -3.5$$

The y-intercept is $(0, -3.5)$.

We plot these points and draw the line.

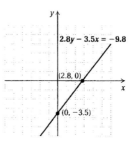

We use a third point as a check. We choose $x = 5$ and solve for y.

$$2.8y - 3.5(5) = -9.8$$
$$2.8y - 17.5 = -9.8$$
$$2.8y = 7.7$$
$$y = 2.75$$

We plot $(5, 2.75)$ and note that it is on the line.

15. $5x + 2y = 7$

To find the x-intercept we let $y = 0$ and solve for x.

$$5x + 2y = 7$$
$$5x + 2 \cdot 0 = 7$$
$$5x = 7$$
$$x = \frac{7}{5}$$

The x-intercept is $\left(\frac{7}{5}, 0\right)$.

To find the y-intercept we let $x = 0$ and solve for y.

$$5x + 2y = 7$$
$$5 \cdot 0 + 2y = 7$$
$$2y = 7$$
$$y = \frac{7}{2}$$

The y-intercept is $\left(0, \frac{7}{2}\right)$.

We plot these points and draw the line.

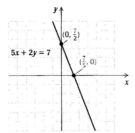

We use a third point as a check. We choose $x = 3$ and solve for y.

$$5 \cdot 3 + 2y = 7$$
$$15 + 2y = 7$$
$$2y = -8$$
$$y = -4$$

We plot $(3, -4)$ and note that it is on the line.

17. $y = \frac{5}{2}x + 1$

First we plot the y-intercept $(0, 1)$. Then we consider the slope $\frac{5}{2}$. Starting at the y-intercept and using the slope, we find another point by moving 5 units up and 2 units to the right. We get to a new point $(2, 6)$.

We can also think of the slope as $\frac{-5}{-2}$. We again start at the y-intercept $(0, 1)$. We move 5 units down and 2 units to the left. We get to another new point $(-2, -4)$. We plot the points and draw the line.

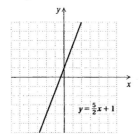

19. $f(x) = -\frac{5}{2}x - 4$

First we plot the y-intercept $(0, -4)$. We can think of the slope as $\frac{-5}{2}$. Starting at the y-intercept and using the slope, we find another point by moving 5 units down and 2 units to the right. We get to a new point $(2, -9)$.

We can also think of the slope as $\frac{5}{-2}$. We again start at the y-intercept $(0, -4)$. We move 5 units up and 2 units to the left. We get to another new point $(-2, 1)$. We plot the points and draw the line.

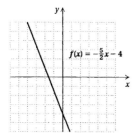

21. $x + 2y = 4$

First we write the equation in slope-intercept form by solving for y.
$$x + 2y = 4$$
$$2y = -x + 4$$
$$\frac{2y}{2} = \frac{-x + 4}{2}$$
$$y = -\frac{1}{2}x + 2$$

Now we plot the y-intercept $(0, 2)$. We can think of the slope as $\frac{-1}{2}$. Starting at the y-intercept and using the slope, we find another point by moving 1 unit down and 2 units to the right. We get to a new point $(2, 1)$.

We can also think of the slope as $\frac{1}{-2}$. We again start at the y-intercept $(0, 2)$. We move 1 unit up and 2 units to

the left. We get to another new point $(-2, 3)$. We plot the points and draw the line.

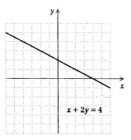

23. $4x - 3y = 12$

First we write the equation in slope-intercept form by solving for y.
$$4x - 3y = 12$$
$$-3y = -4x + 12$$
$$\frac{-3y}{-3} = \frac{-4x + 12}{-3}$$
$$y = \frac{4}{3}x - 4$$

Now we plot the y-intercept $(0, -4)$ and consider the slope $\frac{4}{3}$. Starting at the y-intercept and using the slope, we find another point by moving 4 units up and 3 units to the right. We get to a new point $(3, 0)$. In a similar manner we can move from the point $(3, 0)$ to find another point $(6, 4)$. We plot these points and draw the line.

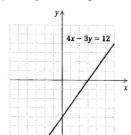

25. $f(x) = \frac{1}{3}x - 4$

First we plot the y-intercept $(0, -4)$. Then we consider the slope $\frac{1}{3}$. Starting at the y-intercept and using the slope, we find another point by moving 1 unit up and 3 units to the right. We get to a new point $(3, -3)$.

We can also think of the slope as $\frac{-1}{-3}$. We again start at the y-intercept $(0, -4)$. We move 1 unit down and 3 units to the left. We get to another new point $(-3, -5)$. We plot these points and draw the line.

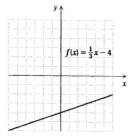

27. $5x + 4 \cdot f(x) = 4$

First we solve for $f(x)$.

$$5x + 4 \cdot f(x) = 4$$
$$4 \cdot f(x) = -5x + 4$$
$$\frac{4 \cdot f(x)}{4} = \frac{-5x + 4}{4}$$
$$f(x) = -\frac{5}{4}x + 1$$

Now we plot the y-intercept $(0, 1)$. We can think of the slope as $\frac{-5}{4}$. Starting at the y-intercept and using the slope, we find another point by moving 5 units down and 4 units to the right. We get to a new point $(4, -4)$.

We can also think of the slope as $\frac{5}{-4}$. We again start at the y-intercept $(0, 1)$. We move 5 units up and 4 units to the left. We get to another new point $(-4, 6)$. We plot these points and draw the line.

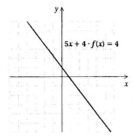

29. $x = 1$

Since y is missing, any number for y will do. Thus all ordered pairs $(1, y)$ are solutions. The graph is parallel to the y-axis.

x	y
1	-2
1	0	← x-intercept
1	3

↑   ↑ ___ Choose any
x must   number for y.
be 1.

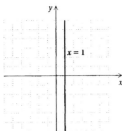

This is a vertical line, so the slope is not defined.

31. $y = -1$

Since x is missing, any number for x will do. Thus all ordered pairs $(x, -1)$ are solutions. The graph is parallel to the x-axis.

x	y
-2	-1
0	-1	← y-intercept
3	-1

↑   ↑ ___ y must be -1.
Choose
any number
for x.

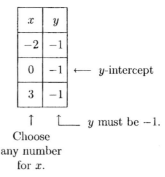

This is a horizontal line, so the slope is 0.

33. $f(x) = -6$

Since x is missing all ordered pairs $(x, 6)$ are solutions. The graph is parallel to the x-axis.

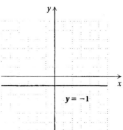

This is a horizontal line, so the slope is 0.

35. $y = 0$

Since x is missing, all ordered pairs $(x, 0)$ are solutions. The graph is the x-axis.

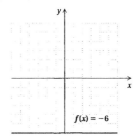

This is a horizontal line, so the slope is 0.

37. $2 \cdot f(x) + 5 = 0$

$$2 \cdot f(x) = -5$$
$$f(x) = -\frac{5}{2}$$

Since x is missing, all ordered pairs $\left(x, -\frac{5}{2}\right)$ are solutions. The graph is parallel to the x-axis.

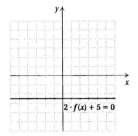

This is a horizontal line, so the slope is 0.

39. $7 - 3x = 4 + 2x$

$7 - 5x = 4$

$-5x = -3$

$x = \dfrac{3}{5}$

Since y is missing, all ordered pairs $\left(\dfrac{3}{5}, y\right)$ are solutions.
The graph is parallel to the y-axis.

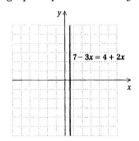

This is a vertical line, so the slope is not defined.

41. We first solve for y and determine the slope of each line.

$x + 6 = y$

$y = x + 6$ Reversing the order

The slope of $y = x + 6$ is 1.

$y - x = -2$

$y = x - 2$

The slope of $y = x - 2$ is 1.

The slopes are the same, and the y-intercepts are different.
The lines are parallel.

43. We first solve for y and determine the slope of each line.

$y + 3 = 5x$

$y = 5x - 3$

The slope of $y = 5x - 3$ is 5.

$3x - y = -2$

$3x + 2 = y$

$y = 3x + 2$ Reversing the order

The slope of $y = 3x + 2$ is 3.

The slopes are not the same; the lines are not parallel.

45. We determine the slope of each line.

The slope of $y = 3x + 9$ is 3.

$2y = 6x - 2$

$y = 3x - 1$

The slope of $y = 3x - 1$ is 3.

The slopes are the same, and the y-intercepts are different.
The lines are parallel.

47. We solve each equation for x.

$12x = 3$ $\qquad\qquad$ $-7x = 10$

$x = \dfrac{1}{4}$ $\qquad\qquad$ $x = -\dfrac{10}{7}$

We have two vertical lines, so they are parallel.

49. We determine the slope of each line.

The slope of $y = 4x - 5$ is 4.

$4y = 8 - x$

$4y = -x + 8$

$y = -\dfrac{1}{4}x + 2$

The slope of $4y = 8 - x$ is $-\dfrac{1}{4}$.

The product of their slopes is $4\left(-\dfrac{1}{4}\right)$, or -1; the lines
are perpendicular.

51. We determine the slope of each line.

$x + 2y = 5$

$2y = -x + 5$

$y = -\dfrac{1}{2}x + \dfrac{5}{2}$

The slope of $x + 2y = 5$ is $-\dfrac{1}{2}$.

$2x + 4y = 8$

$4y = -2x + 8$

$y = -\dfrac{1}{2}x + 2$

The slope of $2x + 4y = 8$ is $-\dfrac{1}{2}$.

The product of their slopes is $\left(-\dfrac{1}{2}\right)\left(-\dfrac{1}{2}\right)$, or $\dfrac{1}{4}$; the
lines are not perpendicular. For the lines to be perpendic-
ular, the product must be -1.

53. We determine the slope of each line.

$2x - 3y = 7$

$-3y = -2x + 7$

$y = \dfrac{2}{3}x - \dfrac{7}{3}$

The slope of $2x - 3y = 7$ is $\dfrac{2}{3}$.

$2y - 3x = 10$

$2y = 3x + 10$

$y = \dfrac{3}{2}x + 5$

The slope of $2y - 3x = 10$ is $\dfrac{3}{2}$.

The product of their slopes is $\dfrac{2}{3} \cdot \dfrac{3}{2} = 1$; the lines are not
perpendicular. For the lines to be perpendicular, the prod-
uct must be -1.

55. Solving the first equation for x and the second for y, we have $x = \frac{3}{2}$ and $y = -2$. The graph of $x = \frac{3}{2}$ is a vertical line, and the graph of $y = -2$ is a horizontal line. Since one line is vertical and the other is horizontal, the lines are perpendicular.

57. Discussion and Writing Exercise

59. 5.3,000,000,000.

      ↑_____|

       10 places

Large number, so the exponent is positive.

$53,000,000,000 = 5.3 \times 10^{10}$

61. 0.01. 8

    └_↑

   2 places

Small number, so the exponent is negative.

$0.018 = 1.8 \times 10^{-2}$

63. Negative exponent, so the number is small.

    0.00002. 13

    ↑_____|

    5 places

$2.13 \times 10^{-5} = 0.0000213$

65. Positive exponent, so the number is large.

    2.0000.

    └____↑

    4 places

$2 \times 10^4 = 20,000$

67. $9x - 15y = 3 \cdot 3x - 3 \cdot 5y = 3(3x - 5y)$

69. $21p - 7pq + 14p = 7p \cdot 3 - 7p \cdot q + 7p \cdot 2$
$$= 7p(3 - q + 2)$$

71. The equation will be of the form $y = b$. Since the line passes through $(-2, 3)$, b must be 3. Thus, we have $y = 3$.

73. Find the slope of each line.
$$5y = ax + 5$$
$$y = \frac{a}{5}x + 1$$

The slope of $5y = ax + 5$ is $\frac{a}{5}$.

$$\frac{1}{4}y = \frac{1}{10}x - 1$$
$$4 \cdot \frac{1}{4}y = 4\left(\frac{1}{10}x - 1\right)$$
$$y = \frac{2}{5}x - 4$$

The slope of $\frac{1}{4}y = \frac{1}{10}x - 1$ is $\frac{2}{5}$.

In order for the graphs to be parallel, their slopes must be the same. (Note that the y-intercepts are different.)
$$\frac{a}{5} = \frac{2}{5}$$
$$a = 2 \qquad \text{Multiplying by 5}$$

75. The y-intercept is $\left(0, \frac{2}{5}\right)$, so the equation is of the form $y = mx + \frac{2}{5}$. We substitute -3 for x and 0 for y in this equation to find m.

$$y = mx + \frac{2}{5}$$
$$0 = m(-3) + \frac{2}{5} \qquad \text{Substituting}$$
$$0 = -3m + \frac{2}{5}$$
$$3m = \frac{2}{5} \qquad \text{Adding } 3m$$
$$m = \frac{2}{15} \qquad \text{Multiplying by } \frac{1}{3}$$

The equation is $y = \frac{2}{15}x + \frac{2}{5}$.

(We could also have found the slope as follows:

$$m = \frac{\frac{2}{5} - 0}{0 - (-3)} = \frac{\frac{2}{5}}{3} = \frac{2}{15})$$

77. All points on the x-axis are pairs of the form $(x, 0)$. Thus any number for x will do and y must be 0. The equation is $y = 0$. This equation is a function because its graph passes the vertical-line test.

79. We substitute 4 for x and 0 for y.
$$y = mx + 3$$
$$0 = m(4) + 3$$
$$-3 = 4m$$
$$-\frac{3}{4} = m$$

81. a) Graph II indicates that 200 mL of fluid was dripped in the first 3 hr, a rate of 200/3 mL/hr. It also indicates that 400 mL of fluid was dripped in the next 3 hr, a rate of 400/3 mL/hr, and that this rate continues until the end of the time period shown. Since the rate of 400/3 mL/hr is double the rate of 200/3 mL/hr, this graph is appropriate for the given situation.

    b) Graph IV indicates that 300 mL of fluid was dripped in the first 2 hr, a rate of 300/2, or 150 mL/hr. In the next 2 hr, 200 mL was dripped. This is a rate of 200/2, or 100 mL/hr. Then 100 mL was dripped in the next 3 hr, a rate of 100/3, or $33\frac{1}{3}$ mL/hr. Finally, in the remaining 2 hr, 0 mL of fluid was dripped, a rate of 0/2, or 0 mL/hr. Since the rate at which the fluid was given decreased as time progressed and eventually became 0, this graph is appropriate for the given situation.

    c) Graph I is the only graph that shows a constant rate for 5 hours, in this case from 3 PM to 8 PM. Thus, it is appropriate for the given situation.

d) Graph III indicates that 100 mL of fluid was dripped in the first 4 hr, a rate of 100/4, or 25 mL/hr. In the next 3 hr, 200 mL was dripped. This is a rate of 200/3, or $66\frac{2}{3}$ mL/hr. Then 100 mL was dripped in the next hour, a rate of 100 mL/hr. In the last hour 200 mL was dripped, a rate of 200 mL/hr. Since the rate at which the fluid was given gradually increased, this graph is appropriate for the given situation.

Exercise Set 12.5

1. We use the slope-intercept equation and substitute -8 for m and 4 for b.
$$y = mx + b$$
$$y = -8x + 4$$

3. We use the slope-intercept equation and substitute 2.3 for m and -1 for b.
$$y = mx + b$$
$$y = 2.3x - 1$$

5. We use the slope-intercept equation and substitute $-\frac{7}{3}$ for m and -5 for b.
$$y = mx + b$$
$$y = -\frac{7}{3}x - 5$$

7. We use the slope-intercept equation and substitute $\frac{2}{3}$ for m and $\frac{5}{8}$ for b.
$$y = mx + b$$
$$y = \frac{2}{3}x + \frac{5}{8}$$

9. Using the point-slope equation:

Substitute 4 for x_1, 3 for y_1, and 5 for m.
$$y - y_1 = m(x - x_1)$$
$$y - 3 = 5(x - 4)$$
$$y - 3 = 5x - 20$$
$$y = 5x - 17$$

Using the slope-intercept equation:

Substitute 4 for x, 3 for y, and 5 for m in $y = mx + b$ and solve for b.
$$y = mx + b$$
$$3 = 5 \cdot 4 + b$$
$$3 = 20 + b$$
$$-17 = b$$

Then we use the equation $y = mx + b$ and substitute 5 for m and -17 for b.
$$y = 5x - 17$$

11. Using the point-slope equation:

Substitute 9 for x_1, 6 for y_1, and -3 for m.
$$y - y_1 = m(x - x_1)$$
$$y - 6 = -3(x - 9)$$
$$y - 6 = -3x + 27$$
$$y = -3x + 33$$

Using the slope-intercept equation:

Substitute 9 for x, 6 for y, and -3 for m in $y = mx + b$ and solve for b.
$$y = mx + b$$
$$6 = -3 \cdot 9 + b$$
$$6 = -27 + b$$
$$33 = b$$

Then we use the equation $y = mx + b$ and substitute -3 for m and 33 for b.
$$y = -3x + 33$$

13. Using the point-slope equation:

Substitute -1 for x_1, -7 for y_1, and 1 for m.
$$y - y_1 = m(x - x_1)$$
$$y - (-7) = 1(x - (-1))$$
$$y + 7 = 1(x + 1)$$
$$y + 7 = x + 1$$
$$y = x - 6$$

Using the slope-intercept equation:

Substitute -1 for x, -7 for y, and 1 for m in $y = mx + b$ and solve for b.
$$y = mx + b$$
$$-7 = 1(-1) + b$$
$$-7 = -1 + b$$
$$-6 = b$$

Then we use the equation $y = mx + b$ and substitute 1 for m and -6 for b.
$$y = 1x - 6, \text{ or } y = x - 6$$

15. Using the point-slope equation:

Substitute 8 for x_1, 0 for y_1, and -2 for m.
$$y - y_1 = m(x - x_1)$$
$$y - 0 = -2(x - 8)$$
$$y = -2x + 16$$

Using the slope-intercept equation:

Substitute 8 for x, 0 for y, and -2 for m in $y = mx + b$ and solve for b.
$$y = mx + b$$
$$0 = -2 \cdot 8 + b$$
$$0 = -16 + b$$
$$16 = b$$

Then we use the equation $y = mx + b$ and substitute -2 for m and 16 for b.
$$y = -2x + 16$$

17. Using the point-slope equation:

Substitute 0 for x_1, -7 for y_1, and 0 for m.

$$y - y_1 = m(x - x_1)$$
$$y - (-7) = 0(x - 0)$$
$$y + 7 = 0$$
$$y = -7$$

Using the slope-intercept equation:

Substitute 0 for x, -7 for y, and 0 for m in $y = mx + b$ and solve for b.

$$y = mx + b$$
$$-7 = 0 \cdot 0 + b$$
$$-7 = b$$

Then we use the equation $y = mx + b$ and substitute 0 for m and -7 for b.

$$y = 0x - 7, \text{ or } y = -7$$

19. Using the point-slope equation:

Substitute 1 for x_1, -2 for y_1, and $\frac{2}{3}$ for m.

$$y - y_1 = m(x - x_1)$$
$$y - (-2) = \frac{2}{3}(x - 1)$$
$$y + 2 = \frac{2}{3}x - \frac{2}{3}$$
$$y = \frac{2}{3}x - \frac{8}{3}$$

Using the slope-intercept equation:

Substitute 1 for x, -2 for y and $\frac{2}{3}$ for m in $y = mx + b$ and solve for b.

$$y = mx + b$$
$$-2 = \frac{2}{3} \cdot 1 + b$$
$$-2 = \frac{2}{3} + b$$
$$-\frac{8}{3} = b$$

Then we use the equation $y = mx + b$ and substitute $\frac{2}{3}$ for m and $-\frac{8}{3}$ for b.

$$y = \frac{2}{3}x - \frac{8}{3}$$

21. First find the slope of the line:

$$m = \frac{6 - 4}{5 - 1} = \frac{2}{4} = \frac{1}{2}$$

Using the point-slope equation:

We choose to use the point $(1, 4)$ and substitute 1 for x_1, 4 for y_1, and $\frac{1}{2}$ for m.

$$y - y_1 = m(x - x_1)$$
$$y - 4 = \frac{1}{2}(x - 1)$$
$$y - 4 = \frac{1}{2}x - \frac{1}{2}$$
$$y = \frac{1}{2}x + \frac{7}{2}$$

Using the slope-intercept equation:

We choose $(1, 4)$ and substitute 1 for x, 4 for y, and $\frac{1}{2}$ for m in $y = mx + b$. Then we solve for b.

$$y = mx + b$$
$$4 = \frac{1}{2} \cdot 1 + b$$
$$4 = \frac{1}{2} + b$$
$$\frac{7}{2} = b$$

Finally, we use the equation $y = mx + b$ and substitute $\frac{1}{2}$ for m and $\frac{7}{2}$ for b.

$$y = \frac{1}{2}x + \frac{7}{2}$$

23. First find the slope of the line:

$$m = \frac{-3 - 2}{-3 - 2} = \frac{-5}{-5} = 1$$

Using the point-slope equation:

We choose to use the point $(2, 2)$ and substitute 2 for x_1, 2 for y_1, and 1 for m.

$$y - y_1 = m(x - x_1)$$
$$y - 2 = 1(x - 2)$$
$$y - 2 = x - 2$$
$$y = x$$

Using the slope-intercept equation:

We choose $(2, 2)$ and substitute 2 for x, 2 for y, and 1 for m in $y = mx + b$. Then we solve for b.

$$y = mx + b$$
$$2 = 1 \cdot 2 + b$$
$$2 = 2 + b$$
$$0 = b$$

Finally, we use the equation $y = mx + b$ and substitute 1 for m and 0 for b.

$$y = 1x + 0, \text{ or } y = x$$

25. First find the slope of the line:

$$m = \frac{0 - 7}{-4 - 0} = \frac{-7}{-4} = \frac{7}{4}$$

Using the point-slope equation:

We choose $(0, 7)$ and substitute 0 for x_1, 7 for y_1, and $\frac{7}{4}$ for m.

$$y - y_1 = m(x - x_1)$$
$$y - 7 = \frac{7}{4}(x - 0)$$
$$y - 7 = \frac{7}{4}x$$
$$y = \frac{7}{4}x + 7$$

Using the slope-intercept equation:

We choose $(0, 7)$ and substitute 0 for x, 7 for y, and $\frac{7}{4}$ for m in $y = mx + b$. Then we solve for b.

$$y = mx + b$$
$$7 = \frac{7}{4} \cdot 0 + b$$
$$7 = b$$

Finally, we use the equation $y = mx + b$ and substitute $\frac{7}{4}$ for m and 7 for b.

$$y = \frac{7}{4}x + 7$$

27. First find the slope of the line:
$$m = \frac{-6 - (-3)}{-4 - (-2)} = \frac{-6 + 3}{-4 + 2} = \frac{-3}{-2} = \frac{3}{2}$$

Using the point-slope equation:

We choose $(-2, -3)$ and substitute -2 for x_1, -3 for y_1, and $\frac{3}{2}$ for m.

$$y - y_1 = m(x - x_1)$$
$$y - (-3) = \frac{3}{2}(x - (-2))$$
$$y + 3 = \frac{3}{2}(x + 2)$$
$$y + 3 = \frac{3}{2}x + 3$$
$$y = \frac{3}{2}x$$

Using the slope-intercept equation:

We choose $(-2, -3)$ and substitute -2 for x, -3 for y, and $\frac{3}{2}$ for m in $y = mx + b$. Then we solve for b.

$$y = mx + b$$
$$-3 = \frac{3}{2}(-2) + b$$
$$-3 = -3 + b$$
$$0 = b$$

Finally, we use the equation $y = mx + b$ and substitute $\frac{3}{2}$ for m and 0 for b.

$$y = \frac{3}{2}x + 0, \text{ or } y = \frac{3}{2}x$$

29. First find the slope of the line:
$$m = \frac{1 - 0}{6 - 0} = \frac{1}{6}$$

Using the point-slope equation:

We choose $(0, 0)$ and substitute 0 for x_1, 0 for y_1, and $\frac{1}{6}$ for m.

$$y - y_1 = m(x - x_1)$$
$$y - 0 = \frac{1}{6}(x - 0)$$
$$y = \frac{1}{6}x$$

Using the slope-intercept equation:

We choose $(0, 0)$ and substitute 0 for x, 0 for y, and $\frac{1}{6}$ for m in $y = mx + b$. Then we solve for b.

$$y = mx + b$$
$$0 = \frac{1}{6} \cdot 0 + b$$
$$0 = b$$

Finally, we use the equation $y = mx + b$ and substitute $\frac{1}{6}$ for m and 0 for b.

$$y = \frac{1}{6}x + 0, \text{ or } y = \frac{1}{6}x$$

31. First find the slope of the line:
$$m = \frac{-\frac{1}{2} - 6}{\frac{1}{4} - \frac{3}{4}} = \frac{-\frac{13}{2}}{-\frac{1}{2}} = 13$$

Using the point-slope equation:

We choose $\left(\frac{3}{4}, 6\right)$ and substitute $\frac{3}{4}$ for x_1, 6 for y_1, and 13 for m.

$$y - y_1 = m(x - x_1)$$
$$y - 6 = 13\left(x - \frac{3}{4}\right)$$
$$y - 6 = 13x - \frac{39}{4}$$
$$y = 13x - \frac{15}{4}$$

Using the slope-intercept equation:

We choose $\left(\frac{3}{4}, 6\right)$ and substitute $\frac{3}{4}$ for x, 6 for y, and 13 for m in $y = mx + b$. Then we solve for b.

$$y = mx + b$$
$$6 = 13 \cdot \frac{3}{4} + b$$
$$6 = \frac{39}{4} + b$$
$$-\frac{15}{4} = b$$

Finally, we use the equation $y = mx + b$ and substitute 13 for m and $-\frac{15}{4}$ for b.

$$y = 13x - \frac{15}{4}$$

33. First solve the equation for y and determine the slope of the given line.

$$x + 2y = 6 \qquad \text{Given line}$$
$$2y = -x + 6$$
$$y = -\frac{1}{2}x + 3$$

The slope of the given line is $-\frac{1}{2}$. The line through $(3, 7)$ must have slope $-\frac{1}{2}$.

Using the point-slope equation:

Substitute 3 for x_1, 7 for y_1, and $-\frac{1}{2}$ for m.

$$y - y_1 = m(x - x_1)$$
$$y - 7 = -\frac{1}{2}(x - 3)$$
$$y - 7 = -\frac{1}{2}x + \frac{3}{2}$$
$$y = -\frac{1}{2}x + \frac{17}{2}$$

Using the slope-intercept equation:

Substitute 3 for x, 7 for y, and $-\frac{1}{2}$ for m and solve for b.

$$y = mx + b$$
$$7 = -\frac{1}{2} \cdot 3 + b$$
$$7 = -\frac{3}{2} + b$$
$$\frac{17}{2} = b$$

Then we use the equation $y = mx + b$ and substitute $-\frac{1}{2}$ for m and $\frac{17}{2}$ for b.

$$y = -\frac{1}{2}x + \frac{17}{2}$$

35. First solve the equation for y and determine the slope of the given line.

$$5x - 7y = 8 \qquad \text{Given line}$$
$$5x - 8 = 7y$$
$$\frac{5}{7}x - \frac{8}{7} = y$$
$$y = \frac{5}{7}x - \frac{8}{7}$$

The slope of the given line is $\frac{5}{7}$. The line through $(2, -1)$ must have slope $\frac{5}{7}$.

Using the point-slope equation:

Substitute 2 for x_1, -1 for y_1, and $\frac{5}{7}$ for m.

$$y - y_1 = m(x - x_1)$$
$$y - (-1) = \frac{5}{7}(x - 2)$$
$$y + 1 = \frac{5}{7}x - \frac{10}{7}$$
$$y = \frac{5}{7}x - \frac{17}{7}$$

Using the slope-intercept equation:

Substitute 2 for x, -1 for y, and $\frac{5}{7}$ for m and solve for b.

$$y = mx + b$$
$$-1 = \frac{5}{7} \cdot 2 + b$$
$$-1 = \frac{10}{7} + b$$
$$-\frac{17}{7} = b$$

Then we use the equation $y = mx + b$ and substitute $\frac{5}{7}$ for m and $-\frac{17}{7}$ for b.

$$y = \frac{5}{7}x - \frac{17}{7}$$

37. First solve the equation for y and determine the slope of the given line.

$$3x - 9y = 2 \qquad \text{Given line}$$
$$3x - 2 = 9y$$
$$\frac{1}{3}x - \frac{2}{9} = y$$

The slope of the given line is $\frac{1}{3}$. The line through $(-6, 2)$ must have slope $\frac{1}{3}$.

Using the point-slope equation:

Substitute -6 for x_1, 2 for y_1, and $\frac{1}{3}$ for m.

$$y - y_1 = m(x - x_1)$$
$$y - 2 = \frac{1}{3}(x - (-6))$$
$$y - 2 = \frac{1}{3}(x + 6)$$
$$y - 2 = \frac{1}{3}x + 2$$
$$y = \frac{1}{3}x + 4$$

Using the slope-intercept equation:

Substitute -6 for x, 2 for y, and $\frac{1}{3}$ for m and solve for b.

$$y = mx + b$$
$$2 = \frac{1}{3}(-6) + b$$
$$2 = -2 + b$$
$$4 = b$$

Then we use the equation $y = mx + b$ and substitute $\frac{1}{3}$ for m and 4 for b.

$$y = \frac{1}{3}x + 4$$

39. First solve the equation for y and determine the slope of the given line.

$$2x + y = -3 \qquad \text{Given line}$$
$$y = -2x - 3$$

The slope of the given line is -2. The slope of the perpendicular line is the opposite of the reciprocal of -2. Thus, the line through $(2, 5)$ must have slope $\frac{1}{2}$.

Using the point-slope equation:

Substitute 2 for x_1, 5 for y_1, and $\frac{1}{2}$ for m.

$$y - y_1 = m(x - x_1)$$
$$y - 5 = \frac{1}{2}(x - 2)$$
$$y - 5 = \frac{1}{2}x - 1$$
$$y = \frac{1}{2}x + 4$$

Using the slope-intercept equation:

Substitute 2 for x, 5 for y, and $\frac{1}{2}$ for m and solve for b.

$$y = mx + b$$
$$5 = \frac{1}{2} \cdot 2 + b$$
$$5 = 1 + b$$
$$4 = b$$

Then we use the equation $y = mx + b$ and substitute $\frac{1}{2}$ for m and 4 for b.

$$y = \frac{1}{2}x + 4$$

41. First solve the equation for y and determine the slope of the given line.

$$3x + 4y = 5 \qquad \text{Given line}$$
$$4y = -3x + 5$$
$$y = -\frac{3}{4}x + \frac{5}{4}$$

The slope of the given line is $-\frac{3}{4}$. The slope of the perpendicular line is the opposite of the reciprocal of $-\frac{3}{4}$. Thus, the line through $(3, -2)$ must have slope $\frac{4}{3}$.

Using the point-slope equation:

Substitute 3 for x_1, -2 for y_1, and $\frac{4}{3}$ for m.

$$y - y_1 = m(x - x_1)$$
$$y - (-2) = \frac{4}{3}(x - 3)$$
$$y + 2 = \frac{4}{3}x - 4$$
$$y = \frac{4}{3}x - 6$$

Using the slope-intercept equation:

Substitute 3 for x, -2 for y, and $\frac{4}{3}$ for m.

$$y = mx + b$$
$$-2 = \frac{4}{3} \cdot 3 + b$$
$$-2 = 4 + b$$
$$-6 = b$$

Then we use the equation $y = mx + b$ and substitute $\frac{4}{3}$ for m and -6 for b.

$$y = \frac{4}{3}x - 6$$

43. First solve the equation for y and determine the slope of the given line.

$$2x + 5y = 7 \qquad \text{Given line}$$
$$5y = -2x + 7$$
$$y = -\frac{2}{5}x + \frac{7}{5}$$

The slope of the given line is $-\frac{2}{5}$. The slope of the perpendicular line is the opposite of the reciprocal of $-\frac{2}{5}$. Thus, the line through $(0, 9)$ must have slope $\frac{5}{2}$.

Using the point-slope equation:

Substitute 0 for x_1, 9 for y_1, and $\frac{5}{2}$ for m.

$$y - y_1 = m(x - x_1)$$
$$y - 9 = \frac{5}{2}(x - 0)$$
$$y - 9 = \frac{5}{2}x$$
$$y = \frac{5}{2}x + 9$$

Using the slope-intercept equation:

Substitute 0 for x, 9 for y, and $\frac{5}{2}$ for m.

$$y = mx + b$$
$$9 = \frac{5}{2} \cdot 0 + b$$
$$9 = b$$

Then we use the equation $y = mx + b$ and substitute $\frac{5}{2}$ for m and 9 for b.

$$y = \frac{5}{2}x + 9$$

45. a) The problem describes a situation in which an hourly fee is charged after an initial assessment of \$85. After 1 hour, the total cost is $\$85 + \$40 \cdot 1$. After 2 hours, the total cost is $\$85 + \$40 \cdot 2$. Then after t hours, the total cost is $C(t) = 85 + 40t$, or $C(t) = 40t + 85$, where $t > 0$.

b) For $C(t) = 40t + 85$, the y-intercept is $(0, 85)$ and the slope, or rate of change, is \$40 per hour. We plot $(0, 85)$ and from there we count up \$40 and to the right 1 hour. This takes us to $(1, 125)$. Then we draw a line through the points, calculating a third value as a check:

$$C(5) = 40 \cdot 5 + 85 = 285$$

c) To find the cost for $6\frac{1}{2}$ hours of moving service, we determine $C(6.5)$:

$$C(6.5) = 40(6.5) + 85 = 345$$

Thus, it would cost \$345 for $6\frac{1}{2}$ hours of moving service.

47. a) The problem describes a situation in which the value of the fax machine decreases at the rate of \$25 per month from an initial value of \$750. After 1 month, the value is \$750 $- \$25 \cdot 1$. After 2 months, the value is \$750 $- \$25 \cdot 2$. Then after t months, the value is $V(t) = 750 - 25t$, where $t \geq 0$.

b) For $V(t) = 750 - 25t$, or $V(t) = -25t + 750$, the y-intercept is $(0, 750)$ and the slope, or rate of change, is $-\$25$ per month. We think of the slope as $\dfrac{-25}{1}$. Plot $(0, 750)$ and from there count down \$25 and to the right 1 month. This takes us to $(1, 725)$. We draw the line through the points, calculating a third value as a check:

$$V(6) = 750 - 25 \cdot 6 = 600$$

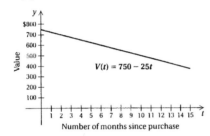

Number of months since purchase

c) To find the value of the machine after 13 months, we determine $V(13)$:

$$V(13) = 750 - 25 \cdot 13 = 425$$

Thus, after 13 months the value of the machine is \$425.

49. a) In 1991, $x = 1991 - 1990 = 1$, so one data point is $(1, 271,000)$. In 2001, $x = 2001 - 1990 = 11$, so the other data point is $(11, 1,430,000)$.

First we find the slope of the line:

$$m = \frac{1,430,000 - 271,000}{11 - 1} = \frac{1,159,000}{10} = 115,900.$$

Using the point-slope equation:

We choose $(1, 271,000)$ and substitute 1 for x_1, 271,000 for y_1, and 115,900 for m.

$$y - y_1 = m(x - x_1)$$
$$y - 271,000 = 115,900(x - 1)$$
$$y - 271,000 = 115,900x - 115,900$$
$$y = 115,900x + 155,100, \text{ or}$$
$$S = 115,900x + 155,100$$

Using the slope-intercept equation:

We choose $(1, 271,000)$ and substitute 1 for x, 271,000 for y, and 115,900 for m in $y = mx + b$.

$$y = mx + b$$
$$271,000 = 115,900 \cdot 1 + b$$
$$271,000 = 115,900 + b$$
$$155,100 = b$$

Finally, we use the equation $y = mx + b$ and substitute 115,900 for m and 155,100 for b.

$$y = 115,900x + 155,100, \text{ or}$$
$$S = 115,900x + 155,100$$

b) In 2005, $x = 2005 - 1990 = 15$. Substitute 15 for x and compute S.

$$S = 115,900(15) + 155,100$$
$$S = 1,738,500 + 155,100$$
$$S = 1,893,600$$

We estimate the average salary in 2005 to be \$1,893,600.

In 2010, $x = 2010 - 1990 = 20$. Substitute 20 for x and compute S.

$$S = 115,900(20) + 155,100$$
$$S = 2,318,000 + 155,100$$
$$S = 2,473,100$$

We estimate the average salary in 2010 to be \$2,473,100.

51. a) We form pairs of the type (t, R) where t is the number of years since 1930 and R is the record. We have two pairs, $(0, 46.8)$ and $(40, 43.8)$. These are two points on the graph of the linear function we are seeking.

First we find the slope:

$$m = \frac{43.8 - 46.8}{40 - 0} = \frac{-3}{40} = -0.075$$

Using the slope and the y-intercept, $(0, 46.8)$ we write the equation of the line.

$$R = -0.075t + 46.8$$
$$R(t) = -0.075t + 46.8 \quad \text{Using function notation}$$

b) 2003 is 73 years since 1930, so to predict the record in 2003, we find $R(73)$:

$$R(73) = -0.075(73) + 46.8$$
$$= 41.325$$

The estimated record is 41.325 seconds in 2003.

2006 is 76 years since 1930, so to predict the record in 2006, we find $R(76)$:

$$R(76) = -0.075(76) + 46.8$$
$$= 41.1$$

The estimated record is 41.1 seconds in 2006.

c) Substitute 40 for $R(t)$ and solve for t:

$$40 = -0.075t + 46.8$$
$$-6.8 = -0.075t$$
$$91 \approx t$$

The record will be 40 seconds about 91 years after 1930, or in 2021.

53. a) We form pairs of the type (t, E) where t is the number of years since 1990 and E is the life expectancy. We have two pairs, $(0, 71.8)$ and $(7, 73.6)$. These are two points on the graph of the linear function we are seeking. First we find the slope.

$$m = \frac{73.6 - 71.8}{7 - 0} = \frac{1.8}{7} = \frac{18}{70} = \frac{9}{35}$$

Using the slope and the y-intercept $(0, 71.8)$, we write the equation of the line.

$$E = \frac{9}{35}t + 71.8$$
$$E(t) = \frac{9}{35}t + 71.8 \quad \text{Using function notation}$$

b) 2007 is 17 years since 1990, so we find $E(17)$:

$$E(17) = \frac{9}{35}(17) + 71.8 \approx 72.6$$

We predict that the life expectancy of males will be about 76.2 years in 2007.

55. Discussion and Writing Exercise

57.
$$\frac{w - t}{t - w} = \frac{w - t}{-(-t + w)}$$
$$= \frac{w - t}{-1(w - t)}$$
$$= -1 \cdot \frac{w - t}{w - t}$$
$$= -1 \cdot 1$$
$$= -1$$

59.
$$\frac{3x^2 + 15x - 72}{6x^2 + 18x - 240} = \frac{3(x^2 + 5x - 24)}{6(x^2 + 3x - 40)}$$
$$= \frac{3(x + 8)(x - 3)}{2 \cdot 3(x + 8)(x - 5)}$$
$$= \frac{\cancel{3}(\cancel{x+8})(x - 3)}{2 \cdot \cancel{3}(\cancel{x+8})(x - 5)}$$
$$= \frac{x - 3}{2(x - 5)}$$

Chapter 13

Systems of Equations

Exercise Set 13.1

---

1. Graph both lines on the same set of axes.

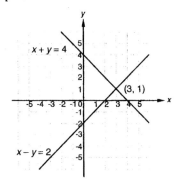

The solution (point of intersection) seems to be the point $(3, 1)$.

Check:

$$\begin{array}{c|c}
x + y = 4 & x - y = 2 \\
\hline
3 + 1 ? 4 & 3 - 1 ? 2 \\
4 \mid \text{TRUE} & 2 \mid \text{TRUE}
\end{array}$$

The solution is $(3, 1)$.

Since the system of equations has a solution it is consistent. Since there is exactly one solution, the equations are independent.

3. Graph both lines on the same set of axes.

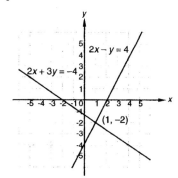

The solution (point of intersection) seems to be the point $(1, -2)$.

Check:

$$\begin{array}{c|c}
2x - y = 4 & 2x + 3y = -4 \\
\hline
2 \cdot 1 - (-2) ? 4 & 2 \cdot 1 + 3(-2) ? -4 \\
2 + 2 & 2 - 6 \\
4 \mid \text{TRUE} & -4 \mid \text{TRUE}
\end{array}$$

The solution is $(1, -2)$.

Since the system of equations has a solution, it is consistent. Since there is exactly one solution, the equations are independent.

5. Graph both lines on the same set of axes.

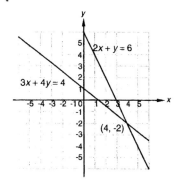

The solution (point of intersection) seems to be the point $(4, -2)$.

Check:

$$\begin{array}{c|c}
2x + y = 6 & 3x + 4y = 4 \\
\hline
2 \cdot 4 + (-2) ? 6 & 3 \cdot 4 + 4(-2) ? 4 \\
8 - 2 & 12 - 8 \\
6 \mid \text{TRUE} & 4 \mid \text{TRUE}
\end{array}$$

The solution is $(4, -2)$.

Since the system of equations has a solution, it is consistent. Since there is exactly one solution, the equations are independent.

7. Graph both lines on the same set of axes.

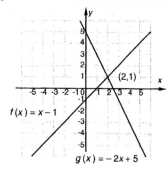

The solution seems to be the point $(2, 1)$.

Check:

$$\begin{array}{c|c}
f(x) = x - 1 & g(x) = -2x + 5 \\
\hline
1 ? 2 - 1 & 1 ? -2 \cdot 2 + 5 \\
1 \quad \text{TRUE} & -4 + 5 \\
& 1 \quad \text{TRUE}
\end{array}$$

The solution is $(2, 1)$.

Since the system of equations has a solution, it is consistent. Since there is exactly one solution, the equations are independent.

9. Graph both lines on the same set of axes.

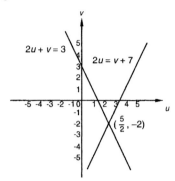

The solution seems to be $\left(\dfrac{5}{2}, -2\right)$.

Check:

$$\begin{array}{c|c} 2u+v=3 & 2u=v+7 \\ \hline 2\cdot\dfrac{5}{2}+(-2)\ ?\ 3 & 2\cdot\dfrac{5}{2}\ ?\ -2+7 \\ 5-2\ \Big|\ & 5\ \Big|\ 5 \qquad \text{TRUE} \\ 3\ \Big|\ \text{TRUE} & \end{array}$$

The solution is $\left(\dfrac{5}{2}, -2\right)$.

Since the system of equations has a solution, it is consistent. Since there is exactly one solution, the equations are independent.

11. Graph both lines on the same set of axes.

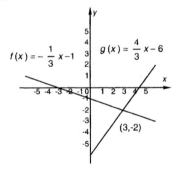

The ordered pair $(3, -2)$ checks in both equations. It is the solution.

Since the system of equations has a solution, it is consistent. Since there is exactly one solution, the equations are independent.

13. Graph both lines on the same set of axes.

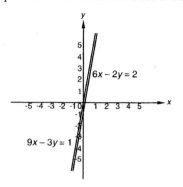

The lines are parallel. There is no solution.

Since the system of equations has no solution, it is inconsistent. Since there is no solution, the equations are independent.

15. Graph both lines on the same set of axes.

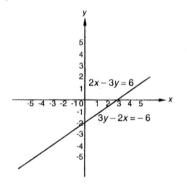

The graphs are the same. Any solution of one of the equations is also a solution of the other. Each equation has an infinite number of solutions. Thus the system of equations has an infinite number of solutions. Since the system of equations has a solution, it is consistent. Since there are infinitely many solutions, the equations are dependent.

17. Graph both lines on the same set of axes.

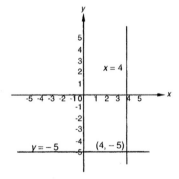

The ordered pair $(4, -5)$ checks in both equations. It is the solution.

Since the system of equations has a solution, it is consistent. Since there is exactly one solution, the equations are independent.

19. Graph both lines on the same set of axes.

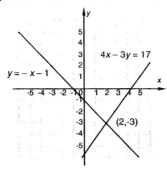

The ordered pair $(2, -3)$ checks in both equations. It is the solution.

Since the system of equations has a solution, it is consistent. Since there is exactly one solution, the equations are independent.

21. Since the system of equations has a solution, it is consistent. Since there is exactly one solution, the equations are independent. The graph of the system consists of a vertical line and a horizontal line, each passing through $(3, 3)$. Thus, system **F** corresponds to this graph.

23. Since the system of equations has a solution, it is consistent. Since there are infinitely many solutions, the equations are dependent. The equations in system B are equivalent, so their graphs are the same. In addition the graph corresponds to the one shown, so system **B** corresponds to this graph.

25. Since the system of equations has no solution, it is inconsistent. Since there is no solution, the equations are independent. The equations in system **D** have the same slope and different y-intercepts and have the graphs shown, so this system corresponds to the given graph.

27. Discussion and Writing Exercise

29. First solve the equation for y and determine the slope of the given line.

$$3x = 5y - 4 \quad \text{Given line}$$
$$3x + 4 = 5y$$
$$\frac{3}{5}x + \frac{4}{5} = y$$

The slope of the given line is $\frac{3}{5}$. The line through $(-4, 2)$ must have slope $\frac{3}{5}$. We find an equation of this new line using the equation $y = mx + b$. We substitute -4 for x, 2 for y, and $\frac{3}{5}$ for m and solve for b.

$$y = mx + b$$
$$2 = \frac{3}{5}(-4) + b$$
$$2 = -\frac{12}{5} + b$$
$$\frac{22}{5} = b$$

Then we use the equation $y = mx + b$ and substitute $\frac{3}{5}$ for m and $\frac{22}{5}$ for b.

$$y = \frac{3}{5}x + \frac{22}{5}$$

31. Graph these equations, solving each equation for y first, if necessary. We get $y = \dfrac{13.78 - 2.18x}{7.81}$ and $y = \dfrac{5.79x - 8.94}{3.45}$. Using the INTERSECT feature, we find that the point of intersection is $(2.23, 1.14)$.

33. Graph both lines on the same set of axes.

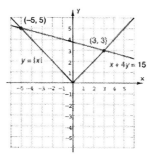

The solutions appear to be $(-5, 5)$ and $(3, 3)$.

Check:

For $(-5, 5)$:

$y = |x|$		$x + 4y = 15$	
$5 \ ? \ |-5|$		$-5 + 4 \cdot 5 \ ? \ 15$	
	5 TRUE	$-5 + 20$	
		15	TRUE

For $(3, 3)$:

$y = |x|$		$x + 4y = 15$	
$3 \ ? \ |3|$		$3 + 4 \cdot 3 \ ? \ 15$	
	3 TRUE	$3 + 12$	
		15	TRUE

Both pairs check. The solutions are $(-5, 5)$ and $(3, 3)$.

Exercise Set 13.2

1. $y = 5 - 4x, \quad (1)$
$\quad 2x - 3y = 13 \quad (2)$

We substitute $5 - 4x$ for y in the second equation and solve for x.

$$2x - 3y = 13 \quad (2)$$
$$2x - 3(5 - 4x) = 13 \quad \text{Substituting}$$
$$2x - 15 + 12x = 13$$
$$14x - 15 = 13$$
$$14x = 28$$
$$x = 2$$

Next we substitute 2 for x in either equation of the original system and solve for y.

$y = 5 - 4x$ (1)

$y = 5 - 4 \cdot 2$ Substituting

$y = 5 - 8$

$y = -3$

We check the ordered pair $(2, -3)$.

$$\frac{y = 5 - 4x}{\begin{array}{c|c} -3 \ ? \ 5 - 4 \cdot 2 & \\ 5 - 8 & \\ -3 & -3 \qquad \text{TRUE} \end{array}}$$

$$\frac{2x - 3y = 13}{\begin{array}{c|c} 2 \cdot 2 - 3(-3) \ ? \ 13 & \\ 4 + 9 & \\ 13 & 13 \quad \text{TRUE} \end{array}}$$

Since $(2, -3)$ checks, it is the solution.

3. $2y + x = 9$,   (1)

$ x = 3y - 3$ (2)

We substitute $3y - 3$ for x in the first equation and solve for y.

$ 2y + x = 9$ (1)

$2y + (3y - 3) = 9$ Substituting

$ 5y - 3 = 9$

$ 5y = 12$

$ y = \dfrac{12}{5}$

Next we substitute $\dfrac{12}{5}$ for y in either equation of the original system and solve for x.

$x = 3y - 3$ (2)

$x = 3 \cdot \dfrac{12}{5} - 3 = \dfrac{36}{5} - \dfrac{15}{5} = \dfrac{21}{5}$

We check the ordered pair $\left(\dfrac{21}{5}, \dfrac{12}{5} \right)$.

$$\frac{2y + x = 9}{\begin{array}{c|c} 2 \cdot \dfrac{12}{5} + \dfrac{21}{15} \ ? \ 9 & \\ \dfrac{24}{5} + \dfrac{21}{5} & \\ \dfrac{45}{5} & \\ 9 & 9 \quad \text{TRUE} \end{array}}$$

$$\frac{x = 3y - 3}{\begin{array}{c|c} \dfrac{21}{5} \ ? \ 3 \cdot \dfrac{12}{5} - 3 & \\ & \dfrac{36}{5} - \dfrac{15}{5} \\ \dfrac{21}{5} & \dfrac{21}{5} \qquad \text{TRUE} \end{array}}$$

Since $\left(\dfrac{21}{5}, \dfrac{12}{5} \right)$ checks, it is the solution.

5. $3s - 4t = 14$,   (1)

$ 5s + t = 8$ (2)

We solve the second equation for t.

$5s + t = 8$ (2)

$ t = 8 - 5s$ (3)

We substitute $8 - 5s$ for t in the first equation and solve for s.

$ 3s - 4t = 14$ (1)

$3s - 4(8 - 5s) = 14$ Substituting

$3s - 32 + 20s = 14$

$23s - 32 = 14$

$23s = 46$

$s = 2$

Next we substitute 2 for s in Equation (1), (2), or (3). It is easiest to use Equation (3) since it is already solved for t.

$t = 8 - 5 \cdot 2 = 8 - 10 = -2$

We check the ordered pair $(2, -2)$.

$$\frac{3s - 4t = 14}{\begin{array}{c|c} 3 \cdot 2 - 4(-2) \ ? \ 14 & \\ 6 + 8 & \\ 14 & 14 \quad \text{TRUE} \end{array}}$$

$$\frac{5s + t = 8}{\begin{array}{c|c} 5 \cdot 2 + (-2) \ ? \ 8 & \\ 10 - 2 & \\ 8 & 8 \quad \text{TRUE} \end{array}}$$

Since $(2, -2)$ checks, it is the solution.

7. $9x - 2y = -6$,   (1)

$7x + 8 = y$ (2)

We substitute $7x + 8$ for y in the first equation and solve for x.

$ 9x - 2y = -6$ (1)

$9x - 2(7x + 8) = -6$ Substituting

$9x - 14x - 16 = -6$

$-5x - 16 = -6$

$-5x = 10$

$x = -2$

Next we substitute -2 for x in either equation of the original system and solve for y.

$$7x + 8 = y \quad (2)$$
$$7(-2) + 8 = y$$
$$-14 + 8 = y$$
$$-6 = y$$

We check the ordered pair $(-2, -6)$.

$$\begin{array}{c|c} 9x - 2y = -6 \\ \hline 9(-2) - 2(-6) \ ? \ -6 \\ -18 + 12 \\ -6 & \text{TRUE} \end{array}$$

$$\begin{array}{c|c} 7x + 8 = y \\ \hline 7(-2) + 8 \ ? \ -6 \\ -14 + 8 \\ -6 & \text{TRUE} \end{array}$$

Since $(-2, -6)$ checks, it is the solution.

9. $-5s + t = 11, \quad (1)$
$4s + 12t = 4 \quad (2)$

We solve the first equation for t.

$$-5s + t = 11 \quad\quad (1)$$
$$t = 5s + 11 \quad (3)$$

We substitute $5s + 11$ for t in the second equation and solve for s.

$$4s + 12t = 4 \quad\quad (2)$$
$$4s + 12(5s + 11) = 4$$
$$4s + 60s + 132 = 4$$
$$64s + 132 = 4$$
$$64s = -128$$
$$s = -2$$

Next we substitute -2 for s in Equation (3).

$$t = 5s + 11 = 5(-2) + 11 = -10 + 11 = 1$$

We check the ordered pair $(-2, 1)$.

$$\begin{array}{c|c} -5s + t = 11 \\ \hline -5(-2) + 1 \ ? \ 11 \\ 10 + 1 \\ 11 & 11 \quad \text{TRUE} \end{array}$$

$$\begin{array}{c|c} 4s + 12t = 4 \\ \hline 4(-2) + 12 \cdot 1 \ ? \ 4 \\ -8 + 12 \\ 4 & 4 \quad \text{TRUE} \end{array}$$

Since $(-2, 1)$ checks, it is the solution.

11. $2x + 2y = 2, \quad (1)$
$3x - y = 1 \quad\quad (2)$

We solve the second equation for y.

$$3x - y = 1 \quad\quad (2)$$
$$-y = -3x + 1$$
$$y = 3x - 1 \quad\quad (3)$$

We substitute $3x - 1$ for y in the first equation and solve for x.

$$2x + 2y = 2 \quad (1)$$
$$2x + 2(3x - 1) = 2$$
$$2x + 6x - 2 = 2$$
$$8x - 2 = 2$$
$$8x = 4$$
$$x = \frac{1}{2}$$

Next we substitute $\frac{1}{2}$ for x in Equation (3).

$$y = 3x - 1 = 3 \cdot \frac{1}{2} - 1 = \frac{3}{2} - 1 = \frac{1}{2}$$

The ordered pair $\left(\frac{1}{2}, \frac{1}{2} \right)$ checks in both equations. It is the solution.

13. $3a - b = 7, \quad (1)$
$2a + 2b = 5 \quad (2)$

We solve the first equation for b.

$$3a - b = 7 \quad\quad (1)$$
$$-b = -3a + 7$$
$$b = 3a - 7 \quad (3)$$

We substitute $3a - 7$ for b in the second equation and solve for a.

$$2a + 2b = 5 \quad\quad (2)$$
$$2a + 2(3a - 7) = 5$$
$$2a + 6a - 14 = 5$$
$$8a - 14 = 5$$
$$8a = 19$$
$$a = \frac{19}{8}$$

We substitute $\frac{19}{8}$ for a in Equation (3).

$$b = 3a - 7 = 3 \cdot \frac{19}{8} - 7 = \frac{57}{8} - \frac{56}{8} = \frac{1}{8}$$

The ordered pair $\left(\frac{19}{8}, \frac{1}{8} \right)$ checks in both equations. It is the solution.

15. $2x - 3 = y \quad (1)$
$y - 2x = 1, \quad (2)$

We substitute $2x - 3$ for y in the second equation and solve for x.

$$y - 2x = 1 \quad (2)$$
$$2x - 3 - 2x = 1 \quad \text{Substituting}$$
$$-3 = 1 \quad \text{Collecting like terms}$$

We have a false equation. Therefore, there is no solution.

17. *Familiarize.* Refer to the drawing in the text. Let $l =$ the length of the court and $w =$ the width. Recall that the perimeter P of a rectangle with length l and width w is given by $P = 2l + 2w$.

Translate.

The perimeter is 120 ft.

$$2l + 2w = 120$$

The length is twice the width.

$$l = 2\cdot w$$

We have system of equations.

$$2l + 2w = 120, \quad (1)$$
$$l = 2w \qquad (2)$$

Solve. We substitute $2w$ for l in Equation (1) and solve for w.

$$2l + 2w = 120$$
$$2 \cdot 2w + 2w = 120$$
$$4w + 2w = 120$$
$$6w = 120$$
$$w = 20$$

Now substitute 20 for w in Equation (2) and find l.

$$l = 2w = 2 \cdot 20 = 40$$

Check. If the length is 40 ft and the width is 20 ft, then the perimeter is $2 \cdot 40 + 2 \cdot 20$, or 120 ft. Also, the length is twice the width. The answer checks.

State. The length of the court is 40 ft, and the width is 20 ft.

19. *Familiarize.* Using the drawing in the text, we let x and y represent the measures of the angles.

Translate.

The sum of the measures is 180°.

$$x + y = 180$$

One angle is 3 times the other less 12°.

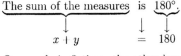

$$x = 3 \cdot y - 12$$

We have a system of equations.

$$x + y = 180, \quad (1)$$
$$x = 3y - 12 \quad (2)$$

Solve. Substitute $3y - 12$ for x in Equation (1) and solve for y.

$$x + y = 180$$
$$(3y - 12) + y = 180$$
$$4y - 12 = 180$$
$$4y = 192$$
$$y = 48$$

Now substitute 48 for y in Equation (2) and find x.

$$x = 3y - 12 = 3 \cdot 48 - 12 = 132$$

Check. The sum of the measures is $48° + 132°$, or 180°. Also, 132° is 12° less than three times 48°. The answer checks.

State. The measures of the angles are 48° and 132°.

21. *Familiarize.* Let $x =$ number of games won and $y =$ number of games tied. The total points earned in x wins is $2x$; the total points earned in y ties is $1 \cdot y$, or y.

Translate.

Points from wins plus points from ties is 60.

$$2x + y = 60$$

Number of wins is 9 more than the number of ties.

$$x = 9 + y$$

We have a system of equations:

$$2x + y = 60,$$
$$x = 9 + y$$

Solve. We solve the system of equations. We use substitution.

$$2(9 + y) + y = 60 \quad \text{Substituting } 9+y \text{ for } x \text{ in (1)}$$
$$18 + 2y + y = 60$$
$$18 + 3y = 60$$
$$3y = 42$$
$$y = 14$$

$$x = 9 + 14 \quad \text{Substituting 14 for } y \text{ in (2)}$$
$$x = 23$$

Check. The number of wins, 23, is 9 more than the number of ties, 14.

Points from wins: $23 \times 2 = 46$

Points from ties: $14 \times 1 = \underline{14}$

                              Total   60

The numbers check.

State. The team had 23 wins and 14 ties.

23. Discussion and Writing Exercise

25. $y = 1.3x - 7$

The equation is in slope-intercept form, $y = mx + b$. The slope is 1.3.

27. $A = \dfrac{pq}{7}$

$$7A = pq \quad \text{Multiplying by 7}$$

$$\frac{7A}{q} = p \quad \text{Dividing by } q$$

29. $2 = m + b, \quad (1) \quad \text{Substituting } (1, 2)$

$4 = -3m + b \quad (2) \quad \text{Substituting } (-3, 4)$

$2 = m + b \quad (1)$

$\underline{-4 = 3m - b} \quad \text{Multiplying (2) by } -1$

$-2 = 4m$

$-\dfrac{1}{2} = m$

Substitute $-\frac{1}{2}$ for m in (1).

$$2 = -\frac{1}{2} + b$$

$$\frac{5}{2} = b$$

Thus, $m = -\frac{1}{2}$ and $b = \frac{5}{2}$.

31. *Familiarize.* Let $l =$ the original length, in inches, and $w =$ the original width, in inches. Then $w - 6 =$ the width after 6 in. is cut off.

Translate.

The original perimeter is 156 in.

$$2l + 2w = 156$$

The length becomes 4 times the new width.

$$l = 4 \cdot (w - 6)$$

We have a system of equations:

$$2l + 2w = 156, \quad (1)$$
$$l = 4 \cdot (w - 6) \quad (2)$$

Solve. Substitute $4(w - 6)$ for l in Equation (1) and solve for w.

$$2l + 2w = 156$$
$$2 \cdot 4(w - 6) + 2w = 156$$
$$8w - 48 + 2w = 156$$
$$10w - 48 = 156$$
$$10w = 204$$
$$w = 20.4$$

Now substitute 20.4 for w in Equation (2) and find l.

$$l = 4 \cdot (w - 6) = 4(20.4 - 6) = 4(14.4) = 57.6$$

Check. The original perimeter is $2(57.6) + 2(20.4)$, or $115.2 + 40.8$, or 156 in. If 6 in. is cut off the width, then the width becomes $20.4 - 6$, or 14.4 in., and the length is 4 times the width, or 57.6 in. The answer checks.

State. The length is 57.6 in., and the width is 20.4 in.

Exercise Set 13.3

1. $x + 3y = 7 \quad (1)$
  $\underline{-x + 4y = 7} \quad (2)$
  $0 + 7y = 14$ Adding
    $7y = 14$
    $y = 2$

Substitute 2 for y in one of the original equations and solve for x.

  $x + 3y = 7$ Equation (1)
  $x + 3 \cdot 2 = 7$ Substituting
  $x + 6 = 7$
  $x = 1$

Check:

$x + 3y = 7$	$-x + 4y = 7$
$1 + 3 \cdot 2 \; ? \; 7$	$-1 + 4 \cdot 2 \; ? \; 7$
$1 + 6$	$-1 + 8$
7 | TRUE	7 | TRUE

Since $(1, 2)$ checks, it is the solution.

3. $9x + 5y = 6 \quad (1)$
  $\underline{2x - 5y = -17} \quad (2)$
  $11x + 0 = -11$ Adding
    $11x = -11$
    $x = -1$

Substitute -1 for x in one of the original equations and solve for y.

  $9x + 5y = 6$ Equation (1)
  $9(-1) + 5y = 6$ Substituting
  $-9 + 5y = 6$
  $5y = 15$
  $y = 3$

We obtain $(-1, 3)$. This checks, so it is the solution.

5. $5x + 3y = 19, \quad (1)$
  $2x - 5y = 11 \quad (2)$

We multiply twice to make two terms become additive inverses.

From (1): $25x + 15y = 95$ Multiplying by 5
From (2): $\underline{6x - 15y = 33}$ Multiplying by 3
    $31x + 0 = 128$ Adding
    $31x = 128$
    $x = \dfrac{128}{31}$

Substitute $\dfrac{128}{31}$ for x in one of the original equations and solve for y.

  $5x + 3y = 19$ Equation (1)
  $5 \cdot \dfrac{128}{31} + 3y = 19$ Substituting
  $\dfrac{640}{31} + 3y = \dfrac{589}{31}$
  $3y = -\dfrac{51}{31}$
  $\dfrac{1}{3} \cdot 3y = \dfrac{1}{3} \cdot \left(-\dfrac{51}{31}\right)$
  $y = -\dfrac{17}{31}$

We obtain $\left(\dfrac{128}{31}, -\dfrac{17}{31}\right)$. This checks, so it is the solution.

7. $5r - 3s = 24, \quad (1)$
  $3r + 5s = 28 \quad (2)$

We multiply twice to make two terms become additive inverses.

From (1): $25r - 15s = 120$ Multiplying by 5

From (2): $\underline{9r + 15s = 84}$ Multiplying by 3

$34r + 0 = 204$ Adding

$34r = 204$

$r = 6$

Substitute 6 for r in one of the original equations and solve for s.

$3r + 5s = 28$ Equation (2)

$3 \cdot 6 + 5s = 28$ Substituting

$18 + 5s = 28$

$5s = 10$

$s = 2$

We obtain $(6, 2)$. This checks, so it is the solution.

9. $0.3x - 0.2y = 4,$

$0.2x + 0.3y = 1$

We first multiply each equation by 10 to clear decimals.

$3x - 2y = 40$ (1)

$2x + 3y = 10$ (2)

We use the multiplication principle with both equations of the resulting system.

From (1): $9x - 6y = 120$ Multiplying by 3

From (2): $\underline{4x + 6y = 20}$ Multiplying by 2

$13x + 0 = 140$ Adding

$13x = 140$

$x = \dfrac{140}{13}$

Substitute $\dfrac{140}{13}$ for x in one of the equations in which the decimals were cleared and solve for y.

$2x + 3y = 10$ Equation (2)

$2 \cdot \dfrac{140}{13} + 3y = 10$ Substituting

$\dfrac{280}{13} + 3y = \dfrac{130}{13}$

$3y = -\dfrac{150}{13}$

$y = -\dfrac{50}{13}$

We obtain $\left(\dfrac{140}{13}, -\dfrac{50}{13}\right)$. This checks, so it is the solution.

11. $\dfrac{1}{2}x + \dfrac{1}{3}y = 4,$

$\dfrac{1}{4}x + \dfrac{1}{3}y = 3$

We first multiply each equation by the LCM of the denominators to clear fractions.

$3x + 2y = 24$ Multiplying by 6

$3x + 4y = 36$ Multiplying by 12

We multiply by -1 on both sides of the first equation and then add.

$-3x - 2y = -24$ Multiplying by -1

$\underline{3x + 4y = 36}$

$0 + 2y = 12$ Adding

$2y = 12$

$y = 6$

Substitute 6 for y in one of the equations in which the fractions were cleared and solve for x.

$3x + 2y = 24$

$3x + 2 \cdot 6 = 24$ Substituting

$3x + 12 = 24$

$3x = 12$

$x = 4$

We obtain $(4, 6)$. This checks, so it is the solution.

13. $\dfrac{2}{5}x + \dfrac{1}{2}y = 2,$

$\dfrac{1}{2}x - \dfrac{1}{6}y = 3$

We first multiply each equation by the LCM of the denominators to clear fractions.

$4x + 5y = 20$ Multiplying by 10

$3x - y = 18$ Multiplying by 6

We multiply by 5 on both sides of the second equation and then add.

$4x + 5y = 20$

$\underline{15x - 5y = 90}$ Multiplying by 5

$19x + 0 = 110$ Adding

$19x = 110$

$x = \dfrac{110}{19}$

Substitute $\dfrac{110}{19}$ for x in one of the equations in which the fractions were cleared and solve for y.

$3x - y = 18$

$3\left(\dfrac{110}{19}\right) - y = 18$ Substituting

$\dfrac{330}{19} - y = \dfrac{342}{19}$

$-y = \dfrac{12}{19}$

$y = -\dfrac{12}{19}$

We obtain $\left(\dfrac{110}{19}, -\dfrac{12}{19}\right)$. This checks, so it is the solution.

15. $2x + 3y = 1,$

$4x + 6y = 2$

Multiply the first equation by -2 and then add.

$-4x - 6y = -2$

$\underline{4x + 6y = 2}$

$0 = 0$ Adding

We have an equation that is true for all numbers x and y. The system is dependent and has an infinite number of solutions.

17. $2x - 4y = 5,$

$2x - 4y = 6$

Multiply the first equation by -1 and then add.

$$-2x + 4y = -5$$
$$\underline{2x - 4y = 6}$$
$$0 = 1$$

We have a false equation. The system has no solution.

19. $5x - 9y = 7,$

$7y - 3x = -5$

We first write the second equation in the form $Ax + By = C$.

$$5x - 9y = 7 \quad (1)$$
$$-3x + 7y = -5 \quad (2)$$

We use the multiplication principle with both equations and then add.

$$15x - 27y = 21 \quad \text{Multiplying by 3}$$
$$\underline{-15x + 35y = -25} \quad \text{Multiplying by 5}$$
$$0 + 8y = -4 \quad \text{Adding}$$
$$8y = -4$$
$$y = -\frac{1}{2}$$

Substitute $-\frac{1}{2}$ for y in one of the original equations and solve for x.

$$5x - 9y = 7 \quad \text{Equation (1)}$$
$$5x - 9\left(-\frac{1}{2}\right) = 7 \quad \text{Substituting}$$
$$5x + \frac{9}{2} = \frac{14}{2}$$
$$5x = \frac{5}{2}$$
$$x = \frac{1}{2}$$

We obtain $\left(\frac{1}{2}, -\frac{1}{2}\right)$. This checks, so it is the solution.

21. $3(a - b) = 15,$

$4a = b + 1$

We first write each equation in the form $Ax + By = C$.

$$3a - 3b = 15 \quad (1)$$
$$4a - b = 1 \quad (2)$$

We multiply by -3 on both sides of the second equation and then add.

$$3a - 3b = 15$$
$$\underline{-12a + 3b = -3} \quad \text{Multiplying by } -3$$
$$-9a + 0 = 12$$
$$-9a = 12$$
$$a = -\frac{12}{9}$$
$$a = -\frac{4}{3}$$

Substitute $-\frac{4}{3}$ for a in either Equation (1) or Equation (2) and solve for b.

$$4a - b = 1 \quad \text{Equation (2)}$$
$$4\left(-\frac{4}{3}\right) - b = 1 \quad \text{Substituting}$$
$$-\frac{16}{3} - b = \frac{3}{3}$$
$$-b = \frac{19}{3}$$
$$b = -\frac{19}{3}$$

We obtain $\left(-\frac{4}{3}, -\frac{19}{3}\right)$. This checks, so it is the solution.

23. $x - \frac{1}{10}y = 100,$

$y - \frac{1}{10}x = -100$

We first write the second equation in the form $Ax + By = C$.

$$x - \frac{1}{10}y = 100$$
$$-\frac{1}{10}x + y = -100$$

Next we multiply each equation by 10 to clear fractions.

$$10x - y = 1000 \quad (1)$$
$$-x + 10y = -1000 \quad (2) \quad \text{Equation (1)}$$

We multiply by 10 on both sides of Equation (1) and then add.

$$100x - 10y = 10,000 \quad \text{Multiplying by 10}$$
$$\underline{-x + 10y = -1000}$$
$$99x + 0 = 9000$$
$$99x = 9000$$
$$x = \frac{9000}{99}$$
$$x = \frac{1000}{11}$$

Substitute $\frac{1000}{11}$ for x in one of the equations in which the fractions were cleared and solve for y.

$$10x - y = 1000 \qquad \text{Equation (1)}$$

$$10\left(\frac{1000}{11}\right) - y = 1000 \qquad \text{Substituting}$$

$$\frac{10,000}{11} - y = \frac{11,000}{11}$$

$$-y = \frac{1000}{11}$$

$$y = -\frac{1000}{11}$$

We obtain $\left(\dfrac{1000}{11}, -\dfrac{1000}{11}\right)$. This checks, so it is the solution.

25. $0.05x + 0.25y = 22,$

$0.15x + 0.05y = 24$

We first multiply each equation by 100 to clear decimals.

$$5x + 25y = 2200 \quad (1)$$

$$15x + 5y = 2400 \quad (2)$$

We multiply by -5 on both sides of the second equation and add.

$$5x + 25y = 2200$$

$$\underline{-75x - 25y = -12,000} \qquad \text{Multiplying by } -5$$

$$-70x + 0 = -9800 \qquad \text{Adding}$$

$$-70x = -9800$$

$$x = \frac{-9800}{-70}$$

$$x = 140$$

Substitute 140 for x in one of the equations in which the decimals were cleared and solve for y.

$$5x + 25y = 2200 \qquad \text{Equation (1)}$$

$$5 \cdot 140 + 25y = 2200 \qquad \text{Substituting}$$

$$700 + 25y = 2200$$

$$25y = 1500$$

$$y = 60$$

We obtain $(140, 60)$. This checks, so it is the solution.

27. Familiarize. Let $l =$ the length of the field and $w =$ the width, in meters. Recall that the perimeter P of a rectangle with length l and width w is given by $P = 2l + 2w$.

Translate.

The perimeter is 340 m.

$$2l + 2w = 340$$

The length is 50 m more than the width.

$$l = 50 + w$$

We have system of equations.

$$2l + 2w = 340, \quad (1)$$

$$l = 50 + w \qquad (2)$$

Solve. First we subtract w on both sides of Equation (2).

$$l = 50 + w$$

$$l - w = 50$$

Now we have

$$2l + 2w = 340, \quad (1)$$

$$l - w = 50. \quad (3)$$

Multiply Equation (3) by 2 and then add.

$$2l + 2w = 340$$

$$\underline{2l - 2w = 100}$$

$$4l = 440$$

$$l = 110$$

Now substitute 110 for l in one of the original equations and solve for w.

$$l = 50 + w \quad (2)$$

$$110 = 50 + w$$

$$60 = w$$

Check. The perimeter is $2 \cdot 110 + 2 \cdot 60$, or $220 + 120$, or 340 m. Also, the length, 110 m, exceeds the width, 60 m, by 50 m. The answer checks.

State. The length is 110 m and the width is 60 m.

29. Familiarize. Using the drawing in the text, we let x and y represent the measures of the angles.

Translate.

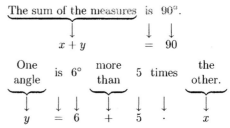

The sum of the measures is $90°$.

$$x + y = 90$$

One angle is $6°$ more than 5 times the other.

$$y = 6 + 5 \cdot x$$

We have system of equations.

$$x + y = 90, \quad (1)$$

$$y = 6 + 5x \quad (2)$$

Solve. First we subtract $5x$ on both sides of Equation (2).

$$y = 6 + 5x$$

$$-5x + y = 6 \qquad (3)$$

Now we have

$$x + y = 90, \quad (1)$$

$$-5x + y = 6. \quad (3)$$

Multiply Equation (1) by 5 and then add.

$$5x + 5y = 450$$

$$\underline{-5x + y = 6}$$

$$6y = 456$$

$$y = 76$$

Substitute 76 for y in one of the original equations and solve for x.

$$x + y = 90 \quad (1)$$

$$x + 76 = 90$$

$$x = 14$$

Check. The sum of the measures is $14° + 76°$, or $90°$. Also, $6°$ more than 5 times the measure of the $14°$ angle is

$5 \cdot 14° + 6° = 70° + 6° = 76°$, the measure of the other angle. The answer checks.

State. The angles measure $14°$ and $76°$.

31. Familiarize. Let $c =$ the number of coach seats and $f =$ the number of first-class seats.

Translate.

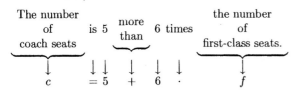

The total number of seats is 152.
$$c + f = 152$$

The number of coach seats is 5 more than 6 times the number of first-class seats.
$$c = 5 + 6 \cdot f$$

We have system of equations.
$$c + f = 152, \quad (1)$$
$$c = 5 + 6f \quad (2)$$

Solve. First we subtract $6f$ on both sides of Equation (2).
$$c = 5 + 6f$$
$$c - 6f = 5 \quad (3)$$

Now we have
$$c + f = 152, \quad (1)$$
$$c - 6f = 5. \quad (3)$$

Multiply Equation (1) by 6 and then add.
$$6c + 6f = 912$$
$$\underline{c - 6f = 5}$$
$$7c \qquad = 917$$
$$c = 131$$

Substitute 131 for c in one of the original equations and solve for f.
$$c + f = 152 \quad (1)$$
$$131 + f = 152$$
$$f = 21$$

Check. The total number of seats is $131 + 21$, or 152. Five more than six times the number of first-class seats is $5 + 6 \cdot 21$, or $5 + 126$, or 131, the number of coach seats. The answer checks.

State. There are 131 coach-class seats and 21 first-class seats.

33. Discussion and Writing Exercise

35. First solve the equation for y and determine the slope of the given line.
$$2x - 7y = 3 \qquad \text{Given line}$$
$$-7y = -2x + 3$$
$$y = \frac{2}{7}x - \frac{3}{7}$$

The slope of the given line is $\frac{2}{7}$. The slope of the perpendicular line is the opposite of the reciprocal of $\frac{2}{7}$. Thus,

the line through $(10, 1)$ must have slope $-\frac{7}{2}$. We find an equation of this new line using the equation $y = mx + b$. We substitute 10 for x, 1 for y, and $-\frac{7}{2}$ for m and solve for b.
$$y = mx + b$$
$$1 = -\frac{7}{2} \cdot 10 + b$$
$$1 = -35 + b$$
$$36 = b$$

Then we use the equation $y = mx + b$ and substitute $-\frac{7}{2}$ for m and 36 for b.
$$y = -\frac{7}{2}x + 36$$

37. $f(x) = 3x^2 - x + 1$
$$f(0) = 3 \cdot 0^2 - 0 + 1 = 0 - 0 + 1 = 1$$

39. $f(x) = 3x^2 - x + 1$
$$f(1) = 3 \cdot 1^2 - 1 + 1 = 3 - 1 + 1 = 3$$

41. $f(x) = 3x^2 - x + 1$
$$f(-2) = 3(-2)^2 - (-2) + 1 = 12 + 2 + 1 = 15$$

43. $f(x) = 3x^2 - x + 1$
$$f(-4) = 3(-4)^2 - (-4) + 1 = 48 + 4 + 1 = 53$$

45. Graph these equations, solving each equation for y first, if necessary. We get $y = \dfrac{3.5x - 106.2}{2.1}$ and $y = \dfrac{-4.1x - 106.28}{16.7}$. Using the INTERSECT feature, we find that the point of intersection is $(23.12, -12.04)$.

47. Substitute -5 for x and -1 for y in the first equation.
$$A(-5) - 7(-1) = -3$$
$$-5A + 7 = -3$$
$$-5A = -10$$
$$A = 2$$

Then substitute -5 for x and -1 for y in the second equation.
$$-5 - B(-1) = -1$$
$$-5 + B = -1$$
$$B = 4$$

We have $A = 2$, $B = 4$.

49. $(0, -3)$ and $\left(-\frac{3}{2}, 6\right)$ are two solutions of $px - qy = -1$. Substitute 0 for x and -3 for y.
$$p \cdot 0 - q \cdot (-3) = -1$$
$$3q = -1$$
$$q = -\frac{1}{3}$$

Substitute $-\frac{3}{2}$ for x and 6 for y.
$$p \cdot \left(-\frac{3}{2}\right) - q \cdot 6 = -1$$
$$-\frac{3}{2}p - 6q = -1$$

Substitute $-\frac{1}{3}$ for q and solve for p.

$$-\frac{3}{2}p - 6 \cdot \left(-\frac{1}{3}\right) = -1$$

$$-\frac{3}{2}p + 2 = -1$$

$$-\frac{3}{2}p = -3$$

$$-\frac{2}{3} \cdot \left(-\frac{3}{2}p\right) = -\frac{2}{3} \cdot (-3)$$

$$p = 2$$

Thus, $p = 2$ and $q = -\frac{1}{3}$.

Exercise Set 13.4

1. **Familiarize.** Let x = the number of less expensive brushes sold and y = the number of more expensive brushes sold.

Translate. We organize the information in a table.

Kind of brush	Less expensive	More expensive	Total
Number sold	x	y	45
Price	\$8.50	\$9.75	
Amount taken in	$8.50x$	$9.75y$	398.75

The "Number sold" row of the table gives us one equation:

$$x + y = 45$$

The "Amount taken in" row gives us a second equation:

$$8.50x + 9.75y = 398.75$$

We have a system of equations:

$$x + y = 45,$$
$$8.50x + 9.75y = 398.75$$

We can multiply the second equation on both sides by 100 to clear the decimals:

$$x + y = 45, \qquad (1)$$
$$850x + 975y = 39,875 \quad (2)$$

Solve. We solve the system of equations using the elimination method. Begin by multiplying Equation (1) by -850.

$$\begin{array}{rl} -850x - 850y = -38,250 & \text{Multiplying (1)} \\ \underline{850x + 975y = 39,875} & \\ 125y = 1625 & \\ y = 13 & \end{array}$$

Substitute 13 for y in (1) and solve for x.

$$x + 13 = 45$$
$$x = 32$$

Check. The number of brushes sold is $32 + 13$, or 45. The amount taken in was $\$8.50(32) + \$9.75(13) = \$272 + \$126.75 = \$398.75$. The answer checks.

State. 32 of the less expensive brushes were sold, and 13 of the more expensive brushes were sold.

3. **Familiarize.** Let h = the number of vials of Humulin Insulin sold and n = the number of vials of Novolin Insulin sold.

Translate. We organize the information in a table.

Brand	Humulin	Novolin	Total
Number sold	h	n	65
Price	\$15.75	\$12.95	
Amount taken in	$15.75h$	$12.95n$	959.35

The "Number sold" row of the table gives us one equation:

$$h + n = 65$$

The "Amount taken in" row gives us a second equation:

$$15.75h + 12.95n = 959.35$$

We have a system of equations:

$$h + n = 65,$$
$$15.75h + 12.95n = 959.35$$

We can multiply the second equation on both sides by 100 to clear the decimals:

$$h + n = 65, \qquad (1)$$
$$1575h + 1295n = 95,935 \quad (2)$$

Solve. We solve the system of equations using the elimination method.

$$\begin{array}{rl} -1295h - 1295n = -84,175 & \text{Multiplying (1)} \\ & \text{by } -1295 \\ \underline{1575h + 1295n = 95,935} & \\ 280h = 11,760 & \\ h = 42 & \end{array}$$

Substitute 42 for h in (1) and solve for n.

$$42 + n = 65$$
$$n = 23$$

Check. A total of $42 + 23$, or 65 vials, was sold. The amount collected was $\$15.75(42) + \$12.95(23) = \$661.50 + \$297.85 = \$959.35$. The answer checks.

State. 42 vials of Humulin Insulin and 23 vials of Novolin Insulin were sold.

5. **Familiarize.** Let x and y represent the number of 30-sec and 60-sec commercials played, respectively. We will convert 10 min to seconds:

$$10 \text{ min} = 10 \times 1 \text{ min} = 10 \times 60 \text{ sec} = 600 \text{ sec}.$$

Translate. We organize the information in a table.

Type	30-sec	60-sec	Total
Number	x	y	12
Time	$30x$	$60y$	600

The "Number" row of the table gives us one equation:

$$x + y = 12$$

The "Time" row gives us a second equation:

$$30x + 60y = 600$$

We have a system of equations:

$$x + y = 12, \qquad (1)$$
$$30x + 60y = 600 \quad (2)$$

Solve. We solve the system of equations using the elimination method.

$$
\begin{array}{ll}
-30x - 30y = -360 & \text{Multiplying (1) by } -30 \\
\underline{30x + 60y = 600} & \\
30y = 240 & \\
y = 8 &
\end{array}
$$

Substitute 8 for y in Equation (1) and solve for x.

$$x + 8 = 12$$
$$x = 4$$

Check. If Rudy plays 4 30-sec and 8 60-sec commercials, then the total number of commercials played is $4 + 8$, or 12. Also, the time for 4 30-sec commercials is $4 \cdot 30$, or 120 sec, and the time for 8 60-sec commercials is $8 \cdot 60$, or 480 sec. Then the total commercial time is $120 + 480$, or 600 sec, or 10 min. The answer checks.

State. Rudy plays 4 30-sec commercials and 8 60-sec commercials.

7. Familiarize. Let x and y represent the number of pounds of the 40% and the 10% mixture to be used. The final mixture contains 25% (10 lb), or 0.25(10 lb), or 2.5 lb of peanuts.

Translate. We organize the information in a table.

	40% mixture	10% mixture	Wedding mixture
Number of pounds	x	y	10
Percent of peanuts	40%	10%	25%
Pounds of peanuts	$0.4x$	$0.1y$	2.5

The first row of the table gives us one equation:

$$x + y = 10$$

The last row gives us a second equation:

$$0.4x + 0.1y = 2.5$$

After clearing decimals, we have the problem translated to a system of equations:

$$x + y = 10, \quad (1)$$
$$4x + y = 25 \quad (2)$$

Solve. We solve the system of equations using the elimination method.

$$
\begin{array}{ll}
-x - y = -10 & \text{Multiplying (1) by } -1 \\
\underline{4x + y = 25} & \\
3x = 15 & \\
x = 5 &
\end{array}
$$

Now substitute 5 for x in Equation (1) and solve for y.

$$5 + y = 10$$
$$y = 5$$

Check. If 5 lb of each mixture is used, the total wedding mixture is $5 + 5$, or 10 lb. The amount of peanuts in the wedding mixture is $0.4(5) + 0.1(5)$, or $2 + 0.5$, or 2.5 lb. The answer checks.

State. 5 lb of each type of mixture should be used.

9. Familiarize. Let $x = $ the number of liters of 25% solution and $y = $ the number of liters of 50% solution to be used. The mixture contains 40%(10 L), or 0.4(10 L) $= 4$ L of acid.

Translate. We organize the information in a table.

	25% solution	50% solution	Mixture
Number of liters	x	y	10
Percent of acid	25%	50%	40%
Amount of acid	$0.25x$	$0.5y$	4 L

We get one equation from the "Number of liters" row of the table.

$$x + y = 10$$

The last row of the table yields a second equation.

$$0.25x + 0.5y = 4$$

After clearing decimals, we have the problem translated to a system of equations:

$$x + y = 10, \quad (1)$$
$$25x + 50y = 400 \quad (2)$$

Solve. We use the elimination method to solve the system of equations.

$$
\begin{array}{ll}
-25x - 25y = -250 & \text{Multiplying (1) by } -25 \\
\underline{25x + 50y = 400} & \\
25y = 150 & \\
y = 6 &
\end{array}
$$

Substitute 6 for y in (1) and solve for x.

$$x + 6 = 10$$
$$x = 4$$

Check. The total amount of the mixture is $4 \text{ lb} + 6 \text{ lb}$, or 10 lb. The amount of acid in the mixture is $0.25(4 \text{ L}) + 0.5(6 \text{ L}) = 1 \text{ L} + 3 \text{ L} = 4$ L. The answer checks.

State. 4 L of the 25% solution and 6 L of the 50% solution should be mixed.

11. Familiarize. Let $x = $ the amount of the 6% loan and $y = $ the amount of the 9% loan. Recall that the formula for simple interest is

$$\text{Interest} = \text{Principal} \cdot \text{Rate} \cdot \text{Time}.$$

Translate. We organize the information in a table.

	6% loan	9% loan	Total
Principal	x	y	$12,000
Interest Rate	6%	9%	
Time	1 yr	1 yr	
Interest	0.06x	0.09y	$855

The "Principal" row of the table gives us one equation:

$$x + y = 12,000$$

The last row of the table yields another equation:

$$0.06x + 0.09y = 855$$

After clearing decimals, we have the problem translated to a system of equations:

$$x + y = 12,000 \quad (1)$$
$$6x + 9y = 85,500 \quad (2)$$

Solve. We use the elimination method to solve the system of equations.

$$
\begin{aligned}
-6x - 6y &= -72,000 \quad \text{Multiplying (1) by } -6 \\
\underline{6x + 9y} &= \underline{85,500} \\
3y &= 13,500 \\
y &= 4500
\end{aligned}
$$

Substitute 4500 for y in (1) and solve for x.

$$x + 4500 = 12,000$$
$$x = 7500$$

Check. The loans total $7500 + $4500, or $12,000. The total interest is $0.06(\$7500) + 0.09(\$4500) = \$450 + \$405 = \$855$. The answer checks.

State. The 6% loan was for $7500, and the 9% loan was for $4500.

13. ***Familiarize.*** From the bar graph we see that whole milk is 4% milk fat, milk for cream cheese is 8% milk fat, and cream is 30% milk fat. Let x = the number of pounds of whole milk and y = the number of pounds of cream to be used. The mixture contains 8%(200 lb), or 0.08(200 lb) = 16 lb of milk fat.

Translate. We organize the information in a table.

	Whole milk	Cream	Mixture
Number of pounds	x	y	200
Percent of milk fat	4%	30%	8%
Amount of milk fat	0.04x	0.3y	16 lb

We get one equation from the " Number of pounds" row of the table:

$$x + y = 200$$

The last row of the table yields a second equation:

$$0.04x + 0.3y = 16$$

After clearing decimals, we have the problem translated to a system of equations:

$$x + y = 200, \quad (1)$$
$$4x + 30y = 1600 \quad (2)$$

Solve. We use the elimination method to solve the system of equations.

$$
\begin{aligned}
-4x - 4y &= -800 \quad \text{Multiplying (1) by } -4 \\
\underline{4x + 30y} &= \underline{1600} \\
26y &= 800 \\
y &= \frac{400}{13}, \text{ or } 30\frac{10}{13}
\end{aligned}
$$

Substitute $\frac{400}{13}$ for y in (1) and solve for x.

$$x + \frac{400}{13} = 200$$
$$x = \frac{2200}{13}, \text{ or } 169\frac{3}{13}$$

Check. The total amount of the mixture is $\frac{2200}{13}$ lb + $\frac{400}{13}$ lb = $\frac{2600}{13}$ lb = 200 lb. The amount of milk fat in the mixture is $0.04\left(\frac{2200}{13} \text{ lb}\right) + 0.3\left(\frac{400}{13} \text{ lb}\right) = \frac{88}{13}$ lb + $\frac{120}{13}$ lb = $\frac{208}{13}$ lb = 16 lb. The answer checks.

State. $169\frac{3}{13}$ lb of whole milk and $30\frac{10}{13}$ lb of cream should be mixed.

15. ***Familiarize.*** Let x = the number of $5 bills and y = the number of $1 bills. The total value of the $5 bills is $5x$, and the total value of the $1 bills is $1 \cdot y$, or y.

Translate.

$$\underbrace{\text{The total number of bills}}_{x + y} \; \underbrace{\text{is}}_{=} \; \underbrace{22.}_{22}$$

$$\underbrace{\text{The total value of the bills}}_{5x + y} \; \underbrace{\text{is}}_{=} \; \underbrace{\$50.}_{50}$$

We have a system of equations:

$$x + y = 22, \quad (1)$$
$$5x + y = 50 \quad (2)$$

Solve. We use the elimination method.

$$
\begin{aligned}
-x - y &= -22 \quad \text{Multiplying (1) by } -1 \\
\underline{5x + y} &= \underline{50} \\
4x &= 28 \\
x &= 7
\end{aligned}
$$

$$7 + y = 22 \quad \text{Substituting 7 for } x \text{ in (1)}$$
$$y = 15$$

Check. Total number of bills: $7 + 15 = 22$

Total value of bills: $\$5 \cdot 7 + \$1 \cdot 15 = \$35 + \$15 = \$50$.

The numbers check.

State. There are 7 $5 bills and 15 $1 bills.

17. Familiarize. We first make a drawing.

Slow train d miles	75 mph	$(t + 2)$ hr
Fast train d miles	125 mph	t hr

From the drawing we see that the distances are the same. Now complete the chart.

$$d = r \cdot t$$

	Distance	Rate	Time	
Slow train	d	75	$t + 2$	$\rightarrow d = 75(t+2)$
Fast train	d	125	t	$\rightarrow d = 125t$

Translate. Using $d = rt$ in each row of the table, we get a system of equations:

$$d = 75(t + 2),$$
$$d = 125t$$

Solve. We solve the system of equations.

$$125t = 75(t + 2) \quad \text{Using substitution}$$
$$125t = 75t + 150$$
$$50t = 150$$
$$t = 3$$

Then $d = 125t = 125 \cdot 3 = 375$

Check. At 125 mph, in 3 hr the fast train will travel $125 \cdot 3 = 375$ mi. At 75 mph, in $3 + 2$, or 5 hr the slow train will travel $75 \cdot 5 = 375$ mi. The numbers check.

State. The trains will meet 375 mi from the station.

19. Familiarize. We first make a drawing. Let $d =$ the distance and $r =$ the speed of the canoe in still water. Then when the canoe travels downstream its speed is $r + 6$, and its speed upstream is $r - 6$. From the drawing we see that the distances are the same.

Downstream, 6 km/h current

d km, $r + 6$, 4 hr

Upstream, 6 km/h current

d km, $r - 6$, 10 hr

Organize the information in a table.

	Distance	Rate	Time
With current	d	$r + 6$	4
Against current	d	$r - 6$	10

Translate. Using $d = rt$ in each row of the table, we get a system of equations:

$$d = 4(r + 6), \qquad d = 4r + 24,$$
$$\text{or}$$
$$d = 10(r - 6) \qquad d = 10r - 60$$

Solve. Solve the system of equations.

$$4r + 24 = 10r - 60 \quad \text{Using substitution}$$
$$24 = 6r - 60$$
$$84 = 6r$$
$$14 = r$$

Check. When $r = 14$, then $r + 6 = 14 + 6 = 20$, and the distance traveled in 4 hr is $4 \cdot 20 = 80$ km. Also, $r - 6 = 14 - 6 = 8$, and the distance traveled in 10 hr is $8 \cdot 10 = 80$ km. The answer checks.

State. The speed of the canoe in still water is 14 km/h.

21. Familiarize. We first make a drawing. Let $d =$ the distance and $t =$ the time at 32 mph. At 4 mph faster, the speed is 36 mph.

32 mph	t hr	d mi
36 mph	$\left(t - \dfrac{1}{2}\right)$ hr	d mi

From the drawing, we see that the distances are the same. List the information in a table.

$$d = r \cdot t$$

	Distance	Rate	Time	
Slower trip	d	32	t	$\rightarrow d = 32t$
Faster trip	d	36	$t - \dfrac{1}{2}$	$\rightarrow d = 36\left(t - \dfrac{1}{2}\right)$

Translate. Using $d = rt$ in each row of the table, we get a system of equations:

$$d = 32t, \qquad (1)$$
$$d = 36\left(t - \frac{1}{2}\right) \quad (2)$$

Solve. We solve the system of equations.

$$32t = 36\left(t - \frac{1}{2}\right) \quad \text{Substituting } 32t \text{ for } d \text{ in (2)}$$
$$32t = 36t - 18$$
$$-4t = -18$$
$$t = \frac{18}{4}, \text{ or } \frac{9}{2}$$

The time at 32 mph is $\dfrac{9}{2}$ hr, and the time at 36 mph is $\dfrac{9}{2} - \dfrac{1}{2}$, or 4 hr.

Check. At 32 mph, in $\dfrac{9}{2}$ hr the salesperson will travel $32 \cdot \dfrac{9}{2}$, or 144 mi. At 36 mph, in 4 hr she will travel $36 \cdot 4$, or 144 mi. Since the distances are the same, the numbers check.

State. The towns are 144 mi apart.

23. Familiarize. We first make a drawing. Let $t =$ the time, $d =$ the distance traveled at 190 km/h, and $780 - d =$ the distance traveled at 200 km/h.

190 km/h t hr t hr   200 km/h

|——————— 780 km ———————|

We list the information in a table.

$$d = r \cdot t$$

	Distance	Rate	Time	
Slower plane	d	190	t	$\rightarrow d = 190t$
Faster plane	$780 - d$	200	t	$\rightarrow 780 - d = 200t$

Translate. Using $d = rt$ in each row of the table, we get a system of equations:

$$d = 190t, \quad (1)$$
$$780 - d = 200t \quad (2)$$

Solve. We solve the system of equations.

$$780 - 190t = 200t \quad \text{Substituting } 190t \text{ for } d \text{ in (2)}$$
$$780 = 390t$$
$$2 = t$$

Check. In 2 hr the slower plane will travel $190 \cdot 2$, or 380 km, and the faster plane will travel $200 \cdot 2$, or 400 km. The sum of the distances is $380 + 400$, or 780 km. The value checks.

State. The planes will meet in 2 hr.

25. Familiarize. We first make a drawing. Let $d =$ the distance traveled at 420 km/h and $t =$ the time traveled. Then $1000 - d =$ the distance traveled at 330 km/h.

d km, 420 km/h, t hr $1000 - d$ km, 330 km/h, t hr

|——————— 1000 km ———————|

We list the information in a table.

$$d = r \cdot t$$

	Distance	Rate	Time	
Faster airplane	d	420	t	$\rightarrow d = 420t$
Slower airplane	$1000 - d$	330	t	$\rightarrow 1000 - d = 330t$

Translate. Using $d = rt$ in each row of the table, we get a system of equations:

$$d = 420t, \quad (1)$$
$$1000 - d = 330t \quad (2)$$

Solve. We use substitution.

$$1000 - 420t = 330t \quad \text{Substituting } 420t \text{ for } d \text{ in (2)}$$
$$1000 = 750t$$
$$\frac{4}{3} = t$$

Check. If $t = \frac{4}{3}$, then $420 \cdot \frac{4}{3} = 560$, the distance traveled by the faster airplane. Also, $330 \cdot \frac{4}{3} = 440$, the distance traveled by the slower plane. The sum of the distances is $560 + 440$, or 1000 km. The values check.

State. The airplanes will meet after $\frac{4}{3}$ hr, or $1\frac{1}{3}$ hr.

27. Familiarize. We make a drawing. Note that the plane's speed traveling toward London is $360 + 50$, or 410 mph, and the speed traveling toward New York City is $360 - 50$, or 310 mph. Also, when the plane is d mi from New York City, it is $3458 - d$ mi from London.

New York City   London
310 mph t hours t hours   410 mph

|——————— 3458 mi ———————|

|——— d ———|——— 3458 mi $-d$ ———|

Organize the information in a table.

	Distance	Rate	Time
Toward NYC	d	310	t
Toward London	$3458 - d$	410	t

Translate. Using $d = rt$ in each row of the table, we get a system of equations:

$$d = 310t, \quad (1)$$
$$3458 - d = 410t \quad (2)$$

Solve. We solve the system of equations.

$$3458 - 310t = 410t \quad \text{Using substitution}$$
$$3458 = 720t$$
$$4.8028 \approx t$$

Substitute 4.8028 for t in (1).

$$d \approx 310(4.8028) \approx 1489$$

Check. If the plane is 1489 mi from New York City, it can return to New York City, flying at 310 mph, in $1489/310 \approx$ 4.8 hr. If the plane is $3458 - 1489$, or 1969 mi from London, it can fly to London, traveling at 410 mph, in $1969/410 \approx$ 4.8 hr. Since the times are the same, the answer checks.

State. The point of no return is about 1489 mi from New York City.

29. Discussion and Writing Exercise

31. $f(x) = 4x - 7$
$$f(0) = 4 \cdot 0 - 7 = 0 - 7 = -7$$

33. $f(x) = 4x - 7$
$$f(1) = 4 \cdot 1 - 7 = 4 - 7 = -3$$

35. $f(x) = 4x - 7$
$$f(-2) = 4(-2) - 7 = -8 - 7 = -15$$

37. $f(x) = 4x - 7$
$$f(-4) = 4(-4) - 7 = -16 - 7 = -23$$

39. $f(x) = 4x - 7$
$$f\left(\frac{3}{4}\right) = 4 \cdot \frac{3}{4} - 7 = 3 - 7 = -4$$

41. $f(x) = 4x - 7$
$$f(-3h) = 4(-3h) - 7 = -12h - 7$$

43. *Familiarize*. Let x = the amount of the original solution that remains after some of the original solution is drained and replaced with pure antifreeze. Let y = the amount of the original solution that is drained and replaced with pure antifreeze.

Translate. We organize the information in a table. Keep in mind that the table contains information regarding the solution *after* some of the original solution is drained and replaced with pure antifreeze.

	Original Solution	Pure Anti-freeze	New Mixture
Amount of solution	x	y	16 L
Percent of antifreeze	30%	100%	50%
Amount of antifreeze in solution	$0.3x$	$1 \cdot y$, or y	$0.5(16)$, or 8

The "Amount of solution" row gives us one equation:
$x + y = 16$

The last row gives us a second equation:
$0.3x + y = 8$

After clearing the decimal we have the following system of equations:

$$x + y = 16, \quad (1)$$
$$3x + 10y = 80 \quad (2)$$

Solve. We use the elimination method.

$$\begin{array}{rl} -3x - 3y = -48 & \text{Multiplying (1) by } -3 \\ \underline{3x + 10y = 80} & \\ 7y = 32 & \\ y = \dfrac{32}{7}, \text{ or } 4\dfrac{4}{7} & \end{array}$$

Although the problem only asks for the amount of pure antifreeze added, we will also find x in order to check.

$$x + 4\frac{4}{7} = 16 \quad \text{Substituting } 4\frac{4}{7} \text{ for } y \text{ in (1)}$$
$$x = 11\frac{3}{7}$$

Check. Total amount of new mixture: $11\frac{3}{7} + 4\frac{4}{7} = $
16 L

Amount of antifreeze in new mixture:
$$0.3\left(11\frac{3}{7}\right) + 4\frac{4}{7} = \frac{3}{10} \cdot \frac{80}{7} + \frac{32}{7} = \frac{56}{7} = 8 \text{ L}$$
The numbers check.

State. Michelle should drain $4\frac{4}{7}$ L of the original solution and replace it with pure antifreeze.

45. *Familiarize*. Let x and y represent the number of city miles and highway miles that were driven, respectively. Then in city driving, $\frac{x}{18}$ gallons of gasoline are used; in highway driving, $\frac{y}{24}$ gallons are used.

Translate. We organize the information in a table.

Type of driving	City	Highway	Total
Number of miles	x	y	465
Gallons of gasoline used	$\frac{x}{18}$	$\frac{y}{24}$	23

The first row of the table gives us one equation:
$x + y = 465$

The second row gives us another equation:
$$\frac{x}{18} + \frac{y}{24} = 23$$

After clearing fractions, we have the following system of equations:

$$x + y = 465, \quad (1)$$
$$24x + 18y = 9936 \quad (2)$$

Solve. We solve the system of equations using the elimination method.

$$\begin{array}{rl} -18x - 18y = -8370 & \text{Multiplying (1) by } -18 \\ \underline{24x + 18y = 9936} & \\ 6x = 1566 & \\ x = 261 & \end{array}$$

Now substitute 261 for x in Equation (1) and solve for y.

$$261 + y = 465$$
$$y = 204$$

Check. The total mileage is $261 + 204$, or 465. In 216 city miles, $261/18$, or 14.5 gal of gasoline are used; in 204 highway miles, $204/24$, or 8.5 gal are used. Then a total of $14.5 + 8.5$ or 23 gal of gasoline are used. The answer checks.

State. 261 miles were driven in the city, and 204 miles were driven on the highway.

47. *Familiarize*. Let x = the number of gallons of pure brown and y = the number of gallons of neutral stain that should be added to the original 0.5 gal. Note that a total of 1 gal of stain needs to be added to bring the amount of stain up to 1.5 gal. The original 0.5 gal of stain contains $20\%(0.5 \text{ gal})$, or $0.2(0.5 \text{ gal}) = 0.1$ gal of brown stain. The final solution contains $60\%(1.5 \text{ gal})$, or $0.6(1.5 \text{ gal}) = 0.9$ gal of brown stain. This is composed of the original 0.1 gal and the x gal that are added.

Translate.

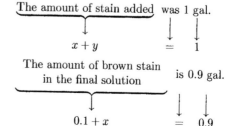

We have a system of equations.

$$x + y = 1, \quad (1)$$
$$0.1 + x = 0.9 \quad (2)$$

Carry out. First we solve (2) for x.

$$0.1 + x = 0.9$$
$$x = 0.8$$

Then substitute 0.8 for x in (1) and solve for y.

$$0.8 + y = 1$$
$$y = 0.2$$

Check. Total amount of stain: $0.5 + 0.8 + 0.2 = 1.5$ gal

Total amount of brown stain: $0.1 + 0.8 = 0.9$ gal

Total amount of neutral stain: $0.8(0.5) + 0.2 = 0.4 + 0.2 = 0.6$ gal $= 0.4(1.5$ gal)

The answer checks.

State. 0.8 gal of pure brown and 0.2 gal of neutral stain should be added.

Exercise Set 13.5

1.
$$x + y + z = 2, \quad (1)$$
$$2x - y + 5z = -5, \quad (2)$$
$$-x + 2y + 2z = 1 \quad (3)$$

Add Equations (1) and (2) to eliminate y:

$$
\begin{array}{l}
x + y + z = 2 \quad (1) \\
\underline{2x - y + 5z = -5 \quad (2)} \\
3x + 6z = -3 \quad (4) \quad \text{Adding}
\end{array}
$$

Use a different pair of equations and eliminate y:

$$
\begin{array}{l}
4x - 2y + 10z = -10 \quad \text{Multiplying (2) by 2} \\
\underline{-x + 2y + 2z = 1 \quad (3)} \\
3x + 12z = -9 \quad (5) \quad \text{Adding}
\end{array}
$$

Now solve the system of Equations (4) and (5).

$$3x + 6z = -3 \quad (4)$$
$$3x + 12z = -9 \quad (5)$$

$$
\begin{array}{l}
-3x - 6z = 3 \quad \text{Multiplying (4) by } -1 \\
\underline{3x + 12z = -9 \quad (5)} \\
 6z = -6 \quad \text{Adding} \\
 z = -1
\end{array}
$$

$$3x + 6(-1) = -3 \quad \text{Substituting } -1 \text{ for } z \text{ in (4)}$$
$$3x - 6 = -3$$
$$3x = 3$$
$$x = 1$$

$$1 + y + (-1) = 2 \quad \text{Substituting 1 for } x \text{ and } -1$$
$$ \text{for } z \text{ in (1)}$$
$$y = 2 \quad \text{Simplifying}$$

We obtain $(1, 2, -1)$. This checks, so it is the solution.

3.
$$2x - y + z = 5, \quad (1)$$
$$6x + 3y - 2z = 10, \quad (2)$$
$$x - 2y + 3z = 5 \quad (3)$$

We start by eliminating z from two different pairs of equations.

$$
\begin{array}{l}
4x - 2y + 2z = 10 \quad \text{Multiplying (1) by 2} \\
\underline{6x + 3y - 2z = 10 \quad (2)} \\
10x + y = 20 \quad (4) \quad \text{Adding}
\end{array}
$$

$$
\begin{array}{l}
-6x + 3y - 3z = -15 \quad \text{Multiplying (1) by } -3 \\
\underline{x - 2y + 3z = 5 \quad (3)} \\
-5x + y = -10 \quad (5) \quad \text{Adding}
\end{array}
$$

Now solve the system of Equations (4) and (5).

$$10x + y = 20 \quad (4)$$
$$
\begin{array}{l}
\underline{5x - y = 10 \quad \text{Multiplying (5) by } -1} \\
15x = 30 \quad \text{Adding} \\
x = 2
\end{array}
$$

$$10 \cdot 2 + y = 20 \quad \text{Substituting 2 for } x \text{ in (4)}$$
$$20 + y = 20$$
$$y = 0$$

$$2 \cdot 2 - 0 + z = 5 \quad \text{Substituting 2 for } x \text{ and 0}$$
$$ \text{for } y \text{ in (1)}$$
$$4 + z = 5$$
$$z = 1$$

We obtain $(2, 0, 1)$. This checks, so it is the solution.

5.
$$2x - 3y + z = 5, \quad (1)$$
$$x + 3y + 8z = 22, \quad (2)$$
$$3x - y + 2z = 12 \quad (3)$$

We start by eliminating y from two different pairs of equations.

$$
\begin{array}{l}
2x - 3y + z = 5 \quad (1) \\
\underline{x + 3y + 8z = 22 \quad (2)} \\
3x + 9z = 27 \quad (4) \quad \text{Adding}
\end{array}
$$

$$
\begin{array}{l}
x + 3y + 8z = 22 \quad (2) \\
\underline{9x - 3y + 6z = 36 \quad \text{Multiplying (3) by 3}} \\
10x + 14z = 58 \quad (5) \quad \text{Adding}
\end{array}
$$

Solve the system of Equations (4) and (5).

$$3x + 9z = 27 \quad (4)$$
$$10x + 14z = 58 \quad (5)$$

$$
\begin{array}{l}
30x + 90z = 270 \quad \text{Multiplying (4) by 10} \\
\underline{-30x - 42z = -174 \quad \text{Multiplying (5) by } -3} \\
 48z = 96 \quad \text{Adding} \\
 z = 2
\end{array}
$$

$$3x + 9 \cdot 2 = 27 \quad \text{Substituting 2 for } z \text{ in (4)}$$
$$3x + 18 = 27$$
$$3x = 9$$
$$x = 3$$

$$2 \cdot 3 - 3y + 2 = 5 \qquad \text{Substituting 3 for } x \text{ and 2}$$
$$\text{for } z \text{ in (1)}$$
$$-3y + 8 = 5$$
$$-3y = -3$$
$$y = 1$$

We obtain $(3, 1, 2)$. This checks, so it is the solution.

7. $3a - 2b + 7c = 13, \quad (1)$
$a + 8b - 6c = -47, \quad (2)$
$7a - 9b - 9c = -3 \quad (3)$

We start by eliminating a from two different pairs of equations.

$$\begin{array}{ll} 3a - 2b + 7c = 13 & (1) \\ -3a - 24b + 18c = 141 & \text{Multiplying (2) by } -3 \\ \hline - 26b + 25c = 154 & (4) \quad \text{Adding} \end{array}$$

$$\begin{array}{ll} -7a - 56b + 42c = 329 & \text{Multiplying (2) by } -7 \\ 7a - 9b - 9c = -3 & (3) \\ \hline - 65b + 33c = 326 & (5) \quad \text{Adding} \end{array}$$

Now solve the system of Equations (4) and (5).

$$-26b + 25c = 154 \quad (4)$$
$$-65b + 33c = 326 \quad (5)$$

$$\begin{array}{ll} -130b + 125c = 770 & \text{Multiplying (4) by 5} \\ 130b - 66c = -652 & \text{Multiplying (5) by } -2 \\ \hline 59c = 118 \\ c = 2 \end{array}$$

$$-26b + 25 \cdot 2 = 154 \qquad \text{Substituting 2 for } c \text{ in (4)}$$
$$-26b + 50 = 154$$
$$-26b = 104$$
$$b = -4$$

$$a + 8(-4) - 6(2) = -47 \qquad \text{Substituting } -4 \text{ for } b \text{ and}$$
$$\text{2 for } c \text{ in (2)}$$
$$a - 32 - 12 = -47$$
$$a - 44 = -47$$
$$a = -3$$

We obtain $(-3, -4, 2)$. This checks, so it is the solution.

9. $2x + 3y + z = 17, \quad (1)$
$x - 3y + 2z = -8, \quad (2)$
$5x - 2y + 3z = 5 \quad (3)$

We start by eliminating y from two different pairs of equations.

$$\begin{array}{ll} 2x + 3y + z = 17 & (1) \\ x - 3y + 2z = -8 & (2) \\ \hline 3x + 3z = 9 & (4) \quad \text{Adding} \end{array}$$

$$\begin{array}{ll} 4x + 6y + 2z = 34 & \text{Multiplying (1) by 2} \\ 15x - 6y + 9z = 15 & \text{Multiplying (3) by 3} \\ \hline 19x + 11z = 49 & (5) \quad \text{Adding} \end{array}$$

Now solve the system of Equations (4) and (5).

$$3x + 3z = 9 \quad (4)$$
$$19x + 11z = 49 \quad (5)$$

$$\begin{array}{ll} 33x + 33z = 99 & \text{Multiplying (4) by 11} \\ -57x - 33z = -147 & \text{Multiplying (5) by } -3 \\ \hline -24x = -48 \\ x = 2 \end{array}$$

$$3 \cdot 2 + 3z = 9 \qquad \text{Substituting 2 for } x \text{ in (4)}$$
$$6 + 3z = 9$$
$$3z = 3$$
$$z = 1$$

$$2 \cdot 2 + 3y + 1 = 17 \qquad \text{Substituting 2 for } x \text{ and 1 for}$$
$$z \text{ in (1)}$$
$$3y + 5 = 17$$
$$3y = 12$$
$$y = 4$$

We obtain $(2, 4, 1)$. This checks, so it is the solution.

11. $2x + y + z = -2, \quad (1)$
$2x - y + 3z = 6, \quad (2)$
$3x - 5y + 4z = 7 \quad (3)$

We start by eliminating y from two different pairs of equations.

$$\begin{array}{ll} 2x + y + z = -2 & (1) \\ 2x - y + 3z = 6 & (2) \\ \hline 4x + 4z = 4 & (4) \quad \text{Adding} \end{array}$$

$$\begin{array}{ll} 10x + 5y + 5z = -10 & \text{Multiplying (1) by 5} \\ 3x - 5y + 4z = 7 & (3) \\ \hline 13x + 9z = -3 & (5) \quad \text{Adding} \end{array}$$

Now solve the system of Equations (4) and (5).

$$4x + 4z = 4 \quad (4)$$
$$13x + 9z = -3 \quad (5)$$

$$\begin{array}{ll} 36x + 36z = 36 & \text{Multiplying (4) by 9} \\ -52x - 36z = 12 & \text{Multiplying (5) by } -4 \\ \hline -16x = 48 & \text{Adding} \\ x = -3 \end{array}$$

$$4(-3) + 4z = 4 \qquad \text{Substituting } -3 \text{ for } x \text{ in (4)}$$
$$-12 + 4z = 4$$
$$4z = 16$$
$$z = 4$$

$$2(-3) + y + 4 = -2 \qquad \text{Substituting } -3 \text{ for } x \text{ and 4}$$
$$\text{for } z \text{ in (1)}$$
$$y - 2 = -2$$
$$y = 0$$

We obtain $(-3, 0, 4)$. This checks, so it is the solution.

13. $x - y + z = 4,$ (1)

$5x + 2y - 3z = 2,$ (2)

$3x - 7y + 4z = 8$ (3)

We start by eliminating z from two different pairs of equations.

$3x - 3y + 3z = 12$ Multiplying (1) by 3

$\underline{5x + 2y - 3z = 2}$ (2)

$8x - y = 14$ (4)   Adding

$-4x + 4y - 4z = -16$ Multiplying (1) by -4

$\underline{3x - 7y + 4z = 8}$ (3)

$-x - 3y = -8$ (5)   Adding

Now solve the system of Equations (4) and (5).

$8x - y = 14$ (4)

$-x - 3y = -8$ (5)

$8x - y = 14$ (4)

$\underline{-8x - 24y = -64}$ Multiplying (5) by 8

$ - 25y = -50$

$y = 2$

$8x - 2 = 14$ Substituting 2 for y in (4)

$8x = 16$

$x = 2$

$2 - 2 + z = 4$ Substituting 2 for x and 2 for

$ y$ in (1)

$z = 4$

We obtain $(2, 2, 4)$. This checks, so it is the solution.

15. $4x - y - z = 4,$ (1)

$2x + y + z = -1,$ (2)

$6x - 3y - 2z = 3$ (3)

We start by eliminating y from two different pairs of equations.

$4x - y - z = 4$ (1)

$\underline{2x + y + z = -1}$ (2)

$6x = 3$ (4)   Adding

At this point we can either continue by eliminating y from a second pair of equations or we can solve (4) for x and substitute that value in a different pair of the original equations to obtain a system of two equations in two variables. We take the second option.

$6x = 3$ (4)

$x = \dfrac{1}{2}$

Substitute $\dfrac{1}{2}$ for x in (1):

$4\left(\dfrac{1}{2}\right) - y - z = 4$

$2 - y - z = 4$

$-y - z = 2$ (5)

Substitute $\dfrac{1}{2}$ for x in (3):

$6\left(\dfrac{1}{2}\right) - 3y - 2z = 3$

$3 - 3y - 2z = 3$

$-3y - 2z = 0$ (6)

Solve the system of Equations (5) and (6).

$2y + 2z = -4$ Multiplying (5) by -2

$\underline{-3y - 2z = 0}$ (6)

$-y = -4$

$y = 4$

$-4 - z = 2$ Substituting 4 for y in (5)

$-z = 6$

$z = -6$

We obtain $\left(\dfrac{1}{2}, 4, -6\right)$. This checks, so it is the solution.

17. $2r + 3s + 12t = 4,$ (1)

$4r - 6s + 6t = 1,$ (2)

$r + s + t = 1$ (3)

We start by eliminating s from two different pairs of equations.

$4r + 6s + 24t = 8$ Multiplying (1) by 2

$\underline{4r - 6s + 6t = 1}$ (2)

$8r + 30t = 9$ (4)   Adding

$4r - 6s + 6t = 1$ (2)

$\underline{6r + 6s + 6t = 6}$ Multiplying (3) by 6

$10r + 12t = 7$ (5)   Adding

Solve the system of Equations (4) and (5).

$40r + 150t = 45$ Multiplying (4) by 5

$\underline{-40r - 48t = -28}$ Multiplying (5) by -4

$102t = 17$

$t = \dfrac{17}{102}$

$t = \dfrac{1}{6}$

$8r + 30\left(\dfrac{1}{6}\right) = 9$ Substituting $\dfrac{1}{6}$ for t in (4)

$8r + 5 = 9$

$8r = 4$

$r = \dfrac{1}{2}$

$\dfrac{1}{2} + s + \dfrac{1}{6} = 1$ Substituting $\dfrac{1}{2}$ for r and

$\phantom{\dfrac{1}{2} + s + \dfrac{1}{6} = 1 \quad} \dfrac{1}{6}$ for t in (3)

$\phantom{\dfrac{1}{2} + }s + \dfrac{2}{3} = 1$

$\phantom{\dfrac{1}{2} + }s = \dfrac{1}{3}$

We obtain $\left(\dfrac{1}{2}, \dfrac{1}{3}, \dfrac{1}{6}\right)$. This checks, so it is the solution.

19. $4a + 9b \qquad = \quad 8, \quad (1)$

$\qquad 8a \qquad + 6c = -1, \quad (2)$

$\qquad\qquad 6b + 6c = -1 \quad (3)$

We will use the elimination method. Note that there is no c in Equation (1). We will use equations (2) and (3) to obtain another equation with no c terms.

$\qquad 8a \qquad + 6c = -1 \quad (2)$

$\underline{\qquad -6b - 6c = \quad 1 \quad \text{Multiplying (3) by } -1}$

$\qquad 8a - 6b \qquad = \quad 0 \quad (4) \quad \text{Adding}$

Now solve the system of Equations (1) and (4).

$\qquad -8a - 18b = -16 \quad \text{Multiplying (1) by } -2$

$\underline{\qquad 8a - \quad 6b = \quad 0}$

$\qquad\qquad -24b = -16$

$\qquad\qquad b = \dfrac{2}{3}$

$8a - 6\left(\dfrac{2}{3}\right) = 0 \quad \text{Substituting } \dfrac{2}{3} \text{ for } b \text{ in (4)}$

$\qquad 8a - 4 = 0$

$\qquad\qquad 8a = 4$

$\qquad\qquad a = \dfrac{1}{2}$

$8\left(\dfrac{1}{2}\right) + 6c = -1 \quad \text{Substituting } \dfrac{1}{2} \text{ for } a \text{ in (2)}$

$\qquad 4 + 6c = -1$

$\qquad\qquad 6c = -5$

$\qquad\qquad c = -\dfrac{5}{6}$

We obtain $\left(\dfrac{1}{2}, \dfrac{2}{3}, -\dfrac{5}{6}\right)$. This checks, so it is the solution.

21. $\quad x + y + z = 57, \quad (1)$

$\quad -2x + y \qquad = \quad 3, \quad (2)$

$\quad x \qquad - z = \quad 6 \quad (3)$

We will use the substitution method. Solve Equations (2) and (3) for y and z, respectively. Then substitute in Equation (1) to solve for x.

$\qquad -2x + y = 3 \qquad \text{Solving (2) for } y$

$\qquad\qquad y = 2x + 3$

$\qquad x - z = 6 \qquad \text{Solving (3) for } z$

$\qquad\qquad -z = -x + 6$

$\qquad\qquad z = x - 6$

$x + (2x + 3) + (x - 6) = 57 \quad \text{Substituting in (1)}$

$\qquad\qquad 4x - 3 = 57$

$\qquad\qquad 4x = 60$

$\qquad\qquad x = 15$

To find y, substitute 15 for x in $y = 2x + 3$:

$y = 2 \cdot 15 + 3 = 33$

To find z, substitute 15 for x in $z = x - 6$:

$z = 15 - 6 = 9$

We obtain $(15, 33, 9)$. This checks, so it is the solution.

23. $\quad r + \quad s \qquad = \quad 5, \quad (1)$

$\qquad 3s + 2t = -1, \quad (2)$

$\quad 4r \qquad + \quad t = \quad 14 \quad (3)$

We will use the elimination method. Note that there is no t in Equation (1). We will use Equations (2) and (3) to obtain another equation with no t terms.

$\qquad 3s + 2t = -1 \quad (2)$

$\underline{\quad -8r \qquad - 2t = -28 \quad \text{Multiplying (3) by } -2}$

$\quad -8r + 3s \qquad = -29 \quad (4) \quad \text{Adding}$

Now solve the system of Equations (1) and (4).

$\qquad r + \quad s = \quad 5 \quad (1)$

$\qquad -8r + 3s = -29 \quad (4)$

$\qquad 8r + \quad 8s = \quad 40 \quad \text{Multiplying (1) by 8}$

$\underline{\quad -8r + \quad 3s = -29 \quad (4)}$

$\qquad\qquad 11s = \quad 11 \quad \text{Adding}$

$\qquad\qquad s = 1$

$r + 1 = 5 \quad \text{Substituting 1 for } s \text{ in (1)}$

$\qquad r = 4$

$4 \cdot 4 + t = 14 \quad \text{Substituting 4 for } r \text{ in (3)}$

$\qquad 16 + t = 14$

$\qquad\qquad t = -2$

We obtain $(4, 1, -2)$. This checks, so it is the solution.

25. Discussion and Writing Exercise

27. $\quad F = 3ab$

$\qquad \dfrac{F}{3b} = a \quad \text{Dividing by } 3b$

29. $\qquad F = \dfrac{1}{2}t(c - d)$

$\qquad 2F = t(c - d) \quad \text{Multiplying by 2}$

$\qquad 2F = tc - td \quad \text{Removing parentheses}$

$\qquad 2F + td = tc \qquad \text{Adding } td$

$\qquad \dfrac{2F + td}{t} = c, \text{ or} \qquad \text{Dividing by } t$

$\qquad \dfrac{2F}{t} + d = c$

31. $\quad Ax - By = c$

$\qquad Ax = By + c \qquad \text{Adding } By$

$\qquad Ax - c = By \qquad \text{Subtracting } c$

$\qquad \dfrac{Ax - c}{B} = y \qquad \text{Dividing by } B$

33. $y = -\dfrac{2}{3}x - \dfrac{5}{4}$

The equation is in slope-intercept form, $y = mx + b$. The slope is $-\dfrac{2}{3}$, and the y-intercept is $\left(0, -\dfrac{5}{4}\right)$.

35. $2x - 5y = 10$

$$-5y = -2x + 10$$

$$-\frac{1}{5}(-5y) = -\frac{1}{5}(-2x + 10)$$

$$y = \frac{2}{5}x - 2$$

The equation is now in slope-intercept form, $y = mx + b$. The slope is $\frac{2}{5}$, and the y-intercept is $(0, -2)$.

37. $w + x + y + z = 2,$ (1)

$w + 2x + 2y + 4z = 1,$ (3)

$w - x + y + z = 6,$ (3)

$w - 3x - y + z = 2$ (4)

Start by eliminating w from three different pairs of equations.

$w + x + y + z = 2$ (1)

$\underline{-w - 2x - 2y - 4z = -1}$ Multiplying (2) by -1

$ -x - y - 3z = 1$ (5) Adding

$w + x + y + z = 2$ (1)

$\underline{-w + x - y - z = -6}$ Multiplying (3) by -1

$ 2x = -4$ (6) Adding

$w + x + y + z = 2$ (1)

$\underline{-w + 3x + y - z = -2}$ Multiplying (4) by -1

$ 4x + 2y = 0$ (7) Adding

We can solve (6) for x:

$$2x = -4$$

$$x = -2$$

Substitute -2 for x in (7):

$$4(-2) + 2y = 0$$

$$-8 + 2y = 0$$

$$2y = 8$$

$$y = 4$$

Substitute -2 for x and 4 for y in (5):

$$-(-2) - 4 - 3z = 1$$

$$-2 - 3z = 1$$

$$-3z = 3$$

$$z = -1$$

Substitute -2 for x, 4 for y, and -1 for z in (1):

$$w - 2 + 4 - 1 = 2$$

$$w + 1 = 2$$

$$w = 1$$

We obtain $(1, -2, 4, -1)$. This checks, so it is the solution.

Exercise Set 13.6

1. *Familiarize*. Let $x =$ the number of 10-oz cups, $y =$ the number of 14-oz cups, and $z =$ the number of 20-oz cups that Reggie filled. Note that five 96-oz pots contain $5 \cdot 96$ oz, or 480 oz of coffee. Also, x 10-oz cups contain a total of $10x$ oz of coffee and bring in $\$1.09x$, y 14-oz cups contain $14y$ oz and bring in $\$1.29y$, and z 20-oz cups contain $20z$ oz and bring in $\$1.49z$.

***Translate*.**

The total number of coffees served was 34.

$$\underbrace{}$$
$$x + y + z \qquad\qquad = \quad 34$$

The total amount of coffee served was 480 oz.

$$\underbrace{} \qquad \underbrace{}$$
$$10x + 14y + 20z \qquad = \qquad 480$$

The total amount collected was $\$44.06$.

$$\underbrace{}$$
$$1.09x + 1.29y + 1.49z \quad = \quad 44.06$$

Now we have a system of equations.

$$x + y + z = 34,$$

$$10x + 14y + 20z = 480,$$

$$1.09x + 1.29y + 1.49z = 43.46$$

***Solve*.** Solving the system we get $(8, 20, 6)$.

***Check*.** The total number of coffees served was $8 + 20 + 6$, or 34. The total amount of coffee served was $10 \cdot 8 + 14 \cdot 20 + 20 \cdot 6 = 80 + 280 + 120 = 480$ oz. The total amount collected was $\$1.09(8) + \$1.29(20) + \$1.49(6) = \$8.72 + \$25.80 + \$8.94 = \$43.46$. The numbers check.

***State*.** Reggie filled 8 10-oz cups, 20 14-oz cups, and 6 20-oz cups.

3. *Familiarize*. We first make a drawing.

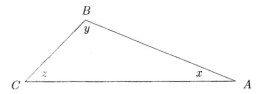

We let x, y, and z represent the measures of angles A, B, and C, respectively. The measures of the angles of a triangle add up to $180°$.

***Translate*.**

The sum of the measures is $180°$.

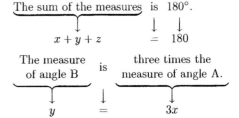

$$x + y + z \qquad = \quad 180$$

The measure of angle B is three times the measure of angle A.

$$y \qquad\qquad = \qquad\qquad 3x$$

The measure of angle C 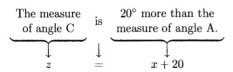 is 20° more than the measure of angle A.

$$z = x + 20$$

We now have a system of equations.

$$x + y + z = 180,$$
$$y = 3x,$$
$$z = x + 20$$

Solve. Solving the system we get $(32, 96, 52)$.

Check. The sum of the measures is $32° + 96° + 52°$, or $180°$. Three times the measure of angle A is $3 \cdot 32°$, or $96°$, the measure of angle B. $20°$ more than the measure of angle A is $32° + 20°$, or $52°$, the measure of angle C. The numbers check.

State. The measures of angles A, B, and C are $32°$, $96°$, and $52°$, respectively.

5. **Familiarize.** Let $x =$ the cost of automatic transmission, $y =$ the cost of power door locks, and $z =$ the cost of air conditioning. The prices of the options are added to the basic price of $12,685.

Translate.

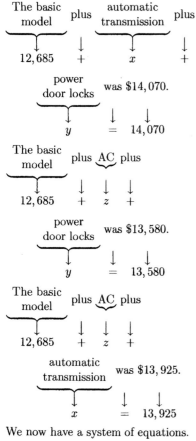

We now have a system of equations.

$$12,685 + x + y = 14,070,$$
$$12,685 + z + y = 13,580,$$
$$12,685 + z + x = 13,925$$

Solve. Solving the system we get $(865, 520, 375)$.

Check. The basic model with automatic transmission and power door locks costs $12,685 + $865 + 520, or $14,070. The basic model with AC and power door locks costs $12,685 + $375 + 520, or $13,580. The basic model with AC and automatic transmission costs $12,685 + $375 + 865, or $13,925. The numbers check.

State. Automatic transmission costs $865, power door locks cost $520, and AC costs $375.

7. **Familiarize.** It helps to organize the information in a table. We let x, y, and z represent the weekly productions of the individual machines.

Machines Working	A	B	C
Weekly Production	x	y	z

Machines Working	A & B	B & C	A, B, & C
Weekly Production	3400	4200	5700

Translate. From the table, we obtain three equations.

$$x + y + z = 5700 \quad \text{(All three machines working)}$$
$$x + y = 3400 \quad \text{(A and B working)}$$
$$y + z = 4200 \quad \text{(B and C working)}$$

Solve. Solving the system we get $(1500, 1900, 2300)$.

Check. The sum of the weekly productions of machines A, B & C is $1500 + 1900 + 2300$, or 5700. The sum of the weekly productions of machines A and B is $1500 + 1900$, or 3400. The sum of the weekly productions of machines B and C is $1900 + 2300$, or 4200. The numbers check.

State. In a week Machine A can polish 1500 lenses, Machine B can polish 1900 lenses, and Machine C can polish 2300 lenses.

9. **Familiarize.** Let $x =$ the amount invested in the first fund, $y =$ the amount invested in the second fund, and $z =$ the amount invested in the third fund. Then the earnings from the investments were $0.1x$, $0.06y$, and $0.15z$.

Translate.

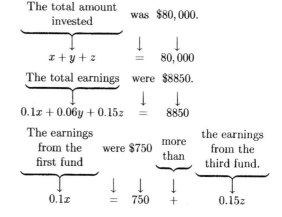

Now we have a system of equations.

$$x + y + z = 80,000$$
$$0.1x + 0.06y + 0.15z = 8850,$$
$$0.1x = 750 + 0.15z$$

Solve. Solving the system we get
$(45,000, 10,000, 25,000)$.

Check. The total investment was $\$45,000+$
$\$10,000 + \$25,000$, or $\$80,000$. The total earnings were
$0.1(\$45,000) + 0.06(10,000) + 0.15(25,000) = \$4500 + \$600 + \$3750 = \$8850$. The earnings from the first fund, $\$4500$, were $\$750$ more than the earnings from the second fund, $\$3750$.

State. $\$45,000$ was invested in the first fund, $\$10,000$ in the second fund, and $\$25,000$ in the third fund.

11. **Familiarize**. Let x, y, and z represent the number of fraternal twin births for Asian-Americans, African-Americans, and Caucasians in the U.S., respectively, out of every 15,400 births.

Translate. Out of every 15,400 births, we have the following statistics:

$\underbrace{\text{The total number of fraternal twin births}}$ is 739.
$\qquad\qquad x + y + z \qquad\qquad = 739$

$\underbrace{\begin{array}{c}\text{The number of}\\\text{fraternal twin}\\\text{births for}\\\text{Asian-Americans}\end{array}}$ is 185 more than $\underbrace{\begin{array}{c}\text{the number}\\\text{for African-}\\\text{Americans.}\end{array}}$
$\qquad x \qquad\quad = 185 \quad + \qquad y$

$\underbrace{\begin{array}{c}\text{The number of}\\\text{fraternal twin}\\\text{births for}\\\text{Asian-Americans}\end{array}}$ is 231 more than $\underbrace{\begin{array}{c}\text{the}\\\text{number for}\\\text{Caucasians.}\end{array}}$
$\qquad x \qquad\quad = 231 \quad + \qquad z$

We have a system of equations.

$$x + y + z = 739,$$
$$x = 185 + y,$$
$$x = 231 + y$$

Solve. Solving the system we get
$(385, 200, 154)$.

Check. The total of the numbers is 739. Also 385 is 185 more than 200, and it is 231 more than 154.

State. Out of every 15,400 births, there are 385 births of fraternal twins for Asian-Americans, 200 for African-Americans, and 154 for Caucasians.

13. **Familiarize**. Let r = the number of servings of roast beef, p = the number of baked potatoes, and b = the number of servings of broccoli. Then r servings of roast beef contain $300r$ Calories, $20r$ g of protein, and no vitamin C. In p baked potatoes there are $100p$ Calories, $5p$ g of protein, and $20p$ mg of vitamin C. And b servings of broccoli contain $50b$

Calories, $5b$ g of protein, and $100b$ mg of vitamin C. The patient requires 800 Calories, 55 g of protein, and 220 mg of vitamin C.

Translate. Write equations for the total number of calories, the total amount of protein, and the total amount of vitamin C.

$$300r + 100p + 50b = 800 \quad (\text{Calories})$$
$$20r + 5p + 5b = 55 \quad (\text{protein})$$
$$20p + 100b = 220 \quad (\text{vitamin C})$$

We now have a system of equations.

Solve. Solving the system we get $(2, 1, 2)$.

Check. Two servings of roast beef provide 600 Calories, 40 g of protein, and no vitamin C. One baked potato provides 100 Calories, 5 g of protein, and 20 mg of vitamin C. And 2 servings of broccoli provide 100 Calories, 10 g of protein, and 200 mg of vitamin C. Together, then, they provide 800 Calories, 55 g of protein, and 220 mg of vitamin C. The values check.

State. The dietician should prepare 2 servings of roast beef, 1 baked potato, and 2 servings of broccoli.

15. **Familiarize**. Let x, y, and z represent the number of par-3, par-4, and par-5 holes, respectively. Then a par golfer shoots $3x$ on the par-3 holes, $4x$ on the par-4 holes, and $5x$ on the par-5 holes.

Translate.

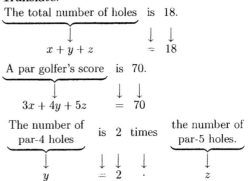

$\underbrace{\text{The total number of holes}}$ is 18.
$\qquad x + y + z \qquad = 18$

$\underbrace{\text{A par golfer's score}}$ is 70.
$\qquad 3x + 4y + 5z \qquad = 70$

$\underbrace{\begin{array}{c}\text{The number of}\\\text{par-4 holes}\end{array}}$ is 2 times $\underbrace{\begin{array}{c}\text{the number of}\\\text{par-5 holes.}\end{array}}$
$\qquad y \qquad = 2 \cdot \qquad z$

We have a system of equations.

$$x + y + z = 18,$$
$$3x + 4y + 5z = 70,$$
$$y = 2z$$

Solve. Solving the system we get $(6, 8, 4)$.

Check. The numbers add up to 18. A par golfer would shoot $3 \cdot 6 + 4 \cdot 8 + 5 \cdot 4$, or 70. The number of par-4 holes, 8, is twice the number of par-5 holes, 4. The numbers check.

State. There are 6 par-3 holes, 8 par-4 holes, and 4 par-5 holes.

17. **Familiarize**. Let x, y, and z represent the number of 2-point field goals, 3-point field goals, and 1-point foul shots made, respectively. The total number of points scored from each of these types of goals is $2x$, $3y$, and z.

Translate.

$\underbrace{\text{The total number of points}}$ was 92.
$\qquad 2x + 3y + z \qquad = 92$

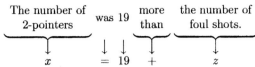

The total number of baskets was 50.

$$x + y + z = 50$$

The number of 2-pointers was 19 more than the number of foul shots.

$$x = 19 + z$$

Now we have a system of equations.

$2x + 3y + z = 92,$

$x + y + z = 50,$

$x = 19 + z$

Solve. Solving the system we get $(32, 5, 13)$.

Check. The total number of points was $2 \cdot 32 + 3 \cdot 5 + 13 = 64 + 15 + 13 = 92$. The number of baskets was $32 + 5 + 13$, or 50. The number of 2-pointers, 32, was 19 more than the number of foul shots, 13. The numbers check.

State. The Knicks made 32 two-point field goals, 5 three-point field goals, and 13 foul shots.

19. Discussion and Writing Exercise

21. Discussion and Writing Exercise

23. The correspondence is not a function, because an input (in fact, both inputs) corresponds to more than one output.

25. We cannot calculate $f(x)$ when the denominator is 0. We set the denominator equal to 0 and solve for x.

$x + 7 = 0$

$x = -7$

The domain is $\{x | x$ is a real number $and \ x \neq -7\}$, or $(-\infty, -7) \cup (-7, \infty)$.

27. Substitute $-\dfrac{3}{5}$ for m and -7 for b in the slope-intercept equation.

$y = mx + b$

$y = -\dfrac{3}{5}x - 7$

29. Familiarize. We first make a drawing with additional labels.

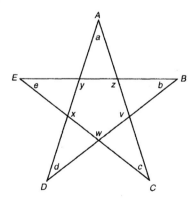

We let a, b, c, d, and e represent the angle measures at the tips of the star. We also label the interior angles of

the pentagon v, w, x, y, and z. We recall the following geometric fact:

The sum of the measures of the interior angles of a polygon of n sides is given by $(n-2)180°$.

Using this fact we know:

1. The sum of the angle measures of a triangle is $(3-2)180°$, or $180°$.

2. The sum of the angle measures of a pentagon is $(5-2)180°$, or $3(180°)$.

Translate. Using fact (1) listed above we obtain a system of 5 equations.

$a + v + d = 180$

$b + w + e = 180$

$c + x + a = 180$

$d + y + b = 180$

$e + z + c = 180$

Solve. Adding we obtain

$2a + 2b + 2c + 2d + 2e + v + w + x + y + z = 5(180)$

$2(a + b + c + d + e) + (v + w + x + y + z) = 5(180)$

Using fact (2) listed above we substitute $3(180)$ for $(v + w + x + y + z)$ and solve for $(a + b + c + d + e)$.

$2(a + b + c + d + e) + 3(180) = 5(180)$

$2(a + b + c + d + e) = 2(180)$

$a + b + c + d + e = 180$

Check. We should repeat the above calculations.

State. The sum of the angle measures at the tips of the star is $180°$.

31. Familiarize. Let $x =$ the one's digit, $y =$ the ten's digit, and $z =$ the hundred's digit. Then the number is represented by $100z + 10y + x$. When the digits are reversed, the resulting number is represented by $100x + 10y + z$.

Translate.

The sum of the digits is 14.

$$x + y + z = 14$$

The ten's digit is 2 more than the one's digit.

$$y = 2 + x$$

The number is the same as the number with the digits reversed.

$$100z + 10y + x = 100x + 10y + z$$

Now we have a system of equations.

$x + y + z = 14,$

$y = 2 + x,$

$100z + 10y + x = 100x + 10y + z$

Solve. Solving the system we get $(4, 6, 4)$.

Check. If the number is 464, then the sum of the digits is $4 + 6 + 4$, or 14. The ten's digit, 6, is 2 more than the one's digit, 4. If the digits are reversed the number is unchanged The result checks.

State. The number is 464.

Exercise Set 13.7

1. $C(x) = 25x + 270,000 \qquad R(x) = 70x$

a) $P(x) = R(x) - C(x)$

$\qquad = 70x - (25x + 270,000)$

$\qquad = 70x - 25x - 270,000$

$\qquad = 45x - 270,000$

b) To find the break-even point we solve the system

$\qquad R(x) = 70x,$

$\qquad C(x) = 25x + 270,000.$

Since both $R(x)$ and $C(x)$ are in dollars and they are equal at the break-even point, we can rewrite the system:

$\qquad d = 70x, \qquad\qquad (1)$

$\qquad d = 25x + 270,000 \quad (2)$

We solve using substitution.

$\qquad 70x = 25x + 270,000$ Substituting $65x$
$\qquad\qquad\qquad\qquad\qquad\qquad$ for d in (2)

$\qquad 45x = 270,000$

$\qquad\qquad x = 6000$

Thus, 6000 units must be produced and sold in order to break even.

The amount taken in is $R(6000) = 70 \cdot 6000 = \$420,000$. Thus, the break-even point is $(6000, \$420,000)$.

3. $C(x) = 10x + 120,000 \qquad R(x) = 60x$

a) $P(x) = R(x) - C(x)$

$\qquad = 60x - (10x + 120,000)$

$\qquad = 60x - 10x - 120,000$

$\qquad = 50x - 120,000$

b) Solve the system

$\qquad R(x) = 60x,$

$\qquad C(x) = 10x + 120,000.$

Since both $R(x)$ and $C(x)$ are in dollars and they are equal at the break-even point, we can rewrite the system:

$\qquad d = 60x, \qquad\qquad (1)$

$\qquad d = 10x + 120,000 \quad (2)$

We solve using substitution.

$\qquad 60x = 10x + 120,000$ Substituting $60x$
$\qquad\qquad\qquad\qquad\qquad\qquad$ for d in (2)

$\qquad 50x = 120,000$

$\qquad\qquad x = 2400$

Thus, 2400 units must be produced and sold in order to break even.

The amount taken in is $R(2400) = 60 \cdot 2400 = \$144,000$. Thus, the break-even point is $(2400, \$144,000)$.

5. $C(x) = 20x + 10,000 \qquad R(x) = 100x$

a) $P(x) = R(x) - C(x)$

$\qquad = 100x - (20x + 10,000)$

$\qquad = 100x - 20x - 10,000$

$\qquad = 80x - 10,000$

b) Solve the system

$\qquad R(x) = 100x,$

$\qquad C(x) = 20x + 10,000.$

Since both $R(x)$ and $C(x)$ are in dollars and they are equal at the break-even point, we can rewrite the system:

$\qquad d = 100x, \qquad\qquad (1)$

$\qquad d = 20x + 10,000 \quad (2)$

We solve using substitution.

$\qquad 100x = 20x + 10,000$ Substituting $100x$
$\qquad\qquad\qquad\qquad\qquad\qquad$ for d in (2)

$\qquad 80x = 10,000$

$\qquad\quad x = 125$

Thus, 125 units must be produced and sold in order to break even.

The amount taken in is $R(125) = 100 \cdot 125 = \$12,500$. Thus, the break-even point is $(125, \$12,500)$.

7. $C(x) = 22x + 16,000 \qquad R(x) = 40x$

a) $P(x) = R(x) - C(x)$

$\qquad = 40x - (22x + 16,000)$

$\qquad = 40x - 22x - 16,000$

$\qquad = 18x - 16,000$

b) Solve the system

$\qquad R(x) = 40x,$

$\qquad C(x) = 22x + 16,000.$

Since both $R(x)$ and $C(x)$ are in dollars and they are equal at the break-even point, we can rewrite the system:

$\qquad d = 40x, \qquad\qquad (1)$

$\qquad d = 22x + 16,000 \quad (2)$

We solve using substitution.

$\qquad 40x = 22x + 16,000$ Substituting $40x$ for
$\qquad\qquad\qquad\qquad\qquad\qquad$ d in (2)

$\qquad 18x = 16,000$

$\qquad\quad x \approx 889$ units

Thus, 889 units must be produced and sold in order to break even.

The amount taken in is $R(889) = 40 \cdot 889 = \$35,560$. Thus, the break-even point is $(889, \$35,560)$.

9. $C(x) = 50x + 195,000 \qquad R(x) = 125x$

a) $P(x) = R(x) - C(x)$

$= 125x - (50x + 195,000)$

$= 125x - 50x - 195,000$

$= 75x - 195,000$

b) Solve the system

$R(x) = 125x,$

$C(x) = 50x + 195,000.$

Since $R(x) = C(x)$ at the break-even point, we can rewrite the system:

$R(x) = 125x, \qquad (1)$

$R(x) = 50x + 195,000 \quad (2)$

We solve using substitution.

$125x = 50x + 195,000$ Substituting $125x$ for $R(x)$ in (2)

$75x = 195,000$

$x = 2600$

To break even 2600 units must be produced and sold.

The amount taken in is $R(2600) = 125 \cdot 2600 = \$325,000$. Thus, the break-even point is $(2600, \$325,000)$.

11. a) $C(x) =$ Fixed costs + Variable costs

$C(x) = 22,500 + 40x,$

where x is the number of lamps produced.

b) Each lamp sells for \$85. The total revenue is 85 times the number of lamps sold. We assume that all lamps produced are sold.

$R(x) = 85x$

c) $P(x) = R(x) - C(x)$

$P(x) = 85x - (22,500 + 40x)$

$= 85x - 22,500 - 40x$

$= 45x - 22,500$

d) $P(3000) = 45(3000) - 22,500$

$= 135,000 - 22,500$

$= 112,500$

The company will realize a profit of \$112,500 when 3000 lamps are produced and sold.

$P(400) = 45(400) - 22,500$

$= 18,000 - 22,500$

$= -4500$

The company will realize a \$4500 loss when 400 lamps are produced and sold.

e) Solve the system

$R(x) = 85x,$

$C(x) = 22,500 + 40x.$

Since both $R(x)$ and $C(x)$ are in dollars and they are equal at the break-even point, we can rewrite the system:

$d = 85x, \qquad (1)$

$d = 22,500 + 40x \quad (2)$

We solve using substitution.

$85x = 22,500 + 40x$ Substituting $85x$ for d in (2)

$45x = 22,500$

$x = 500$

The firm will break even if it produces and sells 500 lamps and takes in a total of $R(500) = 85 \cdot 500 = \$42,500$ in revenue. Thus, the break-even point is $(500, \$42,500)$.

13. a) $C(x) =$ Fixed costs + Variable costs

$C(x) = 16,404 + 6x,$

where x is the number of caps produced, in dozens.

b) Each dozen caps sell for \$18. The total revenue is 18 times the number of caps sold, in dozens. We assume that all caps produced are sold.

$R(x) = 18x$

c) $P(x) = R(x) - C(x)$

$P(x) = 18x - (16,404 + 6x)$

$= 18x - 16,404 - 6x$

$= 12x - 16,404$

d) $P(3000) = 12(3000) - 16,404$

$= 36,000 - 16,404$

$= 19,596$

The company will realize a profit of \$19,596 when 3000 dozen caps are produced and sold.

$P(1000) = 12(1000) - 16,404$

$= 12,000 - 16,404$

$= -4404$

The company will realize a \$4404 loss when 1000 dozen caps are produced and sold.

e) Solve the system

$R(x) = 18x,$

$C(x) = 16,404 + 6x.$

Since both $R(x)$ and $C(x)$ are in dollars and they are equal at the break-even point, we can rewrite the system:

$d = 18x, \qquad (1)$

$d = 16,404 + 6x \quad (2)$

We solve using substitution.

$18x = 16,404 + 6x$ Substituting $18x$ for d in (2)

$12x = 16,404$

$x = 1367$

The firm will break even if it produces and sells 1367 dozen caps and takes in a total of $R(1367) = 18 \cdot 1367 = \$24,606$ in revenue. Thus, the break-even point is (1367,$24,606).

15. $D(p) = 1000 - 10p,$

$\quad S(p) = 230 + p$

Since both demand and supply are quantities, the system can be rewritten:

$\quad q = 1000 - 10p, \quad (1)$

$\quad q = 230 + p \qquad (2)$

Substitute $1000 - 10p$ for q in (2) and solve.

$1000 - 10p = 230 + p$

$\qquad 770 = 11p$

$\qquad 70 = p$

The equilibrium price is $70 per unit. To find the equilibrium quantity we substitute $70 into either $D(p)$ or $S(p)$.

$D(70) = 1000 - 10 \cdot 70 = 1000 - 700 = 300$

The equilibrium quantity is 300 units.

The equilibrium point is ($70, 300$).

17. $D(p) = 760 - 13p,$

$\quad S(p) = 430 + 2p$

Rewrite the system:

$\quad q = 760 - 13p, \quad (1)$

$\quad q = 430 + 2p \qquad (2)$

Substitute $760 - 13p$ for q in (2) and solve.

$760 - 13p = 430 + 2p$

$\qquad 330 = 15p$

$\qquad 22 = p$

The equilibrium price is $22 per unit.

To find the equilibrium quantity we substitute $22 into either $D(p)$ or $S(p)$.

$S(22) = 430 + 2(22) = 430 + 44 = 474$

The equilibrium quantity is 474 units.

The equilibrium point is ($22, 474$).

19. $D(p) = 7500 - 25p,$

$\quad S(p) = 6000 + 5p$

Rewrite the system:

$\quad q = 7500 - 25p, \quad (1)$

$\quad q = 6000 + 5p \qquad (2)$

Substitute $7500 - 25p$ for q in (2) and solve.

$7500 - 25p = 6000 + 5p$

$\qquad 1500 = 30p$

$\qquad 50 = p$

The equilibrium price is $50 per unit.

To find the equilibrium quantity we substitute $50 into either $D(p)$ or $S(p)$.

$D(50) = 7500 - 25(50) = 7500 - 1250 = 6250$

The equilibrium quantity is 6250 units.

The equilibrium point is ($50, 6250$).

21. $D(p) = 1600 - 53p,$

$\quad S(p) = 320 + 75p$

Rewrite the system:

$\quad q = 1600 - 53p, \quad (1)$

$\quad q = 320 + 75p \qquad (2)$

Substitute $1600 - 53p$ for q in (2) and solve.

$1600 - 53p = 320 + 75p$

$\qquad 1280 = 128p$

$\qquad 10 = p$

The equilibrium price is $10 per unit.

To find the equilibrium quantity we substitute $10 into either $D(p)$ or $S(p)$.

$S(10) = 320 + 75(10) = 320 + 750 = 1070$

The equilibrium quantity is 1070 units.

The equilibrium point is ($10, 1070$).

23. Discussion and Writing Exercise

25. $5y - 3x = 8$

$\qquad 5y = 3x + 8$

$\quad \dfrac{1}{5} \cdot 5y = \dfrac{1}{5}(3x + 8)$

$\qquad\quad y = \dfrac{3}{5}x + \dfrac{8}{5}$

The equation is now in slope-intercept form, $y = mx + b$

The slope is $\dfrac{3}{5}$, and the y-intercept is $\left(0, \dfrac{8}{5}\right)$.

27. $\qquad 2y = 3.4x + 98$

$\quad \dfrac{1}{2} \cdot 2y = \dfrac{1}{2}(3.4x + 98)$

$\qquad\quad y = 1.7x + 49$

The equation is now in slope-intercept form, $y = mx + b$.

The slope is 1.7, and the y-intercept is $(0, 49)$.

Chapter 14

More on Inequalities

Exercise Set 14.1

1. $x - 2 \geq 6$

  -4: We substitute and get $-4 - 2 \geq 6$, or $-6 \geq 6$, a false sentence. Therefore, -4 is not a solution.

  0: We substitute and get $0 - 2 \geq 6$, or $-2 \geq 6$, a false sentence. Therefore, 0 is not a solution.

  4: We substitute and get $4 - 2 \geq 6$, or $2 \geq 6$, a false sentence. Therefore, 4 is not a solution.

  8: We substitute and get $8 - 2 \geq 6$, or $6 \geq 6$, a true sentence. Therefore, 8 is a solution.

3. $t - 8 > 2t - 3$

  0: We substitute and get $0 - 8 > 2 \cdot 0 - 3$, or $-8 > -3$, a false sentence. Therefore, 0 is not a solution.

  -8: We substitute and get $-8 - 8 > 2(-8) - 3$, or $-16 > -19$, a true sentence. Therefore, -8 is a solution.

  -9: We substitute and get $-9 - 8 > 2(-9) - 3$, or $-17 > -21$, a true sentence. Therefore, -9 is a solution.

  -3: We substitute and get $-3 - 8 > 2(-3) - 3$, or $-11 > -9$, a false sentence. Therefore, -3 is not a solution.

  $-\dfrac{7}{8}$: We substitute and get $-\dfrac{7}{8} - 8 > 2\left(-\dfrac{7}{8}\right) - 3$, or $-\dfrac{71}{8} > -\dfrac{38}{8}$, a false sentence. Therefore, $-\dfrac{7}{8}$ is not a solution.

5. Interval notation for $\{x | x < 5\}$ is $(-\infty, 5)$.

7. Interval notation for $\{x | -3 \leq x \leq 3\}$ is $[-3, 3]$.

9. Interval notation for the given graph is $(-2, 5)$.

11. Interval notation for the given graph is $(-\sqrt{2}, \infty)$.

13. $\quad x + 2 > 1$

  $x + 2 - 2 > 1 - 2 \qquad$ Subtracting 2

  $\qquad x > -1$

  The solution set is $\{x | x > -1\}$, or $(-1, \infty)$.

15. $\quad y + 3 < 9$

  $y + 3 - 3 < 9 - 3 \qquad$ Subtracting 3

  $\qquad y < 6$

  The solution set is $\{y | y < 6\}$, or $(-\infty, 6)$.

17. $\quad a - 9 \leq -31$

  $a - 9 + 9 \leq -31 + 9 \qquad$ Adding 9

  $\qquad a \leq -22$

  The solution set is $\{a | a \leq -22\}$, or $(-\infty, -22]$.

19. $\quad t + 13 \geq 9$

  $t + 13 - 13 \geq 9 - 13 \qquad$ Subtracting 13

  $\qquad t \geq -4$

  The solution set is $\{t | t \geq -4\}$, or $[-4, \infty)$.

21. $\quad y - 8 > -14$

  $y - 8 + 8 > -14 + 8 \qquad$ Adding 8

  $\qquad y > -6$

  The solution set is $\{y | y > -6\}$, or $(-6, \infty)$.

23. $\quad x - 11 \leq -2$

  $x - 11 + 11 \leq -2 + 11 \qquad$ Adding 11

  $\qquad x \leq 9$

  The solution set is $\{x | x \leq 9\}$, or $(-\infty, 9]$.

25. $8x \geq 24$

  $\dfrac{8x}{8} \geq \dfrac{24}{8} \qquad$ Dividing by 8

  $x \geq 3$

  The solution set is $\{x | x \geq 3\}$, or $[3, \infty)$.

27. $0.3x < -18$

  $\dfrac{0.3x}{0.3} < \dfrac{-18}{0.3} \qquad$ Dividing by 0.3

  $x < -60$

  The solution set is $\{x | x < -60\}$, or $(-\infty, -60)$.

29. $-9x \geq -8.1$

  $\dfrac{-9x}{-9} \leq \dfrac{-8.1}{-9} \qquad$ Dividing by -9 and reversing the inequality symbol

  $x \leq 0.9$

  The solution set is $\{x | x \leq 0.9\}$, or $(-\infty, 0.9]$.

31.
$$-\frac{3}{4}x \ge -\frac{5}{8}$$

$-\frac{4}{3}\left(-\frac{3}{4}x\right) \le -\frac{4}{3}\left(-\frac{5}{8}\right)$ Multiplying by $-\frac{4}{3}$ and reversing the inequality symbol

$$x \le \frac{20}{24}$$

$$x \le \frac{5}{6}$$

The solution set is $\left\{x\middle|x \le \frac{5}{6}\right\}$, or $\left(-\infty, \frac{5}{6}\right]$.

33.
$$2x + 7 < 19$$

$2x + 7 - 7 < 19 - 7$ Subtracting 7

$$2x < 12$$

$\dfrac{2x}{2} < \dfrac{12}{2}$ Dividing by 2

$$x < 6$$

The solution set is $\{x|x < 6\}$, or $(-\infty, 6)$.

35. $5y + 2y \le -21$

$7y \le -21$ Collecting like terms

$\dfrac{7y}{7} \le \dfrac{-21}{7}$ Dividing by 7

$y \le -3$

The solution set is $\{y|y \le -3\}$, or $(-\infty, -3]$.

37.
$$2y - 7 < 5y - 9$$

$-5y + 2y - 7 < -5y + 5y - 9$ Adding $-5y$

$$-3y - 7 < -9$$

$-3y - 7 + 7 < -9 + 7$ Adding 7

$$-3y < -2$$

$\dfrac{-3y}{-3} > \dfrac{-2}{-3}$ Dividing by -3 and reversing the inequality symbol

$$y > \frac{2}{3}$$

The solution set is $\left\{y\middle|y > \frac{2}{3}\right\}$, or $\left(\frac{2}{3}, \infty\right)$.

39.
$$0.4x + 5 \le 1.2x - 4$$

$-1.2x + 0.4x + 5 \le -1.2x + 1.2x - 4$ Adding $-1.2x$

$$-0.8x + 5 \le -4$$

$-0.8x + 5 - 5 \le -4 - 5$ Subtracting 5

$$-0.8x \le -9$$

$\dfrac{-0.8x}{-0.8} \ge \dfrac{-9}{-0.8}$ Dividing by -0.8 and reversing the inequality symbol

$$x \ge 11.25$$

The solution set is $\{x|x \ge 11.25\}$, or $[11.25, \infty)$.

41.
$$5x - \frac{1}{12} \le \frac{5}{12} + 4x$$

$12\left(5x - \dfrac{1}{12}\right) \le 12\left(\dfrac{5}{12} + 4x\right)$ Clearing fractions

$$60x - 1 \le 5 + 48x$$

$60x - 1 - 48x \le 5 + 48x - 48x$ Subtracting $48x$

$$12x - 1 \le 5$$

$12x - 1 + 1 \le 5 + 1$ Adding 1

$$12x \le 6$$

$\dfrac{12x}{12} \le \dfrac{6}{12}$ Dividing by 12

$$x \le \frac{1}{2}$$

The solution set is $\left\{x\middle|x \le \frac{1}{2}\right\}$, or $\left(-\infty, \frac{1}{2}\right]$.

43.
$$4(4y - 3) \ge 9(2y + 7)$$

$16y - 12 \ge 18y + 63$ Removing parentheses

$16y - 12 - 18y \ge 18y + 63 - 18y$ Subtracting $18y$

$$-2y - 12 \ge 63$$

$-2y - 12 + 12 \ge 63 + 12$ Adding 12

$$-2y \ge 75$$

$\dfrac{-2y}{-2} \le \dfrac{75}{-2}$ Dividing by -2 and reversing the inequality symbol

$$y \le -\frac{75}{2}$$

The solution set is $\left\{y\middle|y \le -\frac{75}{2}\right\}$, or $\left(-\infty, -\frac{75}{2}\right]$.

45. $3(2 - 5x) + 2x < 2(4 + 2x)$

$$6 - 15x + 2x < 8 + 4x$$

$6 - 13x < 8 + 4x$ Collecting like terms

$6 - 17x < 8$ Subtracting $4x$

$-17x < 2$ Subtracting 6

$x > -\dfrac{2}{17}$ Dividing by -17 and reversing the inequality symbol

The solution set is $\left\{x\middle|x > -\frac{2}{17}\right\}$, or $\left(-\frac{2}{17}, \infty\right)$.

47. $5[3m - (m + 4)] > -2(m - 4)$

$$5(3m - m - 4) > -2(m - 4)$$

$$5(2m - 4) > -2(m - 4)$$

$$10m - 20 > -2m + 8$$

$12m - 20 > 8$ Adding $2m$

$12m > 28$ Adding 20

$$m > \frac{28}{12}$$

$$m > \frac{7}{3}$$

The solution set is $\left\{m\middle|m > \frac{7}{3}\right\}$, or $\left(\frac{7}{3}, \infty\right)$.

49. $3(r - 6) + 2 > 4(r + 2) - 21$

$\qquad 3r - 18 + 2 > 4r + 8 - 21$

$\qquad 3r - 16 > 4r - 13 \qquad$ Collecting like terms

$\qquad -r - 16 > -13 \qquad$ Subtracting $4r$

$\qquad -r > 3 \qquad$ Adding 16

$\qquad r < -3 \qquad$ Multiplying by -1 and reversing the inequality symbol

The solution set is $\{r | r < -3\}$, or $(-\infty, -3)$.

51. $19 - (2x + 3) \leq 2(x + 3) + x$

$\qquad 19 - 2x - 3 \leq 2x + 6 + x$

$\qquad 16 - 2x \leq 3x + 6 \qquad$ Collecting like terms

$\qquad 16 - 5x \leq 6 \qquad$ Subtracting $3x$

$\qquad -5x \leq -10 \qquad$ Subtracting 16

$\qquad x \geq 2 \qquad$ Dividing by -5 and reversing the inequality symbol

The solution set is $\{x | x \geq 2\}$, or $[2, \infty)$.

53. $\frac{1}{4}(8y + 4) - 17 < -\frac{1}{2}(4y - 8)$

$\qquad 2y + 1 - 17 < -2y + 4$

$\qquad 2y - 16 < -2y + 4 \qquad$ Collecting like terms

$\qquad 4y - 16 < 4 \qquad$ Adding $2y$

$\qquad 4y < 20 \qquad$ Adding 16

$\qquad y < 5$

The solution set is $\{y | y < 5\}$, or $(-\infty, 5)$.

55. $2[4 - 2(3 - x)] - 1 \geq 4[2(4x - 3) + 7] - 25$

$\qquad 2[4 - 6 + 2x] - 1 \geq 4[8x - 6 + 7] - 25$

$\qquad 2[-2 + 2x] - 1 \geq 4[8x + 1] - 25$

$\qquad -4 + 4x - 1 \geq 32x + 4 - 25$

$\qquad 4x - 5 \geq 32x - 21$

$\qquad -28x - 5 \geq -21$

$\qquad -28x \geq -16$

$\qquad x \leq \dfrac{-16}{-28} \qquad$ Dividing by -28 and reversing the inequality symbol

$\qquad x \leq \dfrac{4}{7}$

The solution set is $\left\{x \middle| x \leq \dfrac{4}{7}\right\}$, or $\left(-\infty, \dfrac{4}{7}\right]$.

57. $\frac{4}{5}(7x - 6) < 40$

$\qquad 5 \cdot \frac{4}{5}(7x - 6) < 5 \cdot 40 \qquad$ Clearing the fraction

$\qquad 4(7x - 6) < 200$

$\qquad 28x - 24 < 200$

$\qquad 28x < 224$

$\qquad x < 8$

The solution set is $\{x | x < 8\}$, or $(-\infty, 8)$.

59. $\frac{3}{4}(3 + 2x) + 1 \geq 13$

$\qquad 4\left[\frac{3}{4}(3 + 2x) + 1\right] \geq 4 \cdot 13 \qquad$ Clearing the fraction

$\qquad 3(3 + 2x) + 4 \geq 52$

$\qquad 9 + 6x + 4 \geq 52$

$\qquad 6x + 13 \geq 52$

$\qquad 6x \geq 39$

$\qquad x \geq \dfrac{39}{6}$, or $\dfrac{13}{2}$

The solution set is $\left\{x \middle| x \geq \dfrac{13}{2}\right\}$, or $\left[\dfrac{13}{2}, \infty\right)$.

61. $\frac{3}{4}\left(3x - \frac{1}{2}\right) - \frac{2}{3} < \frac{1}{3}$

$\qquad \dfrac{9x}{4} - \dfrac{3}{8} - \dfrac{2}{3} < \dfrac{1}{3}$

$\qquad 24\left(\dfrac{9x}{4} - \dfrac{3}{8} - \dfrac{2}{3}\right) < 24 \cdot \dfrac{1}{3} \qquad$ Clearing fractions

$\qquad 54x - 9 - 16 < 8$

$\qquad 54x - 25 < 8$

$\qquad 54x < 33$

$\qquad x < \dfrac{33}{54}$, or $\dfrac{11}{18}$

The solution set is $\left\{x \middle| x < \dfrac{11}{18}\right\}$, or $\left(-\infty, \dfrac{11}{18}\right)$.

63. $0.7(3x + 6) \geq 1.1 - (x + 2)$

$\qquad 10[0.7(3x + 6)] \geq 10[1.1 - (x + 2)] \qquad$ Clearing decimals

$\qquad 7(3x + 6) \geq 11 - 10(x + 2)$

$\qquad 21x + 42 \geq 11 - 10x - 20$

$\qquad 21x + 42 \geq -9 - 10x$

$\qquad 31x + 42 \geq -9$

$\qquad 31x \geq -51$

$\qquad x \geq -\dfrac{51}{31}$

The solution set is $\left\{x \middle| x \geq -\dfrac{51}{31}\right\}$, or $\left[-\dfrac{51}{31}, \infty\right)$.

65. $a + (a - 3) \leq (a + 2) - (a + 1)$

$\qquad a + a - 3 \leq a + 2 - a - 1$

$\qquad 2a - 3 \leq 1$

$\qquad 2a \leq 4$

$\qquad a \leq 2$

The solution set is $\{a | a \leq 2\}$, or $(-\infty, 2]$.

67. a) We substitute 214 for W and 73 for H and calculate I.

$\qquad I = \dfrac{704.5W}{H^2}$

$\qquad I = \dfrac{704.5(192)}{73^2}$

$\qquad I \approx 25.38$

b) **Familiarize**. We will use the formula

$I = \dfrac{704.5W}{H^2}$. Recall that $H = 73$ in.

Translate. An index I less than 25 indicates the lowest risk category, so we have

$$I < 25, \text{ or } \frac{704.5W}{H^2} < 25.$$

Now we replace H with 73.

$$\frac{704.5W}{73^2} < 25$$

Solve. We solve the inequality.

$$\frac{704.5W}{73^2} < 25$$

$$\frac{704.5W}{5329} < 25$$

$$704.5W < 133{,}225$$

$$W < 189.1 \qquad \text{Rounding}$$

Check. As a partial check we can substitute a value of W less than 189.1 and a value greater than 189.1 in the formula.

For $W = 189$: $I = \dfrac{704.5(189)}{73^2} \approx 24.97$

For $W = 190$: $I = \dfrac{704.5(190)}{73^2} \approx 25.12$

Since a value of W less than 189.1 gives a body mass index less than 25 and a value of W greater than 189.1 gives an index greater than 25, we have a partial check.

State. Weights of approximately 189.1 lb or less will keep Marv in the lowest risk category. In terms of an inequality we write $\{W | W < \text{(approximately)}\ 189.1 \text{ lb}\}$.

69. Familiarize. List the information in a table. Let $x =$ the score on the fourth test.

Test	Score
Test 1	89
Test 2	92
Test 3	95
Test 4	x
Total	360 or more

Translate. We can easily get an inequality from the table.

$$89 + 92 + 95 + x \geq 360$$

Solve.

$$276 + x \geq 360 \qquad \text{Collecting like terms}$$

$$x \geq 84 \qquad \text{Adding } -276$$

Check. If you get 84 on the fourth test, your total score will be $89 + 92 + 95 + 84$, or 360. Any higher score will also give you an A.

State. A score of 84 or better will give you an A. In terms of an inequality we write $\{x | x \geq 84\}$.

71. Familiarize. Let $v =$ the blue book value of the car. Since the car was not replaced, we know that $9200 does not exceed 80% of the blue book value.

Translate. We write an inequality stating that $9200 does not exceed 80% of the blue book value.

$$9200 \leq 0.8v$$

Solve.

$$9200 \leq 0.8v$$

$$11{,}500 \leq v \qquad \text{Multiplying by } \frac{1}{0.8}$$

Check. We can do a partial check by substituting a value for v greater than 11,500. When $v = 11{,}525$, then 80% of v is $0.8(11{,}525)$, or $9220. This is greater than $9200; that is, $9200 does not exceed this amount. We cannot check all possible values for v, so we stop here.

State. The blue book value of the car is $11,500 or more. In terms of an inequality we write $\{v | v \geq \$11{,}500\}$.

73. Familiarize. We make a table of information.

Plan A: Monthly Income	Plan B: Monthly Income
$400 salary	$610 salary
8% of sales	5% of sales
Total: 400 + 8% of sales	Total: 610 + 5% of sales

Translate. We write an inequality stating that the income from Plan A is greater than the income from Plan B. We let $S =$ gross sales.

$$400 + 8\%S > 610 + 5\%S$$

Solve.

$$400 + 0.08S > 610 + 0.05S$$

$$400 + 0.03S > 610$$

$$0.03S > 210$$

$$S > 7000$$

Check. We calculate for $S = \$7000$ and for some amount greater than $7000 and some amount less than $7000.

Plan A:	Plan B:
$400 + 8\%(7000)$	$610 + 5\%(7000)$
$400 + 0.08(7000)$	$610 + 0.05(7000)$
$400 + 560$	$610 + 350$
$960	$960

When $S = \$7000$, the income from Plan A is equal to the income from Plan B.

Plan A:	Plan B:
$400 + 8\%(8000)$	$610 + 5\%(8000)$
$400 + 0.08(8000)$	$610 + 0.05(8000)$
$400 + 640$	$610 + 400$
$1040	$1010

When $S = \$8000$, the income from Plan A is greater than the income from Plan B.

Plan A: Plan B:

$400 + 8\%(6000)$ $610 + 5\%(6000)$

$400 + 0.08(6000)$ $610 + 0.05(6000)$

$400 + 480$ $610 + 300$

$880 $910

When $S = \$6000$, the income from Plan A is less than the income from Plan B.

State. Plan A is better than Plan B when gross sales are greater than $7000. In terms of an inequality we write $\{S | S > \$7000\}$.

75. *Familiarize.* Let $c =$ the number of checks per month. Then the Anywhere plan will cost $\$0.20c$ per month and the Acu-checking plan will cost $\$2 + \$0.12c$ per month.

Translate. We write an inequality stating that the Acu-checking plan costs less than the Anywhere plan.

$$2 + 0.12c < 0.20c$$

Solve.

$$2 + 0.12c < 0.20c$$
$$2 < 0.08c$$
$$25 < c$$

Check. We can do a partial check by substituting a value for c less than 25 and a value for c greater than 25. When $c = 24$, the Acu-checking plan costs $\$2 + \$0.12(24)$, or $4.88, and the Anywhere plan costs $\$0.20(24)$, or $4.80, so the Anywhere plan is less expensive. When $c = 26$, the Acu-checking plan costs $\$2 + \$0.12(26)$, or $5.12, and the Anywhere plan costs $\$0.20(26)$, or $5.20, so Acu-checking is less expensive. We cannot check all possible values for c, so we stop here.

State. The Acu-checking plan costs less for more than 25 checks per month. In terms of an inequality we write $\{c | c > 25\}$.

77. *Familiarize.* Let $p =$ the number of guests at the wedding party. Then the number of guests in excess of 25 is $p - 25$. The cost under plan A is $30p$, and the cost under plan B is $1300 + 20(p - 25)$.

Translate. We write an inequality stating that plan B costs less than plan A.

$$1300 + 20(p - 25) < 30p$$

Solve. We solve the inequality.

$$1300 + 20(p - 25) < 30p$$
$$1300 + 20p - 500 < 30p$$
$$800 + 20p < 30p$$
$$800 < 10p$$
$$80 < p$$

Check. We calculate for $p = 80$ and for some number less than 80 and some number greater than 80.

Plan A: Plan B:

$30 \cdot 80$ $1300 + 20(80 - 25)$

$2400 $2400

When 80 people attend, plan B costs the same as plan A.

Plan A: Plan B:

$30 \cdot 79$ $1300 + 20(79 - 25)$

$2370 $2380

When fewer than 80 people attend, plan B costs more than plan A.

Plan A: Plan B:

$30 \cdot 81$ $1300 + 20(81 - 25)$

$2430 $2420

When more than 80 people attend, plan B costs less than plan A.

State. For parties of more than 80 people, plan B will cost less. In terms of an inequality we write $\{p | p > 80\}$.

79. *Familiarize.* We want to find the values of s for which $I > 36$.

Translate. $2(s + 10) > 36$

Solve.

$$2s + 20 > 36$$
$$2s > 16$$
$$s > 8$$

Check. For $s = 8$, $I = 2(8 + 10) = 2 \cdot 18 = 36$. Then any U.S. size larger than 8 will give a size larger than 36 in Italy.

State. For U.S. dress sizes larger than 8, dress sizes in Italy will be larger than 36. In terms of an inequality we write $\{s | s > 8\}$.

81. a) Substitute 0 for t and carry out the calculation.

$$N = 0.733(0) + 8.398$$
$$N = 0 + 8.398$$
$$N = 8.398$$

Each person drank 8.398 gal of bottled water in 1990.
Substitute 5 for t and carry out the calculation.

$$N = 0.733(5) + 8.398$$
$$N = 3.665 + 8.398$$
$$N = 12.063$$

Each person drank 12.063 gal of bottled water in 1995.
In 2000, $t = 2000 - 1990 = 10$. Substitute 10 for t and carry out the calculation.

$$N = 0.733(10) + 8.398$$
$$N = 7.33 + 8.398$$
$$N = 15.728$$

Each person will drink 15.728 gal of bottled water in 2000.

b) *Familiarize.* The amount of bottled water that each person drinks t years after 1990 is given by $0.733t + 8.398$.

Translate.

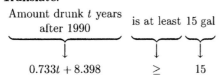

$$0.733t + 8.398 \geq 15$$

Solve. We solve the inequality.

$$0.733t + 8.398 \geq 15$$

$$0.733t \geq 6.602 \qquad \text{Subtracting } 8.398$$

$$t \geq 9 \qquad \text{Rounding}$$

Check. We calculate for 9, for some number less than 9, and for some number greater than 9.

For $t = 9$: $0.733(9) + 8.398 \approx 15$.

For $t = 8$: $0.733(8) + 8.398 \approx 14.3$.

For $t = 10$: $0.733(10) + 8.398 \approx 15.7$.

For a value of t greater than or equal to 9, the number of gallons of bottled water each person drinks is at 15. We cannot check all the possible values of t, so we stop here.

State. Each person will drink at least 15 gal of bottled water 9 or more years after 1990, or for all years after 1999.

83. Discussion and Writing Exercise

85. $(3x - 4)(x + 8) = 3x \cdot x + 3x \cdot 8 - 4 \cdot x - 4 \cdot 8$
$$= 3x^2 + 24x - 4x - 32$$
$$= 3x^2 + 20x - 32$$

87. $(2a - 5)(3a + 11) = 2a \cdot 3a + 2a \cdot 11 - 5 \cdot 3a - 5 \cdot 11$
$$= 6a^2 + 22a - 15a - 55$$
$$= 6a^2 + 7a - 55$$

89. $f(x) = \dfrac{-3}{x + 8}$

Since $\dfrac{-3}{x + 8}$ cannot be calculated when the denominator is 0, we find the x-value that causes $x + 8$ to be 0:

$$x + 8 = 0$$

$$x = -8 \qquad \text{Subtracting 8 on both sides}$$

Thus, -8 is not in the domain of f, while all other real numbers are. The domain of f is

$\{x | x \text{ is a real number } and \ x \neq -8\}$, or

$(-\infty, -8) \cup (-8, \infty)$.

91. $f(x) = |x| - 4$

Since we can calculate $|x| - 4$ for any real number x, the domain is the set of all real numbers.

93. a) *Familiarize.* We will use

$$S = 460 + 94p \quad \text{and} \quad D = 2000 - 60p.$$

Translate. Supply is to exceed demand, so we have

$$S > D, \text{ or}$$

$$460 + 94p > 2000 - 60p.$$

Solve. We solve the inequality.

$$460 + 94p > 2000 - 60p$$

$$460 + 154p > 2000 \qquad \text{Adding } 60p$$

$$154p > 1540 \qquad \text{Subtracting } 460$$

$$p > 10 \qquad \text{Dividing by } 154$$

Check. We calculate for $p = 10$, for some value of p less than 10, and for some value of p greater than 10.

For $p = 10$: $S = 460 + 94 \cdot 10 = 1400$
$D = 2000 - 60 \cdot 10 = 1400$

For $p = 9$: $S = 460 + 94 \cdot 9 = 1306$
$D = 2000 - 60 \cdot 9 = 1460$

For $p = 11$: $S = 460 + 94 \cdot 11 = 1494$
$D = 2000 - 60 \cdot 11 = 1340$

For a value of p greater than 10, supply exceeds demand. We cannot check all possible values of p, so we stop here.

State. Supply exceeds demand for values of p greater than 10. In terms of an inequality we write $\{p | p > 10\}$.

b) We have seen in part (a) that $D = S$ for $p = 10$, $S < D$ for a value of p less than 10, and $S > D$ for a value of p greater than 10. Since we cannot check all possible values of p, we stop here. Supply is less than demand for values of p less than 10. In terms of an inequality we write $\{p | p < 10\}$.

95. True

97. $x + 5 \leq 5 + x$

$$5 \leq 5 \qquad \text{Subtracting } x$$

We get a true inequality, so all real numbers are solutions.

99. $x^2 + 1 > 0$

$x^2 \geq 0$ for all real numbers, so $x^2 + 1 \geq 1 > 0$ for all real numbers.

Exercise Set 14.2

1. $\{9, 10, 11\} \cap \{9, 11, 13\}$

The numbers 9 and 11 are common to the two sets, so the intersection is $\{9, 11\}$.

3. $\{a, b, c, d\} \cap \{b, f, g\}$

Only the letter b is common to the two sets. The intersection is $\{b\}$.

5. $\{9, 10, 11\} \cup \{9, 11, 13\}$

The numbers in either or both sets are 9, 10, 11, and 13, so the union is $\{9, 10, 11, 13\}$.

7. $\{a, b, c, d\} \cup \{b, f, g\}$

The letters in either or both sets are a, b, c, d, f, and g, so the union is $\{a, b, c, d, f, g\}$.

9. $\{2,5,7,9\} \cap \{1,3,4\}$

There are no numbers common to the two sets. The intersection is the empty set, $\emptyset$.

11. $\{3,5,7\} \cup \emptyset$

The numbers in either or both sets are 3, 5, and 7, so the union is $\{3,5,7\}$.

13. $-4 < a$ *and* $a \le 1$ can be written $-4 < a \le 1$. In interval notation we have $(-4,1]$.

The graph is the intersection of the graphs of $a > -4$ and $a \le 1$.

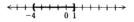

15. We can write $1 < x < 6$ in interval notation as $(1,6)$.

The graph is the intersection of the graphs of $x > 1$ and $x < 6$.

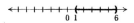

17. $-10 \le 3x + 2$ *and* $3x + 2 < 17$

$\qquad -12 \le 3x \qquad$ *and* $\qquad 3x < 15$

$\qquad -4 \le x \qquad$ *and* $\qquad x < 5$

The solution set is the intersection of the solution sets of the individual inequalities. The numbers common to both sets are those that are greater than or equal to -4 *and* less than 5. Thus the solution set is $\{x| -4 \le x < 5\}$, or $[-4,5)$.

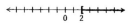

19. $3x + 7 \ge 4 \quad$ *and* $\quad 2x - 5 \ge -1$

$\qquad 3x \ge -3 \;$ *and* $\qquad 2x \ge 4$

$\qquad x \ge -1 \;$ *and* $\qquad x \ge 2$

The solution set is $\{x|x \ge -1\} \cap \{x|x \ge 2\} = \{x|x \ge 2\}$, or $[2, \infty)$.

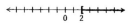

21. $4 - 3x \ge 10 \quad$ *and* $\quad 5x - 2 > 13$

$\qquad -3x \ge 6 \quad$ *and* $\qquad 5x > 15$

$\qquad x \le -2 \;$ *and* $\qquad x > 3$

The solution set is $\{x|x \le -2\} \cap \{x|x > 3\} = \emptyset$.

23. $\qquad -4 < x + 4 < 10$

$-4 - 4 < x + 4 - 4 < 10 - 4 \qquad$ Subtracting 4

$\qquad -8 < x < 6$

The solution set is $\{x| -8 < x < 6\}$, or $(-8,6)$.

25. $\qquad 6 > -x \ge -2$

$\qquad -6 < x \le 2 \qquad$ Multiplying by -1

The solution set is $\{x| -6 < x \le 2\}$, or $(-6,2]$.

27. $\qquad 1 < 3y + 4 \le 19$

$1 - 4 < 3y + 4 - 4 \le 19 - 4 \quad$ Subtracting 4

$\qquad -3 < 3y \le 15$

$\qquad \dfrac{-3}{3} < \dfrac{3y}{3} \le \dfrac{15}{3} \quad$ Dividing by 3

$\qquad -1 < y \le 5$

The solution set is $\{y| -1 < y \le 5\}$, or $(-1,5]$.

29. $\qquad -10 \le 3x - 5 \le -1$

$-10 + 5 \le 3x - 5 + 5 \le -1 + 5 \quad$ Adding 5

$\qquad -5 \le 3x \le 4$

$\qquad \dfrac{-5}{3} \le \dfrac{3x}{3} \le \dfrac{4}{3} \quad$ Dividing by 3

$\qquad -\dfrac{5}{3} \le x \le \dfrac{4}{3}$

The solution set is $\left\{x| -\dfrac{5}{3} \le x \le \dfrac{4}{3}\right\}$, or $\left[-\dfrac{5}{3}, \dfrac{4}{3}\right]$.

31. $\qquad 2 < x + 3 \le 9$

$2 - 3 < x + 3 - 3 \le 9 - 3 \quad$ Subtracting 3

$\qquad -1 < x \le 6$

The solution set is $\{x| -1 < x \le 6\}$, or $(-1,6]$.

33. $\qquad -6 \le 2x - 3 < 6$

$-6 + 3 \le 2x - 3 + 3 < 6 + 3$

$\qquad -3 \le 2x < 9$

$\qquad \dfrac{-3}{2} \le \dfrac{2x}{2} < \dfrac{9}{2}$

$\qquad -\dfrac{3}{2} \le x < \dfrac{9}{2}$

The solution set is $\left\{x| -\dfrac{3}{2} \le x < \dfrac{9}{2}\right\}$, or $\left[-\dfrac{3}{2}, \dfrac{9}{2}\right)$.

35. $\qquad -\dfrac{1}{2} < \dfrac{1}{4}x - 3 \le \dfrac{1}{2}$

$\qquad -\dfrac{1}{2} + 3 < \dfrac{1}{4}x - 3 + 3 \le \dfrac{1}{2} + 3$

$\qquad \dfrac{5}{2} < \dfrac{1}{4}x \le \dfrac{7}{2}$

$\qquad 4 \cdot \dfrac{5}{2} < 4 \cdot \dfrac{1}{4}x \le 4 \cdot \dfrac{7}{2}$

$\qquad 10 < x \le 14$

The solution set is $\{x|10 < x \le 14\}$, or $(10,14]$.

37.
$$-3 < \frac{2x-5}{4} < 8$$
$$4(-3) < 4\left(\frac{2x-5}{4}\right) < 4 \cdot 8$$
$$-12 < 2x - 5 < 32$$
$$-12 + 5 < 2x - 5 + 5 < 32 + 5$$
$$-7 < 2x < 37$$
$$\frac{-7}{2} < \frac{2x}{2} < \frac{37}{2}$$
$$-\frac{7}{2} < x < \frac{37}{2}$$

The solution set is $\left\{x \mid -\frac{7}{2} < x < \frac{37}{2}\right\}$, or $\left(-\frac{7}{2}, \frac{37}{2}\right)$.

39. $x < -2$ *or* $x > 1$ can be written in interval notation as $(-\infty, -2) \cup (1, \infty)$.

The graph is the union of the graphs of $x < -2$ and $x > 1$.

41. $x \leq -3$ *or* $x > 1$ can be written in interval notation as $(-\infty, -3] \cup (1, \infty)$.

The graph is the union of the graphs of $x \leq -3$ and $x > 1$.

43.
$$x + 3 < -2 \qquad or \qquad x + 3 > 2$$
$$x + 3 - 3 < -2 - 3 \quad or \quad x + 3 - 3 > 2 - 3$$
$$x < -5 \qquad or \qquad x > -1$$

The solution set is $\{x \mid x < -5 \text{ or } x > -1\}$, or $(-\infty, -5) \cup (-1, \infty)$.

45.
$$2x - 8 \leq -3 \qquad or \qquad x - 1 \geq 3$$
$$2x - 8 + 8 \leq -3 + 8 \quad or \quad x - 1 + 1 \geq 3 + 1$$
$$2x \leq 5 \qquad or \qquad x \geq 4$$
$$\frac{2x}{2} \leq \frac{5}{2} \qquad or \qquad x \geq 4$$
$$x \leq \frac{5}{2} \qquad or \qquad x \geq 4$$

The solution set is $\left\{x \mid x \leq \frac{5}{2} \text{ or } x \geq 4\right\}$, or $\left(-\infty, \frac{5}{2}\right] \cup [4, \infty)$.

47. $7x + 4 \geq -17$ *or* $6x + 5 \geq -7$
$$7x \geq -21 \quad or \quad 6x \geq -12$$
$$x \geq -3 \quad or \quad x \geq -2$$
The solution set is $\{x \mid x \geq -3\}$, or $[-3, \infty)$.

49.
$$7 > -4x + 5 \qquad or \qquad 10 \leq -4x + 5$$
$$7 - 5 > -4x + 5 - 5 \quad or \quad 10 - 5 \leq -4x + 5 - 5$$
$$2 > -4x \qquad or \qquad 5 \leq -4x$$
$$\frac{2}{-4} < \frac{-4x}{-4} \qquad or \qquad \frac{5}{-4} \geq \frac{-4x}{-4}$$
$$-\frac{1}{2} < x \qquad or \qquad -\frac{5}{4} \geq x$$

The solution set is $\left\{x \mid x \leq -\frac{5}{4} \text{ or } x > -\frac{1}{2}\right\}$, or $\left(-\infty, -\frac{5}{4}\right] \cup \left(-\frac{1}{2}, \infty\right)$.

51. $3x - 7 > -10$ or $5x + 2 \leq 22$
$$3x > -3 \quad or \quad 5x \leq 20$$
$$x > -1 \quad or \quad x \leq 4$$

All real numbers are solutions. In interval notation, the solution set is $(-\infty, \infty)$.

53.
$$-2x - 2 < -6 \qquad or \qquad -2x - 2 > 6$$
$$-2x - 2 + 2 < -6 + 2 \quad or \quad -2x - 2 + 2 > 6 + 2$$
$$-2x < -4 \qquad or \qquad -2x > 8$$
$$\frac{-2x}{-2} > \frac{-4}{-2} \qquad or \qquad \frac{-2x}{-2} < \frac{8}{-2}$$
$$x > 2 \qquad or \qquad x < -4$$

The solution set is $\{x \mid x < -4 \text{ or } x > 2\}$, or $(-\infty, -4) \cup (2, \infty)$.

55.
$$\frac{2}{3}x - 14 < -\frac{5}{6} \qquad or \qquad \frac{2}{3}x - 14 > \frac{5}{6}$$
$$6\left(\frac{2}{3}x - 14\right) < 6\left(-\frac{5}{6}\right) \quad or \quad 6\left(\frac{2}{3}x - 14\right) > 6 \cdot \frac{5}{6}$$
$$4x - 84 < -5 \qquad or \qquad 4x - 84 > 5$$
$$4x - 84 + 84 < -5 + 84 \quad or \quad 4x - 84 + 84 > 5 + 84$$
$$4x < 79 \qquad or \qquad 4x > 89$$
$$\frac{4x}{4} < \frac{79}{4} \qquad or \qquad \frac{4x}{4} > \frac{89}{4}$$
$$x < \frac{79}{4} \qquad or \qquad x > \frac{89}{4}$$

The solution set is $\left\{x \mid x < \frac{79}{4} \text{ or } x > \frac{89}{4}\right\}$, or $\left(-\infty, \frac{79}{4}\right) \cup \left(\frac{89}{4}, \infty\right)$.

57. $\qquad \dfrac{2x-5}{6} \le -3 \qquad or \qquad \dfrac{2x-5}{6} \ge 4$

$6\left(\dfrac{2x-5}{6}\right) \le 6(-3) \quad or \quad 6\left(\dfrac{2x-5}{6}\right) \ge 6 \cdot 4$

$\qquad 2x-5 \le -18 \qquad or \qquad 2x-5 \ge 24$

$2x-5+5 \le -18+5 \quad or \quad 2x-5+5 \ge 24+5$

$\qquad 2x \le -13 \qquad or \qquad 2x \ge 29$

$\qquad \dfrac{2x}{2} \le \dfrac{-13}{2} \qquad or \qquad \dfrac{2x}{2} \ge \dfrac{29}{2}$

$\qquad x \le -\dfrac{13}{2} \qquad or \qquad x \ge \dfrac{29}{2}$

The solution set is $\left\{x \Big| x \le -\dfrac{13}{2} \text{ or } x \ge \dfrac{29}{2}\right\}$, or

$\left(-\infty, -\dfrac{13}{2}\right] \cup \left[\dfrac{29}{2}, \infty\right)$.

59. Familiarize. We will use the formula $P = 1 + \dfrac{d}{33}$.

Translate. We want to find those values of d for which

$$1 \le P \le 7$$

or

$$1 \le 1 + \dfrac{d}{33} \le 7.$$

Solve. We solve the inequality.

$$1 \le 1 + \dfrac{d}{33} \le 7$$

$$0 \le \dfrac{d}{33} \le 6$$

$$0 \le d \le 198$$

Check. We could do a partial check by substituting some values for d in the formula. The result checks.

State. The pressure is at least 1 atm and at most 7 atm for depths d in the set $\{d | 0 \text{ ft} \le d \le 198 \text{ ft}\}$.

61. Familiarize. Let $b =$ the number of beats per minute. Note that $10 \text{ sec} = 10 \text{ sec} \times \dfrac{1 \text{ min}}{60 \text{ sec}} = \dfrac{10}{60} \times \dfrac{\text{sec}}{\text{sec}} \times 1 \text{ min} = \dfrac{1}{6} \text{ min}$. Then in 10 sec, or $\dfrac{1}{6}$ min, the woman should have between $\dfrac{1}{6} \cdot 138$ and $\dfrac{1}{6} \cdot 162$ beats.

Translate. We want to find the value of b for which

$$\dfrac{1}{6} \cdot 138 < b < \dfrac{1}{6} \cdot 162$$

Solve. We solve the inequality.

$$\dfrac{1}{6} \cdot 138 < b < \dfrac{1}{6} \cdot 162$$

$$23 < b < 27$$

Check. If the number of beats in 10 sec, or $\dfrac{1}{6}$ min, is between 23 and 27, then the number of beats per minute is between $6 \cdot 23$ and $6 \cdot 27$, or between 138 and 162. The answer checks.

State. The number of beats should be between 23 and 27.

63. Familiarize. We will use the equation $y = 14.57x + 62.91$.

Translate. Since y is given in millions, we want to find the values of x for which

$$106.62 < y < 194.04$$

or

$$106.62 < 14.57x + 62.91 < 194.04.$$

Solve. We solve the inequality.

$$106.62 < 14.57x + 62.91 < 194.04$$

$$43.71 < 14.57x < 131.13$$

$$3 < x < 9$$

Check. We could do a partial check by substituting some values for x in the equation. The result checks.

State. The number of online shoppers will be between 106,620,000 and 194,040,000 for years x in the set $\{x | 3 \text{ yr} < x < 9 \text{ yr}\}$, where x is the number of years after 2000, or between 2003 and 2009.

65. Discussion and Writing Exercise

67. $\quad 3x - 2y = -7, \quad (1)$

$\quad 2x + 5y = 8 \qquad (2)$

We multiply twice to make two terms become additive inverses.

From (1): $\quad 15x - 10y = -35 \quad$ Multiplying by 5

From (2): $\quad \underline{4x + 10y = 16} \quad$ Multiplying by 2

$\qquad 19x + 0 = -19 \quad$ Adding

$\qquad\qquad 19x = -19$

$\qquad\qquad\quad x = -1$

Substitute -1 for x in one of the original equations and solve for y.

$\quad 2x + 5y = 8 \qquad$ Equation (2)

$2(-1) + 5y = 8 \qquad$ Substituting

$\quad -2 + 5y = 8$

$\qquad\quad 5y = 10$

$\qquad\quad\ y = 2$

We obtain $(-1, 2)$. This checks, so it is the solution.

69. $\quad x + y = 0 \quad (1)$

$\quad \underline{x - y = 8} \quad (2)$

$2x + 0 = 8 \quad$ Adding

$\qquad 2x = 8$

$\qquad\ x = 4$

Substitute 4 for x in one of the original equations and solve for y.

$\quad x + y = 0 \qquad$ Equation (1)

$\quad 4 + y = 0 \qquad$ Substituting

$\qquad\ y = -4$

We obtain $(4, -4)$. This checks, so it is the solution.

71. First find the slope of the line:

$$m = \dfrac{7 - (-1)}{0 - 2} = \dfrac{8}{-2} = -4$$

Now we use the slope and one of the given points to find b. We choose $(0, 7)$ and substitute 0 for x, 7 for y, and -4 for m in $y = mx + b$. Then we solve for b.

$$y = mx + b$$
$$7 = -4 \cdot 0 + b$$
$$7 = b$$

Finally, we use the equation $y = mx + b$ and substitute -4 for m and 7 for b.

$$y = -4x + 7$$

73. $(2a - b)(3a + 5b) = 2a \cdot 3a + 2a \cdot 5b - b \cdot 3a - b \cdot 5b$
$$= 6a^2 + 10ab - 3ab - 5b^2$$
$$= 6a^2 + 7ab - 5b^2$$

75. $(7x - 8)(3x - 5) = 7x \cdot 3x - 7x \cdot 5 - 8 \cdot 3x + 8 \cdot 5$
$$= 21x^2 - 35x - 24x + 40$$
$$= 21x^2 - 59x + 40$$

77.
$$-\frac{2}{15} \le \frac{2}{3}x - \frac{2}{5} \le \frac{2}{15}$$
$$-\frac{2}{15} \le \frac{2}{3}x - \frac{6}{15} \le \frac{2}{15}$$
$$\frac{4}{15} \le \frac{2}{3}x \le \frac{8}{15}$$
$$\frac{3}{2} \cdot \frac{4}{15} \le \frac{3}{2} \cdot \frac{2}{3}x \le \frac{3}{2} \cdot \frac{8}{15}$$
$$\frac{2}{5} \le x \le \frac{4}{5}$$

The solution set is $\left\{ x \middle| \frac{2}{5} \le x \le \frac{4}{5} \right\}$, or $\left[\frac{2}{5}, \frac{4}{5} \right]$.

79.
$$3x < 4 - 5x < 5 + 3x$$
$$0 < 4 - 8x < 5 \qquad \text{Subtracting } 3x$$
$$-4 < -8x < 1$$
$$\frac{1}{2} > x > -\frac{1}{8}$$

The solution set is $\left\{ x \middle| -\frac{1}{8} < x < \frac{1}{2} \right\}$, or $\left(-\frac{1}{8}, \frac{1}{2} \right)$.

81.
$$x + 4 < 2x - 6 \le x + 12$$
$$4 < x - 6 \le 12 \qquad \text{Subtracting } x$$
$$10 < x \le 18$$

The solution set is $\{x | 10 < x \le 18\}$, or $(10, 18]$.

83. If $-b < -a$, then $-1(-b) > -1(-a)$, or $b > a$, or $a < b$. The statement is true.

85. Let $a = 5$, $c = 12$, and $b = 2$. Then $a < c$ and $b < c$, but $a \not< b$. The given statement is false.

87. The numbers in either the set of all rational numbers or the set of all irrational numbers are all real numbers, so the union is all real numbers.

There are no numbers common to the set of all rational numbers and the set of all irrational numbers, so the intersection is $\emptyset$.

Exercise Set 14.3

1. $|9x| = |9| \cdot |x| = 9|x|$

3. $|2x^2| = |2| \cdot |x^2|$
$$= 2|x^2|$$
$$= 2x^2 \qquad \text{Since } x^2 \text{ is never negative}$$

5. $|-2x^2| = |-2| \cdot |x^2|$
$$= 2|x^2|$$
$$= 2x^2 \qquad \text{Since } x^2 \text{ is never negative}$$

7. $|-6y| = |-6| \cdot |y| = 6|y|$

9. $\left| \dfrac{-2}{x} \right| = \dfrac{|-2|}{|x|} = \dfrac{2}{|x|}$

11. $\left| \dfrac{x^2}{-y} \right| = \dfrac{|x^2|}{|-y|}$
$$= \frac{x^2}{|-y|}$$
$$= \frac{x^2}{|y|} \qquad \begin{array}{l} \text{The absolute value of the opposite of} \\ \text{a number is the same as the absolute} \\ \text{value of the number.} \end{array}$$

13. $\left| \dfrac{-8x^2}{2x} \right| = |-4x| = |-4| \cdot |x| = 4|x|$

15. $|-8 - (-46)| = |38| = 38$, or
$$|-46 - (-8)| = |-38| = 38$$

17. $|36 - 17| = |19| = 19$, or
$$|17 - 36| = |-19| = 19$$

19. $|-3.9 - 2.4| = |-6.3| = 6.3$, or
$$|2.4 - (-3.9)| = |6.3| = 6.3$$

21. $|-5 - 0| = |-5| = 5$, or
$$|0 - (-5)| = |5| = 5$$

23. $|x| = 3$
$$x = -3 \ \ or \ \ x = 3 \quad \text{Absolute-value principle}$$

The solution set is $\{-3, 3\}$.

25. $|x| = -3$

The absolute value of a number is always nonnegative. Therefore, the solution set is $\emptyset$.

27. $|q| = 0$

The only number whose absolute value is 0 is 0. The solution set is $\{0\}$.

29. $|x - 3| = 12$
$$x - 3 = -12 \ \ or \ \ x - 3 = 12 \quad \begin{array}{l} \text{Absolute-value} \\ \text{principle} \end{array}$$
$$x = -9 \ \ \ or \ \ \ \ \ \ \ x = 15$$

The solution set is $\{-9, 15\}$.

31. $|2x - 3| = 4$

$2x - 3 = -4 \quad or \quad 2x - 3 = 4 \qquad$ Absolute-value
$\qquad\qquad\qquad\qquad\qquad\qquad\qquad$ principle

$2x = -1 \quad or \qquad 2x = 7$

$x = -\dfrac{1}{2} \quad or \qquad x = \dfrac{7}{2}$

The solution set is $\left\{ -\dfrac{1}{2}, \dfrac{7}{2} \right\}$.

33. $|4x - 9| = 14$

$4x - 9 = -14 \quad or \quad 4x - 9 = 14$

$4x = -5 \quad or \qquad 4x = 23$

$x = -\dfrac{5}{4} \quad or \qquad x = \dfrac{23}{4}$

The solution set is $\left\{ -\dfrac{5}{4}, \dfrac{23}{4} \right\}$.

35. $\qquad |x| + 7 = 18$

$|x| + 7 - 7 = 18 - 7 \qquad$ Subtracting 7

$\qquad |x| = 11$

$x = -11 \quad or \quad x = 11 \quad$ Absolute-value principle

The solution set is $\{-11, 11\}$.

37. $574 = 283 + |t|$

$291 = |t| \qquad$ Subtracting 283

$t = -291 \quad or \quad t = 291 \qquad$ Absolute-value principle

The solution set is $\{-291, 291\}$.

39. $|5x| = 40$

$5x = -40 \quad or \quad 5x = 40$

$x = -8 \quad or \quad x = 8$

The solution set is $\{-8, 8\}$.

41. $|3x| - 4 = 17$

$|3x| = 21 \qquad$ Adding 4

$3x = -21 \quad or \quad 3x = 21$

$x = -7 \quad or \quad x = 7$

The solution set is $\{-7, 7\}$.

43. $7|w| - 3 = 11$

$7|w| = 14 \qquad$ Adding 3

$|w| = 2 \qquad$ Dividing by 7

$w = -2 \quad or \quad w = 2 \qquad$ Absolute-value principle

The solution set is $\{-2, 2\}$.

45. $\left| \dfrac{2x - 1}{3} \right| = 5$

$\dfrac{2x - 1}{3} = -5 \quad or \quad \dfrac{2x - 1}{3} = 5$

$2x - 1 = -15 \quad or \quad 2x - 1 = 15$

$2x = -14 \quad or \qquad 2x = 16$

$x = -7 \quad or \qquad x = 8$

The solution set is $\{-7, 8\}$.

47. $|m + 5| + 9 = 16$

$|m + 5| = 7 \qquad$ Subtracting 9

$m + 5 = -7 \quad or \quad m + 5 = 7$

$m = -12 \quad or \qquad m = 2$

The solution set is $\{-12, 2\}$.

49. $10 - |2x - 1| = 4$

$-|2x - 1| = -6 \qquad$ Subtracting 10

$|2x - 1| = 6 \qquad$ Multiplying by -1

$2x - 1 = -6 \quad or \quad 2x - 1 = 6$

$2x = -5 \quad or \qquad 2x = 7$

$x = -\dfrac{5}{2} \quad or \qquad x = \dfrac{7}{2}$

The solution set is $\left\{ -\dfrac{5}{2}, \dfrac{7}{2} \right\}$.

51. $|3x - 4| = -2$

The absolute value of a number is always nonnegative. The solution set is $\emptyset$.

53. $\left| \dfrac{5}{9} + 3x \right| = \dfrac{1}{6}$

$\dfrac{5}{9} + 3x = -\dfrac{1}{6} \quad or \quad \dfrac{5}{9} + 3x = \dfrac{1}{6}$

$3x = -\dfrac{13}{18} \quad or \qquad 3x = -\dfrac{7}{18}$

$x = -\dfrac{13}{54} \quad or \qquad x = -\dfrac{7}{54}$

The solution set is $\left\{ -\dfrac{13}{54}, -\dfrac{7}{54} \right\}$.

55. $|3x + 4| = |x - 7|$

$3x + 4 = x - 7 \quad or \quad 3x + 4 = -(x - 7)$

$2x + 4 = -7 \quad or \quad 3x + 4 = -x + 7$

$2x = -11 \quad or \quad 4x + 4 = 7$

$x = -\dfrac{11}{2} \quad or \qquad 4x = 3$

$x = -\dfrac{11}{2} \quad or \qquad x = \dfrac{3}{4}$

The solution set is $\left\{ -\dfrac{11}{2}, \dfrac{3}{4} \right\}$.

57. $|x + 3| = |x - 6|$

$x + 3 = x - 6 \quad or \quad x + 3 = -(x - 6)$

$3 = -6 \quad or \quad x + 3 = -x + 6$

$3 = -6 \quad or \qquad 2x = 3$

$3 = -6 \quad or \qquad x = \dfrac{3}{2}$

The first equation has no solution. The second equation has $\dfrac{3}{2}$ as a solution. There is only one solution of the original equation. The solution set is $\left\{ \dfrac{3}{2} \right\}$.

59. $|2a + 4| = |3a - 1|$

$2a + 4 = 3a - 1$ *or* $2a + 4 = -(3a - 1)$

$-a + 4 = -1$ *or* $2a + 4 = -3a + 1$

$-a = -5$ *or* $5a + 4 = 1$

$a = 5$ *or* $5a = -3$

$a = 5$ *or* $a = -\dfrac{3}{5}$

The solution set is $\left\{5, -\dfrac{3}{5}\right\}$.

61. $|y - 3| = |3 - y|$

$y - 3 = 3 - y$ *or* $y - 3 = -(3 - y)$

$2y - 3 = 3$ *or* $y - 3 = -3 + y$

$2y = 6$ *or* $-3 = -3$

$y = 3$ True for all real values of y

All real numbers are solutions.

63. $|5 - p| = |p + 8|$

$5 - p = p + 8$ *or* $5 - p = -(p + 8)$

$5 - 2p = 8$ *or* $5 - p = -p - 8$

$-2p = 3$ *or* $5 = -8$

$p = -\dfrac{3}{2}$ False

The solution set is $\left\{-\dfrac{3}{2}\right\}$.

65. $\left|\dfrac{2x - 3}{6}\right| = \left|\dfrac{4 - 5x}{8}\right|$

$\dfrac{2x - 3}{6} = \dfrac{4 - 5x}{8}$ *or* $\dfrac{2x - 3}{6} = -\left(\dfrac{4 - 5x}{8}\right)$

$24\left(\dfrac{2x - 3}{6}\right) = 24\left(\dfrac{4 - 5x}{8}\right)$ *or* $\dfrac{2x - 3}{6} = \dfrac{-4 + 5x}{8}$

$8x - 12 = 12 - 15x$ *or* $24\left(\dfrac{2x - 3}{6}\right) = 24\left(\dfrac{-4 + 5x}{8}\right)$

$23x - 12 = 12$ *or* $8x - 12 = -12 + 15x$

$23x = 24$ *or* $-7x - 12 = -12$

$x = \dfrac{24}{23}$ *or* $-7x = 0$

$x = 0$

The solution set is $\left\{\dfrac{24}{23}, 0\right\}$.

67. $\left|\dfrac{1}{2}x - 5\right| = \left|\dfrac{1}{4}x + 3\right|$

$\dfrac{1}{2}x - 5 = \dfrac{1}{4}x + 3$ *or* $\dfrac{1}{2}x - 5 = -\left(\dfrac{1}{4}x + 3\right)$

$\dfrac{1}{4}x - 5 = 3$ *or* $\dfrac{1}{2}x - 5 = -\dfrac{1}{4}x - 3$

$\dfrac{1}{4}x = 8$ *or* $\dfrac{3}{4}x - 5 = -3$

$x = 32$ *or* $\dfrac{3}{4}x = 2$

$x = 32$ *or* $x = \dfrac{8}{3}$

The solution set is $\left\{32, \dfrac{8}{3}\right\}$.

69. $|x| < 3$

$-3 < x < 3$

The solution set is $\{x | -3 < x < 3\}$, or $(-3, 3)$.

71. $|x| \geq 2$

$x \leq -2$ or $x \geq 2$

The solution set is $\{x | x \leq -2 \ or \ x \geq 2\}$, or $(-\infty, -2] \cup [2, \infty)$.

73. $|x - 1| < 1$

$-1 < x - 1 < 1$

$0 < x < 2$

The solution set is $\{x | 0 < x < 2\}$, or $(0, 2)$.

75. $5|x + 4| \leq 10$

$|x + 4| \leq 2$ Dividing by 5

$-2 \leq x + 4 \leq 2$

$-6 \leq x \leq -2$ Subtracting 4

The solution set is $\{x | -6 \leq x \leq -2\}$, or $[-6, -2]$.

77. $|2x - 3| \leq 4$

$-4 \leq 2x - 3 \leq 4$

$-1 \leq 2x \leq 7$ Adding 3

$-\dfrac{1}{2} \leq x \leq \dfrac{7}{2}$ Dividing by 2

The solution set is $\left\{x \mid -\dfrac{1}{2} \leq x \leq \dfrac{7}{2}\right\}$, or $\left[-\dfrac{1}{2}, \dfrac{7}{2}\right]$.

79. $|2y - 7| > 10$

$2y - 7 < -10$ *or* $2y - 7 > 10$

$2y < -3$ *or* $2y > 17$ Adding 7

$y < -\dfrac{3}{2}$ *or* $y > \dfrac{17}{2}$ Dividing by 2

The solution set is $\left\{y \mid y < -\dfrac{3}{2} \ or \ y > \dfrac{17}{2}\right\}$, or

$\left(-\infty, -\dfrac{3}{2}\right) \cup \left(\dfrac{17}{2}, \infty\right)$.

81. $|4x - 9| \geq 14$

$\qquad 4x - 9 \leq -14 \quad or \quad 4x - 9 \geq 14$

$\qquad 4x \leq -5 \quad or \qquad 4x \geq 23$

$\qquad x \leq -\dfrac{5}{4} \quad or \qquad x \geq \dfrac{23}{4}$

The solution set is $\left\{ x \middle| x \leq -\dfrac{5}{4} \ or \ x \geq \dfrac{23}{4} \right\}$, or

$\left(-\infty, -\dfrac{5}{4} \right] \cup \left[\dfrac{23}{4}, \infty \right)$.

83. $|y - 3| < 12$

$\qquad -12 < y - 3 < 12$

$\qquad -9 < y < 15 \qquad$ Adding 3

The solution set is $\{ y | -9 < y < 15 \}$, or $(-9, 15)$.

85. $|2x + 3| \leq 4$

$\qquad -4 \leq 2x + 3 \leq 4$

$\qquad -7 \leq 2x \leq 1 \qquad$ Subtracting 3

$\qquad -\dfrac{7}{2} \leq x \leq \dfrac{1}{2} \qquad$ Dividing by 2

The solution set is $\left\{ x \middle| -\dfrac{7}{2} \leq x \leq \dfrac{1}{2} \right\}$, or $\left[-\dfrac{7}{2}, \dfrac{1}{2} \right]$.

87. $|4 - 3y| > 8$

$\qquad 4 - 3y < -8 \quad or \quad 4 - 3y > 8$

$\qquad -3y < -12 \quad or \quad -3y > 4 \qquad$ Subtracting 4

$\qquad y > 4 \qquad or \qquad y < -\dfrac{4}{3} \quad$ Dividing by -3

The solution set is $\left\{ y \middle| y < -\dfrac{4}{3} \ or \ y > 4 \right\}$, or

$\left(-\infty, -\dfrac{4}{3} \right) \cup (4, \infty)$.

89. $|9 - 4x| \geq 14$

$\qquad 9 - 4x \leq -14 \quad or \quad 9 - 4x \geq 14$

$\qquad -4x \leq -23 \quad or \quad -4x \geq 5 \qquad$ Subtracting 9

$\qquad x \geq \dfrac{23}{4} \quad or \qquad x \leq -\dfrac{5}{4} \quad$ Dividing by -4

The solution set is $\left\{ x \middle| x \leq -\dfrac{5}{4} \ or \ x \geq \dfrac{23}{4} \right\}$ or

$\left(-\infty, -\dfrac{5}{4} \right] \cup \left[\dfrac{23}{4}, \infty \right)$.

91. $|3 - 4x| < 21$

$\qquad -21 < 3 - 4x < 21$

$\qquad -24 < -4x < 18 \qquad$ Subtracting 3

$\qquad 6 > x > -\dfrac{9}{2} \qquad$ Dividing by -4 and simplifying

The solution set is $\left\{ x \middle| 6 > x > -\dfrac{9}{2} \right\}$, or

$\left\{ x \middle| -\dfrac{9}{2} < x < 6 \right\}$, or $\left(-\dfrac{9}{2}, 6 \right)$.

93. $\left| \dfrac{1}{2} + 3x \right| \geq 12$

$\qquad \dfrac{1}{2} + 3x \leq -12 \quad or \quad \dfrac{1}{2} + 3x \geq 12$

$\qquad 3x \leq -\dfrac{25}{2} \quad or \qquad 3x \geq \dfrac{23}{2} \qquad$ Subtracting $\dfrac{1}{2}$

$\qquad x \leq -\dfrac{25}{6} \quad or \qquad x \geq \dfrac{23}{6} \qquad$ Dividing by 3

The solution set is $\left\{ x \middle| x \leq -\dfrac{25}{6} \ or \ x \geq \dfrac{23}{6} \right\}$, or

$\left(-\infty, -\dfrac{25}{6} \right] \cup \left[\dfrac{23}{6}, \infty \right)$.

95. $\left| \dfrac{x - 7}{3} \right| < 4$

$\qquad -4 < \dfrac{x - 7}{3} < 4$

$\qquad -12 < x - 7 < 12 \qquad$ Multiplying by 3

$\qquad -5 < x < 19 \qquad$ Adding 7

The solution set is $\{ x | -5 < x < 19 \}$, or $(-5, 19)$.

97. $\left| \dfrac{2 - 5x}{4} \right| \geq \dfrac{2}{3}$

$\qquad \dfrac{2 - 5x}{4} \leq -\dfrac{2}{3} \quad or \quad \dfrac{2 - 5x}{4} \geq \dfrac{2}{3}$

$\qquad 2 - 5x \leq -\dfrac{8}{3} \quad or \quad 2 - 5x \geq \dfrac{8}{3} \qquad$ Multiplying by 4

$\qquad -5x \leq -\dfrac{14}{3} \quad or \quad -5x \geq \dfrac{2}{3} \qquad$ Subtracting 2

$\qquad x \geq \dfrac{14}{15} \quad or \qquad x \leq -\dfrac{2}{15} \quad$ Dividing by -5

The solution set is $\left\{ x \middle| x \leq -\dfrac{2}{15} \ or \ x \geq \dfrac{14}{15} \right\}$, or

$\left(-\infty, -\dfrac{2}{15} \right] \cup \left[\dfrac{14}{15}, \infty \right)$.

99. $|m + 5| + 9 \leq 16$

$\qquad |m + 5| \leq 7 \qquad$ Subtracting 9

$\qquad -7 \leq m + 5 \leq 7$

$\qquad -12 \leq m \leq 2$

The solution set is $\{ m | -12 \leq m \leq 2 \}$, or $[-12, 2]$.

101. $7 - |3 - 2x| \geq 5$

$\qquad -|3 - 2x| \geq -2 \qquad$ Subtracting 7

$\qquad |3 - 2x| \leq 2 \qquad$ Multiplying by -1

$\qquad -2 \leq 3 - 2x \leq 2$

$\qquad -5 \leq -2x \leq -1 \quad$ Subtracting 3

$\qquad \dfrac{5}{2} \geq x \geq \dfrac{1}{2} \qquad$ Dividing by -2

The solution set is $\left\{ x \middle| \dfrac{5}{2} \geq x \geq \dfrac{1}{2} \right\}$, or $\left\{ x \middle| \dfrac{1}{2} \leq x \leq \dfrac{5}{2} \right\}$, or

$\left[\dfrac{1}{2}, \dfrac{5}{2} \right]$.

103. $\left|\dfrac{2x-1}{0.0059}\right| \le 1$

$$-1 \le \frac{2x-1}{0.0059} \le 1$$

$$-0.0059 \le 2x - 1 \le 0.0059$$

$$0.9941 \le 2x \le 1.0059$$

$$0.49705 \le x \le 0.50295$$

The solution set is $\{x|0.49705 \le x \le 0.50295\}$, or $[0.49705, 0.50295]$.

105. Discussion and Writing Exercise

107. $f(x) = |x - 2|$

Since we can calculate $|x - 2|$ for any real number x, the domain is the set of all real numbers.

109. $f(x) = \dfrac{1}{x-5}$

Since $\dfrac{1}{x-5}$ cannot be calculated when the denominator is 0, we find the x-value that causes $x - 5$ to be 0.

$$x - 5 = 0$$

$$x = 5 \quad \text{Adding 5 on both sides}$$

Thus, 5 is not in the domain of f, while all other real numbers are. The domain of f is

$\{x|x \text{ is a real number } and \ x \ne 5\}$, or $(-\infty, 5) \cup (5, \infty)$.

111. $2x + y = 7$,   (1)

$x - 3y = 21$ (2)

Multiply the first equation by 3 and then add.

$6x + 3y = 21$ Multiplying by 3

$\underline{x - 3y = 21}$

$7x + 0 = 42$ Adding

$7x = 42$

$x = 6$

Substitute 6 for x in one of the original equations and solve for y.

$2x + y = 7$ Equation (1)

$2(6) + y = 7$ Substituting

$12 + y = 7$

$y = -5$

We obtain $(6, -5)$. This checks, so it is the solution.

113. $|d - 6 \text{ ft}| \le \dfrac{1}{2}$ ft

$$-\frac{1}{2} \text{ ft} \le d - 6 \text{ ft} \le \frac{1}{2} \text{ ft}$$

$$5\frac{1}{2} \text{ ft} \le d \le 6\frac{1}{2} \text{ ft}$$

The solution set is $\left\{d \,\middle|\, 5\dfrac{1}{2} \text{ ft} \le d \le 6\dfrac{1}{2} \text{ ft}\right\}$.

115. $|x + 5| = x + 5$

From the definition of absolute value, $|x + 5| = x + 5$ only when $x + 5 \ge 0$, or $x \ge -5$. The solution set is $\{x|x \ge -5\}$, or $[-5, \infty)$.

117. $|7x - 2| = x + 4$

From the definition of absolute value, we know $x + 4 \ge 0$, or $x \ge -4$. So we have $x \ge -4$ and

$7x - 2 = x + 4 \quad or \quad 7x - 2 = -(x + 4)$

$6x = 6 \quad\quad or \quad 7x - 2 = -x - 4$

$x = 1 \quad\quad\quad or \quad\quad\quad 8x = -2$

$x = 1 \quad\quad\quad or \quad\quad\quad\quad x = -\dfrac{1}{4}$

The solution set is $\left\{x \,\middle|\, x \ge -4 \ and \ x = 1 \ or \ x = -\dfrac{1}{4}\right\}$, or $\left\{1, -\dfrac{1}{4}\right\}$.

119. $|x - 6| \le -8$

From the definition of absolute value we know that $|x - 6| \ge 0$. Thus $|x - 6| \le -8$ is false for all x. The solution set is $\emptyset$.

121. $|x + 5| > x$

The inequality is true for all $x < 0$ (because absolute value must be nonnegative). The solution set in this case is $\{x|x < 0\}$. If $x = 0$, we have $|0 + 5| > 0$, which is true. The solution set in this case is $\{0\}$. If $x > 0$, we have the following:

$x + 5 < -x \quad or \quad x + 5 > x$

$2x < -5 \quad or \quad\quad\quad 5 > 0$

$x < -\dfrac{5}{2} \quad or \quad\quad\quad 5 > 0$

Although $x > 0$ and $x < -\dfrac{5}{2}$ yields no solution, $x > 0$ and $5 > 0$ (true for all x) yield the solution set $\{x|x > 0\}$ in this case. The solution set for the inequality is $\{x|x < 0\} \cup \{0\} \cup \{x|x > 0\}$, or all real numbers.

123. $|x| \ge 0$

Since the absolute value of a number is always nonnegative, all real numbers are solutions.

125. $-3 < x < 3$ is equivalent to $|x| < 3$.

127. $x \le -6$ or or $x \ge 6$ is equivalent to $|x| \ge 6$.

129. $x < -8 \quad or \quad\quad x > 2$

$x + 3 < -5 \quad or \quad x + 3 > 5$ Adding 3

$|x + 3| > 5$

Exercise Set 14.4

---

1. We use alphabetical order to replace x by -3 and y by 3.

$$\begin{array}{c|c} 3x + y < -5 & \\ \hline 3(-3) + 3 \ ? \ -5 & \\ -9 + 3 & \\ -6 & \text{TRUE} \end{array}$$

Since $-6 < -5$ is true, $(-3, 3)$ is a solution.

3. We use alphabetical order to replace x by 5 and y by 9.

$$\frac{2x - y > -1}{2 \cdot 5 - 9 \; ? \; -1}$$
$$10 - 9 \left|\right.$$
$$1 \left|\right. \quad \text{TRUE}$$

Since $1 > -1$ is true, $(5, 9)$ is a solution.

5. Graph: $y > 2x$

We first graph the line $y = 2x$. We draw the line dashed since the inequality symbol is $>$. To determine which half-plane to shade, test a point not on the line. We try $(1, 1)$ and substitute:

$$\frac{y > 2x}{1 \; ? \; 2 \cdot 1}$$
$$2 \quad \text{FALSE}$$

Since $1 > 2$ is false, $(1, 1)$ is not a solution, nor are any points in the half-plane containing $(1, 1)$. The points in the opposite half-plane are solutions, so we shade that half-plane and obtain the graph.

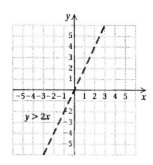

7. Graph: $y < x + 1$

First graph the line $y = x + 1$. Draw it dashed since the inequality symbol is $<$. Test the point $(0, 0)$ to determine if it is a solution.

$$\frac{y < x + 1}{0 \; ? \; 0 + 1}$$
$$1 \quad \text{TRUE}$$

Since $0 < 1$ is true, we shade the half-plane containing $(0, 0)$ and obtain the graph.

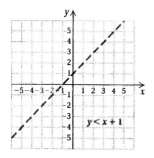

9. Graph: $y > x - 2$

First graph the line $y = x - 2$. Draw a dashed line since the inequality symbol is $>$. Test the point $(0, 0)$ to determine if it is a solution.

$$\frac{y > x - 2}{0 \; ? \; 0 - 2}$$
$$-2 \quad \text{TRUE}$$

Since $0 > -2$ is true, we shade the half-plane containing $(0, 0)$ and obtain the graph.

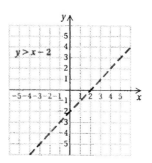

11. Graph: $x + y < 4$

First graph $x + y = 4$. Draw the line dashed since the inequality symbol is $<$. Test the point $(0, 0)$ to determine if it is a solution.

$$\frac{x + y < 4}{0 + 0 \; ? \; 4}$$
$$0 \left|\right. \quad \text{TRUE}$$

Since $0 < 4$ is true, we shade the half-plane containing $(0, 0)$ and obtain the graph.

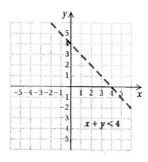

13. Graph: $3x + 4y \leq 12$

We first graph $3x + 4y = 12$. Draw the line solid since the inequality symbol is $\leq$. Test the point $(0, 0)$ to determine if it is a solution.

$$\frac{3x + 4y \leq 12}{3 \cdot 0 + 4 \cdot 0 \; ? \; 12}$$
$$0 \left|\right. \quad \text{TRUE}$$

Since $0 \leq 12$ is true, we shade the half-plane containing $(0, 0)$ and obtain the graph.

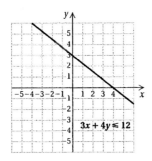

15. Graph: $2y - 3x > 6$

We first graph $2y - 3x = 6$. Draw the line dashed since the inequality symbol is $>$. Test the point $(0, 0)$ to determine if it is a solution.

$$\frac{2y - 3x > 6}{2 \cdot 0 - 3 \cdot 0 \; ? \; 6}$$
$$0 \quad \Big| \quad \text{FALSE}$$

Since $0 > 6$ is false, we shade the half-plane that does not contain $(0, 0)$ and obtain the graph.

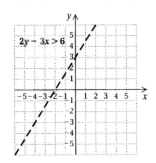

17. Graph: $3x - 2 \le 5x + y$
$$-2 \le 2x + y$$

We first graph $-2 = 2x + y$. Draw the line solid since the inequality symbol is $\le$. Test the point $(0, 0)$ to determine if it is a solution.

$$\frac{-2 \le 2x + y}{-2 \; ? \; 2 \cdot 0 + 0}$$
$$\Big| \; 0 \qquad \text{TRUE}$$

Since $-2 \le 0$ is true, we shade the half-plane containing $(0, 0)$ and obtain the graph.

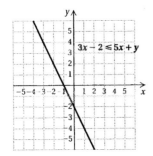

19. Graph: $x < 5$

We first graph $x = 5$. Draw the line dashed since the inequality symbol is $<$. Test the point $(0, 0)$ to determine if it is a solution.

$$\frac{x < 5}{0 \; ? \; 5 \;\; \text{TRUE}}$$

Since $0 < 5$ is true, we shade the half-plane containing $(0, 0)$ and obtain the graph.

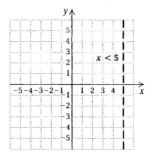

21. Graph: $y > 2$

We first graph $y = 2$. We draw the line dashed since the inequality symbol is $>$. Test the point $(0, 0)$ to determine if it is a solution.

$$\frac{y > 2}{0 \; ? \; 2 \;\; \text{FALSE}}$$

Since $0 > 2$ is false, we shade the half-plane that does not contain $(0, 0)$ and obtain the graph.

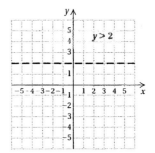

23. Graph: $2x + 3y \le 6$

We first graph $2x + 3y = 6$. We draw the line solid since the inequality symbol is $\le$. Test the point $(0, 0)$ to determine if it is a solution.

$$\frac{2x + 3y \le 6}{2 \cdot 0 + 3 \cdot 0 \; ? \; 6}$$
$$0 \quad \Big| \quad \text{TRUE}$$

Since $0 \le 6$ is true, we shade the half-plane containing $(0, 0)$ and obtain the graph.

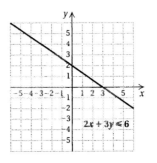

25. The intercepts of the graph of the related equation are $(0, -2)$ and $(3, 0)$, so inequality **F** could be the correct one. Since $(0, 0)$ is in the solution set of this inequality and the half-plane containing $(0, 0)$ is shaded, we know that inequality **F** corresponds to this graph.

27. The intercepts of the graph of the related equation are $(-5, 0)$ and $(0, 3)$, so inequality **B** could be the correct one. Since $(0, 0)$ is in the solution set of this inequality and the half-plane containing $(0, 0)$ is shaded, we know that inequality **B** corresponds to this graph.

29. The intercepts of the graph of the related equation are $(-3, 0)$ and $(0, -3)$, so inequality **C** could be the correct one. Since $(0, 0)$ is not in the solution set of the inequality and the half-plane that does not contain $(0, 0)$ is shaded, we know that inequality **C** corresponds to this graph.

31. Graph: $y \geq x$,

$\qquad y \leq -x + 2$

We graph the lines $y = x$ and $y = -x + 2$, using solid lines. We indicate the region for each inequality by the arrows at the ends of the lines. Note where the regions overlap, and shade the region of solutions.

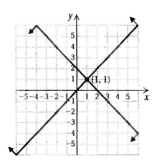

To find the vertex we solve the system of related equations:

$\qquad y = x$,

$\qquad y = -x + 2$

Solving, we obtain the vertex $(1, 1)$.

33. Graph: $y > x$,

$\qquad y < -x + 1$

We graph the lines $y = x$ and $y = -x + 1$, using dashed lines. We indicate the region for each inequality by arrows at the ends of the lines. Note where the regions overlap, and shade the region of solutions.

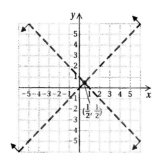

To find the vertex we solve the system of related equations:

$\qquad y = x$,

$\qquad y = -x + 1$

Solving, we obtain the vertex $\left(\dfrac{1}{2}, \dfrac{1}{2}\right)$.

35. Graph: $y \geq -2$,

$\qquad x \geq 1$

We graph the lines $y = -2$ and $x = 1$, using solid lines. We indicate the region for each inequality by arrows. Shade the region where they overlap.

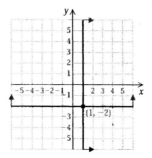

To find the vertex, we solve the system of related equations:

$\qquad y = -2$,

$\qquad x = 1$

Solving, we obtain the vertex $(1, -2)$.

37. Graph: $x \leq 3$,

$\qquad y \geq -3x + 2$

Graph the lines $x = 3$ and $y = -3x + 2$, using solid lines. Indicate the region for each inequality by arrows, and shade the region where they overlap.

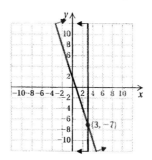

To find the vertex we solve the system of related equations:

$\qquad x = 3$,

$\qquad y = -3x + 2$

Solving, we obtain the vertex $(3, -7)$.

39. Graph: $y \leq 2x + 1$,   (1)

$\qquad y \geq -2x + 1$,   (2)

$\qquad x \leq 2$ (3)

Shade the intersection of the graphs of $y \leq 2x + 1$, $y \geq -2x + 1$, and $x \leq 2$.

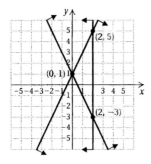

To find the vertices we solve three different systems of equations. From (1) and (2) we obtain the vertex $(0, 1)$. From (1) and (3) we obtain the vertex $(2, 5)$. From (2) and (3) we obtain the vertex $(2, -3)$.

41. Graph: $x + y \leq 1$,
$$x - y \leq 2$$

Graph the lines $x + y = 1$ and $x - y = 2$, using solid lines. Indicate the region for each inequality by arrows, and shade the region where they overlap.

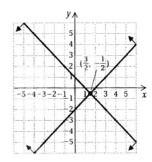

To find the vertex we solve the system of related equations:
$$x + y = 1,$$
$$x - y = 2$$

The vertex is $\left(\frac{3}{2}, -\frac{1}{2}\right)$.

43. Graph: $x + 2y \leq 12$, (1)
$$2x + y \leq 12, \quad (2)$$
$$x \geq 0, \quad (3)$$
$$y \geq 0 \quad (4)$$

Shade the intersection of the graphs of the four inequalities above.

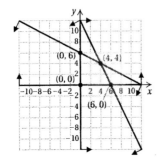

To find the vertices we solve four different systems of equations, as follows:

System of equations	Vertex
From (1) and (2)	$(4, 4)$
From (1) and (3)	$(0, 6)$
From (2) and (4)	$(6, 0)$
From (3) and (4)	$(0, 0)$

45. Graph: $8x + 5y \leq 40$, (1)
$$x + 2y \leq 8, \quad (2)$$
$$x \geq 0, \quad (3)$$
$$y \geq 0 \quad (4)$$

Shade the intersection of the graphs of the four inequalities above.

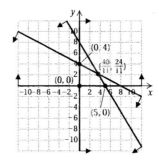

To find the vertices we solve four different systems of equations, as follows:

System of equations	Vertex
From (1) and (2)	$\left(\frac{40}{11}, \frac{24}{11}\right)$
From (1) and (4)	$(5, 0)$
From (2) and (3)	$(0, 4)$
From (3) and (4)	$(0, 0)$

47. Discussion and Writing Exercise

49. First find the slope of the line:
$$m = \frac{-6 - (-2)}{-5 - 3} = \frac{-4}{-8} = \frac{1}{2}$$

Now we use the slope and one of the given points to find b. We choose $(3, -2)$ and substitute 3 for x, -2 for y, and $\frac{1}{2}$ for m in $y = mx + b$. Then we solve for b.
$$y = mx + b$$
$$-2 = \frac{1}{2} \cdot 3 + b$$
$$-2 = \frac{3}{2} + b$$
$$-\frac{7}{2} = b$$

Finally, we use the equation $y = mx + b$ and substitute $\frac{1}{2}$ for m and $-\frac{7}{2}$ for b.
$$y = \frac{1}{2}x - \frac{7}{2}$$

51. First find the slope of the line:
$$m = \frac{6-12}{3-1} = \frac{-6}{2} = -3$$
Now we use the slope and one of the given points to find b. We choose $(3,6)$ and substitute 3 for x, 6 for y, and -3 for m in $y = mx + b$. Then we solve for b.
$$y = mx + b$$
$$6 = -3 \cdot 3 + b$$
$$6 = -9 + b$$
$$15 = b$$
Finally, we use the equation $y = mx + b$ and substitute -3 for m and 15 for b.
$$y = -3x + 15$$

53. First find the slope of the line:
$$m = \frac{5-(-1)}{-5-(-1)} = \frac{6}{-4} = -\frac{3}{2}$$
Now we use the slope and one of the given points to find b. We choose $(-5,5)$ and substitute -5 for x, 5 for y, and $-\frac{3}{2}$ for m in $y = mx + b$. Then we solve for b.
$$y = mx + b$$
$$5 = -\frac{3}{2}(-5) + b$$
$$5 = \frac{15}{2} + b$$
$$-\frac{5}{2} = b$$
Finally, we use the equation $y = mx + b$ and substitute $-\frac{3}{2}$ for m and $-\frac{5}{2}$ for b.
$$y = -\frac{3}{2}x - \frac{5}{2}$$

55. $f(x) = |2 - x|$
$f(0) = |2 - 0| = |2| = 2$

57. $f(x) = |2 - x|$
$f(1) = |2 - 1| = |1| = 1$

59. $f(x) = |2 - x|$
$f(-2) = |2 - (-2)| = |4| = 4$

61. $f(x) = |2 - x|$
$f(-4) = |2 - (-4)| = |6| = 6$

63. Both the width and the height must be positive, but they must be less than 62 in. in order to be checked as luggage, so we have:
$$0 < w \le 62,$$
$$0 < h \le 62$$
The girth is represented by $2w + 2h$ and the length is 62 in. In order to meet postal regulations the sum of the girth and the length cannot exceed 108 in., so we have:
$$62 + 2w + 2h \le 108, \text{ or}$$
$$2w + 2h \le 46, \text{ or}$$
$$w + h \le 23$$

Thus, have a system of inequalities:
$$0 < w \le 62,$$
$$0 < h \le 62,$$
$$w + h \le 23$$

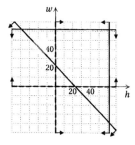

65.

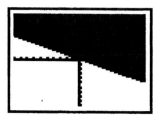

67.

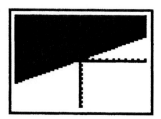

69. Left to the student

Chapter 15

Radical Expressions, Equations, and Functions

1. The square roots of 16 are 4 and -4, because $4^2 = 16$ and $(-4)^2 = 16$.

3. The square roots of 144 are 12 and -12, because $12^2 = 144$ and $(-12)^2 = 144$.

5. The square roots of 400 are 20 and -20, because $20^2 = 400$ and $(-20)^2 = 400$.

7. $-\sqrt{\dfrac{49}{36}} = -\dfrac{7}{6}$ Since $\sqrt{\dfrac{49}{36}} = \dfrac{7}{6}$, $-\sqrt{\dfrac{49}{36}} = -\dfrac{7}{6}$.

9. $\sqrt{196} = 14$ Remember, $\sqrt{}$ indicates the principle square root.

11. $\sqrt{0.0036} = 0.06$

13. $\sqrt{347} \approx 18.628$

15. $\sqrt{\dfrac{285}{74}} \approx 1.962$

17. $9\sqrt{y^2 + 16}$

The radicand is the expression written under the radical sign, $y^2 + 16$.

19. $x^4 y^5 \sqrt{\dfrac{x}{y-1}}$

The radicand is the expression written under the radical sign, $\dfrac{x}{y-1}$.

21. $f(x) = \sqrt{5x - 10}$

$f(6) = \sqrt{5 \cdot 6 - 10} = \sqrt{20} \approx 4.472$

$f(2) = \sqrt{5 \cdot 2 - 10} = \sqrt{0} = 0$

$f(1) = \sqrt{5 \cdot 1 - 10} = \sqrt{-5}$

Since negative numbers do not have real-number square roots, $f(1)$ does not exist as a real number.

$f(-1) = \sqrt{5(-1) - 10} = \sqrt{-15}$

Since negative numbers do not have real-number square roots, $f(-1)$ does not exist as a real number.

23. $g(x) = \sqrt{x^2 - 25}$

$g(-6) = \sqrt{(-6)^2 - 25} = \sqrt{11} \approx 3.317$

$g(3) = \sqrt{3^2 - 25} = \sqrt{-16}$

Since negative numbers do not have real-number square roots, $g(3)$ does not exist as a real number.

$g(6) = \sqrt{6^2 - 25} = \sqrt{11} \approx 3.317$

$g(13) = \sqrt{13^2 - 25} = \sqrt{144} = 12$

25. The domain of $f(x) = \sqrt{5x - 10}$ is the set of all x-values for which $5x - 10 \geq 0$.

$5x - 10 \geq 0$

$5x \geq 10$

$x \geq 2$

The domain is $\{x | x \geq 2\}$, or $[2, \infty)$.

27. $S(x) = 2\sqrt{5x}$

$S(30) = 2\sqrt{5 \cdot 30} = 2\sqrt{150} \approx 24.5$

The speed of a car that left skid marks of length 30 ft was about 24.5 mph.

$S(150) = 2\sqrt{5 \cdot 150} = 2\sqrt{750} \approx 54.8$

The speed of a car that left skid marks of length 150 ft was about 54.8 mph.

29. Graph: $f(x) = 2\sqrt{x}$.

We find some ordered pairs, plot points, and draw the curve.

x	$f(x)$	$(x, f(x))$
0	0	$(0, 0)$
1	2	$(1, 2)$
2	2.8	$(2, 2.8)$
3	3.5	$(3, 3.5)$
4	4	$(4, 4)$
5	4.5	$(5, 4.5)$

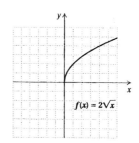

31. Graph: $F(x) = -3\sqrt{x}$.

We find some ordered pairs, plot points, and draw the curve.

x	$f(x)$	$(x, f(x))$
0	0	$(0, 0)$
1	-3	$(1, -3)$
2	-4.2	$(2, -4.2)$
3	-5.2	$(3, -5.2)$
4	-6	$(4, -6)$
5	-6.7	$(5, -6.7)$

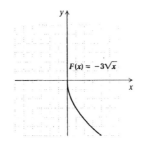

33. Graph: $f(x) = \sqrt{x}$.

We find some ordered pairs, plot points, and draw the curve.

x	$f(x)$	$(x, f(x))$
0	0	$(0, 0)$
1	1	$(1, 1)$
2	1.4	$(2, 1.4)$
3	1.7	$(3, 1.7)$
4	2	$(4, 2)$
5	2.2	$(5, 2.2)$

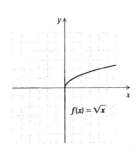

35. Graph: $f(x) = \sqrt{x - 2}$.

We find some ordered pairs, plot points, and draw the curve.

x	$f(x)$	$(x, f(x))$
2	0	$(2, 0)$
3	1	$(3, 1)$
4	1.4	$(4, 1.4)$
5	1.7	$(5, 1.7)$
7	2.2	$(7, 2.2)$
9	2.6	$(9, 2.6)$

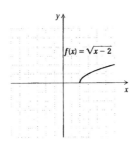

37. Graph: $f(x) = \sqrt{12 - 3x}$.

We find some ordered pairs, plot points, and draw the curve.

x	$f(x)$	$(x, f(x))$
-5	5.2	$(-5, 5.2)$
-3	4.6	$(-3, 4.6)$
-1	3.9	$(-1, 3.9)$
0	3.5	$(0, 3.5)$
2	2.4	$(2, 2.4)$
4	0	$(4, 0)$

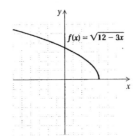

39. Graph: $g(x) = \sqrt{3x + 9}$.

We find some ordered pairs, plot points, and draw the curve.

x	$f(x)$	$(x, f(x))$
-3	0	$(-3, 0)$
-1	2.4	$(-1, 2.4)$
0	3	$(0, 3)$
1	3.5	$(1, 3.5)$
3	4.2	$(3, 4.2)$
5	4.9	$(5, 4.9)$

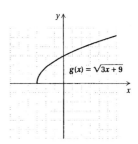

41. $\sqrt{16x^2} = \sqrt{(4x)^2} = |4x| = 4|x|$

(The absolute value is used to ensure that the principal square root is nonnegative.)

43. $\sqrt{(-12c)^2} = |-12c| = |-12| \cdot |c| = 12|c|$

(The absolute value is used to ensure that the principal square root is nonnegative.)

45. $\sqrt{(p + 3)^2} = |p + 3|$

(The absolute value is used to ensure that the principal square root is nonnegative.)

47. $\sqrt{x^2 - 4x + 4} = \sqrt{(x - 2)^2} = |x - 2|$

(The absolute value is used to ensure that the principal square root is nonnegative.)

49. $\sqrt[3]{27} = 3 \quad [3^3 = 27]$

51. $\sqrt[3]{-64x^3} = -4x \quad [(-4x)^3 = -64x^3]$

53. $\sqrt[3]{-216} = -6 \quad [(-6)^3 = -216]$

55. $\sqrt[3]{0.343(x + 1)^3} = 0.7(x + 1)$
$$[(0.7(x + 1))^3 = 0.343(x + 1)^3]$$

57.
$$f(x) = \sqrt[3]{x + 1}$$
$$f(7) = \sqrt[3]{7 + 1} = \sqrt[3]{8} = 2$$
$$f(26) = \sqrt[3]{26 + 1} = \sqrt[3]{27} = 3$$
$$f(-9) = \sqrt[3]{-9 + 1} = \sqrt[3]{-8} = -2$$
$$f(-65) = \sqrt[3]{-65 + 1} = \sqrt[3]{-64} = -4$$

59.
$$f(x) = -\sqrt[3]{3x + 1}$$
$$f(0) = -\sqrt[3]{3 \cdot 0 + 1} = -\sqrt[3]{1} = -1$$
$$f(-7) = -\sqrt[3]{3(-7) + 1} = -\sqrt[3]{-20}, \, or \, \sqrt[3]{20} \approx 2.7144$$
$$f(21) = -\sqrt[3]{3 \cdot 21 + 1} = -\sqrt[3]{64} = -4$$
$$f(333) = -\sqrt[3]{3 \cdot 333 + 1} = -\sqrt[3]{1000} = -10$$

61. $-\sqrt[4]{625} = -5$ Since $5^4 = 625$, then $\sqrt[4]{625} = 5$ and $-\sqrt[4]{625} = -5$.

63. $\sqrt[5]{-1} = -1$ Since $(-1)^5 = -1$

65. $\sqrt[5]{-\dfrac{32}{243}} = -\dfrac{2}{3}$ Since $\left(-\dfrac{2}{3}\right)^5 = -\dfrac{32}{243}$

67. $\sqrt[6]{x^6} = |x|$

The index is even so we use absolute-value notation.

69. $\sqrt[4]{(5a)^4} = |5a| = 5|a|$

The index is even so we use absolute-value notation.

71. $\sqrt[10]{(-6)^{10}} = |-6| = 6$

73. $\sqrt[414]{(a+b)^{414}} = |a+b|$

The index is even so we use absolute-value notation.

75. $\sqrt[7]{y^7} = y$

We do not use absolute-value notation when the index is odd.

77. $\sqrt[5]{(x-2)^5} = x-2$

We do not use absolute-value notation when the index is odd.

79. Discussion and Writing Exercise

81. $x^2 + x - 2 = 0$

$(x+2)(x-1) = 0$ Factoring

$x+2 = 0$ or $x-1 = 0$ Principle of zero products

$x = -2$ or $x = 1$

The solutions are -2 and 1.

83. $4x^2 - 49 = 0$

$(2x+7)(2x-7) = 0$ Factoring

$2x+7 = 0$ or $2x-7 = 0$ Principle of zero products

$2x = -7$ or $2x = 7$

$x = -\dfrac{7}{2}$ or $x = \dfrac{7}{2}$

The solutions are $-\dfrac{7}{2}$ and $\dfrac{7}{2}$.

85. $3x^2 + x = 10$

$3x^2 + x - 10 = 0$

$(3x-5)(x+2) = 0$

$3x-5 = 0$ or $x+2 = 0$

$3x = 5$ or $x = -2$

$x = \dfrac{5}{3}$ or $x = -2$

The solutions are $\dfrac{5}{3}$ and -2.

87. $4x^3 - 20x^2 + 25x = 0$

$x(4x^2 - 20x + 25) = 0$

$x(2x-5)(2x-5) = 0$

$x = 0$ or $2x-5 = 0$ or $2x-5 = 0$

$x = 0$ or $2x = 5$ or $2x = 5$

$x = 0$ or $x = \dfrac{5}{2}$ or $x = \dfrac{5}{2}$

The solutions are 0 and $\dfrac{5}{2}$.

89. $(a^3b^2c^5)^3 = a^{3\cdot3}b^{2\cdot3}c^{5\cdot3} = a^9b^6c^{15}$

91. $f(x) = \dfrac{\sqrt{x+3}}{\sqrt{2-x}}$

In the numerator we must have $x+3 \geq 0$, or $x \geq -3$, and in the denominator we must have $2-x > 0$, or $x < 2$. Thus, we have $x \geq -3$ and $x < 2$, so

Domain of $f = \{x| -3 \leq x < 2\}$, or $[-3, 2)$.

93. From 3 on the x-axis, go up to the graph and across to the y-axis to find $f(3) = \sqrt{3} \approx 1.7$.

From 5 on the x-axis, go up to the graph and across to the y-axis to find $f(5) = \sqrt{5} \approx 2.2$.

From 10 on the x-axis, go up to the graph and across to the y-axis to find $f(10) = \sqrt{10} \approx 3.2$.

95. a) $f(x) = \sqrt[3]{x}$

Domain $= (-\infty, \infty)$; range $= (-\infty, \infty)$

b) $g(x) = \sqrt[3]{4x-5}$

Domain $= (-\infty, \infty)$; range $= (-\infty, \infty)$

c) $q(x) = 2 - \sqrt{x+3}$

Domain $= [-3, \infty)$; range $= (-\infty, 2]$

d) $h(x) = \sqrt[4]{x}$

Domain $= [0, \infty)$; range $= [0, \infty)$

e) $t(x) = \sqrt[4]{x-3}$

Domain $= [3, \infty)$; range $= [0, \infty)$

Exercise Set 15.2

1. $y^{1/7} = \sqrt[7]{y}$

3. $(8)^{1/3} = \sqrt[3]{8} = 2$

5. $(a^3b^3)^{1/5} = \sqrt[5]{a^3b^3}$

7. $16^{3/4} = \sqrt[4]{16^3} = (\sqrt[4]{16})^3 = 2^3 = 8$

9. $49^{3/2} = \sqrt{49^3} = (\sqrt{49})^3 = 7^3 = 343$

11. $\sqrt{17} = 17^{1/2}$

13. $\sqrt[3]{18} = 18^{1/3}$

15. $\sqrt[5]{xy^2z} = (xy^2z)^{1/5}$

17. $(\sqrt{3mn})^3 = (3mn)^{3/2}$

19. $(\sqrt[7]{8x^2y})^5 = (8x^2y)^{5/7}$

21. $27^{-1/3} = \dfrac{1}{27^{1/3}} = \dfrac{1}{\sqrt[3]{27}} = \dfrac{1}{3}$

23. $100^{-3/2} = \dfrac{1}{100^{3/2}} = \dfrac{1}{(\sqrt{100})^3} = \dfrac{1}{10^3} = \dfrac{1}{1000}$

25. $x^{-1/4} = \dfrac{1}{x^{1/4}}$

27. $(2rs)^{-3/4} = \dfrac{1}{(2rs)^{3/4}}$

29. $2a^{3/4}b^{-1/2}c^{2/3} = 2 \cdot a^{3/4} \cdot \dfrac{1}{b^{1/2}} \cdot c^{2/3} = \dfrac{2a^{3/4}c^{2/3}}{b^{1/2}}$

31. $\left(\dfrac{7x}{8yz}\right)^{-3/5} = \left(\dfrac{8yz}{7x}\right)^{3/5}$ $\left(\text{Since } \left(\dfrac{a}{b}\right)^{-n} = \left(\dfrac{b}{a}\right)^n\right)$

33. $\dfrac{1}{x^{-2/3}} = x^{2/3}$

35. $2^{-1/3}x^4 y^{-2/7} = \dfrac{1}{2^{1/3}} \cdot x^4 \cdot \dfrac{1}{y^{2/7}} = \dfrac{x^4}{2^{1/3}y^{2/7}}$

37. $\dfrac{7x}{\sqrt[3]{z}} = \dfrac{7x}{z^{1/3}}$

39. $\dfrac{5a}{3c^{-1/2}} = \dfrac{5a}{3} \cdot c^{1/2} = \dfrac{5ac^{1/2}}{3}$

41. $5^{3/4} \cdot 5^{1/8} = 5^{3/4+1/8} = 5^{6/8+1/8} = 5^{7/8}$

43. $\dfrac{7^{5/8}}{7^{3/8}} = 7^{5/8-3/8} = 7^{2/8} = 7^{1/4}$

45. $\dfrac{4.9^{-1/6}}{4.9^{-2/3}} = 4.9^{-1/6-(-2/3)} = 4.9^{-1/6+4/6} = 4.9^{3/6} = 4.9^{1/2}$

47. $(6^{3/8})^{2/7} = 6^{3/8 \cdot 2/7} = 6^{6/56} = 6^{3/28}$

49. $a^{2/3} \cdot a^{5/4} = a^{2/3+5/4} = a^{8/12+15/12} = a^{23/12}$

51. $(a^{2/3} \cdot b^{5/8})^4 = (a^{2/3})^4 (b^{5/8})^4 = a^{8/3}b^{20/8} = a^{8/3}b^{5/2}$

53. $(x^{2/3})^{-3/7} = x^{2/3(-3/7)} = x^{-2/7} = \dfrac{1}{x^{2/7}}$

55. $\sqrt[6]{a^2} = a^{2/6}$ Converting to exponential notation
   $= a^{1/3}$ Simplifying the exponent
   $= \sqrt[3]{a}$ Returning to radical notation

57. $\sqrt[3]{x^{15}} = x^{15/3}$ Converting to exponential notation
   $= x^5$ Simplifying

59. $\sqrt[6]{x^{-18}} = x^{-18/6}$ Converting to exponential notation
   $= x^{-3}$ Simplifying
   $= \dfrac{1}{x^3}$

61. $(\sqrt[3]{ab})^{15} = (ab)^{15/3}$ Converting to exponential notation
   $= (ab)^5$ Simplifying the exponent
   $= a^5 b^5$ Using the law of exponents

63. $\sqrt[14]{128} = \sqrt[14]{2^7} = 2^{7/14} = 2^{1/2} = \sqrt{2}$

65. $\sqrt[6]{4x^2} = (2^2 x^2)^{1/6} = 2^{2/6}x^{2/6}$
   $= 2^{1/3}x^{1/3} = (2x)^{1/3} = \sqrt[3]{2x}$

67. $\sqrt{x^4 y^6} = (x^4 y^6)^{1/2} = x^{4/2}y^{6/2} = x^2 y^3$

69. $\sqrt[5]{32c^{10}d^{15}} = (2^5 c^{10}d^{15})^{1/5} = 2^{5/5}c^{10/5}d^{15/5}$
   $= 2c^2 d^3$

71. $\sqrt[3]{7} \cdot \sqrt[4]{5} = 7^{1/3} \cdot 5^{1/4} = 7^{4/12} \cdot 5^{3/12} =$
   $(7^4 \cdot 5^3)^{1/12} = \sqrt[12]{7^4 \cdot 5^3}$

73. $\sqrt[4]{5} \cdot \sqrt[5]{7} = 5^{1/4} \cdot 7^{1/5} = 5^{5/20} \cdot 7^{4/20} = (5^5 \cdot 7^4)^{1/20} = \sqrt[20]{5^5 \cdot 7^4}$

75. $\sqrt{x} \sqrt[3]{2x} = x^{1/2} \cdot (2x)^{1/3} = x^{3/6} \cdot (2x)^{2/6} =$
   $[x^3 (2x)^2]^{1/6} = (x^3 \cdot 4x^2)^{1/6} = (4x^5)^{1/6} = \sqrt[6]{4x^5}$

77. $(\sqrt[5]{a^2 b^4})^{15} = (a^2 b^4)^{15/5} = (a^2 b^4)^3 = a^6 b^{12}$

79. $\sqrt[3]{\sqrt[6]{m}} = \sqrt[3]{m^{1/6}} = (m^{1/6})^{1/3} = m^{1/18} = \sqrt[18]{m}$

81. $x^{1/3} \cdot y^{1/4} \cdot z^{1/6} = x^{4/12} \cdot y^{3/12} \cdot z^{2/12} =$
   $(x^4 y^3 z^2)^{1/12} = \sqrt[12]{x^4 y^3 z^2}$

83. $\left(\dfrac{c^{-4/5}d^{5/9}}{c^{3/10}d^{1/6}}\right)^3 = (c^{-4/5-3/10}d^{5/9-1/6})^3 =$
   $(c^{-8/10-3/10}d^{10/18-3/18})^3 = (c^{-11/10}d^{7/18})^3 =$
   $c^{-33/10}d^{7/6} = c^{-99/30}d^{35/30} = (c^{-99}d^{35})^{1/30} =$
   $\left(\dfrac{d^{35}}{c^{99}}\right)^{1/30} = \sqrt[30]{\dfrac{d^{35}}{c^{99}}}$

85. Discussion and Writing Exercise

87. $|7x - 5| = 9$
   $7x - 5 = -9 \quad or \quad 7x - 5 = 9$
   $7x = -4 \quad or \quad\quad 7x = 14$
   $x = -\dfrac{4}{7} \quad or \quad\quad x = 2$
   The solution set is $\left\{-\dfrac{4}{7}, 2\right\}$.

89. $8 - |2x + 5| = -2$
   $-|2x + 5| = -10$
   $|2x + 5| = 10$ Multiplying by -1
   $2x + 5 = -10 \quad or \quad 2x + 5 = 10$
   $2x = -15 \quad or \quad\quad 2x = 5$
   $x = -\dfrac{15}{2} \quad or \quad\quad x = \dfrac{5}{2}$
   The solution set is $\left\{-\dfrac{15}{2}, \dfrac{5}{2}\right\}$.

91.

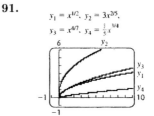

Exercise Set 15.3

1. $\sqrt{24} = \sqrt{4 \cdot 6} = \sqrt{4}\ \sqrt{6} = 2\sqrt{6}$

3. $\sqrt{90} = \sqrt{9 \cdot 10} = \sqrt{9}\ \sqrt{10} = 3\sqrt{10}$

5. $\sqrt[3]{250} = \sqrt[3]{125 \cdot 2} = \sqrt[3]{125}\ \sqrt[3]{2} = 5\sqrt[3]{2}$

7. $\sqrt{180x^4} = \sqrt{36 \cdot 5 \cdot x^4} = \sqrt{36x^4}\ \sqrt{5} = 6x^2\sqrt{5}$

9. $\sqrt[3]{54x^8} = \sqrt[3]{27 \cdot 2 \cdot x^6 \cdot x^2} = \sqrt[3]{27x^6}\ \sqrt[3]{2x^2} = 3x^2\sqrt[3]{2x^2}$

11. $\sqrt[3]{80t^8} = \sqrt[3]{8 \cdot 10 \cdot t^6 \cdot t^2} = \sqrt[3]{8t^6}\ \sqrt[3]{10t^2} = 2t^2\sqrt[3]{10t^2}$

13. $\sqrt[4]{80} = \sqrt[4]{16 \cdot 5} = \sqrt[4]{16}\ \sqrt[4]{5} = 2\sqrt[4]{5}$

15. $\sqrt{32a^2b} = \sqrt{16a^2 \cdot 2b} = \sqrt{16a^2} \cdot \sqrt{2b} = 4a\sqrt{2b}$

17. $\sqrt[4]{243x^8y^{10}} = \sqrt[4]{81x^8y^8 \cdot 3y^2} = \sqrt[4]{81x^8y^8}\,\sqrt[4]{3y^2} = 3x^2y^2\,\sqrt[4]{3y^2}$

19. $\sqrt[5]{96x^7y^{15}} = \sqrt[5]{32x^5y^{15} \cdot 3x^2} = \sqrt[5]{32x^5y^{15}}\,\sqrt[5]{3x^2} = 2xy^3\,\sqrt[5]{3x^2}$

21. $\sqrt{10}\sqrt{5} = \sqrt{10 \cdot 5} = \sqrt{50} = \sqrt{25 \cdot 2} = 5\sqrt{2}$

23. $\sqrt{15}\,\sqrt{6} = \sqrt{15 \cdot 6} = \sqrt{90}$
$= \sqrt{9 \cdot 10} = \sqrt{9}\,\sqrt{10} = 3\sqrt{10}$

25. $\sqrt[3]{2}\,\sqrt[3]{4} = \sqrt[3]{2 \cdot 4} = \sqrt[3]{8} = 2$

27. $\sqrt{45}\,\sqrt{60} = \sqrt{45 \cdot 60} = \sqrt{2700}$
$= \sqrt{900 \cdot 3} = \sqrt{900}\,\sqrt{3} = 30\sqrt{3}$

29. $\sqrt{3x^3}\sqrt{6x^5} = \sqrt{18x^8} = \sqrt{9x^8 \cdot 2} = 3x^4\sqrt{2}$

31. $\sqrt{5b^3}\,\sqrt{10c^4}$
$= \sqrt{5b^3 \cdot 10c^4}$
$= \sqrt{50b^3c^4}$
$= \sqrt{25 \cdot 2 \cdot b^2 \cdot b \cdot c^4}$
$= \sqrt{25b^2c^4}\,\sqrt{2b}$
$= 5bc^2\,\sqrt{2b}$

33. $\sqrt[3]{5a^2}\,\sqrt[3]{2a} = \sqrt[3]{5a^2 \cdot 2a} = \sqrt[3]{10a^3} = \sqrt[3]{a^3 \cdot 10} = a\,\sqrt[3]{10}$

35. $\sqrt[3]{y^4}\,\sqrt[3]{16y^5} = \sqrt[3]{y^4 \cdot 16y^5}$
$= \sqrt[3]{16y^9}$
$= \sqrt[3]{8 \cdot 2 \cdot y^9}$
$= \sqrt[3]{8y^9}\,\sqrt[3]{2}$
$= 2y^3\,\sqrt[3]{2}$

37. $\sqrt[4]{16}\,\sqrt[4]{64} = \sqrt[4]{16 \cdot 64} = \sqrt[4]{1024} = \sqrt[4]{256 \cdot 4} = \sqrt[4]{256}\,\sqrt[4]{4} = 4\sqrt[4]{4}$

39. $\sqrt{12a^3b}\,\sqrt{8a^4b^2} = \sqrt{12a^3b \cdot 8a^4b^2} = \sqrt{96a^7b^3} = \sqrt{16a^6b^2 \cdot 6ab} = \sqrt{16a^6b^2}\,\sqrt{6ab} = 4a^3b\sqrt{6ab}$

41. $\sqrt{2}\,\sqrt[3]{5}$
$= 2^{1/2} \cdot 5^{1/3}$ Converting to exponential notation
$= 2^{3/6} \cdot 5^{2/6}$ Rewriting so that exponents have a common denominator
$= (2^3 \cdot 5^2)^{1/6}$ Using $a^n b^n = (ab)^n$
$= \sqrt[6]{2^3 \cdot 5^2}$ Converting to radical notation
$= \sqrt[6]{8 \cdot 25}$ Simplifying
$= \sqrt[6]{200}$ Multiplying

43. $\sqrt[4]{3}\,\sqrt{2}$
$= 3^{1/4} \cdot 2^{1/2}$ Converting to exponential notation
$= 3^{1/4} \cdot 2^{2/4}$ Rewriting so that exponents have a common denominator
$= (3 \cdot 2^2)^{1/4}$ Using $a^n b^n = (ab)^n$
$= \sqrt[4]{3 \cdot 2^2}$. Converting to radical notation
$= \sqrt[4]{3 \cdot 4}$ Squaring 2
$= \sqrt[4]{12}$ Multiplying

45. $\sqrt{a}\,\sqrt[4]{a^3}$
$= a^{1/2} \cdot a^{3/4}$ Converting to exponential notation
$= a^{5/4}$ Adding exponents
$= a^{1+1/4}$ Writing 5/4 as a mixed number
$= a \cdot a^{1/4}$ Factoring
$= a\sqrt[4]{a}$ Returning to radical notation

47. $\sqrt[5]{b^2}\sqrt{b^3}$
$= b^{2/5} \cdot b^{3/2}$ Converting to exponential notation
$= b^{19/10}$ Adding exponents
$= b^{1+9/10}$ Writing 19/10 as a mixed number
$= b \cdot b^{9/10}$ Factoring
$= b\,\sqrt[10]{b^9}$ Returning to radical notation

49. $\sqrt{xy^3}\,\sqrt[3]{x^2y} = (xy^3)^{1/2}(x^2y)^{1/3}$
$= (xy^3)^{3/6}(x^2y)^{2/6}$
$= [(xy^3)^3(x^2y)^2]^{1/6}$
$= \sqrt[6]{x^3y^9 \cdot x^4y^2}$
$= \sqrt[6]{x^7y^{11}}$
$= \sqrt[6]{x^6y^6 \cdot xy^5}$
$= xy\,\sqrt[6]{xy^5}$

51. $\dfrac{\sqrt{90}}{\sqrt{5}} = \sqrt{\dfrac{90}{5}} = \sqrt{18} = \sqrt{9 \cdot 2} = \sqrt{9}\,\sqrt{2} = 3\sqrt{2}$

53. $\dfrac{\sqrt{35q}}{\sqrt{7q}} = \sqrt{\dfrac{35q}{7q}} = \sqrt{5}$

55. $\dfrac{\sqrt[3]{54}}{\sqrt[3]{2}} = \sqrt[3]{\dfrac{54}{2}} = \sqrt[3]{27} = 3$

57. $\dfrac{\sqrt{56xy^3}}{\sqrt{8x}} = \sqrt{\dfrac{56xy^3}{8x}} = \sqrt{7y^3} = \sqrt{y^2 \cdot 7y} = \sqrt{y^2}\,\sqrt{7y} = y\sqrt{7y}$

59. $\dfrac{\sqrt[3]{96a^4b^2}}{\sqrt[3]{12a^2b}} = \sqrt[3]{\dfrac{96a^4b^2}{12a^2b}} = \sqrt[3]{8a^2b} = \sqrt[3]{8}\,\sqrt[3]{a^2b} = 2\sqrt[3]{a^2b}$

61. $\dfrac{\sqrt{128xy}}{2\sqrt{2}} = \dfrac{1}{2}\dfrac{\sqrt{128xy}}{\sqrt{2}} = \dfrac{1}{2}\sqrt{\dfrac{128xy}{2}} = \dfrac{1}{2}\sqrt{64xy} = \dfrac{1}{2}\sqrt{64}\,\sqrt{xy} = \dfrac{1}{2} \cdot 8\sqrt{xy} = 4\sqrt{xy}$

63. $\dfrac{\sqrt[4]{48x^9y^{13}}}{\sqrt[4]{3xy^5}} = \sqrt[4]{\dfrac{48x^9y^{13}}{3xy^5}} = \sqrt[4]{16x^8y^8} = 2x^2y^2$

65.
$$\dfrac{\sqrt[3]{a}}{\sqrt{a}}$$

$= \dfrac{a^{1/3}}{a^{1/2}}$ Converting to exponential notation

$= a^{1/3-1/2}$ Subtracting exponents

$= a^{2/6-3/6}$

$= a^{-1/6}$

$= \dfrac{1}{a^{1/6}}$

$= \dfrac{1}{\sqrt[6]{a}}$ Converting to radical notation

67.
$$\dfrac{\sqrt[3]{a^2}}{\sqrt[4]{a}}$$

$= \dfrac{a^{2/3}}{a^{1/4}}$ Converting to exponential notation

$= a^{2/3-1/4}$ Subtracting exponents

$= a^{5/12}$ Converting back

$= \sqrt[12]{a^5}$ to radical notation

69.
$$\dfrac{\sqrt[4]{x^2y^3}}{\sqrt[3]{xy}}$$

$= \dfrac{(x^2y^3)^{1/4}}{(xy)^{1/3}}$ Converting to exponential notation

$= \dfrac{x^{2/4}y^{3/4}}{x^{1/3}y^{1/3}}$ Using the power and product rules

$= x^{2/4-1/3}y^{3/4-1/3}$ Subtracting exponents

$= x^{2/12}y^{5/12}$

$= (x^2y^5)^{1/2}$ Converting back to

$= \sqrt[12]{x^2y^5}$ radical notation

71. $\sqrt{\dfrac{25}{36}} = \dfrac{\sqrt{25}}{\sqrt{36}} = \dfrac{5}{6}$

73. $\sqrt{\dfrac{16}{49}} = \dfrac{\sqrt{16}}{\sqrt{49}} = \dfrac{4}{7}$

75. $\sqrt[3]{\dfrac{125}{27}} = \dfrac{\sqrt[3]{125}}{\sqrt[3]{27}} = \dfrac{5}{3}$

77. $\sqrt{\dfrac{49}{y^2}} = \dfrac{\sqrt{49}}{\sqrt{y^2}} = \dfrac{7}{y}$

79. $\sqrt{\dfrac{25y^3}{x^4}} = \dfrac{\sqrt{25y^3}}{\sqrt{x^4}} = \dfrac{\sqrt{25y^2 \cdot y}}{\sqrt{x^4}} = \dfrac{\sqrt{25y^2}\,\sqrt{y}}{\sqrt{x^4}} =$

$\dfrac{5y\sqrt{y}}{x^2}$

81. $\sqrt[3]{\dfrac{27a^4}{8b^3}} = \dfrac{\sqrt[3]{27a^4}}{\sqrt[3]{8b^3}} = \dfrac{\sqrt[3]{27a^3 \cdot a}}{\sqrt[3]{8b^3}} = \dfrac{\sqrt[3]{27a^3}\,\sqrt[3]{a}}{\sqrt[3]{8b^3}} =$

$\dfrac{3a\sqrt[3]{a}}{2b}$

83. $\sqrt[4]{\dfrac{81x^4}{16}} = \dfrac{\sqrt[4]{81x^4}}{\sqrt[4]{16}} = \dfrac{3x}{2}$

85. $\sqrt[5]{\dfrac{32x^8}{y^{10}}} = \dfrac{\sqrt[5]{32x^8}}{\sqrt[5]{y^{10}}} = \dfrac{\sqrt[5]{32 \cdot x^5 \cdot x^3}}{\sqrt[5]{y^{10}}} = \dfrac{\sqrt[5]{32x^5}\,\sqrt[5]{x^3}}{\sqrt[5]{y^{10}}} =$

$\dfrac{2x\sqrt[5]{x^3}}{y^2}$

87. $\sqrt[6]{\dfrac{x^{13}}{y^6z^{12}}} = \dfrac{\sqrt[6]{x^{13}}}{\sqrt[6]{y^6z^{12}}} = \dfrac{\sqrt[6]{x^{12} \cdot x}}{\sqrt[6]{y^6z^{12}}} = \dfrac{\sqrt[6]{x^{12}}\,\sqrt[6]{x}}{\sqrt[6]{y^6z^{12}}} = \dfrac{x^2\sqrt[6]{x}}{yz^2}$

89. Discussion and Writing Exercise

91. *Familiarize.* We will use the formula $d = rt$. When the boat travels downstream, its rate is $14 + 7$, or 21 mph. Its rate traveling upstream is $14 - 7$ or 7 mph.

Translate. We substitute in the formula.

Downstream: $56 = 21t$

Upstream: $56 = 7t$

Solve. We solve the equation.

    Downstream: $56 = 21t$

$$\dfrac{56}{21} = t$$

$$\dfrac{8}{3} = t, \text{ or}$$

$$2\dfrac{2}{3} = t$$

    Upstream: $56 = 7t$

$$8 = t$$

Check. At a rate of 21 mph, in $\dfrac{8}{3}$ hr the boat would travel $21 \cdot \dfrac{8}{3}$, or 56 mi. At a rate of 7 mph, in 8 hr the boat would travel $7 \cdot 8$, or 56 mi. The answer checks.

State. It will take the boat $2\dfrac{2}{3}$ hr to travel 56 mi downstream and 8 hr to travel 56 mi upstream.

93.
$$\dfrac{12x}{x-4} - \dfrac{3x^2}{x+4} = \dfrac{384}{x^2-16}$$

$$\dfrac{12x}{x-4} - \dfrac{3x^2}{x+4} = \dfrac{384}{(x+4)(x-4)},$$
$$\text{LCM is } (x+4)(x-4).$$

$$(x+4)(x-4)\left[\dfrac{12x}{x-4} - \dfrac{3x^2}{x+4}\right] = (x+4)(x-4) \cdot \dfrac{384}{(x+4)(x-4)}$$

$$12x(x+4) - 3x^2(x-4) = 384$$

$$12x^2 + 48x - 3x^3 + 12x^2 = 384$$

$$-3x^3 + 24x^2 + 48x - 384 = 0$$

$$-3(x^3 - 8x^2 - 16x + 128) = 0$$

$$-3[x^2(x-8) - 16(x-8)] = 0$$

$$-3(x-8)(x^2 - 16) = 0$$

$$-3(x-8)(x+4)(x-4) = 0$$

$$x - 8 = 0 \quad or \quad x + 4 = 0 \quad or \quad x - 4 = 0$$

$$x = 8 \quad or \qquad x = -4 \quad or \qquad x = 4$$

Check: For 8:

$$\frac{12x}{x-4} - \frac{3x^2}{x+4} = \frac{384}{x^2-16}$$

$$\frac{12\cdot 8}{8-4} - \frac{3\cdot 8^2}{8+4} \ \overset{?}{=} \ \frac{384}{8^2-16}$$

$$\frac{96}{4} - \frac{192}{12} \ \Big| \ \frac{384}{48}$$

$$24 - 16 \ \Big| \ 8$$

$$8 \ \Big| \qquad \text{TRUE}$$

8 is a solution.

For -4:

$$\frac{12x}{x-4} - \frac{3x^2}{x+4} = \frac{384}{x^2-16}$$

$$\frac{12(-4)}{-4-4} - \frac{3(-4)^2}{-4+4} \ \overset{?}{=} \ \frac{384}{(-4)^2-16}$$

$$\frac{-48}{-8} - \frac{48}{0} \ \Big| \ \frac{384}{16-16} \qquad \text{UNDEFINED}$$

-4 is not a solution.

For 4:

$$\frac{12x}{x-4} - \frac{3x^2}{x+4} = \frac{384}{x^2-16}$$

$$\frac{12\cdot 4}{4-4} - \frac{3\cdot 4^2}{4+4} \ \overset{?}{=} \ \frac{384}{4^2-16}$$

$$\frac{48}{0} - \frac{48}{8} \ \Big| \ \frac{384}{16-16} \qquad \text{UNDEFINED}$$

4 is not a solution.

The solution is 8.

95.
$$\frac{18}{x^2-3x} = \frac{2x}{x-3} - \frac{6}{x}$$

$$\frac{18}{x(x-3)} = \frac{2x}{x-3} - \frac{6}{x},$$

$$\text{LCM is } x(x-3)$$

$$x(x-3)\cdot \frac{18}{x(x-3)} = x(x-3)\left(\frac{2x}{x-3} - \frac{6}{x}\right)$$

$$18 = x(x-3)\cdot \frac{2x}{x-3} - x(x-3)\cdot \frac{6}{x}$$

$$18 = 2x^2 - 6x + 18$$

$$0 = 2x^2 - 6x$$

$$0 = 2x(x-3)$$

$$2x = 0 \ \ or \ \ x-3 = 0$$

$$x = 0 \ \ or \qquad x = 3$$

Each value makes a denominator 0. There is no solution.

97. a) $T = 2\pi\sqrt{\dfrac{65}{980}} \approx 1.62$ sec

b) $T = 2\pi\sqrt{\dfrac{98}{980}} \approx 1.99$ sec

c) $T = 2\pi\sqrt{\dfrac{120}{980}} \approx 2.20$ sec

99. $\dfrac{\sqrt{44x^2y^9z}\,\sqrt{22y^9z^6}}{(\sqrt{11xy^8z^2})^2} = \dfrac{\sqrt{44\cdot 22x^2y^{18}z^7}}{\sqrt{11\cdot 11x^2y^{16}z^4}} =$

$$\sqrt{\frac{44\cdot 22x^2y^{18}z^7}{11\cdot 11x^2y^{16}z^4}} = \sqrt{4\cdot 2y^2z^3} = \sqrt{4y^2z^2\cdot 2z} = 2yz\sqrt{2z}$$

Exercise Set 15.4

1. $7\sqrt{5} + 4\sqrt{5} = (7+4)\sqrt{5}$ Factoring out $\sqrt{5}$

$$= 11\sqrt{5}$$

3. $6\sqrt[3]{7} - 5\sqrt[3]{7} = (6-5)\sqrt[3]{7}$ Factoring out $\sqrt[3]{7}$

$$= \sqrt[3]{7}$$

5. $4\sqrt[3]{y} + 9\sqrt[3]{y} = (4+9)\sqrt[3]{y} = 13\sqrt[3]{y}$

7. $5\sqrt{6} - 9\sqrt{6} - 4\sqrt{6} = (5-9-4)\sqrt{6} = -8\sqrt{6}$

9. $4\sqrt[3]{3} - \sqrt{5} + 2\sqrt[3]{3} + \sqrt{5} =$

$$(4+2)\sqrt[3]{3} + (-1+1)\sqrt{5} = 6\sqrt[3]{3}$$

11. $8\sqrt{27} - 3\sqrt{3} = 8\sqrt{9\cdot 3} - 3\sqrt{3}$ $\Big\}$ Factoring the first radical

$$= 8\sqrt{9}\cdot\sqrt{3} - 3\sqrt{3}$$

$$= 8\cdot 3\sqrt{3} - 3\sqrt{3} \quad \text{Taking the square root}$$

$$= 24\sqrt{3} - 3\sqrt{3}$$

$$= (24-3)\sqrt{3} \quad \text{Factoring out } \sqrt{3}$$

$$= 21\sqrt{3}$$

13. $8\sqrt{45} + 7\sqrt{20} = 8\sqrt{9\cdot 5} + 7\sqrt{4\cdot 5}$ $\Big\}$ Factoring the radicals

$$= 8\sqrt{9}\cdot\sqrt{5} + 7\sqrt{4}\cdot\sqrt{5}$$

$$= 8\cdot 3\sqrt{5} + 7\cdot 2\sqrt{5} \quad \text{Taking the square roots}$$

$$= 24\sqrt{5} + 14\sqrt{5}$$

$$= (24+14)\sqrt{5} \quad \text{Factoring out } \sqrt{5}$$

$$= 38\sqrt{5}$$

15. $18\sqrt{72} + 2\sqrt{98} = 18\sqrt{36\cdot 2} + 2\sqrt{49\cdot 2} =$

$$18\sqrt{36}\cdot\sqrt{2} + 2\sqrt{49}\cdot\sqrt{2} = 18\cdot 6\sqrt{2} + 2\cdot 7\sqrt{2} =$$

$$108\sqrt{2} + 14\sqrt{2} = (108+14)\sqrt{2} = 122\sqrt{2}$$

17. $3\sqrt[3]{16} + \sqrt[3]{54} = 3\sqrt[3]{8\cdot 2} + \sqrt[3]{27\cdot 2} =$

$$3\sqrt[3]{8}\cdot\sqrt[3]{2} + \sqrt[3]{27}\cdot\sqrt[3]{2} = 3\cdot 2\sqrt[3]{2} + 3\sqrt[3]{2} =$$

$$6\sqrt[3]{2} + 3\sqrt[3]{2} = (6+3)\sqrt[3]{2} = 9\sqrt[3]{2}$$

19. $2\sqrt{128} - \sqrt{18} + 4\sqrt{32} =$

$$2\sqrt{64\cdot 2} - \sqrt{9\cdot 2} + 4\sqrt{16\cdot 2} =$$

$$2\sqrt{64}\cdot\sqrt{2} - \sqrt{9}\cdot\sqrt{2} + 4\sqrt{16}\cdot\sqrt{2} =$$

$$2\cdot 8\sqrt{2} - 3\sqrt{2} + 4\cdot 4\sqrt{2} = 16\sqrt{2} - 3\sqrt{2} + 16\sqrt{2} =$$

$$(16 - 3 + 16)\sqrt{2} = 29\sqrt{2}$$

21. $\sqrt{5a} + 2\sqrt{45a^3} = \sqrt{5a} + 2\sqrt{9a^2\cdot 5a} =$

$$\sqrt{5a} + 2\sqrt{9a^2}\cdot\sqrt{5a} = \sqrt{5a} + 2\cdot 3a\sqrt{5a} =$$

$$\sqrt{5a} + 6a\sqrt{5a} = (1 + 6a)\sqrt{5a}$$

23. $\sqrt[3]{24x} - \sqrt[3]{3x^4} = \sqrt[3]{8 \cdot 3x} - \sqrt[3]{x^3 \cdot 3x} =$

$\sqrt[3]{8} \cdot \sqrt[3]{3x} - \sqrt[3]{x^3} \cdot \sqrt[3]{3x} = 2\sqrt[3]{3x} - x\sqrt[3]{3x} =$

$(2 - x)\sqrt[3]{3x}$

25. $5\sqrt[3]{32} - \sqrt[3]{108} + 2\sqrt[3]{256} =$

$5\sqrt[3]{8 \cdot 4} - \sqrt[3]{27 \cdot 4} + 2\sqrt[3]{64 \cdot 4} =$

$5\sqrt[3]{8} \cdot \sqrt[3]{4} - \sqrt[3]{27} \cdot \sqrt[3]{4} + 2\sqrt[3]{64} \cdot \sqrt[3]{4} =$

$5 \cdot 2\sqrt[3]{4} - 3\sqrt[3]{4} + 2 \cdot 4\sqrt[3]{4} = 10\sqrt[3]{4} - 3\sqrt[3]{4} + 8\sqrt[3]{4} =$

$(10 - 3 + 8)\sqrt[3]{4} = 15\sqrt[3]{4}$

27. $\quad \sqrt[3]{6x^4} + \sqrt[3]{48x} - \sqrt[3]{6x}$

$= \sqrt[3]{x^3 \cdot 6x} + \sqrt[3]{8 \cdot 6x} - \sqrt[3]{6x}$

$= \sqrt[3]{x^3} \cdot \sqrt[3]{6x} + \sqrt[3]{8} \cdot \sqrt[3]{6x} - \sqrt[3]{6x}$

$= x\sqrt[3]{6x} + 2\sqrt[3]{6x} - \sqrt[3]{6x}$

$= (x + 2 - 1)\sqrt[3]{6x}$

$= (x + 1)\sqrt[3]{6x}$

29. $\sqrt{4a - 4} + \sqrt{a - 1} = \sqrt{4(a - 1)} + \sqrt{a - 1}$

$= \sqrt{4} \cdot \sqrt{a - 1} + \sqrt{a - 1}$

$= 2\sqrt{a - 1} + \sqrt{a - 1}$

$= (2 + 1)\sqrt{a - 1}$

$= 3\sqrt{a - 1}$

31. $\sqrt{x^3 - x^2} + \sqrt{9x - 9} = \sqrt{x^2(x - 1)} + \sqrt{9(x - 1)}$

$= \sqrt{x^2}\sqrt{x - 1} + \sqrt{9}\sqrt{x - 1}$

$= x\sqrt{x - 1} + 3\sqrt{x - 1}$

$= (x + 3)\sqrt{x - 1}$

33. $\sqrt{5}(4 - 2\sqrt{5}) = \sqrt{5} \cdot 4 - 2(\sqrt{5})^2$ Distributive law

$= 4\sqrt{5} - 2 \cdot 5$

$= 4\sqrt{5} - 10$

35. $\sqrt{3}(\sqrt{2} - \sqrt{7}) = \sqrt{3}\,\sqrt{2} - \sqrt{3}\,\sqrt{7}$ Distributive law

$= \sqrt{6} - \sqrt{21}$

37. $\sqrt{3}(2\sqrt{5} - 3\sqrt{4}) = \sqrt{3}(2\sqrt{5} - 3 \cdot 2) =$

$\sqrt{3} \cdot 2\sqrt{5} - \sqrt{3} \cdot 6 = 2\sqrt{15} - 6\sqrt{3}$

39. $\sqrt[3]{2}(\sqrt[3]{4} - 2\sqrt[3]{32}) = \sqrt[3]{2} \cdot \sqrt[3]{4} - \sqrt[3]{2} \cdot 2\sqrt[3]{32} =$

$\sqrt[3]{8} - 2\sqrt[3]{64} = 2 - 2 \cdot 4 = 2 - 8 = -6$

41. $\sqrt[3]{a}(\sqrt[3]{2a^2} + \sqrt[3]{16a^2}) = \sqrt[3]{a} \cdot \sqrt[3]{2a^2} + \sqrt[3]{a} \cdot \sqrt[3]{16a^2} =$

$\sqrt[3]{2a^3} + \sqrt[3]{16a^3} = \sqrt[3]{a^3 \cdot 2} + \sqrt[3]{8a^3 \cdot 2} = a\sqrt[3]{2} + 2a\sqrt[3]{2} =$

$3a\sqrt[3]{2}$

43. $(\sqrt{3} - \sqrt{2})(\sqrt{3} + \sqrt{2}) = (\sqrt{3})^2 - (\sqrt{2})^2 = 3 - 2 = 1$

45. $(\sqrt{8} + 2\sqrt{5})(\sqrt{8} - 2\sqrt{5}) = (\sqrt{8})^2 - (2\sqrt{5})^2 =$

$8 - 4 \cdot 5 = 8 - 20 = -12$

47. $(7 + \sqrt{5})(7 - \sqrt{5}) = 7^2 - (\sqrt{5})^2 = 49 - 5 = 44$

49. $(2 - \sqrt{3})(2 + \sqrt{3}) = 2^2 - (\sqrt{3})^2 = 4 - 3 = 1$

51. $(\sqrt{8} + \sqrt{5})(\sqrt{8} - \sqrt{5}) = (\sqrt{8})^2 - (\sqrt{5})^2 = 8 - 5 = 3$

53. $(3 + 2\sqrt{7})(3 - 2\sqrt{7}) = 3^2 - (2\sqrt{7})^2 =$

$9 - 4 \cdot 7 = 9 - 28 = -19$

55. $(\sqrt{a} + \sqrt{b})(\sqrt{a} - \sqrt{b}) = (\sqrt{a})^2 - (\sqrt{b})^2 = a - b$

57. $\quad (3 - \sqrt{5})(2 + \sqrt{5})$

$= 3 \cdot 2 + 3\sqrt{5} - 2\sqrt{5} - (\sqrt{5})^2$ Using FOIL

$= 6 + 3\sqrt{5} - 2\sqrt{5} - 5$

$= 1 + \sqrt{5}$ Simplifying

59. $\quad (\sqrt{3} + 1)(2\sqrt{3} + 1)$

$= \sqrt{3} \cdot 2\sqrt{3} + \sqrt{3} \cdot 1 + 1 \cdot 2\sqrt{3} + 1^2$ Using FOIL

$= 2 \cdot 3 + \sqrt{3} + 2\sqrt{3} + 1$

$= 7 + 3\sqrt{3}$ Simplifying

61. $(2\sqrt{7} - 4\sqrt{2})(3\sqrt{7} + 6\sqrt{2}) =$

$2\sqrt{7} \cdot 3\sqrt{7} + 2\sqrt{7} \cdot 6\sqrt{2} - 4\sqrt{2} \cdot 3\sqrt{7} - 4\sqrt{2} \cdot 6\sqrt{2} =$

$6 \cdot 7 + 12\sqrt{14} - 12\sqrt{14} - 24 \cdot 2 =$

$42 + 12\sqrt{14} - 12\sqrt{14} - 48 = -6$

63. $(\sqrt{a} + \sqrt{2})(\sqrt{a} + \sqrt{3}) =$

$(\sqrt{a})^2 + \sqrt{a} \cdot \sqrt{3} + \sqrt{2} \cdot \sqrt{a} + \sqrt{2} \cdot \sqrt{3} =$

$a + \sqrt{3a} + \sqrt{2a} + \sqrt{6}$

65. $(2\sqrt[3]{3} + \sqrt[3]{2})(\sqrt[3]{3} - 2\sqrt[3]{2}) =$

$2\sqrt[3]{3} \cdot \sqrt[3]{3} - 2\sqrt[3]{3} \cdot 2\sqrt[3]{2} + \sqrt[3]{2} \cdot \sqrt[3]{3} - \sqrt[3]{2} \cdot 2\sqrt[3]{2} =$

$2\sqrt[3]{9} - 4\sqrt[3]{6} + \sqrt[3]{6} - 2\sqrt[3]{4} = 2\sqrt[3]{9} - 3\sqrt[3]{6} - 2\sqrt[3]{4}$

67. $(2 + \sqrt{3})^2 = 2^2 + 4\sqrt{3} + (\sqrt{3})^2$ Squaring a binomial

$= 4 + 4\sqrt{3} + 3$

$= 7 + 4\sqrt{3}$

69. $\quad (\sqrt[5]{9} - \sqrt[5]{3})(\sqrt[5]{8} + \sqrt[5]{27})$

$= \sqrt[5]{9} \cdot \sqrt[5]{8} + \sqrt[5]{9} \cdot \sqrt[5]{27} - \sqrt[5]{3} \cdot \sqrt[5]{8} - \sqrt[5]{3} \cdot \sqrt[5]{27}$

Using FOIL

$= \sqrt[5]{72} + \sqrt[5]{243} - \sqrt[5]{24} - \sqrt[5]{81}$

$= \sqrt[5]{72} + 3 - \sqrt[5]{24} - \sqrt[5]{81}$

71. Discussion and Writing Exercise

73. $\dfrac{x^3 + 4x}{x^2 - 16} \div \dfrac{x^2 + 8x + 15}{x^2 + x - 20}$

$= \dfrac{x^3 + 4x}{x^2 - 16} \cdot \dfrac{x^2 + x - 20}{x^2 + 8x + 15}$

$= \dfrac{(x^3 + 4x)(x^2 + x - 20)}{(x^2 - 16)(x^2 + 8x + 15)}$

$= \dfrac{x(x^2 + 4)(x + 5)(x - 4)}{(x + 4)(x - 4)(x + 3)(x + 5)}$

$= \dfrac{x(x^2 + 4)(x + 5)(x - 4)}{(x + 4)(x - 4)(x + 3)(x + 5)}$

$= \dfrac{x(x^2 + 4)}{(x + 4)(x + 3)}$

75. $\dfrac{a^3+8}{a^2-4} \cdot \dfrac{a^2-4a+4}{a^2-2a+4}$

$= \dfrac{(a^3+8)(a^2-4a+4)}{(a^2-4)(a^2-2a+4)}$

$= \dfrac{(a+2)(a^2-2a+4)(a-2)(a-2)}{(a+2)(a-2)(a^2-2a+4)(1)}$

$= \dfrac{(a+2)(a^2-2a+4)(a-2)}{(a+2)(a^2-2a+4)(a-2)} \cdot \dfrac{a-2}{1}$

$= a-2$

77. $\dfrac{x-\dfrac{1}{3}}{x+\dfrac{1}{4}} = \dfrac{x-\dfrac{1}{3}}{x+\dfrac{1}{4}} \cdot \dfrac{12}{12}$

$= \dfrac{\left(x-\dfrac{1}{3}\right)(12)}{\left(x+\dfrac{1}{4}\right)(12)}$

$= \dfrac{12x-4}{12x+3}, \text{ or } \dfrac{4(3x-1)}{3(4x+1)}$

79. $\dfrac{\dfrac{1}{p}-\dfrac{1}{q}}{\dfrac{1}{p^2}-\dfrac{1}{q^2}} = \dfrac{\dfrac{1}{p}-\dfrac{1}{q}}{\dfrac{1}{p^2}-\dfrac{1}{q^2}} \cdot \dfrac{p^2q^2}{p^2q^2}$

$= \dfrac{\left(\dfrac{1}{p}-\dfrac{1}{q}\right)(p^2q^2)}{\left(\dfrac{1}{p^2}-\dfrac{1}{q^2}\right)(p^2q^2)}$

$= \dfrac{p^2q^2 \cdot \dfrac{1}{p} - p^2q^2 \cdot \dfrac{1}{q}}{p^2q^2 \cdot \dfrac{1}{p^2} - p^2q^2 \cdot \dfrac{1}{q^2}}$

$= \dfrac{pq^2 - p^2q}{q^2 - p^2}$

$= \dfrac{pq(q-p)}{(q+p)(q-p)}$

$= \dfrac{pq(q-p)}{(q+p)(q-p)}$

$= \dfrac{pq}{q+p}$

81.

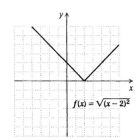

$f(x) = \sqrt{(x-2)^2}$

Since $(x-2)^2$ is nonnegative for all values of x, the domain of f is $\{x|x$ is a real number$\}$, or $(-\infty,\infty)$.

83. $\sqrt{9+3\sqrt{5}}\sqrt{9-3\sqrt{5}} = \sqrt{(9+3\sqrt{5})(9-3\sqrt{5})} =$

$\sqrt{9^2 - (3\sqrt{5})^2} = \sqrt{81-9\cdot5} = \sqrt{81-45} = \sqrt{36} = 6$

85. $(\sqrt{3}+\sqrt{5}-\sqrt{6})^2 = [(\sqrt{3}+\sqrt{5})-\sqrt{6}]^2 =$

$(\sqrt{3}+\sqrt{5})^2 - 2(\sqrt{3}+\sqrt{5})(\sqrt{6}) + (\sqrt{6})^2 =$

$3 + 2\sqrt{15} + 5 - 2\sqrt{18} - 2\sqrt{30} + 6 =$

$14 + 2\sqrt{15} - 2\sqrt{9\cdot2} - 2\sqrt{30} =$

$14 + 2\sqrt{15} - 6\sqrt{2} - 2\sqrt{30}$

87. $(\sqrt[3]{9}-2)(\sqrt[3]{9}+4)$

$= \sqrt[3]{9}\sqrt[3]{9} + 4\sqrt[3]{9} - 2\sqrt[3]{9} - 2\cdot4$

$= \sqrt[3]{81} + 2\sqrt[3]{9} - 8$

$= \sqrt[3]{27\cdot3} + 2\sqrt[3]{9} - 8$

$= 3\sqrt[3]{3} + 2\sqrt[3]{9} - 8$

Exercise Set 15.5

1. $\sqrt{\dfrac{5}{3}} = \sqrt{\dfrac{5}{3}\cdot\dfrac{3}{3}} = \sqrt{\dfrac{15}{9}} = \dfrac{\sqrt{15}}{\sqrt{9}} = \dfrac{\sqrt{15}}{3}$

3. $\sqrt{\dfrac{11}{2}} = \sqrt{\dfrac{11}{2}\cdot\dfrac{2}{2}} = \sqrt{\dfrac{22}{4}} = \dfrac{\sqrt{22}}{\sqrt{4}} = \dfrac{\sqrt{22}}{2}$

5. $\dfrac{2\sqrt{3}}{7\sqrt{5}} = \dfrac{2\sqrt{3}}{7\sqrt{5}}\cdot\dfrac{\sqrt{5}}{\sqrt{5}} = \dfrac{2\sqrt{15}}{7\sqrt{5^2}} = \dfrac{2\sqrt{15}}{7\cdot5} = \dfrac{2\sqrt{15}}{35}$

7. $\sqrt[3]{\dfrac{16}{9}} = \sqrt[3]{\dfrac{16}{9}\cdot\dfrac{3}{3}} = \sqrt[3]{\dfrac{48}{27}} = \dfrac{\sqrt[3]{8\cdot6}}{\sqrt[3]{27}} = \dfrac{2\sqrt[3]{6}}{3}$

9. $\dfrac{\sqrt[3]{3a}}{\sqrt[3]{5c}} = \dfrac{\sqrt[3]{3a}}{\sqrt[3]{5c}}\cdot\dfrac{\sqrt[3]{5^2c^2}}{\sqrt[3]{5^2c^2}} = \dfrac{\sqrt[3]{75ac^2}}{\sqrt[3]{5^3c^3}} = \dfrac{\sqrt[3]{75ac^2}}{5c}$

11. $\dfrac{\sqrt[3]{2y^4}}{\sqrt[3]{6x^4}} = \dfrac{\sqrt[3]{2y^4}}{\sqrt[3]{6x^4}}\cdot\dfrac{\sqrt[3]{6^2x^2}}{\sqrt[3]{6^2x^2}} = \dfrac{\sqrt[3]{72x^2y^4}}{\sqrt[3]{6^3x^6}} = \dfrac{\sqrt[3]{8y^3\cdot9x^2y}}{6x^2} =$

$\dfrac{2y\sqrt[3]{9x^2y}}{6x^2} = \dfrac{y\sqrt[3]{9x^2y}}{3x^2}$

13. $\dfrac{1}{\sqrt[4]{st}} = \dfrac{1}{\sqrt[4]{st}}\cdot\dfrac{\sqrt[4]{s^3t^3}}{\sqrt[4]{s^3t^3}} = \dfrac{\sqrt[4]{s^3t^3}}{\sqrt[4]{s^4t^4}} = \dfrac{\sqrt[4]{s^3t^3}}{st}$

15. $\sqrt{\dfrac{3x}{20}} = \sqrt{\dfrac{3x}{20}\cdot\dfrac{5}{5}} = \sqrt{\dfrac{15x}{100}} = \dfrac{\sqrt{15x}}{\sqrt{100}} = \dfrac{\sqrt{15x}}{10}$

17. $\sqrt[3]{\dfrac{4}{5x^5y^2}} = \sqrt[3]{\dfrac{4}{5x^5y^2}\cdot\dfrac{25xy}{5^2xy}} = \sqrt[3]{\dfrac{100xy}{5^3x^6y^3}} =$

$\dfrac{\sqrt[3]{100xy}}{\sqrt[3]{5^3x^6y^3}} = \dfrac{\sqrt[3]{100xy}}{5x^2y}$

19. $\sqrt[4]{\dfrac{1}{8x^7y^3}} = \sqrt[4]{\dfrac{1}{2^3x^7y^3}\cdot\dfrac{2xy}{2xy}} = \sqrt[4]{\dfrac{2xy}{2^4x^8y^4}} = \dfrac{\sqrt[4]{2xy}}{\sqrt[4]{2^4x^8y^4}} =$

$\dfrac{\sqrt[4]{2xy}}{2x^2y}$

21. $\dfrac{9}{6-\sqrt{10}} = \dfrac{9}{6-\sqrt{10}}\cdot\dfrac{6+\sqrt{10}}{6+\sqrt{10}} = \dfrac{9(6+\sqrt{10})}{6^2-(\sqrt{10})^2} =$

$\dfrac{9(6+\sqrt{10})}{36-10} = \dfrac{54+9\sqrt{10}}{26}$

23. $\dfrac{-4\sqrt{7}}{\sqrt{5}-\sqrt{3}} = \dfrac{-4\sqrt{7}}{\sqrt{5}-\sqrt{3}} \cdot \dfrac{\sqrt{5}+\sqrt{3}}{\sqrt{5}+\sqrt{3}} =$

$\dfrac{-4\sqrt{7}(\sqrt{5}+\sqrt{3})}{(\sqrt{5})^2-(\sqrt{3})^2} = \dfrac{-4\sqrt{7}(\sqrt{5}+\sqrt{3})}{5-3} =$

$\dfrac{-4\sqrt{7}(\sqrt{5}+\sqrt{3})}{2} = -2\sqrt{7}(\sqrt{5}+\sqrt{3}) = -2\sqrt{35}-2\sqrt{21}$

25. $\dfrac{\sqrt{5}-2\sqrt{6}}{\sqrt{3}-4\sqrt{5}} = \dfrac{\sqrt{5}-2\sqrt{6}}{\sqrt{3}-4\sqrt{5}} \cdot \dfrac{\sqrt{3}+4\sqrt{5}}{\sqrt{3}+4\sqrt{5}} =$

$\dfrac{\sqrt{15}+4\cdot5-2\sqrt{18}-8\sqrt{30}}{(\sqrt{3})^2-(4\sqrt{5})^2} =$

$\dfrac{\sqrt{15}+20-2\sqrt{9\cdot2}-8\sqrt{30}}{3-16\cdot5} =$

$\dfrac{\sqrt{15}+20-6\sqrt{2}-8\sqrt{30}}{-77}$, or $-\dfrac{\sqrt{15}+20-6\sqrt{2}-8\sqrt{30}}{77}$

27. $\dfrac{2-\sqrt{a}}{3+\sqrt{a}} = \dfrac{2-\sqrt{a}}{3+\sqrt{a}} \cdot \dfrac{3-\sqrt{a}}{3-\sqrt{a}} = \dfrac{6-2\sqrt{a}-3\sqrt{a}+a}{9-a} =$

$\dfrac{6-5\sqrt{a}+a}{9-a}$

29. $\dfrac{5\sqrt{3}-3\sqrt{2}}{3\sqrt{2}-2\sqrt{3}} = \dfrac{5\sqrt{3}-3\sqrt{2}}{3\sqrt{2}-2\sqrt{3}} \cdot \dfrac{3\sqrt{2}+2\sqrt{3}}{3\sqrt{2}+2\sqrt{3}} =$

$\dfrac{15\sqrt{6}+10\cdot3-9\cdot2-6\sqrt{6}}{9\cdot2-4\cdot3} = \dfrac{12+9\sqrt{6}}{6} =$

$\dfrac{3(4+3\sqrt{6})}{3\cdot2} = \dfrac{4+3\sqrt{6}}{2}$

31. $\dfrac{\sqrt{x}-\sqrt{y}}{\sqrt{x}+\sqrt{y}} = \dfrac{\sqrt{x}-\sqrt{y}}{\sqrt{x}+\sqrt{y}} \cdot \dfrac{\sqrt{x}-\sqrt{y}}{\sqrt{x}-\sqrt{y}} =$

$\dfrac{x-\sqrt{xy}-\sqrt{xy}+y}{x-y} = \dfrac{x-2\sqrt{xy}+y}{x-y}$

33. Discussion and Writing Exercise

35. $\dfrac{1}{2}-\dfrac{1}{3} = \dfrac{5}{t}$, LCM is $6t$

$6t\left(\dfrac{1}{2}-\dfrac{1}{3}\right) = 6t\left(\dfrac{5}{t}\right)$

$3t-2t = 30$

$t = 30$

Check:

$\dfrac{\dfrac{1}{2}-\dfrac{1}{3} = \dfrac{1}{t}}{\begin{array}{c|c} \dfrac{1}{2}-\dfrac{1}{3} \; ? \; \dfrac{5}{30} \\ \dfrac{3}{6}-\dfrac{2}{6} & \dfrac{1}{6} \\ \dfrac{1}{6} & \text{TRUE} \end{array}}$

The solution is 30.

37. $\dfrac{1}{x^3-y^2} \div \dfrac{1}{(x-y)(x^2+xy+y^2)}$

$= \dfrac{1}{(x-y)(x^2+xy+y^2)} \cdot \dfrac{(x-y)(x^2+xy+y^2)}{1}$

$= \dfrac{(x-y)(x^2+xy+y^2)}{(x-y)(x^2+xy+y^2)}$

$= 1$

39. Left to the student

41. $\sqrt{a^2-3} - \dfrac{a^2}{\sqrt{a^2-3}}$

$= \sqrt{a^2-3} - \dfrac{a^2}{\sqrt{a^2-3}} \cdot \dfrac{\sqrt{a^2-3}}{\sqrt{a^2-3}}$

$= \sqrt{a^2-3} - \dfrac{a^2\sqrt{a^2-3}}{a^2-3}$

$= \sqrt{a^2-3} \cdot \dfrac{a^2-3}{a^2-3} - \dfrac{a^2\sqrt{a^2-3}}{a^2-3}$

$= \dfrac{a^2\sqrt{a^2-3}-3\sqrt{a^2-3}-a^2\sqrt{a^2-3}}{a^2-3}$

$= \dfrac{-3\sqrt{a^2-3}}{a^2-3}$, or $-\dfrac{3\sqrt{a^2-3}}{a^2-3}$

Exercise Set 15.6

1. $\sqrt{2x-3} = 4$

$(\sqrt{2x-3})^2 = 4^2$ Principle of powers

$2x-3 = 16$

$2x = 19$

$x = \dfrac{19}{2}$

Check:

$\dfrac{\sqrt{2x-3} = 4}{\begin{array}{c|c} \sqrt{2\cdot\dfrac{19}{2}-3} \; ? \; 4 \\ \sqrt{19-3} \\ \sqrt{16} \\ 4 & \text{TRUE} \end{array}}$

The solution is $\dfrac{19}{2}$.

3. $\sqrt{6x}+1 = 8$

$\sqrt{6x} = 7$ Subtracting to isolate the radical

$(\sqrt{6x})^2 = 7^2$ Principle of powers

$6x = 49$

$x = \dfrac{49}{6}$

Check:

$\dfrac{\sqrt{6x}+1 = 8}{\begin{array}{c|c} \sqrt{6\cdot\dfrac{49}{6}}+1 \; ? \; 8 \\ \sqrt{49}+1 \\ 7+1 \\ 8 & \text{TRUE} \end{array}}$

The solution is $\dfrac{49}{6}$.

5. $\sqrt{y+7} - 4 = 4$

$\sqrt{y+7} = 8$ Adding to isolate the radical

$(\sqrt{y+7})^2 = 8^2$ Principle of powers

$y + 7 = 64$

$y = 57$

Check: $\dfrac{\sqrt{y+7} - 4 = 4}{}$

$\sqrt{57 + 7} - 4 ~?~ 4$

$\sqrt{64} - 4 ~\Big|$

$8 - 4 ~\Big|$

$4 ~\Big|$ TRUE

The solution is 57.

7. $\sqrt{5y+8} = 10$

$(\sqrt{5y+8})^2 = 10^2$ Principle of powers

$5y + 8 = 100$

$5y = 92$

$y = \dfrac{92}{5}$

Check: $\dfrac{\sqrt{5y+8} = 10}{}$

$\sqrt{5 \cdot \dfrac{92}{5} + 8} ~?~ 10$

$\sqrt{92 + 8} ~\Big|$

$\sqrt{100} ~\Big|$

$10 ~\Big|$ TRUE

The solution is $\dfrac{92}{5}$.

9. $\sqrt[3]{x} = -1$

$(\sqrt[3]{x})^3 = (-1)^3$ Principle of powers

$x = -1$

Check: $\dfrac{\sqrt[3]{x} = -1}{}$

$\sqrt[3]{-1} ~?~ -1$

$-1 ~\Big|$ TRUE

The solution is -1.

11. $\sqrt{x+2} = -4$

$(\sqrt{x+2})^2 = (-4)^2$

$x + 2 = 16$

$x = 14$

Check: $\dfrac{\sqrt{x+2} = -4}{}$

$\sqrt{14 + 2} ~?~ -4$

$\sqrt{16} ~\Big|$

$4 ~\Big|$ FALSE

The number 14 does not check. The equation has no solution. We might have observed at the outset that this equation has no solution because the principle square root of a number is never negative.

13. $\sqrt[3]{x+5} = 2$

$(\sqrt[3]{x+5})^3 = 2^3$

$x + 5 = 8$

$x = 3$

Check: $\dfrac{\sqrt[3]{x+5} = 2}{}$

$\sqrt[3]{3 + 5} ~?~ 2$

$\sqrt[3]{8} ~\Big|$

$2 ~\Big|$ TRUE

The solution is 3.

15. $\sqrt[4]{y-3} = 2$

$(\sqrt[4]{y-3})^4 = 2^4$

$y - 3 = 16$

$y = 19$

Check: $\dfrac{\sqrt[4]{y-3} = 2}{}$

$\sqrt[4]{19 - 3} ~?~ 2$

$\sqrt[4]{16} ~\Big|$

$2 ~\Big|$ TRUE

The solution is 19.

17. $\sqrt[3]{6x+9} + 8 = 5$

$\sqrt[3]{6x+9} = -3$

$(\sqrt[3]{6x+9})^3 = (-3)^3$

$6x + 9 = -27$

$6x = -36$

$x = -6$

Check: $\dfrac{\sqrt[3]{6x+9} + 8 = 5}{}$

$\sqrt[3]{6(-6) + 9} + 8 ~?~ 5$

$\sqrt[3]{-27} + 8 ~\Big|$

$-3 + 8 ~\Big|$

$5 ~\Big|$ TRUE

The solution is -6.

19. $8 = \dfrac{1}{\sqrt{x}}$

$8 \cdot \sqrt{x} = \dfrac{1}{\sqrt{x}} \cdot \sqrt{x}$

$8\sqrt{x} = 1$

$(8\sqrt{x})^2 = 1^2$

$64x = 1$

$x = \dfrac{1}{64}$

Check:

$$8 = \frac{1}{\sqrt{x}}$$

$$8 \ ? \ \frac{1}{\sqrt{\dfrac{1}{64}}}$$

$$\frac{1}{\dfrac{1}{8}}$$

$$8 \qquad \text{TRUE}$$

The solution is $\frac{1}{64}$.

21.
$$x - 7 = \sqrt{x - 5}$$
$$(x - 7)^2 = (\sqrt{x - 5})^2$$
$$x^2 - 14x + 49 = x - 5$$
$$x^2 - 15x + 54 = 0$$
$$(x - 6)(x - 9) = 0$$
$$x - 6 = 0 \ \ or \ \ x - 9 = 0$$
$$x = 6 \ \ or \qquad x = 9$$

Check: For 6:
$$x - 7 = \sqrt{x - 5}$$
$$6 - 7 \ ? \ \sqrt{6 - 5}$$
$$-1 \ \Big| \ \sqrt{1}$$
$$-1 \ \Big| \qquad 1 \ \ \text{FALSE}$$

Check: For 9:
$$x - 7 = \sqrt{x - 5}$$
$$9 - 7 \ ? \ \sqrt{9 - 5}$$
$$2 \ \Big| \ \sqrt{4}$$
$$2 \ \Big| \qquad 2 \ \ \text{TRUE}$$

The number 6 does not check, but 9 does. The solution is 9.

23.
$$2\sqrt{x + 1} + 7 = x$$
$$2\sqrt{x + 1} = x - 7$$
$$(2\sqrt{x + 1})^2 = (x - 7)^2$$
$$4(x + 1) = x^2 - 14x + 49$$
$$4x + 4 = x^2 - 14x + 49$$
$$0 = x^2 - 18x + 45$$
$$0 = (x - 3)(x - 15)$$
$$x - 3 = 0 \ \ or \ \ x - 15 = 0$$
$$x = 3 \ \ or \qquad x = 15$$

Check: For 3:
$$2\sqrt{x + 1} + 7 = x$$
$$2\sqrt{3 + 1} + 7 \ ? \ 3$$
$$2\sqrt{4} + 7$$
$$2 \cdot 2 + 7$$
$$4 + 7$$
$$11 \ \Big| \ 3 \ \ \text{FALSE}$$

Check: For 15:
$$2\sqrt{x + 1} + 7 = x$$
$$2\sqrt{15 + 1} + 7 \ ? \ 15$$
$$2\sqrt{16} + 7$$
$$2 \cdot 4 + 7$$
$$8 + 7$$
$$15 \ \Big| \ 15 \ \ \text{TRUE}$$

The number 3 does not check, but 15 does. The solution is 15.

25.
$$3\sqrt{x - 1} - 1 = x$$
$$3\sqrt{x - 1} = x + 1$$
$$(3\sqrt{x - 1})^2 = (x + 1)^2$$
$$9(x - 1) = x^2 + 2x + 1$$
$$9x - 9 = x^2 + 2x + 1$$
$$0 = x^2 - 7x + 10$$
$$0 = (x - 2)(x - 5)$$
$$x - 2 = 0 \ \ or \ \ x - 5 = 0$$
$$x = 2 \ \ or \qquad x = 5$$

Check: For 2:
$$3\sqrt{x - 1} - 1 = x$$
$$3\sqrt{2 - 1} - 1 \ ? \ 2$$
$$3\sqrt{1} - 1$$
$$3 \cdot 1 - 1$$
$$3 - 1$$
$$2 \ \Big| \ 2 \ \ \text{TRUE}$$

Check: For 5:
$$3\sqrt{x - 1} - 1 = x$$
$$3\sqrt{5 - 1} - 1 \ ? \ 5$$
$$3\sqrt{4} - 1$$
$$3 \cdot 2 - 1$$
$$6 - 1$$
$$5 \ \Big| \ 5 \ \ \text{TRUE}$$

Both numbers check. The solutions are 2 and 5.

27.
$$x - 3 = \sqrt{27 - 3x}$$
$$(x - 3)^2 = (\sqrt{27 - 3x})^2$$
$$x^2 - 6x + 9 = 27 - 3x$$
$$x^2 - 3x - 18 = 0$$
$$(x - 6)(x + 3) = 0$$
$$x - 6 = 0 \ \ or \ \ x + 3 = 0$$
$$x = 6 \ \ or \qquad x = -3$$

Check: For 6:
$$x - 3 = \sqrt{27 - 3x}$$
$$6 - 3 \ ? \ \sqrt{27 - 3 \cdot 6}$$
$$3 \ \Big| \ \sqrt{27 - 18}$$
$$\Big| \ \sqrt{9}$$
$$3 \ \Big| \qquad 3 \ \ \text{TRUE}$$

Check: For -3:

$$x - 3 = \sqrt{27 - 3x}$$

$$\begin{array}{c|c} -3 - 3 \ ? & \sqrt{27 - 3(-3)} \\ -6 & \sqrt{27 + 9} \\ & \sqrt{36} \\ -6 & 6 \quad \text{FALSE} \end{array}$$

The number 6 checks but -3 does not. The solution is 6.

29.
$$\sqrt{3y + 1} = \sqrt{2y + 6}$$
$$(\sqrt{3y + 1})^2 = (\sqrt{2y + 6})^2$$
$$3y + 1 = 2y + 6$$
$$y = 5$$

Check:
$$\sqrt{3y + 1} = \sqrt{2y + 6}$$
$$\begin{array}{c|c} \sqrt{3 \cdot 5 + 1} \ ? & \sqrt{2 \cdot 5 + 6} \\ \sqrt{16} & \sqrt{16} \quad \text{TRUE} \end{array}$$

The solution is 5.

31.
$$\sqrt{y - 5} + \sqrt{y} = 5$$
$$\sqrt{y - 5} = 5 - \sqrt{y} \qquad \text{Isolating one radical}$$
$$(\sqrt{y - 5})^2 = (5 - \sqrt{y})^2$$
$$y - 5 = 25 - 10\sqrt{y} + y$$
$$10\sqrt{y} = 30 \qquad \text{Isolating the remaining radical}$$
$$\sqrt{y} = 3 \qquad \text{Dividing by 10}$$
$$(\sqrt{y})^2 = 3^2$$
$$y = 9$$

The number 9 checks, so it is the solution.

33.
$$3 + \sqrt{z - 6} = \sqrt{z + 9}$$
$$(3 + \sqrt{z - 6})^2 = (\sqrt{z + 9})^2$$
$$9 + 6\sqrt{z - 6} + z - 6 = z + 9$$
$$6\sqrt{z - 6} = 6$$
$$\sqrt{z - 6} = 1 \qquad \text{Dividing by 6}$$
$$(\sqrt{z - 6})^2 = 1^2$$
$$z - 6 = 1$$
$$z = 7$$

The number 7 checks, so it is the solution.

35.
$$\sqrt{20 - x} + 8 = \sqrt{9 - x} + 11$$
$$\sqrt{20 - x} = \sqrt{9 - x} + 3 \qquad \text{Isolating one radical}$$
$$(\sqrt{20 - x})^2 = (\sqrt{9 - x} + 3)^2$$
$$20 - x = 9 - x + 6\sqrt{9 - x} + 9$$
$$2 = 6\sqrt{9 - x} \qquad \text{Isolating the remaining radical}$$
$$1 = 3\sqrt{9 - x} \qquad \text{Dividing by 2}$$
$$1^2 = (3\sqrt{9 - x})^2$$
$$1 = 9(9 - x)$$
$$1 = 81 - 9x$$
$$9x = 80$$
$$x = \frac{80}{9}$$

The number $\frac{80}{9}$ checks, so it is the solution.

37.
$$\sqrt{4y + 1} - \sqrt{y - 2} = 3$$
$$\sqrt{4y + 1} = 3 + \sqrt{y - 2} \qquad \text{Isolating one radical}$$
$$(\sqrt{4y + 1})^2 = (3 + \sqrt{y - 2})^2$$
$$4y + 1 = 9 + 6\sqrt{y - 2} + y - 2$$
$$3y - 6 = 6\sqrt{y - 2} \qquad \text{Isolating the remaining radical}$$
$$y - 2 = 2\sqrt{y - 2} \qquad \text{Multiplying by } \frac{1}{3}$$
$$(y - 2)^2 = (2\sqrt{y - 2})^2$$
$$y^2 - 4y + 4 = 4(y - 2)$$
$$y^2 - 4y + 4 = 4y - 8$$
$$y^2 - 8y + 12 = 0$$
$$(y - 6)(y - 2) = 0$$
$$y - 6 = 0 \quad or \quad y - 2 = 0$$
$$y = 6 \quad or \qquad y = 2$$

The numbers 6 and 2 check, so they are the solutions.

39.
$$\sqrt{x + 2} + \sqrt{3x + 4} = 2$$
$$\sqrt{x + 2} = 2 - \sqrt{3x + 4} \qquad \text{Isolating one radical}$$
$$(\sqrt{x + 2})^2 = (2 - \sqrt{3x - 4})^2$$
$$x + 2 = 4 - 4\sqrt{3x + 4} + 3x + 4$$
$$-2x - 6 = -4\sqrt{3x + 4} \qquad \text{Isolating the remaining radical}$$
$$x + 3 = 2\sqrt{3x + 4} \qquad \text{Dividing by 2}$$
$$(x + 3)^2 = (2\sqrt{3x + 4})^2$$
$$x^2 + 6x + 9 = 4(3x + 4)$$
$$x^2 + 6x + 9 = 12x + 16$$
$$x^2 - 6x - 7 = 0$$
$$(x - 7)(x + 1) = 0$$
$$x - 7 = 0 \quad or \quad x + 1 = 0$$
$$x = 7 \quad or \qquad x = -1$$

Check: For 7: $$\sqrt{x+2} + \sqrt{3x+4} = 2$$
$$\sqrt{7+2} + \sqrt{3\cdot 7+4} \; ? \; 2$$
$$\sqrt{9} + \sqrt{25}$$
$$8 \;\Big|\; \text{FALSE}$$

Check: For -1: $$\sqrt{x+2} + \sqrt{3x+4} = 2$$
$$\sqrt{-1+2} + \sqrt{3(-1)+4} \; ? \; 2$$
$$\sqrt{1} + \sqrt{1}$$
$$2 \;\Big|\; \text{TRUE}$$

Since -1 checks but 7 does not, the solution is -1.

41. $\sqrt{3x-5} + \sqrt{2x+3} + 1 = 0$
$$\sqrt{3x-5} + 1 = -\sqrt{2x+3}$$
$$(\sqrt{3x-5} + 1)^2 = (-\sqrt{2x+3})^2$$
$$3x - 5 + 2\sqrt{3x-5} + 1 = 2x+3$$
$$2\sqrt{3x-5} = -x + 7$$
$$(2\sqrt{3x-5})^2 = (-x+7)^2$$
$$4(3x-5) = x^2 - 14x + 49$$
$$12x - 20 = x^2 - 14x + 49$$
$$0 = x^2 - 26x + 69$$
$$0 = (x-23)(x-3)$$
$$x - 23 = 0 \quad or \quad x - 3 = 0$$
$$x = 23 \quad or \quad x = 3$$

Neither number checks. There is no solution. (At the outset we might have observed that there is no solution since the sum on the left side of the equation must be at least 1.)

43. $2\sqrt{t-1} - \sqrt{3t-1} = 0$
$$2\sqrt{t-1} = \sqrt{3t-1}$$
$$(2\sqrt{t-1})^2 = (\sqrt{3t-1})^2$$
$$4(t-1) = 3t-1$$
$$4t - 4 = 3t - 1$$
$$t = 3$$

Since 3 checks, it is the solution.

45. $V = 3.5\sqrt{h}$
$$V = 3.5\sqrt{9000}$$
$$V \approx 332$$

You can see about 332 km to the horizon at an altitude of 9000 m.

47. $V = 3.5\sqrt{h}$
$$50.4 = 3.5\sqrt{h}$$
$$(50.4)^2 = (3.5\sqrt{h})^2$$
$$2540.16 = 12.25h$$
$$207.36 = h$$

The altitude is 207.36 m.

49. $V = 3.5\sqrt{h}$
$$21 = 3.5\sqrt{h}$$
$$21^2 = (3.5\sqrt{h})^2$$
$$441 = 12.25h$$
$$36 = h$$

The height of the steeplejack's eyes is 36 m.

51. $V = 3.5\sqrt{h}$
$$V = 3.5\sqrt{37}$$
$$V \approx 21$$

The sailor can see about 21 km to the horizon.

53. At 55 mph: $$r = 2\sqrt{5L}$$
$$55 = 2\sqrt{5L}$$
$$27.5 = \sqrt{5L}$$
$$(27.5)^2 = (\sqrt{5L})^2$$
$$756.25 = 5L$$
$$151.25 = L$$

At 55 mph, a car will skid 151.25 ft.

At 75 mph: $$r = 2\sqrt{5L}$$
$$75 = 2\sqrt{5L}$$
$$37.5 = \sqrt{5L}$$
$$(37.5)^2 = (\sqrt{5L})^2$$
$$1406.25 = 5L$$
$$281.25 = L$$

At 75 mph, a car will skid 281.25 ft.

55. $$S = 21.9\sqrt{5t + 2457}$$
$$1113 = 21.9\sqrt{5t+2457}$$
$$\frac{1113}{21.9} = \sqrt{5t+2457}$$
$$\left(\frac{1113}{21.9}\right)^2 = (\sqrt{5t+2457})^2$$
$$2583 \approx 5t + 2457$$
$$126 \approx 5t$$
$$25.2 \approx t$$

The temperature was approximately $25°F$.

57.
$$T = 2\pi\sqrt{\frac{L}{32}}$$

$$1 = 2(3.14)\sqrt{\frac{L}{32}}$$

$$1 = 6.28\sqrt{\frac{L}{32}}$$

$$1^2 = \left(6.28\sqrt{\frac{L}{32}}\right)^2$$

$$1 = 39.4384 \cdot \frac{L}{32}$$

$$\frac{32}{39.4384} = L$$

$$0.81 \approx L$$

The pendulum is about 0.81 ft long.

59. Discussion and Writing Exercise

61. **Familiarize.** Let t = the time it will take Julia and George to paint the room working together.

Translate. We use the work principle.

$$\frac{t}{a} + \frac{t}{b} = 1$$

$$\frac{t}{8} + \frac{t}{10} = 1$$

Solve. We first multiply by 40 to clear fractions.

$$40\left(\frac{t}{8} + \frac{t}{10}\right) = 40 \cdot 1$$

$$40 \cdot \frac{t}{8} + 40 \cdot \frac{t}{10} = 40$$

$$5t + 4t = 40$$

$$9t = 40$$

$$t = \frac{40}{9}, \text{ or } 4\frac{4}{9}$$

Check. In $\frac{40}{9}$ hr, Julia does $\frac{40}{9}\left(\frac{1}{8}\right)$, or $\frac{5}{9}$, of the job and George does $\frac{40}{9}\left(\frac{1}{10}\right)$, or $\frac{4}{9}$, of the job. Together they do $\frac{5}{9} + \frac{4}{9}$, or 1 entire job. The answer checks.

State. It will take them $4\frac{4}{9}$ hr to paint the room, working together.

63. **Familiarize.** Let d = the distance the cyclist would travel in 56 days at the same rate.

Translate. We translate to a proportion.

$$\begin{array}{c}\text{Distance} \rightarrow \\ \text{Days} \rightarrow\end{array} \frac{702}{14} = \frac{d}{56} \begin{array}{c}\leftarrow \text{Distance} \\ \leftarrow \text{Days}\end{array}$$

Solve. We equate cross products.

$$\frac{702}{14} = \frac{d}{56}$$

$$702 \cdot 56 = 14 \cdot d$$

$$\frac{702 \cdot 56}{14} = d$$

$$2808 = d$$

Check. We substitute in the proportion and check cross products.

$$\frac{702}{14} = \frac{2808}{56}; \ 702 \cdot 56 = 39,312; \ 14 \cdot 2808 = 39,312$$

The cross products are the same, so the answer checks.

State. The cyclist would have traveled 2808 mi in 56 days.

65.
$$x^2 + 2.8x = 0$$
$$x(x + 2.8) = 0$$
$$x = 0 \ \ or \ \ x + 2.8 = 0$$
$$x = 0 \ \ or \ \ \ \ \ \ \ \ x = -2.8$$

The solutions are 0 and -2.8.

67.
$$x^2 - 64 = 0$$
$$(x + 8)(x - 8) = 0$$
$$x + 8 = 0 \ \ \ or \ \ x - 8 = 0$$
$$x = -8 \ \ or \ \ \ \ \ \ \ x = 8$$

The solutions are -8 and 8.

69. Left to the student

71.
$$\sqrt[3]{\frac{z}{4}} - 10 = 2$$

$$\sqrt[3]{\frac{z}{4}} = 12$$

$$\left(\sqrt[3]{\frac{z}{4}}\right)^3 = 12^3$$

$$\frac{z}{4} = 1728$$

$$z = 6912$$

The number 6912 checks, so it is the solution.

73.
$$\sqrt{\sqrt{y + 49} - \sqrt{y}} = \sqrt{7}$$

$$\left(\sqrt{\sqrt{y + 49} - \sqrt{y}}\right)^2 = (\sqrt{7})^2$$

$$\sqrt{y + 49} - \sqrt{y} = 7$$

$$\sqrt{y + 49} = 7 + \sqrt{y}$$

$$(\sqrt{y + 49})^2 = (7 + \sqrt{y})^2$$

$$y + 49 = 49 + 14\sqrt{y} + y$$

$$0 = 14\sqrt{y}$$

$$0 = \sqrt{y}$$

$$0^2 = (\sqrt{y})^2$$

$$0 = y$$

The number 0 checks and is the solution.

75.
$$\sqrt{\sqrt{x^2 + 9x + 34}} = 2$$
$$\left(\sqrt{\sqrt{x^2 + 9x + 34}}\right)^2 = 2^2$$
$$\sqrt{x^2 + 9x + 34} = 4$$
$$(\sqrt{x^2 + 9x + 34})^2 = 4^2$$
$$x^2 + 9x + 34 = 16$$
$$x^2 + 9x + 18 = 0$$
$$(x + 6)(x + 3) = 0$$
$$x + 6 = 0 \quad \text{or} \quad x + 3 = 0$$
$$x = -6 \quad \text{or} \quad x = -3$$

Both values check. The solutions are -6 and -3.

77.
$$\sqrt{x - 2} - \sqrt{x + 2} + 2 = 0$$
$$\sqrt{x - 2} + 2 = \sqrt{x + 2}$$
$$(\sqrt{x - 2} + 2)^2 = (\sqrt{x + 2})^2$$
$$(x - 2) + 4\sqrt{x - 2} + 4 = x + 2$$
$$4\sqrt{x - 2} = 0$$
$$\sqrt{x - 2} = 0$$
$$(\sqrt{x - 2})^2 = 0^2$$
$$x - 2 = 0$$
$$x = 2$$

The number 2 checks, so it is the solution.

79.
$$\sqrt{a^2 + 30a} = a + \sqrt{5a}$$
$$(\sqrt{a^2 + 30a})^2 = (a + \sqrt{5a})^2$$
$$a^2 + 30a = a^2 + 2a\sqrt{5a} + 5a$$
$$25a = 2a\sqrt{5a}$$
$$(25a)^2 = (2a\sqrt{5a})^2$$
$$625a^2 = 4a^2 \cdot 5a$$
$$625a^2 = 20a^3$$
$$0 = 20a^3 - 625a^2$$
$$0 = 5a^2(4a - 125)$$
$$5a^2 = 0 \quad \text{or} \quad 4a - 125 = 0$$
$$a^2 = 0 \quad \text{or} \quad 4a = 125$$
$$a = 0 \quad \text{or} \quad a = \frac{125}{4}$$

Both values check. The solutions are 0 and $\frac{125}{4}$.

81.
$$\frac{x - 1}{\sqrt{x^2 + 3x + 6}} = \frac{1}{4},$$
$$\text{LCM} = 4\sqrt{x^2 + 3x + 6}$$
$$4\sqrt{x^2 + 3x + 6} \cdot \frac{x - 1}{\sqrt{x^2 + 3x + 6}} = 4\sqrt{x^2 + 3x + 6} \cdot \frac{1}{4}$$
$$4x - 4 = \sqrt{x^2 + 3x + 6}$$
$$16x^2 - 32x + 16 = x^2 + 3x + 6$$
$$\text{Squaring both sides}$$
$$15x^2 - 35x + 10 = 0$$
$$3x^2 - 7x + 2 = 0 \qquad \text{Dividing by 5}$$
$$(3x - 1)(x - 2) = 0$$
$$3x - 1 = 0 \quad \text{or} \quad x - 2 = 0$$
$$3x = 1 \quad \text{or} \quad x = 2$$
$$x = \frac{1}{3} \quad \text{or} \quad x = 2$$

The number 2 checks but $\frac{1}{3}$ does not. The solution is 2.

83.
$$\sqrt{y^2 + 6} + y - 3 = 0$$
$$\sqrt{y^2 + 6} = 3 - y$$
$$(\sqrt{y^2 + 6})^2 = (3 - y)^2$$
$$y^2 + 6 = 9 - 6y + y^2$$
$$-3 = -6y$$
$$\frac{1}{2} = y$$

The number $\frac{1}{2}$ checks and is the solution.

85.
$$\sqrt{y + 1} - \sqrt{2y - 5} = \sqrt{y - 2}$$
$$(\sqrt{y + 1} - \sqrt{2y - 5})^2 = (\sqrt{y - 2})^2$$
$$y + 1 - 2\sqrt{(y + 1)(2y - 5)} + 2y - 5 = y - 2$$
$$-2\sqrt{2y^2 - 3y - 5} = -2y + 2$$
$$\sqrt{2y^2 - 3y - 5} = y - 1$$
$$\text{Dividing by } -2$$
$$(\sqrt{2y^2 - 3y - 5})^2 = (y - 1)^2$$
$$2y^2 - 3y - 5 = y^2 - 2y + 1$$
$$y^2 - y - 6 = 0$$
$$(y - 3)(y + 2) = 0$$
$$y - 3 = 0 \quad \text{or} \quad y + 2 = 0$$
$$y = 3 \quad \text{or} \quad y = -2$$

The number 3 checks but -2 does not. The solution is 3.

Exercise Set 15.7

1. $a = 3, \quad b = 5$

Find c.
$$c^2 = a^2 + b^2 \qquad \text{Pythagorean equation}$$
$$c^2 = 3^2 + 5^2 \qquad \text{Substituting}$$
$$c^2 = 9 + 25$$
$$c^2 = 34$$
$$c = \sqrt{34} \qquad \text{Exact answer}$$
$$c \approx 5.831 \qquad \text{Approximation}$$

3. $a = 15, \quad b = 15$

Find c.

$c^2 = a^2 + b^2$ — Pythagorean equation

$c^2 = 15^2 + 15^2$ — Substituting

$c^2 = 225 + 225$

$c^2 = 450$

$c = \sqrt{450}$ — Exact answer

$c \approx 21.213$ — Approximation

5. $b = 12, \quad c = 13$

Find a.

$a^2 + b^2 = c^2$ — Pythagorean equation

$a^2 + 12^2 = 13^2$ — Substituting

$a^2 + 144 = 169$

$a^2 = 25$

$a = 5$

7. $c = 7, \quad a = \sqrt{6}$

Find b.

$c^2 = a^2 + b^2$ — Pythagorean equation

$7^2 = (\sqrt{6})^2 + b^2$ — Substituting

$49 = 6 + b^2$

$43 = b^2$

$\sqrt{43} = b$ — Exact answer

$6.557 \approx b$ — Approximation

9. $b = 1, \quad c = \sqrt{13}$

Find a.

$a^2 + b^2 = c^2$ — Pythagorean equation

$a^2 + 1^2 = (\sqrt{13})^2$ — Substituting

$a^2 + 1 = 13$

$a^2 = 12$

$a = \sqrt{12}$ — Exact answer

$a \approx 3.464$ — Approximation

11. $a = 1, \quad c = \sqrt{n}$

Find b.

$a^2 + b^2 = c^2$

$1^2 + b^2 = (\sqrt{n})^2$

$1 + b^2 = n$

$b^2 = n - 1$

$b = \sqrt{n-1}$

13. We add labels to the drawing in the text. We let $h =$ the height of the bulge.

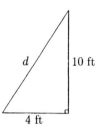

Note that 1 mi = 5280 ft, so 1 mi + 1 ft = 5280 + 1, or 5281 ft.

We use the Pythagorean equation to find h.

$5281^2 = 5280^2 + h^2$

$27,888,961 = 27,878,400 + h^2$

$10,561 = h^2$

$\sqrt{10,561} = h$

$102.767 \approx h$

The bulge is $\sqrt{10,561}$ ft, or about 102.767 ft high.

15. We add some labels to the drawing in the text.

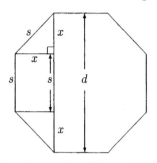

Note that $d = s + 2x$. We use the Pythagorean equation to find x.

$x^2 + x^2 = s^2$

$2x^2 = s^2$

$x^2 = \dfrac{s^2}{2}$

$x = \sqrt{\dfrac{s^2}{2}}$

$x = \dfrac{s}{\sqrt{2}}$

$x = \dfrac{s\sqrt{2}}{2}$ — Rationalizing the denominator

Then $d = s + 2x = s + 2\left(\dfrac{s\sqrt{2}}{2}\right) = s + s\sqrt{2}$.

17. We make a drawing and let $d =$ the length of the guy wire.

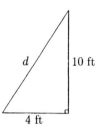

We use the Pythagorean equation to find d.

$d^2 = 4^2 + 10^2$

$d^2 = 16 + 100$

$d^2 = 116$

$d = \sqrt{116}$

$d \approx 10.770$

The wire is $\sqrt{116}$ ft, or about 10.770 ft long.

19. $L = \dfrac{0.000169d^{2.27}}{h}$

$L = \dfrac{0.000169(200)^{2.27}}{4}$

≈ 7.1

The length of the letters should be about 7.1 ft.

21. We make a drawing. Let $x =$ the width of the rectangle. Then $x + 1 =$ the length.

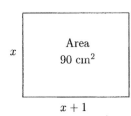

We first find the length and width of the rectangle. Recall the formula for the area of a rectangle, $A = lw$. We substitute 90 for A, $x + 1$ for l, and x for w in this formula and solve for x.

$90 = (x + 1)x$

$90 = x^2 + x$

$0 = x^2 + x - 90$

$0 = (x + 10)(x - 9)$

$x + 10 = 0 \quad$ or $\quad x - 9 = 0$

$x = -10 \quad$ or $\qquad x = 9$

Since the width cannot be negative, we know that the width is 9 cm. Thus the length is 10 cm. (These numbers check since 9 and 10 are consecutive integers and the area of a rectangle with width 9 cm and length 10 cm is $10 \cdot 9$, or 90 cm^2.)

Now we find the length of the diagonal of the rectangle. We make another drawing, letting $d =$ the length of the diagonal.

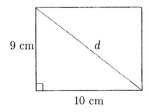

We use the Pythagorean equation to find d.

$d^2 = 9^2 + 10^2$

$d^2 = 81 + 100$

$d^2 = 181$

$d = \sqrt{181}$

$d \approx 13.454$

The length of the diagonal is $\sqrt{181}$ cm, or about 13.454 cm.

23. We use the drawing in the text, replacing w with 16 in.

We use the Pythagorean equation to find h.

$h^2 + 16^2 = 20^2$

$h^2 + 256 = 400$

$h^2 = 144$

$h = 12$

The height is 12 in.

25. We first make a drawing. A point on the x-axis has coordinates $(x, 0)$ and is $|x|$ units from the origin.

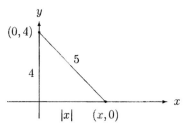

We use the Pythagorean equation to find x.

$4^2 + |x|^2 = 5^2$

$16 + x^2 = 25 \quad |x|^2 = x^2$

$x^2 - 9 = 0 \quad$ Subtracting 25

$(x + 3)(x - 3) = 0$

$x - 3 = 0 \quad$ or $\quad x + 3 = 0$

$x = 3 \quad$ or $\qquad x = -3$

The points are $(3, 0)$ and $(-3, 0)$.

27. We make a drawing, letting $d =$ the distance the wire will run diagonally, disregarding the slack.

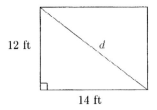

We use the Pythagorean equation to find d.

$d^2 = 12^2 + 14^2$

$d^2 = 144 + 196$

$d^2 = 340$

$d = \sqrt{340}$

Adding 4 ft of slack on each end, we find that a wire of length $\sqrt{340} + 4 + 4$, or $\sqrt{340} + 8$ ft should be purchased. This is approximately 26.439 ft of wire.

29. Referring to the drawing in the text, we let t = the travel. Then we use the Pythagorean equation.

$$t^2 = (17.75)^2 + (10.25)^2$$
$$t^2 = 315.0625 + 105.0625$$
$$t^2 = 420.125$$
$$t = \sqrt{420.125}$$
$$t \approx 20.497$$

The travel is $\sqrt{420.125}$ in., or about 20.497 in.

31. Discussion and Writing Exercise

33. **Familiarize.** Let r = the speed of the Carmel Crawler. Then $r + 14$ = the speed of the Zionsville Flash. We organize the information in a table.

	Distance	Speed	Time
Crawler	230	r	t
Flash	290	$r + 14$	t

Translate. Using the formula $t = d/r$ and noting that the times are the same, we have

$$\frac{230}{r} = \frac{290}{r + 14}.$$

Solve. We first clear fractions by multiplying by the LCM of the denominators, $r(r + 14)$.

$$r(r + 14) \cdot \frac{230}{r} = r(r + 14) \cdot \frac{290}{r + 14}$$
$$230(r + 14) = 290r$$
$$230r + 3220 = 290r$$
$$3220 = 60r$$
$$\frac{161}{3} = r, \text{ or}$$
$$53\frac{2}{3} = r$$

If $r = 53\frac{2}{3}$, then $r + 14 = 67\frac{2}{3}$.

Check. At $53\frac{2}{3}$, or $\frac{161}{3}$ mph, the Crawler travels 230 mi in $\frac{230}{161/3}$, or about 4.3 hr. At $67\frac{2}{3}$, or $\frac{203}{3}$ mph, the Flash travels 290 mi in $\frac{290}{203/3}$, or about 4.3 hr. Since the times are the same, the answer checks.

State. The Carmel Crawler's speed is $53\frac{2}{3}$ mph, and the Zionsville Flash's speed is $67\frac{2}{3}$ mph.

35. $2x^2 + 11x - 21 = 0$
$$(2x - 3)(x + 7) = 0$$
$$2x - 3 = 0 \quad or \quad x + 7 = 0$$
$$x = \frac{3}{2} \quad or \quad x = -7$$

The solutions are $\frac{3}{2}$ and -7.

37.
$$\frac{x + 2}{x + 3} = \frac{x - 4}{x - 5},$$
LCM is $(x + 3)(x - 5)$
$$(x + 3)(x - 5) \cdot \frac{x + 2}{x + 3} = (x + 3)(x - 5) \cdot \frac{x - 4}{x - 5}$$
$$(x - 5)(x + 2) = (x + 3)(x - 4)$$
$$x^2 - 3x - 10 = x^2 - x - 12$$
$$-2x = -2$$
$$x = 1$$

The number 1 checks, so it is the solution.

39. $\frac{x - 5}{x - 7} = \frac{4}{3}$, LCM is $3(x - 7)$
$$3(x - 7) \cdot \frac{x - 5}{x - 7} = 3(x - 7) \cdot \frac{4}{3}$$
$$3(x - 5) = 4(x - 7)$$
$$3x - 15 = 4x - 28$$
$$13 = x$$

The number 13 checks and is the solution.

41.

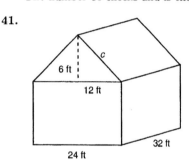

$$c^2 = 6^2 + 12^2 = 36 + 144 = 180$$
$$c = \sqrt{180} \text{ ft}$$

Area of the roof = $2 \cdot \sqrt{180} \cdot 32 = 64\sqrt{180}$ ft^2

Number of packets = $\dfrac{64\sqrt{180}}{33\frac{1}{3}} \approx 26$

Kit should buy 26 packets of shingles.

43.

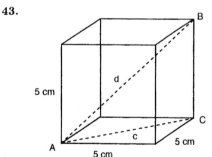

First find the length of a diagonal of the base of the cube. It is the hypotenuse of an isosceles right triangle with $a = 5$ cm. Then $c = a\sqrt{2} = 5\sqrt{2}$ cm.

Triangle ABC is a right triangle with legs of $5\sqrt{2}$ cm and 5 cm and hypotenuse d. Use the Pythagorean equation to find d, the length of the diagonal that connects two opposite corners of the cube.

$d^2 = (5\sqrt{2})^2 + 5^2$

$d^2 = 25 \cdot 2 + 25$

$d^2 = 50 + 25$

$d^2 = 75$

$d = \sqrt{75}$

Exact answer: $d = \sqrt{75}$ cm

Exercise Set 15.8

1. $\sqrt{-35} = \sqrt{-1 \cdot 35} = \sqrt{-1} \cdot \sqrt{35} = i\sqrt{35}$, or $\sqrt{35}i$

3. $\sqrt{-16} = \sqrt{-1 \cdot 16} = \sqrt{-1} \cdot \sqrt{16} = i \cdot 4 = 4i$

5. $-\sqrt{-12} = -\sqrt{-1 \cdot 12} = -\sqrt{-1} \cdot \sqrt{12} = -i \cdot 2\sqrt{3} = -2\sqrt{3}i$, or $-2i\sqrt{3}$

7. $\sqrt{-3} = \sqrt{-1 \cdot 3} = \sqrt{-1} \cdot \sqrt{3} = i\sqrt{3}$, or $\sqrt{3}i$

9. $\sqrt{-81} = \sqrt{-1 \cdot 81} = \sqrt{-1} \cdot \sqrt{81} = i \cdot 9 = 9i$

11. $\sqrt{-98} = \sqrt{-1 \cdot 98} = \sqrt{-1} \cdot \sqrt{98} = i \cdot 7\sqrt{2} = 7\sqrt{2}i$, or $7i\sqrt{2}$

13. $-\sqrt{-49} = -\sqrt{-1 \cdot 49} = -\sqrt{-1} \cdot \sqrt{49} = -i \cdot 7 = -7i$

15. $4 - \sqrt{-60} = 4 - \sqrt{-1 \cdot 60} = 4 - \sqrt{-1} \cdot \sqrt{60} = 4 - i \cdot 2\sqrt{15} = 4 - 2\sqrt{15}i$, or $4 - 2i\sqrt{15}$

17. $\sqrt{-4} + \sqrt{-12} = \sqrt{-1 \cdot 4} + \sqrt{-1 \cdot 12} = \sqrt{-1} \cdot \sqrt{4} + \sqrt{-1} \cdot \sqrt{12} = i \cdot 2 + i \cdot 2\sqrt{3} = (2 + 2\sqrt{3})i$

19. $(7 + 2i) + (5 - 6i)$
$= (7 + 5) + (2 - 6)i$ Collecting like terms
$= 12 - 4i$

21. $(4 - 3i) + (5 - 2i)$
$= (4 + 5) + (-3 - 2)i$ Collecting like terms
$= 9 - 5i$

23. $(9 - i) + (-2 + 5i) = (9 - 2) + (-1 + 5)i$
$= 7 + 4i$

25. $(6 - i) - (10 + 3i) = (6 - 10) + (-1 - 3)i$
$= -4 - 4i$

27. $(4 - 2i) - (5 - 3i) = (4 - 5) + [-2 - (-3)]i$
$= -1 + i$

29. $(9 + 5i) - (-2 - i) = [9 - (-2)] + [5 - (-1)]i$
$= 11 + 6i$

31. $\sqrt{-36} \cdot \sqrt{-9} = \sqrt{-1} \cdot \sqrt{36} \cdot \sqrt{-1} \cdot \sqrt{9}$
$= i \cdot 6 \cdot i \cdot 3$
$= i^2 \cdot 18$
$= -1 \cdot 18 \qquad i^2 = -1$
$= -18$

33. $\sqrt{-7} \cdot \sqrt{-2} = \sqrt{-1} \cdot \sqrt{7} \cdot \sqrt{-1} \cdot \sqrt{2}$
$= i \cdot \sqrt{7} \cdot i \cdot \sqrt{2}$
$= i^2(\sqrt{14})$
$= -1(\sqrt{14}) \qquad i^2 = -1$
$= -\sqrt{14}$

35. $-3i \cdot 7i = -21 \cdot i^2$
$= -21(-1) \qquad i^2 = -1$
$= 21$

37. $-3i(-8 - 2i) = -3i(-8) - 3i(-2i)$
$= 24i + 6i^2$
$= 24i + 6(-1) \qquad i^2 = -1$
$= 24i - 6$
$= -6 + 24i$

39. $(3 + 2i)(1 + i)$
$= 3 + 3i + 2i + 2i^2$ Using FOIL
$= 3 + 3i + 2i - 2 \qquad i^2 = -1$
$= 1 + 5i$

41. $(2 + 3i)(6 - 2i)$
$= 12 - 4i + 18i - 6i^2$ Using FOIL
$= 12 - 4i + 18i + 6 \qquad i^2 = -1$
$= 18 + 14i$

43. $(6 - 5i)(3 + 4i) = 18 + 24i - 15i - 20i^2$
$= 18 + 24i - 15i + 20$
$= 38 + 9i$

45. $(7 - 2i)(2 - 6i) = 14 - 42i - 4i + 12i^2$
$= 14 - 42i - 4i - 12$
$= 2 - 46i$

47. $(3 - 2i)^2 = 3^2 - 2 \cdot 3 \cdot 2i + (2i)^2$ Squaring a binomial
$= 9 - 12i + 4i^2$
$= 9 - 12i - 4 \qquad i^2 = -1$
$= 5 - 12i$

49. $(1 + 5i)^2$
$= 1^2 + 2 \cdot 1 \cdot 5i + (5i)^2$ Squaring a binomial
$= 1 + 10i + 25i^2$
$= 1 + 10i - 25 \qquad i^2 = -1$
$= -24 + 10i$

51. $(-2 + 3i)^2 = 4 - 12i + 9i^2 = 4 - 12i - 9 = -5 - 12i$

53. $i^7 = i^6 \cdot i = (i^2)^3 \cdot i = (-1)^3 \cdot i = -1 \cdot i = -i$

55. $i^{24} = (i^2)^{12} = (-1)^{12} = 1$

57. $i^{42} = (i^2)^{21} = (-1)^{21} = -1$

59. $i^9 = (i^2)^4 \cdot i = (-1)^4 \cdot i = 1 \cdot i = i$

61. $i^6 = (i^2)^3 = (-1)^3 = -1$

63. $(5i)^3 = 5^3 \cdot i^3 = 125 \cdot i^2 \cdot i = 125(-1)(i) = -125i$

65. $7 + i^4 = 7 + (i^2)^2 = 7 + (-1)^2 = 7 + 1 = 8$

67. $i^{28} - 23i = (i^2)^{14} - 23i = (-1)^{14} - 23i = 1 - 23i$

69. $i^2 + i^4 = -1 + (i^2)^2 = -1 + (-1)^2 = -1 + 1 = 0$

71. $i^5 + i^7 = i^4 \cdot i + i^6 \cdot i = (i^2)^2 \cdot i + (i^2)^3 \cdot i =$
$(-1)^2 \cdot i + (-1)^3 \cdot i = 1 \cdot i + (-1)i = i - i = 0$

73. $1 + i + i^2 + i^3 + i^4 = 1 + i + i^2 + i^2 \cdot i + (i^2)^2$
$$= 1 + i + (-1) + (-1) \cdot i + (-1)^2$$
$$= 1 + i - 1 - i + 1$$
$$= 1$$

75. $5 - \sqrt{-64} = 5 - \sqrt{-1} \cdot \sqrt{64} = 5 - i \cdot 8 = 5 - 8i$

77. $\dfrac{8 - \sqrt{-24}}{4} = \dfrac{8 - \sqrt{-1} \cdot \sqrt{24}}{4} = \dfrac{8 - i \cdot 2\sqrt{6}}{4} =$
$\dfrac{2(4 - i\sqrt{6})}{2 \cdot 2} = \dfrac{\cancel{2}(4 - i\sqrt{6})}{\cancel{2} \cdot 2} = \dfrac{4 - i\sqrt{6}}{2} = 2 - \dfrac{\sqrt{6}}{2}i$

79. $\dfrac{4 + 3i}{3 - i} = \dfrac{4 + 3i}{3 - i} \cdot \dfrac{3 + i}{3 + i}$
$$= \dfrac{(4 + 3i)(3 + i)}{(3 - i)(3 + i)}$$
$$= \dfrac{12 + 4i + 9i + 3i^2}{9 - i^2}$$
$$= \dfrac{12 + 13i - 3}{9 - (-1)}$$
$$= \dfrac{9 + 13i}{10}$$
$$= \dfrac{9}{10} + \dfrac{13}{10}i$$

81. $\dfrac{3 - 2i}{2 + 3i} = \dfrac{3 - 2i}{2 + 3i} \cdot \dfrac{2 - 3i}{2 - 3i}$
$$= \dfrac{(3 - 2i)(2 - 3i)}{(2 + 3i)(2 - 3i)}$$
$$= \dfrac{6 - 9i - 4i + 6i^2}{4 - 9i^2}$$
$$= \dfrac{6 - 13i - 6}{4 - 9(-1)}$$
$$= \dfrac{-13i}{13}$$
$$= -i$$

83. $\dfrac{8 - 3i}{7i} = \dfrac{8 - 3i}{7i} \cdot \dfrac{-7i}{-7i}$
$$= \dfrac{-56i + 21i^2}{-49i^2}$$
$$= \dfrac{-21 - 56i}{49}$$
$$= -\dfrac{21}{49} - \dfrac{56}{49}i$$
$$= -\dfrac{3}{7} - \dfrac{8}{7}i$$

85. $\dfrac{4}{3 + i} = \dfrac{4}{3 + i} \cdot \dfrac{3 - i}{3 - i}$
$$= \dfrac{12 - 4i}{9 - i^2}$$
$$= \dfrac{12 - 4i}{9 - (-1)}$$
$$= \dfrac{12 - 4i}{10}$$
$$= \dfrac{12}{10} - \dfrac{4}{10}i$$
$$= \dfrac{6}{5} - \dfrac{2}{5}i$$

87. $\dfrac{2i}{5 - 4i} = \dfrac{2i}{5 - 4i} \cdot \dfrac{5 + 4i}{5 + 4i}$
$$= \dfrac{10i + 8i^2}{25 - 16i^2}$$
$$= \dfrac{10i + 8(-1)}{25 - 16(-1)}$$
$$= \dfrac{-8 + 10i}{41}$$
$$= -\dfrac{8}{41} + \dfrac{10}{41}i$$

89. $\dfrac{4}{3i} = \dfrac{4}{3i} \cdot \dfrac{-3i}{-3i}$
$$= \dfrac{-12i}{-9i^2}$$
$$= \dfrac{-12i}{-9(-1)}$$
$$= \dfrac{-12i}{9}$$
$$= -\dfrac{4}{3}i$$

91. $\dfrac{9 - 4i}{8i} = \dfrac{2 - 4i}{8i} \cdot \dfrac{-8i}{-8i}$
$$= \dfrac{-16i + 32i^2}{-64i^2}$$
$$= \dfrac{-16i + 32(-1)}{-64(-1)}$$
$$= \dfrac{-32 - 16i}{64}$$
$$= -\dfrac{32}{64} - \dfrac{16}{64}i$$
$$= -\dfrac{1}{2} - \dfrac{1}{4}i$$

93.
$$\frac{6+3i}{6-3i} = \frac{6+3i}{6-3i} \cdot \frac{6+3i}{6+3i}$$
$$= \frac{36+18i+18i+9i^2}{36-9i^2}$$
$$= \frac{36+36i-9}{36-9(-1)}$$
$$= \frac{27+36i}{45}$$
$$= \frac{27}{45} + \frac{36}{45}i$$
$$= \frac{3}{5} + \frac{4}{5}i$$

95. Substitute $1-2i$ for x in the equation.
$$x^2 - 2x + 5 = 0$$

$(1-2i)^2 - 2(1-2i) + 5$? 0	
$1 - 4i + 4i^2 - 2 + 4i + 5$	
$1 - 4i - 4 - 2 + 4i + 5$	
	0 $\quad$ TRUE

$1 - 2i$ is a solution.

97. Substitute $2+i$ for x in the equation.
$$x^2 - 4x - 5 = 0$$

$(2+i)^2 - 4(2+i) - 5$? 0	
$4 + 4i + i^2 - 8 - 4i - 5$	
$4 + 4i - 1 - 8 - 4i - 5$	
	-10 $\quad$ FALSE

$2 + i$ is not a solution.

99. Discussion and Writing Exercise

101.
$$\frac{196}{x^2 - 7x + 49} - \frac{2x}{x+7} = \frac{2058}{x^3 + 343}$$

Note: $x^3 + 343 = (x+7)(x^2 - 7x + 49)$.

The LCM $= (x+7)(x^2 - 7x + 49)$.

$$(x+7)(x^2 - 7x + 49)\left(\frac{196}{x^2 - 7x + 49} - \frac{2x}{x+7}\right) =$$
$$(x+7)(x^2 - 7x + 49) \cdot \frac{2058}{x^3 + 343}$$
$$196(x+7) - 2x(x^2 - 7x + 49) = 2058$$
$$196x + 1372 - 2x^3 + 14x^2 - 98x = 2058$$
$$98x - 686 - 2x^3 + 14x^2 = 0$$
$$49x - 343 - x^3 + 7x^2 = 0 \quad \text{Dividing by 2}$$
$$49(x-7) - x^2(x-7) = 0$$
$$(49 - x^2)(x-7) = 0$$
$$(7-x)(7+x)(x-7) = 0$$

$7 - x = 0 \quad or \quad 7 + x = 0 \quad or \quad x - 7 = 0$
$7 = x \quad or \quad x = -7 \quad or \quad x = 7$

Only 7 checks. It is the solution.

103. $|3x + 7| = 22$
$$3x + 7 = -22 \quad or \quad 3x + 7 = 22$$
$$3x = -29 \quad or \quad 3x = 15$$
$$x = -\frac{29}{3} \quad or \quad x = 5$$

The solutions are $-\frac{29}{3}$ and 5.

105. $|3x + 7| \geq 22$
$$3x + 7 \leq -22 \quad or \quad 3x + 7 \geq 22$$
$$3x \leq -29 \quad or \quad 3x \geq 15$$
$$x \leq -\frac{29}{3} \quad or \quad x \geq 5$$

The solution set is $\left\{x \middle| x \leq -\frac{29}{3} \; or \; x \geq 5\right\}$, or $\left(-\infty, -\frac{29}{3}\right] \cup [5, \infty)$.

107. $g(2i) = \dfrac{(2i)^4 - (2i)^2}{2i - 1} = \dfrac{16i^4 - 4i^2}{-1 + 2i} = \dfrac{20}{-1 + 2i} =$
$$\frac{20}{-1 + 2i} \cdot \frac{-1 - 2i}{-1 - 2i} = \frac{-20 - 40i}{5} = -4 - 8i;$$

$g(i+1) = \dfrac{(i+1)^4 - (i+1)^2}{(i+1) - 1} =$
$$\frac{(i+1)^2[(i+1)^2 - 1]}{i} = \frac{2i(2i-1)}{i} = 2(2i - 1) =$$
$$-2 + 4i;$$

$g(2i-1) = \dfrac{(2i-1)^4 - (2i-1)^2}{(2i-1) - 1} =$
$$\frac{(2i-1)^2[(2i-1)^2 - 1]}{2i - 2} = \frac{(-3-4i)(-4-4i)}{-2 + 2i} =$$
$$\frac{(-3-4i)(-2-2i)}{-1 + i} = \frac{-2 + 14i}{-1 + i} =$$
$$\frac{-2 + 14i}{-1 + i} \cdot \frac{-1 - i}{-1 - i} = \frac{16 - 12i}{2} = 8 - 6i$$

109. $\dfrac{1}{8}\left(-24 - \sqrt{-1024}\right) = \dfrac{1}{8}(-24 - 32i) = -3 - 4i$

111. $7\sqrt{-64} - 9\sqrt{-256} = 7 \cdot 8i - 9 \cdot 16i = 56i - 144i = -88i$

113. $(1-i)^3(1+i)^3 =$
$$(1-i)(1+i) \cdot (1-i)(1+i) \cdot (1-i)(1+i) =$$
$$(1 - i^2)(1 - i^2)(1 - i^2) = (1+1)(1+1)(1+1) =$$
$$2 \cdot 2 \cdot 2 = 8$$

115. $\dfrac{6}{1 + \dfrac{3}{i}} = \dfrac{6}{\dfrac{i+3}{i}} = \dfrac{6i}{i+3} = \dfrac{6i}{i+3} \cdot \dfrac{-i+3}{-i+3} =$
$$\frac{-6i^2 + 18i}{-i^2 + 9} = \frac{6 + 18i}{10} = \frac{6}{10} + \frac{18}{10}i = \frac{3}{5} + \frac{9}{5}i$$

117. $\dfrac{i - i^{38}}{1 + i} = \dfrac{i - (i^2)^{19}}{1 + i} = \dfrac{i - (-1)^{19}}{1 + i} = \dfrac{i - (-1)}{1 + i} =$
$$\frac{i + 1}{1 + i} = 1$$

Chapter 16

Quadratic Equations and Functions

1. a) $6x^2 = 30$

$x^2 = 5$ Dividing by 6

$x = \sqrt{5}$ *or* $x = -\sqrt{5}$ Principle of square roots

Check: $\qquad \dfrac{6x^2 = 30}{}$

$6(\pm\sqrt{5})\ ?\ 30$

$6 \cdot 5$

30 | TRUE

The solutions are $\sqrt{5}$ and $-\sqrt{5}$, or $\pm\sqrt{5}$.

b) The real-number solutions of the equation $6x^2 = 30$ are the first coordinates of the x-intercepts of the graph of $f(x) = 6x^2 - 30$. Thus, the x-intercepts are $(-\sqrt{5}, 0)$ and $(\sqrt{5}, 0)$.

3. a) $9x^2 + 25 = 0$

$9x^2 = -25$ Subtracting 25

$x^2 = -\dfrac{25}{9}$ Dividing by 9

$x = \sqrt{-\dfrac{25}{9}}$ *or* $x = -\sqrt{-\dfrac{25}{9}}$ Principle of square roots

$x = \dfrac{5}{3}i$ *or* $x = -\dfrac{5}{3}i$ Simplifying

Check: $\qquad \dfrac{9x^2 + 25 = 0}{}$

$9\left(\pm\dfrac{5}{3}i\right) + 25\ ?\ 0$

$9\left(-\dfrac{25}{9}\right) + 25$

$-25 + 25$

0 | TRUE

The solutions are $\dfrac{5}{3}i$ and $-\dfrac{5}{3}i$, or $\pm\dfrac{5}{3}i$.

b) Since the equation $9x^2 + 25 = 0$ has no real-number solutions, the graph of $f(x) = 9x^2 + 25$ has no x-intercepts.

5. $2x^2 - 3 = 0$

$2x^2 = 3$

$x^2 = \dfrac{3}{2}$

$x = \sqrt{\dfrac{3}{2}}$ *or* $x = -\sqrt{\dfrac{3}{2}}$ Principle of square roots

$x = \sqrt{\dfrac{3}{2} \cdot \dfrac{2}{2}}$ *or* $x = -\sqrt{\dfrac{3}{2} \cdot \dfrac{2}{2}}$ Rationalizing denominators

$x = \dfrac{\sqrt{6}}{2}$ *or* $x = -\dfrac{\sqrt{6}}{2}$

Check: $\qquad \dfrac{2x^2 - 3 = 0}{}$

$2\left(\pm\dfrac{\sqrt{6}}{2}\right)^2 - 3\ ?\ 0$

$2 \cdot \dfrac{6}{4} - 3$

$3 - 3$

0 | TRUE

The solutions are $\dfrac{\sqrt{6}}{2}$ and $-\dfrac{\sqrt{6}}{2}$, or $\pm\dfrac{\sqrt{6}}{2}$. Using a calculator, we find that the solutions are approximately ± 1.225.

7. $(x + 2)^2 = 49$

$x + 2 = 7$ *or* $x + 2 = -7$ Principle of square roots

$x = 5$ *or* $\qquad x = -9$

The solutions are 5 and -9.

9. $(x - 4)^2 = 16$

$x - 4 = 4$ *or* $x - 4 = -4$ Principle of square roots

$x = 8$ *or* $\qquad x = 0$

The solutions are 8 and 0.

11. $(x - 11)^2 = 7$

$x - 11 = \sqrt{7}$ *or* $x - 11 = -\sqrt{7}$

$x = 11 + \sqrt{7}$ *or* $\qquad x = 11 - \sqrt{7}$

The solutions are $11 + \sqrt{7}$ and $11 - \sqrt{7}$, or $11 \pm \sqrt{7}$. Using a calculator, we find that the solutions are approximately 8.354 and 13.646.

13. $(x - 7)^2 = -4$

$x - 7 = \sqrt{-4}$ *or* $x - 7 = -\sqrt{-4}$

$x - 7 = 2i$ *or* $x - 7 = -2i$

$x = 7 + 2i$ *or* $\qquad x = 7 - 2i$

The solutions are $7 + 2i$ and $7 - 2i$, or $7 \pm 2i$.

15. $(x - 9)^2 = 81$

$x - 9 = 9$ *or* $x - 9 = -9$

$x = 18$ *or* $\qquad x = 0$

The solutions are 18 and 0.

17. $\left(x - \dfrac{3}{2}\right)^2 = \dfrac{7}{2}$

$x - \dfrac{3}{2} = \sqrt{\dfrac{7}{2}}$ or $x - \dfrac{3}{2} = -\sqrt{\dfrac{7}{2}}$

$x - \dfrac{3}{2} = \sqrt{\dfrac{7}{2} \cdot \dfrac{2}{2}}$ or $x - \dfrac{3}{2} = -\sqrt{\dfrac{7}{2} \cdot \dfrac{2}{2}}$

$x - \dfrac{3}{2} = \dfrac{\sqrt{14}}{2}$ or $x - \dfrac{3}{2} = -\dfrac{\sqrt{14}}{2}$

$x = \dfrac{3}{2} + \dfrac{\sqrt{14}}{2}$ or $x = \dfrac{3}{2} - \dfrac{\sqrt{14}}{2}$

$x = \dfrac{3 + \sqrt{14}}{2}$ or $x = \dfrac{3 - \sqrt{14}}{2}$

The solutions are $\dfrac{3 + \sqrt{14}}{2}$ and $\dfrac{3 - \sqrt{14}}{2}$, or $\dfrac{3 \pm \sqrt{14}}{2}$. Using a calculator, we find that the solutions are approximately -0.371 and 3.371.

19. $x^2 + 6x + 9 = 64$

$(x + 3)^2 = 64$

$x + 3 = 8$ or $x + 3 = -8$

$x = 5$ or $x = -11$

The solutions are 5 and -11.

21. $y^2 - 14y + 49 = 4$

$(y - 7)^2 = 4$

$y - 7 = 2$ or $y - 7 = -2$

$y = 9$ or $y = 5$

The solutions are 9 and 5.

23. $x^2 + 4x \quad = 2$ Original equation

$x^2 + 4x + 4 = 2 + 4$ Adding 4: $\left(\dfrac{4}{2}\right)^2 = 2^2 = 4$

$(x + 2)^2 = 6$

$x + 2 = \sqrt{6}$ or $x + 2 = -\sqrt{6}$ Principle of square roots

$x = -2 + \sqrt{6}$ or $x = -2 - \sqrt{6}$

The solutions are $-2 \pm \sqrt{6}$.

25. $x^2 - 22x \quad = 11$ Original equation

$x^2 - 22x + 121 = 11 + 121$ Adding 121: $\left(\dfrac{-22}{2}\right)^2 =$

$(x - 11)^2 = 132$ $(-11)^2 = 121$

$x - 11 = \sqrt{132}$ or $x - 11 = -\sqrt{132}$

$x - 11 = 2\sqrt{33}$ or $x - 11 = -2\sqrt{33}$

$x = 11 + 2\sqrt{33}$ or $x = 11 - 2\sqrt{33}$

The solutions are $11 \pm 2\sqrt{33}$.

27. $x^2 + x \quad = 1$

$x^2 + x + \dfrac{1}{4} = 1 + \dfrac{1}{4}$ Adding $\dfrac{1}{4}$: $\left(\dfrac{1}{2}\right)^2 = \dfrac{1}{4}$

$\left(x + \dfrac{1}{2}\right)^2 = \dfrac{5}{4}$

$x + \dfrac{1}{2} = \dfrac{\sqrt{5}}{2}$ or $x + \dfrac{1}{2} = -\dfrac{\sqrt{5}}{2}$

$x = \dfrac{-1 + \sqrt{5}}{2}$ or $x = \dfrac{-1 - \sqrt{5}}{2}$

The solutions are $\dfrac{-1 \pm \sqrt{5}}{2}$.

29. $t^2 - 5t \quad = 7$

$t^2 - 5t + \dfrac{25}{4} = 7 + \dfrac{25}{4}$ Adding $\dfrac{25}{4}$: $\left(\dfrac{-5}{2}\right)^2 = \dfrac{25}{4}$

$\left(t - \dfrac{5}{2}\right)^2 = \dfrac{53}{4}$

$t - \dfrac{5}{2} = \dfrac{\sqrt{53}}{2}$ or $t - \dfrac{5}{2} = -\dfrac{\sqrt{53}}{2}$

$t = \dfrac{5 + \sqrt{53}}{2}$ or $t = \dfrac{5 - \sqrt{53}}{2}$

The solutions are $\dfrac{5 \pm \sqrt{53}}{2}$.

31. $x^2 + \dfrac{3}{2}x \quad = 3$

$x^2 + \dfrac{3}{2}x + \dfrac{9}{16} = 3 + \dfrac{9}{16}$ $\left(\dfrac{1}{2} \cdot \dfrac{3}{2}\right)^2 = \left(\dfrac{3}{4}\right)^2 = \dfrac{9}{16}$

$\left(x + \dfrac{3}{4}\right)^2 = \dfrac{57}{16}$

$x + \dfrac{3}{4} = \dfrac{\sqrt{57}}{4}$ or $x + \dfrac{3}{4} = -\dfrac{\sqrt{57}}{4}$

$x = \dfrac{-3 + \sqrt{57}}{4}$ or $x = \dfrac{-3 - \sqrt{57}}{4}$

The solutions are $\dfrac{-3 \pm \sqrt{57}}{4}$.

33. $m^2 - \dfrac{9}{2}m \quad = \dfrac{3}{2}$ Original equation

$m^2 - \dfrac{9}{2}m + \dfrac{81}{16} = \dfrac{3}{2} + \dfrac{81}{16}$ $\left[\dfrac{1}{2}\left(-\dfrac{9}{2}\right)\right]^2 = \left(-\dfrac{9}{4}\right)^2 = \dfrac{81}{16}$

$\left(m - \dfrac{9}{4}\right)^2 = \dfrac{105}{16}$

$m - \dfrac{9}{4} = \dfrac{\sqrt{105}}{4}$ or $m - \dfrac{9}{4} = -\dfrac{\sqrt{105}}{4}$

$m = \dfrac{9 + \sqrt{105}}{4}$ or $m = \dfrac{9 - \sqrt{105}}{4}$

The solutions are $\dfrac{9 \pm \sqrt{105}}{4}$.

35. $x^2 + 6x - 16 = 0$

$x^2 + 6x \quad = 16$ Adding 16

$x^2 + 6x + 9 = 16 + 9$ $\left(\dfrac{6}{2}\right)^2 = 3^2 = 9$

$(x + 3)^2 = 25$

$x + 3 = 5$ or $x + 3 = -5$

$x = 2$ or $x = -8$

The solutions are 2 and -8.

37. $x^2 + 22x + 102 = 0$

$x^2 + 22x \qquad = -102 \qquad$ Subtracting 102

$x^2 + 22x + 121 = -102 + 121 \qquad \left(\dfrac{22}{2}\right)^2 = 11^2 = 121$

$(x + 11)^2 = 19$

$x + 11 = \sqrt{19} \qquad or \quad x + 11 = -\sqrt{19}$

$\qquad x = -11 + \sqrt{19} \quad or \qquad x = -11 - \sqrt{19}$

The solutions are $-11 \pm \sqrt{19}$.

39. $x^2 - 10x - 4 = 0$

$x^2 - 10x \qquad = 4 \qquad$ Adding 4

$x^2 - 10x + 25 = 4 + 25 \qquad \left(\dfrac{-10}{2}\right)^2 = (-5)^2 = 25$

$(x - 5)^2 = 29$

$x - 5 = \sqrt{29} \qquad or \quad x - 5 = -\sqrt{29}$

$\qquad x = 5 + \sqrt{29} \quad or \qquad x = 5 - \sqrt{29}$

The solutions are $5 \pm \sqrt{29}$.

41. a) $x^2 + 7x - 2 = 0$

$x^2 + 7x \qquad = 2 \qquad$ Adding 2

$x^2 + 7x + \dfrac{49}{4} = 2 + \dfrac{49}{4} \qquad \left(\dfrac{7}{2}\right)^2 = \dfrac{49}{4}$

$\left(x + \dfrac{7}{2}\right)^2 = \dfrac{57}{4}$

$x + \dfrac{7}{2} = \dfrac{\sqrt{57}}{2} \qquad or \quad x + \dfrac{7}{2} = -\dfrac{\sqrt{57}}{2}$

$\qquad x = \dfrac{-7 + \sqrt{57}}{2} \quad or \qquad x = \dfrac{-7 - \sqrt{57}}{2}$

The solutions are $\dfrac{-7 \pm \sqrt{57}}{2}$.

b) The real-number solutions of the equation $x^2 + 7x - 2 = 0$ are the first coordinates of the x-intercepts of the graph of $f(x) = x^2 + 7x - 2$. Thus, the x-intercepts are $\left(\dfrac{-7 - \sqrt{57}}{2}, 0\right)$ and $\left(\dfrac{-7 + \sqrt{57}}{2}, 0\right)$.

43. a) $\quad 2x^2 - 5x + 8 = 0$

$\dfrac{1}{2}(2x^2 - 5x + 8) = \dfrac{1}{2} \cdot 0 \qquad$ Multiplying by $\dfrac{1}{2}$ to make the x^2-coefficient 1

$x^2 - \dfrac{5}{2}x + \quad 4 = 0$

$x^2 - \dfrac{5}{2}x \qquad = -4 \qquad$ Subtracting 4

$x^2 - \dfrac{5}{2}x + \dfrac{25}{16} = -4 + \dfrac{25}{16}$

$\left[\dfrac{1}{2}\left(-\dfrac{5}{2}\right)\right]^2 = \left(-\dfrac{5}{4}\right)^2 = \dfrac{25}{16}$

$\left(x - \dfrac{5}{4}\right)^2 = -\dfrac{64}{16} + \dfrac{25}{16}$

$\left(x - \dfrac{5}{4}\right)^2 = -\dfrac{39}{16}$

$x - \dfrac{5}{4} = \sqrt{-\dfrac{39}{16}} \qquad or \quad x - \dfrac{5}{4} = -\sqrt{-\dfrac{39}{16}}$

$x - \dfrac{5}{4} = i\sqrt{\dfrac{39}{16}} \qquad or \quad x - \dfrac{5}{4} = -i\sqrt{\dfrac{39}{16}}$

$x = \dfrac{5}{4} + i\dfrac{\sqrt{39}}{4} \qquad or \qquad x = \dfrac{5}{4} - i\dfrac{\sqrt{39}}{4}$

The solutions are $\dfrac{5}{4} \pm i\dfrac{\sqrt{39}}{4}$.

b) Since the equation $2x^2 - 5x + 8 = 0$ has no real-number solutions, the graph of $f(x) = 2x^2 - 5x + 8$ has no x-intercepts.

45. $x^2 - \dfrac{3}{2}x - \dfrac{1}{2} = 0$

$x^2 - \dfrac{3}{2}x \qquad = \dfrac{1}{2}$

$x^2 - \dfrac{3}{2}x + \dfrac{9}{16} = \dfrac{1}{2} + \dfrac{9}{16} \qquad \left[\dfrac{1}{2}\left(-\dfrac{3}{2}\right)\right]^2 = \left(-\dfrac{3}{4}\right)^2 = \dfrac{9}{16}$

$\left(x - \dfrac{3}{4}\right)^2 = \dfrac{17}{16}$

$x - \dfrac{3}{4} = \dfrac{\sqrt{17}}{4} \qquad or \quad x - \dfrac{3}{4} = -\dfrac{\sqrt{17}}{4}$

$\qquad x = \dfrac{3 + \sqrt{17}}{4} \quad or \qquad x = \dfrac{3 - \sqrt{17}}{4}$

The solutions are $\dfrac{3 \pm \sqrt{17}}{4}$.

47. $2x^2 - 3x - 17 = 0$

$\frac{1}{2}(2x^2 - 3x - 17) = \frac{1}{2} \cdot 0$ Multiplying by $\frac{1}{2}$ to make the x^2-coefficient 1

$x^2 - \frac{3}{2}x - \frac{17}{2} = 0$

$x^2 - \frac{3}{2}x \quad = \frac{17}{2}$ Adding $\frac{17}{2}$

$x^2 - \frac{3}{2}x + \frac{9}{16} = \frac{17}{2} + \frac{9}{16}$ $\left[\frac{1}{2}\left(-\frac{3}{2}\right)\right]^2 =$

$\left(-\frac{3}{4}\right)^2 = \frac{9}{16}$

$\left(x - \frac{3}{4}\right)^2 = \frac{145}{16}$

$x - \frac{3}{4} = \frac{\sqrt{145}}{4}$ or $x - \frac{3}{4} = -\frac{\sqrt{145}}{4}$

$x = \frac{3 + \sqrt{145}}{4}$ or $x = \frac{3 - \sqrt{145}}{4}$

The solutions are $\frac{3 \pm \sqrt{145}}{4}$.

49. $3x^2 - 4x - 1 = 0$

$\frac{1}{3}(3x^2 - 4x - 1) = \frac{1}{3} \cdot 0$ Multiplying to make the x^2-coefficient 1

$x^2 - \frac{4}{3}x - \frac{1}{3} = 0$

$x^2 - \frac{4}{3}x \quad = \frac{1}{3}$ Adding $\frac{1}{3}$

$x^2 - \frac{4}{3}x + \frac{4}{9} = \frac{1}{3} + \frac{4}{9}$ $\left[\frac{1}{2}\left(-\frac{4}{3}\right)\right]^2 =$

$\left(-\frac{2}{3}\right)^2 = \frac{4}{9}$

$\left(x - \frac{2}{3}\right)^2 = \frac{7}{9}$

$x - \frac{2}{3} = \frac{\sqrt{7}}{3}$ or $x - \frac{2}{3} = -\frac{\sqrt{7}}{3}$

$x = \frac{2 + \sqrt{7}}{3}$ or $x = \frac{2 - \sqrt{7}}{3}$

The solutions are $\frac{2 \pm \sqrt{7}}{3}$.

51. $x^2 + x + 2 = 0$

$x^2 + x \quad = -2$ Subtracting 2

$x^2 + x + \frac{1}{4} = -2 + \frac{1}{4}$ $\left(\frac{1}{2}\right)^2 = \frac{1}{4}$

$\left(x + \frac{1}{2}\right)^2 = -\frac{7}{4}$

$x + \frac{1}{2} = \sqrt{-\frac{7}{4}}$ or $x + \frac{1}{2} = -\sqrt{-\frac{7}{4}}$

$x + \frac{1}{2} = i\sqrt{\frac{7}{4}}$ or $x + \frac{1}{2} = -i\sqrt{\frac{7}{4}}$

$x = -\frac{1}{2} + i\frac{\sqrt{7}}{2}$ or $x = -\frac{1}{2} - i\frac{\sqrt{7}}{2}$

The solutions are $-\frac{1}{2} \pm i\frac{\sqrt{7}}{2}$.

53. $x^2 - 4x + 13 = 0$

$x^2 - 4x \quad = -13$ Subtracting 13

$x^2 - 4x + 4 = -13 + 4$ $\left(\frac{-4}{2}\right)^2 = (-2)^2 = 4$

$(x - 2)^2 = -9$

$x - 2 = \sqrt{-9}$ or $x - 2 = -\sqrt{-9}$

$x - 2 = 3i$ or $x - 2 = -3i$

$x = 2 + 3i$ or $x = 2 - 3i$

The solutions are $2 \pm 3i$.

55. $V = 48T^2$

$36 = 48T^2$ Substituting 36 for V

$\frac{36}{48} = T^2$ Solving for T^2

$0.75 = T^2$

$\sqrt{0.75} = T$

$0.866 \approx T$

The hang time is 0.866 sec.

57. $s(t) = 16t^2$

$850 = 16t^2$ Substituting 850 for $s(t)$

$\frac{850}{16} = t^2$ Solving for t^2

$53.125 = t^2$

$\sqrt{53.125} = t$

$7.3 \approx t$

It will take about 7.3 sec for an object to fall from the top.

59. $s(t) = 16t^2$

$745 = 16t^2$ Substituting 745 for $s(t)$

$\frac{745}{16} = t^2$ Solving for t^2

$46.5625 = t^2$

$\sqrt{46.5625} = t$

$6.8 \approx t$

It will take about 6.8 sec for an object to fall from the top.

61. Discussion and Writing Exercise

63. a) First find the slope.

$m = \frac{410 - 275}{5 - 1} = \frac{135}{4} = 33.75$

Now find the function. We substitute in the point-slope equation, using the point $(1, 275)$.

$T - T_1 = m(t - t_1)$

$T - 275 = 33.75(t - 1)$

$T - 275 = 33.75t - 33.75$

$T = 33.75t + 241.25$

Using the function notation we have $T(t) = 33.75t + 241.25$.

b) Since $2005 - 1995 = 10$, the year 2005 is 10 years since 1995. We find $T(10)$.

$$T(t) = 33.75t + 241.25$$
$$T(10) = 33.75(10) + 241.25$$
$$= 337.5 + 241.25$$
$$= 578.75$$

In 2005 about 578.75 thousand, or 578,750 people, will visit a doctor for tattoo removal.

c) Since $1,085,000 = 1085$ thousand, we substitute 1085 for $T(t)$ and solve for t.

$$T(t) = 33.75t + 241.25$$
$$1085 = 33.75t + 241.25$$
$$843.75 = 33.75t$$
$$25 = t$$

1,085,000 people will visit a doctor for tattoo removal 25 yr after 1995, or in 2020.

65. Graph $f(x) = 5 - 2x$

We find some ordered pairs $(x, f(x))$, plot them, and draw the graph.

x	$f(x)$	$(x, f(x))$
0	5	$(0, 5)$
1	3	$(1, 3)$
3	-1	$(3, -1)$
5	-5	$(5, -5)$

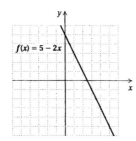

67. Graph $f(x) = |5 - 2x|$

We find some ordered pairs $(x, f(x))$, plot them, and draw the graph.

x	$f(x)$	$(x, f(x))$
0	5	$(0, 5)$
1	3	$(1, 3)$
$\frac{5}{2}$	0	$\left(\frac{5}{2}, 0\right)$
3	1	$(3, 1)$
5	5	$(5, 5)$

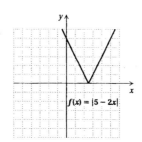

69. $\sqrt{\dfrac{2}{5}} = \sqrt{\dfrac{2}{5} \cdot \dfrac{5}{5}} = \sqrt{\dfrac{10}{25}} = \dfrac{\sqrt{10}}{\sqrt{25}} = \dfrac{\sqrt{10}}{5}$

71. Left to the student

73. In order for $x^2 + bx + 64$ to be a trinomial square, the following must be true:

$$\left(\frac{b}{2}\right)^2 = 64$$
$$\frac{b^2}{4} = 64$$
$$b^2 = 256$$
$$b = 16 \quad or \quad b = -16$$

75. $x(2x^2 + 9x - 56)(3x + 10) = 0$

$$x(2x - 7)(x + 8)(3x + 10) = 0$$
$$x=0 \ or \ 2x - 7=0 \ or \ x + 8=0 \ or \ 3x + 10=0$$
$$x=0 \ or \qquad x=\frac{7}{2} \ or \qquad x=-8 \ or \qquad x=-\frac{10}{3}$$

The solutions are -8, $-\dfrac{10}{3}$, 0, and $\dfrac{7}{2}$.

Exercise Set 16.2

1. $x^2 + 6x + 4 = 0$

$a = 1, \ b = 6, \ c = 4$

$$x = \frac{-b \pm \sqrt{b^2 - 4ac}}{2a}$$
$$x = \frac{-6 \pm \sqrt{6^2 - 4 \cdot 1 \cdot 4}}{2 \cdot 1} = \frac{-6 \pm \sqrt{36 - 16}}{2}$$
$$x = \frac{-6 \pm \sqrt{20}}{2} = \frac{-6 \pm 2\sqrt{5}}{2}$$
$$x = \frac{2(-3 \pm \sqrt{5})}{2} = -3 \pm \sqrt{5}$$

The solutions are $-3 + \sqrt{5}$ and $-3 - \sqrt{5}$.

3. $\qquad 3p^2 = -8p - 1$

$3p^2 + 8p + 1 = 0 \qquad$ Finding standard form

$a = 3, \ b = 8, \ c = 1$

$$p = \frac{-b \pm \sqrt{b^2 - 4ac}}{2a}$$
$$p = \frac{-8 \pm \sqrt{8^2 - 4 \cdot 3 \cdot 1}}{2 \cdot 3} = \frac{-8 \pm \sqrt{64 - 12}}{6}$$
$$x = \frac{-8 \pm \sqrt{52}}{6} = \frac{-8 \pm 2\sqrt{13}}{6}$$
$$x = \frac{2(-4 \pm \sqrt{13})}{2 \cdot 3} = \frac{-4 \pm \sqrt{13}}{3}$$

The solutions are $\dfrac{-4 + \sqrt{13}}{3}$ and $\dfrac{-4 - \sqrt{13}}{3}$.

5. $x^2 - x + 1 = 0$

$a = 1, \ b = -1, \ c = 1$

$$x = \frac{-(-1) \pm \sqrt{(-1)^2 - 4 \cdot 1 \cdot 1}}{2 \cdot 1} = \frac{1 \pm \sqrt{1 - 4}}{2}$$
$$x = \frac{1 \pm \sqrt{-3}}{2} = \frac{1 \pm i\sqrt{3}}{2} = \frac{1}{2} \pm i\frac{\sqrt{3}}{2}$$

The solutions are $\dfrac{1}{2} + i\dfrac{\sqrt{3}}{2}$ and $\dfrac{1}{2} - i\dfrac{\sqrt{3}}{2}$.

7. $\qquad x^2 + 13 = 4x$

$x^2 - 4x + 13 = 0 \qquad$ Finding standard form

$a = 1, \ b = -4, \ c = 13$

$$x = \frac{-(-4) \pm \sqrt{(-4)^2 - 4 \cdot 1 \cdot 13}}{2 \cdot 1} = \frac{4 \pm \sqrt{16 - 52}}{2}$$
$$x = \frac{4 \pm \sqrt{-36}}{2} = \frac{4 \pm 6i}{2} = 2 \pm 3i$$

The solutions are $2 + 3i$ and $2 - 3i$.

9. $r^2 + 3r = 8$

$r^2 + 3r - 8 = 0$ Finding standard form

$a = 1,\ b = 3,\ c = -8$

$r = \dfrac{-3 \pm \sqrt{3^2 - 4 \cdot 1 \cdot (-8)}}{2 \cdot 1} = \dfrac{-3 \pm \sqrt{9 + 32}}{2}$

$r = \dfrac{-3 \pm \sqrt{41}}{2}$

The solutions are $\dfrac{-3 + \sqrt{41}}{2}$ and $\dfrac{-3 - \sqrt{41}}{2}$.

11. $1 + \dfrac{2}{x} + \dfrac{5}{x^2} = 0$

$x^2 + 2x + 5 = 0$ Multiplying by x^2, the LCM
                 of the denominators

$a = 1,\ b = 2,\ c = 5$

$x = \dfrac{-2 \pm \sqrt{2^2 - 4 \cdot 1 \cdot 5}}{2 \cdot 1} = \dfrac{-2 \pm \sqrt{4 - 20}}{2}$

$x = \dfrac{-2 \pm \sqrt{-16}}{2} = \dfrac{-2 \pm 4i}{2} = -1 \pm 2i$

The solutions are $-1 + 2i$ and $-1 - 2i$.

13. a) $3x + x(x - 2) = 0$

$3x + x^2 - 2x = 0$

$x^2 + x = 0$

$x(x + 1) = 0$

$x = 0 \ \ or \ \ x + 1 = 0$

$x = 0 \ \ or \ \ \ \ \ \ \ x = -1$

The solutions are 0 and -1.

b) The solutions of the equation $3x + x(x - 2) = 0$ are
the first coordinates of the x-intercepts of the graph
of $f(x) = 3x + x(x - 2)$. Thus, the x-intercepts are
$(-1, 0)$ and $(0, 0)$.

15. a) $11x^2 - 3x - 5 = 0$

$a = 11,\ b = -3,\ c = -5$

$x = \dfrac{-(-3) \pm \sqrt{(-3)^2 - 4 \cdot 11 \cdot (-5)}}{2 \cdot 11}$

$x = \dfrac{3 \pm \sqrt{9 + 220}}{22} = \dfrac{3 \pm \sqrt{229}}{22}$

The solutions are $\dfrac{3 + \sqrt{229}}{22}$ and $\dfrac{3 - \sqrt{229}}{22}$.

b) The solutions of the equation $11x^2 - 3x - 5 =$
0 are the first coordinates of the x-intercepts of
the graph of $f(x) = 11x^2 - 3x - 5$. Thus,
the x-intercepts are $\left(\dfrac{3 - \sqrt{229}}{22}, 0\right)$ and
$\left(\dfrac{3 + \sqrt{229}}{22}, 0\right)$.

17. a) $25x^2 - 20x + 4 = 0$

$(5x - 2)(5x - 2) = 0$

$5x - 2 = 0 \ \ or \ \ 5x - 2 = 0$

$5x = 2 \ \ or \ \ \ \ \ \ \ 5x = 2$

$x = \dfrac{2}{5} \ \ or \ \ \ \ \ \ \ x = \dfrac{2}{5}$

The solution is $\dfrac{2}{5}$.

b) The solution of the equation $25x^2 - 20x + 4 = 0$
is the first coordinate of the x-intercept of $f(x) =$
$25x^2 - 20x + 4$. Thus, the x-intercept is $\left(\dfrac{2}{5}, 0\right)$.

19. $4x(x - 2) - 5x(x - 1) = 2$

$4x^2 - 8x - 5x^2 + 5x = 2$ Removing parentheses

$-x^2 - 3x = 2$

$-x^2 - 3x - 2 = 0$

$x^2 + 3x + 2 = 0$ Multiplying by -1

$(x + 2)(x + 1) = 0$

$x + 2 = 0 \ \ or \ \ x + 1 = 0$

$x = -2 \ or \ \ \ \ \ \ \ x = -1$

The solutions are -2 and -1.

21. $14(x - 4) - (x + 2) = (x + 2)(x - 4)$

$14x - 56 - x - 2 = x^2 - 2x - 8$

$13x - 58 = x^2 - 2x - 8$

$0 = x^2 - 15x + 50$

$0 = (x - 10)(x - 5)$

$x - 10 = 0 \ \ or \ \ x - 5 = 0$

$x = 10 \ or \ \ \ \ \ \ \ x = 5$

The solutions are 10 and 5.

23. $5x^2 = 17x - 2$

$5x^2 - 17x + 2 = 0$

$a = 5,\ b = -17,\ c = 2$

$x = \dfrac{-(-17) \pm \sqrt{(-17)^2 - 4 \cdot 5 \cdot 2}}{2 \cdot 5}$

$x = \dfrac{17 \pm \sqrt{289 - 40}}{10} = \dfrac{17 \pm \sqrt{249}}{10}$

The solutions are $\dfrac{17 + \sqrt{249}}{10}$ and $\dfrac{17 - \sqrt{249}}{10}$.

25. $x^2 + 5 = 4x$

$x^2 - 4x + 5 = 0$

$a = 1,\ b = -4,\ c = 5$

$x = \dfrac{-(-4) \pm \sqrt{(-4)^2 - 4 \cdot 1 \cdot 5}}{2 \cdot 1} = \dfrac{4 \pm \sqrt{16 - 20}}{2}$

$x = \dfrac{4 \pm \sqrt{-4}}{2} = \dfrac{4 \pm 2i}{2} = 2 \pm i$

The solutions are $2 + i$ and $2 - i$.

27.
$$x + \frac{1}{x} = \frac{13}{6}, \text{ LCM is } 6x$$
$$6x\left(x + \frac{1}{x}\right) = 6x \cdot \frac{13}{6}$$
$$6x^2 + 6 = 13x$$
$$6x^2 - 13x + 6 = 0$$
$$(2x - 3)(3x - 2) = 0$$
$$2x - 3 = 0 \quad or \quad 3x - 2 = 0$$
$$2x = 3 \quad or \quad 3x = 2$$
$$x = \frac{3}{2} \quad or \quad x = \frac{2}{3}$$
The solutions are $\frac{3}{2}$ and $\frac{\cdot 2}{3}$.

29.
$$\frac{1}{y} + \frac{1}{y + 2} = \frac{1}{3}, \text{ LCM is } 3y(y + 2)$$
$$3y(y + 2)\left(\frac{1}{y} + \frac{1}{y + 2}\right) = 3y(y + 2) \cdot \frac{1}{3}$$
$$3(y + 2) + 3y = y(y + 2)$$
$$3y + 6 + 3y = y^2 + 2y$$
$$6y + 6 = y^2 + 2y$$
$$0 = y^2 - 4y - 6$$
$$a = 1, \ b = -4, \ c = -6$$
$$y = \frac{-(-4) \pm \sqrt{(-4)^2 - 4 \cdot 1 \cdot (-6)}}{2 \cdot 1} = \frac{4 \pm \sqrt{16 + 24}}{2}$$
$$y = \frac{4 \pm \sqrt{40}}{2} = \frac{4 \pm 2\sqrt{10}}{2}$$
$$y = \frac{2(2 \pm \sqrt{10})}{2 \cdot 1} = 2 \pm \sqrt{10}$$
The solutions are $2 + \sqrt{10}$ and $2 - \sqrt{10}$.

31.
$$(2t - 3)^2 + 17t = 15$$
$$4t^2 - 12t + 9 + 17t = 15$$
$$4t^2 + 5t - 6 = 0$$
$$(4t - 3)(t + 2) = 0$$
$$4t - 3 = 0 \quad or \quad t + 2 = 0$$
$$t = \frac{3}{4} \quad or \quad t = -2$$
The solutions are $\frac{3}{4}$ and -2.

33.
$$(x - 2)^2 + (x + 1)^2 = 0$$
$$x^2 - 4x + 4 + x^2 + 2x + 1 = 0$$
$$2x^2 - 2x + 5 = 0$$
$$a = 2, \ b = -2, \ c = 5$$
$$x = \frac{-(-2) \pm \sqrt{(-2)^2 - 4 \cdot 2 \cdot 5}}{2 \cdot 2} = \frac{2 \pm \sqrt{4 - 40}}{4}$$
$$x = \frac{2 \pm \sqrt{-36}}{4} = \frac{2 \pm 6i}{4}$$
$$x = \frac{2(1 \pm 3i)}{2 \cdot 2} = \frac{1 \pm 3i}{2} = \frac{1}{2} \pm \frac{3}{2}i$$
The solutions are $\frac{1}{2} + \frac{3}{2}i$ and $\frac{1}{2} - \frac{3}{2}i$.

35.
$$x^3 - 1 = 0$$
$$(x - 1)(x^2 + x + 1) = 0$$
$$x - 1 = 0 \quad or \quad x^2 + x + 1 = 0$$
$$x = 1 \quad or \quad x = \frac{-1 \pm \sqrt{1^2 - 4 \cdot 1 \cdot 1}}{2 \cdot 1}$$
$$x = 1 \quad or \quad x = \frac{-1 \pm \sqrt{-3}}{2}$$
$$x = 1 \quad or \quad x = \frac{-1 \pm i\sqrt{3}}{2} = -\frac{1}{2} \pm i\frac{\sqrt{3}}{2}$$
The solutions are 1, $-\frac{1}{2} + i\frac{\sqrt{3}}{2}$, and $-\frac{1}{2} - i\frac{\sqrt{3}}{2}$.

37. $x^2 + 6x + 4 = 0$
$$a = 1, \ b = 6, \ c = 4$$
$$x = \frac{-6 \pm \sqrt{6^2 - 4 \cdot 1 \cdot 4}}{2 \cdot 1} = \frac{-6 \pm \sqrt{36 - 16}}{2}$$
$$x = \frac{-6 \pm \sqrt{20}}{2} = \frac{-6 \pm \sqrt{4 \cdot 5}}{2}$$
$$x = \frac{-6 \pm 2\sqrt{5}}{2} = \frac{2(-3 \pm \sqrt{5})}{2}$$
$$x = -3 \pm \sqrt{5}$$
We can use a calculator to approximate the solutions:
$-3 + \sqrt{5} \approx -0.764; -3 - \sqrt{5} \approx -5.236$
The solutions are $-3 + \sqrt{5}$ and $-3 - \sqrt{5}$, or approximately -0.764 and -5.236.

39. $x^2 - 6x + 4 = 0$
$$a = 1, \ b = -6, \ c = 4$$
$$x = \frac{-(-6) \pm \sqrt{(-6)^2 - 4 \cdot 1 \cdot 4}}{2 \cdot 1} = \frac{6 \pm \sqrt{36 - 16}}{2}$$
$$x = \frac{6 \pm \sqrt{20}}{2} = \frac{6 \pm \sqrt{4 \cdot 5}}{2}$$
$$x = \frac{6 \pm 2\sqrt{5}}{2} = \frac{2(3 \pm \sqrt{5})}{2}$$
$$x = 3 \pm \sqrt{5}$$
We can use a calculator to approximate the solutions:
$3 + \sqrt{5} \approx 5.236; 3 - \sqrt{5} \approx 0.764$
The solutions are $3 + \sqrt{5}$ and $3 - \sqrt{5}$, or approximately 5.236 and 0.764.

41. $2x^2 - 3x - 7 = 0$
$$a = 2, \ b = -3, \ c = -7$$
$$x = \frac{-(-3) \pm \sqrt{(-3)^2 - 4 \cdot 2 \cdot (-7)}}{2 \cdot 2} = \frac{3 \pm \sqrt{9 + 56}}{4}$$
$$x = \frac{3 \pm \sqrt{65}}{4}$$
We can use a calculator to approximate the solutions:
$\frac{3 + \sqrt{65}}{4} \approx 2.766; \frac{3 - \sqrt{65}}{4} \approx -1.266$
The solutions are $\frac{3 + \sqrt{65}}{4}$ and $\frac{3 - \sqrt{65}}{4}$, or approximately 2.766 and -1.266.

43.
$$5x^2 = 3 + 8x$$
$$5x^2 - 8x - 3 = 0$$
$$a = 5, \ b = -8, \ c = -3$$
$$x = \frac{-(-8) \pm \sqrt{(-8)^2 - 4 \cdot 5 \cdot (-3)}}{2 \cdot 5} = \frac{8 \pm \sqrt{64 + 60}}{10}$$
$$x = \frac{8 \pm \sqrt{124}}{10} = \frac{8 \pm \sqrt{4 \cdot 31}}{10}$$
$$x = \frac{8 \pm 2\sqrt{31}}{10} = \frac{2(4 \pm \sqrt{31})}{2 \cdot 5}$$
$$x = \frac{4 \pm \sqrt{31}}{5}$$

We can use a calculator to approximate the solutions:
$$\frac{4 + \sqrt{31}}{5} \approx 1.914; \ \frac{4 - \sqrt{31}}{5} \approx -0.314$$

The solutions are $\dfrac{4 + \sqrt{31}}{5}$ and $\dfrac{4 - \sqrt{31}}{5}$, or approximately 1.914 and -0.314.

45. Discussion and Writing Exercise

47.
$$x = \sqrt{x + 2}$$
$$x^2 = (\sqrt{x + 2})^2 \quad \text{Principle of powers}$$
$$x^2 = x + 2$$
$$x^2 - x - 2 = 0$$
$$(x - 2)(x + 1) = 0$$
$$x - 2 = 0 \ \ or \ \ x + 1 = 0$$
$$x = 2 \ \ or \ \ \ \ \ x = -1$$

The number 2 checks but -1 does not, so the solution is 2.

49.
$$\sqrt{x + 2} = \sqrt{2x - 8}$$
$$(\sqrt{x + 2})^2 = (\sqrt{2x + 8})^2 \quad \text{Principle of powers}$$
$$x + 2 = 2x - 8$$
$$2 = x - 8$$
$$10 = x$$

The number 10 checks, so it is the solution.

51. $\sqrt{x + 5} = -7$

Since the square root of a number must be nonnegative, this equation has no solution.

53.
$$\sqrt[3]{4x - 7} = 2$$
$$(\sqrt[3]{4x - 7})^3 = 2^3 \quad \text{Principle of powers}$$
$$4x - 7 = 8$$
$$4x = 15$$
$$x = \frac{15}{4}$$

The number $\dfrac{15}{4}$ checks, so it is the solution.

55. The solutions of $2.2x^2 + 0.5x - 1 = 0$ are approximately -0.797 and 0.570.

57. $2x^2 - x - \sqrt{5} = 0$
$$a = 2, \ b = -1, \ c = -\sqrt{5}$$
$$x = \frac{-(-1) \pm \sqrt{(-1)^2 - 4 \cdot 2 \cdot (-\sqrt{5})}}{2 \cdot 2} = \frac{1 \pm \sqrt{1 + 8\sqrt{5}}}{4}$$

The solutions are $\dfrac{1 + \sqrt{1 + 8\sqrt{5}}}{4}$ and $\dfrac{1 - \sqrt{1 + 8\sqrt{5}}}{4}$.

59. $ix^2 - x - 1 = 0$
$$a = i, \ b = -1, \ c = -1$$
$$x = \frac{-(-1) \pm \sqrt{(-1)^2 - 4 \cdot i \cdot (-1)}}{2 \cdot i} = \frac{1 \pm \sqrt{1 + 4i}}{2i}$$
$$x = \frac{1 \pm \sqrt{1 + 4i}}{2i} \cdot \frac{i}{i} = \frac{i \pm i\sqrt{1 + 4i}}{2i^2} = \frac{i \pm i\sqrt{1 + 4i}}{-2}$$
$$x = \frac{-i \pm i\sqrt{1 + 4i}}{2}$$

The solutions are $\dfrac{-i + i\sqrt{1 + 4i}}{2}$ and $\dfrac{-i - i\sqrt{1 + 4i}}{2}$.

61.
$$\frac{x}{x + 1} = 4 + \frac{1}{3x^2 - 3}$$
$$\frac{x}{x + 1} = 4 + \frac{1}{3(x + 1)(x - 1)},$$
$$\text{LCM is } 3x(x + 1)(x - 1)$$
$$3(x + 1)(x - 1) \cdot \frac{x}{x + 1} =$$
$$3(x + 1)(x - 1)\left(4 + \frac{1}{3(x + 1)(x - 1)}\right)$$
$$3x(x - 1) = 12(x + 1)(x - 1) + 1$$
$$3x^2 - 3x = 12x^2 - 12 + 1$$
$$0 = 9x^2 + 3x - 11$$
$$a = 9, \ b = 3, \ c = -11$$
$$x = \frac{-3 \pm \sqrt{3^2 - 4 \cdot 9 \cdot (-11)}}{2 \cdot 9} = \frac{-3 \pm \sqrt{9 + 396}}{18}$$
$$x = \frac{-3 \pm \sqrt{405}}{18} = \frac{-3 \pm 9\sqrt{5}}{18}$$
$$x = \frac{3(-1 \pm 3\sqrt{5})}{3 \cdot 6} = \frac{-1 \pm 3\sqrt{5}}{6}$$

The solutions are $\dfrac{-1 + 3\sqrt{5}}{6}$ and $\dfrac{-1 - 3\sqrt{5}}{6}$.

63. Replace $f(x)$ with 13.
$$13 = (x - 3)^2$$
$$\pm\sqrt{13} = x - 3$$
$$3 \pm \sqrt{13} = x$$

The solutions are $3 + \sqrt{13}$ and $3 - \sqrt{13}$.

Exercise Set 16.3

1. *Familiarize*. We make a drawing and label it. We let $x =$ the length of the rectangle. Then $x - 7 =$ the width.

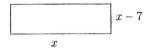

Translate. We use the formula for the area of a rectangle.

$$A = lw$$
$$18 = x(x - 7) \quad \text{Substituting}$$

Solve. We solve the equation.

$$18 = x^2 - 7x$$
$$0 = x^2 - 7x - 18$$
$$0 = (x - 9)(x + 2)$$
$$x - 9 = 0 \quad or \quad x + 2 = 0$$
$$x = 9 \quad or \quad x = -2$$

Check. We only check 9 since the length cannot be negative. If $x = 9$, then $x - 7 = 9 - 7$, or 2, and the area is $9 \cdot 2$, or 18 ft². The value checks.

State. The length of 9 ft, and the width is 2 ft.

3. Familiarize. We make a drawing and label it. We let $x =$ the width of the rectangle. Then $2x =$ the length.

Translate.

$$A = lw$$
$$162 = 2x \cdot x \quad \text{Substituting}$$

Solve. We solve the equation.

$$162 = 2x^2$$
$$81 = x^2$$
$$\pm 9 = x$$

Check. We only check 9 since the width cannot be negative. If $x = 9$, then $2x = 2 \cdot 9$, or 18, and the area is $18 \cdot 9$, or 162 yd². The value checks.

State. The length is 18 yd, and the width is 9 yd.

5. Familiarize. Let h represent the height of the sail. Then $h - 9$ represents the base. Recall that the formula for the area of a triangle is $A = \frac{1}{2} \times$ base $\times$ height.

Translate. The area is 56 m². We substitute in the formula.

$$\frac{1}{2}(h - 9)h = 56$$

Solve. We solve the equation:

$$\frac{1}{2}(h - 9)h = 56$$
$$(h - 9)h = 112 \quad \text{Multiplying by 2}$$
$$h^2 - 9h = 112$$
$$h^2 - 9h - 112 = 0$$
$$(h - 16)(h + 7) = 0$$
$$h - 16 = 0 \quad or \quad h + 7 = 0$$
$$h = 16 \quad or \quad h = -7$$

Check. We check only 16, since height cannot be negative. If the height is 16 m, the base is $16 - 9$, or 7 m, and the area is $\frac{1}{2} \cdot 7 \cdot 16$, or 56 m². We have a solution.

State. The height is 16 m, and the base is 7 m.

7. Familiarize. Let h represent the height of the sail. Then $h - 8$ represents the base. Recall that the formula for the area of a triangle is $A = \frac{1}{2} \times$ base $\times$ height.

Translate. The area 56 ft². We substitute n the formula.

$$\frac{1}{2}(h - 8)h = 56$$

Solve. We solve the equation.

$$\frac{1}{2}(h - 8)h = 56$$
$$(h - 8)h = 112 \quad \text{Multiplying by 2}$$
$$h^2 - 8h = 112$$
$$h^2 - 8h - 112 = 0$$

We use the quadratic formula.

$$h = \frac{-b \pm \sqrt{b^2 - 4ac}}{2a}$$
$$= \frac{-(-8) \pm \sqrt{(-8)^2 - 4 \cdot 1 \cdot (-112)}}{2 \cdot 1}$$
$$= \frac{8 \pm \sqrt{64 + 448}}{2} = \frac{8 \pm \sqrt{512}}{2}$$
$$= \frac{8 \pm \sqrt{256 \cdot 2}}{2} = \frac{8 \pm 16\sqrt{2}}{2}$$
$$= \frac{8(1 \pm 2\sqrt{2})}{2} = 4(1 \pm 2\sqrt{2})$$
$$= 4 \pm 8\sqrt{2}$$

Check. The number $4 - 8\sqrt{2}$ is negative. We do not check it since the height cannot be negative. If the height is $4 + 8\sqrt{2}$ ft, then the base is $4 + 8\sqrt{2} - 8$, or $-4 + 8\sqrt{2}$ ft, and the area is $\frac{1}{2}(-4 + 8\sqrt{2})(4 + 8\sqrt{2})$, or $\frac{1}{2}(-16 + 128)$, or $\frac{1}{2}(112)$, or 56 ft². The answer checks.

State. The base of the sail is $-4 + 8\sqrt{2}$ ft, and the height is $4 + 8\sqrt{2}$ ft.

9. Familiarize. We make a drawing and label it. We let $x =$ the width of the frame.

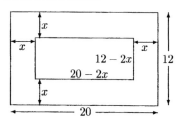

The length and width of the picture that shows are represented by $20 - 2x$ and $12 - 2x$. The area of the picture that shows is 84 cm².

Translate. Using the formula for the area of a rectangle, $A = l \cdot w$, we have

$$84 = (20 - 2x)(12 - 2x).$$

Solve. We solve the equation.

$$84 = (20 - 2x)(12 - 2x)$$
$$84 = 240 - 64x + 4x^2$$
$$0 = 156 - 64x + 4x^2$$
$$0 = 4x^2 - 64x + 156$$
$$0 = x^2 - 16x + 39 \qquad \text{Dividing by 4}$$
$$0 = (x - 13)(x - 3)$$
$$x - 13 = 0 \quad or \quad x - 3 = 0$$
$$x = 13 \quad or \qquad x = 3$$

Check. We see that 13 is not a solution, because when $x = 13$, then $20 - 2x = -6$ and $12 - 2x = -14$ and the dimensions of the frame cannot be negative. We check 3. When $x = 3$, then $20 - 2x = 14$ and $12 - 2x = 6$ and $14 \cdot 6 = 84$, the area of the picture that shows. The number 3 checks.

State. The width of the frame is 3 cm.

11. **Familiarize**. Using the labels on the drawing in the text, we let x and $x + 2$ represent the lengths of the legs of the right triangle.

Translate. We use the Pythagorean equation.

$$a^2 + b^2 = c^2$$
$$x^2 + (x + 2)^2 = 10^2 \qquad \text{Substituting}$$

Solve. We solve the equation.

$$x^2 + x^2 + 4x + 4 = 100$$
$$2x^2 + 4x + 4 = 100$$
$$2x^2 + 4x - 96 = 0$$
$$x^2 + 2x - 48 = 0 \qquad \text{Dividing by 2}$$
$$(x + 8)(x - 6) = 0$$
$$x + 8 = 0 \quad or \quad x - 6 = 0$$
$$x = -8 \quad or \qquad x = 6$$

Check. We only check 6 since the length of a leg cannot be negative. When $x = 6$, then $x + 2 = 8$, and $6^2 + 8^2 = 100 = 10^2$. The number 6 checks.

State. The lengths of the legs are 6 ft and 8 ft.

13. **Familiarize**. The page numbers on facing pages are consecutive integers. Let $x =$ the number on the left-hand page. Then $x + 1 =$ the number on the right-hand page.

Translate.

$$\underbrace{\text{The product of the page numbers}}_{\displaystyle x(x+1)} \quad \underset{=}{\text{is}} \quad \underset{812}{\text{812.}}$$

Solve. We solve the equation.

$$x^2 + x = 812$$
$$x^2 + x - 812 = 0$$
$$(x + 29)(x - 28) = 0$$
$$x + 29 = 0 \quad or \quad x - 28 = 0$$
$$x = -29 \quad or \qquad x = 28$$

Check. We only check 28 since a page number cannot be negative. If $x = 28$, then $x + 1 = 29$ and $28 \cdot 29 = 812$. The number 28 checks.

State. The page numbers are 28 and 29.

15. **Familiarize**. We make a drawing and label it. We let $x =$ the length and $x - 4 =$ the width.

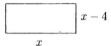

Translate. We use the formula for the area of a rectangle.

$$A = lw$$
$$10 = x(x - 4) \qquad \text{Substituting}$$

Solve. We solve the equation.

$$10 = x^2 - 4x$$
$$0 = x^2 - 4x - 10$$
$$x = \frac{-b \pm \sqrt{b^2 - 4ac}}{2a} = \frac{-(-4) \pm \sqrt{(-4)^2 - 4 \cdot 1 \cdot (-10)}}{2 \cdot 1}$$
$$x = \frac{4 \pm \sqrt{16 + 40}}{2} = \frac{4 \pm \sqrt{56}}{2} = \frac{4 \pm \sqrt{4 \cdot 14}}{2}$$
$$x = \frac{4 \pm 2\sqrt{14}}{2} = 2 \pm \sqrt{14}$$

Check. We only need to check $2 + \sqrt{14}$ since $2 - \sqrt{14}$ is negative and the length cannot be negative. If $x = 2 + \sqrt{14}$, then $x - 4 = (2 + \sqrt{14}) - 4$, or $\sqrt{14} - 2$. Using a calculator we find that the length is $2 + \sqrt{14} \approx 5.742$ ft and the width is $\sqrt{14} - 2 \approx 1.742$ ft, and $(5.742)(1.742) = 10.003 \approx 10$. Our result checks.

State. The length is $2 + \sqrt{14}$ ft ≈ 5.742 ft; the width is $\sqrt{14} - 2$ ft ≈ 1.742 ft.

17. **Familiarize**. We make a drawing and label it. We let $x =$ the width of the margin.

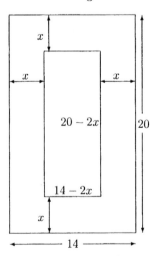

The length and width of the printed text are represented by $20 - 2x$ and $14 - 2x$. The area of the printed text is 100 in^2.

Translate. We use the formula for the area of a rectangle.

$$A = lw$$
$$100 = (20 - 2x)(14 - 2x)$$

Solve. We solve the equation.

$$100 = 280 - 68x + 4x^2$$
$$0 = 4x^2 - 68x + 180$$
$$0 = x^2 - 17x + 45 \qquad \text{Dividing by 4}$$
$$x = \frac{-b \pm \sqrt{b^2 - 4ac}}{2a} = \frac{-(-17) \pm \sqrt{(-17)^2 - 4 \cdot 1 \cdot 45}}{2 \cdot 1}$$
$$x = \frac{17 \pm \sqrt{289 - 180}}{2} = \frac{17 \pm \sqrt{109}}{2}$$
$$x \approx 13.720 \quad or \quad x \approx 3.280$$

Check. If $x \approx 13.720$, then $20 - 2x \approx -7.440$ and $14 - 2x \approx -13.440$. Since the width of the margin cannot be negative, 13.720 is not a solution. If $x \approx 3.280$, then $20 - 2x \approx 13.440$ and $14 - 2x \approx 7.440$ and $(13.440)(7.440) = 99.99 \approx 100$. The number $\frac{17 - \sqrt{109}}{2} \approx 3.280$ checks.

State. The width of the margin is $\frac{17 - \sqrt{109}}{2}$ in. ≈ 3.280 in.

19. *Familiarize*. We make a drawing. We let $x =$ the length of the shorter leg and $x + 14 =$ the length of the longer leg.

Translate. We use the Pythagorean equation.

$$a^2 + b^2 = c^2$$
$$x^2 + (x + 14)^2 = 24^2 \qquad \text{Substituting}$$

Solve. We solve the equation.

$$x^2 + x^2 + 28x + 196 = 576$$
$$2x^2 + 28x - 380 = 0$$
$$x^2 + 14x - 190 = 0 \qquad \text{Dividing by 2}$$
$$x = \frac{-b \pm \sqrt{b^2 - 4ac}}{2a} = \frac{-14 \pm \sqrt{14^2 - 4 \cdot 1 \cdot (-190)}}{2 \cdot 1}$$
$$x = \frac{-14 \pm \sqrt{196 + 760}}{2} = \frac{-14 \pm \sqrt{956}}{2} = \frac{-14 \pm \sqrt{4 \cdot 239}}{2}$$
$$x = \frac{-14 \pm 2\sqrt{239}}{2} = -7 \pm \sqrt{239}$$
$$x \approx 8.460 \quad or \quad x \approx -22.460$$

Check. Since the length of a leg cannot be negative, we only need to check 8.460. If $x = -7 + \sqrt{239} \approx 8.460$, then $x + 14 = -7 + \sqrt{239} + 14 = 7 + \sqrt{239} \approx 22.460$ and $(8.460)^2 + (22.460)^2 = 576.0232 \approx 576 = 24^2$. The number $-7 + \sqrt{239} \approx 8.460$ checks.

State. The lengths of the legs are $-7 + \sqrt{239}$ ft ≈ 8.460 ft and $7 + \sqrt{239}$ ft ≈ 22.460 ft.

21. *Familiarize*. We first make a drawing, labeling it with the known and unknown information. We can also organize the information in a table. We let r represent the speed and t the time for the first part of the trip.

Trip	Distance	Speed	Time
1st part	120	r	t
2nd part	100	$r - 10$	$4 - t$

Translate. Using $r = \frac{d}{t}$, we get two equations from the table, $r = \frac{120}{t}$ and $r - 10 = \frac{100}{4 - t}$.

Solve. We substitute $\frac{120}{t}$ for r in the second equation and solve for t.

$$\frac{120}{t} - 10 = \frac{100}{4 - t}, \text{ LCD is } t(4 - t)$$
$$t(4 - t)\left(\frac{120}{t} - 10\right) = t(4 - t) \cdot \frac{100}{4 - t}$$
$$120(4 - t) - 10t(4 - t) = 100t$$
$$480 - 120t - 40t + 10t^2 = 100t$$
$$10t^2 - 260t + 480 = 0 \qquad \text{Standard form}$$
$$t^2 - 26t + 48 = 0 \qquad \text{Multiplying by } \frac{1}{10}$$
$$(t - 2)(t - 24) = 0$$
$$t = 2 \quad or \quad t = 24$$

Check. Since the time cannot be negative (If $t = 24$, $4 - t = -20$.), we check only 2 hr. If $t = 2$, then $4 - t = 2$. The speed of the first part is $\frac{120}{2}$, or 60 mph. The speed of the second part is $\frac{100}{2}$, or 50 mph. The speed of the second part is 10 mph slower than the first part. The value checks.

State. The speed of the first part was 60 mph, and the speed of the second part was 50 mph.

23. *Familiarize*. We first make a drawing. We also organize the information in a table. We let $r =$ the speed and $t =$ the time of the slower trip.

Trip	Distance	Speed	Time
Slower	200	r	t
Faster	200	$r + 10$	$t - 1$

Translate. Using $t = d/r$, we get two equations from the table:

$$t = \frac{200}{r} \quad \text{and} \quad t - 1 = \frac{200}{r + 10}$$

Solve. We substitute $\frac{200}{r}$ for t in the second equation and solve for r.

$$\frac{200}{r} - 1 = \frac{200}{r + 10}, \text{ LCD is } r(r + 10)$$

$$r(r + 10)\left(\frac{200}{r} - 1\right) = r(r + 10) \cdot \frac{200}{r + 10}$$

$$200(r + 10) - r(r + 10) = 200r$$

$$200r + 2000 - r^2 - 10r = 200r$$

$$0 = r^2 + 10r - 2000$$

$$0 = (r + 50)(r - 40)$$

$$r = -50 \quad or \quad r = 40$$

Check. Since negative speed has no meaning in this problem, we check only 40. If $r = 40$, then the time for the slower trip is $\frac{200}{40}$, or 5 hours. If $r = 40$, then $r + 10 = 50$ and the time for the faster trip is $\frac{200}{50}$, or 4 hours. This is 1 hour less time than the slower trip took, so we have an answer to the problem.

State. The speed is 40 mph.

25. **Familiarize.** We make a drawing and then organize the information in a table. We let $r =$ the speed and $t =$ the time of the Cessna.

600 mi r mph t hr

1000 mi $r + 50$ mph $t + 1$ hr

Plane	Distance	Speed	Time
Cessna	600	r	t
Beechcraft	1000	$r + 50$	$t + 1$

Translate. Using $t = d/r$, we get two equations from the table:

$$t = \frac{600}{r} \quad \text{and} \quad t + 1 = \frac{1000}{r + 50}$$

Solve. We substitute $\frac{600}{r}$ for t in the second equation and solve for r.

$$\frac{600}{r} + 1 = \frac{1000}{r + 50},$$
$$\text{LCD is } r(r + 50)$$

$$r(r + 50)\left(\frac{600}{r} + 1\right) = r(r + 50) \cdot \frac{1000}{r + 50}$$

$$600(r + 50) + r(r + 50) = 1000r$$

$$600r + 30,000 + r^2 + 50r = 1000r$$

$$r^2 - 350r + 30,000 = 0$$

$$(r - 150)(r - 200) = 0$$

$$r = 150 \quad or \quad r = 200$$

Check. If $r = 150$, then the Cessna's time is $\frac{600}{150}$, or 4 hr and the Beechcraft's time is $\frac{1000}{150 + 50}$, or $\frac{1000}{200}$, or 5 hr. If $r = 200$, then the Cessna's time is $\frac{600}{200}$, or 3 hr and the Beechcraft's time is $\frac{1000}{200 + 50}$, or $\frac{1000}{250}$, or 4 hr. Since the Beechcraft's time is 1 hr longer in each case, both values check. There are two solutions.

State. The speed of the Cessna is 150 mph and the speed of the Beechcraft is 200 mph; or the speed of the Cessna is 200 mph and the speed of the Beechcraft is 250 mph.

27. **Familiarize.** We make a drawing and then organize the information in a table. We let r represent the speed and t the time of the trip to Hillsboro.

            Hillsboro

40 mi r mph t hr

40 mi $r - 6$ mph $14 - t$ hr

Trip	Distance	Speed	Time
To Hillsboro	40	r	t
Return	40	$r - 6$	$14 - t$

Translate. Using $t = \frac{d}{r}$, we get two equations from the table,

$$t = \frac{40}{r} \quad \text{and} \quad 14 - t = \frac{40}{r - 6}.$$

Solve. We substitute $\frac{40}{r}$ for t in the second equation and solve for r.

$$14 - \frac{40}{r} = \frac{40}{r - 6},$$
$$\text{LCD is } r(r - 6)$$

$$r(r - 6)\left(14 - \frac{40}{r}\right) = r(r - 6) \cdot \frac{40}{r - 6}$$

$$14r(r - 6) - 40(r - 6) = 40r$$

$$14r^2 - 84r - 40r + 240 = 40r$$

$$14r^2 - 164r + 240 = 0$$

$$7r^2 - 82r + 120 = 0$$

$$(7r - 12)(r - 10) = 0$$

$$r = \frac{12}{7} \quad or \quad r = 10$$

Check. Since negative speed has no meaning in this problem (If $r = \frac{12}{7}$, then $r - 6 = -\frac{30}{7}$.), we check only 10 mph. If $r = 10$, then the time of the trip to Hillsboro is $\frac{40}{10}$, or 4 hr. The speed of the return trip is $10 - 6$, or 4 mph, and the time is $\frac{40}{4}$, or 10 hr. The total time for the round trip is 4 hr + 10 hr, or 14 hr. The value checks.

State. Naoki's speed on the trip to Hillsboro was 10 mph and it was 4 mph on the return trip.

29. **Familiarize.** We make a drawing and organize the information in a table. Let r represent the speed of the barge in still water, and let t represent the time of the trip upriver.

24 mi $r - 4$ mph t hr   Upriver

Downriver   24 mi $r + 4$ mph $5 - t$ hr

Trip	Distance	Speed	Time
Upriver	24	$r - 4$	t
Downriver	24	$r + 4$	$5 - t$

Translate. Using $t = \dfrac{d}{r}$, we get two equations from the table,

$$t = \frac{24}{r - 4} \text{ and } 5 - t = \frac{24}{r + 4}.$$

Solve. We substitute $\dfrac{24}{r - 4}$ for t in the second equation and solve for r.

$$5 - \frac{24}{r - 4} = \frac{24}{r+4},$$

LCD is $(r-4)(r+4)$

$$(r - 4)(r + 4)\left(5 - \frac{24}{r - 4}\right) = (r - 4)(r + 4) \cdot \frac{24}{r + 4}$$

$$5(r - 4)(r + 4) - 24(r + 4) = 24(r - 4)$$

$$5r^2 - 80 - 24r - 96 = 24r - 96$$

$$5r^2 - 48r - 80 = 0$$

We use the quadratic formula.

$$r = \frac{-(-48) \pm \sqrt{(-48)^2 - 4 \cdot 5 \cdot (-80)}}{2 \cdot 5}$$

$$r = \frac{48 \pm \sqrt{3904}}{10}$$

$$r \approx 11 \ \ or \ \ r \approx -1.5$$

Check. Since negative speed has no meaning in this problem, we check only 11 mph. If $r \approx 11$, then the speed upriver is about $11 - 4$, or 7 mph, and the time is about $\dfrac{24}{7}$, or 3.4 hr. The speed downriver is about $11 + 4$, or 15 mph, and the time is about $\dfrac{24}{15}$, or 1.6 hr. The total time of the round trip is $3.4 + 1.6$, or 5 hr. The value checks.

State. The barge must be able to travel about 11 mph in still water.

31.
$$A = 6s^2$$

$$\frac{A}{6} = s^2 \qquad \text{Dividing by 6}$$

$$\sqrt{\frac{A}{6}} = s \qquad \text{Taking the positive square root}$$

33.
$$F = \frac{Gm_1m_2}{r^2}$$

$$Fr^2 = Gm_1m_2 \qquad \text{Multiplying by } r^2$$

$$r^2 = \frac{Gm_1m_2}{F} \qquad \text{Dividing by } F$$

$$r = \sqrt{\frac{Gm_1m_2}{F}} \qquad \text{Taking the positive square root}$$

35.
$$E = mc^2$$

$$\frac{E}{m} = c^2 \qquad \text{Dividing by } m$$

$$\sqrt{\frac{E}{m}} = c \qquad \text{Taking the square root}$$

37.
$$a^2 + b^2 = c^2$$

$$b^2 = c^2 - a^2 \qquad \text{Subtracting } a^2$$

$$b = \sqrt{c^2 - a^2} \qquad \text{Taking the square root}$$

39.
$$N = \frac{k^2 - 3k}{2}$$

$$2N = k^2 - 3k$$

$$0 = k^2 - 3k - 2N \qquad \text{Standard form}$$

$$a = 1, \ b = -3, \ c = -2N$$

$$k = \frac{-(-3) \pm \sqrt{(-3)^3 - 4 \cdot 1 \cdot (-2N)}}{2 \cdot 1} \qquad \begin{array}{l}\text{Using the} \\ \text{quadratic formula}\end{array}$$

$$k = \frac{3 \pm \sqrt{9 + 8N}}{2}$$

Since taking the negative square root would result in a negative answer, we take the positive one.

$$k = \frac{3 + \sqrt{9 + 8N}}{2}$$

41.
$$A = 2\pi r^2 + 2\pi rh$$

$$0 = 2\pi r^2 + 2\pi rh - A \qquad \text{Standard form}$$

$$a = 2\pi, \ b = 2\pi h, \ c = -A$$

$$r = \frac{-2\pi h \pm \sqrt{(2\pi h)^2 - 4 \cdot 2\pi \cdot (-A)}}{2 \cdot 2\pi} \qquad \begin{array}{l}\text{Using the} \\ \text{quadratic formula}\end{array}$$

$$r = \frac{-2\pi h \pm \sqrt{4\pi^2 h^2 + 8\pi A}}{4\pi}$$

$$r = \frac{-2\pi h \pm 2\sqrt{\pi^2 h^2 + 2\pi A}}{4\pi}$$

$$r = \frac{-\pi h \pm \sqrt{\pi^2 h^2 + 2\pi A}}{2\pi}$$

Since taking the negative square root would result in a negative answer, we take the positive one.

$$r = \frac{-\pi h + \sqrt{\pi^2 h^2 + 2\pi A}}{2\pi}$$

43.
$$T = 2\pi\sqrt{\frac{L}{g}}$$

$$\frac{T}{2\pi} = \sqrt{\frac{L}{g}} \qquad \text{Dividing by } 2\pi$$

$$\frac{T^2}{4\pi^2} = \frac{L}{g} \qquad \text{Squaring}$$

$$gT^2 = 4\pi^2 L \qquad \text{Multiplying by } 4\pi^2 g$$

$$g = \frac{4\pi^2 L}{T^2} \qquad \text{Dividing by } T^2$$

45.
$$I = \frac{704.5W}{H^2}$$

$$H^2 I = 704.5W \qquad \text{Multiplying by } H^2$$

$$H^2 = \frac{704.5W}{I} \qquad \text{Dividing by } I$$

$$H = \sqrt{\frac{704.5W}{I}}$$

47.

$$m = \frac{m_0}{\sqrt{1 - \dfrac{v^2}{c^2}}}$$

$$m^2 = \frac{m_0^2}{1 - \dfrac{v^2}{c^2}} \qquad \text{Principle of powers}$$

$$m^2\left(1 - \frac{v^2}{c^2}\right) = m_0^2$$

$$m^2 - \frac{m^2 v^2}{c^2} = m_0^2$$

$$m^2 - m_0^2 = \frac{m^2 v^2}{c^2}$$

$$c^2(m^2 - m_0^2) = m^2 v^2$$

$$\frac{c^2(m^2 - m_0^2)}{m^2} = v^2$$

$$\sqrt{\frac{c^2(m^2 - m_0^2)}{m^2}} = v$$

$$\frac{c\sqrt{m^2 - m_0^2}}{m} = v$$

49. Discussion and Writing Exercise

51.

$$\frac{1}{x-1} + \frac{1}{x^2 - 3x + 2}$$

$$= \frac{1}{x-1} + \frac{1}{(x-1)(x-2)}, \quad \text{LCD is } (x-1)(x-2)$$

$$= \frac{1}{x-1} \cdot \frac{x-2}{x-2} + \frac{1}{(x-1)(x-2)}$$

$$= \frac{x-2}{(x-1)(x-2)} + \frac{1}{(x-1)(x-2)}$$

$$= \frac{x-2+1}{(x-1)(x-2)}$$

$$= \frac{x-1}{(x-1)(x-2)}$$

$$= \frac{(x-1)\cdot 1}{(x-1)(x-2)}$$

$$= \frac{1}{x-2}$$

53.

$$\frac{2}{x+3} - \frac{x}{x-1} + \frac{x^2+2}{x^2+2x-3}$$

$$= \frac{2}{x+3} - \frac{x}{x-1} + \frac{x^2+2}{(x+3)(x-1)},$$
$$\text{LCD is } (x+3)(x-1)$$

$$= \frac{2}{x+3} \cdot \frac{x-1}{x-1} - \frac{x}{x-1} \cdot \frac{x+3}{x+3} + \frac{x^2+2}{(x+3)(x-1)}$$

$$= \frac{2(x-1)}{(x+3)(x-1)} - \frac{x(x+3)}{(x-1)(x+3)} + \frac{x^2+2}{(x+3)(x-1)}$$

$$= \frac{2(x-1) - x(x+3) + x^2 + 2}{(x+3)(x-1)}$$

$$= \frac{2x - 2 - x^2 - 3x + x^2 + 2}{(x+3)(x-1)}$$

$$= \frac{-x}{(x+3)(x-1)}$$

55. $\sqrt{-20} = \sqrt{-1 \cdot 4 \cdot 5} = i \cdot 2 \cdot \sqrt{5} = 2\sqrt{5}i$, or $2i\sqrt{5}$

57.

$$\frac{\dfrac{4}{a^2 b}}{\dfrac{3}{a} - \dfrac{4}{b^2}} = \frac{\dfrac{4}{a^2 b}}{\dfrac{3}{a} - \dfrac{4}{b^2}} \cdot \frac{a^2 b^2}{a^2 b^2}$$

$$= \frac{\dfrac{4}{a^2 b} \cdot a^2 b^2}{\dfrac{3}{a} \cdot a^2 b^2 - \dfrac{4}{b^2} \cdot a^2 b^2}$$

$$= \frac{4b}{3ab^2 - 4a^2}, \text{ or } \frac{4b}{a(3b^2 - 4a)}$$

59.

$$\frac{1}{a-1} = a + 1$$

$$\frac{1}{a-1} \cdot a - 1 = (a+1)(a-1)$$

$$1 = a^2 - 1$$

$$2 = a^2$$

$$\pm\sqrt{2} = a$$

61. Let s represent a length of a side of the cube, let S represent the surface area of the cube, and let A represent the surface area of the sphere. Then the diameter of the sphere is s, so the radius r is $s/2$. From Exercise 32, we know, $A = 4\pi r^2$, so when $r = s/2$ we have $A = 4\pi\left(\dfrac{s}{2}\right)^2 = 4\pi \cdot \dfrac{s^2}{4} = \pi s^2$. From the formula for the surface area of a cube (See Exercise 31.) we know that $S = 6s^2$, so $\dfrac{S}{6} = s^2$ and then $A = \pi \cdot \dfrac{S}{6}$, or $A(S) = \dfrac{\pi S}{6}$.

63.

$$\frac{w}{l} = \frac{l}{w+l}$$

$$l(w+l) \cdot \frac{w}{l} = l(w+l) \cdot \frac{l}{w+l}$$

$$w(w+l) = l^2$$

$$w^2 + lw = l^2$$

$$0 = l^2 - lw - w^2$$

Use the quadratic formula with $a = 1$, $b = -w$, and $c = -w^2$.

$$l = \frac{-(-w) \pm \sqrt{(-w)^2 - 4 \cdot 1 \cdot (-w^2)}}{2 \cdot 1}$$

$$l = \frac{w \pm \sqrt{w^2 + 4w^2}}{2} = \frac{w \pm \sqrt{5w^2}}{2}$$

$$l = \frac{w \pm w\sqrt{5}}{2}$$

Since $\dfrac{w - w\sqrt{5}}{2}$ is negative we use the positive square root:

$$l = \frac{w + w\sqrt{5}}{2}$$

Exercise Set 16.4

1. $x^2 - 8x + 16 = 0$

$a = 1$, $b = -8$, $c = 16$

We compute the discriminant.

$$b^2 - 4ac = (-8)^2 - 4 \cdot 1 \cdot 16$$
$$= 64 - 64$$
$$= 0$$

Since $b^2 - 4ac = 0$, there is just one solution, and it is a real number.

3. $x^2 + 1 = 0$

$a = 1$, $b = 0$, $c = 1$

We compute the discriminant.

$$b^2 - 4ac = 0^2 - 4 \cdot 1 \cdot 1$$
$$= -4$$

Since $b^2 - 4ac < 0$, there are two nonreal solutions.

5. $x^2 - 6 = 0$

$a = 1$, $b = 0$, $c = -6$

We compute the discriminant.

$$b^2 - 4ac = 0^2 - 4 \cdot 1 \cdot (-6)$$
$$= 24$$

Since $b^2 - 4ac > 0$, there are two real solutions.

7. $4x^2 - 12x + 9 = 0$

$a = 4$, $b = -12$, $c = 9$

We compute the discriminant.

$$b^2 - 4ac = (-12)^2 - 4 \cdot 4 \cdot 9$$
$$= 144 - 144$$
$$= 0$$

Since $b^2 - 4ac = 0$, there is just one solution, and it is a real number.

9. $x^2 - 2x + 4 = 0$

$a = 1$, $b = -2$, $c = 4$

We compute the discriminant.

$$b^2 - 4ac = (-2)^2 - 4 \cdot 1 \cdot 4$$
$$= 4 - 16$$
$$= -12$$

Since $b^2 - 4ac < 0$, there are two nonreal solutions.

11. $9t^2 - 3t = 0$

$a = 9$, $b = -3$, $c = 0$

We compute the discriminant.

$$b^2 - 4ac = (-3)^2 - 4 \cdot 9 \cdot 0$$
$$= 9 - 0$$
$$= 9$$

Since $b^2 - 4ac > 0$, there are two real solutions.

13. $y^2 = \frac{1}{2}y + \frac{3}{5}$

$y^2 - \frac{1}{2}y - \frac{3}{5} = 0$ Standard form

$a = 1$, $b = -\frac{1}{2}$, $c = -\frac{3}{5}$

We compute the discriminant.

$$b^2 - 4ac = \left(-\frac{1}{2}\right)^2 - 4 \cdot 1 \cdot \left(-\frac{3}{5}\right)$$
$$= \frac{1}{4} + \frac{12}{5}$$
$$= \frac{53}{20}$$

Since $b^2 - 4ac > 0$, there are two real solutions.

15. $4x^2 - 4\sqrt{3}x + 3 = 0$

$a = 4$, $b = -4\sqrt{3}$, $c = 3$

We compute the discriminant.

$$b^2 - 4ac = (-4\sqrt{3})^2 - 4 \cdot 4 \cdot 3$$
$$= 48 - 48$$
$$= 0$$

Since $b^2 - 4ac = 0$, there is just one solution, and it is a real number.

17. The solutions are -4 and 4.

$x = -4$ *or* $x = 4$

$x + 4 = 0$ *or* $x - 4 = 0$

$(x + 4)(x - 4) = 0$ Principle of zero products

$x^2 - 16 = 0$ $(A + B)(A - B) = A^2 - B^2$

19. The solutions are -2 and -7.

$x = -2$ *or* $x = -7$

$x + 2 = 0$ *or* $x + 7 = 0$

$(x + 2)(x + 7) = 0$ Principle of zero products

$x^2 + 9x + 14 = 0$ FOIL

21. The only solution is 8. It must be a double solution.

$x = 8$ *or* $x = 8$

$x - 8 = 0$ *or* $x - 8 = 0$

$(x - 8)(x - 8) = 0$ Principle of zero products

$x^2 - 16x + 64 = 0$ $(A - B)^2 = A^2 - 2AB + B^2$

23. The solutions are $-\frac{2}{5}$ and $\frac{6}{5}$.

$x = -\frac{2}{5}$ *or* $x = \frac{6}{5}$

$x + \frac{2}{5} = 0$ *or* $x - \frac{6}{5} = 0$

$5x + 2 = 0$ *or* $5x - 6 = 0$ Clearing fractions

$(5x + 2)(5x - 6) = 0$ Principle of zero products

$25x^2 - 20x - 12 = 0$ FOIL

25. The solutions are $\frac{k}{3}$ and $\frac{m}{4}$.

$x = \frac{k}{3}$ *or* $x = \frac{m}{4}$

$x - \frac{k}{3} = 0$ *or* $x - \frac{m}{4} = 0$

$3x - k = 0$ *or* $4x - m = 0$ Clearing fractions

$$(3x - k)(4x - m) = 0 \quad \text{Principle of zero products}$$

$$12x^2 - 3mx - 4kx + km = 0 \quad \text{FOIL}$$

$$12x^2 - (3m + 4k)x + km = 0 \quad \text{Collecting like terms}$$

27. The solutions are $-\sqrt{3}$ and $2\sqrt{3}$.

$$x = -\sqrt{3} \quad or \quad x = 2\sqrt{3}$$

$$x + \sqrt{3} = 0 \quad or \quad x - 2\sqrt{3} = 0$$

$$(x + \sqrt{3})(x - 2\sqrt{3}) = 0 \quad \text{Principle of zero products}$$

$$x^2 - 2\sqrt{3}x + \sqrt{3}x - 2(\sqrt{3})^2 = 0 \quad \text{FOIL}$$

$$x^2 - \sqrt{3}x - 6 = 0$$

29. $x^4 - 6x^2 + 9 = 0$

Let $u = x^2$ and think of x^4 as $(x^2)^2$.

$$u^2 - 6u + 9 = 0 \quad \text{Substituting } u \text{ for } x^2$$

$$(u - 3)(u - 3) = 0$$

$$u - 3 = 0 \quad or \quad u - 3 = 0$$

$$u = 3 \quad or \quad u = 3$$

Now we substitute x^2 for u and solve the equation:

$$x^2 = 3$$

$$x = \pm\sqrt{3}$$

Both $\sqrt{3}$ and $-\sqrt{3}$ check. They are the solutions.

31. $x - 10\sqrt{x} + 9 = 0$

Let $u = \sqrt{x}$ and think of x as $(\sqrt{x})^2$.

$$u^2 - 10u + 9 = 0 \quad \text{Substituting } u \text{ for } \sqrt{x}$$

$$(u - 9)(u - 1) = 0$$

$$u - 9 = 0 \quad or \quad u - 1 = 0$$

$$u = 9 \quad or \quad u = 1$$

Now we substitute $\sqrt{x}$ for u and solve these equations:

$$\sqrt{x} = 9 \quad or \quad \sqrt{x} = 1$$

$$x = 81 \quad or \quad x = 1$$

The numbers 81 and 1 both check. They are the solutions.

33. $(x^2 - 6x) - 2(x^2 - 6x) - 35 = 0$

Let $u = x^2 - 6x$.

$$u^2 - 2u - 35 = 0 \quad \text{Substituting } u \text{ for } x^2 - 6x$$

$$(u - 7)(u + 5) = 0$$

$$u - 7 = 0 \quad or \quad u + 5 = 0$$

$$u = 7 \quad or \quad u = -5$$

Now we substitute $x^2 - 6x$ for u and solve these equations:

$$x^2 - 6x = 7 \quad or \quad x^2 - 6x = -5$$

$$x^2 - 6x - 7 = 0 \quad or \quad x^2 - 6x + 5 = 0$$

$$(x - 7)(x + 1) = 0 \quad or \quad (x - 5)(x - 1) = 0$$

$$x = 7 \text{ or } x = -1 \text{ or } x = 5 \text{ or } x = 1$$

The numbers -1, 1, 5, and 7 check. They are the solutions.

35. $x^{-2} - 5^{-1} - 36 = 0$

Let $u = x^{-1}$.

$$u^2 - 5u - 36 = 0 \quad \text{Substituting } u \text{ for } x^{-1}$$

$$(u - 9)(u + 4) = 0$$

$$u - 9 = 0 \quad or \quad u + 4 = 0$$

$$u = 9 \quad or \quad u = -4$$

Now we substitute x^{-1} for u and solve these equations:

$$x^{-1} = 9 \quad or \quad x^{-1} = -4$$

$$\frac{1}{x} = 9 \quad or \quad \frac{1}{x} = -4$$

$$\frac{1}{9} = x \quad or \quad -\frac{1}{4} = x$$

Both $\frac{1}{9}$ and $-\frac{1}{4}$ check. They are the solutions.

37. $(1 + \sqrt{x})^2 + (1 + \sqrt{x}) - 6 = 0$

Let $u = 1 + \sqrt{x}$.

$$u^2 + u - 6 = 0 \quad \text{Substituting } u \text{ for } 1 + \sqrt{x}$$

$$(u + 3)(u - 2) = 0$$

$$u + 3 = 0 \quad or \quad u - 2 = 0$$

$$u = -3 \quad or \quad u = 2$$

$$1 + \sqrt{x} = -3 \quad or \quad 1 + \sqrt{x} = 2 \quad \text{Substituting } 1 + \sqrt{x} \text{ for } u$$

$$\sqrt{x} = -4 \quad or \quad \sqrt{x} = 1$$

$$\text{No solution} \quad\quad\quad x = 1$$

The number 1 checks. It is the solution.

39. $(y^2 - 5y)^2 - 2(y^2 - 5y) - 24 = 0$

Let $u = y^2 - 5y$.

$$u^2 - 2u - 24 = 0 \quad \text{Substituting } u \text{ for } y^2 - 5y$$

$$(u - 6)(u + 4) = 0$$

$$u - 6 = 0 \quad or \quad u + 4 = 0$$

$$u = 6 \quad or \quad u = -4$$

$$y^2 - 5y = 6 \quad or \quad y^2 - 5y = -4$$

Substituting $y^2 - 5y$ for u

$$y^2 - 5y - 6 = 0 \quad or \quad y^2 - 5y + 4 = 0$$

$$(y - 6)(y + 1) = 0 \quad or \quad (y - 4)(y - 1) = 0$$

$$y = 6 \text{ or } y = -1 \text{ or } y = 4 \text{ or } y = 1$$

The numbers -1, 1, 4, and 6 check. They are the solutions.

41. $t^4 - 6t^2 - 4 = 0$

Let $u = t^2$.

$u^2 - 6u - 4 = 0$ Substituting u for t^2

$u = \dfrac{-(-6) \pm \sqrt{(-6)^2 - 4 \cdot 1 \cdot (-4)}}{2 \cdot 1}$

$u = \dfrac{6 \pm \sqrt{52}}{2} = \dfrac{6 \pm 2\sqrt{13}}{2}$

$u = 3 \pm \sqrt{13}$

Now we substitute t^2 for u and solve these equations:

$t^2 = 3 + \sqrt{13}$ or $t^2 = 3 - \sqrt{13}$

$t = \pm\sqrt{3 + \sqrt{13}}$ or $t = \pm\sqrt{3 - \sqrt{13}}$

All four numbers check. They are the solutions.

43. $2x^{-2} + x^{-1} - 1 = 0$

Let $u = x^{-1}$.

$2u^2 + u - 1 = 0$ Substituting u for x^{-1}

$(2u - 1)(u + 1) = 0$

$2u - 1 = 0$ or $u + 1 = 0$

$2u = 1$ or $u = -1$

$u = \dfrac{1}{2}$ or $u = -1$

$x^{-1} = \dfrac{1}{2}$ or $x^{-1} = -1$ Substituting x^{-1}
                              for u

$\dfrac{1}{x} = \dfrac{1}{2}$ or $\dfrac{1}{x} = -1$

$x = 2$ or $x = -1$

Both 2 and -1 check. They are the solutions.

45. $6x^4 - 19x^2 + 15 = 0$

Let $u = x^2$.

$6u^2 - 19u + 15 = 0$ Substituting u for x^2

$(3u - 5)(2u - 3) = 0$

$3u - 5 = 0$ or $2u - 3 = 0$

$3u = 5$ or $2u = 3$

$u = \dfrac{5}{3}$ or $u = \dfrac{3}{2}$

$x^2 = \dfrac{5}{3}$ or $x^2 = \dfrac{3}{2}$ Substituting x^2
                                for u

$x = \pm\sqrt{\dfrac{5}{3}}$ or $x = \pm\sqrt{\dfrac{3}{2}}$

$x = \pm\dfrac{\sqrt{15}}{3}$ or $x = \pm\dfrac{\sqrt{6}}{2}$
                       Rationalizing denominators

All four numbers check. They are the solutions.

47. $x^{2/3} - 4x^{1/3} - 5 = 0$

Let $u = x^{1/3}$.

$u^2 - 4u - 5 = 0$ Substituting u for $x^{1/3}$

$(u - 5)(u + 1) = 0$

$u - 5 = 0$ or $u + 1 = 0$

$u = 5$ or $u = -1$

$x^{1/3} = 5$ or $x^{1/3} = -1$ Substituting $x^{1/3}$
                                 for u

$(x^{1/3})^3 = 5^3$ or $(x^{1/3})^3 = (-1)^3$ Principle
                                      of powers

$x = 125$ or $x = -1$

Both 125 and -1 check. They are the solutions.

49. $\left(\dfrac{x - 4}{x + 1}\right)^2 - 2\left(\dfrac{x - 4}{x + 1}\right) - 35 = 0$

Let $u = \dfrac{x - 4}{x + 1}$.

$u^2 - 2u - 35 = 0$ Substituting u for $\dfrac{x - 4}{x + 1}$

$(u - 7)(u + 5) = 0$

$u - 7 = 0$ or $u + 5 = 0$

$u = 7$ or $u = -5$

$\dfrac{x - 4}{x + 1} = 7$ or $\dfrac{x - 4}{x + 1} = -5$ Substituting
                                      $\dfrac{x - 4}{x + 1}$ for u

$x - 4 = 7(x + 1)$ or $x - 4 = -5(x + 1)$

$x - 4 = 7x + 7$ or $x - 4 = -5x - 5$

$-6x = 11$ or $6x = -1$

$x = -\dfrac{11}{6}$ or $x = -\dfrac{1}{6}$

Both $-\dfrac{11}{6}$ and $-\dfrac{1}{6}$ check. They are the solutions.

51. $9\left(\dfrac{x + 2}{x + 3}\right)^2 - 6\left(\dfrac{x + 2}{x + 3}\right) + 1 = 0$

Let $u = \dfrac{x + 2}{x + 3}$.

$9u^2 - 6u + 1 = 0$ Substituting u for $\dfrac{x + 2}{x + 3}$

$(3u - 1)(3u - 1) = 0$

$3u - 1 = 0$ or $3u - 1 = 0$

$3u = 1$ or $3u = 1$

$u = \dfrac{1}{3}$ or $u = \dfrac{1}{3}$

Now we substitute $\dfrac{x + 2}{x + 3}$ for u and solve the equation:

$\dfrac{x + 2}{x + 3} = \dfrac{1}{3}$

$3(x + 2) = x + 3$ Multiplying by $3(x + 3)$

$3x + 6 = x + 3$

$2x = -3$

$x = -\dfrac{3}{2}$

The number $-\dfrac{3}{2}$ checks. It is the solution.

53. $\left(\dfrac{x^2-2}{x}\right)^2 - 7\left(\dfrac{x^2-2}{x}\right) - 18 = 0$

Let $u = \dfrac{x^2-2}{x}$.

$u^2 - 7u - 18 = 0$　　Substituting u for $\dfrac{x^2-2}{x}$

$(u-9)(u+2) = 0$

$\quad u - 9 = 0 \quad or \quad\quad u + 2 = 0$

$\quad\quad u = 9 \quad or \quad\quad\quad u = -2$

$\quad \dfrac{x^2-2}{x} = 9 \quad or \quad\quad \dfrac{x^2-2}{x} = -2$

$\quad\quad\quad\quad\quad\quad$ Substituting $\dfrac{x^2-2}{x}$ for u

$\quad x^2 - 2 = 9x \quad or \quad\quad x^2 - 2 = -2x$

$x^2 - 9x - 2 = 0 \quad or \quad x^2 + 2x - 2 = 0$

$\quad x = \dfrac{-(-9) \pm \sqrt{(-9)^2 - 4 \cdot 1 \cdot (-2)}}{2 \cdot 1}$

$\quad x = \dfrac{9 \pm \sqrt{89}}{2}$

or

$\quad x = \dfrac{-2 \pm \sqrt{2^2 - 4 \cdot 1 \cdot (-2)}}{2 \cdot 1} = \dfrac{-2 \pm \sqrt{12}}{2}$

$\quad x = \dfrac{-2 \pm 2\sqrt{3}}{2} = -1 \pm \sqrt{3}$

All four numbers check. They are the solutions.

55. The x-intercepts occur where $f(x) = 0$. Thus, we must have $5x + 13\sqrt{x} - 6 = 0$.

Let $u = \sqrt{x}$.

$\quad 5u^2 + 13u - 6 = 0$　　Substituting

$\quad (5u - 2)(u + 3) = 0$

$u = \dfrac{2}{5} \; or \; u = -3$

Now replace u with $\sqrt{x}$ and solve these equations:

$\quad \sqrt{x} = \dfrac{2}{5} \quad or \quad \sqrt{x} = -3$

$\quad\quad x = \dfrac{4}{25} \quad\quad$ No solution

The number $\dfrac{4}{25}$ checks. Thus, the x-intercept is $\left(\dfrac{4}{25}, 0\right)$.

57. The x-intercepts occur where $f(x) = 0$. Thus, we must have $(x^2 - 3x)^2 - 10(x^2 - 3x) + 24 = 0$.

Let $u = x^2 - 3x$.

$\quad u^2 - 10u + 24 = 0$　　Substituting

$\quad (u - 6)(u - 4) = 0$

$u = 6 \; or \; u = 4$

Now replace u with $x^2 - 3x$ and solve these equations:

$\quad x^2 - 3x = 6 \quad or \quad\quad x^2 - 3x = 4$

$x^2 - 3x - 6 = 0 \quad or \quad x^2 - 3x - 4 = 0$

$x = \dfrac{-(-3) \pm \sqrt{(-3)^2 - 4(1)(-6)}}{2 \cdot 1} \; or$

$\quad\quad\quad\quad (x - 4)(x + 1) = 0$

$x = \dfrac{3 \pm \sqrt{33}}{2} \; or \; x = 4 \; or \; x = -1$

All four numbers check. Thus, the x-intercepts are $\left(\dfrac{3+\sqrt{33}}{2}, 0\right)$, $\left(\dfrac{3-\sqrt{33}}{2}, 0\right)$, $(4, 0)$, and $(-1, 0)$.

59. Discussion and Writing Exercise

61. *Familiarize*. Let x = the number of pounds of Kenyan coffee and y = the number of pounds of Peruvian coffee in the mixture. We organize the information in a table.

Type of Coffee	Kenyan	Peruvian	Mixture
Price per pound	$6.75	$11.25	$8.55
Number of pounds	x	y	50
Total cost	$6.75x$	$11.25y$	8.55×50, or $427.50

Translate. From the last two rows of the table we get a system of equations.

$\quad x + y = 50,$

$\quad 6.75x + 11.25y = 427.50$

Solve. Solving the system of equations, we get $(30, 20)$.

Check. The total number of pounds in the mixture is $30 + 20$, or 50. The total cost of the mixture is $6.75(30) + 11.25(20) = 427.50$. The values check.

State. The mixture should consist of 30 lb of Kenyan coffee and 20 lb of Peruvian coffee.

63. $\sqrt{8x} \cdot \sqrt{2x} = \sqrt{8x \cdot 2x} = \sqrt{16x^2} = \sqrt{(4x)^2} = 4x$

65. $\sqrt[4]{9a^2} \cdot \sqrt[4]{18a^3} = \sqrt[4]{9a^2 \cdot 18a^3} =$
$\sqrt[4]{3 \cdot 3 \cdot a^2 \cdot 3 \cdot 3 \cdot 2 \cdot a^2 \cdot a} = \sqrt[4]{3^4 a^4 \cdot 2a} = \sqrt[4]{3^4} \sqrt[4]{a^4} \sqrt[4]{2a} = 3a\sqrt[4]{2a}$

67. Graph $f(x) = -\dfrac{3}{5}x + 4$.

Choose some values for x, find the corresponding values of $f(x)$, plot the points $(x, f(x))$, and draw the graph.

x	$f(x)$	$(x, f(x))$
-5	7	$(-5, 7)$
0	4	$(0, 4)$
5	1	$(5, 1)$

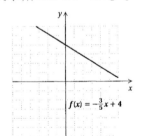

69. Graph $y = 4$.

The graph of $y = 4$ is a horizontal line with y-intercept $(0, 4)$.

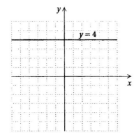

71. Left to the student

73. a) $kx^2 - 2x + k = 0$; one solution is -3

We first find k by substituting -3 for x.

$$k(-3)^2 - 2(-3) + k = 0$$
$$9k + 6 + k = 0$$
$$10k = -6$$
$$k = -\frac{6}{10}$$
$$k = -\frac{3}{5}$$

b) $-\frac{3}{5}x^2 - 2x + \left(-\frac{3}{5}\right) = 0$ Substituting $-\frac{3}{5}$ for k

$$3x^2 + 10x + 3 = 0$$ Multiplying by -5
$$(3x + 1)(x + 3) = 0$$
$$3x + 1 = 0 \quad or \quad x + 3 = 0$$
$$3x = -1 \quad or \quad x = -3$$
$$x = -\frac{1}{3} \quad or \quad x = -3$$

The other solution is $-\frac{1}{3}$.

75. For $ax^2 + bx + c = 0$, $-\frac{b}{a}$ is the sum of the solutions and $\frac{c}{a}$ is the product of the solutions. Thus $-\frac{b}{a} = \sqrt{3}$ and $\frac{c}{a} = 8$.

$$ax^2 + bx + c = 0$$
$$x^2 + \frac{b}{a}x + \frac{c}{a} = 0$$ Multiplying by $\frac{1}{a}$
$$x^2 - \left(-\frac{b}{a}\right)x + \frac{c}{a} = 0$$
$$x^2 - \sqrt{3}x + 8 = 0$$ Substituting $\sqrt{3}$ for $-\frac{b}{a}$ and 8 for $\frac{c}{a}$

77. The graph includes the points $(-3, 0)$, $(0, -3)$, and $(1, 0)$. Substituting in $y = ax^2 + bx + c$, we have three equations.

$$0 = 9a - 3b + c,$$
$$-3 = \qquad\quad c,$$
$$0 = a + b + c$$

The solution of this system of equations is $a = 1$, $b = 2$, $c = -3$.

79. $\dfrac{x}{x - 1} - 6\sqrt{\dfrac{x}{x - 1}} - 40 = 0$

Let $u = \sqrt{\dfrac{x}{x - 1}}$.

$$u^2 - 6u - 40 = 0 \quad \text{Substituting for } \sqrt{\dfrac{x}{x - 1}}$$
$$(u - 10)(u + 4) = 0$$
$$u = 10 \qquad or \qquad u = -4$$
$$\sqrt{\dfrac{x}{x - 1}} = 10 \quad or \quad \sqrt{\dfrac{x}{x - 1}} = -4$$

Substituting for u

$$\dfrac{x}{x - 1} = 100 \qquad or \qquad \text{No solution}$$
$$x = 100x - 100 \quad \text{Multiplying by } (x - 1)$$
$$100 = 99x$$
$$\dfrac{100}{99} = x$$

This number checks. It is the solution.

81. $\sqrt{x - 3} - \sqrt[4]{x - 3} = 12$

$$(x - 3)^{1/2} - (x - 3)^{1/4} - 12 = 0$$

Let $u = (x - 3)^{1/4}$.

$$u^2 - u - 12 = 0 \quad \text{Substituting for } (x - 3)^{1/4}$$
$$(u - 4)(u + 3) = 0$$
$$u = 4 \qquad or \qquad u = -3$$
$$(x - 3)^{1/4} = 4 \quad or \quad (x - 3)^{1/4} = -3$$

Substituting for u

$$x - 3 = 4^4 \quad or \qquad \text{No solution}$$
$$x - 3 = 256$$
$$x = 259$$

This number checks. It is the solution.

83. $x^6 - 28x^3 + 27 = 0$

Let $u = x^3$.

$$u^2 - 28u + 27 = 0 \quad \text{Substituting for } x^3$$
$$(u - 27)(u - 1) = 0$$
$$u = 27 \quad or \quad u = 1$$
$$x^3 = 27 \quad or \quad x^3 = 1 \quad \text{Substituting for } u$$
$$x = 3 \quad or \quad x = 1$$

Both 3 and 1 check. They are the solutions.

Exercise Set 16.5

1. $f(x) = 4x^2$

$f(x) = 4x^2$ is of the form $f(x) = ax^2$. Thus we know that the vertex is $(0,0)$ and $x = 0$ is the line of symmetry.

We know that $f(x) = 0$ when $x = 0$ since the vertex is $(0,0)$.

For $x = 1$, $f(x) = 4x^2 = 4 \cdot 1^2 = 4$.

For $x = -1$, $f(x) = 4x^2 = 4 \cdot (-1)^2 = 4$.

For $x = 2$, $f(x) = 4x^2 = 4 \cdot 2^2 = 16$.

For $x = -2$, $f(x) = 4x^2 = 4 \cdot (-2)^2 = 16$.

We complete the table.

x	$f(x)$	
0	0	← Vertex
1	4	
2	16	
−1	4	
−2	16	

We plot the ordered pairs $(x, f(x))$ from the table and connect them with a smooth curve.

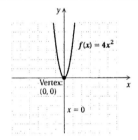

3. $f(x) = \dfrac{1}{3}x^2$ is of the form $f(x) = ax^2$. Thus we know that the vertex is $(0,0)$ and $x = 0$ is the line of symmetry. We choose some numbers for x and find the corresponding values for $f(x)$. Then we plot the ordered pairs $(x, f(x))$ and connect them with a smooth curve.

For $x = 3$, $f(x) = \dfrac{1}{3}x^2 = \dfrac{1}{3} \cdot 3^2 = 3$.

For $x = -3$, $f(x) = \dfrac{1}{3}x^2 = \dfrac{1}{3} \cdot (-3)^2 = 3$.

For $x = 6$, $f(x) = \dfrac{1}{3}x^2 = \dfrac{1}{3} \cdot 6^2 = 12$.

For $x = -6$, $f(x) = \dfrac{1}{3}x^2 = \dfrac{1}{3} \cdot (-6)^2 = 12$.

x	$f(x)$	
0	0	← Vertex
1	$\dfrac{1}{3}$	
2	$\dfrac{4}{3}$	
−1	$\dfrac{1}{3}$	
−2	$\dfrac{4}{3}$	

5. $f(x) = -\dfrac{1}{2}x^2$ is of the form $f(x) = ax^2$. Thus we know that the vertex is $(0,0)$ and $x = 0$ is the line of symmetry. We choose some numbers for x and find the corresponding values for $f(x)$. Then we plot the ordered pairs $(x, f(x))$ and connect them with a smooth curve.

For $x = 2$, $f(x) = -\dfrac{1}{2}x^2 = -\dfrac{1}{2} \cdot 2^2 = -2$.

For $x = -2$, $f(x) = -\dfrac{1}{2}x^2 = -\dfrac{1}{2} \cdot (-2)^2 = -2$.

For $x = 4$, $f(x) = -\dfrac{1}{2}x^2 = -\dfrac{1}{2} \cdot 4^2 = -8$.

For $x = -4$, $f(x) = -\dfrac{1}{2}x^2 = -\dfrac{1}{2} \cdot (-4)^2 = -8$.

x	$f(x)$	
0	0	← Vertex
2	−2	
−2	−2	
4	−8	
−4	−8	

7. $f(x) = -4x^2$ is of the form $f(x) = ax^2$. Thus we know that the vertex is $(0,0)$ and $x = 0$ is the line of symmetry. We choose some numbers for x and find the corresponding values for $f(x)$. Then we plot the ordered pairs $(x, f(x))$ and connect them with a smooth curve.

For $x = 1$, $f(x) = -4x^2 = -4 \cdot 1^2 = -4$.

For $x = -1$, $f(x) = -4x^2 = -4 \cdot (-1)^2 = -4$.

For $x = 2$, $f(x) = -4x^2 = -4 \cdot 2^2 = -16$.

For $x = -2$, $f(x) = -4x^2 = -4 \cdot (-2)^2 = -16$.

x	$f(x)$	
0	0	← Vertex
1	−4	
−1	−4	
2	−16	
−2	−16	

9. $f(x) = (x + 3)^2 = [x - (-3)]^2$ is of the form $f(x) = a(x - h)^2$.

Thus we know that the vertex is $(-3, 0)$ and $x = -3$ is the line of symmetry. We choose some numbers for x and find the corresponding values for $f(x)$. Then we plot the ordered pairs $(x, f(x))$ and connect them with a smooth curve.

x	$f(x)$	
−3	0	← Vertex
−2	1	
−1	4	
−4	1	
−5	4	

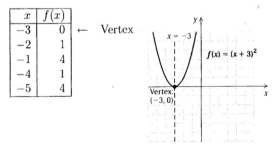

11. Graph: $f(x) = 2(x-4)^2$

We choose some values of x and compute $f(x)$. Then we plot these ordered pairs and connect them with a smooth curve.

x	$f(x)$
4	0
5	2
3	2
6	8
2	8

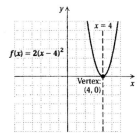

The graph of $f(x) = 2(x-4)^2$ looks like the graph of $f(x) = 2x^2$ except that it is translated 4 units to the right. The vertex is $(4, 0)$, and the line of symmetry is $x = 4$.

13. Graph: $f(x) = -2(x+2)^2$

We choose some values of x and compute $f(x)$. Then we plot these ordered pairs and connect them with a smooth curve.

x	$f(x)$
1	-18
0	-8
-1	-2
-2	0
-3	-2
-4	-8

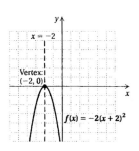

We can express the equation in the equivalent form $f(x) = -2[x-(-2)]^2$. Then we know that the graph looks like the graph of $f(x) = -2x^2$ translated 2 units to the left. The vertex is $(-2, 0)$, and the line of symmetry is $x = -2$.

15. Graph: $f(x) = 3(x-1)^2$

We choose some values of x and compute $f(x)$. Then we plot these ordered pairs and connect them with a smooth curve.

x	$f(x)$
1	0
2	3
0	3
3	12
-1	12

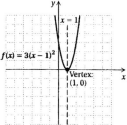

The graph of $f(x) = 3(x-1)^2$ looks like the graph of $f(x) = 3x^2$ except that it is translated 1 unit to the right. The vertex is $(1, 0)$, and the line of symmetry is $x = 1$.

17. Graph: $f(x) = -\dfrac{3}{2}(x+2)^2$

We choose some values of x and compute $f(x)$. Then we plot these ordered pairs and connect them with a smooth curve.

x	$f(x)$
-4	-6
-2	0
0	-6
2	-24

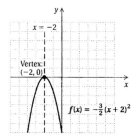

We can express the equation in the equivalent form $f(x) = -\dfrac{3}{2}[x-(-2)]^2$. Then we know that the graph looks like the graph of $f(x) = -\dfrac{3}{2}x^2$ translated 2 units to the left. The vertex is $(-2, 0)$, and the line of symmetry is $x = -2$.

19. Graph: $f(x) = (x-3)^2 + 1$

We choose some values of x and compute $f(x)$. Then we plot these ordered pairs and connect them with a smooth curve.

x	$f(x)$
3	1
4	2
2	2
5	5
1	5

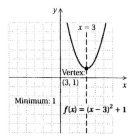

The graph of $f(x) = (x-3)^2 + 1$ looks like the graph of $f(x) = x^2$ except that it is translated 3 units right and 1 unit up. The vertex is $(3, 1)$, and the line of symmetry is $x = 3$. The equation is of the form $f(x) = a(x-h)^2 + k$ with $a = 1$. Since $1 > 0$, we know that 1 is the minimum value.

21. Graph: $f(x) = -3(x+4)^2 + 1$
$$f(x) = -3[x-(-4)]^2 + 1$$

We choose some values of x and compute $f(x)$. Then we plot these ordered pairs and connect them with a smooth curve.

x	$f(x)$
-4	1
$-3\frac{1}{2}$	$\frac{1}{4}$
$-4\frac{1}{2}$	$\frac{1}{4}$
-3	-2
-5	-2
-2	-11
-6	-11

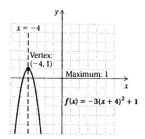

The graph of $f(x) = -3(x+4)^2 + 1$ looks like the graph of $f(x) = 3x^2$ except that it is translated 4 units left and 1 unit up and opens downward. The vertex is $(-4, 1)$, and the line of symmetry is $x = -4$. Since $-3 < 0$, we know that 1 is the maximum value.

23. Graph: $f(x) = \dfrac{1}{2}(x+1)^2 + 4$

$$f(x) = \dfrac{1}{2}[x - (-1)]^2 + 4$$

We choose some values of x and compute $f(x)$. Then we plot these ordered pairs and connect them with a smooth curve.

x	$f(x)$
1	6
2	$8\frac{1}{2}$
0	$4\frac{1}{2}$
-1	4
-2	$4\frac{1}{2}$
-3	6

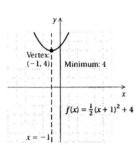

The graph of $f(x) = \dfrac{1}{2}(x+1)^2 + 4$ looks like the graph of $f(x) = \dfrac{1}{2}x^2$ except that it is translated 1 unit left and 4 units up. The vertex is $(-1, 4)$, and the line of symmetry is $x = -1$. Since $\dfrac{1}{2} > 0$, we know that 4 is the minimum value.

25. Graph: $f(x) = -(x+1)^2 - 2$
$$f(x) = -[x - (-1)]^2 + (-2)$$

We choose some values of x and compute $f(x)$. Then we plot these ordered pairs and connect them with a smooth curve.

x	$f(x)$
-1	-2
0	-3
-2	-3
1	-6
-3	-6

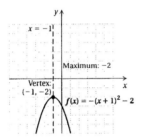

The graph of $f(x) = -(x+1)^2 - 2$ looks like the graph of $f(x) = x^2$ except that it is translated 1 unit left and 2 units down and opens downward. The vertex is $(-1, -2)$, and the line of symmetry is $x = -1$. Since $-1 < 0$, we know that -2 is the maximum value.

27. Discussion and Writing Exercise

29.
$$x - 5 = \sqrt{x+7}$$
$$(x-5)^2 = (\sqrt{x+7})^2 \quad \text{Principle of powers}$$
$$x^2 - 10x + 25 = x + 7$$
$$x^2 - 11x + 18 = 0$$
$$(x-9)(x-2) = 0$$
$$x - 9 = 0 \quad or \quad x - 2 = 0$$
$$x = 9 \quad or \qquad x = 2$$

Check: For 9:

$$\begin{array}{c|c} x - 5 = \sqrt{x+7} \\ \hline 9 - 5 \; ? \; \sqrt{9+7} \\ 4 \; \Big| \; \sqrt{16} \\ \Big| \; 4 \qquad \text{TRUE} \end{array}$$

For 2:

$$\begin{array}{c|c} x - 5 = \sqrt{x+7} \\ \hline 2 - 5 \; ? \; \sqrt{2+7} \\ -3 \; \Big| \; \sqrt{9} \\ \Big| \; 3 \qquad \text{FALSE} \end{array}$$

Only 9 checks. It is the solution.

31. $\sqrt{x+4} = -11$

The equation has no solution, because the principal square root of a number is always nonnegative.

33. Left to the student

Exercise Set 16.6

1. $f(x) = x^2 - 2x - 3 = (x^2 - 2x) - 3$

We complete the square inside the parentheses. We take half the x-coefficient and square it.

$$\dfrac{1}{2} \cdot (-2) = -1 \text{ and } (-1)^2 = 1$$

Then we add $1 - 1$ inside the parentheses.

$$\begin{aligned} f(x) &= (x^2 - 2x + 1 - 1) - 3 \\ &= (x^2 - 2x + 1) - 1 - 3 \\ &= (x-1)^2 - 4 \\ &= (x-1)^2 + (-4) \end{aligned}$$

Vertex: $(1, -4)$

Line of symmetry: $x = 1$

The coefficient of x^2 is 1, which is positive, so the graph opens up. This tells us that -4 is a minimum.

We plot a few points and draw the curve.

x	$f(x)$
1	-4
2	-3
0	-3
3	0
-1	0
4	5
-2	5

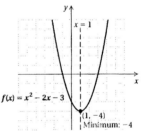

3. $f(x) = -x^2 - 4x - 2 = -(x^2 + 4x) - 2$

We complete the square inside the parentheses. We take half the x-coefficient and square it.

$$\dfrac{1}{2} \cdot 4 = 2 \text{ and } 2^2 = 4$$

Then we add $4 - 4$ inside the parentheses.

$$f(x) = -(x^2 + 4x + 4 - 4) - 2$$
$$= -(x^2 + 4x + 4) + (-1)(-4) - 2$$
$$= -(x + 2)^2 + 4 - 2$$
$$= -(x + 2)^2 + 2$$
$$= -[x - (-2)]^2 + 2$$

Vertex: $(-2, 2)$

Line of symmetry: $x = -2$

The coefficient of x^2 is -1, which is negative, so the graph opens down. This tells us that 2 is a maximum.

We plot a few points and draw the curve.

x	$f(x)$
-2	2
-4	-2
-3	1
-1	1
0	-2

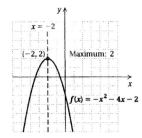

5. $f(x) = 3x^2 - 24x + 50 = 3(x^2 - 8x) + 50$

We complete the square inside the parentheses. We take half the x-coefficient and square it.

$$\frac{1}{2} \cdot (-8) = -4 \text{ and } (-4)^2 = 16$$

Then we add $16 - 16$ inside the parentheses.

$$f(x) = 3(x^2 - 8x + 16 - 16) + 50$$
$$= 3(x^2 - 8x + 16) - 48 + 50$$
$$= 3(x - 4)^2 + 2$$

Vertex: $(4, 2)$

Line of symmetry: $x = 4$

The coefficient of x^2 is 3, which is positive, so the graph opens up. This tells us that 2 is a minimum.

We plot a few points and draw the curve.

x	$f(x)$
4	2
5	5
3	5
6	14
2	14

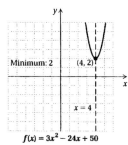

7. $f(x) = -2x^2 - 2x + 3 = -2(x^2 + x) + 3$

We complete the square inside the parentheses. We take half the x-coefficient and square it.

$$\frac{1}{2} \cdot 1 = \frac{1}{2} \text{ and } \left(\frac{1}{2}\right)^2 = \frac{1}{4}$$

Then we add $\frac{1}{4} - \frac{1}{4}$ inside the parentheses.

$$f(x) = -2\left(x^2 + x + \frac{1}{4} - \frac{1}{4}\right) + 3$$
$$= -2\left(x^2 + x + \frac{1}{4}\right) + (-2)\left(-\frac{1}{4}\right) + 3$$
$$= -2\left(x + \frac{1}{2}\right)^2 + \frac{1}{2} + 3$$
$$= -2\left(x + \frac{1}{2}\right)^2 + \frac{7}{2}$$
$$= -2\left[x - \left(-\frac{1}{2}\right)\right]^2 + \frac{7}{2}$$

Vertex: $\left(-\frac{1}{2}, \frac{7}{2}\right)$

Line of symmetry: $x = -\frac{1}{2}$

The coefficient of x^2 is -2, which is negative, so the graph opens down. This tells us that $\frac{7}{2}$ is a maximum.

We plot a few points and draw the curve.

x	$f(x)$
$-\frac{1}{2}$	$\frac{7}{2}$
-2	-1
-1	3
0	3
1	-1

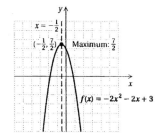

9. $f(x) = 5 - x^2 = -x^2 + 5 = -(x - 0)^2 + 5$

Vertex: $(0, 5)$

Line of symmetry: $x = 0$

The coefficient of x^2 is -1, which is negative, so the graph opens down. This tells us that 5 is a maximum.

We plot a few points and draw the curve.

x	$f(x)$
0	5
1	4
-1	4
2	1
-2	1
3	-4
-3	-4

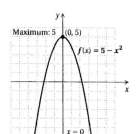

11. $f(x) = 2x^2 + 5x - 2 = 2\left(x^2 + \frac{5}{2}x\right) - 2$

We complete the square inside the parentheses. We take half the x-coefficient and square it.

$$\frac{1}{2} \cdot \frac{5}{2} = \frac{5}{4} \text{ and } \left(\frac{5}{4}\right)^2 = \frac{25}{16}$$

Then we add $\frac{25}{16} - \frac{25}{16}$ inside the parentheses.

$$f(x) = 2\left(x^2 + \frac{5}{2}x + \frac{25}{16} - \frac{25}{16}\right) - 2$$

$$= 2\left(x^2 + \frac{5}{2}x + \frac{25}{16}\right) + 2\left(-\frac{25}{16}\right) - 2$$

$$= 2\left(x + \frac{5}{4}\right)^2 - \frac{25}{8} - 2$$

$$= 2\left(x + \frac{5}{4}\right)^2 - \frac{41}{8}$$

$$= 2\left[x - \left(-\frac{5}{4}\right)\right]^2 + \left(-\frac{41}{8}\right)$$

Vertex: $\left(-\dfrac{5}{4}, -\dfrac{41}{8}\right)$

Line of symmetry: $x = -\dfrac{5}{4}$

The coefficient of x^2 is 2, which is positive, so the graph opens up. This tells us that $-\dfrac{41}{8}$ is a minimum.

We plot a few points and draw the curve.

x	$f(x)$
$-\dfrac{5}{4}$	$-\dfrac{41}{8}$
-3	1
-2	-4
-1	-5
0	-2
1	5

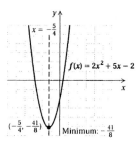

13. $f(x) = x^2 - 6x + 1$

The y-intercept is $(0, f(0))$. Since $f(0) = 0^2 - 6 \cdot 0 + 1 = 1$, the y-intercept is $(0, 1)$.

To find the x-intercepts, we solve $x^2 - 6x + 1 = 0$. Using the quadratic formula gives us $x = 3 \pm 2\sqrt{2}$.

Thus, the x-intercepts are $(3 - 2\sqrt{2}, 0)$ and $(3 + 2\sqrt{2}, 0)$, or approximately $(0.172, 0)$ and $(5.828, 0)$.

15. $f(x) = -x^2 + x + 20$

The y-intercept is $(0, f(0))$. Since $f(0) = -0^2 + 0 + 20 = 20$, the y-intercept is $(0, 20)$.

To find the x-intercepts, we solve $-x^2 + x + 20 = 0$. Factoring and using the principle of zero products gives us $x = -4$ or $x = 5$. Thus, the x-intercepts are $(-4, 0)$ and $(5, 0)$.

17. $f(x) = 4x^2 + 12x + 9$

The y-intercept is $(0, f(0))$. Since $f(0) = 4 \cdot 0^2 + 12 \cdot 0 + 9 = 9$, the y-intercept is $(0, 9)$.

To find the x-intercepts, we solve $4x^2 + 12x + 9 = 0$. Factoring and using the principle of zero products gives us $x = -\dfrac{3}{2}$. Thus, the x-intercept is $\left(-\dfrac{3}{2}, 0\right)$.

19. $f(x) = 4x^2 - x + 8$

The y-intercept is $(0, f(0))$. Since $f(0) = 4 \cdot 0^2 - 0 + 8 = 8$, the y-intercept is $(0, 8)$.

To find the x-intercepts, we solve $4x^2 - x + 8 = 0$. Using the quadratic formula gives us $x = \dfrac{1 \pm i\sqrt{127}}{8}$. Since there are no real-number solutions, there are no x-intercepts.

21. Discussion and Writing Exercise

23. a)
$$D = kw$$
$$420 = k \cdot 28$$
$$\frac{420}{28} = k$$
$$15 = k$$

The equation of variation is $D = 15w$.

b) We substitute 42 for w and compute D.
$$D = 15w$$
$$D = 15 \cdot 42$$
$$D = 630$$

630 mg would be recommended for a child who weighs 42 kg.

25.
$$y = \frac{k}{x}$$
$$125 = \frac{k}{2}$$
$$250 = k \qquad \text{Variation constant}$$
$$y = \frac{250}{x} \qquad \text{Equation of variation}$$

27.
$$y = kx$$
$$125 = k \cdot 2$$
$$\frac{125}{2} = k \qquad \text{Variation constant}$$
$$y = \frac{125}{2}x \qquad \text{Equation of variation}$$

29. a) Minimum: -6.954

b) Maximum: 7.014

31. $f(x) = |x^2 - 1|$

We plot some points and draw the curve. Note that it will lie entirely on or above the x-axis since absolute value is never negative.

x	$f(x)$
-3	8
-2	3
-1	0
0	1
1	0
2	3
3	8

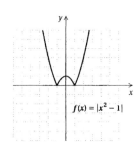

33. $f(x) = |x^2 - 3x - 4|$

We plot some points and draw the curve. Note that it will lie entirely on or above the x-axis since absolute value is never negative.

x	$f(x)$
-4	24
-3	14
-2	6
-1	0
0	4
1	6
2	6
3	4
4	0
5	6
6	14

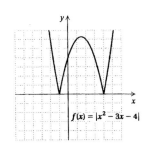

$f(x) = |x^2 - 3x - 4|$

35. The horizontal distance from $(-1, 0)$ to $(3, -5)$ is $|3-(-1)|$, or 4, so by symmetry the other x-intercept is $(3 + 4, 0)$, or $(7, 0)$. Substituting the three ordered pairs $(-1, 0)$, $(3, -5)$, and $(7, 0)$ in the equation $f(x) = ax^2 + bx + c$ yields a system of equations:

$$0 = a(-1)^2 + b(-1) + c,$$
$$-5 = a \cdot 3^2 + b \cdot 3 + c,$$
$$0 = a \cdot 7^2 + b \cdot 7 + c$$

or

$$0 = a - b + c,$$
$$-5 = 9a + 3b + c,$$
$$0 = 49a + 7b + c$$

The solution of this system of equations is $\left(\dfrac{5}{16}, -\dfrac{15}{8}, -\dfrac{35}{16}\right)$, so $f(x) = \dfrac{5}{16}x^2 - \dfrac{15}{8}x - \dfrac{35}{16}$.

37. $f(x) = \dfrac{x^2}{8} + \dfrac{x}{4} - \dfrac{3}{8}$

The x-coordinate of the vertex is $-b/2a$:

$$-\frac{b}{2a} = -\frac{\dfrac{1}{4}}{2 \cdot \dfrac{1}{8}} = -\frac{\dfrac{1}{4}}{\dfrac{1}{4}} = -1$$

The second coordinate is $f(-1)$:

$$f(-1) = \frac{(-1)^2}{8} + \frac{-1}{4} - \frac{3}{8}$$
$$= \frac{1}{8} - \frac{1}{4} - \frac{3}{8}$$
$$= -\frac{1}{2}$$

The vertex is $\left(-1, -\dfrac{1}{2}\right)$.

The line of symmetry is $x = -1$.

The coefficient of x^2 is $\dfrac{1}{8}$, which is positive, so the graph opens up. This tells us that $-\dfrac{1}{2}$ is a minimum.

We plot some points and draw the graph.

x	$f(x)$
-5	$\dfrac{3}{2}$
-3	0
-1	$-\dfrac{1}{2}$
0	$-\dfrac{3}{8}$
1	0
3	$\dfrac{3}{2}$
5	4

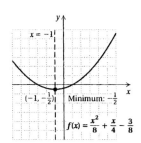

$x = -1$

$\left(-1, -\dfrac{1}{2}\right)$ Minimum: $-\dfrac{1}{2}$

$f(x) = \dfrac{x^2}{8} + \dfrac{x}{4} - \dfrac{3}{8}$

39. Graph $y_1 = x^2 - 4x + 2$ and $y_2 = 2 + x$ and use INTERSECT to find the points of intersection. They are $(0, 2)$ and $(5, 7)$.

Exercise Set 16.7

1. *Familiarize.* Referring to the drawing in the text, we let l = the length of the atrium and w = the width. Then the perimeter of each floor is $2l + 2w$, and the area is $l \cdot w$.

Translate. Using the formula for perimeter we have:

$$2l + 2w = 720$$
$$2l = 720 - 2w$$
$$l = \frac{720 - 2w}{2}$$
$$l = 360 - w$$

Substituting $360 - w$ for l in the formula for area, we get a quadratic function.

$$A = lw = (360 - w)w = 360w - w^2 = -w^2 + 360w$$

Carry out. We complete the square in order to find the vertex of the quadratic function.

$$A = -w^2 + 360w$$
$$= -(w^2 - 360w)$$
$$= -(w^2 - 360w + 32,400 - 32,400)$$
$$= -(w^2 - 360w + 32,400) + (-1)(-32,400)$$
$$= -(w - 180)^2 + 32,400$$

The vertex is $(180, 32,400)$. The coefficient of w^2 is negative, so the graph of a function is a parabola that opens down. This tells us that the function has a maximum value and that value occurs when $w = 180$. When $w = 180$, $l = 360 - w = 360 - 180 = 180$.

Check. We could find the value of the function for some values of w less than 180 and for some values greater than 180, determining that the maximum value we found, 32,400, is larger than these function values. We could also use the graph of the function to check the maximum value. Our answer checks.

State. Floors with dimensions 180 ft by 180 ft will allow an atrium with maximum area.

3. **Familiarize.** Let x represent the height of the file and y represent the width. We make a drawing.

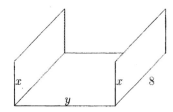

Translate. We have two equations.

$2x + y = 14$

$V = 8xy$

Solve the first equation for y.

$y = 14 - 2x$

Substitute for y in the second equation.

$V = 8x(14 - 2x)$

$V = -16x^2 + 112x$

Carry out. Completing the square, we get

$$V = -16\left(x - \frac{7}{2}\right)^2 + 196.$$

The maximum function value of 196 occurs when $x = \frac{7}{2}$. When $x = \frac{7}{2}$, $y = 14 - 2 \cdot \frac{7}{2} = 7$.

Check. Check a function value for x less than $\frac{7}{2}$ and for x greater than $\frac{7}{2}$.

$$V(3) = -16 \cdot 3^2 + 112 \cdot 3 = 192$$
$$V(4) = -16 \cdot 4^2 + 112 \cdot 4 = 192$$

Since 196 is greater than these numbers, it looks as though we have a maximum.

We could also use the graph of the function to check the maximum value.

State. The file should be $\frac{7}{2}$ in., or 3.5 in., tall.

5. **Familiarize and Translate.** We want to find the value of x for which $C(x) = 0.1x^2 - 0.7x + 2.425$ is a minimum.

Carry out. We complete the square.

$C(x) = 0.1(x^2 - 7x + 12.25) + 2.425 - 1.225$

$C(x) = 0.1(x - 3.5)^2 + 1.2$

The minimum function value of 1.2 occurs when $x = 3.5$.

Check. Check a function value for x less than 3.5 and for x greater than 3.5.

$C(3) = 0.1(3)^2 - 0.7(3) + 2.425 = 1.225$

$C(4) = 0.1(4)^2 - 0.7(4) + 2.425 = 1.225$

Since 1.2 is less than these numbers, it looks as though we have a minimum.

We could also use the graph of the function to check the minimum value.

State. The shop should build 3.5 hundred, or 350 bicycles.

7. **Familiarize.** We make a drawing and label it.

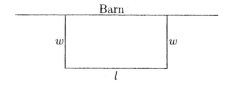

Translate. We have two equations.

$l + 2w = 40$

$A = lw$

Solve the first equation for l.

$l = 40 - 2w$

Substitute for l in the second equation.

$A = (40 - 2w)w = 40w - 2w^2$

$\quad = -2w^2 + 40w$

Carry out. Completing the square, we get

$A = -2(w - 10)^2 + 200$

The maximum function value is 200. It occurs when $w = 10$. When $w = 10$, $l = 40 - 2 \cdot 10 = 20$.

Check. Check a function value for w less than 10 and for w greater than 10.

$$A(9) = -2 \cdot 9^2 + 40 \cdot 9 = 198$$
$$A(11) = -2 \cdot 11^2 + 40 \cdot 11 = 198$$

Since 200 is greater than these numbers, it looks as though we have a maximum. We could also use the graph of the function to check the maximum value.

State. The maximum area of 200 ft^2 will occur when the dimensions are 10 ft by 20 ft.

9. **Familiarize and Translate.** We are given the function $N(x) = -0.4x^2 + 9x + 11$.

Carry out. To find the value of x for which $N(x)$ is a maximum, we first find $-\dfrac{b}{2a}$:

$$-\frac{b}{2a} = -\frac{9}{2(-0.4)} = 11.25$$

Now we find the maximum value of the function $N(11.25)$:

$N(11.25) = -0.4(11.25)^2 + 9(11.25) + 11 = 61.625$

Check. We can go over the calculations again. We could also solve the problem again by completing the square. The answer checks.

State. Daily ticket sales will peak 11 days after the concert was announced. About 62 tickets will be sold that day.

11. Find the total profit:

$P(x) = R(x) - C(x)$

$P(x) = (1000x - x^2) - (3000 + 20x)$

$P(x) = -x^2 + 980x - 3000$

To find the maximum value of the total profit and the value of x at which it occurs we complete the square:

$$P(x) = -(x^2 - 980x) - 3000$$
$$= -(x^2 - 980x + 240,100 - 240,100) - 3000$$
$$= -(x^2 - 980x + 240,100) - (-240,100) - 3000$$
$$= -(x - 490)^2 + 237,100$$

The maximum profit of \$237,100 occurs at $x = 490$.

13. Familiarize. Let x and y represent the numbers.

Translate. The sum of the numbers is 22, so we have $x + y = 22$. Solving for y, we get $y = 22 - x$. The product of the numbers is xy. Substituting $22 - x$ for y in the product, we get a quadratic function:

$$P = xy = x(22 - x) = 22x - x^2 = -x^2 + 22x$$

Carry out. The coefficient of x^2 is negative, so the graph of the function is a parabola that opens down and a maximum exists. We complete the square in order to find the vertex of the quadratic function.

$$P = -x^2 + 22x$$
$$= -(x^2 - 22x)$$
$$= -(x^2 - 22x + 121 - 121)$$
$$= -(x^2 - 22x + 121) + (-1)(-121)$$
$$= -(x - 11)^2 + 121$$

The vertex is $(11, 121)$. This tells us that the maximum product is 121. The maximum occurs when $x = 11$. Note that when $x = 11$, $y = 22 - x = 22 - 11 = 11$, so the numbers that yield the maximum product are 11 and 11.

Check. We could find the value of the function for some values of x less than 11 and for some greater than 11, determining that the maximum value we found is larger than these function values. We could also use the graph of the function to check the maximum value. Our answer checks.

State. The maximum product is 121. The numbers 11 and 11 yield this product.

15. Familiarize. Let x and y represent the numbers.

Translate. The difference of the numbers is 4, so we have $x - y = 4$. Solve for x, we get $x = y + 4$. The product of the numbers is xy. Substituting $y + 4$ for x in the product, we get a quadratic function:

$$P = xy = (y + 4)y = y^2 + 4y$$

Carry out. The coefficient of y^2 is positive, so the graph of the function opens up and a minimum exists. We complete the square in order to find the vertex of the quadratic function.

$$P = y^2 + 4y$$
$$= y^2 + 4y + 4 - 4$$
$$= (y + 2)^2 - 4$$
$$= [y - (-2)]^2 + (-4)$$

The vertex is $(-2, 4)$. This tells us that the minimum product is -4. The minimum occurs when $y = -2$. Note that when $y = -2$, $x = y + 4 = -2 + 4 = 2$, so the numbers that yield the minimum product are 2 and -2.

Check. We could find the value of the function for some values of y less than -2 and for some greater than -2,

determining that the minimum value we found is smaller than these function values. We could also use the graph of the function to check the minimum value. Our answer checks.

State. The minimum product is -4. The numbers 2 and -2 yield this product.

17. Familiarize. We let x and y represent the two numbers, and we let P represent their product.

Translate. We have two equations.
$$x + y = -12,$$
$$P = xy$$

Solve the first equation for y.
$$y = -12 - x$$

Substitute for y in the second equation.
$$P = x(-12 - x) = -12x - x^2$$
$$= -x^2 - 12x$$

Carry out. Completing the square, we get
$$P = -(x + 6)^2 + 36$$

The maximum function value is 36. It occurs when $x = -6$. When $x = -6$, $y = -12 - (-6)$, or -6.

Check. Check a function value for x less than -6 and for x greater than -6.
$$P(-7) = -(-7)^2 - 12(-7) = 35$$
$$P(-5) = -(-5)^2 - 12(-5) = 35$$

Since 36 is greater than these numbers, it looks as though we have a maximum.

We could also use the graph of the function to check the maximum value.

State. The maximum product of 36 occurs for the numbers -6 and -6.

19. The data seem to fit a linear function $f(x) = mx + b$.

21. The data fall and then rise in a curved manner fitting a quadratic function $f(x) = ax^2 + bx + c$, $a > 0$.

23. The data fall, then rise, then fall again so they do not fit a linear or a quadratic function but might fit a polynomial function that is neither quadratic nor linear.

25. The data rise and then fall in a curved manner fitting a quadratic function $f(x) = ax^2 + bx + c$, $a < 0$.

27. We look for a function of the form $f(x) = ax^2 + bx + c$. Substituting the data points, we get
$$4 = a(1)^2 + b(1) + c,$$
$$-2 = a(-1)^2 + b(-1) + c,$$
$$13 = a(2)^2 + b(2) + c,$$
or
$$4 = a + b + c,$$
$$-2 = a - b + c,$$
$$13 = 4a + 2b + c.$$

Solving this system, we get
$$a = 2, \; b = 3, \text{ and } c = -1.$$

Therefore the function we are looking for is

$$f(x) = 2x^2 + 3x - 1.$$

29. We look for a function of the form $f(x) = ax^2 + bx + c$. Substituting the data points, we get

$$0 = a(2)^2 + b(2) + c,$$
$$3 = a(4)^2 + b(4) + c,$$
$$-5 = a(12)^2 + b(12) + c,$$

or

$$0 = 4a + 2b + c,$$
$$3 = 16a + 4b + c,$$
$$-5 = 144a + 12b + c.$$

Solving this system, we get

$$a = -\frac{1}{4}, \ b = 3, \ c = -5.$$

Therefore the function we are looking for is

$$f(x) = -\frac{1}{4}x^2 + 3x - 5.$$

31. a) We look for a function of the form $A(s) = as^2 + bs + c$, where $A(s)$ represents the number of nighttime accidents (for every 200 million km) and s represents the travel speed (in km/h). We substitute the given values of s and $A(s)$.

$$400 = a(60)^2 + b(60) + c,$$
$$250 = a(80)^2 + b(80) + c,$$
$$250 = a(100)^2 + b(100) + c,$$

or

$$400 = 3600a + 60b + c,$$
$$250 = 6400a + 80b + c,$$
$$250 = 10,000a + 100b + c.$$

Solving the system of equations, we get

$$a = \frac{3}{16}, \ b = -\frac{135}{4}, \ c = 1750.$$

Thus, the function $A(s) = \frac{3}{16}s^2 - \frac{135}{4}s + 1750$ fits the data.

b) Find $A(50)$.

$$A(50) = \frac{3}{16}(50)^2 - \frac{135}{4}(50) + 1750 = 531.25$$

About 531 accidents per 200,000,000 km driven occur at 50 km/h.

33. *Familiarize.* Think of a coordinate system placed on the drawing in the text with the origin at the point where the arrow is released. Then three points on the arrow's parabolic path are $(0,0)$, $(63, 27)$, and $(126, 0)$. We look for a function of the form $h(d) = ad^2 + bd + c$, where $h(d)$ represents the arrow's height and d represents the distance the arrow has traveled horizontally.

Translate. We substitute the values given above for d and $h(d)$.

$$0 = a \cdot 0^2 + b \cdot 0 + c,$$
$$27 = a \cdot 63^2 + b \cdot 63 + c,$$
$$0 = a \cdot 126^2 + b \cdot 126 + c$$

or

$$0 = c,$$
$$27 = 3969a + 63b + c,$$
$$0 = 15,876a + 126b + c$$

Solve. Solving the system of equations, we get

$$a = -\frac{1}{147} \approx -0.0068, \ b = \frac{6}{7} \approx 0.8571, \text{ and } c = 0.$$

Check. Recheck the calculations.

State. The function $h(d) = -\frac{1}{147}d^2 + \frac{6}{7}d \approx -0.0068d^2 + 0.8571d$ expresses the arrow's height as a function of the distance it has traveled horizontally.

35. Discussion and Writing Exercise

37. $\sqrt[4]{5x^3y^5}\sqrt[4]{125x^2y^3} = \sqrt[4]{625x^5y^8} = \sqrt[4]{625x^4y^8 \cdot x} = \sqrt[4]{625x^4y^8} \cdot \sqrt[4]{x} = 5xy^2\sqrt[4]{x}$

39.
$$\sqrt{4x-4} = \sqrt{x+4} + 1$$
$$(\sqrt{4x-4})^2 = (\sqrt{x+4}+1)^2$$
$$4x - 4 = x + 4 + 2\sqrt{x+4} + 1$$
$$3x - 9 = 2\sqrt{x+4}$$
$$(3x-9)^2 = (2\sqrt{x+4})^2$$
$$9x^2 - 54x + 81 = 4(x+4)$$
$$9x^2 - 54x + 81 = 4x + 16$$
$$9x^2 - 58x + 65 = 0$$
$$(9x-13)(x-5) = 0$$
$$9x - 13 = 0 \quad or \quad x - 5 = 0$$
$$9x = 13 \quad or \quad x = 5$$
$$x = \frac{13}{9} \quad or \quad x = 5$$

Only 5 checks. It is the solution.

41. $-35 = \sqrt{2x+5}$

The equation has no solution, because the principle square root of a number is always nonnegative.

43. We will let x represent the number of years since 1997 and y represent the gross profit, in billions of dollars. Enter the data points (x, y) in STAT lists and use the quartic regression feature to find the equation $y = -0.290x^4 + 2.699x^3 - 8.306x^2 + 9.190x + 12.235$, where x is the number of years after 1997 and y is in billions of dollars.

Exercise Set 16.8

1. $(x-6)(x+2) > 0$

The solutions of $(x-6)(x+2) = 0$ are 6 and -2. They divide the real-number line into three intervals as shown:

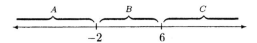

We try test numbers in each interval.

A: Test -3, $y = (-3 - 6)(-3 + 2) = 9 > 0$

B: Test 0, $y = (0 - 6)(0 + 2) = -12 < 0$

C: Test 7, $y = (7 - 6)(7 + 2) = 9 > 0$

The expression is positive for all values of x in intervals A and C. The solution set is $\{x | x < -2 \text{ or } x > 6\}$, or $(-\infty, -2) \cup (6, \infty)$.

From the graph in the text we see that the value of $(x - 6)(x + 2)$ is positive to the left of -2 and to the right of 6. This verifies the answer we found algebraically.

3. $4 - x^2 \geq 0$

$(2 + x)(2 - x) \geq 0$

The solutions of $(2 + x)(2 - x) = 0$ are -2 and 2. They divide the real-number line into three intervals as shown.

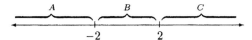

We try test numbers in each interval.

A: Test -3, $y = 4 - (-3)^2 = -5 < 0$

B: Test 0, $y = 4 - 0^2 = 4 > 0$

C: Test 3, $y = 4 - 3^2 = -5 < 0$

The expression is positive for values of x in interval B. Since the inequality symbol is $\geq$ we also include the intercepts. The solution set is $\{x | -2 \leq x \leq 2\}$, or $[-2, 2]$.

From the graph we see that $4 - x^2 \geq 0$ at the intercepts and between them. This verifies the answer we found algebraically.

5. $3(x + 1)(x - 4) \leq 0$

The solutions of $3(x + 1)(x - 4) = 0$ are -1 and 4. They divide the real-number line into three intervals as shown:

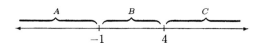

We try test numbers in each interval.

A: Test -2, $y = 3(-2 + 1)(-2 - 4) = 18 > 0$

B: Test 0, $y = 3(0 + 1)(0 - 4) = -12 < 0$

C: Test 5, $y = 3(5 + 1)(5 - 4) = 18 > 0$

The expression is negative for all numbers in interval B. The inequality symbol is $\leq$, so we need to include the intercepts. The solution set is $\{x | -1 \leq x \leq 4\}$, or $[-1, 4]$.

7. $x^2 - x - 2 < 0$

$(x + 1)(x - 2) < 0$ Factoring

The solutions of $(x + 1)(x - 2) = 0$ are -1 and 2. They divide the real-number line into three intervals as shown:

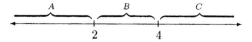

We try test numbers in each interval.

A: Test -2, $y = (-2 + 1)(-2 - 2) = 4 > 0$

B: Test 0, $y = (0 + 1)(0 - 2) = -2 < 0$

C: Test 3, $y = (3 + 1)(3 - 2) = 4 > 0$

The expression is negative for all numbers in interval B. The solution set is $\{x | -1 < x < 2\}$, or $(-1, 2)$.

9. $x^2 - 2x + 1 \geq 0$

$(x - 1)^2 \geq 0$

The solution of $(x - 1)^2 = 0$ is 1. For all real-number values of x except 1, $(x - 1)^2$ will be positive. Thus the solution set is $\{x | x \text{ is a real number}\}$, or $(-\infty, \infty)$.

11. $x^2 + 8 < 6x$

$x^2 - 6x + 8 < 0$

$(x - 4)(x - 2) < 0$

The solutions of $(x - 4)(x - 2) = 0$ are 4 and 2. They divide the real-number line into three intervals as shown:

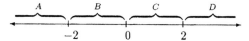

We try test numbers in each interval.

A: Test 0, $y = (0 - 4)(0 - 2) = 8 > 0$

B: Test 3, $y = (3 - 4)(3 - 2) = -1 < 0$

C: Test 5, $y = (5 - 4)(5 - 2) = 3 > 0$

The expression is negative for all numbers in interval B. The solution set is $\{x | 2 < x < 4\}$, or $(2, 4)$.

13. $3x(x + 2)(x - 2) < 0$

The solutions of $3x(x + 2)(x - 2) = 0$ are 0, -2, and 2. They divide the real-number line into four intervals as shown:

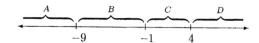

We try test numbers in each interval.

A: Test -3, $y = 3(-3)(-3 + 2)(-3 - 2) = -45 < 0$

B: Test -1, $y = 3(-1)(-1 + 2)(-1 - 2) = 9 > 0$

C: Test 1, $y = 3(1)(1 + 2)(1 - 2) = -9 < 0$

D: Test 3, $y = 3(3)(3 + 2)(3 - 2) = 45 > 0$

The expression is negative for all numbers in intervals A and C. The solution set is $\{x | x < -2 \text{ or } 0 < x < 2\}$, or $(-\infty, -2) \cup (0, 2)$.

15. $(x + 9)(x - 4)(x + 1) > 0$

The solutions of $(x + 9)(x - 4)(x + 1) = 0$ are -9, 4, and -1. They divide the real-number line into four intervals as shown:

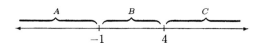

We try test numbers in each interval.

A: Test -10, $y = (-10+9)(-10-4)(-10+1) = -126 < 0$

B: Test -2, $y = (-2+9)(-2-4)(-2+1) = 42 > 0$

C: Test 0, $y = (0+9)(0-4)(0+1) = -36 < 0$

D: Test 5, $y = (5+9)(5-4)(5+1) = 84 > 0$

The expression is positive for all values of x in intervals B and D. The solution set is $\{x| -9 < x < -1 \ or \ x > 4\}$, or $(-9, -1) \cup (4, \infty)$.

17. $(x+3)(x+2)(x-1) < 0$

The solutions of $(x+3)(x+2)(x-1) = 0$ are -3, -2, and 1. They divide the real-number line into four intervals as shown:

We try test numbers in each interval.

A: Test -4, $y = (-4+3)(-4+2)(-4-1) = -10 < 0$

B: Test $-\dfrac{5}{2}$, $y = \left(-\dfrac{5}{2}+3\right)\left(-\dfrac{5}{2}+2\right)\left(-\dfrac{5}{2}-1\right) = \dfrac{7}{8} > 0$

C: Test 0, $y = (0+3)(0+2)(0-1) = -6 < 0$

D: Test 2, $y = (2+3)(2+2)(2-1) = 20 > 0$

The expression is negative for all numbers in intervals A and C. The solution set is $\{x|x < -3 \ or \ -2 < x < 1\}$, or $(-\infty, -3) \cup (-2, 1)$.

19. $\dfrac{1}{x-6} < 0$

We write the related equation by changing the $<$ symbol to $=$:

$$\frac{1}{x-6} = 0$$

We solve the related equation.

$$(x-6) \cdot \frac{1}{x-6} = (x-6) \cdot 0$$

$$1 = 0$$

We get a false equation, so the related equation has no solution.

Next we find the numbers for which the rational expression is undefined by setting the denominator equal to 0 and solving:

$$x - 6 = 0$$

$$x = 6$$

We use 6 to divide the number line into two intervals as shown:

We try test numbers in each interval.

A: Test 0,
$$\frac{1}{x-6} < 0$$
$$\frac{1}{0-6} \ ? \ 0$$
$$-\frac{1}{6} \ \bigg| \ \text{TRUE}$$

The number 0 is a solution of the inequality, so the interval A is part of the solution set.

B: Test 7,
$$\frac{1}{x-6} < 0$$
$$\frac{1}{7-6} \ ? \ 0$$
$$1 \ \bigg| \ \text{FALSE}$$

The number 7 is not a solution of the inequality, so the interval B is not part of the solution set. The solution set is $\{x|x < 6\}$, or $(-\infty, 6)$.

21. $\dfrac{x+1}{x-3} > 0$

Solve the related equation.

$$\frac{x+1}{x-3} = 0$$

$$x + 1 = 0$$

$$x = -1$$

Find the numbers for which the rational expression is undefined.

$$x - 3 = 0$$

$$x = 3$$

Use the numbers -1 and 3 to divide the number line into intervals as shown:

Try test numbers in each interval.

A: Test -2,
$$\frac{x+1}{x-3} > 0$$
$$\frac{-2+1}{-2-3} \ ? \ 0$$
$$\frac{-1}{-5} \ \bigg|$$
$$\frac{1}{5} \ \bigg| \ \text{TRUE}$$

The number -2 is a solution of the inequality, so the interval A is part of the solution set.

B: Test 0,
$$\frac{x+1}{x-3} > 0$$
$$\frac{0+1}{0-3} \ ? \ 0$$
$$-\frac{1}{3} \ \bigg| \ \text{FALSE}$$

The number 0 is not a solution of the inequality, so the interval B is not part of the solution set.

C: Test 4,
$$\frac{x+1}{x-3} > 0$$
$$\frac{4+1}{4-3} \ ? \ 0$$
$$\frac{5}{1} \ \bigg|$$
$$5 \ \bigg| \ \text{TRUE}$$

The number 4 is a solution of the inequality, so the interval C is part of the solution set. The solution set is

$\{x \mid x < -1 \ or \ x > 3\}$, or $(-\infty, -1) \cup (3, \infty)$.

23. $\dfrac{3x + 2}{x - 3} \leq 0$

Solve the related equation.

$$\frac{3x + 2}{x - 3} = 0$$

$$3x + 2 = 0$$

$$3x = -2$$

$$x = -\frac{2}{3}$$

Find the numbers for which the rational expression is undefined.

$$x - 3 = 0$$

$$x = 3$$

Use the numbers $-\dfrac{2}{3}$ and 3 to divide the number line into intervals as shown:

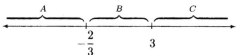

Try test numbers in each interval.

A: Test -1,
$$\frac{3x + 2}{x - 3} \leq 0$$
$$\frac{3(-1) + 2}{-1 - 3} \ ? \ 0$$
$$\frac{-1}{-4}$$
$$\frac{1}{4} \quad \Big| \quad \text{FALSE}$$

The number -1 is not a solution of the inequality, so the interval A is not part of the solution set.

B: Test 0,
$$\frac{3x + 2}{x - 3} \leq 0$$
$$\frac{3 \cdot 0 + 2}{0 - 3} \ ? \ 0$$
$$\frac{2}{-3}$$
$$-\frac{2}{3} \quad \Big| \quad \text{TRUE}$$

The number 0 is a solution of the inequality, so the interval B is part of the solution set.

C: Test 4,
$$\frac{3x + 2}{x - 3} \leq 0$$
$$\frac{3 \cdot 4 + 2}{4 - 3} \ ? \ 0$$
$$14 \quad \Big| \quad \text{FALSE}$$

The number 4 is not a solution of the inequality, so the interval C is not part of the solution set. The solution set includes the interval B. The number $-\dfrac{2}{3}$ is also included

since the inequality symbol is $\leq$ and $-\dfrac{2}{3}$ is the solution of the related equation. The number 3 is not included because the rational expression is undefined for 3. The solution set is $\left\{ x \mid -\dfrac{2}{3} \leq x < 3 \right\}$, or $\left[-\dfrac{2}{3}, 3 \right)$.

25. $\dfrac{x - 1}{x - 2} > 3$

Solve the related equation.

$$\frac{x - 1}{x - 2} = 3$$

$$x - 1 = 3(x - 2)$$

$$x - 1 = 3x - 6$$

$$5 = 2x$$

$$\frac{5}{2} = x$$

Find the numbers for which the rational expression is undefined.

$$x - 2 = 0$$

$$x = 2$$

Use the numbers $\dfrac{5}{2}$ and 2 to divide the number line into intervals as shown:

Try test numbers in each interval.

A: Test 0,
$$\frac{x - 1}{x - 2} > 3$$
$$\frac{0 - 1}{0 - 2} \ ? \ 3$$
$$\frac{1}{2} \quad \Big| \quad \text{FALSE}$$

The number 0 is not a solution of the inequality, so the interval A is not part of the solution set.

B: Test $\dfrac{9}{4}$,
$$\frac{x - 1}{x - 2} > 3$$
$$\frac{\dfrac{9}{4} - 1}{\dfrac{9}{4} - 2} \ ? \ 3$$
$$\frac{\dfrac{5}{4}}{\dfrac{1}{4}}$$
$$5 \quad \Big| \quad \text{TRUE}$$

The number $\dfrac{9}{4}$ is a solution of the inequality, so the interval B is part of the solution set.

C: Test 3,
$$\frac{x - 1}{x - 2} > 3$$
$$\frac{3 - 1}{3 - 2} \ ? \ 3$$
$$2 \quad \Big| \quad \text{FALSE}$$

The number 3 is not a solution of the inequality, so the interval C is not part of the solution set. The solution set is $\left\{x \middle| 2 < x < \dfrac{5}{2}\right\}$, or $\left(2, \dfrac{5}{2}\right)$.

27. $\dfrac{(x-2)(x+1)}{x-5} < 0$

Solve the related equation.

$$\frac{(x-2)(x+1)}{x-5} = 0$$
$$(x-2)(x+1) = 0$$
$$x = 2 \text{ or } x = -1$$

Find the numbers for which the rational expression is undefined.

$$x - 5 = 0$$
$$x = 5$$

Use the numbers 2, -1, and 5 to divide the number line into intervals as shown:

Try test numbers in each interval.

A: Test -2,
$$\frac{(x-2)(x+1)}{x-5} < 0$$
$$\frac{(-2-2)(-2+1)}{-2-5} \ ? \ 0$$
$$\frac{-4(-1)}{-7}$$
$$-\frac{4}{7} \ \bigg| \ \text{TRUE}$$

Interval A is part of the solution set.

B: Test 0,
$$\frac{(x-2)(x+1)}{x-5} < 0$$
$$\frac{(0-2)(0+1)}{0-5} \ ? \ 0$$
$$\frac{-2 \cdot 1}{-5}$$
$$\frac{2}{5} \ \bigg| \ \text{FALSE}$$

Interval B is not part of the solution set.

C: Test 3,
$$\frac{(x-2)(x+1)}{x-5} < 0$$
$$\frac{(3-2)(3+1)}{3-5} \ ? \ 0$$
$$\frac{1 \cdot 4}{-2}$$
$$-2 \ \bigg| \ \text{TRUE}$$

Interval C is part of the solution set.

D: Test 6,

$$\frac{(x-2)(x+1)}{x-5} < 0$$
$$\frac{(6-2)(6+1)}{6-5} \ ? \ 0$$
$$\frac{4 \cdot 7}{1}$$
$$28 \ \bigg| \ \text{FALSE}$$

Interval D is not part of the solution set.

The solution set is $\{x | x < -1 \ or \ 2 < x < 5\}$, or $(-\infty, -1) \cup (2, 5)$.

29. $\dfrac{x+3}{x} \le 0$

Solve the related equation.

$$\frac{x+3}{x} = 0$$
$$x + 3 = 0$$
$$x = -3$$

Find the numbers for which the rational expression is undefined.

$$x = 0$$

Use the numbers -3 and 0 to divide the number line into intervals as shown:

Try test numbers in each interval.

A: Test -4,
$$\frac{x+3}{x} \le 0$$
$$\frac{-4+3}{-4} \ ? \ 0$$
$$\frac{1}{4} \ \bigg| \ \text{FALSE}$$

Interval A is not part of the solution set.

B: Test -1,
$$\frac{x+3}{x} \le 0$$
$$\frac{-1+3}{-1} \ ? \ 0$$
$$-2 \ | \ \text{TRUE}$$

Interval B is part of the solution set.

C: Test 1,
$$\frac{x+3}{x} \le 0$$
$$\frac{1+3}{1} \ ? \ 0$$
$$4 \ | \ \text{FALSE}$$

Interval C is not part of the solution set.

The solution set includes the interval B. The number -3 is also included since the inequality symbol is $\le$ and -3 is a solution of the related equation. The number 0 is not included because the rational expression is undefined for 0. The solution set is $\{x | -3 \le x < 0\}$, or $[-3, 0)$.

31. $\dfrac{x}{x-1} > 2$

Solve the related equation.

$$\frac{x}{x-1} = 2$$

$$x = 2x - 2$$

$$2 = x$$

Find the numbers for which the rational expression is undefined.

$$x - 1 = 0$$

$$x = 1$$

Use the numbers 1 and 2 to divide the number line into intervals as shown:

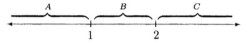

Try test numbers in each interval.

A: Test 0,

$$\frac{x}{x-1} > 2$$

$$\frac{0}{0-1} \;?\; 2$$

$$0 \;\Big|\; \text{FALSE}$$

Interval A is not part of the solution set.

B: Test $\dfrac{3}{2}$,

$$\frac{x}{x-1} > 2$$

$$\frac{\frac{3}{2}}{\frac{3}{2}-1} \;?\; 2$$

$$\frac{\frac{3}{2}}{\frac{1}{2}}$$

$$3 \;\Big|\; \text{TRUE}$$

Interval B is part of the solution set.

C: Test 3,

$$\frac{x}{x-1} > 2$$

$$\frac{3}{3-1} \;?\; 2$$

$$\frac{3}{2} \;\Big|\; \text{FALSE}$$

Interval C is not part of the solution set.

The solution set is $\{x \mid 1 < x < 2\}$, or $(1, 2)$.

33. $\dfrac{x-1}{(x-3)(x+4)} < 0$

Solve the related equation.

$$\frac{x-1}{(x-3)(x+4)} = 0$$

$$x - 1 = 0$$

$$x = 1$$

Find the numbers for which the rational expression is undefined.

$$(x-3)(x+4) = 0$$

$$x = 3 \text{ or } x = -4$$

Use the numbers 1, 3, and −4 to divide the number line into intervals as shown:

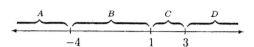

Try test numbers in each interval.

A: Test −5,

$$\frac{x-1}{(x-3)(x+4)} < 0$$

$$\frac{-5-1}{(-5-3)(-5+4)} \;?\; 0$$

$$\frac{-6}{-8(-1)}$$

$$-\frac{3}{4} \;\Big|\; \text{TRUE}$$

Interval A is part of the solution set.

B: Test 0,

$$\frac{x-1}{(x-3)(x+4)} < 0$$

$$\frac{0-1}{(0-3)(0+4)} \;?\; 0$$

$$\frac{-1}{-3 \cdot 4}$$

$$\frac{1}{12} \;\Big|\; \text{FALSE}$$

Interval B is not part of the solution set.

C: Test 2,

$$\frac{x-1}{(x-3)(x+4)} < 0$$

$$\frac{2-1}{(2-3)(2+4)} \;?\; 0$$

$$\frac{1}{-1 \cdot 6}$$

$$-\frac{1}{6} \;\Big|\; \text{TRUE}$$

Interval C is part of the solution set.

D: Test 4,

$$\frac{x-1}{(x-3)(x+4)} < 0$$

$$\frac{4-1}{(4-3)(4+4)} \;?\; 0$$

$$\frac{3}{1 \cdot 8}$$

$$\frac{3}{8} \;\Big|\; \text{FALSE}$$

Interval D is not part of the solution set.

The solution set is $\{x \mid x < -4 \text{ or } 1 < x < 3\}$, or $(-\infty, -4) \cup (1, 3)$.

35. $3 < \dfrac{1}{x}$

Solve the related equation.

$$3 = \frac{1}{x}$$

$$x = \frac{1}{3}$$

Find the numbers for which the rational expression is undefined.

$$x = 0$$

Use the numbers 0 and $\dfrac{1}{3}$ to divide the number line into intervals as shown:

Try test numbers in each interval.

A: Test -1,

$$3 < \frac{1}{x}$$

$$3 \; ? \; \frac{1}{-1}$$

$$\Big|\; -1 \qquad \text{FALSE}$$

Interval A is not part of the solution set.

B: Test $\dfrac{1}{6}$,

$$3 < \frac{1}{x}$$

$$3 \; ? \; \frac{1}{\frac{1}{6}}$$

$$\Big|\; 6 \qquad \text{TRUE}$$

Interval B is part of the solution set.

C: Test 1,

$$3 < \frac{1}{x}$$

$$3 \; ? \; \frac{1}{1}$$

$$\Big|\; 1 \qquad \text{FALSE}$$

Interval C is not part of the solution set.

The solution set is $\left\{ x \Big| 0 < x < \dfrac{1}{3} \right\}$, or $\left(0, \dfrac{1}{3}\right)$.

37. $\dfrac{(x-1)(x+2)}{(x+3)(x-4)} > 0$

Solve the related equation.

$$\frac{(x-1)(x+2)}{(x+3)(x-4)} = 0$$

$$(x-1)(x+2) = 0$$

$$x = 1 \text{ or } x = -2$$

Find the numbers for which the rational expression is undefined.

$$(x+3)(x-4) = 0$$

$$x = -3 \text{ or } x = 4$$

Use the numbers 1, -2, -3, and 4 to divide the number line into intervals as shown:

Try test numbers in each interval.

A: Test -4,

$$\frac{(x-1)(x+2)}{(x+3)(x-4)} > 0$$

$$\frac{(-4-1)(-4+2)}{(-4+3)(-4-4)} \; ? \; 0$$

$$\frac{-5(-2)}{-1(-8)}$$

$$\frac{5}{4} \;\Big|\; \text{TRUE}$$

Interval A is part of the solution set.

B: Test $-\dfrac{5}{2}$,

$$\frac{(x-1)(x+2)}{(x+3)(x-4)} > 0$$

$$\frac{\left(-\frac{5}{2}-1\right)\left(-\frac{5}{2}+2\right)}{\left(-\frac{5}{2}+3\right)\left(-\frac{5}{2}-4\right)} \; ? \; 0$$

$$\frac{-\frac{7}{2}\left(-\frac{1}{2}\right)}{\frac{1}{2}\left(-\frac{13}{2}\right)}$$

$$-\frac{7}{13} \;\Big|\; \text{FALSE}$$

Interval B is not part of the solution set.

C: Test 1,

$$\frac{(x-1)(x+2)}{(x+3)(x-4)} > 0$$

$$\frac{(0-1)(0+2)}{(0+3)(0-4)} \; ? \; 0$$

$$\frac{-1 \cdot 2}{3(-4)}$$

$$\frac{1}{6} \;\Big|\; \text{TRUE}$$

Interval C is part of the solution set.

D: Test 2,

$$\frac{(x-1)(x+2)}{(x+3)(x-4)} > 0$$

$$\frac{(2-1)(2+2)}{(2+3)(2-4)} \; ? \; 0$$

$$\frac{1 \cdot 4}{5(-2)}$$

$$-\frac{2}{5} \;\Big|\; \text{FALSE}$$

Interval D is not part of the solution set.

E: Test 5,
$$\frac{(x-1)(x+2)}{(x+3)(x-4)} > 0$$

$$\frac{(5-1)(5+2)}{(5+3)(5-4)} \; ? \; 0$$

$$\frac{4 \cdot 7}{8 \cdot 1}$$

$$\frac{7}{2} \quad \bigg| \quad \text{TRUE}$$

Interval *E* is part of the solution set.

The solution set is $\{x | x < -3 \ or \ -2 < x < 1 \ or \ x > 4\}$, or $(-\infty, -3) \cup (-2, 1) \cup (4, \infty)$.

39. Discussion and Writing Exercise

41. $\sqrt[3]{\dfrac{125}{27}} = \dfrac{\sqrt[3]{125}}{\sqrt[3]{27}} = \dfrac{5}{3}$

43. $\sqrt{\dfrac{16a^3}{b^4}} = \dfrac{\sqrt{16a^3}}{\sqrt{b^4}} = \dfrac{\sqrt{16a^2 \cdot a}}{\sqrt{b^4}} = \dfrac{\sqrt{16a^2}\sqrt{a}}{\sqrt{b^4}} = \dfrac{4a}{b^2}\sqrt{a}$

45. $3\sqrt{8} - 5\sqrt{2} = 3\sqrt{4 \cdot 2} - 5\sqrt{2}$
$\qquad = 3\sqrt{4}\sqrt{2} - 5\sqrt{2}$
$\qquad = 3 \cdot 2\sqrt{2} - 5\sqrt{2}$
$\qquad = 6\sqrt{2} - 5\sqrt{2}$
$\qquad = \sqrt{2}$

47. $5\sqrt[3]{16a^4} + 7\sqrt[3]{2a} = 5\sqrt[3]{8a^3 \cdot 2a} + 7\sqrt[3]{2a}$
$\qquad = 5\sqrt[3]{8a^3}\sqrt[3]{2a} + 7\sqrt[3]{2a}$
$\qquad = 5 \cdot 2a\sqrt[3]{2a} + 7\sqrt[3]{2a}$
$\qquad = 10a\sqrt[3]{2a} + 7\sqrt[3]{2a}$
$\qquad = (10a + 7)\sqrt[3]{2a}$

49. For Exercise 11, graph $y_1 = x^2 + 8$ and $y_2 = 6x$. Then determine the values of x for which the graph of y_1 lies below the graph of y_2.

For Exercise 22, graph $y_1 = \dfrac{x-2}{x+5}$ and $y_2 = 0$. Then determine the values of x for which the graph of y_1 lies below the graph of y_2. Since the graph of $y_2 = 0$ is the x-axis, this could also be done by graphing $y_1 = \dfrac{x-2}{x+5}$ and determining the values of x for which the graph of y_1 lies below the x-axis.

For Exercise 25, graph $y_1 = \dfrac{x-1}{x-2}$ and $y_2 = 3$. Then determine the values of x for which the graph of y_1 lies above the graph of y_2.

51. $\qquad x^2 - 2x \le 2$
$x^2 - 2x - 2 \le 0$

The solutions of $x^2 - 2x - 2 = 0$ are found using the quadratic formula. They are $1 \pm \sqrt{3}$, or about 2.7 and -0.7. These numbers divide the number line into three intervals as shown:

$$\overset{A}{\underbrace{\qquad\qquad}} \quad \overset{B}{\underbrace{\qquad\qquad}} \quad \overset{C}{\underbrace{\qquad\qquad}}$$
$$1 - \sqrt{3} \qquad 1 + \sqrt{3}$$

We try test numbers in each interval.

A: Test -1, $y = (-1)^2 - 2(-1) - 2 = 1 > 0$

B: Test 0, $y = 0^2 - 2 \cdot 0 - 2 = -2 < 0$

C: Test 3, $y = 3^2 - 2 \cdot 3 - 2 = 1 > 0$

The expression is negative for all values of x in interval *B*. The inequality symbol is $\le$, so we must also include the intercepts. The solution set is $\{x | 1 - \sqrt{3} \le x \le 1 + \sqrt{3}\}$, or $[1 - \sqrt{3}, 1 + \sqrt{3}]$.

53. $\qquad x^4 + 2x^2 > 0$
$x^2(x^2 + 2) > 0$

$x^2 > 0$ for all $x \ne 0$, and $x^2 + 2 > 0$ for all values of x. Then $x^2(x^2 + 2) > 0$ for all $x \ne 0$. The solution set is $\{x | x \ne 0\}$, or the set of all real numbers except 0, or $(-\infty, 0) \cup (0, \infty)$.

55. $\left| \dfrac{x+2}{x-1} \right| < 3$

$-3 < \dfrac{x+2}{x-1} < 3$

We rewrite the inequality using "and."

$-3 < \dfrac{x+2}{x-1} \ and \ \dfrac{x+2}{x-1} < 3$

We will solve each inequality and then find the intersection of their solution sets.

Solve: $-3 < \dfrac{x+2}{x-1}$

Solve the related equation.

$$-3 = \frac{x+2}{x-1}$$
$$-3x + 3 = x + 2$$
$$1 = 4x$$
$$\frac{1}{4} = x$$

Find the numbers for which the rational expression is undefined.

$$x - 1 = 0$$
$$x = 1$$

Use the numbers $\dfrac{1}{4}$ and 1 to divide the number line into intervals as shown:

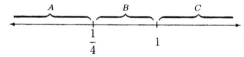

Try test numbers in each interval.

A: Test 0,
$$-3 < \frac{x+2}{x-1}$$

$$-3 \; ? \; \frac{0+2}{0-1}$$

$$\bigg| \; -2 \quad \text{TRUE}$$

Interval *A* is part of the solution set.

B: Test $\frac{1}{2}$,

$$-3 < \frac{x+2}{x-1}$$

$$-3 \ ? \ \frac{\frac{1}{2}+2}{\frac{1}{2}-1}$$

$$\left| \ \frac{\frac{5}{2}}{-\frac{1}{2}} \right.$$

$$\left| \ -5 \quad \text{FALSE} \right.$$

Interval B is not part of the solution set.

C: Test 2,

$$-3 < \frac{x+2}{x-1}$$

$$-3 \ ? \ \frac{2+2}{2-1}$$

$$\left| \ 4 \quad \text{TRUE} \right.$$

Interval C is part of the solution set.

The solution set of $-3 < \frac{x+2}{x-1}$ is $\left\{ x \middle| x < \frac{1}{4} \ or \ x > 1 \right\}$, or $\left(-\infty, \frac{1}{4} \right) \cup (1, \infty)$.

Solve: $\frac{x+2}{x-1} > 3$

Solve the related equation.

$$\frac{x+2}{x-1} = 3$$

$$x + 2 = 3x - 3$$

$$5 = 2x$$

$$\frac{5}{2} = x$$

From our work above we know that the rational expression is undefined for 1.

Use the numbers $\frac{5}{2}$ and 1 to divide the number line into intervals as shown:

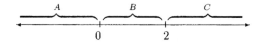

Try test numbers in each interval.

A: Test 0,

$$\frac{x+2}{x-1} < 3$$

$$\frac{0+2}{0-1} \ ? \ 3$$

$$-2 \ \middle| \quad \text{TRUE}$$

Interval A is part of the solution set.

B: Test 2,

$$\frac{x+2}{x-1} < 3$$

$$\frac{2+2}{2-1} \ ? \ 3$$

$$4 \ \middle| \quad \text{FALSE}$$

Interval B is not part of the solution set.

C: Test 3,

$$\frac{x+2}{x-1} < 3$$

$$\frac{3+2}{3-1} \ ? \ 3$$

$$\frac{5}{2} \ \middle| \quad \text{TRUE}$$

Interval C is part of the solution set.

The solution set of $\frac{x+2}{x-1} < 3$ is $\left\{ x \middle| x < 1 \ or \ x > \frac{5}{2} \right\}$, or $(-\infty, 1) \cup \left(\frac{5}{2}, \infty \right)$.

The solution set of the original inequality is

$$\left\{ x \middle| x < \frac{1}{4} \ or \ x > 1 \right\} \cap \left\{ x \middle| x < 1 \ or \ x > \frac{5}{2} \right\}, \ \text{or}$$

$$\left\{ x \middle| x < \frac{1}{4} \ or \ x > \frac{5}{2} \right\}, \ \text{or} \ \left(-\infty, \frac{1}{4} \right) \cup \left(\frac{5}{2}, \infty \right).$$

57. a) Solve: $-16t^2 + 32t + 1920 > 1920$

$$-16t^2 + 32t > 0$$

$$t^2 - 2t < 0$$

$$t(t-2) < 0$$

The solutions of $t(t-2) = 0$ are 0 and 2. They divide the number line into three intervals as shown:

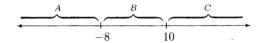

Try test numbers in each interval.

A: Test -1, $y = -1(-1-2) = 3 > 0$

B: Test 1, $y = 1(1-2) = -1 < 0$

C: Test 3, $y = 3(3-2) = 3 > 0$

The expression is negative for all values of t in interval B. The solution set is $\{t | 0 < t < 2\}$, or $(0, 2)$.

b) Solve: $-16t^2 + 32t + 1920 < 640$

$$-16t^2 + 32t + 1280 < 0$$

$$t^2 - 2t - 80 > 0$$

$$(t-10)(t+8) > 0$$

The solutions of $(t-10)(t+8) = 0$ are 10 and -8. They divide the number line into three intervals as shown:

Try test numbers in each interval.

A: Test -10, $y = (-10-10)(-10+8) = 40 > 0$

B: Test 0, $y = (0-10)(0+8) = -80 < 0$

C: Test 20, $y = (20-10)(20+8) = 80 = 280 > 0$

The expression is positive for all values of t in intervals A and C. However, since negative values of t have no meaning in this problem, we disregard interval A. Thus, the solution set is $\{t | t > 10\}$, or $(10, \infty)$.

Chapter 17

Exponential and Logarithmic Functions

1. Graph: $f(x) = 2^x$

We compute some function values and keep the results in a table.

$f(0) = 2^0 = 1$

$f(1) = 2^1 = 2$

$f(2) = 2^2 = 4$

$f(-1) = 2^{-1} = \dfrac{1}{2^1} = \dfrac{1}{2}$

$f(-2) = 2^{-2} = \dfrac{1}{2^2} = \dfrac{1}{4}$

x	$f(x)$
0	1
1	2
2	4
3	8
-1	$\dfrac{1}{2}$
-2	$\dfrac{1}{4}$
-3	$\dfrac{1}{8}$

Next we plot these points and connect them with a smooth curve.

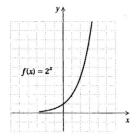

3. Graph: $f(x) = 5^x$

We compute some function values and keep the results in a table.

$f(0) = 5^0 = 1$

$f(1) = 5^1 = 5$

$f(2) = 5^2 = 25$

$f(-1) = 5^{-1} = \dfrac{1}{5^1} = \dfrac{1}{5}$

$f(-2) = 5^{-2} = \dfrac{1}{5^2} = \dfrac{1}{25}$

x	$f(x)$
0	1
1	5
2	25
-1	$\dfrac{1}{5}$
-2	$\dfrac{1}{25}$

Next we plot these points and connect them with a smooth curve.

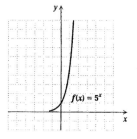

5. Graph: $f(x) = 2^{x+1}$

We compute some function values and keep the results in a table.

$f(0) = 2^{0+1} = 2^1 = 2$

$f(-1) = 2^{-1+1} = 2^0 = 1$

$f(-2) = 2^{-2+1} = 2^{-1} = \dfrac{1}{2^1} = \dfrac{1}{2}$

$f(-3) = 2^{-3+1} = 2^{-2} = \dfrac{1}{2^2} = \dfrac{1}{4}$

$f(1) = 2^{1+1} = 2^2 = 4$

$f(2) = 2^{2+1} = 2^3 = 8$

x	$f(x)$
0	2
-1	1
-2	$\dfrac{1}{2}$
-3	$\dfrac{1}{4}$
1	4
2	8

Next we plot these points and connect them with a smooth curve.

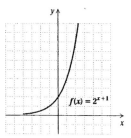

7. Graph: $f(x) = 3^{x-2}$

We compute some function values and keep the results in a table.

$f(0) = 3^{0-2} = 3^{-2} = \dfrac{1}{3^2} = \dfrac{1}{9}$

$f(1) = 3^{1-2} = 3^{-1} = \dfrac{1}{3^1} = \dfrac{1}{3}$

$f(2) = 3^{2-2} = 3^0 = 1$

$f(3) = 3^{3-2} = 3^1 = 3$

$f(4) = 3^{4-2} = 3^2 = 9$

$f(-1) = 3^{-1-2} = 3^{-3} = \dfrac{1}{3^3} = \dfrac{1}{27}$

$f(-2) = 3^{-2-2} = 3^{-4} = \dfrac{1}{3^4} = \dfrac{1}{81}$

x	$f(x)$
0	$\dfrac{1}{9}$
1	$\dfrac{1}{3}$
2	1
3	3
4	9
-1	$\dfrac{1}{27}$
-2	$\dfrac{1}{81}$

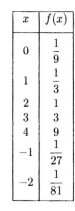

Next we plot these points and connect them with a smooth curve.

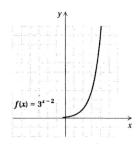

$f(x) = 3^{x-2}$

9. Graph: $f(x) = 2^x - 3$

We construct a table of values. Then we plot the points and connect them with a smooth curve.

$f(0) = 2^0 - 3 = 1 - 3 = -2$

$f(1) = 2^1 - 3 = 2 - 3 = -1$

$f(2) = 2^2 - 3 = 4 - 3 = 1$

$f(3) = 2^3 - 3 = 8 - 3 = 5$

$f(-1) = 2^{-1} - 3 = \dfrac{1}{2} - 3 = -\dfrac{5}{2}$

$f(-2) = 2^{-2} - 3 = \dfrac{1}{4} - 3 = -\dfrac{11}{4}$

x	$f(x)$
0	-2
1	-1
2	1
3	5
-1	$-\dfrac{5}{2}$
-2	$-\dfrac{11}{4}$

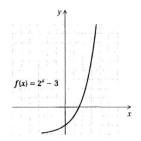

$f(x) = 2^x - 3$

11. Graph: $f(x) = 5^{x+3}$

We construct a table of values. Then we plot the points and connect them with a smooth curve.

$f(0) = 5^{0+3} = 5^3 = 125$

$f(-1) = 5^{-1+3} = 5^2 = 25$

$f(-2) = 5^{-2+3} = 5^1 = 5$

$f(-3) = 5^{-3+3} = 5^0 = 1$

$f(-4) = 5^{-4+3} = 5^{-1} = \dfrac{1}{5}$

$f(-5) = 5^{-5+3} = 5^{-2} = \dfrac{1}{25}$

x	$f(x)$
0	125
-1	25
-2	5
-3	1
-4	$\dfrac{1}{5}$
-5	$\dfrac{1}{25}$

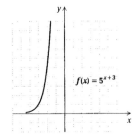

$f(x) = 5^{x+3}$

13. Graph: $f(x) = \left(\dfrac{1}{2}\right)^x$

We construct a table of values. Then we plot the points and connect them with a smooth curve.

$f(0) = \left(\dfrac{1}{2}\right)^0 = 1$

$f(1) = \left(\dfrac{1}{2}\right)^1 = \dfrac{1}{2}$

$f(2) = \left(\dfrac{1}{2}\right)^2 = \dfrac{1}{4}$

$f(3) = \left(\dfrac{1}{2}\right)^3 = \dfrac{1}{8}$

$f(-1) = \left(\dfrac{1}{2}\right)^{-1} = \dfrac{1}{\left(\dfrac{1}{2}\right)^1} = \dfrac{1}{\dfrac{1}{2}} = 2$

$f(-2) = \left(\dfrac{1}{2}\right)^{-2} = \dfrac{1}{\left(\dfrac{1}{2}\right)^2} = \dfrac{1}{\dfrac{1}{4}} = 4$

$f(-3) = \left(\dfrac{1}{2}\right)^{-3} = \dfrac{1}{\left(\dfrac{1}{2}\right)^3} = \dfrac{1}{\dfrac{1}{8}} = 8$

x	$f(x)$
0	1
1	$\dfrac{1}{2}$
2	$\dfrac{1}{4}$
3	$\dfrac{1}{8}$
-1	2
-2	4
-3	8

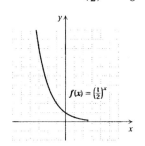

$f(x) = \left(\dfrac{1}{2}\right)^x$

15. Graph: $f(x) = \left(\dfrac{1}{5}\right)^x$

We construct a table of values. Then we plot the points and connect them with a smooth curve.

$f(0) = \left(\dfrac{1}{5}\right)^0 = 1$

$f(1) = \left(\dfrac{1}{5}\right)^1 = \dfrac{1}{5}$

$f(2) = \left(\dfrac{1}{5}\right)^2 = \dfrac{1}{25}$

$f(-1) = \left(\dfrac{1}{5}\right)^{-1} = \dfrac{1}{\dfrac{1}{5}} = 5$

$f(-2) = \left(\dfrac{1}{5}\right)^{-2} = \dfrac{1}{\dfrac{1}{25}} = 25$

x	$f(x)$
0	1
1	$\dfrac{1}{5}$
2	$\dfrac{1}{25}$
-1	5
-2	25

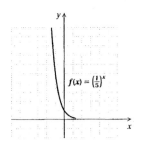

$f(x) = \left(\dfrac{1}{5}\right)^x$

17. Graph: $f(x) = 2^{2x-1}$

We construct a table of values. Then we plot the points and connect them with a smooth curve.

$f(0) = 2^{2 \cdot 0 - 1} = 2^{-1} = \frac{1}{2}$

$f(1) = 2^{2 \cdot 1 - 1} = 2^1 = 2$

$f(2) = 2^{2 \cdot 2 - 1} = 2^3 = 8$

$f(-1) = 2^{2(-1)-1} = 2^{-3} = \frac{1}{8}$

$f(-2) = 2^{2(-2)-1} = 2^{-5} = \frac{1}{32}$

x	$f(x)$
0	$\frac{1}{2}$
1	2
2	8
-1	$\frac{1}{8}$
-2	$\frac{1}{32}$

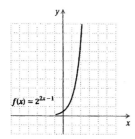

$f(x) = 2^{2x-1}$

19. Graph: $x = 2^y$

We can find ordered pairs by choosing values for y and then computing values for x.

For $y = 0$, $x = 2^0 = 1$.

For $y = 1$, $x = 2^1 = 2$.

For $y = 2$, $x = 2^2 = 4$.

For $y = 3$, $x = 2^3 = 8$.

For $y = -1$, $x = 2^{-1} = \frac{1}{2^1} = \frac{1}{2}$.

For $y = -2$, $x = 2^{-2} = \frac{1}{2^2} = \frac{1}{4}$.

For $y = -3$, $x = 2^{-3} = \frac{1}{2^3} = \frac{1}{8}$.

x	y
1	0
2	1
4	2
8	3
$\frac{1}{2}$	-1
$\frac{1}{4}$	-2
$\frac{1}{8}$	-3

(1) Choose values for y.

(2) Compute values for x.

We plot these points and connect them with a smooth curve.

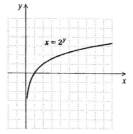

$x = 2^y$

21. Graph: $x = \left(\frac{1}{2}\right)^y$

We can find ordered pairs by choosing values for y and then computing values for x. Then we plot these points and connect them with a smooth curve.

For $y = 0$, $x = \left(\frac{1}{2}\right)^0 = 1$.

For $y = 1$, $x = \left(\frac{1}{2}\right)^1 = \frac{1}{2}$.

For $y = 2$, $x = \left(\frac{1}{2}\right)^2 = \frac{1}{4}$.

For $y = 3$, $x = \left(\frac{1}{2}\right)^3 = \frac{1}{8}$.

For $y = -1$, $x = \left(\frac{1}{2}\right)^{-1} = \frac{1}{\frac{1}{2}} = 2$.

For $y = -2$, $x = \left(\frac{1}{2}\right)^{-2} = \frac{1}{\frac{1}{4}} = 4$.

For $y = -3$, $x = \left(\frac{1}{2}\right)^{-3} = \frac{1}{\frac{1}{8}} = 8$.

x	y
1	0
$\frac{1}{2}$	1
$\frac{1}{4}$	2
$\frac{1}{8}$	3
2	-1
4	-2
8	-3

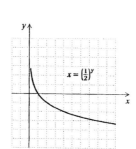

$x = \left(\frac{1}{2}\right)^y$

23. Graph: $x = 5^y$

We can find ordered pairs by choosing values for y and then computing values for x. Then we plot these points and connect them with a smooth curve.

For $y = 0$, $x = 5^0 = 1$.

For $y = 1$, $x = 5^1 = 5$.

For $y = 2$, $x = 5^2 = 25$.

For $y = -1$, $x = 5^{-1} = \frac{1}{5}$.

For $y = -2$, $x = 5^{-2} = \frac{1}{25}$.

x	y
1	0
5	1
25	2
$\frac{1}{5}$	-1
$\frac{1}{25}$	-2

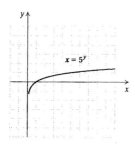

25. Graph $y = 2^x$ (see Exercise 1) and $x = 2^y$ (see Exercise 19) using the same set of axes.

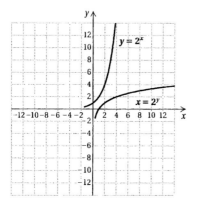

27. a) We substitute \$50,000 for P and 6%, or 0.06, for r in the formula $A = P(1 + r)^t$:
$$A(t) = \$50,000(1 + 0.06)^t = \$50,000(1.06)^t$$

b) $A(0) = \$50,000(1.06)^0 = \$50,000$

$A(1) = \$50,000(1.06)^1 = \$53,000$

$A(2) = \$50,000(1.06)^2 = \$56,180$

$A(4) = \$50,000(1.06)^4 \approx \$63,123.85$

$A(8) = \$50,000(1.06)^8 \approx \$76,692.40$

$A(10) = \$50,000(1.06)^{10} \approx \$89,542.38$

$A(20) = \$50,000(1.06)^{20} \approx \$160,356.77$

c)

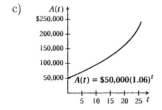

29. $V(t) = 4000(1.22)^t$

a) As the exercise states, $t = 0$ corresponds to 1958.
$V(0) = 4000(1.22)^0 = \$4000$

$1970 - 1958 = 12$, so $t = 12$ corresponds to 1970.
$V(12) = 4000(1.22)^{12} \approx \$43,489$

$1980 - 1958 = 22$, so $t = 22$ corresponds to 1980.
$V(22) = 4000(1.22)^{22} \approx \$317,670$

$1990 - 1958 = 32$, so $t = 32$ corresponds to 1990.
$V(32) = 4000(1.22)^{32} \approx \$2,320,463$

$1998 - 1958 = 40$, so $t = 40$ corresponds to 1998.
$V(40) = 4000(1.22)^{40} \approx \$11,388,151$

b) $1999 - 1958 = 41$, so $t = 41$ corresponds to 1999.
$V(41) = 4000(1.22)^{41} \approx \$13,893,544$

c) $2001 - 1958 = 43$, so $t = 43$ corresponds to 2001.
$V(43) = 4000(1.22)^{43} \approx \$20,679,151$

d)

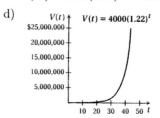

31. a) In 1930, $t = 1930 - 1900 = 30$.
$$P(t) = 150(0.960)^t$$
$$P(30) = 150(0.960)^{30}$$
$$\approx 44.079$$

In 1930, about 44.079 thousand, or 44,079, humpback whales were alive.
In 1960, $t = 1960 - 1900 = 60$.
$$P(t) = 150(0.960)^t$$
$$P(60) = 150(0.960)^{60}$$
$$\approx 12.953$$

In 1960, about 12.953 thousand, or 12,953, humpback whales were alive.

b) Plot the points found in part (a), (30, 44,079) and (60, 12,953) and additional points as needed and graph the function.

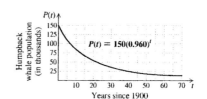

33. $N(t) = 3000(2)^{t/20}$

a) $N(10) = 3000(2)^{10/20} \approx 4243$

There will be approximately 4243 bacteria after 10 min.

$N(20) = 3000(2)^{20/20} = 6000$

There will be 6000 bacteria after 20 min.

$N(30) = 3000(2)^{30/20} \approx 8485$

There will be approximately 8485 bacteria after 30 min.

$N(40) = 3000(2)^{40/20} = 12,000$

There will be 12,000 bacteria after 40 min.

$N(60) = 3000(2)^{60/20} = 24,000$

There will be 24,000 bacteria after 60 min.

b) We use the function values computed in part (a) to draw the graph. Other values can also be computed if needed.

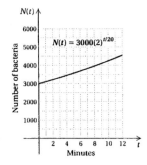

35. Discussion and Writing Exercise

37. $x^{-5} \cdot x^3 = x^{-5+3} = x^{-2} = \dfrac{1}{x^2}$

39. $9^0 = 1$ (For any nonzero number a, $a^0 = 1$.)

41. $\left(\dfrac{2}{3}\right)^1 = \dfrac{2}{3}$ (For any number a, $a^1 = a$.)

43. $\dfrac{x^{-3}}{x^4} = x^{-3-4} = x^{-7} = \dfrac{1}{x^7}$

45. $\dfrac{x}{x^0} = x^{1-0} = x^1 = x$

(This exercise could also be done as follows:

$\dfrac{x}{x^0} = \dfrac{x}{1} = x$.)

47. $(5^{\sqrt{2}})^{2\sqrt{2}} = 5^{\sqrt{2} \cdot 2\sqrt{2}} = 5^4$, or 625

49. Graph: $y = 2^x + 2^{-x}$

Construct a table of values, thinking of y as $f(x)$. Then plot these points and connect them with a curve.

$f(0) = 2^0 + 2^{-0} = 1 + 1 = 2$

$f(1) = 2^1 + 2^{-1} = 2 + \dfrac{1}{2} = 2\dfrac{1}{2}$

$f(2) = 2^2 + 2^{-2} = 4 + \dfrac{1}{4} = 4\dfrac{1}{4}$

$f(3) = 2^3 + 2^{-3} = 8 + \dfrac{1}{8} = 8\dfrac{1}{8}$

$f(-1) = 2^{-1} + 2^{-(-1)} = \dfrac{1}{2} + 2 = 2\dfrac{1}{2}$

$f(-2) = 2^{-2} + 2^{-(-2)} = \dfrac{1}{4} + 4 = 4\dfrac{1}{4}$

$f(-3) = 2^{-3} + 2^{-(-3)} = \dfrac{1}{8} + 8 = 8\dfrac{1}{8}$

x	y, or $f(x)$
0	2
1	$2\dfrac{1}{2}$
2	$4\dfrac{1}{4}$
3	$8\dfrac{1}{8}$
−1	$2\dfrac{1}{2}$
−2	$4\dfrac{1}{4}$
−3	$8\dfrac{1}{8}$

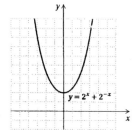

51. $y = \left| \left(\dfrac{1}{2}\right)^x - 1 \right|$

Construct a table of values, thinking of y as $f(x)$. Then plot these points and connect them with a curve.

$f(-4) = \left| \left(\dfrac{1}{2}\right)^{-4} - 1 \right| = |16 - 1| = |15| = 15$

$f(-2) = \left| \left(\dfrac{1}{2}\right)^{-2} - 1 \right| = |4 - 1| = |3| = 3$

$f(-1) = \left| \left(\dfrac{1}{2}\right)^{-1} - 1 \right| = |2 - 1| = |1| = 1$

$f(0) = \left| \left(\dfrac{1}{2}\right)^0 - 1 \right| = |1 - 1| = |0| = 0$

$f(1) = \left| \left(\dfrac{1}{2}\right)^1 - 1 \right| = \left| \dfrac{1}{2} - 1 \right| = \left| -\dfrac{1}{2} \right| = \dfrac{1}{2}$

$f(2) = \left| \left(\dfrac{1}{2}\right)^2 - 1 \right| = \left| \dfrac{1}{4} - 1 \right| = \left| -\dfrac{3}{4} \right| = \dfrac{3}{4}$

$f(3) = \left| \left(\dfrac{1}{2}\right)^3 - 1 \right| = \left| \dfrac{1}{8} - 1 \right| = \left| -\dfrac{7}{8} \right| = \dfrac{7}{8}$

x	y, or $f(x)$
−4	15
−2	3
−1	1
0	0
1	$\dfrac{1}{2}$
2	$\dfrac{3}{4}$
3	$\dfrac{7}{8}$

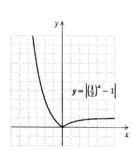

53. Construct a table of values for each equation and then draw the graphs on the same set of axes.

For $y = 3^{-(x-1)}$:

x	y
−3	81
−2	27
−1	9
0	3
1	1
2	$\dfrac{1}{3}$
3	$\dfrac{1}{9}$
4	$\dfrac{1}{27}$

For $x = 3^{-(y-1)}$:

x	y
81	−3
27	−2
9	−1
3	0
1	1
$\dfrac{1}{3}$	2
$\dfrac{1}{9}$	3
$\dfrac{1}{27}$	4

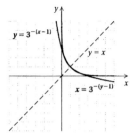

55. Left to the student

Exercise Set 17.2

---

1. To find the inverse of the given relation we interchange the first and second coordinates of each ordered pair. The inverse of the relation is $\{(2,1), (-3,6), (-5,-3)\}$.

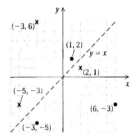

3. We interchange x and y to obtain an equation of the inverse of the relation. It is $x = 2y + 6$. The x-values in the first table become the y-values in the second table. We have

x	y
4	−1
6	0
8	1
10	2
12	3

We graph the original relation and its inverse. Since there is no horizontal line that crosses the graph more than once, the function is one-to-one.

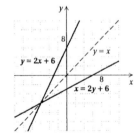

5. The graph of $f(x) = x - 5$ is shown below. Since no horizontal line crosses the graph more than once, the function is one-to-one.

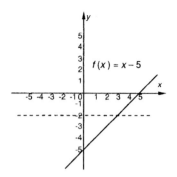

7. The graph of $f(x) = x^2 - 2$ is shown below. There are many horizontal lines that cross the graph more than once, so the function is not one-to-one.

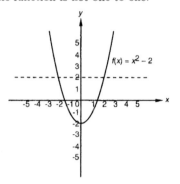

9. The graph of $g(x) = |x| - 3$ is shown below. There are many horizontal lines that cross the graph more than once, so the function is not one-to-one.

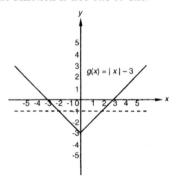

11. The graph of $g(x) = 3^x$ is shown below. Since no horizontal line crosses the graph more than once, the function is one-to-one.

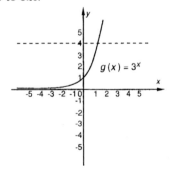

13. The graph of $f(x) = 5x - 2$ is shown below. It passes the horizontal-line test, so it is one-to-one.

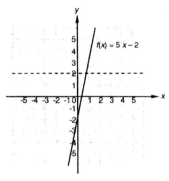

We find a formula for the inverse.

1. Replace $f(x)$ by y : $y = 5x - 2$
2. Interchange x and y : $x = 5y - 2$
3. Solve for y : $x + 2 = 5y$

$$\frac{x+2}{5} = y$$

4. Replace y by $f^{-1}(x)$: $f^{-1}(x) = \dfrac{x+2}{5}$

15. The graph of $f(x) = \dfrac{-2}{x}$ is shown below. It passes the horizontal-line test, so it is one-to-one.

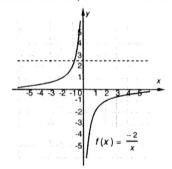

We find a formula for the inverse.

1. Replace $f(x)$ by y : $y = \dfrac{-2}{x}$
2. Interchange x and y : $x = \dfrac{-2}{y}$
3. Solve for y : $y = \dfrac{-2}{x}$
4. Replace y by $f^{-1}(x)$: $f^{-1}(x) = \dfrac{-2}{x}$

17. The graph of $f(x) = \dfrac{4}{3}x + 7$ is shown below. It passes the horizontal line test, so it is one-to-one.

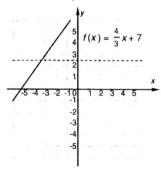

We find a formula for the inverse.

1. Replace $f(x)$ by y: $y = \dfrac{4}{3}x + 7$
2. Interchange x and y: $x = \dfrac{4}{3}y + 7$
3. Solve for y: $x - 7 = \dfrac{4}{3}y$

$$\frac{3}{4}(x - 7) = y$$

4. Replace y by $f^{(-1)}(x)$: $f^{-1}(x) = \dfrac{3}{4}(x - 7)$

19. The graph of $f(x) = \dfrac{2}{x+5}$ is shown below. It passes the horizontal line test, so it is one-to-one.

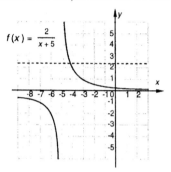

We find a formula for the inverse.

1. Replace $f(x)$ by y : $y = \dfrac{2}{x+5}$
2. Interchange x and y : $x = \dfrac{2}{y+5}$
3. Solve for y : $x(y + 5) = 2$

$$y + 5 = \frac{2}{x}$$

$$y = \frac{2}{x} - 5$$

4. Replace y by $f^{-1}(x)$: $f^{-1}(x) = \dfrac{2}{x} - 5$

21. The graph of $f(x) = 5$ is shown below. The horizontal line $y = 5$ crosses the graph more than once, so the function is not one-to-one.

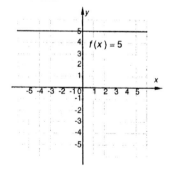

23. The graph of $f(x) = \dfrac{2x+1}{5x+3}$ is shown below. It passes the horizontal line test, so it is one-to-one.

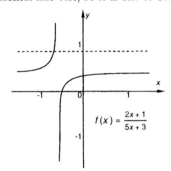

$$f(x) = \frac{2x+1}{5x+3}$$

We find a formula for the inverse.

1. Replace $f(x)$ by y: $y = \dfrac{2x+1}{5x+3}$

2. Interchange x and y: $x = \dfrac{2y+1}{5y+3}$

3. Solve for y: $5xy + 3x = 2y + 1$

$$5xy - 2y = 1 - 3x$$

$$y(5x - 2) = 1 - 3x$$

$$y = \frac{1-3x}{5x-2}$$

4. Replace y by $f^{-1}(x)$: $f^{-1}(x) = \dfrac{1-3x}{5x-2}$

25. The graph of $f(x) = x^3 - 1$ is shown below. It passes the horizontal line test, so it is one-to-one.

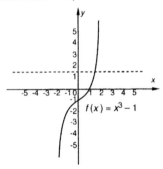

$$f(x) = x^3 - 1$$

1. Replace $f(x)$ by y: $y = x^3 - 1$

2. Interchange x and y: $x = y^3 - 1$

3. Solve for y: $x + 1 = y^3$

$$\sqrt[3]{x+1} = y$$

4. Replace y by $f^{-1}(x)$: $f^{-1}(x) = \sqrt[3]{x+1}$

27. The graph of $f(x) = \sqrt[3]{x}$ is shown below. It passes the horizontal line test, so it is one-to-one.

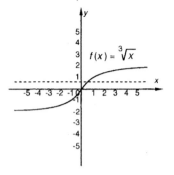

$$f(x) = \sqrt[3]{x}$$

1. Replace $f(x)$ by y: $y = \sqrt[3]{x}$

2. Interchange x and y: $x = \sqrt[3]{y}$

3. Solve for y: $x^3 = y$

4. Replace y by $f^{-1}(x)$: $f^{-1}(x) = x^3$

29. We first graph $f(x) = \dfrac{1}{2}x - 3$. The graph of f^{-1} can be obtained by reflecting the graph of f across the line $y = x$.

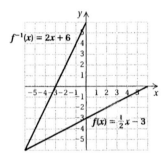

$$f^{-1}(x) = 2x + 6$$

$$f(x) = \frac{1}{2}x - 3$$

31. We first graph $f(x) = x^3$. The graph of f^{-1} can be obtained by reflecting the graph of f across the line $y = x$.

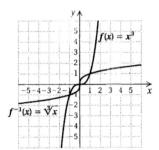

$$f(x) = x^3$$

$$f^{-1}(x) = \sqrt[3]{x}$$

33. $f \circ g(x) = f(g(x)) = f(6 - 4x) = 2(6 - 4x) - 3 =$
$$12 - 8x - 3 = -8x + 9$$

$g \circ f(x) = g(f(x)) = g(2x - 3) = 6 - 4(2x - 3) =$
$$6 - 8x + 12 = -8x + 18$$

35. $f \circ g(x) = f(g(x)) = f(2x - 1) = 3(2x - 1)^2 + 2 =$
$$3(4x^2 - 4x + 1) + 2 = 12x^2 - 12x + 3 + 2 =$$
$$12x^2 - 12x + 5$$

$g \circ f(x) = g(f(x)) = g(3x^2 + 2) = 2(3x^2 + 2) - 1 =$
$$6x^2 + 4 - 1 = 6x^2 + 3$$

37. $f \circ g(x) = f(g(x)) = f\left(\dfrac{2}{x}\right) = 4\left(\dfrac{2}{x}\right)^2 - 1 =$

$\qquad 4\left(\dfrac{4}{x^2}\right) - 1 = \dfrac{16}{x^2} - 1$

$g \circ f(x) = g(f(x)) = g(4x^2 - 1) = \dfrac{2}{4x^2 - 1}$

39. $f \circ g(x) = f(g(x)) = f(x^2 - 5) = (x^2 - 5)^2 + 5 =$

$\qquad x^4 - 10x^2 + 25 + 5 = x^4 - 10x^2 + 30$

$g \circ f(x) = g(f(x)) = g(x^2 + 5) = (x^2 + 5)^2 - 5 =$

$\qquad x^4 + 10x^2 + 25 - 5 = x^4 + 10x^2 + 20$

41. $h(x) = (5 - 3x)^2$

This is $5 - 3x$ raised to the second power, so the two most obvious functions are $f(x) = x^2$ and $g(x) = 5 - 3x$.

43. $h(x) = \sqrt{5x + 2}$

This is the square root of $5x + 2$, so the two most obvious functions are $f(x) = \sqrt{x}$ and $g(x) = 5x + 2$.

45. $h(x) = \dfrac{1}{x - 1}$

This is the reciprocal of $x - 1$, so the two most obvious functions are $f(x) = \dfrac{1}{x}$ and $g(x) = x - 1$.

47. $h(x) = \dfrac{1}{\sqrt{7x + 2}}$

This is the reciprocal of the square root of $7x + 2$. Two functions that can be used are $f(x) = \dfrac{1}{\sqrt{x}}$ and $g(x) = 7x + 2$.

49. $h(x) = (\sqrt{x} + 5)^4$

This is $\sqrt{x} + 5$ raised to the fourth power, so the two most obvious functions are $f(x) = x^4$ and $g(x) = \sqrt{x} + 5$.

51. We check to see that $f^{-1} \circ f(x) = x$ and $f \circ f^{-1}(x) = x$.

$f^{-1} \circ f(x) = f^{-1}(f(x)) = f^{-1}\left(\dfrac{4}{5}x\right) =$

$\dfrac{5}{4} \cdot \dfrac{4}{5}x = x$

$f \circ f^{-1}(x) = f(f^{-1}(x)) = f\left(\dfrac{5}{4}x\right) =$

$\dfrac{4}{5} \cdot \dfrac{5}{4}x = x$

53. We check to see that $f^{-1} \circ f(x) = x$ and $f \circ f^{-1}(x) = x$.

$f^{-1} \circ f(x) = f^{-1}(f(x)) = f^{-1}\left(\dfrac{1 - x}{x}\right) =$

$\dfrac{1}{\dfrac{1 - x}{x} + 1} = \dfrac{1}{\dfrac{1 - x}{x} + 1} \cdot \dfrac{x}{x} = \dfrac{x}{1 - x + x} =$

$\dfrac{x}{1} = x$

$f \circ f^{-1}(x) = f(f^{-1}(x)) = f\left(\dfrac{1}{x + 1}\right) =$

$\dfrac{1 - \dfrac{1}{x + 1}}{\dfrac{1}{x + 1}} = \dfrac{1 - \dfrac{1}{x + 1}}{\dfrac{1}{x + 1}} \cdot \dfrac{x + 1}{x + 1} =$

$\dfrac{x + 1 - 1}{1} = \dfrac{x}{1} = x$

55. The function $f(x) = 3x$ multiplies an input by 3, so the inverse would divide an input by 3. We have $f^{-1}(x) = \dfrac{x}{3}$.

Now we check to see that $f^{-1} \circ f(x) = x$ and $f \circ f^{-1}(x) = x$.

$f^{-1} \circ f(x) = f^{-1}(f(x)) = f^{-1}(3x) = \dfrac{3x}{3} = x$

$f \circ f^{-1}(x) = f(f^{-1}(x)) = f\left(\dfrac{x}{3}\right) = 3 \cdot \dfrac{x}{3} = x$

The inverse is correct.

57. The function $f(x) = -x$ takes the opposite of an input so the inverse would also take the opposite of an input. We have $f^{-1}(x) = -x$.

Now we check to see that $f^{-1} \circ f(x) = x$ and $f \circ f^{-1}(x) = x$.

$f^{-1} \circ f(x) = f^{-1}(f(x)) = f^{-1}(-x) = -(-x) = x$

$f \circ f^{-1}(x) = f(f^{-1}(x)) = f(-x) = -(-x) = x$

The inverse is correct.

59. The function $f(x) = \sqrt[3]{x - 5}$ subtracts 5 from an input and then takes the cube root of the difference, so the inverse would cube an input and then add 5. We have $f^{-1}(x) = x^3 + 5$.

Now we check to see that $f^{-1} \circ f(x) = x$ and $f \circ f^{-1}(x) = x$.

$f^{-1} \circ f(x) = f^{-1}(f(x)) = f^{-1}(\sqrt[3]{x - 5}) = (\sqrt[3]{x - 5})^3 + 5 = x - 5 + 5 = x$

$f \circ f^{-1}(x) = f(f^{-1}(x)) = f(x^3 + 5) = \sqrt[3]{x^3 + 5 - 5} = \sqrt[3]{x^3} = x$

The inverse is correct.

61. a) $f(8) = 8 + 32 = 40$

Size 40 in France corresponds to size 8 in the U.S.

$f(10) = 10 + 32 = 42$

Size 42 in France corresponds to size 10 in the U.S.

$f(14) = 14 + 32 = 46$

Size 46 in France corresponds to size 14 in the U.S.

$f(18) = 18 + 32 = 50$

Size 50 in France corresponds to size 18 in the U.S.

b) The graph of $f(x) = x + 32$ is shown below. It passes the horizontal-line test, so the function is one-to-one and, hence, has an inverse that is a function.

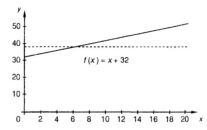

We now find a formula for the inverse.

1. Replace $f(x)$ by y: $y = x + 32$

2. Interchange x and y: $x = y + 32$

3. Solve for y: $x - 32 = y$

4. Replace y by $f^{-1}(x)$: $f^{-1}(x) = x - 32$

c) $f^{-1}(40) = 40 - 32 = 8$

Size 8 in the U.S. corresponds to size 40 in France.

$f^{-1}(42) = 42 - 32 = 10$

Size 10 in the U.S. corresponds to size 42 in France.

$f^{-1}(46) = 46 - 32 = 14$

Size 14 in the U.S. corresponds to size 46 in France.

$f^{-1}(50) = 50 - 32 = 18$

Size 18 in the U.S. corresponds to size 50 in France.

63. Discussion and Writing Exercise

65. $\sqrt[6]{a^2} = a^{2/6} = a^{1/3} = \sqrt[3]{a}$

67. $\sqrt{a^4b^6} = (a^4b^6)^{1/2} = a^2b^3$

69. $\sqrt[8]{81} = (3^4)^{1/8} = 3^{1/2} = \sqrt{3}$

71. $\sqrt[12]{64x^6y^6} = (2^6x^6y^6)^{1/12} = 2^{1/2}x^{1/2}y^{1/2} = (2xy)^{1/2} = \sqrt{2xy}$

73. $\sqrt[5]{32a^{15}b^{40}} = (2^5a^{15}b^{40})^{1/5} = 2a^3b^8$

75. $\sqrt[4]{81a^8b^8} = (3^4a^8b^8)^{1/4} = 3a^2b^2$

77. Graph the functions in a square window and determine whether one is a reflection of the other across the line $y = x$. The graphs show that these functions are not inverses of each other.

79. Graph the functions in a square window and determine whether one is a reflection of the other across the line $y = x$. The graphs show that these functions are inverses of each other.

81. (1) C; (2) A; (3) B; (4) D

83. Reflect the graph of f across the line $y = x$.

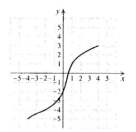

85. $f(x) = \dfrac{1}{2}x + 3$, $g(x) = 2x - 6$

Since $(f \circ g)(x) = x$ and $(g \circ f)(x) = x$, the functions are inverse.

Exercise Set 17.3

1. Graph: $f(x) = \log_2 x$

The equation $f(x) = y = \log_2 x$ is equivalent to $2^y = x$. We can find ordered pairs by choosing values for y and computing the corresponding x-values.

For $y = 0$, $x = 2^0 = 1$.

For $y = 1$, $x = 2^1 = 2$.

For $y = 2$, $x = 2^2 = 4$.

For $y = 3$, $x = 2^3 = 8$.

For $y = -1$, $x = 2^{-1} = \dfrac{1}{2}$.

For $y = -2$, $x = 2^{-2} = \dfrac{1}{4}$.

x, or 2^y	y
1	0
2	1
4	2
8	3
$\dfrac{1}{2}$	-1
$\dfrac{1}{4}$	-2
$\dfrac{1}{8}$	-3

⌐ (1) Select y.

└ (2) Compute x.

We plot the set of ordered pairs and connect the points with a smooth curve.

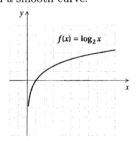

3. Graph: $f(x) = \log_{1/3} x$

The equation $f(x) = y = \log_{1/3} x$ is equivalent to $\left(\frac{1}{3}\right)^y = x$. We can find ordered pairs by choosing values for y and computing the corresponding x-values.

For $y = 0$, $x = \left(\frac{1}{3}\right)^0 = 1$.

For $y = 1$, $x = \left(\frac{1}{3}\right)^1 = \frac{1}{3}$.

For $y = 2$, $x = \left(\frac{1}{3}\right)^2 = \frac{1}{9}$.

For $y = -1$, $x = \left(\frac{1}{3}\right)^{-1} = 3$.

For $y = -2$, $x = \left(\frac{1}{3}\right)^{-2} = 9$.

x, or $\left(\frac{1}{3}\right)^y$	y
1	0
$\frac{1}{3}$	1
$\frac{1}{9}$	2
3	-1
9	-2

We plot the set of ordered pairs and connect the points with a smooth curve.

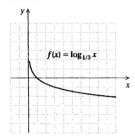

5. Graph $f(x) = 3^x$ (see Exercise Set 10.1, Exercise 2) and $f^{-1}(x) = \log_3 x$ on the same set of axes. We can obtain the graph of f^{-1} by reflecting the graph of f across the line $y = x$.

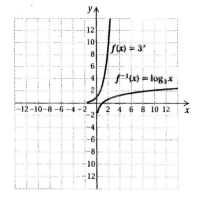

7. The exponent is the logarithm.

$$10^3 = 1000 \Rightarrow 3 = \log_{10} 1000$$

The base remains the same.

9. The exponent is the logarithm.

$$5^{-3} = \frac{1}{125} \Rightarrow -3 = \log_5 \frac{1}{125}$$

The base remains the same.

11. $8^{1/3} = 2 \Rightarrow \frac{1}{3} = \log_8 2$

13. $10^{0.3010} = 2 \Rightarrow 0.3010 = \log_{10} 2$

15. $e^2 = t \Rightarrow 2 = \log_e t$

17. $Q^t = x \Rightarrow t = \log_Q x$

19. $e^2 = 7.3891 \Rightarrow 2 = \log_e 7.3891$

21. $e^{-2} = 0.1353 \Rightarrow -2 = \log_e 0.1353$

23. The logarithm is the exponent.

$$w = \log_4 10 \Rightarrow 4^w = 10$$

The base remains the same.

25. The logarithm is the exponent.

$$\log_6 36 = 2 \Rightarrow 6^2 = 36$$

The base remains the same.

27. $\log_{10} 0.01 = -2 \Rightarrow 10^{-2} = 0.01$

29. $\log_{10} 8 = 0.9031 \Rightarrow 10^{0.9031} = 8$

31. $\log_e 100 = 4.6052 \Rightarrow e^{4.6052} = 100$

33. $\log_t Q = k \Rightarrow t^k = Q$

35. $\log_3 x = 2$

$\quad 3^2 = x \qquad$ Converting to an exponential equation

$\quad 9 = x \qquad$ Computing 3^2

37. $\log_x 16 = 2$

$\quad x^2 = 16 \qquad$ Converting to an exponential equation

$x = 4$ or $x = -4 \qquad$ Principle of square roots

$\log_4 16 = 2$ because $4^2 = 16$. Thus, 4 is a solution. Since all logarithm bases must be positive, $\log_{-4} 16$ is not defined and -4 is not a solution.

39. $\log_2 16 = x$

$\quad 2^x = 16 \qquad$ Converting to an exponential equation

$\quad 2^x = 2^4$

$\quad x = 4 \qquad$ The exponents are the same.

41. $\log_3 27 = x$

$\quad 3^x = 27 \qquad$ Converting to an exponential equation

$\quad 3^x = 3^3$

$\quad x = 3 \qquad$ The exponents are the same.

43. $\log_x 25 = 1$

$\quad x^1 = 25 \qquad$ Converting to an exponential equation

$\quad x = 25$

45. $\log_3 x = 0$

$3^0 = x$ Converting to an exponential equation

$1 = x$

47. $\log_2 x = -1$

$2^{-1} = x$ Converting to an exponential equation

$\dfrac{1}{2} = x$ Simplifying

49. $\log_8 x = \dfrac{1}{3}$

$8^{1/3} = x$

$2 = x$

51. Let $\log_{10} 100 = x$. Then

$10^x = 100$

$10^x = 10^2$

$x = 2$

Thus, $\log_{10} 100 = 2$.

53. Let $\log_{10} 0.1 = x$. Then

$10^x = 0.1 = \dfrac{1}{10}$

$10^x = 10^{-1}$

$x = -1$

Thus, $\log_{10} 0.1 = -1$.

55. Let $\log_{10} 1 = x$. Then

$10^x = 1$

$10^x = 10^0$ $(10^0 = 1)$

$x = 0$

Thus, $\log_{10} 1 = 0$.

57. Let $\log_5 625 = x$. Then

$5^x = 625$

$5^x = 5^4$

$x = 4$

Thus, $\log_5 625 = 4$.

59. Think of the meaning of $\log_7 49$. It is the exponent to which you raise 7 to get 49. That exponent is 2. Therefore, $\log_7 49 = 2$.

61. Think of the meaning of $\log_2 8$. It is the exponent to which you raise 2 to get 8. That exponent is 3. Therefore, $\log_2 8 = 3$.

63. Let $\log_9 \dfrac{1}{81} = x$. Then

$9^x = \dfrac{1}{81}$

$9^x = 9^{-2}$

$x = -2$

Thus, $\log_9 81 = -2$.

65. Let $\log_8 1 = x$. Then

$8^x = 1$

$8^x = 8^0$ $(8^0 = 1)$

$x = 0$

Thus, $\log_8 1 = 0$.

67. Let $\log_e e = x$. Then

$e^x = c$

$e^x = e^1$

$x = 1$

Thus, $\log_e e = 1$.

69. Let $\log_{27} 9 = x$. Then

$27^x = 9$

$(3^3)^x = 3^2$

$3^{3x} = 3^2$

$3x = 2$

$x = \dfrac{2}{3}$

Thus, $\log_{27} 9 = \dfrac{2}{3}$.

71. 4.8970

73. -0.1739

75. Does not exist

77. 0.9464

79. $6 = 10^{0.7782}$; $84 = 10^{1.9243}$; $987,606 = 10^{5.9946}$; $0.00987606 = 10^{-2.0054}$; $98,760.6 = 10^{4.9946}$; $70,000,000 = 10^{7.8451}$; $7000 = 10^{3.8451}$

81. Discussion and Writing Exercise

83. $f(x) = 4 - x^2$

a) The x-coordinate of the vertex is $-\dfrac{b}{2a} = -\dfrac{0}{-2} = 0$.

The y-coordinate is $f(0) = 4 - 0^2 = 4$.

The vertex is $(0, 4)$.

b) The line of symmetry is $x = 0$.

c) Since the coefficient of x^2 is negative, the graph opens down and, hence, has a maximum. The maximum value is 4.

d) We find some points on each side of the vertex and use them to draw the graph.

x	y
0	4
-1	3
1	3
-2	0
2	0
-3	-5
3	-5

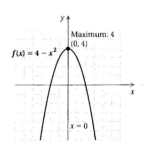

85. $f(x) = -2(x-1)^2 - 3$

  a) The equation is in the form $f(x) = a(x-h)^2 + k$ so we know that the vertex is (h, k), or $(1, -3)$.

  b) The line of symmetry is $x = 1$.

  c) Since the coefficient of x^2 is negative, the graph opens down and, hence, has a maximum. The maximum value is -3.

  d) We find some points on each side of the vertex and use them to draw the graph.

x	y
0	-5
2	-5
-1	-11
3	-11

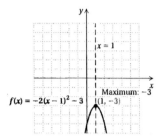

87.
$$E = mc^2$$
$$\frac{E}{m} = c^2 \qquad \text{Dividing by } m$$
$$\sqrt{\frac{E}{m}} = c \qquad \text{Taking the positive square root}$$

89.
$$A = \sqrt{3ab}$$
$$A^2 = (\sqrt{3ab})^2 \qquad \text{Squaring both sides}$$
$$A^2 = 3ab$$
$$\frac{A^2}{3a} = b \qquad \text{Dividing by } 3a$$

91. a) We substitute in the equation $y = ax^2 + bx + c$ and get a system of equations.

  $31 = a(20)^2 + b(20) + c$, or $31 = 400a + 20b + c$;

  $34 = a(24)^2 + b(24) + c$, or $34 = 576a + 24b + c$;

  $22 = a(34)^2 + b(34) + c$, or $22 = 1156a + 34b + c$

  The solution of the system of equation is

  $\left(-\dfrac{39}{280}, \dfrac{963}{140}, -\dfrac{356}{7}\right)$, so the quadratic function that fits the data is

  $f(x) = -\dfrac{39}{280}x^2 + \dfrac{963}{140}x - \dfrac{356}{7}$.

  b) $f(30) = -\dfrac{39}{280}(30)^2 + \dfrac{963}{140}(30) - \dfrac{356}{7} \approx 30$

  About 30% of drivers will be involved in accidents at age 30.

  $f(37) = -\dfrac{39}{280}(37)^2 + \dfrac{963}{140}(37) - \dfrac{356}{7} \approx 13$

  About 13% of drivers will be involved in accidents at age 37.

93. Graph: $f(x) = \log_3 |x+1|$

x	$f(x)$
0	0
2	1
8	2
-2	0
-4	1
-9	2

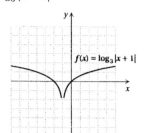

95.
$$\log_{125} x = \frac{2}{3}$$
$$125^{2/3} = x$$
$$(5^3)^{2/3} = x$$
$$5^2 = x$$
$$25 = x$$

97.
$$\log_8(2x+1) = -1$$
$$8^{-1} = 2x+1$$
$$\frac{1}{8} = 2x+1$$
$$1 = 16x + 8 \qquad \text{Multiplying by 8}$$
$$-7 = 16x$$
$$-\frac{7}{16} = x$$

99. Let $\log_{1/4} \dfrac{1}{64} = x$. Then
$$\left(\frac{1}{4}\right)^x = \frac{1}{64}$$
$$\left(\frac{1}{4}\right)^x = \left(\frac{1}{4}\right)^3$$
$$x = 3$$

Thus, $\log_{1/4} \dfrac{1}{64} = 3$.

101.
$$\log_{10}(\log_4(\log_3 81))$$
$$= \log_{10}(\log_4 4) \qquad (\log_3 81 = 4)$$
$$= \log_{10} 1 \qquad (\log_4 4 = 1)$$
$$= 0$$

103. Let $\log_{1/5} 25 = x$. Then
$$\left(\frac{1}{5}\right)^x = 25$$
$$(5^{-1})^x = 25$$
$$5^{-x} = 5^2$$
$$-x = 2$$
$$x = -2$$

Thus, $\log_{1/5} 25 = -2$.

Exercise Set 17.4

1. $\log_2 (32 \cdot 8) = \log_2 32 + \log_2 8$ Property 1

3. $\log_4 (64 \cdot 16) = \log_4 64 + \log_4 16$ Property 1

5. $\log_a Qx = \log_a Q + \log_a x$ Property 1

7. $\log_b 3 + \log_b 84 = \log_b (3 \cdot 84)$ Property 1
$$= \log_b 252$$

9. $\log_c K + \log_c y = \log_c K \cdot y$ Property 1
$$= \log_c Ky$$

11. $\log_c y^4 = 4 \log_c y$ Property 2

13. $\log_b t^6 = 6 \log_b t$ Property 2

15. $\log_b C^{-3} = -3 \log_b C$ Property 2

17. $\log_a \dfrac{67}{5} = \log_a 67 - \log_a 5$ Property 3

19. $\log_b \dfrac{2}{5} = \log_b 2 - \log_b 5$ Property 3

21. $\log_c 22 - \log_c 3 = \log_c \dfrac{22}{3}$ Property 3

23. $\log_a x^2 y^3 z$
$$= \log_a x^2 + \log_a y^3 + \log_b z \quad \text{Property 1}$$
$$= 2 \log_a x + 3 \log_a y + \log_a z \quad \text{Property 2}$$

25. $\log_b \dfrac{xy^2}{z^3}$
$$= \log_b xy^2 - \log_b z^3 \quad\quad\quad \text{Property 3}$$
$$= \log_b x + \log_b y^2 - \log_b z^3 \quad \text{Property 1}$$
$$= \log_b x + 2 \log_b y - 3 \log_b z \quad \text{Property 2}$$

27. $\log_c \sqrt[3]{\dfrac{x^4}{y^3 z^2}}$
$$= \log_c \left(\dfrac{x^4}{y^3 z^2} \right)^{1/3}$$
$$= \frac{1}{3} \log_c \dfrac{x^4}{y^3 z^2} \quad\quad\quad\quad \text{Property 2}$$
$$= \frac{1}{3} (\log_c x^4 - \log_c y^3 z^2) \quad\quad \text{Property 3}$$
$$= \frac{1}{3} [\log_c x^4 - (\log_c y^3 + \log_c z^2)] \quad \text{Property 1}$$
$$= \frac{1}{3} (\log_c x^4 - \log_c y^3 - \log_c z^2) \quad \begin{array}{l}\text{Removing}\\\text{parentheses}\end{array}$$
$$= \frac{1}{3} (4 \log_c x - 3 \log_c y - 2 \log_c z) \quad \text{Property 2}$$
$$= \frac{4}{3} \log_c x - \log_c y - \frac{2}{3} \log_c z$$

29. $\log_a \sqrt[4]{\dfrac{m^8 n^{12}}{a^3 b^5}}$
$$= \log_a \left(\dfrac{m^8 n^{12}}{a^3 b^5} \right)^{1/4}$$
$$= \frac{1}{4} \log_a \dfrac{m^8 n^{12}}{a^3 b^5} \quad\quad\quad \text{Property 2}$$
$$= \frac{1}{4} (\log_a m^8 n^{12} - \log_a a^3 b^5) \quad\quad \text{Property 3}$$
$$= \frac{1}{4} [\log_a m^8 + \log_a n^{12} - (\log_a a^3 + \log_a b^5)] \text{ Property 1}$$
$$= \frac{1}{4} (\log_a m^8 + \log_a n^{12} - \log_a a^3 - \log_a b^5)$$
$$\quad\quad\quad\quad\quad\quad\quad\quad\quad\quad \text{Removing parentheses}$$
$$= \frac{1}{4} (\log_a m^8 + \log_a n^{12} - 3 - \log_a b^5) \quad \text{Property 4}$$
$$= \frac{1}{4} (8 \log_a m + 12 \log_a n - 3 - 5 \log_a b) \quad \text{Property 2}$$
$$= 2 \log_a m + 3 \log_a n - \frac{3}{4} - \frac{5}{4} \log_a b$$

31. $\dfrac{2}{3} \log_a x - \dfrac{1}{2} \log_a y$
$$= \log_a x^{2/3} - \log_a y^{1/2} \quad \text{Property 2}$$
$$= \log_a \dfrac{x^{2/3}}{y^{1/2}}, \text{ or} \quad\quad\quad \text{Property 3}$$
$$\log_a \dfrac{\sqrt[3]{x^2}}{\sqrt{y}}$$

33. $\log_a 2x + 3(\log_a x - \log_a y)$
$$= \log_a 2x + 3 \log_a x - 3 \log_a y$$
$$= \log_a 2x + \log_a x^3 - \log_a y^3 \quad \text{Property 2}$$
$$= \log_a 2x^4 - \log_a y^3 \quad\quad\quad\quad \text{Property 1}$$
$$= \log_a \dfrac{2x^4}{y^3} \quad\quad\quad\quad\quad\quad \text{Property 3}$$

35. $\log_a \dfrac{a}{\sqrt{x}} - \log_a \sqrt{ax}$
$$= \log_a ax^{-1/2} - \log_a a^{1/2} x^{1/2}$$
$$= \log_a \dfrac{ax^{-1/2}}{a^{1/2} x^{1/2}} \quad\quad\quad \text{Property 3}$$
$$= \log_a \dfrac{a^{1/2}}{x}, \text{ or}$$
$$\log_a \dfrac{\sqrt{a}}{x}$$

37. $\log_b 15 = \log_b (3 \cdot 5)$
$$= \log_b 3 + \log_b 5 \quad \text{Property 1}$$
$$= 1.099 + 1.609$$
$$= 2.708$$

39. $\log_b \dfrac{5}{3} = \log_b 5 - \log_b 3$ Property 3
$$= 1.609 - 1.099$$
$$= 0.51$$

41. $\log_b \dfrac{1}{5} = \log_b 1 - \log_b 5$ Property 3
$= 0 - 1.609$ $(\log_b 1 = 0)$
$= -1.609$

43. $\log_b \sqrt{b^3} = \log_b b^{3/2} = \dfrac{3}{2}$ Property 4

45. $\log_b 5b = \log_b 5 + \log_b b$ Property 1
$= 1.609 + 1$ $(\log_b b = 1)$
$= 2.609$

47. $\log_e e^t = t$ Property 4

49. $\log_p p^5 = 5$ Property 4

51. $\log_2 2^7 = x$
$7 = x$ Property 4

53. $\log_e e^x = -7$
$x = -7$ Property 4

55. Discussion and Writing Exercise

57. $i^{29} = i^{28} \cdot i = (i^2)^{14} \cdot i = (-1)^{14} \cdot i = 1 \cdot i = i$

59. $(2+i)(2-i) = 4 - i^2 = 4 - (-1) = 4 + 1 = 5$

61. $(7 - 8i) - (-16 + 10i) = 7 - 8i + 16 - 10i = 23 - 18i$

63. $(8 + 3i)(-5 - 2i) = -40 - 16i - 15i - 6i^2 =$
$-40 - 16i - 15i + 6 = -34 - 31i$

65. Enter $y_1 = \log\ x^2$ and $y_2 = (\log\ x)(\log\ x)$ and show that the graphs are different and that the y-values in a table of values are not the same.

67. $\log_a\ (x^8 - y^8) - \log_a\ (x^2 + y^2)$
$= \log_a \dfrac{x^8 - y^8}{x^2 + y^2}$ Property 3
$= \log_a \dfrac{(x^4 + y^4)(x^2 + y^2)(x+y)(x-y)}{x^2 + y^2}$ Factoring
$= \log_a\ [(x^4 + y^4)(x+y)(x-y)]$ Simplifying
$= \log_a\ (x^6 - x^4y^2 + x^2y^4 - y^6)$ Multiplying

69. $\log_a\ \sqrt{1 - s^2}$
$= \log_a\ (1 - s^2)^{1/2}$
$= \dfrac{1}{2}\ \log_a\ (1 - s^2)$
$= \dfrac{1}{2}\ \log_a\ [(1 - s)(1 + s)]$
$= \dfrac{1}{2}\ \log_a\ (1 - s) + \dfrac{1}{2}\ \log_a\ (1 + s)$

71. False. For example, let $a = 10$, $P = 100$, and $Q = 10$.
$\dfrac{\log 100}{\log 10} = \dfrac{2}{1} = 2$, but
$\log \dfrac{100}{10} = \log 10 = 1$.

73. True, by Property 1

75. False. For example, let $a = 2$, $P = 1$, and $Q = 1$.
$\log_2(1 + 1) = \log_2 2 = 1$, but
$\log_2 1 + \log_2 1 = 0 + 0 = 0$.

Exercise Set 17.5

1. 0.6931

3. 4.1271

5. 8.3814

7. -5.0832

9. -1.6094

11. Does not exist

13. -1.7455

15. 1

17. 15.0293

19. 0.0305

21. 109.9472

23. 5

25. We will use common logarithms for the conversion. Let $a = 10$, $b = 6$, and $M = 100$ and substitute in the change-of-base formula.
$$\log_b M = \frac{\log_a M}{\log_a b}$$
$$\log_6 100 = \frac{\log_{10} 100}{\log_{10} 6}$$
$$\approx \frac{2}{0.7782}$$
$$\approx 2.5702$$

27. We will use common logarithms for the conversion. Let $a = 10$, $b = 2$, and $M = 100$ and substitute in the change-of-base formula.
$$\log_2 100 = \frac{\log_{10} 100}{\log_{10} 2}$$
$$\approx \frac{2}{0.3010}$$
$$\approx 6.6439$$

29. We will use natural logarithms for the conversion. Let $a = e$, $b = 7$, and $M = 65$ and substitute in the change-of-base formula.
$$\log_7 65 = \frac{\ln 65}{\ln 7}$$
$$\approx \frac{4.1744}{1.9459}$$
$$\approx 2.1452$$

31. We will use natural logarithms for the conversion. Let $a = e$, $b = 0.5$, and $M = 5$ and substitute in the change-of-base formula.

$$\log_{0.5} 5 = \frac{\ln 5}{\ln 0.5}$$
$$\approx \frac{1.6094}{-0.6931}$$
$$\approx -2.3219$$

33. We will use common logarithms for the conversion. Let $a = 10$, $b = 2$, and $M = 0.2$ and substitute in the change-of-base formula.

$$\log_2 0.2 = \frac{\log_{10} 0.2}{\log_{10} 2}$$
$$\approx \frac{-0.6990}{0.3010}$$
$$\approx -2.3219$$

35. We will use natural logarithms for the conversion. Let $a = e$, $b = \pi$, and $M = 200$.

$$\log_\pi 200 = \frac{\ln 200}{\ln \pi}$$
$$\approx \frac{5.2983}{1.1447}$$
$$\approx 4.6285$$

If $\ln 200$ and $\ln \pi$ are not rounded before the division is performed, the result is 4.6284.

37. Graph: $f(x) = e^x$

We find some function values with a calculator. We use these values to plot points and draw the graph.

x	e^x
0	1
1	2.7
2	7.4
3	20.1
−1	0.4
−2	0.1
−3	0.05

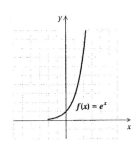

39. Graph: $f(x) = e^{-0.5x}$

We find some function values, plot points, and draw the graph.

x	$e^{-0.5x}$
0	1
1	0.61
2	0.37
−1	1.65
−2	2.72
−3	4.48
−4	7.39

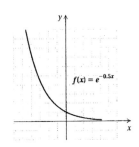

41. Graph: $f(x) = e^{x-1}$

We find some function values, plot points, and draw the graph.

x	e^{x-1}
0	0.4
1	1
2	2.7
3	7.4
4	20.1
−1	0.1
−2	0.05

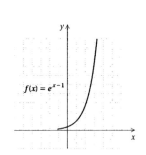

43. Graph: $f(x) = e^{x+2}$

We find some function values, plot points, and draw the graph.

x	e^{x+2}
1	20.1
0	7.4
−2	1
−3	0.4
−4	0.1

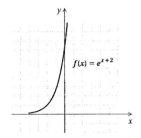

45. Graph: $f(x) = e^x - 1$

We find some function values, plot points, and draw the graph.

x	$e^x - 1$
0	0
1	1.72
2	6.39
3	19.09
−1	−0.63
−2	−0.86
−4	−0.98

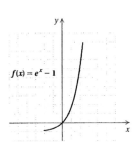

47. Graph: $f(x) = \ln(x + 2)$

We find some function values, plot points, and draw the graph.

x	$\ln(x + 2)$
0	0.69
1	1.10
2	1.39
3	1.61
−0.5	0.41
−1	0
−1.5	−0.69

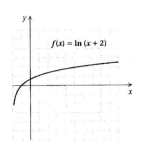

49. Graph: $f(x) = \ln(x-3)$

We find some function values, plot points, and draw the graph.

x	$\ln(x-3)$
3	Undefined
4	0
5	0.69
6	1.10
8	1.61
10	1.95

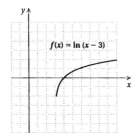

51. Graph: $f(x) = 2 \ln x$

x	$2 \ln x$
0.5	-1.4
1	0
2	1.4
3	2.2
4	2.8
5	3.2
6	3.6

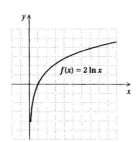

53. Graph: $f(x) = \dfrac{1}{2} \ln x + 1$

x	$\dfrac{1}{2} \ln x + 1$
1	1
2	1.35
3	1.55
4	1.69
6	1.90

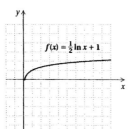

55. Graph: $f(x) = |\ln x|$

x	$\ln x$
$\dfrac{1}{4}$	1.4
$\dfrac{1}{2}$	0.7
1	0
3	1.1
5	1.6

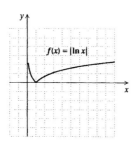

57. Discussion and Writing Exercise

59. $x^{1/2} - 6x^{1/4} + 8 = 0$

Let $u = x^{1/4}$.

$u^2 - 6u + 8 = 0$ Substituting

$(u-4)(u-2) = 0$

$u = 4$ or $u = 2$

$x^{1/4} = 4$ or $x^{1/4} = 2$

$x = 256$ or $x = 16$ Raising both sides to the fourth power

Both numbers check. The solutions are 256 and 16.

61. $x - 18\sqrt{x} + 77 = 0$

Let $u = \sqrt{x}$.

$u^2 - 18u + 77 = 0$ Substituting

$(u-7)(u-11) = 0$

$u = 7$ or $u = 11$

$\sqrt{x} = 7$ or $\sqrt{x} = 11$

$x = 49$ or $x = 121$ Squaring both sides

Both numbers check. The solutions are 49 and 121.

63. Domain: $(-\infty, \infty)$, range: $[0, \infty)$

65. Domain: $(-\infty, \infty)$, range: $(-\infty, 100)$

67. $f(x)$ can be calculated for positive values of $2x - 5$. We have:

$2x - 5 > 0$

$2x > 5$

$x > \dfrac{5}{2}$

The domain is $\left\{ x \middle| x > \dfrac{5}{2} \right\}$, or $\left(\dfrac{5}{2}, \infty \right)$.

Exercise Set 17.6

1. $2^x = 8$

$2^x = 2^3$

$x = 3$ The exponents are the same.

3. $4^x = 256$

$4^x = 4^4$

$x = 4$ The exponents are the same.

5. $2^{2x} = 32$

$2^{2x} = 2^5$

$2x = 5$

$x = \dfrac{5}{2}$

7. $3^{5x} = 27$

$3^{5x} = 3^3$

$5x = 3$

$x = \dfrac{3}{5}$

9. $2^x = 11$

$\log 2^x = \log 11$ Taking the common logarithm on both sides

$x \log 2 = \log 11$ Property 2

$x = \dfrac{\log 11}{\log 2}$

$x \approx 3.4594$

11. $2^x = 43$

$\log 2^x = \log 43$ Taking the common logarithm on both sides

$x \log 2 = \log 43$ Property 2

$x = \dfrac{\log 43}{\log 2}$

$x \approx 5.4263$

13. $5^{4x-7} = 125$

$5^{4x-7} = 5^3$

$4x - 7 = 3$ The exponents are the same.

$4x = 10$

$x = \dfrac{10}{4}$, or $\dfrac{5}{2}$

15. $3^{x^2} \cdot 3^{4x} = \dfrac{1}{27}$

$3^{x^2+4x} = 3^{-3}$

$x^2 + 4x = -3$

$x^2 + 4x + 3 = 0$

$(x+3)(x+1) = 0$

$x = -3 \text{ or } x = -1$

17. $4^x = 8$

$(2^2)^x = 2^3$

$2^{2x} = 2^3$

$2x = 3$ The exponents are the same.

$x = \dfrac{3}{2}$

19. $e^t = 100$

$\ln e^t = \ln 100$ Taking ln on both sides

$t = \ln 100$ Property 4

$t \approx 4.6052$ Using a calculator

21. $e^{-t} = 0.1$

$\ln e^{-t} = \ln 0.1$ Taking ln on both sides

$-t = \ln 0.1$ Property 4

$-t \approx -2.3026$

$t \approx 2.3026$

23. $e^{-0.02t} = 0.06$

$\ln e^{-0.02t} = \ln 0.06$ Taking ln on both sides

$-0.02t = \ln 0.06$ Property 4

$t = \dfrac{\ln 0.06}{-0.02}$

$t \approx \dfrac{-2.8134}{-0.02}$

$t \approx 140.6705$

25. $2^x = 3^{x-1}$

$\log 2^x = \log 3^{x-1}$

$x \log 2 = (x-1) \log 3$

$x \log 2 = x \log 3 - \log 3$

$\log 3 = x \log 3 - x \log 2$

$\log 3 = x(\log 3 - \log 2)$

$\dfrac{\log 3}{\log 3 - \log 2} = x$

$\dfrac{0.4771}{0.4771 - 0.3010} \approx x$

$2.7095 \approx x$

27. $(3.6)^x = 62$

$\log (3.6)^x = \log 62$

$x \log 3.6 = \log 62$

$x = \dfrac{\log 62}{\log 3.6}$

$x \approx 3.2220$

29. $\log_4 x = 4$

$x = 4^4$ Writing an equivalent exponential equation

$x = 256$

31. $\log_2 x = -5$

$x = 2^{-5}$ Writing an equivalent exponential equation

$x = \dfrac{1}{32}$

33. $\log x = 1$ The base is 10.

$x = 10^1$

$x = 10$

35. $\log x = -2$ The base is 10.

$x = 10^{-2}$

$x = \dfrac{1}{100}$

37. $\ln x = 2$

$x = e^2 \approx 7.3891$

39. $\ln x = -1$

$x = e^{-1}$

$x = \dfrac{1}{e} \approx 0.3679$

41. $\log_3 (2x+1) = 5$

$2x + 1 = 3^5$ Writing an equivalent exponential equation

$2x + 1 = 243$

$2x = 242$

$x = 121$

43. $\log x + \log (x - 9) = 1$ The base is 10.

$\log_{10} [x(x - 9)] = 1$ Property 1

$x(x - 9) = 10^1$

$x^2 - 9x = 10$

$x^2 - 9x - 10 = 0$

$(x - 10)(x + 1) = 0$

$x = 10 \text{ or } x = -1$

Check: For 10:

$$\frac{\log x + \log (x - 9) = 1}{\log 10 + \log (10 - 9) \; ? \; 1}$$
$$\log 10 + \log 1 \;\Big|$$
$$1 + 0 \;\Big|$$
$$1 \;\Big| \quad \text{TRUE}$$

For -1:

$$\frac{\log x + \log (x - 9) = 1}{\log(-1) + \log (-1 - 9) \; ? \; 1} \quad \text{FALSE}$$

The number -1 does not check, because negative numbers do not have logarithms. The solution is 10.

45. $\log x - \log (x + 3) = -1$ The base is 10.

$\log_{10} \dfrac{x}{x + 3} = -1$ Property 3

$\dfrac{x}{x + 3} = 10^{-1}$

$\dfrac{x}{x + 3} = \dfrac{1}{10}$

$10x = x + 3$

$9x = 3$

$x = \dfrac{1}{3}$

The answer checks. The solution is $\dfrac{1}{3}$.

47. $\log_2 (x + 1) + \log_2 (x - 1) = 3$

$\log_2 [(x + 1)(x - 1)] = 3$ Property 1

$(x + 1)(x - 1) = 2^3$

$x^2 - 1 = 8$

$x^2 = 9$

$x = \pm 3$

The number 3 checks, but -3 does not. The solution is 3.

49. $\log_4 (x + 6) - \log_4 x = 2$

$\log_4 \dfrac{x + 6}{x} = 2$ Property 3

$\dfrac{x + 6}{x} = 4^2$

$\dfrac{x + 6}{x} = 16$

$x + 6 = 16x$

$6 = 15x$

$\dfrac{2}{5} = x$

The answer checks. The solution is $\dfrac{2}{5}$.

51. $\log_4 (x + 3) + \log_4 (x - 3) = 2$

$\log_4 [(x + 3)(x - 3)] = 2$ Property 1

$(x + 3)(x - 3) = 4^2$

$x^2 - 9 = 16$

$x^2 = 25$

$x = \pm 5$

The number 5 checks, but -5 does not. The solution is 5.

53. $\log_3 (2x - 6) - \log_3 (x + 4) = 2$

$\log_3 \dfrac{2x - 6}{x + 4} = 2$ Property 3

$\dfrac{2x - 6}{x + 4} = 3^2$

$\dfrac{2x - 6}{x + 4} = 9$

$2x - 6 = 9x + 36$
$\quad\quad\quad$ Multiplying by $(x + 4)$

$-42 = 7x$

$-6 = x$

Check:

$$\frac{\log_3 (2x - 6) - \log_3 (x + 4) = 2}{\log_3 [2(-6) - 6] - \log_3 (-6 + 4) \; ? \; 2}$$
$$\log_3 (-18) - \log_3 (-2) \;\Big| \quad \text{FALSE}$$

The number -6 does not check, because negative numbers do not have logarithms. There is no solution.

55. Discussion and Writing Exercise

57. $\quad\quad\quad x^4 + 400 = 104x^2$

$x^4 - 104x^2 + 400 = 0$

Let $u = x^2$.

$u^2 - 104u + 400 = 0$

$(u - 100)(u - 4) = 0$

$u = 100 \quad or \quad u = 4$

$x^2 = 100 \quad or \quad x^2 = 4$ Replacing u with x^2

$x = \pm 10 \quad or \quad x = \pm 2$

The solutions are ± 10 and ± 2.

59.
$$(x^2 + 5x)^2 + 2(x^2 + 5x) = 24$$
$$(x^2 + 5x)^2 + 2(x^2 + 5x) - 24 = 0$$
Let $u = x^2 + 5x$.
$$u^2 + 2u - 24 = 0$$
$$(u + 6)(u - 4) = 0$$
$$u = -6 \quad or \quad u = 4$$
$$x^2 + 5x = 6 \quad or \quad x^2 + 5x = 4$$
Replacing u with $x^2 + 5x$
$$x^2 + 5x + 6 = 0 \quad or \quad x^2 + 5x - 4 = 0$$
$$(x+3)(x+2) = 0 \quad or \quad x = \frac{-5 \pm \sqrt{5^2 - 4 \cdot 1(-4)}}{2 \cdot 1}$$
$$x = -3 \; or \; x = -2 \; or \quad x = \frac{-5 \pm \sqrt{41}}{2}$$
The solutions are -3, -2, and $\dfrac{-5 \pm \sqrt{41}}{2}$.

61. $(125x^3 y^{-2} z^6)^{-2/3} =$
$$(5^3)^{-2/3}(x^3)^{-2/3}(y^{-2})^{-2/3}(z^6)^{-2/3} =$$
$$5^{-2} x^{-2} y^{4/3} z^{-4} = \frac{1}{25} x^{-2} y^{4/3} z^{-4}, \text{ or}$$
$$\frac{y^{4/3}}{25 x^2 z^4}$$

63. Find the first coordinate of the point of intersection of $y_1 = \ln x$ and $y_2 = \log x$. The value of x for which the natural logarithm of x is the same as the common logarithm of x is 1.

65. a) 0.3770
   b) -1.9617
   c) 0.9036
   d) -1.5318

67.
$$2^{2x} + 128 = 24 \cdot 2^x$$
$$2^{2x} - 24 \cdot 2^x + 128 = 0$$
Let $u = 2^x$.
$$u^2 - 24u + 128 = 0$$
$$(u - 8)(u - 16) = 0$$
$$u = 8 \quad or \quad u = 16$$
$$2^x = 8 \quad or \quad 2^x = 16 \quad \text{Replacing } u \text{ with } 2^x$$
$$2^x = 2^3 \quad or \quad 2^x = 2^4$$
$$x = 3 \quad or \quad x = 4$$
The solutions are 3 and 4.

69.
$$8^x = 16^{3x+9}$$
$$(2^3)^x = (2^4)^{3x+9}$$
$$2^{3x} = 2^{12x+36}$$
$$3x = 12x + 36$$
$$-36 = 9x$$
$$-4 = x$$

71.
$$\log_6 (\log_2 x) = 0$$
$$\log_2 x = 6^0$$
$$\log_2 x = 1$$
$$x = 2^1$$
$$x = 2$$

73.
$$\log_5 \sqrt{x^2 - 9} = 1$$
$$\sqrt{x^2 - 9} = 5^1$$
$$x^2 - 9 = 25 \qquad \text{Squaring both sides}$$
$$x^2 = 34$$
$$x = \pm\sqrt{34}$$
Both numbers check. The solutions are $\pm\sqrt{34}$.

75.
$$\log (\log x) = 5 \qquad \text{The base is 10.}$$
$$\log x = 10^5$$
$$\log x = 100,000$$
$$x = 10^{100,000}$$
The number checks. The solution is $10^{100,000}$.

77.
$$\log x^2 = (\log x)^2$$
$$2 \log x = (\log x)^2$$
$$0 = (\log x)^2 - 2 \log x$$
Let $u = \log x$.
$$0 = u^2 - 2u$$
$$0 = u(u - 2)$$
$$u = 0 \quad or \quad u = 2$$
$$\log x = 0 \quad or \quad \log x = 2$$
$$x = 10^0 \quad or \quad x = 10^2$$
$$x = 1 \quad or \quad x = 100$$
Both numbers check. The solutions are 1 and 100.

79.
$$\log_a a^{x^2+4x} = 21$$
$$x^2 + 4x = 21 \qquad \text{Property 4}$$
$$x^2 + 4x - 21 = 0$$
$$(x + 7)(x - 3) = 0$$
$$x = -7 \; or \; x = 3$$
Both numbers check. The solutions are $= -7$ and 3.

81. $3^{2x} - 8 \cdot 3^x + 15 = 0$
Let $u = 3^x$ and substitute.
$$u^2 - 8u + 15 = 0$$
$$(u - 5)(u - 3) = 0$$
$$u = 5 \quad or \quad u = 3$$
$$3^x = 5 \quad or \quad 3^x = 3 \qquad \text{Substituting } 3^x \text{ for } u$$
$$\log 3^x = \log 5 \quad or \quad 3^x = 3^1$$
$$x \log 3 = \log 5 \quad or \quad x = 1$$
$$x = \frac{\log 5}{\log 3} \quad or \quad x = 1, \text{ or}$$
$$x \approx 1.4650 \quad or \quad x = 1$$

Both numbers check. Note that we can also express $\dfrac{\log 5}{\log 3}$ as $\log_3 5$ using the change-of-base formula.

Exercise Set 17.7

1. $L = 10 \cdot \log \dfrac{I}{I_0}$

$= 10 \cdot \log \dfrac{3.2 \times 10^{-3}}{10^{-12}}$ Substituting

$= 10 \cdot \log(3.2 \times 10^9)$

$\approx 10(9.5)$

≈ 95

The sound level is about 95 dB.

3. $\qquad L = 10 \cdot \log \dfrac{I}{I_0}$

$105 = 10 \cdot \log \dfrac{I}{10^{-12}}$

$10.5 = \log \dfrac{I}{10^{-12}}$

$10.5 = \log I - \log 10^{-12}$ $(\log 10^a = a)$

$10.5 = \log I - (-12)$

$10.5 = \log I + 12$

$-1.5 = \log I$

$10^{-1.5} = I$ Converting to an exponential equation

$3.2 \times 10^{-2} \approx I$

The intensity of the sound is $10^{-1.5}$ W/m², or about 3.2×10^{-2} W/m².

5. $\text{pH} = -\log[H^+]$

$= -\log[1.6 \times 10^{-7}]$

$\approx -(-6.795880)$

≈ 6.8

The pH of milk is about 6.8.

7. $\qquad \text{pH} = -\log[H^+]$

$7.8 = -\log[H^+]$

$-7.8 = \log[H^+]$

$10^{-7.8} = [H^+]$

$1.58 \times 10^{-8} \approx [H^+]$

The hydrogen ion concentration is about 1.58×10^{-8} moles per liter.

9. $\quad 3,251,876 = 3251.876$ thousands

$w(P) = 0.37 \ln P + 0.05$

$w(3251.876) = 0.37 \ln 3251.876 + 0.05$

≈ 3.04 ft/sec

11. $\qquad 311,121 = 311.121$ thousands

$w(P) = 0.37 \ln P + 0.05$

$w(311.121) = 0.37 \ln 311.121 + 0.05$

≈ 2.17 ft/sec

13. $N(t) = 3^t$

a) $N(5) = 3^5 = 243$ people

b) $\qquad 6,200,000,000 = 3^t$

$\ln 6,200,000,000 = \ln 3^t$

$\ln 6,200,000,000 = t \ln 3$

$\dfrac{\ln 6,200,000,000}{\ln 3} = t$

$20.5 \approx t$

The acts of kindness will reach the entire world in about 20.5 months.

c) $\qquad 2 = 3^t$

$\ln 2 = \ln 3^t$

$\ln 2 = t \ln 3$

$\dfrac{\ln 2}{\ln 3} = t$

$0.6 \approx t$

The doubling time is about 0.6 month.

15. a) In 2005, $t = 2005 - 2000$, or 5.

$C(5) = 11,054(1.06)^5 \approx \$14,793$

b) $\qquad 21,000 = 11,054(1.06)^t$

$\dfrac{21,000}{11,054} = 1.06^t$

$\log \dfrac{21,000}{11,054} = \log 1.06^t$

$\log \dfrac{21,000}{11,054} = t \log 1.06$

$\dfrac{\log \dfrac{21,000}{11,054}}{\log 1.06} = t$

$11 \approx t$

The cost will be \$21,000 11 yr after 2000, or in 2011.

c) $\qquad 22,108 = 11,054(1.06)^t$

$2 = (1.06)^t$

$\log 2 = \log(1.06)^t$

$\log 2 = t \log 1.06$

$\dfrac{\log 2}{\log 1.06} = t$

$11.9 \approx t$

The doubling time is about 11.9 years.

17. a) $P(t) = 6e^{0.015t}$, where $P(t)$ is in billions and t is the number of years after 1998.

b) In 2010, $t = 2010 - 1998$, or 12.

$P(12) = 6e^{0.015(12)} \approx 7.2$

In 2010, the population will be about 7.2 billion.

c)
$$10 = 6e^{0.015t}$$
$$\frac{5}{3} = e^{0.015t}$$
$$\ln\left(\frac{5}{3}\right) = \ln e^{0.015t}$$
$$\ln\left(\frac{5}{3}\right) = 0.015t$$
$$\frac{\ln\left(\frac{5}{3}\right)}{0.015} = t$$
$$34 \approx t$$

The population will be 10 billion 34 years after 1998, or in 2032.

d)
$$12 = 6e^{0.015t}$$
$$2 = e^{0.015t}$$
$$\ln 2 = \ln e^{0.015t}$$
$$\ln 2 = 0.015t$$
$$\frac{\ln 2}{0.015} = t$$
$$46.2 \approx t$$

The doubling time is about 46.2 yr.

19. a) $P(t) = P_0 e^{0.06t}$

b) To find the balance after one year, replace P_0 with 5000 and t with 1. We find $P(1)$:

$$P(1) = 5000 e^{0.06(1)} \approx \$5309.18$$

To find the balance after 2 years, replace P_0 with 5000 and t with 2. We find $P(2)$:

$$P(2) = 5000 e^{0.06(2)} \approx \$5637.48$$

To find the balance after 10 years, replace P_0 with 5000 and t with 10. We find $P(10)$: $P(10) = 5000 e^{0.06(10)} \approx \$9110.59$

c) To find the doubling time, replace P_0 with 5000 and $P(t)$ with 10,000 and solve for t.

$$10,000 = 5000 e^{0.06t}$$
$$2 = e^{0.06t}$$

$\ln 2 = \ln e^{0.06t}$ Taking the natural logarithm on both sides

$\ln 2 = 0.06t$ Finding the logarithm of the base to a power

$$\frac{\ln 2}{0.06} = t$$
$$11.6 \approx t$$

The investment will double in about 11.6 years.

21. a)
$$P(t) = P_0 e^{kt}$$
$$1,563,282 = 852,737 e^{k \cdot 10}$$
$$\frac{1,563,282}{852,737} = e^{10k}$$
$$\ln \frac{1,563,282}{852,737} = \ln e^{10k}$$
$$\ln \frac{1,563,282}{852,737} = 10k$$
$$\frac{\ln \dfrac{1,563,282}{852,737}}{10} = k$$
$$0.061 \approx k$$

The exponential growth rate is 0.061, or 6.1%.

The exponential growth function is $P(t) = 852,737 e^{0.061t}$, where t is the number of years after 1990.

b) In 2010, $t = 2010 - 1990$, or 20.
$$P(20) = 852,737 e^{0.061(20)}$$
$$= 852,737 e^{1.22}$$
$$\approx 2,888,380$$

In 2010, the population of Las Vegas will be about 2,888,380.

c)
$$8,000,000 = 852,737 e^{0.061t}$$
$$\frac{8,000,000}{852,737} = e^{0.061t}$$
$$\ln \frac{8,000,000}{852,737} = \ln e^{0.061t}$$
$$\ln \frac{8,000,000}{852,737} = 0.061t$$
$$\frac{\ln \dfrac{8,000,000}{852,737}}{0.061} = t$$
$$37 \approx t$$

The population will reach 8,000,000 about 37 yr after 1990, or in 2027.

23. If the scrolls had lost 22.3% of their carbon-14 from an initial amount P_0, then 77.7%(P_0) is the amount present. To find the age t of the scrolls, we substitute 77.7%(P_0), or $0.777P_0$, for $P(t)$ in the carbon-14 decay function and solve for t.

$$P(t) = P_0 e^{-0.00012t}$$
$$0.777P_0 = P_0 e^{-0.00012t}$$
$$0.777 = e^{-0.00012t}$$
$$\ln 0.777 = \ln e^{-0.00012t}$$
$$-0.2523 \approx -0.00012t$$
$$t \approx \frac{-0.2523}{-0.00012} \approx 2103$$

The scrolls are about 2103 years old.

25. The function $P(t) = P_0 e^{-kt}$, $k > 0$, can be used to model decay. For iodine-131, $k = 9.6\%$, or 0.096. To find the half-life we substitute 0.096 for k and $\frac{1}{2}P_0$ for $P(t)$, and solve for t.

$$\frac{1}{2}P_0 = P_0 e^{-0.096t}, \text{ or } \frac{1}{2} = e^{-0.096t}$$

$$\ln\frac{1}{2} = \ln e^{-0.096t} = -0.096t$$

$$t = \frac{\ln 0.5}{-0.096} \approx \frac{-0.6931}{-0.096} \approx 7.2 \text{ days}$$

27. The function $P(t) = P_0 e^{-kt}$, $k > 0$, can be used to model decay. We substitute $\frac{1}{2}P_0$ for $P(t)$ and 1 for t and solve for the decay rate k.

$$\frac{1}{2}P_0 = P_0 e^{-k\cdot 1}$$

$$\frac{1}{2} = e^{-k}$$

$$\ln\frac{1}{2} = \ln e^{-k}$$

$$-0.693 \approx -k$$

$$0.693 \approx k$$

The decay rate is 0.693, or 69.3% per year.

29. a) We use the exponential decay equation $W(t) = W_0 e^{-kt}$, where t is the number of years after 1996 and $W(t)$ is in millions of tons. In 1996, at $t = 0$, 17.5 million tons of yard waste were discarded. We substitute 17.5 for W_0.

$$W(t) = 17.5e^{-kt}.$$

To find the exponential decay rate k, observe that 2 years after 1996, in 1998, 14.5 million tons of yard waste were discarded. We substitute 2 for t and 14.5 for $W(t)$.

$$14.5 = 17.5e^{-k\cdot 2}$$

$$0.8286 \approx e^{-2k}$$

$$\ln 0.8286 \approx \ln e^{-2k}$$

$$\ln 0.8286 \approx -2k$$

$$\frac{\ln 0.8286}{-2} \approx k$$

$$0.094 \approx k$$

Then we have $W(t) = 17.5e^{-0.094t}$, where t is the number of years after 1996 and $W(t)$ is in millions of tons.

b) In 2010, $t = 2010 - 1996 = 14$.

$$W(14) = 17.5e^{-0.094(14)}$$

$$= 17.5e^{-1.316}$$

$$\approx 4.7$$

In 2010, about 4.7 million tons of yard waste were discarded.

c) 1 ton is equivalent to 0.000001 million tons.

$$0.000001 = 17.5e^{-0.094t}$$

$$5.71 \times 10^{-8} \approx e^{-0.094t}$$

$$\ln(5.71 \times 10^{-8}) \approx \ln e^{-0.094t}$$

$$\ln(5.71 \times 10^{-8}) \approx -0.094t$$

$$\frac{\ln(5.71 \times 10^{-8})}{-0.094} \approx t$$

$$177 \approx t$$

Only one ton of yard waste will be discarded about 177 years after 1996, or in 2173.

31. a) We start with the exponential growth equation

$$V(t) = V_0 e^{kt}, \text{ where } t \text{ is the number of years after 1991.}$$

Substituting 451,000 for V_0, we have

$$V(t) = 451,000 e^{kt}.$$

To find the exponential growth rate k, observe that the card sold for \$640,500 in 1996, or 5 years after 1991. We substitute and solve for k.

$$V(5) = 451,000 e^{k\cdot 5}$$

$$640,500 = 451,000 e^{5k}$$

$$1.42 = e^{5k}$$

$$\ln 1.42 = \ln e^{5k}$$

$$\ln 1.42 = 5k$$

$$\frac{\ln 1.42}{5} = k$$

$$0.07 \approx k$$

Thus the exponential growth function is $V(t) = 451,000 e^{0.07t}$, where t is the number of years after 1991.

b) In 2005, $t = 2005 - 1991$, or 14.

$$V(14) = 451,000 e^{0.07(14)} \approx 1,201,670$$

The card's value in 2005 will be about \$1,201,670.

c) Substitute \$902,000 for $V(t)$ and solve for t.

$$902,000 = 451,000 e^{0.07t}$$

$$2 = e^{0.07t}$$

$$\ln 2 = \ln e^{0.07t}$$

$$\ln 2 = 0.07t$$

$$\frac{\ln 2}{0.07} = t$$

$$9.9 \approx t$$

The doubling time is about 9.9 years.

d) Substitute $\$1,000,000$ for $V(t)$ and solve for t.

$$1,000,000 = 451,000\, e^{0.07t}$$

$$2.217 \approx e^{0.07t}$$

$$\ln 2.217 \approx \ln e^{0.07t}$$

$$\ln 2.217 \approx 0.07t$$

$$\frac{\ln 2.217}{0.07} \approx t$$

$$11.4 \approx t$$

The value of the card will be $\$1,000,000$ in $1991 + 11$, or 2002.

e) In 2001, $t = 2001 - 1991$, or 10.

$$V(10) = 451,000e^{0.07(10)}$$

$$= 451,000e^{0.7}$$

$$\approx 908,202$$

The function estimates that the card's value in 2001 would be about $\$908,202$. According to this, the card would not be a good buy at $\$1.1$ million.

33. a)

$$P(t) = P_0 e^{-kt}$$

$$2,242,798 = 2,394,811 e^{-k \cdot 10}$$

$$\frac{2,242,798}{2,394,811} = e^{-10k}$$

$$\ln \frac{2,242,798}{2,394,811} = \ln e^{-10k}$$

$$\ln \frac{2,242,798}{2,394,811} = -10k$$

$$\frac{\ln \dfrac{2,242,798}{2,394,811}}{-10} = k$$

$$0.007 \approx k$$

The exponential decay rate is 0.007, or 0.7%. The exponential decay function is $P(t) = 2,394,811e^{-0.007t}$, where t is the number of years after 1990.

b) In 2010, $t = 2010 - 1990$, or 20.

$$P(20) = 2,394,811e^{-0.007(20)}$$

$$= 2,394,811e^{-0.14}$$

$$\approx 2,081,949$$

In 2010, the population of Pittsburgh will be about 2,081,949.

c)

$$1,000,000 = 2,394,811e^{-0.007t}$$

$$\frac{1,000,000}{2,394,811} = e^{-0.007t}$$

$$\ln \frac{1,000,000}{2,394,811} = \ln e^{-0.007t}$$

$$\ln \frac{1,000,000}{2,394,811} = -0.007t$$

$$\frac{\ln \dfrac{1,000,000}{2,394,811}}{-0.007} = t$$

$$125 \approx t$$

The population will decline to 1 million about 125 yr after 1990, or in 2115.

35. Discussion and Writing Exercise

37. $i^{46} = (i^2)^{23} = (-1)^{23} = -1$

39. $i^{53} = (i^2)^{26} \cdot i = (-1)^{26} \cdot i = i$

41. $i^{14} + i^{15} = (i^2)^7 + (i^2)^7 \cdot i = (-1)^7 + (-1)^7 \cdot i = -1 - i$

43.

$$\frac{8-i}{8+i} = \frac{8-i}{8+i} \cdot \frac{8-i}{8-i}$$

$$= \frac{64 - 16i + i^2}{64 - i^2}$$

$$= \frac{64 - 16i - 1}{64 - (-1)}$$

$$= \frac{63 - 16i}{65}$$

$$= \frac{63}{65} - \frac{16}{65}i$$

45. $(5 - 4i)(5 + 4i) = 25 - 16i^2 = 25 + 16 = 41$

47. $-0.937,\ 1.078,\ 58.770$

49. $-0.767,\ 2,\ 4$

51. We will use the exponential growth equation $V(t) = V_0 e^{kt}$, where t is the number of years after 2001 and $V(t)$ is in millions of dollars. We substitute 21 for $V(t)$, 0.05 for k, and 9 for t and solve for V_0.

$$21 = V_0 e^{0.05(9)}$$

$$21 = V_0 e^{0.45}$$

$$\frac{21}{e^{0.45}} = V_0$$

$$13.4 \approx V_0$$

George Steinbrenner needs to invest $\$13.4$ million at 5% interest compounded continuously in order to have $\$21$ million to pay Derek Jeter in 2010.